EMS PRESS

EMS Series of Congress Reports

The *EMS Series of Congress Reports* publishes volumes originating from conferences or seminars focusing on any field of pure or applied mathematics. The individual volumes include an introduction into their subject and review of the contributions in this context. Articles are required to undergo a refereeing process and are accepted only if they contain a survey or significant results not published elsewhere in the literature.

Previously published in this series:

Trends in Representation Theory of Algebras and Related Topics, *A. Skowroński (Ed.)*
K-Theory and Noncommutative Geometry, *G. Cortiñas et al. (Eds.)*
Classification of Algebraic Varieties, *C. Faber, G. van der Geer, E. Looijenga (Eds.)*
Surveys in Stochastic Processes, *J. Blath, P. Imkeller, S. Rœlly (Eds.)*
Representations of Algebras and Related Topics, *A. Skowroński, K. Yamagata (Eds.)*
Contributions to Algebraic Geometry. Impanga Lecture Notes, *P. Pragacz (Ed.)*
Geometry and Arithmetic, *C. Faber, G. Farkas, R. de Jong (Eds.)*
Derived Categories in Algebraic Geometry. Toyko 2011, *Y. Kawamata (Ed.)*
Advances in Representation Theory of Algebras, *D. J. Benson, H. Krause, A. Skowroński (Eds.)*
Valuation Theory in Interaction, *A. Campillo, F.-V. Kuhlmann, B. Teissier (Eds.)*
Representation Theory – Current Trends and Perspectives, *H. Krause et al. (Eds.)*
Functional Analysis and Operator Theory for Quantum Physics. The Pavel Exner Anniversary Volume, *J. Dittrich, H. Kovařík, A. Laptev (Eds.)*
Schubert Varieties, Equivariant Cohomology and Characteristic Classes, *J. Buczyński, M. Michałek, E. Postinghel (Eds.)*
Non-Linear Partial Differential Equations, Mathematical Physics, and Stochastic Analysis, *F. Gesztesy et al. (Eds.)*
Spectral Structures and Topological Methods in Mathematics, *M. Baake, F. Götze, W. Hoffmann (Eds.)*
t-Motives: Hodge Structures, Transcendence and Other Motivic Aspects, *G. Böckle et al. (Eds.)*
Probabilistic Structures in Evolution, *E. Baake, A. Wakolbinger (Eds.)*

Partial Differential Equations, Spectral Theory, and Mathematical Physics

The Ari Laptev Anniversary Volume

Edited by

Pavel Exner
Rupert L. Frank
Fritz Gesztesy
Helge Holden
Timo Weidl

EMS PRESS

Editors:

Pavel Exner
Doppler Institute for Mathematical Physics and Applied Mathematics, Czech Technical University, Břehová 7, 11519 Prague, Czech Republic; and Department of Theoretical Physics, Nuclear Physics Institute, Czech Academy of Sciences, 25068 Řež, Czech Republic
E-mail: exner@ujf.cas.cz

Rupert L. Frank
Department of Mathematics, California Institute of Technology, Pasadena, CA 91125, USA; and Mathematisches Institut, Ludwig-Maximilians-Universität München, Theresienstraße 39, 80333 München, Germany
E-mail: rlfrank@caltech.edu

Fritz Gesztesy
Department of Mathematics, Baylor University, Sid Richardson Bldg., 1410 S. 4th Street, Waco, TX 76706, USA
E-mail: fritz_gesztesy@baylor.edu

Helge Holden
Department of Mathematical Sciences, Norwegian University of Science and Technology, Alfred Getz vei 1, 7491 Trondheim, Norway
E-mail: helge.holden@ntnu.no

Timo Weidl
Institut für Analysis, Dynamik und Modellierung, Fakultät für Mathematik und Physik, Universität Stuttgart, Pfaffenwaldring 57, 70569 Stuttgart, Germany
E-mail: weidl@mathematik.uni-stuttgart.de

2020 Mathematics Subject Classification: 34L15, 34L40, 35J10, 35P05, 35P15, 35P20, 44A10, 47A10, 47A63, 47B28, 47F10, 81Q10

Keywords: Friedrichs inequality, Hardy inequality, Lieb–Thirring inequality, Feshbach–Schur map, scattering theory, Wehrl-type entropy inequalities, electron density estimates, stability of matter, Bose–Einstein condensation, heat kernel estimates, Euler–Bardina equations, nonlinear Schrödinger equation, Brezis–Nirenberg problem, d-bar problem, Bogoliubov theory, wave packet evolution

ISBN 978-3-98547-007-5

Bibliographic information published by the Deutsche Nationalbibliothek
The Deutsche Nationalbibliothek lists this publication in the Deutsche Nationalbibliografie; detailed bibliographic data are available on the Internet at http://dnb.dnb.de.

Published by EMS Press, an imprint of the

European Mathematical Society – EMS – Publishing House GmbH
Institut für Mathematik
Technische Universität Berlin
Straße des 17. Juni 136
10623 Berlin, Germany

https://ems.press

Typeset using the authors' LaTeX sources: WisSat Publishing + Consulting GmbH, Fürstenwalde, Germany
Printing and binding: Beltz Bad Langensalza GmbH, Bad Langensalza, Germany
♾ Printed on acid free paper
9 8 7 6 5 4 3 2 1

Preface

The present volume is dedicated to Ari Laptev at the happy occasion of his seventieth birthday.

Ari Laptev was born on August 10, 1950, in the city then called Leningrad, USSR. The place where he studied gave him the opportunity to join one of the best schools of mathematical physics; he defended his PhD in 1978 prepared under the supervision of Mikhael Solomyak and subsequently kept working at Leningrad University.

From the outset one might naively expect the beginning of a bright – and smooth – career. However, Ari's life took a very indirect path toward his ultimate profession. In 1982 he lost his job for reasons that nowadays someone half his age would have difficulties to comprehend. He had met Marilyn, his future wife, abroad and that is why the Soviet regime forced him to make his living as a construction worker in the following five years, who could do mathematics in his spare time only. Fortunately for Ari, the end of the system which put him into this plight was nearing. Another break came in 1987 when he left Russia for Sweden; in the new environment his multiple talents began to flourish. He became a lecturer at the Linköping university, followed by a move to KTH Stockholm, where he became professor in 1999; since 2007 he has held a professorship at the Imperial College, London.

Forging new paths in the territory of mathematics was always Ari's first priority. In particular, he has made fundamental contributions to the area of spectral theory for Schrödinger-type operators. One of his main focuses was the question of sharp constants and, using the elegant idea of "lifting in dimension," he has verified the Pólya conjecture for product domains and, in joint work with one of his students, proved the optimal version of the Lieb–Thirring inequality in higher dimensions for a large range of exponents. He has also written a series of influential papers on Hardy inequalities in various (geometrical) settings, including magnetic field situations. Recently, he has been one of the driving forces in the emerging field of eigenvalue inequalities for Schrödinger operators with complex-valued potentials. His papers in these and other areas continue to have a profound impact on the field.

Ari's research work has been appreciated in many ways. Among notable awards, let us mention the Royal Society Wolfson Research Merit Award he received in 2007. He was elected member of the Royal Swedish Academy of Sciences in 2012, and fellow of the European Academy of Sciences in 2020.

The impact of one's research work on the profession is made even more lasting when the talent is combined with the ability to pass the torch. Ari is an excellent educator as hundreds of young people who attended his courses will attest. He supervised over twenty PhD students, most at the KTH in Stockholm and later at Imperial College. Several of them have remained in academia and continue their research in the general area of mathematical physics in the spirit of their teacher.

Besides research and education, Ari served the community at large in numerous other ways, and we mention here only the most important of those achievements.

After chairing the Swedish Mathematical Society in 2001–2003, and successfully organizing the 4th European Mathematical Congress in Stockholm in 2004, he was elected president of the European Mathematical Society for the period of 2007–2010. In this context his unique ability to interact with people from many different countries came to full fruition. European collaboration is not always easy, but Ari naturally has great success in bringing people together. Moreover, even after the end of his term, the EMS could still rely on him in many matters of great importance.

He also had a fundamental impact on the life of the mathematical community through his exemplary work as director of the Mittag-Leffler Institute for the period of 2011–2018, especially, as he brought new life to this century-old institution. For instance, he was able to secure funds to build a new lecture hall, and provide a more secure storage facility for the old and precious book collection. Furthermore, the apartments got a much needed refurbishing. (Since the director also had a background in carpentry, it was not easy to fool him during the construction work!) This modernization has further solidified the position of the Mittag-Leffler Institute as a leading research center worldwide. During that period he was also Editor-in-Chief of Acta Mathematica. Simultaneously, he provided editorial service for a number of additional journals, in particular, he was one of the principal founders of the Journal of Spectral Theory (JST).

This brief portrait of Ari's life would not be complete if we did not return to mention his family, his wife Marilyn, the meeting with whom forty years ago caused a sharp turn in his destiny, and his children, Ekaterina, Eugenia, and Ivan, who are very proud of their father.

The present volume collects contributions from Ari's colleagues and collaborators resonating his varied scientific interests. They include, in short, topics such as Friedrichs, Hardy, and Lieb–Thirring inequalities, Feshbach–Schur maps and perturbation theory, eigenvalue bounds and asymptotics, scattering theory and orthogonal polynomials, stability of matter, electron density estimates, Bose–Einstein condensation, Wehrl-type entropy inequalities, Bogoliubov theory, wave packet evolution, heat kernel estimates, homogenization, d-bar problems, Brezis–Nirenberg problems, NLS in magnetic fields, classical discriminants, 2D Euler–Bardina equation, and Ari's fundamental role in starting JST.

Presenting this collection, we wish Ari good health and numerous fruitful years to come in mathematics and in life in general.

February 2021

Pavel Exner
Rupert Frank
Fritz Gesztesy
Helge Holden
Timo Weidl

Contents

A non-existence result for a generalized radial Brezis–Nirenberg problem

Rafael D. Benguria and Soledad Benguria

To Ari Laptev on the occasion of his 70th birthday

We develop a new method for estimating the region of the spectral parameter of a generalized Brezis–Nirenberg problem for which there are no, non-trivial, smooth solutions. This new method combines the standard Rellich–Pohozaev argument with a "Hardy-type" inequality for bounded domains.

1 Introduction

As pointed by us in [5], virial theorems have played, for a long time, a key role in the localization of linear and nonlinear eigenvalues. In the spectral theory of Schrödinger Operators, the virial theorem has been widely used to prove the absence of positive eigenvalues for various multiparticle quantum systems (see, e.g., [2, 18, 20]). In 1983, Brezis and Nirenberg [9] considered the existence and non-existence of solutions of the nonlinear equation

$$-\Delta u = \lambda u + |u|^{q-1}u,$$

defined on a bounded, smooth domain of $\mathbb{R}^n$, $n > 2$, with Dirichlet boundary conditions, where $q = (n+2)/(n-2)$ is the critical Sobolev exponent. Here, $\lambda \in \mathbb{R}$. In particular, they used a virial theorem, namely the Pohozaev identity [16], to prove the non-existence of regular solutions when the domain is star-shaped, for any $\lambda \leq 0$, in any $n > 2$. After the classical paper [9] of Brezis and Nirenberg, many authors have considered extensions of this problem in different settings. In particular, the Brezis–Nirenberg (BN) problem has been studied on bounded, smooth, domains of the hyperbolic space $\mathbb{H}^n$ (see, e.g., [3, 6, 12, 19]), where one replaces the Laplacian by the Laplace–Beltrami operator in $\mathbb{H}^n$. Stapelkamp [19] proved the analog of the above mentioned non-existence result of Brezis–Nirenberg in $\mathbb{H}^n$. Namely, she proved that

Keywords: Brezis–Nirenberg problem, hyperbolic space, non-existence of solutions, Pohozaev identity, Hardy inequality
2020 Mathematics Subject Classification: Primary 35XX; secondary 35B33, 35A24, 35J25, 35J60

there are no regular solutions of the BN problem for bounded, smooth, star-shaped domains in $\mathbb{H}^n$ ($n > 2$), if $\lambda \leq \frac{n(n-2)}{4}$.

The purpose of this manuscript is to develop a new method for estimating the region of the spectral parameter of a generalized Brezis–Nirenberg problem, for which there are no, non-trivial, smooth solutions. This new method combines the standard Rellich–Pohozaev argument with a "Hardy-type" inequality for bounded domains. The estimates we get are better than the usual estimates for low dimensions.

Here we consider a generalized radial Brezis–Nirenberg problem, which is given through the following boundary value problem. Given $R > 0$, we are interested in estimating the region of the spectral parameter λ for which there are no non-trivial smooth (more precisely $u \in C^2([0, R])$) solutions of

$$\begin{cases} -u''(\theta) - (n-1)\dfrac{a'(\theta)}{a(\theta)}u'(\theta) = \lambda u(\theta) + |u(\theta)|^{q-1}u(\theta), \\ \qquad\qquad u'(0) = u(R) = 0, \end{cases} \tag{1.1}$$

where $q = \frac{n+2}{n-2}$ is the critical Sobolev constant, $n > 2$, and $a \in C^3([0, R])$ satisfies

(A1) $a(0) = 0$,

(A2) $a'(\theta) > 0$ for all $\theta \in (0, R)$,

(A3) there exists $\omega \geq 0$ such that $a''(\theta) \geq \omega a(\theta)$ for all $\theta \in (0, R)$.

Actually, the boundary value problem given by (1.1) is the radial reduction of a Brezis–Nirenberg problem defined in an n-dimensional manifold M characterized by the conformal Laplacian

$$\Delta_M(\cdot) = p^{-n}\,\mathrm{div}(p^{n-2}\nabla(\cdot)),$$

where p is given in terms of a through

$$p = \frac{a(\theta)}{r} = \frac{1}{r'}. \tag{1.2}$$

Here θ is the radial geodesic coordinate in M, conformal to the Euclidean space $\mathbb{R}^n$, $x \in \mathbb{R}^n$ and $r = |x|$. There is a one-to-one correspondence between the Euclidean radial variable r and θ, and here $r' = \frac{dr}{d\theta}$. In fact, integrating (1.2), one can write $r(\theta)$ in terms of a as

$$r(\theta) = \exp \int_{\theta_0}^{\theta} \frac{1}{a(s)}\,ds, \tag{1.3}$$

and the conformal factor p, through (1.2), as $\frac{a}{r}$. Hence, one can look at (1.1) as the radial case (in geodesic coordinates) of the Brezis–Nirenberg problem

$$-\Delta_M(u) = \lambda u(\theta) + |u(\theta)|^{q-1}u(\theta) \tag{1.4}$$

on a geodesic ball of radius R of the manifold M with Dirichlet boundary conditions (see the beginning of Section 2 for details). In the case $M = \mathbb{H}^n$, $a(\theta) = \sinh(\theta)$ and, from (1.2) and (1.3) one gets $p(r) = \frac{2}{1-r^2}$.

Moreover, as we did in [4,5], we can change the parameter n, which in principle is an integer (i.e., the dimension of the manifold M) by a positive real number d using the procedure we discuss below. What we do is to replace the conformal Laplacian Δ_M in (1.4) by a particular form of a weighted Laplacian, or if one prefers, the *drift Laplacian*. The interest on weighted Laplacians originated in the early 1980s for different reasons coming from physics, geometry, and probability. Depending on the context, a weighted Laplacian is often called the *Witten Laplacian* (after [21]) or the *Bakry–Émery Laplacian* (after [1]). During the past decade there has been a growing interest in studying the spectral properties of weighted Laplacians or *drift Laplacians* (see, e.g., [7,8,10,15]). As described in [10], a Bakry–Émery manifold, denoted by the triple (M, g, ϕ) is a complete Riemannian manifold (M, g) together with some function $\phi \in C^2(M)$, where the measure on M is the weighted measure $\exp(-\phi)\, dV_g$. The Bakry–Émery Laplacian Δ_ϕ associated with such a manifold is given by

$$\Delta_\phi = \Delta_M - \nabla_M \phi \cdot \nabla_M$$

which is self-adjoint with respect to the inner product associated with the weighted measure. Here Δ_M is the standard Laplace–Beltrami operator and ∇_M is the gradient operator on the Bakry–Émery manifold. Weighted Laplacians were also introduced, in a different context, by Chavel and Feldman [11] in the early nineties.

Typically, in the Bakry–Émery Laplacian, the potential ϕ is smooth, and so is the drift term. However, singular drifts have also been considered in the literature. In fluid mechanics a weighted Laplacian with a singular drift is rather common, but typically the drift is divergence free, in other words, away from the singularities the potential ϕ is harmonic. More recently heat kernels with singular drifts have been considered (see, e.g., [13, 14]). In [14] a singular drift is considered with a (singular) potential of the form $\phi(\theta) = |\theta|^{-\alpha}$ with $\alpha > 0$. In that case the singular drift has, generically, a nonzero divergence. In our case, when we consider the Brezis–Nirenberg problem for the weighted Laplacian in M (an n-dimensional manifold with $n \geq 3$, the singular drift derives from a potential of the form $\phi = \delta \log(a)$. Notice that in this case the weighted measure, $\exp(-\phi)\, dV_g$ becomes $a^{-\delta}\, dV_g$. Thus the weighted measure can be thought of as the Lebesgue measure on a space of an effective *fractional* dimension $d \equiv n - \delta$, a fact that we will use intensively in the proofs of our theorems. Here, $(4 - n) < \delta < \frac{n-2}{2}$ (see [4]).

The standard procedure to determine the region of the parameter λ for which there are no solutions to (1.1) is to write the equation in terms of a *conformal* second-order operator and then use the Rellich–Pohozaev technique [16, 17]. See, e.g., [6, 19] for the determination of the range of the parameter λ for which there are no solutions of the equivalent of (1.1) in the case of $\mathbb{H}^n$, i.e., when $a(\theta) = \sinh(\theta)$. Applying this procedure, we can prove the following.

Theorem 1.1. *Problem* (1.1) *has no non-trivial solution if*

$$\lambda \leq \mu^*(n, R) \equiv \frac{n-2}{4} \inf_{0<\theta<R} \left\{ (n-1)\frac{a''}{a} + \frac{a'''}{a'} \right\}.$$

Remark. We make the following observations.

(1) In the case of geodesic balls in $\mathbb{H}^n$, $n \geq 2$, which corresponds to having the function $a(\theta) = \sinh(\theta)$ in (1.1) we recover the result of [19, Theorem 1 (ii)]. In that case there are no solutions of (1.1) if

$$\lambda \leq \mu^*(n, R) = \frac{n(n-2)}{4}.$$

(2) If one restricts to the case of positive solutions, one can prove non-existence when $\lambda \leq 1 + \frac{\pi^2}{16 \operatorname{arctanh}^2 R}$ for $n = 3$ (see [19, Theorem 2]). See also [6, Theorem 1.1] for the analogous result in the hyperbolic case for $2 \leq n \leq 4$.

(3) Also, if we restrict to the case of positive solutions, by a standard argument one does not have solutions if $\lambda \geq \lambda_1(n, R)$, where $\lambda_1(n, R)$ is the first eigenvalue of the operator

$$-\frac{d^2}{d\theta^2} - (n-1)\frac{a'(\theta)}{a(\theta)}\frac{d}{d\theta}$$

on $(0, R)$, with boundary conditions $u'(0) = u(R) = 0$.

As pointed out above, the purpose of this manuscript is to develop a different approach to prove non-existence of solutions. In fact, we develop a scheme that combines the use of the Rellich–Pohozaev technique together with the use of some "Hardy-type" inequalities on bounded domains. Using this new scheme we can prove the following.

Theorem 1.2. *Problem* (1.1) *has no non-trivial solution if*

$$\lambda \leq \lambda^*(n, R) \equiv \frac{n(n-1)}{4} C,$$

where

$$C = \min\left\{\frac{D + 2\omega}{2(n+2)}, \frac{D}{4}\right\} \tag{1.5}$$

and

$$D = \inf_{0<\theta<R}\left\{(2n-3)\frac{a''}{a} + \frac{a'''}{a'}\right\}. \tag{1.6}$$

The rest of the manuscript is organized as follows. In Section 2 we express the problem (1.1) in terms of an appropriate operator reminiscent of the conformal Laplacian. Once we have that, we apply the standard Rellich–Pohozaev technique [16, 17] to prove Theorem 1.1. In Section 3 we set forward our new approach. We start with the problem (1.1) stated as such. This, in the $\mathbb{H}^n$ case, would correspond to studying radial solutions of the BN problem for geodesic balls, in terms of geodesic coordinates centered at the center of the balls. In that setting we use a combination of the Rellich–Pohozaev technique [16, 17] together with a "Hardy-type" inequality to prove Theorem 1.2. Finally, in Section 4 we illustrate and compare the bounds obtained in Theorems 1.1 and 1.2 through a particular example. The bound obtained through our new technique is better than the standard bound for a whole region of the parameter space (n, R), in particular for all R when $n \leq 4$.

2 Non-existence of solutions, via the Pohozaev virial identity, using the conformal Laplacian

A key step towards proving the lower bound $\lambda > \frac{n(n-2)}{4}$ for the existence of solutions to the Brezis–Nirenberg problem on a bounded domain in the Hyperbolic space $\mathbb{H}^n$ is to use the so called "stereographic" projection. By using the stereographic projection one writes the Laplace–Beltrami $\Delta_{\mathbb{H}^n}$ in terms of the conformal Laplacian in Euclidean coordinates. In fact, one can write

$$\Delta_{\mathbb{H}^n} u = p^{-n} \operatorname{div}(p^{n-2} \nabla u),$$

where $p = \frac{2}{1-|x|^2}$. Here we mimic what is done in $\mathbb{H}^n$ for our generalized operator

$$\frac{d^2}{d\theta^2} + (n-1)\frac{a'(\theta)}{a(\theta)}\frac{d}{d\theta} \tag{2.1}$$

and write this as the radial part of a conformal Laplacian

$$p^{-n} \operatorname{div}(p^{n-2} \nabla(\cdot))$$

that defines the operator (2.1), written in Euclidean coordinates, for some appropriate radial function $p(r)$. Given the function $a(\theta)$ that defines our generalized operator (2.1), our first goal is to determine the conformal factor $p(r)$ in our case. In order to do that, we need to make the change of variables $\theta \to r(\theta)$, with $r = |x|$ in such a way that

$$H(u) \equiv u'' + (n-1)\frac{a'(\theta)}{a(\theta)} u' = p^{-n} \operatorname{div}(p^{n-2} \nabla u) \equiv L(u). \tag{2.2}$$

Since here we have radial symmetry, it follows that

$$L(u) = p^{-n} r^{1-n} \frac{d}{dr}\left(r^{n-1} p^{n-2} \frac{du}{dr}\right).$$

Recall that if $\vec{F}$ is of the form $\vec{F} = F(r)\hat{r}$, where $r = |x|$ and $\hat{r} = \nabla|x|$,

$$\operatorname{div} \vec{F} = \nabla \cdot \vec{F} = \frac{1}{r^{n-1}} \frac{d}{dr}(r^{n-1} F(r)).$$

If we denote $\dot{u} = \frac{du}{dr}$, we can write $L(u)$ as

$$L(u) = \frac{1}{p^2}\left[\ddot{u} + \left(\frac{n-1}{r} + (n-2)\frac{\dot{p}}{p}\right)\dot{u}\right]. \tag{2.3}$$

On the other hand, if we denote $u' = \frac{du}{d\theta}$ and use the chain rule, we have

$$u' = \dot{u}\frac{dr}{d\theta}$$

and

$$u'' = \ddot{u}\left(\frac{dr}{d\theta}\right)^2 + \dot{u}\frac{d^2r}{d\theta^2}.$$

Hence,

$$H(u) = \ddot{u}\left(\frac{dr}{d\theta}\right)^2 + \left(\frac{d^2r}{d\theta^2} + (n-1)\frac{a'(\theta)}{a(\theta)}\frac{dr}{d\theta}\right)\dot{u}. \tag{2.4}$$

Replacing H, given by (2.4), and L, given by (2.3) in (2.2), and comparing coefficients, we conclude at once that

$$\left(\frac{dr}{d\theta}\right) = \frac{1}{p}, \tag{2.5}$$

and also that

$$\frac{d^2r}{d\theta^2} + (n-1)\frac{a'(\theta)}{a(\theta)}\frac{dr}{d\theta} = \frac{1}{p^2}\left(\frac{n-1}{r} + (n-2)\frac{\dot{p}}{p}\right). \tag{2.6}$$

Differentiating (2.5) with respect to θ and again using the chain rule, we have

$$\frac{d^2r}{d\theta^2} = -\frac{\dot{p}}{p^2}\frac{dr}{d\theta} = -\frac{\dot{p}}{p^3}. \tag{2.7}$$

Using (2.5) and (2.7) in (2.6), we conclude

$$\frac{a'}{a} = \frac{1}{pr} + \frac{\dot{p}}{p^2} \equiv B(r). \tag{2.8}$$

For the change of variables $\theta \to r(\theta)$ to be well defined we need $p > 0$. From (2.8) and (2.5) is simple to obtain $r(\theta)$ and $p(\theta)$. In fact, $\dot{p} = p'\frac{d\theta}{dr} = p'p$. Since we also have $\frac{1}{p} = r'$, from (2.8) we get

$$\frac{a'}{a} = \frac{r'}{r} + \frac{p'}{p},$$

which can be immediately integrated to yield

$$a = pr \tag{2.9}$$

(here we have chosen an irrelevant integrating constant to be 1). However, $p = \frac{1}{r'}$. So, from (2.9) we have

$$a(\theta) = \frac{r}{r'},$$

which can be integrated to yield

$$r(\theta) = \exp \int_{\theta_0}^{\theta} \frac{1}{a(s)}\, ds.$$

Proof of Theorem 1.1. Once we have done the transformation from our original variable θ to r, via the relation with the conformal Laplacian, we are ready to apply the standard Rellich–Pohozaev [16, 17] argument. This is done with the purpose of determining the region of λ for which there are no non-trivial solutions of the BN problem. For $n \geq 3$ the equation we consider is

$$-L(u) = -p^{-n} r^{1-n} \frac{d}{dr}\left(r^{n-1} p^{n-2} \frac{du}{dr}\right) = \lambda u + |u|^{\frac{4}{n-2}} u. \tag{2.10}$$

We now introduce the change $u \to v$ given by

$$u = p^{1-\frac{n}{2}} v. \tag{2.11}$$

After some long and straightforward computations one can rewrite (2.10) as

$$-\ddot{v} - \frac{n-1}{r}\dot{v} + Vv = \lambda p^2 v + |v|^{\frac{4}{n-2}} v, \tag{2.12}$$

where the *potential* V is given by

$$V = (n-2)\left[\frac{1}{2}\frac{\ddot{p}}{p} + \frac{1}{4}(n-4)\left(\frac{\dot{p}}{p}\right)^2 + \frac{1}{2}(n-1)\frac{\dot{p}}{rp}\right].$$

By using (2.8), the potential V can be expressed in terms of B (introduced in (2.8)) as

$$V = (n-2)\left[\frac{n}{4}p^2 B^2 + \frac{1}{2}p\dot{B} - \frac{n-2}{4r^2}\right]. \tag{2.13}$$

From (2.11), (2.5), and the boundary conditions for u it follows that $\dot{v}(0) = 0$ and $v(\tilde{R}) = 0$, where $\tilde{R} = r(R)$. If we multiply (2.12) by v, we obtain

$$-\frac{1}{r^{n-1}}\frac{d}{dr}(r^{n-1} v\dot{v}) + \dot{v}^2 + Vv^2 = \lambda v^2 + |v|^{\frac{4}{n-2}} v^2. \tag{2.14}$$

Now we proceed with the Rellich–Pohozaev argument applied to equation (2.12). Multiply (2.12) by $r\dot{v}$ (here we are using that $\frac{rd}{dr}$ is the generator of radial dilations). Among other quantities we need the identity,

$$r\dot{v}Vv = \frac{1}{r^{n-1}}\frac{d}{dr}\left(r^n \frac{v^2}{2} V\right) - \frac{n}{2}v^2 V - r\frac{v^2}{2}\dot{V}. \tag{2.15}$$

Using (2.13) and the fact that

$$\frac{\dot{p}}{p} + \frac{1}{r} = pB, \tag{2.16}$$

after some simple computations we conclude that

$$\frac{1}{2}r\dot{V} + V = \frac{n-2}{4} rBp^3 T, \tag{2.17}$$

where T is given, in terms of the function B, by

$$T \equiv nB^2 + (n+1)\frac{\dot{B}}{p} + \frac{\dot{B}}{B}\frac{1}{p^2 r} + \frac{\ddot{B}}{Bp^2}. \tag{2.18}$$

This particular expression involving the potential V as well as the definition of T will prove to be useful later.

Moreover, we need the identity

$$r\dot{v}\ddot{v} + (n-1)\dot{v}^2 = \frac{n-2}{2}\dot{v}^2 + \frac{1}{r^{n-1}}\frac{d}{dr}\left(\frac{1}{2}r^n\dot{v}^2\right). \tag{2.19}$$

The last identity we need to have is the following:

$$\begin{aligned} r\dot{v}\left(\lambda v p^2 + |v|^{\frac{4}{n-2}}v\right) = {} & \frac{1}{r^{n-1}}\frac{d}{dr}\left(r^n\left(\frac{\lambda}{2}p^2v^2 + \frac{n-2}{2n}|v|^{\frac{4}{n-2}}v^2\right)\right) \\ & - \lambda\frac{v^2}{2}np^2 - \lambda r v^2 p\dot{p} - \frac{n-2}{2}|v|^{\frac{4}{n-2}}v^2. \end{aligned} \tag{2.20}$$

With all these identities at hand we can recapitulate. Multiplying (2.12) by $r\dot{v}$ and using (2.15), (2.19), and (2.20), we get

$$\begin{aligned} & \frac{1}{r^{n-1}}\frac{d}{dr}\left(r^n\left(\frac{1}{2}v^2V - \frac{1}{2}\dot{v}^2 - \frac{\lambda}{2}p^2v^2 - \frac{n-2}{2n}|v|^{\frac{4}{n-2}}v^2\right)\right) \\ & = \frac{n-2}{2}\dot{v}^2 + \frac{n}{2}v^2V + r\frac{v^2}{2}\dot{V} - \lambda\frac{v^2}{2}np^2 \\ & \quad - \lambda r v^2 p\dot{p} - \frac{n-2}{2}|v|^{\frac{4}{n-2}}v^2. \end{aligned} \tag{2.21}$$

Multiplying (2.14) by $\frac{n-2}{2}$ and adding (2.21) to the result, we obtain

$$\begin{aligned} & \frac{1}{r^{n-1}}\frac{d}{dr}\left(r^n\left(\frac{1}{2}v^2V - \frac{1}{2}\dot{v}^2 - \frac{n-2}{2}\frac{v\dot{v}}{r} - \frac{\lambda}{2}p^2v^2 - \frac{n-2}{2n}|v|^{\frac{4}{n-2}}v^2\right)\right) \\ & = v^2\left(V + \frac{1}{2}r\dot{V}\right) - \lambda v^2p^2r\left(\frac{\dot{p}}{p} + \frac{1}{r}\right). \end{aligned} \tag{2.22}$$

Finally, multiplying (2.22) by r^{n-1} and integrating the result from 0 to $\tilde{R} \equiv r(R)$, we obtain the following *virial* identity:

$$\frac{1}{2}\tilde{R}^n\dot{v}(\tilde{R})^2 = \int_0^{\tilde{R}} v^2p^3B\left[\lambda - \frac{n-2}{4}T\right]r^n\,dr. \tag{2.23}$$

To obtain (2.23), we used the boundary conditions on v, in particular that $v(\tilde{R}) = 0$. We also used (2.16) and (2.17) in order to express the right side of (2.23) in terms of B and T.

Using the chain rule and the definition of p, we have

$$\dot{B} = B'p \quad \text{and} \quad \ddot{B} = B''p^2 + B'\dot{p}.$$

Replacing these two expressions in definition (2.18) of T, we get

$$\begin{aligned} T &= nB^2 + (n+1)B' + \frac{B''}{B} + \frac{B'}{B}\left(\frac{\dot{p}}{p^2} + \frac{1}{rp}\right) \\ &= nB^2 + (n+1)B' + \frac{B''}{B} + B', \end{aligned} \tag{2.24}$$

where we used (2.16) to obtain the second equality. Using the fact that $B = \frac{a'}{a}$, after some algebra starting from (2.24), we finally conclude that

$$T(r(\theta)) = (n-1)\frac{a''(\theta)}{a(\theta)} + \frac{a'''(\theta)}{a'(\theta)}.$$

Inserting this expression for T back into the *virial* identity (2.23), we have

$$\frac{1}{2}\tilde{R}^n \dot{v}(\tilde{R})^2 = \int_0^{\tilde{R}} v^2 p^3 B\left[\lambda - \frac{n-2}{4}\left((n-1)\frac{a''}{a} + \frac{a'''}{a'}\right)\right] r^n \, dr. \tag{2.25}$$

Notice here that $\dot{v}(\tilde{R})$ cannot vanish. In fact, we have $v(\tilde{R}) = 0$, so if $\dot{v}(\tilde{R})$ vanishes, it would follow from (2.12) and unique continuation that v would be identically zero in $[0, \tilde{R}]$. Since the right side of (2.25) is positive, we conclude that if

$$\lambda \le \mu^*(n, R) \equiv \frac{n-2}{4} \inf_{\theta \in [0,R]} \left((n-1)\frac{a''}{a} + \frac{a'''}{a'}\right), \tag{2.26}$$

then there is no non-trivial solution of (1.1).

This concludes the proof of our Theorem 1.1. ■

3 Non-existence of solutions using the Rellich–Pohozaev argument and a "Hardy-type" inequality

In this section we use a combination of the Rellich–Pohozaev technique together with a "Hardy-type" inequality for bounded intervals in order to prove Theorem 1.2. We begin with the following lemma.

Lemma 3.1. *Let $u \in C^2([0, R])$ be a solution of* (1.1), *and let*

$$G(\theta) = \int_0^{\theta} a^{n-1}(s)\, ds \quad \text{and} \quad S(\theta) = \frac{G(\theta)a'(\theta)}{a(\theta)} - \frac{a(\theta)^{n-1}}{n}.$$

Then

$$\lambda \ge \frac{n(n-1)}{4}\left(\frac{\int_0^R u'^2(\theta)S(\theta)\, d\theta}{\int_0^R \frac{G(\theta)^2 u'(\theta)^2}{G'(\theta)}\, d\theta}\right).$$

Proof. Suppose $u \in C^2([0,R)]$ is a solution of (1.1). Multiplying (1.1) by ua^{n-1} and integrating, we obtain, after integrating the first term by parts,

$$\int_0^R u'^2 a^{n-1}\, d\theta = \lambda \int_0^R u^2 a^{n-1}\, d\theta + \int_0^R |u|^{q+1} a^{n-1}\, d\theta. \tag{3.1}$$

On the other hand, multiplying equation (1.1) by $u'G$, where $G(\theta) = \int_0^\theta a^{n-1}(s)\, ds$ we obtain

$$\begin{aligned} -\int_0^R \left(\frac{u'^2}{2}\right)' G\, d\theta - (n-1)\int_0^R u'^2 \frac{Ga'}{a}\, d\theta \\ = \lambda \int_0^R \left(\frac{u^2}{2}\right)' G\, d\theta + \int_0^R \left(\frac{|u|^{q+1}}{q+1}\right)' G\, d\theta. \end{aligned} \tag{3.2}$$

After integrating by parts, and since $G'(\theta) = a^{n-1}(\theta)$ and $G(0) = 0$, equation (3.2) reads

$$\begin{aligned} -\frac{u'(R)^2 G(R)}{2} + \int_0^R u'^2 \frac{a^{n-1}}{2}\, d\theta - (n-1)\int_0^R u'^2 \frac{Ga'}{a}\, d\theta \\ = -\lambda \int_0^R u^2 \frac{a^{n-1}}{2}\, d\theta - \int_0^R \frac{|u|^{q+1}}{q+1} a^{n-1}\, d\theta. \end{aligned} \tag{3.3}$$

Now, solving for the term in $|u|^{q+1}$ in equation (3.1) and substituting this term in equation (3.3), we obtain

$$\begin{aligned} \int_0^R u'^2 \left(\frac{a^{n-1}}{2} + \frac{a^{n-1}}{p+1} - (n-1)\frac{Ga'}{a}\right) d\theta \\ + \int_0^R u^2 a^{n-1}\left(\frac{\lambda}{2} - \frac{\lambda}{q+1}\right) d\theta = \frac{u'(R)^2 G(R)}{2}. \end{aligned}$$

However, since $\frac{1}{2} + \frac{1}{q+1} = \frac{n-1}{n}$ and $\frac{1}{2} - \frac{1}{q+1} = \frac{1}{n}$, we can write

$$\frac{\lambda}{n}\int_0^R u^2 a^{n-1}\, d\theta = \frac{u'(R)^2 G(R)}{2} + (n-1)\int_0^R u'^2 \left(\frac{Ga'}{a} - \frac{a^{n-1}}{n}\right) d\theta.$$

By hypothesis, if $\theta > 0$, then $a(\theta) > 0$ (which follows from **(A1)** and **(A2)**). Therefore,

$$\lambda \geq \frac{n(n-1)\int_0^R u'^2 \left(\frac{Ga'}{a} - \frac{a^{n-1}}{n}\right) d\theta}{\int_0^R u^2 a^{n-1}\, d\theta}. \tag{3.4}$$

Now let $S(\theta) = \frac{Ga'}{a} - \frac{a^{n-1}}{n}$, the coefficient of u'^2 of the integral in the numerator. Then $S \geq 0$ if $\theta > 0$. In fact, let $m(\theta) = Ga' - \frac{a^n}{n}$. Since $G(0) = 0$, one has that $m(0) = 0$. Also, since $G'(\theta) = a^{n-1}(\theta)$, we have $m'(\theta) = Ga''$. In particular, since by hypothesis $a'' \geq \omega a \geq 0$, we have $m'(\theta) \geq 0$. It follows that $m \geq 0$ for all $\theta \in (0,R)$, and since a is positive in this range, $S \geq 0$.

We will now use a "Hardy-type" inequality to rewrite the integral in the denominator in terms of u'^2. Integrating by parts, we can write

$$\int_0^R u^2 G' \, d\theta = -2\int_0^R uu'G \, d\theta = -2\int_0^R (ua^{\frac{n-1}{2}})(Gu'a^{\frac{1-n}{2}}) \, d\theta.$$

By Cauchy–Schwarz, it follows that

$$\left(\int_0^R u^2 a^{n-1} \, d\theta\right)^2 < 4\int_0^R u^2 a^{n-1} \, d\theta \int_0^R \frac{G^2 u'^2}{a^{n-1}} \, d\theta, \tag{3.5}$$

where the conditions on u require the above inequality to be strict (in fact, the case of equality in (3.5) would be attained if u is proportional to a power of G, but G is strictly positive for $\theta > 0$ and u has to vanish at $\theta = R$).

Thus,

$$\int_0^R u^2 a^{n-1} \, d\theta < 4\int_0^R \frac{G^2 u'^2}{a^{n-1}} \, d\theta. \tag{3.6}$$

Hence, by using that $a^{n-1}(\theta) = G'(\theta)$, it follows from equations (3.4) and (3.6) that

$$\lambda > \frac{n(n-1)}{4}\left(\frac{\int_0^R u'(\theta)^2 S(\theta) \, d\theta}{\int_0^R \frac{G(\theta)^2 u'(\theta)^2}{G'(\theta)} \, d\theta}\right),$$

which proves the lemma. ■

Lemma 3.2. *We have $S(\theta) \geq C\frac{G^2(\theta)}{G'(\theta)}$, where C is given by equation* (1.5).

Proof. Let $f(\theta) = S(\theta)G'(\theta) - CG(\theta)^2$. We need to show that $f \geq 0$. As before, we write $S(\theta) = \frac{m(\theta)}{a(\theta)}$, with $m(\theta) = G(\theta)a'(\theta) - \frac{a(\theta)^n}{n} \geq 0$. Then

$$f(\theta) = a^{n-2}(\theta)m(\theta) - CG(\theta)^2.$$

Since $a(0) = G(0) = 0$, it follows that $f(0) = 0$. Thus, it suffices to show that $f' \geq 0$.

We have $f'(\theta) = a^{n-3}(\theta)g(\theta)$, where

$$g(\theta) \equiv (n-2)a'(\theta)m(\theta) + G(\theta)a(\theta)(a''(\theta) - 2Ca(\theta)). \tag{3.7}$$

Notice that $g(0) = 0$ so, in order to prove that $g(\theta) \geq 0$, it suffices to show that $g'(\theta) \geq 0$. Differentiating (3.7), we can write

$$g'(\theta) = (2n-3)Ga'a'' + \left(\frac{2}{n}\right)a^n a'' - 2Ca^{n+1} - 4CGaa' + Gaa'''.$$

Since by hypothesis $a \geq 0$ and $a' \geq 0$, it follows from equation (1.6) that

$$Daa' \leq (2n-3)a'a'' + aa''',$$

so we can write

$$g'(\theta) \geq Gaa'(D - 4C) + \frac{2a^n a''}{n} - 2Ca^{n+1}.$$

Furthermore, since by hypothesis $a'' \geq \omega a$, it follows that

$$g'(\theta) \geq Gaa'(D - 4C) + \frac{a^{n+1}}{n}(2\omega - 2Cn).$$

However, since $m \geq 0$, we have $Ga' \geq \frac{a^n}{n}$. In particular, if $C \leq \frac{D}{4}$, then

$$g'(\theta) \geq \frac{a^{n+1}}{n}(D - 4C + 2\omega - 2Cn).$$

It follows that $g' \geq 0$ provided that $C \leq \frac{D+2\omega}{2(n+2)}$.

Now, by choosing

$$C = \min\left\{\frac{D + 2\omega}{2(n+2)}, \frac{D}{4}\right\},$$

it follows that $S(\theta) \geq C\frac{G^2(\theta)}{G'(\theta)}$, which proves the lemma. ∎

It follows from Lemmas 3.1 and 3.2 that if u is a solution of (1.1), then

$$\lambda > \frac{n(n-1)}{4}\left(\frac{\int_0^R S(\theta)u'(\theta)^2 \,\mathrm{d}\theta}{\int_0^R \frac{G(\theta)^2 u'(\theta)^2}{G'(\theta)} \,\mathrm{d}\theta}\right) \geq \frac{Cn(n-1)}{4}.$$

Hence, we conclude that if

$$\lambda \leq \lambda^*(n, R) \equiv \frac{Cn(n-1)}{4},$$

then problem (1.1) has no non-trivial solution. This concludes the proof of Theorem 1.2.

4 An illustrative example

To compare the bounds $\mu^*(n, R)$ and $\lambda^*(n, R)$ embodied in Theorems 1.1 and 1.2 above, it is instructive to work a specific example as an application. Consider

$$a(\theta) = \theta e^\theta.$$

Then $a'(\theta) = e^\theta(1+\theta)$, $a''(\theta) = e^\theta(2+\theta)$, and $a'''(\theta) = e^\theta(3+\theta)$. Define

$$f(\theta) \equiv (n-1)\frac{a''}{a} + \frac{a'''}{a'} = \left(1 + \frac{2}{\theta}\right)(n-1) + \frac{3+\theta}{1+\theta}.$$

The function $f(\theta)$ is strictly decreasing, therefore

$$\inf_{0<\theta<R} f(\theta) = f(R) = n + 2\frac{n(1+R) - 1}{R(1+R)}.$$

Then, according to (2.26), problem (1.1) with $a(\theta) = \theta e^\theta$ does not have a solution if

$$\lambda \leq \mu^*(n, R) \equiv \frac{1}{4}(n-2)\left(n + 2\frac{n(1+R)-1}{R(1+R)}\right). \tag{4.1}$$

On the other hand, define

$$g(\theta) \equiv (2n-3)\frac{a''}{a} + \frac{a'''}{a'} = \left(1 + \frac{2}{\theta}\right)(2n-3) + \frac{3+\theta}{1+\theta}.$$

Again, the function $g(\theta)$ is strictly decreasing, therefore

$$D \equiv \inf_{0<\theta<R} g(\theta) = g(R) = \left(1 + \frac{2}{R}\right)(2n-3) + \frac{3+R}{1+R}.$$

Also, we have

$$\omega \equiv \inf_{0<\theta<R} \frac{a''}{a} = \inf_{0<\theta<R} \left(1 + \frac{2}{\theta}\right) = 1 + \frac{2}{R}.$$

Thus, we have on the one hand

$$D + 2\omega = \left(1 + \frac{2}{R}\right)(2n-1) + \frac{3+R}{1+R},$$

and on the other hand

$$\frac{D}{4} = \frac{1}{4}\left\{\left(1 + \frac{2}{R}\right)(2n-3) + \frac{3+R}{1+R}\right\}.$$

If we introduce $s = (1+R)(2+R)$, after some simple calculations we conclude that

$$\frac{D + 2\omega}{2n+4} \leq \frac{D}{4}$$

provided

$$n \geq \hat{n}(s) = \frac{1}{2}\left[\left(1 + \frac{1}{s}\right) + \sqrt{\left(1 + \frac{1}{s}\right)^2 + 8}\,\right].$$

Notice that $\hat{n}(s)$ is a decreasing function of s such that

$$\hat{n}(2) = \frac{1}{4}(3 + \sqrt{41}) = 2.35078\ldots \quad \text{and} \quad \lim_{s\to\infty} \hat{n}(s) = 2.$$

Hence, if $n \geq \hat{n}((1+R)(2+R))$, then

$$\lambda^*(n, R) = \frac{n(n-1)}{8(n+2)}\left\{\left(1 + \frac{2}{R}\right)(2n-1) + \frac{3+R}{1+R}\right\}. \tag{4.2}$$

Let us compare our new type of bound $\lambda^*(n, R)$ with the more standard one $\mu^*(n, R)$ for this example. In order to do that, we need to determine for which values of n and R,

$$\lambda^*(n, R) \geq \mu^*(n, R). \tag{4.3}$$

Using (4.1) and (4.2), one can check that (4.3) holds if and only if

$$\begin{aligned} F(n,R) &\equiv 2(R^2+3R)(4-n)n + 2(8-n)(n-1) \\ &= 2\{(s-2)(4-n)n + (8-n)(n-1)\} \geq 0. \end{aligned} \tag{4.4}$$

In general, this holds for all $n \geq 4$, independently of the values R. As a function of $s = (R+1)(R+2)$, (4.4) holds provided

$$n \geq \tilde{n}(s) = \frac{1}{2(s-1)}\big[(4s+1) + \sqrt{16s^2 - 24s + 33}\,\big].$$

It is not hard to show that $\tilde{n}(s)$ is a decreasing function of s for all $s > 1$. In fact, $\tilde{n}$ decreases from 8, for $R = 0$ (i.e., $s = 2$) to 4 when $R \to \infty$ (i.e., $s \to \infty$). Moreover, it is a simple exercise, which we leave to the reader, to prove that $\frac{D}{4} \geq \mu^*(n,R)$, so in the case $2 \leq n < \hat{n}(s)$, we have $\lambda^*(n,R) > \mu^*(n,R)$ for all R. Hence, in the (n,R) parameter space, our bound is better provided that $2 \leq n < \tilde{n}(s)$, where $s = (R+1)(R+2)$.

Remark. For the radial hyperbolic case, i.e., for problem (1.1) with $a(\theta) = \sinh(\theta)$, it follows from Theorems 1.1 and 1.2 that

$$\mu^*(n,R) = \frac{n(n-2)}{4},$$

whereas

$$\lambda^*(n,R) = \frac{n^2(n-1)}{4(n+2)},$$

which are independent of R. In that case $\lambda^*(n,R)$ is better than $\mu^*(n,R)$ for all $2 \leq n < 4$, and all $R > 0$. This case was reported by us in [5].

Acknowledgements. We thank the anonymous referee for many suggestions that helped us to improve the manuscript. The work of Rafael D. Benguria has been supported by Fondecyt (Chile) Projects #116-0586 and #120-1055.

Bibliography

[1] D. Bakry and M. Émery, Diffusions hypercontractives. In *Séminaire de probabilités, XIX, 1983/84*, pp. 177–206, Lecture Notes in Math. 1123, Springer, Berlin, 1985

[2] E. Balslev, Absence of positive eigenvalues of Schrödinger operators. *Arch. Ration. Mech. Anal.* **59** (1975), 343–357

[3] C. Bandle and Y. Kabeya, On the positive, "radial" solutions of a semilinear elliptic equation in $\mathbb{H}^N$. *Adv. Nonlinear Anal.* **1** (2012), 1–25

[4] R. D. Benguria and S. Benguria, The Brezis–Nirenberg problem for the Laplacian with a singular drift in $\mathbb{R}^n$ and $\mathbb{S}^n$. *Nonlinear Anal.* **157** (2017), 189–211

[5] R. D. Benguria and S. Benguria, An improved bound for the non-existence of radial solutions of the Brezis–Nirenberg problem in $\mathbb{H}^n$. In *Functional analysis and operator theory for quantum physics*, pp. 153–160, EMS Ser. Congr. Rep., European Mathematical Society, Zürich, 2017

[6] S. Benguria, The solution gap of the Brezis–Nirenberg problem on the hyperbolic space. *Monatsh. Math.* **181** (2016), 537–559

[7] H. Berestycki, F. Hamel and N. Nadirashvili, Elliptic eigenvalue problems with large drift and applications to nonlinear propagation phenomena. *Comm. Math. Phys.* **253** (2005), 451–480

[8] K. Bogdan and T. Jakubowski, Estimates of heat kernel of fractional Laplacian perturbed by gradient operators. *Comm. Math. Phys.* **271** (2007), 179–198

[9] H. Brézis and L. Nirenberg, Positive solutions of nonlinear elliptic equations involving critical Sobolev exponents. *Comm. Pure Appl. Math.* **36** (1983), 437–477

[10] N. Charalambous, Z. Lu and J. Rowlett, Eigenvalue estimates on Bakry-Émery manifolds. In *Elliptic and parabolic equations*, pp. 45–61, Springer Proc. Math. Stat. 119, Springer, Cham, 2015

[11] I. Chavel and E. A. Feldman, Modified isoperimetric constants, and large time heat diffusion in Riemannian manifolds. *Duke Math. J.* **64** (1991), 473–499

[12] D. Ganguly and S. Kunnath, Sign changing solutions of the Brezis–Nirenberg problem in the hyperbolic space. *Calc. Var. Partial Differential Equations* **50** (2014), 69–91

[13] A. Grigor'yan, Heat kernels on weighted manifolds and applications. In *The ubiquitous heat kernel*, pp. 93–191, Contemp. Math. 398, American Mathematical Society, Providence, 2006

[14] A. Grigor'yan, S. Ouyang and M. Röckner, Heat kernel estimates for an operator with a singular drift and isoperimetric inequalities. *J. Reine Angew. Math.* **736** (2018), 1–31

[15] F. Hamel, N. Nadirashvili and E. Russ, Rearrangement inequalities and applications to isoperimetric problems for eigenvalues. *Ann. of Math. (2)* **174** (2011), 647–755

[16] S. I. Pohožaev, On the eigenfunctions of the equation $\Delta u + \lambda f(u) = 0$. *Dokl. Akad. Nauk SSSR* **165** (1965), 36–39

[17] F. Rellich, Darstellung der Eigenwerte von $\Delta u + \lambda u = 0$ durch ein Randintegral. *Math. Z.* **46** (1940), 635–636

[18] B. Simon, Absence of positive eigenvalues in a class of multiparticle quantum systems. *Math. Ann.* **207** (1974), 133–138

[19] S. Stapelkamp, The Brézis–Nirenberg problem on $\mathbb{H}^n$. Existence and uniqueness of solutions. In *Elliptic and parabolic problems (Rolduc/Gaeta, 2001)*, pp. 283–290, World Scientific Publishing, River Edge, 2002

[20] J. Weidmann, The virial theorem and its application to the spectral theory of Schrödinger operators. *Bull. Amer. Math. Soc.* **73** (1967), 452–456

[21] E. Witten, Supersymmetry and Morse theory. *J. Differential Geometry* **17** (1982), 661–692 (1983)

Friedrichs-type inequalities in arbitrary domains

Andrea Cianchi and Vladimir Maz'ya

To Ari Laptev on the occasion of his 70th birthday

A collection of first- and second-order inequalities for Sobolev functions, involving optimal norms, in arbitrary domains in the Euclidean space is offered. The norms in the relevant Sobolev spaces depend on the highest-order derivatives of functions and on their traces on the boundary of the domain. This allows for constants in the inequalities which are independent of the geometry of the domain. Sobolev spaces of vector-valued functions, defined in terms of their symmetric gradient, are also considered. The results presented rely on a general theory developed in our earlier papers [4] and [5].

1 Introduction

Almost one century ago, in his paper [8] on boundary value and eigenvalue problems in the elasticity theory of plates, K. Friedrichs showed that, given a bounded open set $\Omega \subset \mathbb{R}^2$ whose boundary $\partial\Omega$ satisfies a very mild regularity condition, there exists a constant $C = C(\Omega)$ such that

$$\|u\|_{L^2(\Omega)} \leq C\big(\|\nabla u\|_{L^2(\Omega)} + \|u\|_{L^2(\partial\Omega)}\big) \tag{1.1}$$

for every sufficiently smooth function $u : \overline{\Omega} \to \mathbb{R}$. Here, $\|\cdot\|_{L^2(\partial\Omega)}$ stands for the L^2-norm on $\partial\Omega$ with respect to the $(n-1)$-dimensional Hausdorff measure. As noticed in the same paper, an application of inequality (1.1) to the first-order derivatives of u yields

$$\|\nabla u\|_{L^2(\Omega)} \leq C\big(\|\nabla^2 u\|_{L^2(\Omega)} + \|\nabla u\|_{L^2(\partial\Omega)}\big) \tag{1.2}$$

for some constant $C = C(\Omega)$.

Inequality (1.1) can be regarded as a forerunner of inequalities in Sobolev-type spaces involving trace norms over the boundary of the underlying domain. The latter

Keywords: Sobolev inequalities, irregular domains, boundary traces, symmetric gradient, Lorentz norms, Orlicz norms
2020 Mathematics Subject Classification: 46E35, 46E30

play a role in a variety of problems in the theory of partial differential equations, including the analysis of solutions to elliptic equations subject to Robin boundary conditions. Criteria for the validity of inequalities of the form

$$\|u\|_{L^q(\Omega)} \le C_1 \|\nabla u\|_{L^p(\Omega)} + C_2 \|u\|_{L^r(\partial\Omega)}, \tag{1.3}$$

for some exponents $p, q, r \ge 1$, in terms of isoperimetric or isocapacitary-type inequalities relative to Ω, can be found in [19, Section 6.11]. As pointed out in the same section, from those criteria necessary and sufficient conditions on Ω follow for the solvability of the problem

$$\begin{cases} -\mathrm{div}(A(x)\nabla u) = 0 & \text{in } \Omega, \\ A(x)\nabla u \cdot \nu + \lambda u = \varphi & \text{on } \partial\Omega. \end{cases} \tag{1.4}$$

Here, A is an $n \times n$-matrix-valued function in $L^\infty(\Omega)$ such that $A(x)$ is symmetric and positive definite, with lowest eigenvalue uniformly bounded away from 0, the function $\varphi \in L^\sigma(\partial\Omega)$ for some $\sigma \in (1, 2]$, ν is the outward normal unit vector on $\partial\Omega$ and $\lambda > 0$. Criteria for the discreteness of the spectrum of the operator in the first equation in (1.4) are also provided.

Remarkably, no regularity on Ω is needed for inequality (1.3) to hold, provided that the exponents $p, q, r \ge 1$ are suitably related. If $1 \le p < n$, $r = \frac{p(n-1)}{n-p}$ and $q = \frac{pn}{n-p}$, then inequality (1.3) holds for any open set $\Omega \subset \mathbb{R}^n$, with constants C_1 and C_2 independent of Ω. In this case, the optimal constants C_1 and C_2 are also known. They appeared in 1960 in the paper [17] for $p = 1$. In this case the inequality reads

$$\|u\|_{L^{\frac{n}{n-1}}(\Omega)} \le \frac{\Gamma(1+\frac{n}{2})^{\frac{1}{n}}}{n\sqrt{\pi}} \big(\|\nabla u\|_{L^1(\Omega)} + \|u\|_{L^1(\partial\Omega)}\big). \tag{1.5}$$

Inequality (1.5) is in fact equivalent to the classical isoperimetric inequality in $\mathbb{R}^n$, as shown in [17]. An application of this inequality to a suitable power of $|u|$ easily implies a parallel inequality for $1 < p < n$. The optimal constants for these values of p have much more recently been exhibited in [12], via mass-transportation techniques. More generally, the inequality in (1.3) holds for any open set Ω with finite Lebesgue measure $\mathcal{L}^n(\Omega)$, provided that $1 \le r \le \frac{p(n-1)}{n-p}$ and $q = \frac{rn}{n-1}$, the latter exponent being the largest possible if no additional assumption is imposed on Ω – see [19, Corollary 6.11.2 and subsequent Example]. Related results in the space of functions of bounded variation can be found in [24].

In our paper [4] a theory of Friedrichs-type inequalities, for any-order Sobolev spaces, in arbitrary domains is proposed, which allows for more general Borel measures in Ω on the left-hand sides and more general norms of u and of its derivatives. The idea pursued in that paper is that domain regularity can be replaced by appropriate information on boundary traces in Sobolev-type inequalities. The punctum is that the constants appearing in the inequalities in question are independent of the geometry of Ω. This is a critical feature even when smooth domains are considered, and can be of use, for instance, in approximation arguments.

The present note centers on applications of results of [4] to first- and second-order inequalities for specific classes of norms. The dependence of the constants appearing in most of the relevant inequalities is also exhibited. Compact embeddings are proposed as well.

The first-order inequalities offered here have the form

$$\|u\|_{Y(\Omega,\mu)} \leq C_1\|\nabla u\|_{X(\Omega)} + C_2\|u\|_{Z(\partial\Omega)}, \tag{1.6}$$

where $\|\cdot\|_{X(\Omega)}$ and $\|\cdot\|_{Z(\partial\Omega)}$ are Banach function norms on Ω and $\partial\Omega$ with respect to the Lebesgue measure and the $(n-1)$-dimensional Hausdorff measure, respectively, and $\|\cdot\|_{Y(\Omega,\mu)}$ is a Banach function norm with respect to an α-upper Ahlfors regular measure μ, with $\alpha \in (n-1, n]$. The norms that will come into play in our discussion are of Lebesgue, Lorentz, Lorentz–Zygmund and Orlicz type.

The second-order inequalities to be presented disclose further new traits. A major novelty with respect to (1.2), and to other customary inequalities, is that the boundary norms only depend on the trace of u on $\partial\Omega$ and not on that of ∇u. Indeed, our second-order inequalities for u read

$$\|u\|_{Y(\Omega,\mu)} \leq C_1\|\nabla^2 u\|_{X(\Omega)} + C_2\|g_u\|_{U(\partial\Omega)} + C_3\|u\|_{Z(\partial\Omega)}, \tag{1.7}$$

where $\|\cdot\|_{X(\Omega)}$, $\|\cdot\|_{Z(\partial\Omega)}$ and $\|\cdot\|_{Y(\Omega,\mu)}$ are as above, $\|\cdot\|_{U(\partial\Omega)}$ is a Banach function norm on $\partial\Omega$, and g_u denotes any upper gradient, in the sense of Hajłasz [10], of the trace of u over $\partial\Omega$. The function g_u can be regarded as a surrogate for the tangential gradient of u over $\partial\Omega$ when the latter is not smooth, and hence it is not endowed with a differential structure.

Let us emphasize that, although inequalities of the form (1.7) are known to hold in regular domains even without the middle term on the right-hand side, such a term cannot be dispensed with in highly irregular domains, such as the one in Figure 1.1. We refer to the paper [4, Example 7.1] for an example with this regard, and for more examples demonstrating the sharpness of our results under various respects.

Bounds for first-order derivatives can also be provided in the form

$$\|\nabla u\|_{Y(\Omega,\mu)} \leq C_1\|\nabla^2 u\|_{X(\Omega)} + C_2\|g_u\|_{U(\partial\Omega)}.$$

This inequality will be shown to hold, in particular, when $\mu = \mathcal{L}^n$, Ω is such that $\mathcal{L}^n(\Omega) < \infty$, $\mathcal{H}^{n-1}(\partial\Omega) < \infty$, and $X = Y = Z = L^2$. Hence, it augments inequality (1.2) in that it only involves the boundary trace of u, through g_u, instead of the trace of ∇u, on the right-hand side.

Finally, we report on companion inequalities from [5] for vector-valued functions $\mathbf{u} : \Omega \to \mathbb{R}^n$, where the full matrix-valued gradient $\nabla\mathbf{u}$ is replaced by its sole symmetric part, denoted by $\mathcal{E}\mathbf{u}$ and called symmetric gradient of $\mathbf{u}$. Namely,

$$\mathcal{E}\mathbf{u} = \frac{1}{2}(\nabla\mathbf{u} + (\nabla\mathbf{u})^T),$$

where $(\nabla\mathbf{u})^T$ stands for the transpose matrix of $\nabla\mathbf{u}$. The symmetric gradient enters various mathematical models for physical phenomena. Instances in this connection are

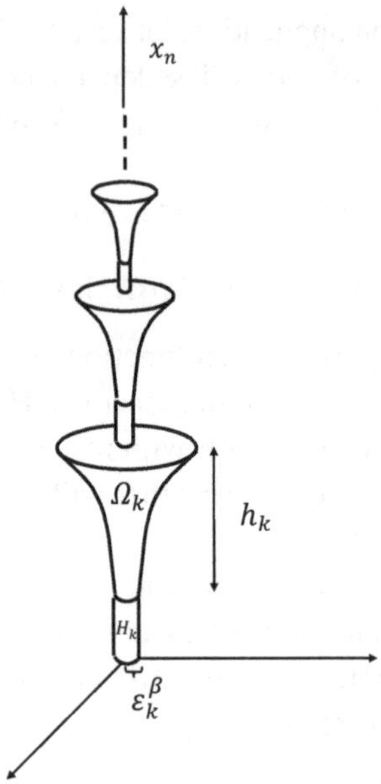

Figure 1.1. A domain showing the optimality of inequality (1.7).

provided by the theory of non-Newtonian fluids, where the function $\mathbf{u}$ represents the velocity of a fluid [1, 6, 9, 14–16], and the theories of plasticity and nonlinear elasticity, where $\mathbf{u}$ stands for the displacement of a body [9, 11, 25].

The approach developed in [4] and [5] rests upon pointwise bounds for functions and their derivatives in terms of classical Riesz potentials of their highest-order derivatives, plus integral operators of a new kind. These bounds, and ensuing estimates in rearrangement form and reduction principles to one-dimensional inequalities, are reproduced here in the cases needed for our purposes. The general reduction principles of use in the proofs of the theorems displayed in Sections 3 and 4, dealing with first- and second-order inequalities, respectively, can be found in [4]. A few special instances of the theorems in question are already explicitly stated and proved in that paper. To give an idea of how the reduction principles can be exploited, in Section 6 we also reproduce a sketch of the proofs of a couple of results on second-order inequalities. Proofs of the results concerning symmetric gradients, which are the subject of Section 5, can be accomplished along the same lines, starting from the reduction principles of [5]. The latter paper also contains statements and proofs of some special cases of these results.

2 Background and notations

Let $\mathcal{R}$ be a measure space endowed with a σ-finite, non-atomic measure m. The decreasing rearrangement $\phi_m^* : [0, \infty) \to [0, \infty]$ of an m-measurable function $\phi : \mathcal{R} \to \mathbb{R}$

is defined as

$$\phi_m^*(t) = \inf\{\tau \geq 0 : m(\{|\phi| > \tau\}) \leq t\} \quad \text{for } t \in [0, \infty).$$

A rearrangement-invariant space $X(\mathcal{R}, m)$ on $(\mathcal{R}, m)$ is a Banach function space (in the sense of Luxemburg) equipped with a norm $\|\cdot\|_{X(\mathcal{R},m)}$, called a rearrangement-invariant norm, such that

$$\|\phi\|_{X(\mathcal{R},m)} = \|\psi\|_{X(\mathcal{R},m)} \quad \text{whenever } \phi_m^* = \psi_m^*.$$

Every rearrangement-invariant space $X(\mathcal{R}, m)$ admits a representation space $\overline{X}(0, \infty)$, namely another rearrangement-invariant space on $(0, \infty)$, endowed with the Lebesgue measure, such that

$$\|\phi\|_{X(\mathcal{R},m)} = \|\phi_m^*\|_{\overline{X}(0,\infty)} \quad \text{for every } \phi \in X(\mathcal{R}, m).$$

A basic example of rearrangement-invariant space is the Lebesgue space $L^p(\mathcal{R}, m)$, defined as usual for $p \in [1, \infty]$.

A generalization of the Lebesgue spaces is constituted by the Lorentz spaces. Assume that either $1 < p < \infty$ and $1 \leq \sigma \leq \infty$, or $p = \sigma = 1$, or $p = \sigma = \infty$. Then the functional given by

$$\|\phi\|_{L^{p,\sigma}(\mathcal{R},m)} = \left\| t^{\frac{1}{p}-\frac{1}{\sigma}} \phi_m^*(t) \right\|_{L^\sigma(0,m(\mathcal{R}))}$$

for an m-measurable function ϕ on $\mathcal{R}$ is (equivalent) to a rearrangement-invariant norm. Here, and in what follows, we use the convention that $\frac{1}{\infty} = 0$. Observe that $L^{p,p}(\mathcal{R}, m) = L^p(\mathcal{R}, m)$ for $p \in [1, \infty]$. Moreover, $L^{p,\sigma_1}(\mathcal{R}, m) \subsetneq L^{p,\sigma_2}(\mathcal{R}, m)$ if $\sigma_1 < \sigma_2$.

The Lorentz spaces admit, in their turn, a further extension, on finite measure spaces, provided by the Lorentz–Zygmund spaces. Assume that $1 \leq p, q \leq \infty$ and $\theta \in \mathbb{R}$, and that $m(\mathcal{R}) < \infty$. The Lorentz–Zygmund functional is defined as

$$\|\phi\|_{L^{p,\sigma;\theta}(\mathcal{R},m)} = \left\| t^{\frac{1}{p}-\frac{1}{\sigma}} \log^\theta\Big(1 + \frac{m(\mathcal{R})}{t}\Big) \phi_m^*(t) \right\|_{L^\sigma(0,m(\mathcal{R}))}$$

for an m-measurable function ϕ on $\mathcal{R}$. For suitable choices of the parameters p, σ, θ, this functional is equivalent to a rearrangement-invariant norm. A description of the admissible choices of these parameters and of the main properties of Lorentz–Zygmund spaces can be found in [7, 21] and [23, Chapter 9].

The Orlicz spaces generalize the Lebesgue spaces in a different direction. Let $A : [0, \infty) \to [0, \infty]$ be a Young function, namely a left-continuous convex function which is neither identically equal to 0 nor to ∞. Then the Orlicz space $L^A(\mathcal{R}, m)$ associated with A is the rearrangement-invariant space equipped with the Luxemburg norm given by

$$\|\phi\|_{L^A(\mathcal{R},m)} = \inf\Big\{\lambda > 0 : \int_{\mathcal{R}} A\Big(\frac{|\phi|}{\lambda}\Big)\, dm \leq 1\Big\}.$$

The Lebesgue space $L^p(\mathcal{R}, m)$ is recovered with the choice

$$A(t) = \begin{cases} t^p & \text{if } p \in [1, \infty), \\ \infty\chi_{(1,\infty)}(t) & \text{if } p = \infty, \end{cases}$$

where χ_E denotes the characteristic function of a set E. If $m(\mathcal{R}) < \infty$ and $\gamma > 0$, we denote by $\exp L^\gamma(\mathcal{R}, m)$ the Orlicz space built upon a Young function $A(t)$ equivalent to $e^{t^\gamma} - 1$ near infinity. Notice that the space $\exp L^\gamma(\mathcal{R}, m)$ belongs to the family of Lorentz–Zygmund spaces as well. Indeed,

$$\exp L^\gamma(\mathcal{R}, m) = L^{\infty,\infty;-\frac{1}{\gamma}}(\mathcal{R}, m), \tag{2.1}$$

up to equivalent norms, depending on γ and on $m(\mathcal{R})$.

Throughout this paper, $\mathcal{R}$ will be either an open subset Ω of $\mathbb{R}^n$, or its topological boundary $\partial\Omega$. The Lebesgue measure in $\mathbb{R}^n$ is denoted by $\mathcal{L}^n$. The notation $\mathcal{H}^{n-1}$ is employed for the $(n-1)$-dimensional Hausdorff measure restricted to $\partial\Omega$. The measure will be omitted in the notation of rearrangement-invariant spaces on Ω or on $\partial\Omega$ equipped with these standard measures.

Upper Ahlfors regular measures, also called Frostman measures, on Ω will also considered. Recall that an α-upper Ahlfors regular measure μ on Ω, with $\alpha \in (0, n]$, is a Borel measure such that

$$\mu(B_r(x) \cap \Omega) \le C_\mu r^\alpha \quad \text{for } x \in \Omega \text{ and } r > 0, \tag{2.2}$$

for some constant $C_\mu > 0$. Here, $B_r(x)$ denotes the ball, centered at x, with radius r.

The spaces of continuous functions in Ω and in $\overline{\Omega}$ are denoted by $C(\Omega)$ and $C(\overline{\Omega})$, respectively. The notation $C_b(\overline{\Omega})$ is adopted for the space of functions in $C(\overline{\Omega})$ with bounded support. The restriction to $\partial\Omega$ of a function $u \in C(\overline{\Omega})$ is denoted by $u_{\partial\Omega}$. We shall simply write u instead of $u_{\partial\Omega}$ in the notation of norms of $u_{\partial\Omega}$ over $\partial\Omega$.

For each $x \in \Omega$, we set

$$(\partial\Omega)_x = \{y \in \partial\Omega : (1-t)x + ty \subset \Omega \text{ for every } t \in (0,1)\}.$$

This is the largest subset of $\partial\Omega$ which can be "seen" from x. One can show that $(\partial\Omega)_x$ is a Borel set. Next, we define the set

$$(\Omega \times \mathbb{S}^{n-1})_0 = \{(x, \vartheta) \in \Omega \times \mathbb{S}^{n-1} : x + t\vartheta \in \partial\Omega \text{ for some } t > 0\}.$$

Clearly, $(\Omega \times \mathbb{S}^{n-1})_0 = \Omega \times \mathbb{S}^{n-1}$ if Ω is bounded. Let $\zeta : (\Omega \times \mathbb{S}^{n-1})_0 \to \mathbb{R}^n$ be the function defined as

$$\zeta(x, \vartheta) = x + t\vartheta, \quad \text{where } t \text{ is such that } x + t\vartheta \in (\partial\Omega)_x.$$

In other words, $\zeta(x, \vartheta)$ is the first point of intersection of the half-line $\{x + t\vartheta : t > 0\}$ with $\partial\Omega$.

Given a function $\varphi : \partial\Omega \to \mathbb{R}$, with bounded support, we adopt the convention that $\varphi(\zeta(x,\vartheta))$ is defined for every $(x,\vartheta) \in \Omega \times \mathbb{S}^{n-1}$, on extending it by 0 on $(\Omega \times \mathbb{S}^{n-1}) \setminus (\Omega \times \mathbb{S}^{n-1})_0$; namely, we set

$$\varphi(\zeta(x,\vartheta)) = 0 \quad \text{if } (x,\vartheta) \in (\Omega \times \mathbb{S}^{n-1}) \setminus (\Omega \times \mathbb{S}^{n-1})_0. \tag{2.3}$$

3 First-order inequalities

In this section we are concerned with inequalities for first-order Sobolev spaces.

Given a rearrangement-invariant space $X(\Omega)$ on an open set $\Omega \subset \mathbb{R}^n$, we denote by $V^1X(\Omega)$ the first-order homogeneous Sobolev-type space defined as

$$V^1X(\Omega) = \{u : u \text{ is weakly differentiable in } \Omega, \text{ and } |\nabla u| \in X(\Omega)\}. \tag{3.1}$$

Notice that no assumption on the integrability of u is made in the definition of $V^1X(\Omega)$.

In the following statements, norms of Sobolev functions and of their derivatives with respect to an α-upper Ahlfors regular measures μ appear. The relevant functions have to interpreted in the sense of traces with respect to μ. These traces are well defined, thanks to standard (local) Sobolev inequalities with measures, owing to the assumption that $\alpha \in (n-1, n]$ in (2.2). An analogous convention applies to the integral operators that enter our discussion.

3.1 First-order Friedrichs-type inequalities

We begin with inequalities involving classical Sobolev and boundary trace spaces built upon Lebesgue norms. The target spaces are also of Lebesgue type, except for a borderline case, where Orlicz spaces of exponential type naturally come into play.

Theorem 3.1 (First-order inequalities). *Let Ω be any open set in $\mathbb{R}^n$, $n \geq 2$, with $\mathcal{L}^n(\Omega) < \infty$ and $\mathcal{H}^{n-1}(\partial\Omega) < \infty$. Assume that μ is a Borel measure in Ω fulfilling* (2.2) *for some $\alpha \in (n-1, n]$ and for some $C_\mu > 0$, and such that $\mu(\Omega) < \infty$.*

(i) *Let $1 < p < n$, $r > 1$, and*

$$q = \min\left\{\frac{p\alpha}{n-p}, \frac{r\alpha}{n-1}\right\}. \tag{3.2}$$

Then there exists a constant $C = C(n, p, r, \alpha, C_\mu)$ such that

$$\begin{aligned}\|u\|_{L^q(\Omega,\mu)} \leq C\Big(&\max\{\mu(\Omega)^{\frac{n}{\alpha}}, \mathcal{L}^n(\Omega)\}^{\frac{\alpha}{qn} - \frac{n-p}{pn}} \|\nabla u\|_{L^p(\Omega)} \\ &+ \max\{\mu(\Omega)^{\frac{n-1}{\alpha}}, \mathcal{H}^{n-1}(\partial\Omega)\}^{\frac{\alpha}{q(n-1)} - \frac{1}{r}} \|u\|_{L^r(\partial\Omega)}\Big)\end{aligned} \tag{3.3}$$

for every $u \in V^1L^p(\Omega) \cap C_b(\overline{\Omega})$.

(ii) *Let $\beta > 0$, and*

$$\gamma = \min\{n', \beta\}. \tag{3.4}$$

Then there exists a constant $C = C(n, \beta, \alpha, C_\mu, \mathcal{L}^n(\Omega), \mathcal{H}^{n-1}(\partial\Omega), \mu(\Omega))$ *such that*

$$\|u\|_{\exp L^\gamma(\Omega,\mu)} \le C\big(\|\nabla u\|_{L^n(\Omega)} + \|u\|_{\exp L^\beta(\partial\Omega)}\big) \tag{3.5}$$

for every $u \in V^1L^n(\Omega) \cap C_b(\overline{\Omega})$.

(iii) *Let* $p > n$. *Then there exists a constant* $C = C(n, p)$ *such that*

$$\|u\|_{L^\infty(\Omega)} \le C\big(\mathcal{L}^n(\Omega)^{\frac{1}{n}-\frac{1}{p}}\|\nabla u\|_{L^p(\Omega)} + \|u\|_{L^\infty(\partial\Omega)}\big)$$

for every $u \in V^1L^p(\Omega) \cap C_b(\overline{\Omega})$.

Remark. Clearly, the inequalities of Theorem 3.1 continue to hold for functions in the closure of the space $V^1L^p(\Omega) \cap C_b(\overline{\Omega})$ with respect to the norms appearing on their right-hand sides. An analogous remark holds for all our inequalities.

Remark. If $q = \frac{p\alpha}{n-p}$, the exponent of the coefficient multiplying the gradient norm in inequality (3.3) vanishes, and, in fact, the assumption $\mathcal{L}^n(\Omega) < \infty$ can be dropped. On the other hand, if $q = \frac{r\alpha}{n-1}$, then the exponent of the coefficient multiplying the boundary norm in inequality (3.3) vanishes, and the assumption $\mathcal{H}^{n-1}(\partial\Omega) < \infty$ can be removed. Analogous weakenings of the assumptions on Ω are admissible in the inequalities stated below, whenever the constants involved turn out to be independent of $\mathcal{L}^n(\Omega)$ or $\mathcal{H}^{n-1}(\partial\Omega)$.

Part (i) of Theorem 3.1 extends a version of the Sobolev inequality for measures, on regular domains [19, Theorem 1.4.5]. It also augments the results mentioned in Section 1 on inequality (1.3) about general domains, which are confined to norms on the left-hand side with respect to the Lebesgue measure. Part (ii) generalizes the Yudovich–Pohozaev–Trudinger inequality to possibly irregular domains. Moreover, it enhances, under some respect, a result of [13], where optimal constants are exhibited, but for a weaker exponential norm, and just with the Lebesgue measure, on the left-hand side.

The following theorem provides us with a compactness result for subcritical norms on the left-hand side.

Theorem 3.2 (Compact embeddings). *Let* Ω *be any open set in* $\mathbb{R}^n$, $n \ge 2$, *with* $\mathcal{L}^n(\Omega) < \infty$ *and* $\mathcal{H}^{n-1}(\partial\Omega) < \infty$. *Let* μ *be a Borel measure in* Ω *fulfilling* (2.2) *for some* $\alpha \in (n-1, n]$ *and for some* $C_\mu > 0$, *and such that* $\mu(\Omega) < \infty$.

(i) *Let* $1 < p < n$ *and* $r > 1$. *Assume that*

$$1 \le q < \min\left\{\frac{p\alpha}{n-p}, \frac{r\alpha}{n-1}\right\}.$$

Then any bounded sequence $\{u_k\}$ *in the space* $V^1L^p(\Omega) \cap C_b(\overline{\Omega})$, *endowed with the norm appearing on the right-hand side of inequality* (3.3), *has a Cauchy subsequence in* $L^q(\Omega, \mu)$.

(ii) *Assume that*

$$0 < \gamma < \min\{n', \beta\}.$$

Then any bounded sequence $\{u_k\}$ *in the space* $V^{1,n}(\Omega) \cap C_b(\overline{\Omega})$, *endowed with the norm appearing on the right-hand side of inequality* (3.5), *has a Cauchy subsequence in* $\exp L^\gamma(\Omega, \mu)$.

The next result concerns inequalities for functions whose gradient belongs to a Lorentz space $L^{p,\sigma}(\Omega)$. Since $L^{p,p}(\Omega) = L^p(\Omega)$, it extends and improves Theorem 3.1. In fact, it augments the conclusions of parts (i) and (ii) of Theorem 3.1 also in this special case when $\sigma = p$, inasmuch as it provides us with strictly smaller target spaces in the respective inequalities.

Theorem 3.3 (First-order Lorentz–Sobolev inequalities). *Let* Ω *be any open set in* $\mathbb{R}^n$, $n \geq 2$, *with* $\mathcal{L}^n(\Omega) < \infty$ *and* $\mathcal{H}^{n-1}(\partial\Omega) < \infty$. *Let* μ *be a Borel measure in* Ω *fulfilling* (2.2) *for some* $\alpha \in (n-1, n]$ *and for some* $C_\mu > 0$, *and such that* $\mu(\Omega) < \infty$.

(i) *Assume that* $1 < p < n$, $r > 1$, *and* $1 \leq \sigma, \varrho \leq \infty$. *Let* q *be defined as in* (3.2) *and let*

$$\eta \geq \begin{cases} \sigma & \text{if } q = \frac{p\alpha}{n-p}, \\ \varrho & \text{if } q = \frac{r\alpha}{n-1}. \end{cases}$$

Then there exists a constant $C = C(n, p, r, \sigma, \varrho, \alpha, C_\mu)$ *such that*

$$\|u\|_{L^{q,\eta}(\Omega,\mu)} \leq C\Big(\max\{\mu(\Omega)^{\frac{n}{\alpha}}, \mathcal{L}^n(\Omega)\}^{\frac{\alpha}{qn}-\frac{n-p}{pn}} \|\nabla u\|_{L^{p,\sigma}(\Omega)} + \max\{\mu(\Omega)^{\frac{n-1}{\alpha}}, \mathcal{H}^{n-1}(\partial\Omega)\}^{\frac{\alpha}{q(n-1)}-\frac{1}{r}} \|u\|_{L^{r,\varrho}(\partial\Omega)}\Big)$$

for every $u \in V^1L^{p,q}(\Omega) \cap C_b(\overline{\Omega})$.

(ii) *Assume that* $\sigma, \varrho > 1$ *and* $\varsigma < -\frac{1}{\varrho}$. *Let*

$$\eta \geq \max\{\sigma, \varrho\} \quad \text{and} \quad \theta \leq \min\left\{-1 + \frac{1}{\sigma} - \frac{1}{\eta}, \varsigma + \frac{1}{\varrho} - \frac{1}{\eta}\right\}.$$

Then there exists $C = C(n, \eta, \sigma, \varrho, \theta, \varsigma, \alpha, C_\mu, \mathcal{L}^n(\Omega), \mu(\Omega), \mathcal{H}^{n-1}(\partial\Omega))$ *such that*

$$\|u\|_{L^{\infty,\eta;\theta}(\Omega,\mu)} \leq C\big(\|\nabla u\|_{L^{n,\sigma}(\Omega)} + \|u\|_{L^{\infty,\varrho;\varsigma}(\partial\Omega)}\big)$$

for every $u \in V^1L^{n,\sigma}(\Omega) \cap C_b(\overline{\Omega})$.

(iii) *Assume that either* $p = n$ *and* $\sigma = 1$, *or* $p > n$ *and* $1 \leq \sigma \leq \infty$. *Then there exists a constant* $C = C(n, p, \sigma, \alpha, C_\mu)$ *such that*

$$\|u\|_{L^\infty(\Omega)} \leq C\big(\mathcal{L}^n(\Omega)^{\frac{1}{n}-\frac{1}{p}} \|\nabla u\|_{L^{p,\sigma}(\Omega)} + \|u\|_{L^\infty(\partial\Omega)}\big)$$

for every $u \in V^1L^{p,\sigma}(\Omega) \cap C_b(\overline{\Omega})$.

Part (i) of Theorem 3.3 is a counterpart in the present setting of well-known results of [20, 22]. The borderline situation considered in part (ii) corresponds to a classical optimal integrability result which follows from a capacitary inequality of [18] – see [19, Inequality (2.3.14)].

3.2 First-order pointwise estimates and reduction principle

The point of departure of our method is the following pointwise estimate involving an unconventional integral operator.

Theorem 3.4 (First-order pointwise estimate). *Let Ω be any open set in $\mathbb{R}^n$, $n \geq 2$. Then there exists a constant $C = C(n)$ such that*

$$|u(x)| \leq C\left(\int_\Omega \frac{|\nabla u(y)|}{|x-y|^{n-1}}\,dy + \int_{\mathbb{S}^{n-1}} |u_{\partial\Omega}(\zeta(x,\vartheta))|\,d\mathcal{H}^{n-1}(\vartheta)\right) \tag{3.6}$$

for $x \in \Omega$ and for every $u \in V^1L^1(\Omega) \cap C_b(\overline{\Omega})$. Here, convention (2.3) *is adopted.*

Sharp endpoint continuity properties of the integral operators on the right-hand side of inequality (3.6), combined with interpolation techniques, yield the bound, in rearrangement form, stated in the next theorem.

Theorem 3.5 (First-order rearrangement estimate). *Let Ω be any open set in $\mathbb{R}^n$, $n \geq 2$. Assume that μ is a Borel measure in Ω fulfilling* (2.2) *for some $\alpha \in (n-1, n]$ and for some $C_\mu > 0$. Then there exist constants $c = c(n)$ and $C = C(n, \alpha, C_\mu)$ such that*

$$\begin{aligned} u_\mu^*(ct) \leq C\Bigg(& t^{-\frac{n-1}{\alpha}} \int_0^{t^{\frac{n}{\alpha}}} |\nabla u|_{\mathcal{L}^n}^*(\rho)\,d\rho + \int_{t^{\frac{n}{\alpha}}}^\infty \rho^{-\frac{n-1}{n}} |\nabla u|_{\mathcal{L}^n}^*(\rho)\,d\rho \\ &+ t^{-\frac{n-1}{\alpha}} \int_0^{t^{\frac{n-1}{\alpha}}} (u_{\partial\Omega})_{\mathcal{H}^{n-1}}^*(\rho)\,d\rho\Bigg) \end{aligned} \tag{3.7}$$

for $t > 0$ and for every $u \in V^1L^1(\Omega) \cap C_b(\overline{\Omega})$.

The rearrangement estimate (3.7) enables us to reduce inequalities of the form (1.6) to considerably simpler one-dimensional Hardy-type inequalities in the corresponding representation norms.

Theorem 3.6 (First-order reduction principle). *Let Ω be any open set in $\mathbb{R}^n$, $n \geq 2$. Assume that μ is a measure in Ω fulfilling* (2.2) *for some $\alpha \in (n-1, n]$ and for some constant C_μ. Let $X(\Omega)$, $Y(\Omega, \mu)$ and $Z(\partial\Omega)$ be rearrangement-invariant spaces such that*

$$\begin{aligned} \Bigg\| \chi_{(0,\mu(\Omega))}(t)\Bigg(& t^{-\frac{n-1}{\alpha}} \int_0^{t^{\frac{n}{\alpha}}} \chi_{(0,\mathcal{L}^n(\Omega))}(\rho) f(\rho)\,d\rho \\ &+ \int_{t^{\frac{n}{\alpha}}}^\infty \rho^{-\frac{n-1}{n}} \chi_{(0,\mathcal{L}^n(\Omega))}(\rho) f(\rho)\,d\rho\Bigg)\Bigg\|_{\overline{Y}(0,\infty)} \\ \leq C_1 \|\chi_{(0,\mathcal{L}^n(\Omega))} f\|_{\overline{X}(0,\infty)}, \end{aligned} \tag{3.8}$$

and

$$\left\| \chi_{(0,\mu(\Omega))}(t)t^{-\frac{n-1}{\alpha}} \int_0^{t^{\frac{n-1}{\alpha}}} \chi_{(0,\mathcal{H}^{n-1}(\partial\Omega))}(\rho) f(\rho)\, d\rho \right\|_{\overline{Y}(0,\infty)}$$
$$\leq C_2 \|\chi_{(0,\mathcal{H}^{n-1}(\partial\Omega))} f\|_{\overline{Z}(0,\infty)} \tag{3.9}$$

for some constants C_1, C_2, and for every non-increasing function $f : [0,\infty) \to [0,\infty)$. Then

$$\|u\|_{Y(\Omega,\mu)} \leq C'\big(C_1\|\nabla u\|_{X(\Omega)} + C_2\|u\|_{Z(\partial\Omega)}\big)$$

for some constant $C' = C'(n)$ and for every $u \in V^1X(\Omega) \cap C_b(\overline{\Omega})$.

Remark. One can verify that the expression appearing in each of the norms on the left-hand sides of inequalities (3.8) and (3.9) is a nonnegative non-increasing function of t. Hence, it agrees with its decreasing rearrangement. This observation can be of use when dealing with rearrangement-invariant spaces $Y(\Omega,\mu)$, such as Lorentz and Lorentz–Zygmund spaces, whose norms are defined in terms of rearrangements.

4 Second-order inequalities

The second-order homogeneous Sobolev space built upon a rearrangement-invariant space $X(\Omega)$ is defined as

$$V^2X(\Omega) = \{u : u \text{ is twice-weakly differentiable in } \Omega, \text{ and } |\nabla^2 u| \in X(\Omega)\}.$$

Here, $\nabla^2 u$ denotes the matrix of all second-order derivatives of u.

As mentioned in Section 1, upper gradients of the restriction to $\partial\Omega$ of trial functions come into play in our second-order inequalities. Let $\varphi : \partial\Omega \to \mathbb{R}$ be a measurable function. A Borel function $g_\varphi : \partial\Omega \to \mathbb{R}$ is called an upper gradient for φ in the sense of Hajłasz [10] if

$$|\varphi(x) - \varphi(y)| \leq |x-y|(g_\varphi(x) + g_\varphi(y)) \quad \text{for } \mathcal{H}^{n-1}\text{-a.e. } x, y \in \partial\Omega.$$

Given a rearrangement-invariant space $X(\partial\Omega)$ on $\partial\Omega$, we denote by $\mathcal{V}^1X(\partial\Omega)$ the space of functions φ that admit an upper gradient in $X(\partial\Omega)$. It is easily verified that $\mathcal{V}^1X(\partial\Omega)$ is a linear space. Furthermore, the functional defined as

$$\|\varphi\|_{\mathcal{V}^1X(\partial\Omega)} = \inf \|g_\varphi\|_{X(\partial\Omega)},$$

where the infimum is extended over all upper gradients g_φ of φ in $X(\partial\Omega)$, is a seminorm on the space $\mathcal{V}^1X(\partial\Omega)$.

Our estimates for u involving Lebesgue norms of $\nabla^2 u$ are the content of the following theorem.

Theorem 4.1 (Second-order inequalities for u). *Let Ω be any open set in $\mathbb{R}^n$, $n \geq 3$, with $\mathcal{L}^n(\Omega) < \infty$ and $\mathcal{H}^{n-1}(\partial\Omega) < \infty$. Assume that μ is a Borel measure in Ω fulfilling* (2.2) *for some $\alpha \in (n-1, n]$ and for some $C_\mu > 0$, and such that $\mu(\Omega) < \infty$.*

(i) *Let $1 < p < \frac{n}{2}$, $1 < s < n-1$, $r > 1$, and let*

$$q = \min\left\{\frac{p\alpha}{n-2p}, \frac{s\alpha}{n-1-s}, \frac{r\alpha}{n-1}\right\}. \tag{4.1}$$

Then there exists a constant $C = C(n, p, r, s, \alpha, C_\mu)$ such that

$$\begin{aligned} \|u\|_{L^q(\Omega,\mu)} \leq C\Big(&\max\{\mu(\Omega)^{\frac{n}{\alpha}}, \mathcal{L}^n(\Omega)\}^{\frac{\alpha}{qn}-\frac{n-2p}{pn}}\|\nabla^2 u\|_{L^p(\Omega)} \\ &+ \max\{\mu(\Omega)^{\frac{n-1}{\alpha}}, \mathcal{H}^{n-1}(\partial\Omega)\}^{\frac{\alpha}{q(n-1)}-\frac{n-1-s}{s(n-1)}}\|u\|_{\mathcal{V}^1L^s(\partial\Omega)} \\ &+ \max\{\mu(\Omega)^{\frac{n-1}{\alpha}}, \mathcal{H}^{n-1}(\partial\Omega)\}^{\frac{\alpha}{q(n-1)}-\frac{1}{r}}\|u\|_{L^r(\partial\Omega)}\Big) \end{aligned} \tag{4.2}$$

for every $u \in V^2L^p(\Omega) \cap \mathcal{V}^1L^s(\partial\Omega) \cap C_b(\overline{\Omega})$.

(ii) *Let $\beta > 0$, $s > n-1$, and let*

$$\gamma = \min\left\{\frac{n}{n-2}, \beta\right\}.$$

Then there exists a constant $C = C(n, \beta, \alpha, C_\mu, \mathcal{L}^n(\Omega), \mathcal{H}^{n-1}(\Omega), \mu(\Omega))$ such that

$$\|u\|_{\exp L^\gamma(\Omega,\mu)} \leq C\big(\|\nabla^2 u\|_{L^{\frac{n}{2}}(\Omega)} + \|u\|_{\mathcal{V}^1L^s(\partial\Omega)} + \|u\|_{\exp L^\beta(\partial\Omega)}\big) \tag{4.3}$$

for every $u \in V^2L^n(\Omega) \cap \mathcal{V}^1L^s(\partial\Omega) \cap C_b(\overline{\Omega})$.

(iii) *Let $p > \frac{n}{2}$ and $s > n-1$. Then there exists a constant $C = C(n, p, s)$ such that*

$$\begin{aligned} \|u\|_{L^\infty(\Omega)} \leq C\big(&\mathcal{L}^n(\Omega)^{\frac{2}{n}-\frac{1}{p}}\|\nabla^2 u\|_{L^p(\Omega)} \\ &+ \mathcal{H}^{n-1}(\partial\Omega)^{\frac{1}{n-1}-\frac{1}{s}}\|u\|_{\mathcal{V}^1L^s(\partial\Omega)} + \|u\|_{L^\infty(\partial\Omega)}\big) \end{aligned} \tag{4.4}$$

for every $u \in V^2L^p(\Omega) \cap \mathcal{V}^1L^s(\partial\Omega) \cap C_b(\overline{\Omega})$

Remark. In the doubly borderline case when $p = \frac{n}{2}$ and $s = n-1$, part (ii) of Theorem 4.1 admits the following variant. Let $\beta > 0$ and

$$\gamma = \min\left\{\frac{n-1}{n-2}, \beta\right\}.$$

Then there exists a constant $C = C(n, \beta, \alpha, C_\mu, \mathcal{L}^n(\Omega), \mathcal{H}^{n-1}(\Omega), \mu(\Omega))$ such that

$$\|u\|_{\exp L^\gamma(\Omega,\mu)} \leq C\big(\|\nabla^2 u\|_{L^{\frac{n}{2}}(\Omega)} + \|u\|_{\mathcal{V}^1L^{n-1}(\partial\Omega)} + \|u\|_{\exp L^\beta(\partial\Omega)}\big)$$

for every $u \in V^2L^n(\Omega) \cap \mathcal{V}^1L^{n-1}(\partial\Omega) \cap C_b(\overline{\Omega})$.

The compactness of embeddings associated with the norms defined via the right-hand sides of the inequalities of Theorem 4.1 is discussed in the next result.

Theorem 4.2 (Second-order compact embeddings). *Let Ω be any open set in $\mathbb{R}^n$, $n \geq 3$, with $\mathcal{L}^n(\Omega) < \infty$ and $\mathcal{H}^{n-1}(\partial\Omega) < \infty$. Let μ be a Borel measure in Ω fulfilling* (2.2) *for some $\alpha \in (n-1, n]$ and for some $C_\mu > 0$, and such that $\mu(\Omega) < \infty$.*

(i) *Let $1 < p < \frac{n}{2}$, $1 < s < n-1$ and $r > 1$. Assume that*

$$1 \leq q < \min\left\{\frac{p\alpha}{n-2p}, \frac{s\alpha}{n-1-s}, \frac{r\alpha}{n-1}\right\}. \tag{4.5}$$

Then any bounded sequence $\{u_k\}$ in the space $V^2L^p(\Omega) \cap \mathcal{V}^1L^s(\partial\Omega) \cap C_b(\overline{\Omega})$, endowed with the norm appearing on the right-hand side of (4.2), *has a Cauchy subsequence in $L^q(\Omega, \mu)$.*

(ii) *Let $p = \frac{n}{2}$ and $s > n-1$. Assume that*

$$0 < \gamma < \min\left\{\frac{n}{n-2}, \beta\right\}. \tag{4.6}$$

Then any bounded sequence $\{u_k\}$ in the space $V^2L^n(\Omega) \cap \mathcal{V}^1L^s(\partial\Omega) \cap C_b(\overline{\Omega})$, endowed with the norm appearing on the right-hand side of inequality (4.3), *has a Cauchy subsequence in $\exp L^\gamma(\Omega, \mu)$.*

Theorem 4.1 admits a counterpart, dealing with bounds for ∇u instead of u, which reads as follows.

Theorem 4.3 (Second-order inequality for ∇u). *Let Ω be any open set in $\mathbb{R}^n$, $n \geq 2$, with $\mathcal{L}^n(\Omega) < \infty$ and $\mathcal{H}^{n-1}(\partial\Omega) < \infty$. Assume that μ is a Borel measure in Ω fulfilling* (2.2) *for some $\alpha \in (n-1, n]$ and for some $C_\mu > 0$, and such that $\mu(\Omega) < \infty$.*

(i) *Let $1 < p < n$, let $r > 1$, and let q be defined as in* (3.2). *Then there exists a constant $C = C(n, p, r, \alpha, C_\mu)$ such that*

$$\|\nabla u\|_{L^q(\Omega,\mu)} \leq C\Big(\max\{\mu(\Omega)^{\frac{n}{\alpha}}, \mathcal{L}^n(\Omega)\}^{\frac{\alpha}{qn} - \frac{n-p}{pn}} \|\nabla^2 u\|_{L^p(\Omega)} + \max\{\mu(\Omega)^{\frac{n}{\alpha}}, \mathcal{H}^{n-1}(\partial\Omega)\}^{\frac{\alpha}{q(n-1)} - \frac{1}{r}} \|u\|_{\mathcal{V}^1L^r(\partial\Omega)}\Big)$$

for every $u \in V^2L^p(\Omega) \cap \mathcal{V}^1L^r(\partial\Omega) \cap C_b(\overline{\Omega})$.

(ii) *Let $\beta > 0$, and let γ be defined as in equation* (3.4). *Then there exists a constant $C = C(n, \beta, \alpha, C_\mu, \mathcal{L}^n(\Omega), \mathcal{H}^{n-1}(\partial\Omega), \mu(\Omega))$ such that*

$$\|\nabla u\|_{\exp L^\gamma(\Omega,\mu)} \leq C\big(\|\nabla^2 u\|_{L^n(\Omega)} + \|u\|_{\mathcal{V}^1 \exp L^\beta(\partial\Omega)}\big)$$

for every $u \in V^2L^n(\Omega) \cap \mathcal{V}^1 \exp L^\beta(\partial\Omega) \cap C_b(\overline{\Omega})$.

(iii) *Let $p > n$. Then there exists a constant $C = C(n, p)$ such that*

$$\|\nabla u\|_{L^\infty(\Omega)} \leq C\big(\mathcal{L}^n(\Omega)^{\frac{1}{n} - \frac{1}{p}} \|\nabla^2 u\|_{L^p(\Omega)} + \|u\|_{\mathcal{V}^1L^\infty(\partial\Omega)}\big)$$

for every $u \in V^2L^p(\Omega) \cap \mathcal{V}^1L^\infty(\partial\Omega) \cap C_b(\overline{\Omega})$.

Inequalities for functions in second-order Lorentz-Sobolev spaces, with improved target spaces, can also be derived. The conclusions about bounds for u are collected in Theorem 4.4. An analogue concerning estimates for ∇u holds, and calls into play the same norms as in Theorem 3.3. Its statement is omitted for brevity.

Theorem 4.4 (Second-order Lorentz–Sobolev inequalities). *Let Ω be any open set in $\mathbb{R}^n$, $n \geq 3$, with $\mathcal{L}^n(\Omega) < \infty$ and $\mathcal{H}^{n-1}(\partial\Omega) < \infty$. Assume that μ is a Borel measure in Ω fulfilling (2.2) for some $\alpha \in (n-1, n]$ and for some $C_\mu > 0$, and such that $\mu(\Omega) < \infty$.*

(i) *Let $1 < p < \frac{n}{2}$, $1 < s < n-1$, $r > 1$, $1 \leq \sigma, \upsilon, \varrho \leq \infty$. Let q be defined as in (4.1) and let*

$$\eta \geq \begin{cases} \sigma & \text{if } q = \frac{p\alpha}{n-2p}, \\ \upsilon & \text{if } q = \frac{s\alpha}{n-1-s}, \\ \varrho & \text{if } q = \frac{r\alpha}{n-1}. \end{cases}$$

Then there exists a constant $C = C(n, p, s, r, \eta, \sigma, \upsilon, \alpha, \varrho, C_\mu)$ such that

$$\begin{aligned}\|u\|_{L^{q,\eta}(\Omega,\mu)} \leq C\Big(&\max\{\mu(\Omega)^{\frac{n}{\alpha}}, \mathcal{L}^n(\Omega)\}^{\frac{\alpha}{qn}-\frac{n-2p}{pn}} \|\nabla^2 u\|_{L^{p,\sigma}(\Omega)} \\ &+ \max\{\mu(\Omega)^{\frac{n-1}{\alpha}}, \mathcal{H}^{n-1}(\partial\Omega)\}^{\frac{\alpha}{q(n-1)}-\frac{n-1-s}{s(n-1)}} \|u\|_{\mathcal{V}^1L^{s,\upsilon}(\partial\Omega)} \\ &+ \max\{\mu(\Omega)^{\frac{n-1}{\alpha}}, \mathcal{H}^{n-1}(\partial\Omega)\}^{\frac{\alpha}{q(n-1)}-\frac{1}{r}} \|u\|_{L^{r,\varrho}(\partial\Omega)}\Big)\end{aligned}$$

for every $u \in V^2L^{p,\sigma}(\Omega) \cap \mathcal{V}^1L^{s,\upsilon}(\partial\Omega) \cap C_b(\overline{\Omega})$.

(ii) *Assume that $\sigma, \upsilon, \varrho > 1$ and $\varsigma < -\frac{1}{\varrho}$. Let*

$$\eta \geq \max\{\sigma, \upsilon, \varrho\},$$

and

$$\theta \leq \min\Big\{-1 + \frac{1}{\sigma} - \frac{1}{\eta}, -1 + \frac{1}{\upsilon} - \frac{1}{\eta}, \varsigma + \frac{1}{\varrho} - \frac{1}{\eta}\Big\}.$$

Then there is $C = C(n, \eta, \sigma, \varrho, \theta, \upsilon, \varsigma, \alpha, C_\mu, \mathcal{L}^n(\Omega), \mu(\Omega), \mathcal{H}^{n-1}(\partial\Omega))$ such that

$$\begin{aligned}\|u\|_{L^{\infty,\eta;\theta}(\Omega,\mu)} \leq C\big(&\|\nabla^2 u\|_{L^{\frac{n}{2},\sigma}(\Omega)} + \|u\|_{\mathcal{V}^1L^{n-1,\upsilon}(\partial\Omega)} \\ &+ \|u\|_{L^{\infty,\varrho;\varsigma}(\partial\Omega)}\big)\end{aligned}$$

for every $u \in V^2L^{\frac{n}{2},\sigma}(\Omega) \cap \mathcal{V}^1L^{n-1,\upsilon}(\partial\Omega) \cap C_b(\overline{\Omega})$.

(iii) *Assume that either $p = \frac{n}{2}$ and $\sigma = 1$, or $p > n$ and $1 \leq \sigma \leq \infty$, and either $s = n-1$ and $\upsilon = 1$, or $s > n-1$ and $1 \leq \upsilon \leq \infty$. Then there exists a constant $C = C(n, p, \sigma, s, \upsilon, \alpha, C_\mu)$ such that*

$$\begin{aligned}\|u\|_{L^\infty(\Omega)} \leq C\big(&\mathcal{L}^n(\Omega)^{\frac{2}{n}-\frac{1}{p}} \|\nabla^2 u\|_{L^{p,\sigma}(\Omega)} \\ &+ \mathcal{H}^{n-1}(\partial\Omega)^{\frac{1}{n-1}-\frac{1}{s}} \|u\|_{\mathcal{V}^1L^{s,\upsilon}(\partial\Omega)} + \|u\|_{L^\infty(\partial\Omega)}\big)\end{aligned}$$

for every $u \in V^2L^{p,\sigma}(\Omega) \cap \mathcal{V}^1L^{s,\upsilon}(\partial\Omega) \cap C_b(\overline{\Omega})$.

4.1 Second-order pointwise estimates and reduction principle

In this subsection we collect second-order versions of the results of Section 3.2, namely pointwise estimates, rearrangement-estimates, and reduction principles, for both u and ∇u.

Theorem 4.5 (Second-order pointwise estimates). *Let Ω be any open set in $\mathbb{R}^n$.*

(i) *Assume that $n \geq 3$. There exists a constant $C = C(n)$ such that*

$$|u(x)| \leq C\Big(\int_\Omega \frac{|\nabla^2 u(y)|}{|x-y|^{n-2}}\,dy + \int_\Omega \int_{\mathbb{S}^{n-1}} \frac{g_u(\zeta(y,\vartheta))}{|x-y|^{n-1}}\,d\mathcal{H}^{n-1}(\vartheta)\,dy + \int_{\mathbb{S}^{n-1}} |u_{\partial\Omega}(\zeta(x,\vartheta))|\,d\mathcal{H}^{n-1}(\vartheta)\Big) \tag{4.7}$$

for $x \in \Omega$ and for every $u \in V^2L^1(\Omega) \cap \mathcal{V}^1L^1(\partial\Omega) \cap C_b(\overline{\Omega})$.

(ii) *Assume that $n \geq 2$. There exists a constant $C = C(n)$ such that*

$$|\nabla u(x)| \leq C\Big(\int_\Omega \frac{|\nabla^2 u(y)|}{|x-y|^{n-1}}\,dy + \int_{\mathbb{S}^{n-1}} g_u(\zeta(x,\vartheta))\,d\mathcal{H}^{n-1}(\vartheta)\Big) \tag{4.8}$$

for a.e. $x \in \Omega$ and for every $u \in V^2L^1(\Omega) \cap \mathcal{V}^1L^1(\partial\Omega) \cap C_b(\overline{\Omega})$.

In inequalities (4.7) and (4.8), g_u denotes any Hajłasz gradient of $u_{\partial\Omega}$, and convention (2.3) is adopted.

Theorem 4.6 (Second-order rearrangement estimates). *Let Ω be any open set in $\mathbb{R}^n$. Assume that μ is a Borel measure in Ω fulfilling (2.2) for some $\alpha \in (n-1, n]$ and for some $C_\mu > 0$.*

(i) *Assume that $n \geq 3$. There exist constants $c = c(n)$ and $C = C(n, \alpha, C_\mu)$ such that*

$$\begin{aligned} u_\mu^*(ct) \leq C\Big(& t^{-\frac{n-2}{\alpha}} \int_0^{t^{\frac{n}{\alpha}}} |\nabla^2 u|_{\mathcal{L}^n}^*(\rho)\,d\rho \\ &+ \int_{t^{\frac{n}{\alpha}}}^\infty r^{-\frac{n-2}{n}} |\nabla^2 u|_{\mathcal{L}^n}^*(\rho)\,d\rho \\ &+ t^{-\frac{n-2}{\alpha}} \int_0^{t^{\frac{n-1}{\alpha}}} (g_u)_{\mathcal{H}^{n-1}}^*(\rho)\,d\rho \\ &+ \int_{t^{\frac{n-1}{\alpha}}}^\infty \rho^{-\frac{n-2}{n-1}} (g_u)_{\mathcal{H}^{n-1}}^*(\rho)\,d\rho \\ &+ t^{-\frac{n-1}{\alpha}} \int_0^{t^{\frac{n-1}{\alpha}}} (u_{\partial\Omega})_{\mathcal{H}^{n-1}}^*(\rho)\,d\rho\Big) \end{aligned} \tag{4.9}$$

for $t > 0$ and for every $u \in V^2L^1(\Omega) \cap \mathcal{V}^1L^1(\partial\Omega) \cap C_b(\overline{\Omega})$.

(ii) *Assume that $n \geq 2$. There exist constants $c = c(n)$ and $C = C(n, \alpha, C_\mu)$ such that*

$$|\nabla u|^*_\mu(ct) \leq C\biggl(t^{-\frac{n-1}{\alpha}} \int_0^{t^{\frac{n}{\alpha}}} |\nabla^2 u|^*_{\mathcal{L}^n}(\rho)\,d\rho + \int_{t^{\frac{n}{\alpha}}}^\infty \rho^{-\frac{n-1}{n}} |\nabla^2 u|^*_{\mathcal{L}^n}(\rho)\,d\rho + t^{-\frac{n-1}{\alpha}} \int_0^{t^{\frac{n-1}{\alpha}}} (g_u)^*_{\mathcal{H}^{n-1}}(\rho)\,d\rho\biggr) \tag{4.10}$$

for $t > 0$ and for every $u \in V^2L^1(\Omega) \cap \mathcal{V}^1L^1(\partial\Omega) \cap C_b(\overline{\Omega})$.

In inequalities (4.9) *and* (4.10), *g_u denotes any Hajłasz gradient of $u_{\partial\Omega}$.*

Theorem 4.7 (Second-order reduction principles). *Let Ω be any open set in $\mathbb{R}^n$. Assume that μ is a measure in Ω fulfilling* (2.2) *for some $\alpha \in (n-1, n]$, and for some constant C_μ.*

(i) *Assume that $n \geq 3$. Let $X(\Omega)$, $Y(\Omega,\mu)$, $U(\partial\Omega)$ and $Z(\partial\Omega)$ be rearrangement-invariant spaces such that*

$$\biggl\|\chi_{(0,\mu(\Omega))}(t)\biggl(t^{-\frac{n-2}{\alpha}} \int_0^{t^{\frac{n}{\alpha}}} \chi_{(0,\mathcal{L}^n(\Omega))}(\rho) f(\rho)\,d\rho + \int_{t^{\frac{n}{\alpha}}}^\infty \rho^{-\frac{n-2}{n}} \chi_{(0,\mathcal{L}^n(\Omega))}(\rho) f(\rho)\,d\rho\biggr)\biggr\|_{\overline{Y}(0,\infty)} \leq C_1 \|\chi_{(0,\mathcal{L}^n(\Omega))} f\|_{\overline{X}(0,\infty)}, \tag{4.11}$$

$$\biggl\|\chi_{(0,\mu(\Omega))}(t)\biggl(t^{-\frac{n-2}{\alpha}} \int_0^{t^{\frac{n-1}{\alpha}}} \chi_{(0,\mathcal{H}^{n-1}(\partial\Omega))}(\rho) f(\rho)\,d\rho + \int_{t^{\frac{n-1}{\alpha}}}^\infty \rho^{-\frac{n-2}{n-1}} \chi_{(0,\mathcal{H}^{n-1}(\partial\Omega))}(\rho) f(\rho)\,d\rho\biggr)\biggr\|_{\overline{Y}(0,\infty)} \leq C_2 \|\chi_{(0,\mathcal{H}^{n-1}(\partial\Omega))} f\|_{\overline{U}(0,\infty)} \tag{4.12}$$

$$\biggl\|\chi_{(0,\mu(\Omega))}(t) t^{-\frac{n-1}{\alpha}} \int_0^{t^{\frac{n-1}{\alpha}}} \chi_{(0,\mathcal{H}^{n-1}(\partial\Omega))}(\rho) f(\rho)\,d\rho\biggr\|_{\overline{Y}(0,\infty)} \leq C_3 \|\chi_{(0,\mathcal{H}^{n-1}(\partial\Omega))} f\|_{\overline{Z}(0,\infty)} \tag{4.13}$$

for some positive constants C_1, C_2 and C_3, and for every non-increasing function $f : [0,\infty) \to [0,\infty)$. Then

$$\|u\|_{Y(\Omega,\mu)} \leq C'\bigl(C_1\|\nabla^2 u\|_{X(\Omega)} + C_2\|u\|_{\mathcal{V}^1U(\partial\Omega)} + C_3\|u\|_{Z(\partial\Omega)}\bigr)$$

for some constant $C' = C'(n)$ and for every $u \in V^2X(\Omega) \cap \mathcal{V}^1U(\partial\Omega) \cap C_b(\overline{\Omega})$.

(ii) *Assume that $n \geq 2$. Let $X(\Omega)$, $Y(\Omega,\mu)$ and $Z(\partial\Omega)$ be rearrangement-invariant spaces such that inequalities* (3.8) *and* (3.9) *hold. Then*

$$\|\nabla u\|_{Y(\Omega,\mu)} \leq C'\bigl(C_1\|\nabla^2 u\|_{X(\Omega)} + C_2\|u\|_{\mathcal{V}^1Z(\partial\Omega)}\bigr)$$

for some constant $C' = C'(n)$ and for every $u \in V^2X(\Omega) \cap \mathcal{V}^1Z(\partial\Omega) \cap C_b(\overline{\Omega})$.

Remark. The expression appearing in each of the norms on the left-hand sides of inequalities (4.11)–(4.13) is a nonnegative non-increasing function of s. Therefore, it agrees with its decreasing rearrangement.

5 Inequalities for the symmetric gradient

In this section we deal with Friedrichs-type inequalities, in the spirit of those presented in Section 1, involving just the symmetric gradient $\mathcal{E}\mathbf{u}$ of functions $\mathbf{u} : \Omega \to \mathbb{R}^n$.

In analogy with (3.1), given a rearrangement-invariant space $X(\Omega)$, we define the symmetric gradient Sobolev space as

$$E^1X(\Omega) = \{\mathbf{u} \in L^1_{\mathrm{loc}}(\Omega) : |\mathcal{E}\mathbf{u}| \in X(\Omega)\}.$$

Interestingly, it turns out that the norms in the relevant symmetric gradient inequalities are exactly the same as those entering their full gradient counterparts. As an example, we reproduce here the basic conclusions concerning functions in the space $E^1L^p(\Omega)$, with $p \in (1, \infty]$. A related result can also be found in the recent paper [3].

Theorem 5.1 (Symmetric gradient inequalities). *Let Ω be any open set in $\mathbb{R}^n$, $n \geq 2$, with $\mathcal{L}^n(\Omega) < \infty$ and $\mathcal{H}^{n-1}(\partial\Omega) < \infty$. Assume that μ is a Borel measure in Ω fulfilling* (2.2) *for some $\alpha \in (n-1, n]$ and for some $C_\mu > 0$, and such that $\mu(\Omega) < \infty$.*

(i) *Let $1 < p < n$, let $r > 1$, and let q be defined as in* (3.2). *Then there exists a constant $C = C(n, p, r, \alpha, C_\mu)$ such that*

$$\|\mathbf{u}\|_{L^q(\Omega,\mu)} \leq C\Big(\max\{\mu(\Omega)^{\frac{n}{\alpha}}, \mathcal{L}^n(\Omega)\}^{\frac{\alpha}{qn}-\frac{n-p}{pn}} \|\mathcal{E}\mathbf{u}\|_{L^p(\Omega)} + \max\{\mu(\Omega)^{\frac{n-1}{\alpha}}, \mathcal{H}^{n-1}(\partial\Omega)\}^{\frac{\alpha}{q(n-1)}-\frac{1}{r}} \|\mathbf{u}\|_{L^r(\partial\Omega)}\Big)$$

for every $\mathbf{u} \in E^1L^p(\Omega) \cap C_b(\overline{\Omega})$.

(ii) *Let $\beta > 0$, and let γ be defined as in equation* (3.4). *Then there exists a constant $C = C(n, \beta, \alpha, C_\mu, \mathcal{L}^n(\Omega), \mathcal{H}^{n-1}(\Omega), \mu(\Omega))$ such that*

$$\|\mathbf{u}\|_{\exp L^\gamma(\Omega,\mu)} \leq C\big(\|\mathcal{E}\mathbf{u}\|_{L^n(\Omega)} + \|\mathbf{u}\|_{\exp L^\beta(\partial\Omega)}\big)$$

for every $\mathbf{u} \in E^1L^n(\Omega) \cap C_b(\overline{\Omega})$.

(iii) *Let $p > n$. Then there exists a constant $C = C(n, p)$ such that*

$$\|\mathbf{u}\|_{L^\infty(\Omega)} \leq C\big(\mathcal{L}^n(\Omega)^{\frac{1}{n}-\frac{1}{p}} \|\mathcal{E}\mathbf{u}\|_{L^p(\Omega)} + \|\mathbf{u}\|_{L^\infty(\partial\Omega)}\big)$$

for every $\mathbf{u} \in E^1L^p(\Omega) \cap C_b(\overline{\Omega})$.

5.1 Symmetric-gradient pointwise estimates and reduction principle

Our approach to Friedrichs inequalities in the spaces $E^1X(\Omega)$ is grounded on pointwise and rearrangement bounds in terms of the symmetric gradient. They parallel those presented in Section 3.2 for the full gradient and are exposed in the next two theorems.

Theorem 5.2 (Symmetric gradient pointwise estimate). *Let Ω be any open set in $\mathbb{R}^n$, $n \geq 2$. Then there exists a constant $C = C(n)$ such that*

$$|\mathbf{u}(x)| \leq C\Big(\int_\Omega \frac{|\mathcal{E}\mathbf{u}(y)|}{|x-y|^{n-1}}\,dy + \int_{\mathbb{S}^{n-1}} |\mathbf{u}_{\partial\Omega}(\zeta(x,\vartheta))|\,d\mathcal{H}^{n-1}(\vartheta)\Big)$$

for $x \in \Omega$ and for every $\mathbf{u} \in E^1L^1(\Omega) \cap C_b(\overline{\Omega})$. Here, convention (2.3) *is adopted.*

Theorem 5.3 (Symmetric gradient rearrangement estimate). *Let Ω be any open set in $\mathbb{R}^n$, $n \geq 2$. Assume that μ is a Borel measure in Ω fulfilling* (2.2) *for some $\alpha \in (n-1, n]$ and for some $C_\mu > 0$. Then there exist some constants $c = c(n)$ and $C = C(n, \alpha, C_\mu)$ such that*

$$|\mathbf{u}|_\mu^*(ct) \leq C\Big(t^{-\frac{n-1}{\alpha}} \int_0^{t^{\frac{n}{\alpha}}} |\mathcal{E}\mathbf{u}|_{\mathcal{L}^n}^*(\rho)\,d\rho + \int_{t^{\frac{n}{\alpha}}}^\infty \rho^{-\frac{n-1}{n}} |\mathcal{E}\mathbf{u}|_{\mathcal{L}^n}^*(\rho)\,d\rho$$
$$+ t^{-\frac{n-1}{\alpha}} \int_0^{t^{\frac{n-1}{\alpha}}} |\mathbf{u}_{\partial\Omega}|_{\mathcal{H}^{n-1}}^*(\rho)\,d\rho\Big)$$

for $t > 0$ and for every $\mathbf{u} \in E^1L^1(\Omega) \cap C_b(\overline{\Omega})$.

As a consequence of Theorem 5.3, a reduction principle for symmetric gradient inequalities takes exactly the same form as that stated in Theorem 3.6.

Theorem 5.4 (Symmetric gradient reduction principle). *Let Ω be any open set in $\mathbb{R}^n$, $n \geq 2$. Assume that μ is a measure in Ω fulfilling* (2.2) *for some $\alpha \in (n-1, n]$, and for some constant C_μ. Let $X(\Omega)$, $Y(\Omega,\mu)$ and $Z(\partial\Omega)$ be rearrangement-invariant spaces such that inequalities* (3.8)–(3.9) *hold for some constants C_1 and C_2. Then*

$$\|\mathbf{u}\|_{Y(\Omega,\mu)} \leq C'\Big(C_1\|\mathcal{E}\mathbf{u}\|_{X(\Omega)} + C_2\|\mathbf{u}\|_{Z(\partial\Omega)}\Big) \tag{5.1}$$

for some constant $C' = C'(n)$ and for every $\mathbf{u} \in E^1X(\Omega) \cap C_b(\overline{\Omega})$.

6 Sketches of proofs

We conclude by outlining the proofs of Theorems 4.1 and 4.2. This should help the reader grasp methods to derive inequalities and compactness results via our reduction principles. For proofs of the latter we refer to the papers [4, 5].

Proof of Theorem 4.1. (i) By Theorem 4.7 and a change of variables, inequality (4.2) will follow if we show that

$$\Big\|t^{-\frac{n-2}{n}+(\frac{\alpha}{n}-1)\frac{1}{q}} \int_0^t f(\rho)\,d\rho\Big\|_{L^q(0,\ell_1)}$$
$$+ \Big\|t^{(\frac{\alpha}{n}-1)\frac{1}{q}} \int_t^{\ell_1} \rho^{-\frac{n-2}{n}} f(\rho)\,d\rho\Big\|_{L^q(0,\ell_1)} \leq C_1\|f\|_{L^p(0,\ell_1)} \tag{6.1}$$

for every non-increasing function $f : (0, \ell_1) \to [0, \infty)$, where

$$\ell_1 = \max\{\mu(\Omega)^{\frac{n}{\alpha}}, \mathcal{L}^n(\Omega)\}, \quad C_1 = c\ell_1^{\frac{\alpha}{qn} - \frac{n-2p}{pn}}$$

and $c = c(n, p, r, s, \alpha, C_\mu)$;

$$\begin{aligned} &\left\| t^{-\frac{n-2}{n-1} + (\frac{\alpha}{n-1} - 1)\frac{1}{q}} \int_0^t f(\rho)\, d\rho \right\|_{L^q(0,\ell_2)} \\ &\quad + \left\| t^{(\frac{\alpha}{n-1} - 1)\frac{1}{q}} \int_t^{\ell_2} \rho^{-\frac{n-2}{n-1}} f(\rho)\, d\rho \right\|_{L^q(0,\ell_2)} \le C_2 \|f\|_{L^s(0,\ell_2)} \end{aligned} \tag{6.2}$$

for every non-increasing function $f : (0, \ell_2) \to [0, \infty)$, where

$$\ell_2 = \max\{\mu(\Omega)^{\frac{n-1}{\alpha}}, \mathcal{H}^{n-1}(\partial\Omega)\}, \quad C_2 = c\ell_2^{\frac{\alpha}{q(n-1)} - \frac{n-1-s}{s(n-1)}}$$

and $c = c(n, p, r, s, \alpha, C_\mu)$;

$$\left\| t^{-1 + (\frac{\alpha}{n-1} - 1)\frac{1}{q}} \int_0^t f(\rho)\, d\rho \right\|_{L^q(0,\ell_2)} \le C_3 \|f\|_{L^r(0,\ell_2)} \tag{6.3}$$

for every non-increasing function $f : (0, \ell_2) \to [0, \infty)$, where

$$C_3 = c\ell_2^{\frac{\alpha}{q(n-1)} - \frac{1}{r}}$$

and $c = c(n, p, r, s, \alpha, C_\mu)$.

Inequalities (6.1)–(6.3) can be established via standard criteria for one-dimensional Hardy-type inequalities – see e.g. [19, Section 1.3.2].

(ii) Owing to Theorem 4.7, the subsequent remark, equation (2.1), and a change of variables, the proof of inequality (4.3) is reduced to showing that

$$\begin{aligned} &\left\| \left(t^{-\frac{n-2}{n}} \int_0^t f(\rho)\, d\rho \right) \log^{-\frac{1}{\gamma}} \left(1 + \frac{\ell_1}{t^{\frac{\alpha}{n}}} \right) \right\|_{L^\infty(0,\ell_1)} \\ &\quad + \left\| \left(\int_t^{\ell_1} \rho^{-\frac{n-2}{n}} f(\rho)\, d\rho \right) \log^{-\frac{1}{\gamma}} \left(1 + \frac{\ell_1}{t^{\frac{\alpha}{n}}} \right) \right\|_{L^\infty(0,\ell_1)} \le C_1 \|f\|_{L^{\frac{n}{2}}(0,\ell_1)} \end{aligned} \tag{6.4}$$

for every non-increasing function $f : (0, \ell_1) \to [0, \infty)$ and for some positive constant $C_1 = C_1(n, \alpha, C_\mu, \mu(\Omega), \mathcal{L}^n(\Omega))$;

$$\begin{aligned} &\left\| \left(t^{-\frac{n-2}{n-1}} \int_0^t f(\rho)\, d\rho \right) \log^{-\frac{1}{\gamma}} \left(1 + \frac{\ell_2}{t^{\frac{\alpha}{n-1}}} \right) \right\|_{L^\infty(0,\ell_2)} \\ &\quad + \left\| \left(\int_t^{\ell_2} r^{-\frac{n-2}{n-1}} f(r) dr \right) \log^{-\frac{1}{\gamma}} \left(1 + \frac{\ell_2}{t^{\frac{\alpha}{n-1}}} \right) \right\|_{L^\infty(0,\ell_2)} \le C_2 \|f\|_{L^s(0,\ell_2)} \end{aligned} \tag{6.5}$$

for every non-increasing function $f : (0, \ell_2) \to [0, \infty)$ and for some positive constant $C_2 = C_2(n, \alpha, s, C_\mu, \mu(\Omega), \mathcal{H}^{n-1}(\partial\Omega))$;

$$\left\| \left(t^{-1} \int_0^t f(\rho)\, d\rho \right) \log^{-\frac{1}{\gamma}} \left(1 + \frac{\ell_2}{t^{\frac{\alpha}{n-1}}} \right) \right\|_{L^\infty(0,\ell_2)} \\ \leq C_3 \left\| f(t) \log^{-\frac{1}{\gamma}} \left(1 + \frac{\ell_2}{t^{\frac{\alpha}{n-1}}} \right) \right\|_{L^\infty(0,\ell_2)} \tag{6.6}$$

for every non-increasing function $f : (0, \ell_2) \to [0, \infty)$ and for some positive constant $C_3 = C_3(n, r, \alpha, C_\mu, \mu(\Omega), \mathcal{H}^{n-1}(\partial\Omega))$.

Inequalities (6.4)–(6.6) follow via the criteria for weighted one-dimensional Hardy-type inequalities mentioned above.

(iii) The proof of inequality (4.4) is analogous to that of inequality (4.3). One has just to set $\mu = \mathcal{L}^n$, $\alpha = n$, $q = \infty$ in inequalities (6.1)–(6.3), and derive the correct dependence of the constants C_1, C_2 and C_3 from appropriate results for Hardy-type inequalities [19, Section 1.3.2]. ■

Proof of Theorem 4.2. (i) Fix any $\varepsilon > 0$. Then, there exists a compact set $K \subset \Omega$ such that $\mu(\Omega \setminus K) < \varepsilon$. Let $\xi \in C_0^\infty(\Omega)$ be such that $0 \leq \xi \leq 1$, $\xi = 1$ in K. Thus, $K \subset \operatorname{supp}(\xi)$, the support of ξ, and hence $\mu(\operatorname{supp}(1-\xi)) \leq \mu(\Omega \setminus K) < \varepsilon$. Let Ω' be an open set, with a smooth boundary, such that $\operatorname{supp}(\xi) \subset \Omega' \subset \Omega$. Let $\{u_k\}$ be a bounded sequence in the space $V^2L^p(\Omega) \cap \mathcal{V}^1L^s(\partial\Omega) \cap C_b(\overline{\Omega})$. Thus, owing to an application of Theorem 4.1, with $\mu = \mathcal{L}^n$, such a sequence is also bounded in the standard Sobolev space $W^{2,p}(\Omega')$. A weighted version of Rellich's compactness theorem [19, Theorem 1.4.6/1] ensures that $\{u_k\}$ has a Cauchy subsequence, still denoted by $\{u_k\}$, in $L^q(\Omega', \mu)$. As a consequence, there exists $k_0 \in \mathbb{N}$ such that

$$\|u_k - u_j\|_{L^q(\Omega',\mu)} < \varepsilon \tag{6.7}$$

if $k, j > k_0$. Now, denote by $\widehat{q}$ the minimum in equation (4.5). By Hölder's inequality,

$$\|(1-\xi)(u_k - u_j)\|_{L^q(\Omega,\mu)} \leq \|u_k - u_j\|_{L^{\widehat{q}}(\Omega,\mu)} \mu(\operatorname{supp}(1-\xi))^{\frac{1}{q} - \frac{1}{\widehat{q}}} \\ \leq C \varepsilon^{\frac{1}{q} - \frac{1}{\widehat{q}}} \tag{6.8}$$

for some constant C independent of k and j. Inequalities (6.7) and (6.8) imply that

$$\|u_k - u_j\|_{L^q(\Omega,\mu)} \leq \|u_k - u_j\|_{L^q(\Omega',\mu)} + \|(1-\xi)(u_k - u_j)\|_{L^q(\Omega,\mu)} \\ \leq \varepsilon + C \varepsilon^{\frac{1}{q} - \frac{1}{\widehat{q}}} \tag{6.9}$$

if $k, j > k_0$. Owing to the arbitrariness of ε, inequality (6.9) tells us that $\{u_k\}$ is a Cauchy sequence in $L^q(\Omega, \mu)$.

(ii) Given $\varepsilon > 0$, let K, Ω, ξ as above, and let $\{u_k\}$ be a bounded sequence in the space $V^2L^{\frac{n}{2}}(\Omega) \cap \mathcal{V}^1L^s(\partial\Omega) \cap C_b(\overline{\Omega})$. By Theorem 4.1, with $\mu = \mathcal{L}^n$, the sequence $\{u_k\}$ is bounded in $W^{2,\frac{n}{2}}(\Omega')$ as well. From [2, Theorem 5.3, part (ii)] we hence deduce that $\{u_k\}$ has a Cauchy subsequence, still denoted by $\{u_k\}$, in the space $\exp L^{\gamma}(\Omega', \mu)$. Note that, as observed in [2], the theorem in question, although stated for the space $W_0^{2,\frac{n}{2}}(\Omega')$, continues to hold for $W^{2,\frac{n}{2}}(\Omega')$ if Ω' is regular enough. Therefore there exists $k_0 \in \mathbb{N}$ such that

$$\|u_k - u_j\|_{\exp L^{\gamma}(\Omega',\mu)} < \varepsilon \tag{6.10}$$

if $k, j > k_0$. Call $\widehat{\gamma}$ the minimum in equation (4.6). On making use of a version of Hölder's inequality in Orlicz spaces one can deduce that

$$\begin{aligned} &\|(1-\xi)(u_k - u_j)\|_{\exp L^{\gamma}(\Omega',\mu)} \\ &\leq C \|u_k - u_j\|_{\exp L^{\widehat{\gamma}}(\Omega',\mu)} \log^{\frac{1}{\gamma}-\frac{1}{\widehat{\gamma}}}\left(1 + \frac{1}{\mu(\operatorname{supp}(1-\xi))}\right) \\ &\leq C' \log^{\frac{1}{\gamma}-\frac{1}{\widehat{\gamma}}}\left(1 + \frac{1}{\varepsilon}\right) \end{aligned} \tag{6.11}$$

for some constants C and C' independent of k and j. Owing to the arbitrariness of ε, the inequalities in (6.10) and (6.11) tell us that $\{u_k\}$ is a Cauchy sequence in the space $\exp L^{\gamma}(\Omega', \mu)$. ■

Acknowledgements. The research of A. Cianchi was partly supported by Research Project 2201758MTR2 of the Italian Ministry of University and Research (MIUR) Prin 2017 "Direct and inverse problems for partial differential equations: Theoretical aspects and applications" and by GNAMPA of the Italian INdAM – National Institute of High Mathematics. V. Maz'ya was supported by RUDN University Strategic Academic Leadership Program.

Bibliography

[1] G. Astarita and G. Marucci, *Principles of non-Newtonian fluid mechanics*. McGraw–Hill, London, 1974

[2] P. Cavaliere, Z.Mihula, Compactness of Sobolev-type embeddings with measures. *Comm. Contemp. Math.*, to appear

[3] N. V. Chemetov and A. L. Mazzucato, Embeddings for the space LD^p_γ on sets of finite perimeter. *Proc. Roy. Soc. Edinburgh Sect. A* **150** (2020), 2442–2461

[4] A. Cianchi and V. Maz'ya, Sobolev inequalities in arbitrary domains. *Adv. Math.* **293** (2016), 644–696

[5] A. Cianchi and V. G. Maz'ya, Sobolev inequalities for the symmetric gradient in arbitrary domains. *Nonlinear Anal.* **194** (2020), Article ID 111515

[6] G. Duvaut and J.-L. Lions, *Inequalities in mechanics and physics*. Springer, Berlin, 1976

[7] W. D. Evans, B. Opic and L. Pick, Interpolation of operators on scales of generalized Lorentz–Zygmund spaces. *Math. Nachr.* **182** (1996), 127–181

[8] K. Friedrichs, Die Randwert-und Eigenwertprobleme aus der Theorie der elastischen Platten. (Anwendung der direkten Methoden der Variationsrechnung). *Math. Ann.* **98** (1928), 205–247

[9] M. Fuchs and G. Seregin, *Variational methods for problems from plasticity theory and for generalized Newtonian fluids*. Lecture Notes in Math. 1749, Springer, Berlin, 2000

[10] P. Hajłasz, Sobolev spaces on an arbitrary metric space. *Potential Anal.* **5** (1996), 403–415

[11] R. V. Kohn, *New estimates for deformations in terms of their strains*. Ph.D. thesis, Princeton University, Princeton, 1979

[12] F. Maggi and C. Villani, Balls have the worst best Sobolev inequalities. *J. Geom. Anal.* **15** (2005), 83–121

[13] F. Maggi and C. Villani, Balls have the worst best Sobolev inequalities. II. Variants and extensions. *Calc. Var. Partial Differential Equations* **31** (2008), 47–74

[14] J. Málek, J. Nečas, M. Rokyta and M. Ružička, *Weak and measure valued solutions to evolutionary PDEs*. Chapman & Hall, London, 1996

[15] J. Málek and K. R. Rajagopal, Mathematical issues concerning the Navier–Stokes equations and some of its generalizations. In *Evolutionary equations. Vol. II*, edited by C. Dafermos and E. Feireisl, pp. 371–459, Handb. Differ. Equ., Elsevier/North-Holland, Amsterdam, 2005

[16] J. Málek and K. R. Rajagopal, Compressible generalized Newtonian fluids. *Z. Angew. Math. Phys.* **61** (2010), 1097–1110

[17] V. G. Maz'ya, Classes of regions and imbedding theorems for function spaces. *Dokl. Akad. Nauk. SSSR* **133** (1960), 527–530 (in Russian); English transl. *Soviet Math. Dokl.* **1** (1960), 882–885

[18] V. G. Maz'ya, Certain integral inequalities for functions of many variables. *Probl. Math. Anal. Leningrad Univ.* **3** (1972), 33–68 (in Russian); English transl. *J. Soviet Math.* **1** (1973), 205–234

[19] V. Maz'ya, *Sobolev spaces with applications to elliptic partial differential equations*. Augmented edn., Grundlehren Math. Wiss. 342, Springer, Heidelberg, 2011

[20] R. O'Neil, Convolution operators and $L(p,q)$ spaces. *Duke Math. J.* **30** (1963), 129–142

[21] B. Opic and L. Pick, On generalized Lorentz–Zygmund spaces. *Math. Inequal. Appl.* **2** (1999), 391–467

[22] J. Peetre, Espaces d'interpolation et théorème de Soboleff. *Ann. Inst. Fourier (Grenoble)* **16** (1966), 279–317

[23] L. Pick, A. Kufner, O. John and S. Fučík, *Function spaces. Vol. 1*. Extended edn., De Gruyter Ser. Nonlinear Anal. Appl. 14, Walter de Gruyter, Berlin, 2013

[24] L. Rondi, A Friedrichs–Maz'ya inequality for functions of bounded variation. *Math. Nachr.* **290** (2017), 1830–1839

[25] R. Temam, *Problèmes mathématiques en plasticité*. Méth. Math. Inform. 12, Gauthier-Villars, Montrouge, 1983

Ari Laptev and the Journal of Spectral Theory

E. Brian Davies

In July 2005 Ari Laptev and I both went to a conference in Matrei in the Austrian Tyrol, organized by Thomas Hoffmann-Ostenhof. In one of our conversations the subject turned to an idea that Ari had been discussing with others for some time – the formation of a new journal to deal with topics in mathematical physics and analysis, in which we all had a common interest. I was very enthusiastic and pointed out that there was currently no journal dealing with spectral theory. Indeed it was not even recognized as a major discipline by Mathematical Reviews.

Ari was clearly interested in this idea and some time later asked me whether I would be willing to become the founding editor of a new Journal of Spectral Theory under the auspices of the European Mathematical Society. I knew that this would involve a lot of work, and immediately said that I would not have the time to do this, because I had just been chosen as the next President of the London Mathematical Society, which I expected to involve a large amount of work for two years starting in November 2007. (I had no idea how much, but that is another story.) Ari, however, had other ideas and said that he was very relaxed about such a delay until after my term of office had ended, because he thought that I was the ideal person to manage such a venture. Anyone who has met Ari knows how charming, and irresistible, he can be when he wants to achieve some goal (the funding of the Mittag-Leffler Institute is another example) and I was not able to find any other excuse on the spot.

There was, however, another issue. As forthcoming President of the LMS, a long established mathematical publisher, it would look very odd if I were to choose to launch a new journal with another body. I took the opportunity to discuss this with officials in the LMS, but it turned out that they were fully occupied at that time digesting a number of other journals with which they had recently signed agreements. They did not want to take further risks at that time. The European Mathematical Society (EMS) was, however, enthusiastic to proceed and seemed not to be concerned about the inevitable early losses that a new journal was bound to make, so the matter was decided.

Ari promised that he would deal with all negotiations with the EMS publisher, so my role would be limited to the academic side, i.e. choosing the Editors, deciding the processes to be followed when papers were submitted and adjudicating on issues that might arise. I am not sure whether Ari knew that I had been Editor of the Quarterly Journal of Mathematics during most of the 1970s, so I had a fair amount of experience of the care needed when dealing with unhappy authors. I made it clear that all important decisions would be taken after consultation.

The first task was clearly to choose a Board of Editors. The JST would only succeed if the members of the Board had a high enough status and I decided to approach Barry Simon for advice. Ari put forward a list of suggestions and Barry provided more. I wrote to most of them myself, emphasizing who had recommended them and also making it clear that their task would be limited to choosing the referees and deciding what academic decisions should be taken. I would handle everything else and provide advice as needed.

Ari and I were in no doubt about the success of the new journal. The high international standing of the editors more or less guaranteed that, but it happened more rapidly than we expected. Between Volume 1 in 2011 and Volume 9 in 2019 its size increased from 460 to 1521 pages. I was constantly surprised at the willingness of the EMS to increase the number of pages, but Ari did not bother me about the financial/publishing management of the EMS. I knew from my experience with the LMS that the financial health of journals could not be taken for granted, but I never found out how much the physical growth of the JST depended on Ari's persuasiveness. He used to wave aside my enquiries about such matters as non-problems. Maybe so, for him! Whenever I needed to consult him on some editorial matter he always replied quickly and positively, as did Barry.

The only crisis, if it deserves to be called that, came when I realized in the autumn of 2014 that I would have to retire from my position as Chief Editor within a few months. I was very pleased when Ari and Barry both approved my suggestion for Fritz Gesztesy as the next Chief Editor. Barry agreed to approach Fritz directly, and "made him an offer he could not refuse", so he accepted the inevitable. We had several months during which we were able to discuss matters as they arose, and I retired confident that the JST was in good hands. Since then it has gone from strength to strength, the only problem being the steadily increasing number of high quality papers submitted to it. This is a tribute to the entire Board of Editors, but particularly to Fritz and to Ari, who initiated the entire venture.

Critical magnetic field for 2D magnetic Dirac–Coulomb operators and Hardy inequalities

Jean Dolbeault, Maria J. Esteban and Michael Loss

It is with great pleasure that we dedicate this paper to Ari on the occasion of his 70th birthday, because of his strong interest in Hardy inequalities and his pioneering work in this area

This paper is devoted to the study of the two-dimensional Dirac–Coulomb operator in presence of an Aharonov–Bohm external magnetic potential. We characterize the highest intensity of the magnetic field for which a two-dimensional magnetic Hardy inequality holds. Up to this critical magnetic field, the operator admits a distinguished self-adjoint extension and there is a notion of ground state energy, defined as the lowest eigenvalue in the gap of the continuous spectrum.

1 Introduction and main results

Aharonov–Bohm magnetic fields describe an idealized situation where a solenoid interacts with a charged quantum mechanical particle and it is customary to take this to be a particle without spin. It was realized by Laptev and Weidl [43] that in such a situation a Hardy inequality holds, provided that the flux of the solenoid is not an integer multiple of 2π. This result was extended in [5] to the nonlinear case yielding a sharp Caffarelli–Kohn–Nirenberg-type inequality. The current paper is devoted to *Hardy inequalities* but for spinor valued functions in two dimensions. It has been known for some time that Hardy-type inequalities yield information about the bound state problem for the three-dimensional Dirac–Coulomb equation [18]. It is therefore natural to ask whether this relationship continues to hold for the two-dimensional Dirac–Coulomb problem and whether it continues to hold in presence of a solenoid. Let us emphasize that we take the Coulomb potential to be $\frac{1}{r}$ and not $-\ln r$. It turns

---

Keywords: Aharonov–Bohm magnetic potential, magnetic Dirac operator, Coulomb potential, critical magnetic field, self-adjoint operators, eigenvalues, ground state energy, Hardy inequality, Wirtinger derivatives, Pauli operator

2020 Mathematics Subject Classification: Primary 81Q10; secondary 46N50, 81Q05, 47A75

out that a number of spectral properties such as the eigenvalues and eigenfunctions can be worked out in an elementary fashion. Another point of interest is that the approach to self-adjoint extensions through Hardy inequalities as pioneered in [27, 28] works also for this case. In addition to working out the *ground state energy* which falls into the gap, we also get a *critical magnetic field* for which the ground state energy hits the value 0 and beyond which the operator cannot be further defined, on the basis of pure energy considerations, as a self-adjoint operator. As the field strength approaches this critical value, the slope of the eigenvalue as a function of the field strength tends to negative infinity. We also exhibit the corresponding eigenfunction for all magnetic field strengths below the critical value.

An additional bonus is that all the quantities are given explicitly which allows to investigate in a simple fashion the non-relativistic limit. The presence of the magnetic field in the Dirac equation manifests itself entirely through the vector potential. Formally the Pauli equation should be related with as a non-relativistic limit and involve the magnetic field. The problem, as mentioned above, is that the magnetic field is a delta function at the origin, i.e., it appears as a point interaction. Such situations were investigated before in [24] where self-adjoint extensions for the Pauli Hamiltonian involving magnetic point interactions are constructed.

Let $\boldsymbol{\sigma} = (\sigma_i)_{i=1,2,3}$ be the Pauli-matrices defined by

$$\sigma_1 := \begin{pmatrix} 0 & 1 \\ 1 & 0 \end{pmatrix}, \quad \sigma_2 := \begin{pmatrix} 0 & -i \\ i & 0 \end{pmatrix}, \quad \sigma_3 := \begin{pmatrix} 1 & 0 \\ 0 & -1 \end{pmatrix}.$$

Throughout this paper, let us consider on $\mathbb{R}^2$ the Aharonov–Bohm magnetic potential

$$\mathbf{A}_a(x) := \frac{a}{\rho^2} \begin{pmatrix} -x_2 \\ x_1 \end{pmatrix}, \quad \text{where } \rho = |x| = \sqrt{x_1^2 + x_2^2}.$$

Here a is a real parameter and (x_1, x_2) are standard Cartesian coordinates of $x \in \mathbb{R}^2$. The magnetic gradient is defined as

$$\nabla_a := \nabla - i\mathbf{A}_a.$$

In atomic units ($m = \hbar = c = 1$) the two-dimensional magnetic Dirac operator can be written as

$$H_a := \sigma_3 - i\boldsymbol{\sigma} \cdot \nabla_a = \begin{pmatrix} 1 & D_a \\ D_a^* & -1 \end{pmatrix},$$

where $z = x_1 + i x_2$ and $\bar{z} = x_1 - i x_2$. Here

$$D_a := -2i\,\partial_z + i a \frac{\bar{z}}{|z|^2} \quad \text{and} \quad D_a^* := -2i\,\partial_{\bar{z}} - i a \frac{z}{|z|^2},$$

where

$$\partial_z := \frac{1}{2}(\partial_1 - i\,\partial_2) \quad \text{and} \quad \partial_{\bar{z}} := \frac{1}{2}(\partial_1 + i\,\partial_2)$$

are the *Wirtinger derivatives*. In polar coordinates such that

$$\rho = |x| = |z| \quad \text{and} \quad \theta = \arctan\left(\frac{x_2}{x_1}\right)$$

so that $z = \rho e^{i\theta}$, we have

$$2\partial_z = e^{-i\theta}\left(\partial_\rho - \frac{i}{\rho}\partial_\theta\right),$$
$$2\partial_{\bar z} = e^{i\theta}\left(\partial_\rho + \frac{i}{\rho}\partial_\theta\right)$$

and

$$D_a = -ie^{-i\theta}\left(\partial_\rho - \frac{i}{\rho}\partial_\theta - \frac{a}{\rho}\right),$$
$$D_a^* = -ie^{i\theta}\left(\partial_\rho + \frac{i}{\rho}\partial_\theta + \frac{a}{\rho}\right).$$

For self-adjointness properties of H_a, we refer to [9, 31, 55].

For any $\nu \in (0, \frac{1}{2}]$, let us consider the *magnetic Dirac–Coulomb operator*

$$H_{\nu,a} := H_a - \frac{\nu}{|x|}.$$

Our purpose is to establish the range of a for which $H_{\nu,a}$ is self-adjoint for a well chosen domain. According to the approach in [27], the self-adjointness is based on the inequality

$$\int_{\mathbb{R}^2}\left(\frac{|D_a^*\varphi|^2}{1+\lambda+\frac{\nu}{|x|}} + \left(1-\lambda-\frac{\nu}{|x|}\right)|\varphi|^2\right)dx \geq 0, \quad \varphi \in C_c^\infty(\mathbb{R}^2\setminus\{0\}, \mathbb{C}) \tag{1.1}$$

for some $\lambda \in (-1, 1)$. For every $\nu \in (0, \frac{1}{2}]$, let us define the *critical magnetic field* by

$$\mathsf{a}(\nu) := \sup\big\{\mathsf{a} > 0 : \text{there is } \lambda \in (-1,1) \text{ such that (1.1) holds true for any} a \in [0, \mathsf{a})\big\}.$$

Our first result deals only with (1.1). Let us define the function

$$\mathsf{c}(s) := \frac{1}{2} - s.$$

Theorem 1.1 (Hardy inequality for the magnetic Dirac–Coulomb operator). *For any $\nu \in (0, \frac{1}{2})$, we have*

$$\mathsf{a}(\nu) = \mathsf{c}(\nu)$$

and (1.1) *holds true for any $a \in (0, \mathsf{a}(\nu)]$ and $\lambda \in (-1, \lambda_{\nu,a}]$, where*

$$\lambda_{\nu,a} := \frac{\sqrt{\mathsf{c}(a)^2 - \nu^2}}{\mathsf{c}(a)}. \tag{1.2}$$

Notice that $a \leq \mathsf{c}(\nu)$ is equivalent to $\nu \leq \mathsf{c}(a)$. Under this condition, the quadratic form

$$\varphi \mapsto \mathcal{Q}_{\nu,a,\lambda}(\varphi) := \int_{\mathbb{R}^2} \left(\frac{|D_a^* \varphi|^2}{1 + \lambda + \frac{\nu}{|x|}} + \left(1 - \lambda - \frac{\nu}{|x|}\right) |\varphi|^2 \right) dx \tag{1.3}$$

is nonnegative on $C_c^\infty(\mathbb{R}^2 \setminus \{0\}, \mathbb{C})$ for any $\lambda \in (-1, \lambda_{\nu,a}]$. Since $\mathcal{Q}_{\nu,a,\lambda}$ is associated with a symmetric operator, it is closable. The results of [27] can be adapted as follows.

Theorem 1.2 (Self-adjointness of the magnetic Dirac–Coulomb operator). *Suppose $\nu \in (0, \frac{1}{2}]$. For any $a \in (0, \mathsf{a}(\nu)]$, the quadratic form $\mathcal{Q}_{\nu,a,\lambda}$ is closable and its form domain is a Hilbert space $\mathcal{F}_{\nu,a}$ which does not depend on $\lambda \in (-1, 1)$. Let*

$$\mathcal{D}_{\nu,a} := \{\psi = (\varphi, \chi)^\top \in L^2(\mathbb{R}^2, \mathbb{C}^2) : \varphi \in \mathcal{F}_{\nu,a},\ H_{\nu,a}\psi \in L^2(\mathbb{R}^2, \mathbb{C}^2)\},$$

where $H_{\nu,a}\psi$ is understood in the sense of distributions. Then the operator $H_{\nu,a}$ with domain $\mathcal{D}_{\nu,a}$ is self-adjoint. Additionally, if $a < \mathsf{a}(\nu)$, then $\mathcal{F}_{\nu,a}$ does not depend on ν and

$$\mathcal{F}_{\nu,a} = \left\{ \varphi \in L^2(\mathbb{R}^2, \mathbb{C}) \cap H^1_{\mathrm{loc}}(\mathbb{R}^2 \setminus \{0\}, \mathbb{C}) : \sqrt{\frac{|x|}{1 + |x|}} D_a^* \varphi \in L^2(\mathbb{R}^2, \mathbb{C}) \right\}. \tag{1.4}$$

In other words, the domain $\mathcal{D}_{\nu,a}$ of $H_{\nu,a}$ is the space of spinors

$$\psi = \begin{pmatrix} \varphi \\ \chi \end{pmatrix} = (\varphi, \chi)^\top \quad \text{with } \varphi \in \mathcal{F}_{\nu,a} \text{ and } \chi \in L^2(\mathbb{R}^2, \mathbb{C})$$

such that

$$D_a^* \varphi - \left(1 + \frac{\nu}{|x|}\right) \chi \in L^2(\mathbb{R}^2, \mathbb{C}),$$
$$D_a \chi + \left(1 - \frac{\nu}{|x|}\right) \varphi \in L^2(\mathbb{R}^2, \mathbb{C}).$$

Here $D_a^* \varphi$, $\frac{\varphi}{|x|}$, $D_a \chi$, and $\frac{\chi}{|x|}$ are interpreted in the sense of distributions. The operator $H_{\nu,a}$ acts on the two components of the spinor ψ and we shall say that φ is the *upper component* and χ the *lower component*. The claim of Theorem 1.2 is that $H_{\nu,a}$ with domain $\mathcal{D}_{\nu,a}$ is self-adjoint if the field a is at most equal to the critical magnetic field $\mathsf{a}(\nu)$. This critical field manifests itself yet in another way. We recall (see for instance [57, Theorem 4.7]) that the essential spectrum of $H_{\nu,a}$ is $(-\infty, -1] \cup [1, +\infty)$. Even if the operator $H_{\nu,a}$ is not bounded from below, there is a notion of *ground state*, which also makes sense in the non-relativistic limit (see Section 5).

Theorem 1.3 (Ground state energy of the magnetic Dirac operator). *Let $\nu \in (0, \frac{1}{2}]$ and $a \in (0, \mathsf{a}(\nu)]$. Then $\lambda_{\nu,a}$ is the lowest eigenvalue in $(-1, 1)$ of the operator $H_{\nu,a}$ with domain $\mathcal{D}_{\nu,a}$.*

In the subcritical and critical range, the ground state energy $\lambda_{\nu,a}$ of the magnetic Dirac–Coulomb operator is given by (1.2) and the ground state itself can be computed: see Proposition 4.1. The Hardy inequality (1.1) is our key tool in the analysis of the magnetic Dirac–Coulomb operator. This deserves an explanation. If $\psi \in \mathcal{D}_{\nu,a}$ is an eigenfunction of $H_{\nu,a}$ associated with an eigenvalue $\lambda \in (-1,1)$, then $H_{\nu,a}\psi = \lambda\psi$ can be rewritten as a system for the upper and lower components φ and χ, namely

$$D_a\chi + \Big(1 - \lambda - \frac{\nu}{|x|}\Big)\varphi = 0, \quad D_a^*\varphi - \Big(1 + \lambda + \frac{\nu}{|x|}\Big)\chi = 0. \tag{1.5}$$

By eliminating χ from the second equation, we find that φ is a critical point of the quadratic form $\mathcal{Q}_{\nu,a,\lambda}$ defined by (1.3) which moreover realizes the equality case in (1.1). We aim at characterizing $\lambda_{\nu,a}$ as the minimum of a variational problem, which further justifies why we call it a *ground state* energy.

Our strategy is to prove (1.1) directly using the Aharonov–Casher transformation

$$\psi(x) = \begin{pmatrix} |x|^{-a}\eta_1(x) \\ |x|^{a}\eta_2(x) \end{pmatrix}, \quad x \in \mathbb{R}^2, \tag{1.6}$$

with $\eta_i : \mathbb{R}^2 \to \mathbb{C}$, for $i = 1, 2$. System (1.5) amounts to

$$\begin{aligned} \eta_1 - 2i\rho^{2a}\partial_z\eta_2 - \frac{\nu}{\rho}\eta_1 &= \lambda\eta_1, \\ -2i\rho^{-2a}\partial_{\bar z}\eta_1 - \frac{\nu}{\rho}\eta_2 - \eta_2 &= \lambda\eta_2. \end{aligned} \tag{1.7}$$

As for (1.5), using the second equation, we can eliminate the lower component and obtain

$$\eta_2 = -2i\frac{\rho^{-2a}\partial_{\bar z}\eta_1}{1 + \lambda + \frac{\nu}{\rho}}. \tag{1.8}$$

On the space

$$\mathcal{G}_{\nu,a} := \{\eta \in L^2(\mathbb{R}^2, \mathbb{C}; |x|^{-2a}\,dx) : |x|^{-a}\eta(x) \in \mathcal{F}_{\nu,a}\},$$

let us define the counterpart of $\mathcal{Q}_{\nu,a,\lambda}$ as in (1.3), that is,

$$J(\eta,\nu,a,\mu) := \int_{\mathbb{R}^2}\Big(\frac{4|\partial_{\bar z}\eta|^2}{1 + \mu + \frac{\nu}{|x|}} + \Big(1 - \mu - \frac{\nu}{|x|}\Big)|\eta|^2\Big)\frac{dx}{|x|^{2a}}.$$

The map $\eta \mapsto J(\eta,\nu,a,\lambda)$ is differentiable and we read from (1.7) that η_1 is a critical point after eliminating η_2 using (1.8).

For a given function $\eta \in \mathcal{G}_{\nu,a}$, if $J(\eta,\nu,a,0)$ is finite, then the function defined by $\mu \mapsto J(\eta,\nu,a,\mu)$ is well defined and monotone non-increasing on $(-1,+\infty)$, with $\lim_{\mu\to+\infty} J(\eta,\nu,a,\mu) = -\infty$. As a consequence, the equation $J(\eta,\nu,a,\mu) = 0$ has one and only one solution in $(-1,+\infty)$ if, for instance, $\lim_{\mu\to(-1)_+} J(\eta,\nu,a,\mu) > 0$. Let us denote this solution by $\lambda_\star(\eta,\nu,a)$, so that

$$J(\eta,\nu,a,\lambda_\star(\eta,\nu,a)) = 0.$$

By convention, we take $\lambda_\star(\eta, \nu, a) = +\infty$ if the equation has no solution in $(-1, +\infty)$. The main technical estimate and the key result for proving Theorems 1.1, 1.2 and 1.3 goes as follows.

Proposition 1.4 (Variational characterization). *For any* $\nu \in (0, \frac{1}{2}]$ *and* $a \in [0, \mathsf{c}(\nu)]$,

$$\min_{\eta \in \mathcal{G}_{\nu,a}} \lambda_\star(\eta, \nu, a) = \lambda_{\nu,a}, \tag{1.9}$$

where $\lambda_{\nu,a}$ *is defined by* (1.2), *and the equality case is achieved, up to multiplication by a constant, by*

$$\eta_\star(x) = |x|^{\sqrt{\mathsf{c}(a)^2 - \nu^2} - \mathsf{c}(a)} e^{-\frac{\nu}{\mathsf{c}(a)}|x|}, \quad x \in \mathbb{R}^2 \setminus \{0\}.$$

Proposition 1.4 is at the core of the paper. Let us notice that $\lambda_{\nu,a}$ is the largest $\lambda > -1$ such that $J(\eta, \nu, a, \lambda) \ge 0$ for all $\eta \in \mathcal{G}_{\nu,a}$. This is a *Hardy-type inequality* which deserves some additional considerations. A simple consequence of Proposition 1.4 is indeed the fact that

$$J(\eta, \nu, a, \mu) \ge 0, \quad \eta \in C^\infty_{\mathrm{c}}(\mathbb{R}^2 \setminus \{0\}, \mathbb{C}), \tag{1.10}$$

for any $\mu \in (-1, \lambda_{\nu,a}]$. Inequality (1.10) provides us with the interesting inequality

$$\int_{\mathbb{R}^2} \left(\frac{4|\partial_{\bar z} \eta|^2}{1 + \lambda_{\nu,a} + \frac{\nu}{|x|}} + \left(1 - \lambda_{\nu,a} - \frac{\nu}{|x|}\right)|\eta|^2 \right) \frac{dx}{|x|^{2a}} \ge 0, \tag{1.11}$$

for all $\eta \in C^\infty_{\mathrm{c}}(\mathbb{R}^2 \setminus \{0\}, \mathbb{C})$ in the case $\mu = \lambda_{\nu,a}$. Using (1.6), we already prove (1.1) written for $\lambda = \lambda_{\nu,a}$, namely

$$\int_{\mathbb{R}^2} \left(\frac{|D_a^* \varphi|^2}{1 + \lambda_{\nu,a} + \frac{\nu}{|x|}} + \left(1 - \lambda_{\nu,a} - \frac{\nu}{|x|}\right)|\varphi|^2 \right) dx \ge 0 \tag{1.12}$$

for all $\varphi \in C^\infty_{\mathrm{c}}(\mathbb{R}^2 \setminus \{0\}, \mathbb{C})$.

It is an essential property of the Aharonov–Bohm magnetic field that (1.6) transforms the quadratic form associated with (1.12) into (1.11), which is a weighted inequality, without magnetic field. In terms of scalings, one has to think of a Hardy-type inequality for a Dirac–Coulomb operator in a non-integer dimension $2 - 2a$. By taking the limit of the inequality $J(\eta, \nu, a, \mu) \ge 0$ as $\mu \to (-1)_+$, we obtain

$$\int_{\mathbb{R}^2} \left(\frac{4}{\nu}|x|^{1-2a}|\partial_{\bar z}\eta|^2 + 2\frac{|\eta|^2}{|x|^{2a}} \right) dx \ge \nu \int_{\mathbb{R}^2} \frac{|\eta|^2}{|x|^{2a+1}}\, dx$$

for all $\eta \in C^\infty_{\mathrm{c}}(\mathbb{R}^2 \setminus \{0\}, \mathbb{C})$ under the assumption $a \in (0, \frac{1}{2})$ and $\nu \in (0, \mathsf{c}(a))$. Using scalings, we can get rid of the non-homogeneous term and find by taking the limit as $\nu \to \mathsf{c}(a)$ that

$$\int_{\mathbb{R}^2} |x|^{1-2a}|\partial_{\bar z}\eta|^2\, dx \ge \frac{1}{4}\mathsf{c}(a)^2 \int_{\mathbb{R}^2} \frac{|\eta|^2}{|x|^{2a+1}}\, dx$$

for all $\eta \in C_c^\infty(\mathbb{R}^2 \setminus \{0\}, \mathbb{C})$. These weighted magnetic Hardy inequalities are part of a larger family of inequalities.

Lemma 1.5 (Hardy inequalities for Wirtinger derivatives). *Let $\beta \in \mathbb{R}$. Then we have the two following inequalities (with optimal constants) for all $\eta \in C_c^\infty(\mathbb{R}^2 \setminus \{0\}, \mathbb{C})$:*

$$\int_{\mathbb{R}^2} |x|^{-2\beta} |\partial_{\bar{z}} \eta|^2 \, dx \geq \frac{1}{4} \min_{\ell \in \mathbb{Z}} (\beta - \ell)^2 \int_{\mathbb{R}^2} |x|^{-2\beta-2} |\eta|^2 \, dx, \tag{1.13}$$

$$\int_{\mathbb{R}^2} |x|^{2\beta} |\partial_z \eta|^2 \, dx \geq \frac{1}{4} \min_{\ell \in \mathbb{Z}} (\beta - \ell)^2 \int_{\mathbb{R}^2} |x|^{2\beta-2} |\eta|^2 \, dx. \tag{1.14}$$

As a simple consequence of Lemma 1.5, we also obtain a family of Hardy inequalities for spinors corresponding to the *magnetic Pauli operator* $-i\boldsymbol{\sigma} \cdot \nabla_a$ (see [24]).

Corollary 1.6 (Hardy inequalities for the magnetic Pauli operator). *Let $\zeta \in \mathbb{R}$ and $a \in [0, \frac{1}{2}]$. Then*

$$\int_{\mathbb{R}^2} |x|^\zeta |(\boldsymbol{\sigma} \cdot \nabla_a)\psi|^2 \, dx \geq C_{a,\zeta} \int_{\mathbb{R}^2} |x|^{\zeta-2} |\psi|^2 \, dx \tag{1.15}$$

for all $\psi \in C_c^\infty(\mathbb{R}^2 \setminus \{0\}, \mathbb{C}^2)$ and the optimal constant in inequality (1.15) *is given by*

$$C_{a,\zeta} = \min_{\pm, \ell \in \mathbb{Z}} \left(a \pm \frac{\zeta}{2} - \ell \right)^2.$$

A straightforward consequence of the expression of $C_{a,\zeta}$ is that, for all $\zeta \in \mathbb{R}$, the function $a \mapsto C_{a,\zeta}$ is 1-periodic. More inequalities of interest are listed in Appendices A.2 and A.3.

Let us give an overview of the literature. In absence of magnetic field, the computation of the eigenvalues of the hydrogen atom in the setting of the Dirac equation goes back to [12,35]. As noted in [45], an explicit computation of the spectrum can be done using Laguerre polynomials. By the transformation (1.6), which replaces a problem with magnetic field by an equivalent problem with weights, we obtain a similar algebra, which would allow us to adapt the computations of [23]. As we are interested only in the *ground state energy*, we use a simpler approach based on Hardy-type inequalities for the upper component of the Dirac operator: the inequality follows from a simple *completion of a square* as in [17, Remark, p. 9]. Notice that (1.6) can be seen as a special case of the transformation introduced by Aharonov and Casher in [1] and used to define the Pauli operator for some measure valued magnetic fields (see [24, inequality (3)]). Although rather elementary, the computation in dimension two of the ground state and the ground state energy for the magnetic Dirac–Coulomb operator in presence of the Aharonov–Bohm magnetic field is, as far as we know, original.

As noted in [26], the *two-dimensional case* is relevant in solid state physics for the study of graphene: the low-energy electronic excitations are modeled by a massless two-dimensional Dirac equation [11] but the study of strained graphene involves a massive Dirac operator [59]. Magnetic impurities are known to play a role. The coupling with

Aharonov–Bohm magnetic fields raises interesting mathematical questions which have probably interesting consequences from the experimental point of view.

Our interest for *Hardy-type inequalities* in the presence of an Aharonov–Bohm magnetic field goes back to the inequality proved by Laptev and Weidl in [43]. In the perspective of Schrödinger operators, such an inequality in dimension two is somewhat surprising, because it is well known that there is no such inequality without magnetic field. It is therefore natural to investigate whether there is a relativistic counterpart, which is our purpose in this paper, as well as to consider the non-relativistic limit. The link between ground states for Dirac operators and Hardy inequalities is known for instance from [18, p. 222] and has been exploited in works like [6, 15, 17]. Here we have a new example which is particularly interesting as the ground state is explicit and its upper component realizes the equality case in the corresponding Hardy inequality.

The continuous spectrum of $H_{\nu,a}$ depends neither on ν nor on a: see [57, Theorem 4.7] and [37]. If $a = 0$, see [34] for all $\nu > 0$ and also [40, 63] and [2, 3, 48]. The case $a \neq 0$ is less standard but does not raise additional difficulties. It is well known that the lower part of the continuous spectrum prevents $H_{\nu,a}$ to be semi-bounded from below in any reasonable sense. In order to characterize the eigenvalues in the gap, one has to address more subtle *variational principles* than standard Rayleigh quotients. The first min-max formulae based on a decomposition into an upper and a lower component of the free Dirac operator perturbed by an electrostatic potential with Coulomb-type singularity at the origin were proposed by Talman in [56] and Datta and Devaiah in [13]. From the mathematical point of view, *min-max methods* for the characterization of eigenvalues go back to [18, 19, 29, 36]. See also [16, 46, 47] for more recent results and especially [51] which provides a comprehensive overview. Such a variational approach provides numerical schemes for computing Dirac eigenvalues, which have been studied in [10, 20, 22, 42, 44, 64]. The symmetry of the electron case and the *positron* case is inherent to the Dirac equation: see for instance [54, Chapter 11] for a review of the classical invariances associated with the Dirac operator. This symmetry has been reformulated from a variational point of view in [21]. In the present paper, this is exploited in Appendix A.3 for obtaining a dual family of Hardy-type inequalities, which is formally summarized by the transformation $(\nu, a, \lambda) \mapsto (-\nu, -a, -\lambda)$, $D_a \mapsto D_a^*$ and $\partial_{\bar{z}} \mapsto \partial_z$. Non-relativistic limits are from the early papers [12, 35] a natural question and have been studied more recently, for instance, in the context of the Dirac-Fock model in [25, 30, 50].

A delicate issue for Dirac operators with singular Coulomb potentials defined on smooth functions is the determination of a self-adjoint extension. According to [37, Theorem 2.1], there are three different regimes of the Dirac–Coulomb operator on $\mathbb{R}^3$. If $0 < \nu \leq \frac{\sqrt{3}}{2}$, the operator is essentially self-adjoint, with finite potential and kinetic energies. In the interval $\frac{\sqrt{3}}{2} < \nu < 1$, besides other self-adjoint extensions, distinguished self-adjoint extensions can be singled out (see [38, 52, 53]) for which either the potential energy (see [60–62]) or the kinetic energy (see [48, 49]) are finite. All these extensions were proved to be equivalent by Klaus and Wüst in [39] and coincide with the distinguished extension of [27]. In the critical case $\nu = 1$,

Hardy–Dirac inequalities lead to a distinguished self-adjoint extension with finite total energy: see [4, 27, 28]. For $\nu > 1$, the operator enjoys a family of self-adjoint extensions, but standard finiteness of the energies fail according to [8, 58, 63].

On $\mathbb{R}^2$, the self-adjointness properties of the operator H_a (without Coulomb potential) have been studied for instance in [9, 31, 55]. Without magnetic field, self-adjoint extensions preserving gaps have been studied in [7, 41] and it is shown there that the decomposition used in the variational approach of [51, Theorem 1.1] enters in the framework of Krein's criterion (see [51, Remark 1.3]). All self-adjoint extensions of the Dirac–Coulomb operator for $\nu < 1$ are classified in [33] with corresponding eigenvalues obtained in [32]. As noted in [51], methods similar to those of [28] are applied in [14] to a two-body Dirac operator with Coulomb interaction (without spectral gap). The characterization of the domain of the extension of the operator is of course a very natural question. This is studied in detail in [26], including the two-dimensional case but without magnetic field (also see references therein). Some of the results in this paper are natural extensions of the considerations in [26] for the case without magnetic field.

This paper is organized as follows. Section 2 is devoted to a direct proof of the homogeneous Hardy-like inequalities of Lemma 1.5 and Corollary 1.6. In Section 3, we prove some inhomogeneous Hardy-like inequalities which allow us to study a variational problem leading to the proof of Proposition 1.4 and Theorem 1.1. This takes us to the identification of the *critical magnetic field* under which all our results hold true. This also proves the Hardy-type inequality (1.11) and its consequences (also see Corollary 3.2 and Appendix A.2). Section 4 is devoted to the self-adjointness properties of $H_{\nu,a}$ and to the identification of the *ground state*. An explicit expression is found, which is explained by the role played by Laguerre polynomials in the computation of the spectrum of $H_{\nu,a}$ (see Appendix A.1). The non-relativistic limit is discussed in Section 5. The inequalities corresponding to the positron case are collected in Appendix A.

2 Homogeneous Hardy-like inequalities

In this section, we prove the inequalities of Lemma 1.5 and Corollary 1.6, which are independent of our other results, but rely on the completion of squares and on (1.6). These proofs are a useful introduction to the main results of this paper.

Proof of Lemma 1.5. Let us start with the proof of inequality (1.14). For any $\ell \in \mathbb{Z}$, if $\eta(\rho,\theta) = e^{i\ell\theta}\phi(\rho)$, then

$$\begin{aligned} 4\int_{\mathbb{R}^2} |x|^{2\beta}|\partial_z\eta|^2\,dx &= \int_{\mathbb{R}^2} \rho^{2\beta}\left|\left(\partial_\rho - i\frac{\partial_\theta}{\rho}\right)e^{i\ell\theta}\phi\right|^2 dx \\ &= 2\pi\int_0^{+\infty} \rho^{2\beta+1}\left|\left(\partial_\rho + \frac{\ell}{\rho}\right)\phi\right|^2 d\rho \end{aligned}$$

and

$$\int_{\mathbb{R}^2} |x|^{2\beta-2}|\eta|^2\,dx = 2\pi \int_0^{+\infty} \rho^{2\beta-1}|\phi|^2\,d\rho.$$

With an expansion of the square and an integration by parts with respect to ρ, we obtain

$$\begin{aligned}\int_0^{+\infty} \rho^{2\beta+1}\left|\partial_\rho\phi + \frac{\kappa}{\rho}\phi\right|^2 d\rho &= \int_0^{+\infty} \rho^{2\beta+1}|\partial_\rho\phi|^2\,d\rho \\ &\quad + \kappa(\kappa-2\beta)\int_0^{+\infty} \rho^{2\beta-1}|\phi|^2\,d\rho\end{aligned}$$

for any $\kappa \in \mathbb{R}$. Applied with $\kappa = \beta$ and with $\kappa = \ell$, this proves that

$$\int_0^{+\infty} \rho^{2\beta+1}|\partial_\rho\phi|^2\,d\rho \geq \beta^2 \int_0^{+\infty} \rho^{2\beta-1}|\phi|^2\,d\rho,$$

$$\begin{aligned}\int_0^{+\infty} \rho^{2\beta+1}\left|\partial_\rho\phi + \frac{\ell}{\rho}\phi\right|^2 d\rho &= \int_0^{+\infty} \rho^{2\beta+1}|\partial_\rho\phi|^2\,d\rho \\ &\quad + \ell(\ell-2\beta)\int_0^{+\infty} \rho^{2\beta-1}|\phi|^2\,d\rho \\ &\geq (\beta-\ell)^2 \int_0^{+\infty} \rho^{2\beta-1}|\phi|^2\,d\rho.\end{aligned}$$

Altogether, we have

$$\int_{\mathbb{R}^2} |x|^{2\beta}|\partial_z\eta|^2\,dx \geq \frac{1}{4}(\beta-\ell)^2 \int_{\mathbb{R}^2} |x|^{2\beta-2}|\eta|^2\,dx.$$

Inequality (1.14) for a general $\eta \in C_c^\infty(\mathbb{R}^2 \setminus \{0\}, \mathbb{C})$ is then a consequence of a decomposition in Fourier modes. The proof of (1.13) is exactly the same, up to the sign changes $\beta \mapsto -\beta$ and $\ell \mapsto -\ell$. ∎

Proof of Corollary 1.6. Using (1.6), inequality (1.15) is equivalent to

$$\begin{aligned}&4\int_{\mathbb{R}^2} \left(|x|^{\zeta-2a}|\partial_{\bar z}\eta_1|^2 + |x|^{\zeta+2a}|\partial_z\eta_2|^2\right) dx \\ &\geq C_{a,\zeta}\int_{\mathbb{R}^2} \left(|x|^{\zeta-2-2a}|\eta_1|^2 + |x|^{\zeta-2+2a}|\eta_2|^2\right) dx\end{aligned}$$

and the result follows by applying (1.13) and (1.14) to η_1 and η_2 with respectively $\eta = a - \frac{\zeta}{2}$ and $\eta = a + \frac{\zeta}{2}$. ∎

3 The minimization problem

The goal of this section is to prove Proposition 1.4 and Theorem 1.1. It is centered on the variational problem (1.9) and its consequences on the definition of the *critical magnetic field*.

Lemma 3.1. *Assume that $\nu \in (0, \frac{1}{2}]$ and $a \in [0, \mathsf{c}(\nu)]$. Then*

$$\mu = \frac{\mathsf{c}(a) - \sqrt{\mathsf{c}(a)^2 - \nu^2}}{\nu} \tag{3.1}$$

is the smallest value of $\mu \in \mathbb{R}$ such that the inequality

$$\int_0^{+\infty} \frac{|\phi'|^2}{\nu + \rho} \rho^{2-2a} \, d\rho + \mu^2 \int_0^{+\infty} |\phi|^2 \rho^{1-2a} \, d\rho \geq \nu \int_0^{+\infty} |\phi|^2 \rho^{-2a} \, d\rho$$

holds for any function $\phi \in C^1((0, +\infty), \mathbb{C})$. Moreover, the equality case with μ given by (3.1) *is achieved by*

$$\phi_\star(\rho) = \rho^{-\mu\nu} e^{-\mu\rho}, \quad \rho \in (0, +\infty).$$

Proof. An expansion of the square and an integration by parts show that

$$\begin{aligned}
0 &\leq \int_0^{+\infty} \frac{|\rho\phi' + \mu(\nu + \rho)\phi|^2}{\nu + \rho} \rho^{-2a} \, d\rho \\
&= \int_0^{+\infty} \frac{\rho^2 |\phi'|^2}{\nu + \rho} \rho^{-2a} \, d\rho + \mu \int_0^{+\infty} (|\phi|^2)' \rho^{1-2a} \, d\rho \\
&\quad + \mu^2 \int_0^{+\infty} (\nu + \rho) |\phi|^2 \rho^{-2a} \, d\rho \\
&= \int_0^{+\infty} \frac{|\phi'|^2}{\nu + \rho} \rho^{2-2a} \, d\rho + \mu^2 \int_0^{+\infty} |\phi|^2 \rho^{1-2a} \, d\rho \\
&\quad + (\nu\mu^2 - (1 - 2a)\mu) \int_0^{+\infty} |\phi|^2 \rho^{-2a} \, d\rho.
\end{aligned}$$

The conclusion holds after observing that $\nu\mu^2 - (1 - 2a)\mu = -\nu$ and solving, in the equality case, the equation

$$\rho\phi' + \mu(\nu + \rho)\phi = 0, \quad \rho \in (0, +\infty).$$

The lemma is proved. ∎

Next we apply Lemma 3.1 to the variational problem (1.9).

Proof of Proposition 1.4. For any function ϕ smooth enough and any $\ell \in \mathbb{Z} \setminus \{0\}$, we have

$$\int_0^{+\infty} \frac{|\rho\phi' - \ell\phi|^2}{(1 + \lambda)\rho + \nu} \frac{d\rho}{\rho^{2a}} \geq \int_0^{+\infty} \frac{\rho^2 |\phi'|^2}{(1 + \lambda)\rho + \nu} \frac{d\rho}{\rho^{2a}}$$

because an integration by parts shows that

$$\begin{aligned}
&\int_0^{+\infty} \frac{\ell^2 |\phi|^2 - 2\rho\ell\phi\phi'}{(1 + \lambda)\rho + \nu} \frac{d\rho}{\rho^{2a}} \\
&= \int_0^{+\infty} \frac{\ell(\ell - 2a)(1 + \lambda)\rho + \ell(\ell + 1 - 2a)\nu}{((1 + \lambda)\rho + \nu)^2} |\phi|^2 \frac{d\rho}{\rho^{2a}}
\end{aligned}$$

and $\ell(\ell-2a)\geq 0$ and $\ell(\ell+1-2a)\geq 0$. Using polar coordinates and a decomposition in Fourier modes,

$$\eta(x)=\sum_{\ell\in\mathbb{Z}}\eta_\ell(\rho)e^{i\ell\theta}, \tag{3.2}$$

we obtain

$$J(\eta,\nu,a,\lambda)=2\pi\sum_{\ell\in\mathbb{Z}}\int_0^{+\infty}\left(\frac{|\rho\eta_\ell'-\ell\eta_\ell|^2}{(1+\lambda)\rho+\nu}+((1-\lambda)\rho-\nu)|\eta_\ell|^2\right)\frac{d\rho}{\rho^{2a}}. \tag{3.3}$$

As a consequence, in order to minimize $\lambda_\star(\eta,\nu,a)$, it is enough to consider only the $\ell=0$ mode, i.e., minimize on radial functions.

Let us consider the change of variables $\rho\mapsto(1+\lambda)\rho$ and write

$$\eta(\rho)=\phi((1+\lambda)\rho).$$

We observe that the largest value of $\lambda>-1$ for which $J(\eta,\nu,a,\lambda)\geq 0$ for any $\eta\in\mathcal{G}_{\nu,a}$ is the largest value of $\lambda>-1$ for which the inequality

$$\begin{aligned}\int_0^{+\infty}\frac{|\phi'|^2}{\nu+\rho}\rho^{2-2a}\,d\rho+\frac{1-\lambda}{1+\lambda}\int_0^{+\infty}|\phi|^2\rho^{1-2a}\,d\rho\\ -\nu\int_0^{+\infty}|\phi|^2\rho^{-2a}\,d\rho\geq 0\end{aligned} \tag{3.4}$$

holds for any ϕ. With the notation of Lemma 3.1,

$$\frac{1-\lambda}{1+\lambda}=\mu^2\iff\lambda=\frac{1-\mu^2}{1+\mu^2},$$

that is,

$$\lambda=\frac{1}{\mathsf{c}(a)}\sqrt{\mathsf{c}(a)^2-\nu^2}=\lambda_{\nu,a}$$

according to (1.2). Equality in (3.4) is obtained, up to multiplication by a constant, with $\eta_\star(\rho)=\phi_\star((1+\lambda)\rho)$. ∎

The above proof deserves an observation. In the proof of Lemma 3.1, we solve $\nu\mu^2-(1-2a)\mu+\nu=0$. It turns out that for any $\nu<\mathsf{c}(a)$, this equation has two roots, $\mu_\pm==\frac{1}{\nu}(-\mathsf{c}(a)\pm\sqrt{\mathsf{c}(a)^2-\nu^2})$, and the inequality is true for any $\mu\in[\mu_-,\mu_+]$. In the proof of Proposition 1.4, we solve $\frac{1-\lambda}{1+\lambda}=\mu^2$ and we look for the largest value of λ for which $J(\eta,\nu,a,\lambda)\geq 0$ for any η, which is the reason why we pick the value of λ corresponding to $\mu=\mu_+$. See Appendix A.3 for similar considerations in the positron case.

Using the Aharonov–Casher transformation

$$\varphi(x)=|x|^{-a}\eta(x),\quad x\in\mathbb{R}^2. \tag{3.5}$$

which transforms a function $\eta\in\mathcal{G}_{\nu,a}$ into a function $\varphi\in\mathcal{F}_{\nu,a}$, we can rephrase the result of Proposition 1.4 as follows.

Corollary 3.2. *Under the conditions* $\nu \in (0, \frac{1}{2}]$ *and* $a \in [0, \mathsf{c}(\nu)]$, *the largest value of* $\lambda > -1$ *for which* $\mathcal{Q}_{\nu,a,\lambda}$ *as defined in* (1.3) *is a nonnegative quadratic form on* $\mathcal{F}_{\nu,a}$ *is* $\lambda_{\nu,a}$. *Additionally, if* $a < \mathsf{c}(\nu)$, *the equality case in* $\mathcal{Q}_{\nu,a,\lambda_{\nu,a}}(\varphi) \geq 0$ *is achieved if and only if* $\varphi = \varphi_\star$ *up to a multiplicative constant, where*

$$\varphi_\star(\rho, \theta) = \rho^{\sqrt{\mathsf{c}(a)^2-\nu^2}-\frac{1}{2}} e^{-\frac{\nu}{\mathsf{c}(a)}\rho}, \quad (\rho, \theta) \in \mathbb{R}^+ \times [0, 2\pi). \tag{3.6}$$

However, we read from (1.2) that $\lim_{a\to\mathsf{c}(\nu)_-} \lambda_{\nu,a} = 0 > -1$. As we shall see next, as soon as $a \in (\mathsf{c}(\nu), \frac{1}{2})$ for some $\nu \in (0, \frac{1}{2}]$, there is no $\lambda \in (-1, 1)$ such that the Hardy-like inequality (1.1) holds true anymore.

Proposition 3.3. *Let* $\nu \in (0, \frac{1}{2}]$ *and* $a \in (\mathsf{c}(\nu), \frac{1}{2})$. *Then*

$$\inf_{\eta \in \mathcal{G}_{\nu,a}} \lambda_\star(\eta, \nu, a) \leq -1$$

and for any $\lambda \in (-1, 1)$, *there is some* $\varphi \in \varphi \in L^2(\mathbb{R}^2, \mathbb{C}) \cap H^1_{\mathrm{loc}}(\mathbb{R}^2 \setminus \{0\}, \mathbb{C})$ *such that*

$$\sqrt{\frac{|x|}{1+|x|}} D_a^* \varphi \in L^2(\mathbb{R}^2, \mathbb{C})$$

and $\mathcal{Q}_{\nu,a,\lambda}(\varphi) < 0$.

Proof. Let $\epsilon > 0$, $\mu \in [-1, 1)$ and, for any $\rho > 0$ and consider a function $\eta_\epsilon \in \mathcal{G}_{\nu,a}$ such that

$$\eta_\epsilon(x) = |x|^{a-\frac{1}{2}} e^{-|x|} \quad \text{if } |x| \geq \epsilon$$

and

$$|\nabla \eta_\epsilon(x)| \leq \epsilon^{a-\frac{3}{2}} \quad \text{if } |x| \leq \epsilon.$$

A computation shows the existence of two positive constants C_1 and C_2 such that

$$J(\eta_\epsilon, \nu, a, \mu) \leq J(\eta_\epsilon, \nu, a, -1) \leq C_1 |\log \epsilon| \left(\frac{\mathsf{c}(a)^2}{\nu} - \nu \right) + C_2 < 0$$

for ϵ small enough, so that $\lambda_\star(\eta_\epsilon, \nu, a) \leq -1$. Using the transformation (3.5), this also proves that $\mathcal{Q}_{\nu,a,\lambda}$ achieves a negative value. ∎

Proof of Theorem 1.1. We know from Corollary 3.2 that $\mathsf{a}(\nu) \geq \mathsf{c}(\nu)$ and from Proposition 3.3 that $\mathsf{a}(\nu) \leq \mathsf{c}(\nu)$. Since $\lambda \mapsto \mathcal{Q}_{\nu,a,\lambda}(\varphi)$ as defined in (1.3) is monotone non-increasing, inequality (1.12) holds true for any $\lambda \in [-1, \lambda_{\nu,a})]$. ∎

4 The 2D magnetic Dirac–Coulomb operator with an Aharonov–Bohm magnetic field

This section is devoted to the self-adjointness of the operator $H_{\nu,a}$ with domain $\mathcal{D}_{\nu,a}$ when $\nu \in (0, \frac{1}{2}]$ and $a \in (0, \mathsf{c}(\nu)]$, that is, Theorem 1.2. In the same range of parameters, we also identify $\lambda_{\nu,a}$ as the *ground-state* of $H_{\nu,a}$ (Theorem 1.3).

Proof of Theorem 1.2. We first deal with abstract results when $a \le \mathsf{a}(\nu)$ before characterizing $\mathcal{F}_{\nu,a}$ in the subcritical range $a < \mathsf{a}(\nu)$.

Critical and subcritical cases, $a \le \mathsf{a}(\nu)$: Domain and self-adjointness. We follow the method of [28] for dealing with non-magnetic three-dimensional Dirac–Coulomb operators in [27, 51]. This method applies almost without change and we provide only a sketch of the proof.

By Theorem 1.1, the quadratic form $\mathcal{Q}_{\nu,a,\lambda}$ defined by (1.3) is nonnegative on $C_c^\infty(\mathbb{R}^2 \setminus \{0\}, \mathbb{C})$ for any $\lambda \in (-1, \lambda_{\nu,a}]$. Let us define the norms $\|\cdot\|$ and $\|\cdot\|_\lambda$ by

$$\|\varphi\|^2 := \|\varphi\|^2_{L^2(\mathbb{R}^2,\mathbb{C})} + \left\| \sqrt{\frac{|x|}{1+|x|}}\, D_a^* \varphi \right\|^2_{L^2(\mathbb{R}^2,\mathbb{C})}$$

and

$$\|\varphi\|^2_\lambda := \|\varphi\|^2_{L^2(\mathbb{R}^2,\mathbb{C})} + \left\| \frac{D_a^*\varphi}{\sqrt{1+\lambda+\frac{\nu}{|x|}}} \right\|^2_{L^2(\mathbb{R}^2,\mathbb{C})}.$$

For the same reasons as in [18, Lemma 2.1] or [51, Lemma 5], the norms $\|\cdot\|$ and $\|\cdot\|_\lambda$ are equivalent for any $\lambda \in (-1,1)$ and the operator $\varphi \mapsto D_a^*\varphi/(1+\lambda+\nu|x|^{-1})$ on $C_c^\infty(\mathbb{R}^2 \setminus \{0\}, \mathbb{C})$ is closable with respect to $\|\cdot\|_\lambda$, with a domain that does not depend on λ. By arguing as in [18, Lemma 2.1] and [51, Lemma 7], there is a constant $\kappa_\lambda \ge 1$ such that $\mathcal{Q}_{\nu,a,\lambda}(\varphi) + \kappa_\lambda \|\varphi\|^2_\lambda$ is equivalent to $\mathcal{Q}_{\nu,a,0}(\varphi) + \|\varphi\|_0^2$ for all $\lambda \in (-1,1)$. These quadratic forms are nonnegative and closable with form domain $\mathcal{F}_{\nu,a} \subset L^2(\mathbb{R}^2,\mathbb{C})$ as defined in Theorem 1.2. Moreover, $\mathcal{F}_{\nu,a}$ does not depend on λ because of the equivalence of the quadratic forms. On $\mathcal{D}_{\nu,a}$, the operator $H_{\nu,a}$ such that

$$H_{\nu,a}\psi = \begin{pmatrix} D_a\chi + \left(1-\frac{\nu}{|x|}\right)\varphi \\ D_a^*\varphi - \left(1+\frac{\nu}{|x|}\right)\chi \end{pmatrix}, \quad \psi = \begin{pmatrix} \varphi \\ \chi \end{pmatrix} \in \mathcal{D}_{\nu,a},$$

is self-adjoint, for the same reasons as in [27], because for all $\lambda \in (-1, \lambda_{\nu,a})$, the operator $H_{\nu,a} - \lambda$ is symmetric and it is a bijection from $\mathcal{D}_{\nu,a}$ onto $L^2(\mathbb{R}^2, \mathbb{C}^2)$.

Subcritical range $a < \mathsf{a}(\nu)$. The space $\mathcal{F}_{\nu,a}$ endowed with the norm $\|\cdot\|$ is given by (1.4). Let us prove it. With $C = \min\{1+\lambda, \nu\}$, we have

$$\begin{aligned} \mathcal{Q}_{\nu,a,\lambda}(\varphi) &\le \frac{1}{C}\int_{\mathbb{R}^2} \frac{|x|}{1+|x|} |D_a^*\varphi|^2\,dx + 2\int_{\mathbb{R}^2} |\varphi|^2\,dx \\ &\le \max\{C^{-1}, 2\}\|\varphi\|^2. \end{aligned}$$

for any $\varphi \in C_c^\infty(\mathbb{R}^2 \setminus \{0\}, \mathbb{C})$. The reverse inequality can be proved as follows. For any $\lambda \in (-1, \lambda_{\nu,a})$, let

$$t = \frac{\mathsf{c}(a)}{\nu}\sqrt{1-\lambda^2}$$

and notice that $t > 1$. This choice of t is made such that $\lambda = \lambda_{t\nu,a}$. Let us choose some $\varphi \in C_c^\infty(\mathbb{R}^2 \setminus \{0\}, \mathbb{C})$ and consider $\phi(x) = \varphi(t^{-1}x)$. A simple change of variables shows that

$$\int_{\mathbb{R}^2}\left(\frac{1}{t^2}\frac{|D_a^*\varphi|^2}{1+\lambda+\frac{\nu}{|x|}}-\frac{\nu}{|x|}|\varphi|^2\right)dx=\frac{1}{t^2}\int_{\mathbb{R}^2}\left(\frac{|D_a^*\phi|^2}{1+\lambda+\frac{t\nu}{|x|}}-\frac{t\nu}{|x|}|\phi|^2\right)dx$$
$$\geq\frac{\lambda-1}{t^2}\int_{\mathbb{R}^2}|\phi|^2\,dx=(\lambda-1)\int_{\mathbb{R}^2}|\varphi|^2\,dx,$$

where the inequality is obtained by writing $\mathcal{Q}_{t\nu,a,\lambda}(\phi)\geq 0$. As a consequence, we have

$$\mathcal{Q}_{\nu,a,\lambda}(\varphi)\geq\left(1-\frac{1}{t^2}\right)\frac{1}{\max\{1+\lambda,\nu\}}\int_{\mathbb{R}^2}\frac{|x|}{1+|x|}|D_a^*\varphi|^2\,dx,$$

which concludes the proof. ∎

Proof of Theorem 1.3. As noted in the introduction, if $\lambda\in(-1,1)$ is an eigenvalue of $H_{\nu,a}$, its upper component is, after the Aharonov–Casher transformation (3.5), a critical point of $\eta\mapsto J(\eta,\nu,a,\lambda)$, so that the ground state energy, i.e., the lowest eigenvalue of $H_{\nu,a}$ in $(-1,1)$, is larger or equal than $\lambda_{\nu,a}$. We have equality if $\eta_1=\eta_\star$ determines an eigenfunction of $H_{\nu,a}$ through (1.6) and (1.8). This is straightforward in the subcritical range as $\eta_\star\in\mathcal{G}_{\nu,a}$ if $a<\mathsf{a}(\nu)$ by Theorem 1.2.

The critical case $a=\mathsf{a}(\nu)$ is more subtle as we have no explicit characterization of $\mathcal{F}_{\nu,a}$ or, equivalently, $\mathcal{G}_{\nu,a}$. We have indeed to prove that $\eta_\star\in\mathcal{G}_{\nu,a}$ if $a=\mathsf{a}(\nu)$. For all $\epsilon\in(0,1]$, let us define the truncation function

$$\theta_\epsilon:=\begin{cases}0 & \text{if } 0\leq\rho\leq\frac{\epsilon}{2},\\ \frac{1}{2}\left(1+\sin\left(\frac{2\pi}{\epsilon}\rho-\frac{3\pi}{2}\right)\right) & \text{if } \frac{\epsilon}{2}<\rho\leq\epsilon,\\ 1 & \text{if } \epsilon\leq\rho\leq\frac{1}{\epsilon},\\ \frac{1}{2}\left(1+\sin\left(\rho+\frac{\pi}{2}-\frac{1}{\epsilon}\right)\right) & \text{if } \frac{1}{\epsilon}\leq\rho\leq\frac{1}{\epsilon}+\pi,\\ 0 & \text{if } \rho\geq\frac{1}{\epsilon}+\pi.\end{cases}$$

With $\varphi_\star$ defined by (3.6), the functions $\varphi_\epsilon:=\varphi_\star\theta_\epsilon$ are equal to 0 in a neighborhood of the origin and converge almost everywhere to $\varphi_\star$ as $\epsilon\to 0_+$, with

$$\limsup_{\epsilon\to 0_+}\mathcal{Q}_{\nu,a,0}(\varphi_\epsilon)<+\infty.$$

But these functions are not in $C^\infty(\mathbb{R}^2\setminus\{0\},\mathbb{C})$. To end the proof we regularize them using a convolution product. ∎

An eigenfunction associated with $\lambda_{\nu,a}$ is obtained as a sub-product of our method.

Proposition 4.1. *Let $\nu\in(0,\frac{1}{2}]$. For any $a\in(0,\mathsf{c}(\nu)]$, the eigenspace of $H_{\nu,a}$ associated with $\lambda_{\nu,a}$ is generated by the spinor*

$$\psi_{\nu,a}(\rho,\theta)=\begin{pmatrix}1\\ i e^{i\theta}\sqrt{\frac{1-\lambda_{\nu,a}}{1+\lambda_{\nu,a}}}\end{pmatrix}\rho^{\sqrt{\mathsf{c}(a)^2-\nu^2}-\frac{1}{2}}e^{-\frac{\nu}{\mathsf{c}(a)}\rho},\quad(\rho,\theta)\in\mathbb{R}^+\times[0,2\pi).$$

Proof. The result follows from $\varphi = \varphi_\star$ as in Corollary 3.2 for the upper component and from (1.5) for the lower component. ∎

Let us conclude this section by some comments. In Proposition 1.4 and in Corollary 3.2, we have $J(\eta_\star, \nu, a, \lambda_{\nu,a}) = 0$ and $\mathcal{Q}_{\nu,a,\lambda_{\nu,a}}(\varphi_\star) = 0$ if $a = \mathsf{c}(\nu)$, but we have to pay attention to the fact that the various terms in the integrals are not all individually integrable. In the proof of Theorem 1.3, this is reflected by the fact that we have to use a truncation argument. The characterization of the space $\mathcal{F}_{\nu,a}$ in the critical case $a = \mathsf{a}(\nu)$ is more technical than in the subcritical regime and we will not do it here. For similar computations without magnetic field in dimensions 2 and 3, see [26, Section 1.5 and Appendix A.3].

5 Non-relativistic limit

In this section, we discuss the non-relativistic limit of the ground state spinor and of the ground state energy of the magnetic Dirac–Coulomb operator on $\mathbb{R}^2$ with the Aharonov–Bohm magnetic field $\mathbf{A}_a$. Let us introduce the *speed of light* c in the operator and consider

$$H^c_{\nu,a} := c^2\sigma_3 - i\,c\,\sigma\cdot\nabla_a - \frac{\nu}{|x|}\,\mathrm{Id}\,.$$

Up to this point, we considered atomic units and took $c = 1$, $H_{\nu,a} = H^1_{\nu,a}$. Here we consider the limit as $c \to +\infty$.

If ψ_c is an eigenfunction of $H^c_{\nu,a}$ with eigenvalue λ_c, that is, $H^c_{\nu,a}\psi_c = \lambda_c\psi_c$, then $\Psi_c(x) := \psi_c(\frac{x}{c})$ solves $H_{\frac{\nu}{c},a}\Psi_c = \frac{\lambda_c}{c^2}\Psi_c$. As a consequence of Proposition 4.1,

$$\psi_c = \begin{pmatrix}\varphi_c\\ \chi_c\end{pmatrix}$$

where

$$\begin{cases}\varphi_c(\rho,\theta) = \rho^{\sqrt{\mathsf{c}(a)^2 - \frac{\nu^2}{c^2}} - \frac12}\, e^{-\frac{\nu}{\mathsf{c}(a)}\rho}\,,\\[2mm] \chi_c(\rho,\theta) = i\,e^{i\theta}\sqrt{\dfrac{c^2-\lambda_c}{c^2+\lambda_c}}\,\varphi_c(\rho,\theta)\end{cases}\qquad (\rho,\theta)\in\mathbb{R}^+\times[0,2\pi),$$

with

$$\lambda_c = c^2\sqrt{1 - \frac{\nu^2}{\mathsf{c}(a)^2c^2}}$$

so that, by passing to the limit as $c \to +\infty$, we obtain

$$\lim_{c\to+\infty}(\lambda_c - c^2) = -\frac{\nu^2}{2\mathsf{c}(a)^2}$$

and

$$\varphi_c \to \rho^{-a} e^{-\frac{\nu}{\mathrm{c}(a)}\rho}, \quad \chi_c \to 0,$$

in $H^1_{\mathrm{loc}}(\mathbb{R}^2 \setminus \{0\}, \mathbb{C})$.

We recall that $\mathrm{c}(a) = \frac{1}{2} - a$. The eigenvalue problem written as a system is

$$c D_a \chi_c + \left(c^2 - \frac{\nu}{\rho}\right)\varphi_c = \lambda_c \varphi_c, \quad c D_a^* \varphi_c - \left(c^2 + \frac{\nu}{\rho}\right)\chi_c = \lambda_c \chi_c.$$

After eliminating the lower component using

$$\chi_c = \frac{c D_a^* \varphi_c}{\lambda_c + c^2 + \frac{\nu}{\rho}},$$

we obtain for the upper component the equation

$$D_a\left(\frac{c^2 D_a^* \varphi_c}{\lambda_c + c^2 + \frac{\nu}{\rho}}\right) - \frac{\nu}{\rho}\varphi_c = (\lambda_c - c^2)\varphi_c$$

and notice that this is consistent with the fact that the limiting solution solves the equation

$$D_a D_a^* \varphi - \frac{\nu}{\rho}\varphi = -\frac{\nu^2}{2\mathrm{c}(a)^2}\varphi$$

in the sense of distributions. On $C^0(\mathbb{R}^2, \mathbb{C}) \cap C_c^2(\mathbb{R}^2 \setminus \{0\}, \mathbb{C})$, an elementary computation shows that $D_a D_a^* = -\nabla_a^2 - \mathbf{B}$, where the magnetic field is $\mathbf{B} = 2\pi a \delta_0$, corresponding to a Dirac mass at the origin. This operator is similar to the Pauli operator for measure valued magnetic fields studied by Erdoes and Vougalter in [24] using the Aharonov–Casher transformation (3.5). If we define the quadratic form

$$q(\varphi, \varphi') := \int_{\mathbb{R}^2} \partial_{\bar z}(|x|^a \varphi)\overline{\partial_{\bar z}(|x|^a \varphi')}|x|^{-2a}\,dx,$$

we can follow [24, Theorem 2.5] and define $D_a D_a^*$ as the Friedrichs extension on $L^2(\mathbb{R}^2, \mathbb{C})$ of the unique self-adjoint operator associated with q, with domain

$$\left\{\varphi \in L^2(\mathbb{R}^2, \mathbb{C}) : q(\varphi, \varphi) < +\infty,\ q(\varphi, \cdot) \in L^2(\mathbb{R}^2, \mathbb{C})'\right\}.$$

A Appendix

A.1 The ground state and Laguerre polynomials

Based on (3.2) and (3.3), we can provide an alternative computation of the optimal function $\eta_\star$ in Proposition 1.4.

As a consequence of the properties of the Laguerre polynomials (see [45]), for any $\ell \in \mathbb{Z}$, solutions of

$$-\left(\frac{\rho\phi' - \ell\phi}{(1+\lambda)\rho + \nu}\right)' - (\ell + 1 - 2a)\frac{\phi' - \frac{\ell}{\rho}\phi}{(1+\lambda)\rho + \nu} + \left(1 - \lambda - \frac{\nu}{\rho}\right)\phi = 0 \quad \text{(A.1)}$$

are generated by the functions $\phi(\rho) = \rho^A e^{-B\rho}$ with either

$$A = a - \frac{1}{2} - \frac{\mathsf{c}_\ell(a)}{|\mathsf{c}_\ell(a)|}\sqrt{\mathsf{c}_\ell(a)^2 - \nu^2}, \quad B = \frac{\nu}{\mathsf{c}_\ell(a)}, \quad \lambda = -\frac{\sqrt{\mathsf{c}_\ell(a)^2 - \nu^2}}{|\mathsf{c}_\ell(a)|},$$

or

$$A = a - \frac{1}{2} + \frac{\mathsf{c}_\ell(a)}{|\mathsf{c}_\ell(a)|}\sqrt{\mathsf{c}_\ell(a)^2 - \nu^2}, \quad B = \frac{\nu}{\mathsf{c}_\ell(a)}, \quad \lambda = \frac{\sqrt{\mathsf{c}_\ell(a)^2 - \nu^2}}{|\mathsf{c}_\ell(a)|},$$

where $\mathsf{c}_\ell(a) := \frac{1}{2} + \ell - a$. However, the integrability of $\rho \mapsto \rho^{2-2a}|\phi'(\rho)|^2$ in a positive neighborhood of $\rho = 0$ selects the second one, with $\ell \geq 0$. For any $\ell \in \mathbb{N}$, let $\phi_\ell(\rho) := \kappa_\ell \rho^{A_\ell(a)} e^{-B_\ell(a)\rho}$ with

$$\begin{aligned} A_\ell(a) &= a - \frac{1}{2} + \frac{\mathsf{c}_\ell(a)}{|\mathsf{c}_\ell(a)|}\sqrt{\mathsf{c}_\ell(a)^2 - \nu^2}, \\ B_\ell(a) &= \frac{\nu}{\mathsf{c}_\ell(a)}, \\ \lambda_\ell(a) &= \frac{\sqrt{\mathsf{c}_\ell(a)^2 - \nu^2}}{|\mathsf{c}_\ell(a)|}. \end{aligned}$$

With this choice, we find that $J(\eta_\star, \nu, a, \lambda_0(a)) \geq 0$ with equality if and only if $\kappa_\ell = 0$ for any $\ell \geq 1$. On the other hand, $\eta_\star$ is a critical point of J, so that ϕ_ℓ solves (A.1) with $\lambda = \lambda_\star(\eta_\star, \nu, a)$. Altogether we conclude that $\lambda_\star(\eta_\star, \nu, a) = \lambda_0(a)$.

A.2 Special cases of Hardy-type inequalities

Here we list some special cases of Hardy-type inequalities related with the Dirac–Coulomb operator, with or without magnetic fields, which are of interest by themselves. This list of inequalities complements the inequalities of Section 1 (after the statement of Proposition 1.4).

For any $\eta \in C^\infty_{\mathrm{c}}(\mathbb{R}^2 \setminus \{0\}, \mathbb{C})$, inequality (1.11) becomes

$$\int_{\mathbb{R}^2} \left(\frac{2|x|}{2|x|+1} \left|(\partial_1 + i\partial_2)\eta\right|^2 + |\eta|^2 \right) dx \geq \frac{1}{2} \int_{\mathbb{R}^2} \frac{|\eta|^2}{|x|}\, dx$$

for $\nu = \frac{1}{2}$ and $a = 0$, and

$$\int_{\mathbb{R}^2} \left(\frac{|x|}{|x|+\nu} \left|(\partial_1 + i\partial_2)\eta\right|^2 + |\eta|^2 \right) |x|^{2\nu-1}\, dx \geq \nu \int_{\mathbb{R}^2} |\eta|^2 |x|^{2\nu-2}\, dx$$

for $\nu \in (0, \frac{1}{2})$, $a = \mathsf{c}(\nu) > 0$ while, in that case, a scaling also shows that

$$\frac{1}{\nu} \int_{\mathbb{R}^2} |x|^{2\nu} \left|(\partial_1 + i\partial_2)\eta\right|^2 |x|^{2\nu}\, dx \geq \nu \int_{\mathbb{R}^2} |\eta|^2 |x|^{2\nu-2}\, dx, \quad \eta \in C^\infty_{\mathrm{c}}(\mathbb{R}^2 \setminus \{0\}, \mathbb{C}).$$

In the case $a = \mathsf{c}(\nu)$ and $\nu \in (0, \frac{1}{2})$, (1.12) becomes

$$\int_{\mathbb{R}^2} \left(\frac{|x|}{|x|+\nu} |D^*_{\mathsf{c}(\nu)}\varphi|^2 + |\varphi|^2 \right) dx \geq \nu \int_{\mathbb{R}^2} \frac{|\varphi|^2}{|x|^2}\, dx, \quad \varphi \in C^\infty_{\mathrm{c}}(\mathbb{R}^2 \setminus \{0\}, \mathbb{C}),$$

and it is interesting to relate this last inequality with the homogeneous case of (1.15), which can be written as

$$\int_{\mathbb{R}^2} \frac{|x|}{\nu} |(\sigma \cdot \nabla_{\mathrm{c}(\nu)})\psi|^2 \, dx \geq \nu \int_{\mathbb{R}^2} \frac{|\psi|^2}{|x|} \, dx, \quad \psi \in C_{\mathrm{c}}^\infty(\mathbb{R}^2 \setminus \{0\}, \mathbb{C}^2).$$

A.3 The positron case

In the positron case with positively charged singularity, the eigenvalue problem

$$H_{a,-\nu}\psi = \lambda\psi$$

is transformed using (1.6) into the system

$$\begin{aligned} \eta_1 - 2i\rho^{2a}\partial_z\eta_2 + \frac{\nu}{\rho}\eta_1 &= \lambda\eta_1, \\ -2i\rho^{-2a}\partial_{\bar{z}}\eta_1 + \frac{\nu}{\rho}\eta_2 - \eta_2 &= \lambda\eta_2. \end{aligned}$$

Using the first equation, we can eliminate the upper component

$$\eta_1 = -2i \frac{\rho^{2a}\partial_z\eta_2}{\lambda - 1 - \frac{\nu}{\rho}}$$

and obtain the equation

$$-4\rho^{-2a}\partial_{\bar{z}}\Big(\frac{\rho^{2a}\partial_z\eta_2}{\lambda - 1 - \frac{\nu}{\rho}}\Big) - \Big(1 + \lambda - \frac{\nu}{\rho}\Big)\eta_2 = 0,$$

which is a critical point of

$$J^{(+)}(\eta, \nu, a, \lambda) := \int_{\mathbb{R}^2} \Big(\frac{4|\partial_z\eta|^2}{1 - \lambda + \frac{\nu}{|x|}} + \Big(1 + \lambda - \frac{\nu}{|x|}\Big)|\eta|^2 \Big)|x|^{2a} \, dx.$$

As a function of λ, $J^{(+)}(\eta, \nu, a, \lambda)$ is monotone increasing and the same method as in the proof of Proposition 1.4 applies up to the change $\lambda \mapsto -\lambda$. This is why in the computation of the roots of

$$\nu\mu^2 - (1 - 2a)\mu + \nu = 0,$$

we choose $\mu_- = -\frac{1}{\nu}(\mathrm{c}(a) + \sqrt{\mathrm{c}(a)^2 - \nu^2})$ and obtain by solving $\frac{1+\lambda}{1-\lambda} = \mu_-^2$ that the optimal value for λ is

$$\lambda = -\frac{\sqrt{\mathrm{c}(a)^2 - \nu^2}}{\mathrm{c}(a)} = -\lambda_{\nu,a}.$$

The result of Theorem 1.1 becomes

$$\int_{\mathbb{R}^2} \Big(\frac{|D_a\chi|^2}{1 + \lambda_{\nu,a} + \frac{\nu}{|x|}} + \Big(1 - \lambda_{\nu,a} - \frac{\nu}{|x|}\Big)|\chi|^2 \Big) \, dx \geq 0, \quad \chi \in C_{\mathrm{c}}^\infty(\mathbb{R}^2 \setminus \{0\}, \mathbb{C}),$$

with equality if χ is the lower component of $\psi_{-\nu,a}$ in Proposition 4.1. This also means that (1.11) is replaced by

$$\int_{\mathbb{R}^2}\left(\frac{|\partial_z\eta|^2}{1+\lambda_{\nu,a}+\frac{\nu}{|x|}}+\left(1-\lambda_{\nu,a}-\frac{\nu}{|x|}\right)|\eta|^2\right)|x|^{2a}\,dx\geq 0, \quad \eta\in C_{\mathrm{c}}^\infty(\mathbb{R}^2\setminus\{0\},\mathbb{C}),$$

with same consequences: under the assumption $a\in(0,\frac{1}{2}]$ and $\nu\in(0,\mathsf{c}(a)]$, we obtain

$$\int_{\mathbb{R}^2}\left(\frac{4}{\nu}|x||\partial_z\eta|^2+2|\eta|^2\right)|x|^{2a}\,dx\geq\nu\int_{\mathbb{R}^2}|\eta|^2|x|^{2a-1}\,dx, \quad \eta\in C_{\mathrm{c}}^\infty(\mathbb{R}^2\setminus\{0\},\mathbb{C}),$$

and

$$\int_{\mathbb{R}^2}|\partial_z\eta|^2|x|^{2a+1}\,dx\geq\frac{1}{4}\mathsf{c}(a)^2\int_{\mathbb{R}^2}|\eta|^2|x|^{2a-1}\,dx, \quad \eta\in C_{\mathrm{c}}^\infty(\mathbb{R}^2\setminus\{0\},\mathbb{C}).$$

Acknowledgements. This research has been partially supported by the projects *EFI*, contract ANR-17-CE40-0030 (Jean Dolbeault) and *molQED* (Maria J. Esteban) of the French National Research Agency (ANR) and by the NSF grant DMS-1856645 (Michael Loss).

Bibliography

[1] Y. Aharonov and A. Casher, Ground state of a spin-$\frac{1}{2}$ charged particle in a two-dimensional magnetic field. *Phys. Rev. A (3)* **19** (1979), 2461–2462

[2] M. Arai, On essential selfadjointness, distinguished selfadjoint extension and essential spectrum of Dirac operators with matrix valued potentials. *Publ. Res. Inst. Math. Sci.* **19** (1983), 33–57

[3] M. Arai and O. Yamada, Essential selfadjointness and invariance of the essential spectrum for Dirac operators. *Publ. Res. Inst. Math. Sci.* **18** (1982), 973–985

[4] N. Arrizabalaga, Distinguished self-adjoint extensions of Dirac operators via Hardy–Dirac inequalities. *J. Math. Phys.* **52** (2011), Article ID 092301

[5] D. Bonheure, J. Dolbeault, M. J. Esteban, A. Laptev and M. Loss, Symmetry results in two-dimensional inequalities for Aharonov–Bohm magnetic fields. *Comm. Math. Phys.* **375** (2020), 2071–2087

[6] R. Bosi, J. Dolbeault and M. J. Esteban, Estimates for the optimal constants in multipolar Hardy inequalities for Schrödinger and Dirac operators. *Commun. Pure Appl. Anal.* **7** (2008), 533–562

[7] J. F. Brasche and H. Neidhardt, Some remarks on Kreĭn's extension theory. *Math. Nachr.* **165** (1994), 159–181

[8] C. Burnap, H. Brysk and P. F. Zweifel, Dirac Hamiltonian for strong Coulomb fields. *Nuovo Cimento B (11)* **64** (1981), 407–419

[9] F. Cacciafesta and L. Fanelli, Dispersive estimates for the Dirac equation in an Aharonov–Bohm field. *J. Differential Equations* **263** (2017), 4382–4399

[10] A. Cáceres and C. Doran, Minimax determination of the energy spectrum of the Dirac equation in a Schwarzschild background. *Phys. Rev. A (3)* **72** (2005), Article ID 022103

[11] A. H. Castro Neto, F. Guinea, N. M. R. Peres, K. S. Novoselov and A. K. Geim, The electronic properties of graphene. *Rev. Mod. Phys.* **81** (2009), 109–162

[12] C. G. Darwin, The wave equations of the electron. *Proc. Royal Soc. Lond. Ser. A* **118** (1928), 654–680

[13] S. N. Datta and G. Devaiah, The minimax technique in relativistic Hartree–Fock calculations. *Pramana* **30** (1988), 387–405

[14] D.-A. Deckert and M. Oelker, Distinguished self-adjoint extension of the two-body Dirac operator with Coulomb interaction. *Ann. Henri Poincaré* **20** (2019), 2407–2445

[15] J. Dolbeault, M. J. Esteban, J. Duoandikoetxea and L. Vega, Hardy-type estimates for Dirac operators. *Ann. Sci. Éc. Norm. Supér. (4)* **40** (2007), 885–900

[16] J. Dolbeault, M. J. Esteban and M. Loss, Relativistic hydrogenic atoms in strong magnetic fields. *Ann. Henri Poincaré* **8** (2007), 749–779

[17] J. Dolbeault, M. J. Esteban, M. Loss and L. Vega, An analytical proof of Hardy-like inequalities related to the Dirac operator. *J. Funct. Anal.* **216** (2004), 1–21

[18] J. Dolbeault, M. J. Esteban and E. Séré, On the eigenvalues of operators with gaps. Application to Dirac operators. *J. Funct. Anal.* **174** (2000), 208–226

[19] J. Dolbeault, M. J. Esteban and E. Séré, Variational characterization for eigenvalues of Dirac operators. *Calc. Var. Partial Differential Equations* **10** (2000), 321–347

[20] J. Dolbeault, M. J. Esteban and E. Séré, A variational method for relativistic computations in atomic and molecular physics. *Int. J. Quantum Chem.* **93** (2003), 149–155

[21] J. Dolbeault, M. J. Esteban and E. Séré, General results on the eigenvalues of operators with gaps, arising from both ends of the gaps. Application to Dirac operators. *J. Eur. Math. Soc. (JEMS)* **8** (2006), 243–251

[22] J. Dolbeault, M. J. Esteban, E. Séré and M. Vanbreugel, Minimization methods for the one-particle Dirac equation. *Phys. Rev. Lett.* **85** (2000), 4020–4023

[23] S.-H. Dong and Z.-Q. Ma, Exact solutions to the Dirac equation with a Coulomb potential in $2 + 1$ dimensions. *Phys. Lett. A* **312** (2003), 78–83

[24] L. Erdős and V. Vougalter, Pauli operator and Aharonov–Casher theorem for measure valued magnetic fields. *Comm. Math. Phys.* **225** (2002), 399–421

[25] M. J. Esteban, M. Lewin and A. Savin, Symmetry breaking of relativistic multiconfiguration methods in the nonrelativistic limit. *Nonlinearity* **23** (2010), 767–791

[26] M. J. Esteban, M. Lewin and E. Séré, Domains for Dirac–Coulomb min-max levels. *Rev. Mat. Iberoam.* **35** (2019), 877–924

[27] M. J. Esteban and M. Loss, Self-adjointness for Dirac operators via Hardy–Dirac inequalities. *J. Math. Phys.* **48** (2007), Article ID 112107

[28] M. J. Esteban and M. Loss, Self-adjointness via partial Hardy-like inequalities. In *Mathematical results in quantum mechanics*, pp. 41–47, World Scientific Publishing, Hackensack, 2008

[29] M. J. Esteban and E. Séré, Existence and multiplicity of solutions for linear and nonlinear Dirac problems. In *Partial differential equations and their applications (Toronto, ON, 1995)*, pp. 107–118, CRM Proc. Lecture Notes 12, American Mathematical Society, Providence, 1997

[30] M. J. Esteban and E. Séré, Nonrelativistic limit of the Dirac–Fock equations. *Ann. Henri Poincaré* **2** (2001), 941–961

[31] H. Falomir and P. A. G. Pisani, Hamiltonian self-adjoint extensions for $(2+1)$-dimensional Dirac particles. *J. Phys. A* **34** (2001), 4143–4154

[32] M. Gallone and A. Michelangeli, Discrete spectra for critical Dirac–Coulomb Hamiltonians. *J. Math. Phys.* **59** (2018), Article ID 062108

[33] M. Gallone and A. Michelangeli, Self-adjoint realisations of the Dirac–Coulomb Hamiltonian for heavy nuclei. *Anal. Math. Phys.* **9** (2019), 585–616

[34] V. Georgescu and M. Măntoiu, On the spectral theory of singular Dirac type Hamiltonians. *J. Operator Theory* **46** (2001), 289–321

[35] W. Gordon, Die Energieniveaus des Wasserstoffatoms nach der Diracschen Quantentheorie des Elektrons. *Z. Physik* **48** (1928), 11–14

[36] M. Griesemer and H. Siedentop, A minimax principle for the eigenvalues in spectral gaps. *J. Lond. Math. Soc. (2)* **60** (1999), 490–500

[37] H. Hogreve, The overcritical Dirac–Coulomb operator. *J. Phys. A* **46** (2013), Article ID 025301

[38] H. Kalf, U.-W. Schmincke, J. Walter and R. Wüst, On the spectral theory of Schrödinger and Dirac operators with strongly singular potentials. In *Spectral theory and differential equations (Proc. Sympos., Dundee, 1974; dedicated to Konrad Jörgens)*, pp. 182–226, Lecture Notes in Math. 448, Springer, Berlin, 1975

[39] M. Klaus and R. Wüst, Characterization and uniqueness of distinguished selfadjoint extensions of Dirac operators. *Comm. Math. Phys.* **64** (1978/79), 171–176

[40] M. Klaus and R. Wüst, Spectral properties of Dirac operators with singular potentials. *J. Math. Anal. Appl.* **72** (1979), 206–214

[41] M. Krein, The theory of self-adjoint extensions of semi-bounded Hermitian transformations and its applications. I. *Rec. Math. [Mat. Sbornik] N.S.* **20(62)** (1947), 431–495

[42] O. Kullie, D. Kolb and A. Rutkowski, Two-spinor fully relativistic finite-element (FEM) solution of the two-center Coulomb problem. *Chem. Phys. Lett.* **383** (2004), 215–221

[43] A. Laptev and T. Weidl, Hardy inequalities for magnetic Dirichlet forms. In *Mathematical results in quantum mechanics (Prague, 1998)*, pp. 299–305, Oper. Theory Adv. Appl. 108, Birkhäuser, Basel, 1999

[44] E. van Lenthe, E. Baerends and J. Snijders, Relativistic regular two-component Hamiltonians. *J. Chem. Phys.* **99** (1993), 4597–4610

[45] J. Mawhin and A. Ronveaux, Schrödinger and Dirac equations for the hydrogen atom, and Laguerre polynomials. *Arch. Hist. Exact Sci.* **64** (2010), 429–460

[46] S. Morozov and D. Müller, On the minimax principle for Coulomb–Dirac operators. *Math. Z.* **280** (2015), 733–747

[47] D. Müller, Minimax principles, Hardy–Dirac inequalities, and operator cores for two and three dimensional Coulomb–Dirac operators. *Doc. Math.* **21** (2016), 1151–1169

[48] G. Nenciu, Self-adjointness and invariance of the essential spectrum for Dirac operators defined as quadratic forms. *Comm. Math. Phys.* **48** (1976), 235–247

[49] G. Nenciu, Distinguished self-adjoint extension for Dirac operator with potential dominated by multicenter Coulomb potentials. *Helv. Phys. Acta* **50** (1977), 1–3

[50] H. Ounaies, Perturbation method for a class of nonlinear Dirac equations. *Differential Integral Equations* **13** (2000), 707–720

[51] L. Schimmer, J. P. Solovej and S. Tokus, Friedrichs extension and min-max principle for operators with a gap. *Ann. Henri Poincaré* **21** (2020), 327–357

[52] U.-W. Schmincke, Distinguished selfadjoint extensions of Dirac operators. *Math. Z.* **129** (1972), 335–349

[53] U.-W. Schmincke, Essential selfadjointness of Dirac operators with a strongly singular potential. *Math. Z.* **126** (1972), 71–81

[54] F. Schwabl, Symmetries and further properties of the Dirac equation. In *Advanced texts in physics*, pp. 209–244, Springer, Berlin, 2004

[55] Y. A. Sitenko, Self-adjointness of the two-dimensional massless Dirac Hamiltonian and vacuum polarization effects in the background of a singular magnetic vortex. *Ann. Physics* **282** (2000), 167–217

[56] J. D. Talman, Minimax principle for the Dirac equation. *Phys. Rev. Lett.* **57** (1986), 1091–1094

[57] B. Thaller, *The Dirac equation*. Texts Monogr. Phys., Springer, Berlin, 1992

[58] B. L. Voronov, D. M. Gitman and I. V. Tyutin, The Dirac Hamiltonian with a superstrong Coulomb field. *Theoret. Math. Phys.* **150** (2007), 34–72

[59] M. A. H. Vozmediano, M. I. Katsnelson and F. Guinea, Gauge fields in graphene. *Phys. Rep.* **496** (2010), 109–148

[60] R. Wüst, A convergence theorem for selfadjoint operators applicable to Dirac operators with cutoff potentials. *Math. Z.* **131** (1973), 339–349

[61] R. Wüst, Distinguished self-adjoint extensions of Dirac operators constructed by means of cut-off potentials. *Math. Z.* **141** (1975), 93–98

[62] R. Wüst, Dirac operations with strongly singular potentials. Distinguished self-adjoint extensions constructed with a spectral gap theorem and cut-off potentials. *Math. Z.* **152** (1977), 259–271

[63] J. Xia, On the contribution of the Coulomb singularity of arbitrary charge to the Dirac Hamiltonian. *Trans. Amer. Math. Soc.* **351** (1999), 1989–2023

[64] H. Zhang, O. Kullie and D. Kolb, Minimax LCAO approach to the relativistic two-centre Coulomb problem and its finite element (FEM) spectrum. *J. Phys. B* **37** (2004), 905–916

The Feshbach–Schur map and perturbation theory

Geneviève Dusson, Israel Michael Sigal and Benjamin Stamm

To Ari with friendship and admiration

This paper deals with perturbation theory for discrete spectra of linear operators. To simplify exposition, we consider here self-adjoint operators. This theory is based on the Feshbach–Schur map and it has advantages with respect to the standard perturbation theory in three aspects: (a) it readily produces rigorous estimates on eigenvalues and eigenfunctions with explicit constants; (b) it is compact and elementary (it uses properties of norms and the fundamental theorem of algebra about solutions of polynomial equations); and (c) it is based on a self-contained formulation of a fixed point problem for the eigenvalues and eigenfunctions, allowing for easy iterations. We apply our abstract results to obtain rigorous bounds on the ground states of Helium-type ions.

1 Set-up and result

The eigenvalue perturbation theory is a major tool in mathematical physics and applied mathematics. In the present form, it goes back to Rayleigh and Schrödinger and became a robust mathematical field in the works of Kato and Rellich. It was extended to quantum resonances by Simon, see [2,5,6,11–15] for books and a book-size review.

A different approach to the eigenvalue perturbation problem going back to works of Feshbach and based on the Feshbach–Schur map was introduced in [1] and extended in [3,4].

In this paper, we develop further this approach proposing a self-contained theory in a form of a fixed point problem for the eigenvalues and eigenfunctions. It is more compact and direct than the traditional one and, as we show elsewhere, extends to the non-linear eigenvalue problem.

We show that this approach leads naturally to bounds on the eigenvalues and eigenfunctions with explicit constants, which we use in an estimation of the ground state energies of the Helium-type ions.

Keywords: Perturbation theory, spectrum, Feshbach–Schur map, Schrödinger operator, atomic systems, Helium-type ions, ground state

2020 Mathematics Subject Classification: Primary 47A55, 35P15, 81Q15; secondary 47A75, 35J10

The approach can handle tougher perturbations, non-isolated eigenvalues (see [1, 4]) and continuous spectra as well as discrete ones. In this paper, we restrict ourselves to the latter. Namely, we address the eigenvalue perturbation problem for operators on a Hilbert space of the form

$$H = H_0 + W, \tag{1.1}$$

where H_0 is an operator with some isolated eigenvalues and W is an operator, small relative to H_0 in an appropriate norm. The goal is to show that H has eigenvalues near those of H_0 and estimate these eigenvalues.

Specifically, with $\|\cdot\|$ standing for the vector and operator norms in the underlying Hilbert space, we assume that

(A) H_0 is a self-adjoint, non-negative operator ($H_0 \geq \beta > 0$).

(B) W is symmetric and form-bounded with respect to H_0, in the sense that

$$\|H_0^{-\frac{1}{2}} W H_0^{-\frac{1}{2}}\| < \infty,$$

(C) H_0 has an isolated eigenvalue $\lambda_0 > 0$ of a finite multiplicity, m.

Here $\|\cdot\|$ is the operator norm and H_0^{-s}, $0 < s < 1$, is defined either by the spectral theory or by the explicit formula

$$H_0^{-s} := c \int_0^\infty (H_0 + \omega)^{-1} \frac{d\omega}{\omega^s},$$

where $c := [\int_0^\infty (1+\omega)^{-1} \frac{d\omega}{\omega^s}]^{-1}$.

It turns out to be useful in the proofs below to use the following form-norm:

$$\|W\|_{H_0} := \|H_0^{-\frac{1}{2}} W H_0^{-\frac{1}{2}}\|.$$

Let P be the orthogonal projection onto the span of the eigenfunctions of H_0 corresponding to the eigenvalue λ_0 and let $P^\perp := \mathbf{1} - P$. Let γ_0 be the distance of λ_0 to the rest of the spectrum of H_0, and $\lambda_* := \lambda_0 + \gamma_0$. In what follows, we often deal with the expression

$$\Phi(W) := \frac{\lambda_0 \lambda_* \|P^\perp W P\|_{H_0}^2}{\gamma_0}. \tag{1.2}$$

The following theorem proven in Section 2 is the main result of this paper.

Main Theorem 1.1. *Let assumptions* (A)–(C) *be satisfied and assume that, for some* $0 < b < 1$ *and* $0 < a < 1 - b$,

$$\|P^\perp W P^\perp\|_{H_0} \leq \frac{b\gamma_0}{\lambda_*}, \tag{1.3}$$

$$\|PWP\| + k\Phi(W) < a\gamma_0, \tag{1.4}$$

$$k\Phi(W) < \frac{1}{2}(a\gamma_0 - \|PWP\|), \tag{1.5}$$

where $k := \frac{1}{1-a-b}$. *Then the spectrum of the operator* H *near* λ_0 *consists of isolated eigenvalues* λ_i *of the total multiplicity* m *satisfying, together with their normalized eigenfunctions* φ_i, *the following estimates:*

$$|\lambda_i - \lambda_0| \leq \|PWP\| + k\Phi(W), \tag{1.6}$$

$$\|\varphi_i - \varphi_{0i}\| \leq k\sqrt{\frac{\Phi(W)}{\gamma_0}}, \tag{1.7}$$

where φ_{0i} *are appropriate eigenfunctions of* H_0 *corresponding to the eigenvalue* λ_0.

Remark 1 (Comparison with [3]). A similar result was already proven in [3]. Here, the theory is made self-contained and formulated as the fixed point problem and the bounds are tightened.

Remark 2 (Conditions on W). The fine tuning conditions (1.3)–(1.5) on W is used in application of this theorem to atomic systems in Section 3.

Note that because of the elementary estimate

$$\Phi(W) \leq \frac{1}{\gamma_0}\|PWP^{\perp}WP\|,$$

see (3.11) below, the computation of $\|PWP\|$ and $\Phi(W)$ reduces to computing the largest eigenvalues of the simple $m \times m$-matrices $PWP, -PWP$ and $PWP^{\perp}WP$.

Remark 3 (Higher-order estimates and non-degenerate λ_0). In fact, one can estimate λ_i (as well as the eigenfunctions φ_i) to an arbitrary order in $\frac{1}{\gamma_0}\|P^{\perp}WP\|_{H_0}$. As a demonstration, we derive (after the proof of Theorem 1.1) the second-order estimate of the eigenvalue in the rank-one ($m = 1$) case

$$|\lambda_1 - \lambda_0 - \langle W\rangle| \leq k\Phi(W), \tag{1.8}$$

where φ_0 is the eigenfunction of H_0 corresponding to λ_0, $\langle W\rangle := \langle\varphi_0, W\varphi_0\rangle$.

For degenerate λ_0, we would like to prove a similar bound on $|\lambda_i - \lambda_0 - \mu_i|$, where μ_i is the corresponding eigenvalue of the $m \times m$-matrix PWP. Here, we have a partial result (proven at the end of the next section) for the lowest eigenvalue λ_1 of H:

$$\lambda_0 + \mu_{\min} - k\Phi(W) \leq \lambda_1 \leq \lambda_0 + \mu_{\min}, \tag{1.9}$$

where $\mu_{\min}$ denotes the smallest eigenvalue of the matrix PWP.

Remark 4 (Non-self-adjoint H). With the sacrifice of the explicit constants in (1.6) and (1.7) (mostly coming from (2.5)), the self-adjointness assumption on H can be removed. However, for the problem of quantum resonances, one can still obtain explicit estimates.

In the rest of this section H is an abstract operator not necessarily self-adjoint or of the form (1.1). Our approach is grounded in the following theorem (see [4, Theorem 11.1]).

Theorem 1.2. *Let H be an operator on a Hilbert space and P and $P^\perp$, a pair of projections such that $P + P^\perp = 1$. Assume $H^\perp := P^\perp H P^\perp$ is invertible on* Ran $P^\perp$ *and the expression*

$$F_P(H) := P(H - HR^\perp H)P, \tag{1.10}$$

where $R^\perp := P^\perp (H^\perp)^{-1} P^\perp$, defines a bounded operator. Then F_P, considered as a map on the space of operators, is isospectral in the following sense:

(a) $\lambda \in \sigma(H)$ *if and only if* $0 \in \sigma(F_P(H - \lambda))$,

(b) $H\psi = \lambda\psi$ *if and only if* $F_P(H - \lambda)\varphi = 0$,

(c) $\dim \mathrm{Null}(H - \lambda) = \dim \mathrm{Null}\, F_P(H - \lambda)$,

(d) ψ *and* φ *in* (b) *are related as* $\varphi = P\psi$ *and* $\psi = Q_P(\lambda)\varphi$, *where*

$$Q_P(\lambda) := P - P^\perp (H^\perp - \lambda)^{-1} P^\perp H P. \tag{1.11}$$

Finally, $F_P(H)^* = F_{P^*}(H^*)$ and therefore, if H and P are self-adjoint, then so is $F_P(H)$.

A proof of this theorem is elementary and short; it can be found in [1, Section IV.1, pp. 346–348] and [4, Appendix 11.6, pp. 123–125].

The map F_P on the space of operators, is called the *Feshbach–Schur map*. The relation $\psi = Q_P(\lambda)\varphi$ allows us to reconstruct the full eigenfunction from the projected one. By statement (a), we have:

Corollary 1.3. *Assume there is an open set $\Lambda \subset \mathbb{C}$ such that $H - \lambda$, $\lambda \in \Lambda$, is in the domain of the map F_P, i.e., $F_P(H - \lambda)$ is well defined. Define the operator-family*

$$H(\lambda) := F_P(H - \lambda) + \lambda P,$$

and let $\nu_i(\lambda)$, for λ in Λ, denote its eigenvalues counted with multiplicities. Then the eigenvalues of H in Λ are in one-to-one correspondence with the solutions of the equations

$$\nu_i(\lambda) = \lambda. \tag{1.12}$$

Concentrating on the eigenvalue problem, Corollary 1.3 shows that the original problem

$$H\psi = \lambda\psi \tag{1.13}$$

is mapped into the equivalent eigenvalue problem

$$H(\lambda)\varphi = \lambda\varphi, \tag{1.14}$$

non-linear in the spectral parameter λ, but on the smaller space Ran P.

Since the projection P is of a finite rank, the original eigenvalue problem (1.13) is mapped into an equivalent lower-dimensional/finite-dimensional one, (1.14). Of course, we have to pay a price for this: at one step we have to solve a one-dimensional fixed point problem that can be equivalently seen as a non-linear eigenvalue problem and invert an operator in Ran $P^\perp$.

We call this approach the *Feshbach–Schur map method*, or *FSM method*, for short. It is rather compact, as one easily see skimming through this paper and entirely elementary.

We call $H(\lambda)$ the *effective Hamiltonian* (matrix) and write as

$$H(\lambda) = PHP + U(\lambda), \tag{1.15}$$

with the self-adjoint *effective interaction*, or a Schur complement, $U(\lambda)$, defined as

$$U(\lambda) := -PHP^{\perp}(H^{\perp} - \lambda)^{-1}P^{\perp}HP. \tag{1.16}$$

It is shown in Lemma 2.1 below that (1.16) defines a bounded operator family.

We mention here some additional properties discussed in [3].

Proposition 1.4. *Let H be self-adjoint, let Λ be the same as in Corollary* 1.3, *let m be the rank of P and let the eigenvalues of $H(\lambda)$, $\lambda \in \Lambda \cap \mathbb{R}$, be labeled in the order of their increase counting their multiplicities so that*

$$\nu_1(\lambda) \leq \cdots \leq \nu_m(\lambda). \tag{1.17}$$

Then we have:

(a) *a solution of the equation $\nu_i(\lambda) = \lambda$, for $\lambda \in \Lambda \cap \mathbb{R}$, is the i-th eigenvalue, λ_i, of H and vice versa,*

(b) *ν_i is differentiable in λ and $\nu_i'(\lambda) \leq 0$, for $\lambda \in \Lambda \cap \mathbb{R}$.*

Proof. (a) By (1.17), $\lambda_i = \nu_i(\lambda_i) < \nu_{i+1}(\lambda) = \lambda_{i+1}$ (except for the level crossings), which proves the result. For (b), by the explicit formula

$$U'(\lambda) := -PHP^{\perp}(H^{\perp} - \lambda)^{-2}P^{\perp}HP \leq 0,$$

we have $U'(\lambda) \leq 0$, which by the Hellmann–Feynman theorem (cf. (2.19)) implies $\nu_i'(\lambda) \leq 0$. ■

Remark 5 (Perturbation expansion). In the context of Hamiltonians of form (1.1) satisfying assumptions (A)–(C), the FSM leads to a perturbation expansion to an arbitrary order. Indeed, in this case, P is the orthogonal projection onto $\text{Null}(H_0 - \lambda_0)$, $P^{\perp} := \mathbf{1} - P$ and $U(\lambda)$ can be written as

$$U(\lambda) := -PWP^{\perp}(H^{\perp} - \lambda)^{-1}P^{\perp}WP.$$

Now, using the notation $A^{\perp} := P^{\perp}AP^{\perp}|_{\text{Ran}\, P^{\perp}}$ and expanding

$$(H^{\perp} - \lambda)^{-1} = (H_0^{\perp} + W^{\perp} - \lambda)^{-1}$$

in $W^{\perp}$ and at the same time iterating fixed point equation (1.12), we generate a perturbation expansion for eigenvalues H to an arbitrary order (see also Remark 2 above).

The paper is organized as follows. In Section 2, we present the proof of the main result, Theorem 1.1. Section 3 uses this result to obtain bounds of the ground-state energy of Helium-type ions.

2 Perturbation estimates

We want to use Theorem 1.2 (b)–(c) to reduce the original eigenvalue problem to a simpler one. In this section, we assume that conditions (A)–(C) of Section 1 are satisfied. Recall that γ_0 is the distance of λ_0 to the rest of the spectrum of H_0 and the expression $\Phi(W)$ is defined in (1.2). First, we prove that the operator $F_P(H-\lambda)$ is well-defined for $\lambda \in \Omega$, where, with a the same as in Theorem 1.1,

$$\Omega := \{z \in \mathbb{C} : |z-\lambda_0| \le a\gamma_0\}. \tag{2.1}$$

Recall that P denotes the orthogonal projection onto $\mathrm{Null}(H_0-\lambda_0)$ and $P^\perp := \mathbf{1}-P$. Denote

$$H^\perp := P^\perp H P^\perp|_{\mathrm{Ran}\,P^\perp} \quad \text{and} \quad R^\perp(\lambda) := P^\perp(H^\perp-\lambda)^{-1}P^\perp$$

and recall $k = \frac{1}{1-a-b}$. We have:

Lemma 2.1. *Recall* $\lambda_* := \lambda_0 + \gamma_0$ *and assume* (1.3). *Then, for* $\lambda \in \Omega$, *the following statements hold:*

(a) *The operator* $H^\perp - \lambda$ *is invertible on* $\mathrm{Ran}\,P^\perp$.

(b) *The inverse* $R^\perp(\lambda) := P^\perp(H^\perp-\lambda)^{-1}P^\perp$ *defines a bounded, analytic operator-family.*

(c) *The expression*

$$U(\lambda) := -PHR^\perp(\lambda)HP \tag{2.2}$$

defines a finite-rank, analytic operator-family, bounded as

$$\|U(\lambda)\| \le k\Phi(W). \tag{2.3}$$

(d) $U(\lambda)$ *is symmetric for any* $\lambda \in \Omega \cap \mathbb{R}$ *and therefore* $H(\lambda)$ *is self-adjoint as well.*

Proof. With the notation $A^\perp := P^\perp A P^\perp|_{\mathrm{Ran}\,P^\perp}$, we write

$$H^\perp = H_0^\perp + W^\perp.$$

To prove (a), we let $H_\lambda := H_0^\perp - \lambda$ and use the factorization $H_\lambda = |H_\lambda|^{\frac12} V_\lambda |H_\lambda|^{\frac12}$, where V_λ is a unitary operator, and use that, for $\lambda \in \Omega$, the operator H_λ is invertible and therefore we have the identity

$$H^\perp - \lambda = |H_\lambda|^{\frac12}[V_\lambda + K_\lambda]|H_\lambda|^{\frac12}, \tag{2.4}$$

where $K_\lambda := |H_\lambda|^{-\frac12} W^\perp |H_\lambda|^{-\frac12}$. For $\lambda \in \Omega$, we have $\|H_0^\perp |H_\lambda|^{-1} P^\perp\| = f(\lambda)$,

where

$$f(\lambda) := \sup\left\{\left|\frac{s}{s-\lambda}\right| : s \geq 0,\ |s-\lambda_0| \geq \gamma_0\right\}.$$

Assuming $|\lambda - \lambda_0| \leq a\gamma_0$ and using

$$\left|\frac{s}{s-\lambda}\right| \leq 1 + \left|\frac{\lambda}{s-\lambda}\right| \leq 1 + \frac{\lambda_0 + a\gamma_0}{(1-a)\gamma_0} = \frac{\lambda_*}{(1-a)\gamma_0},$$

we obtain

$$f(\lambda) \leq \frac{\lambda_*}{(1-a)\gamma_0}.$$

Since $H_0^{\frac{1}{2}}|H_\lambda|^{-\frac{1}{2}}P^\perp = (H_0|H_\lambda|^{-1}|_{\mathrm{Ran}\,P^\perp})^{\frac{1}{2}}P^\perp$, we have for $\lambda \in \Omega$,

$$\|H_0^{\frac{1}{2}}|H_\lambda|^{-\frac{1}{2}}P^\perp\| \leq \left[\frac{\lambda_*}{(1-a)\gamma_0}\right]^{\frac{1}{2}}, \tag{2.5}$$

which implies in particular that $\|K_\lambda\| \leq \frac{\lambda_*}{(1-a)\gamma_0}\|W^\perp\|_{H_0}$. By assumption (1.3), i.e., $\|W^\perp\|_{H_0} \leq \frac{b\gamma_0}{\lambda_*}$, we have

$$\|K_\lambda\| \leq \frac{b}{1-a}. \tag{2.6}$$

Since $1 - a > b$, by (2.4), the operator $H^\perp - \lambda$ is invertible and its inverse is analytic in $\lambda \in \Omega$, which proves (a) and (b).

We show that statement (c) is also satisfied. Since $PH_0 = H_0P$ and $PP^\perp = 0$, we have

$$PHP^\perp = PWP^\perp, \quad P^\perp HP = P^\perp WP.$$

These relations and definition (2.2) yield

$$U(\lambda) = -PWR^\perp(\lambda)WP. \tag{2.7}$$

Inverting (2.4) on Ran $P^\perp$ and recalling the notation $R^\perp(\lambda) := P^\perp(H^\perp - \lambda)^{-1}P^\perp$ gives

$$R^\perp(\lambda) = |H_\lambda|^{-\frac{1}{2}}P^\perp[V_\lambda + K_\lambda]^{-1}P^\perp|H_\lambda|^{-\frac{1}{2}}. \tag{2.8}$$

Now, using identity (2.8), estimate (2.5) and (2.6), we find, for $\lambda \in \Omega$,

$$\|H_0^{\frac{1}{2}}R^\perp(\lambda)H_0^{\frac{1}{2}}\| \leq \frac{k\lambda_*}{\gamma_0}. \tag{2.9}$$

Furthermore, by the eigen-equation $H_0P = \lambda_0P$, we have

$$\|H_0^{\frac{1}{2}}P\|^2 = \lambda_0. \tag{2.10}$$

Using expression (2.7) and estimates (2.9) and (2.10), we arrive at inequality (2.3). The analyticity follows from (2.7) and the analyticity of $R^\perp(\lambda)$.

For (d), since H_0, W and P are self-adjoint, so are $U(\lambda)$, for any $\lambda \in \Omega \cap \mathbb{R}$, and, since $U(\lambda)$ is bounded, $H(\lambda)$ is self-adjoint as well. ■

Proof of Theorem 1.1. Let Ω be given by equation (2.1). Recall that, by Lemma 2.1 and Theorem 1.2, the $m \times m$ matrix-family $H(\lambda) := F_P(H-\lambda) + \lambda P$, with F_P given in (1.10), is well defined, for each $\lambda \in \Omega$, and can be written as (1.15). Since $PHP = \lambda_0 P + PWP$, equation (1.15) can be rewritten as

$$H(\lambda) = \lambda_0 P + PWP + U(\lambda). \tag{2.11}$$

Equations (2.3) and (2.11) imply the inequality

$$\|H(\lambda) - \lambda_0 P - PWP\| \le k\Phi(W). \tag{2.12}$$

By a fact from Linear Algebra, for each $\lambda \in \Omega \cap \mathbb{R}$, the total multiplicity of the eigenvalues of the $m \times m$ self-adjoint matrix $H(\lambda)$ is m.

Denote by $\nu_i(\lambda), i = 1, 2, \ldots, m$, the eigenvalues of $H(\lambda)$, *repeated according to their multiplicities*. Equation (2.12) yields

$$|\nu_i(\lambda) - \lambda_0| \le \|PWP\| + k\Phi(W). \tag{2.13}$$

Indeed, let $P_i(\lambda)$ be the orthogonal projection onto $\mathrm{Null}(H(\lambda) - \nu_i(\lambda))$. Then

$$H(\lambda)P_i(\lambda) = \nu_i(\lambda)P_i(\lambda),$$

which, due to (2.11) and $P_i(\lambda)P = P_i(\lambda)$, can be rewritten as

$$(\nu_i(\lambda) - \lambda_0)P_i(\lambda) = (PWP + U(\lambda))P_i(\lambda).$$

Equating the operator norms of both sides of this equation and using (2.12) and $\|P_i(\lambda)\| = 1$ gives (2.13).

By Corollary 1.3, the eigenvalues of H in the interval $\Omega \cap \mathbb{R}$ are in one-to-one correspondence with the solutions of the equations

$$\nu_i(\lambda) = \lambda \tag{2.14}$$

in $\Omega \cap \mathbb{R}$. If this equation has a solution, then, due to (2.13), this solution would satisfy (1.6). Thus, we address (2.14). Let

$$\Omega' := \{z \in \mathbb{R} : |z - \lambda_0| \le r\gamma_0\}, \tag{2.15}$$

with

$$r := \frac{1}{\gamma_0}(\|PWP\| + k\Phi(W)).$$

By our assumption (1.4), $r < a < 1 - b$.

Recall that by the definition, a branch point is a point at which the multiplicity of one of the eigenvalue families (branches) changes. One could think on a branch point as a point where two or more distinct eigenvalue branches intersect. Our next result shows that the eigenvalue branches of $H(\lambda)$ could be chosen in a differentiable way and estimates their derivatives.

Proposition 2.2. *The following statements hold.*

(i) *The eigenvalues* $\nu_i(\lambda)$ *of* $H(\lambda)$ *and the corresponding eigenfunctions can be chosen to be differentiable for* $\lambda \in \Omega'$.

(ii) *The derivatives* $\nu_i'(\lambda)$, $\lambda \in \Omega'$, *are bounded as*

$$|\nu_i'(\lambda)| \leq \frac{k}{a-r}\frac{\Phi(W)}{\gamma_0}. \tag{2.16}$$

(iii) $\nu_i(\lambda)$ *maps the interval* Ω' *into itself.*

Consequently, since by (1.5) *the right-hand side of* (2.16) *is* < 1, *the equations* $\nu_i(\lambda) = \lambda$ *have unique solutions in* Ω'.

Proof. Proof of (i) for simple λ_0. For a simple eigenvalue λ_0, P is a rank-one projection on the space spanned by the eigenvector φ_0 of H_0 corresponding to the eigenvalue λ_0 and therefore equation (2.11) implies that $H(\lambda) = \nu(\lambda)P$, with

$$\nu(\lambda) := \lambda_0 + \langle \varphi_0, (W + U(\lambda))\varphi_0 \rangle.$$

This and Lemma 2.1 show that the eigenvalue $\nu(\lambda)$ is analytic.

Proof of (i) for degenerate λ_0. We pick an *arbitrary* point μ in Ω' and let $P_{\mu i}$ be the orthogonal projections onto the eigenspaces of $H(\mu)$ corresponding to the eigenvalues $\nu_i(\mu)$, i.e., for a fixed i, $P_{\mu i}$ projects on the spans of all eigenvectors with the same eigenvalue $\nu_i(\mu)$. We now show that, in a neighborhood of μ, the eigenfunctions $\chi_i(\lambda)$ can be chosen in a differentiable way. We introduce the system of equations for the eigenvalues $\nu_i(\lambda)$ and corresponding eigenfunctions $\chi_i(\lambda)$:

$$\nu_i(\lambda) = \frac{\langle \chi_i(\lambda), H(\lambda)\chi_i(\lambda)\rangle}{\|\chi_i(\lambda)\|^2}, \tag{2.17}$$

$$\chi_i(\lambda) = \chi_i(\mu) - R^{\perp}(\nu_i(\lambda), \lambda)W_\lambda \chi_i(\mu), \tag{2.18}$$

where

$$R^{\perp}(\nu, \lambda) := P_{\mu i}^{\perp}(P_{\mu i}^{\perp} H(\lambda) P_{\mu i}^{\perp} - \nu)^{-1} P_{\mu i}^{\perp}, \quad W_\lambda := U(\lambda) - U(\mu).$$

The expression on the right side of (2.18) is the Q-operator for $H(\lambda)$ and $P_{\mu i}$ defined according to (1.11) and applied to $\chi_i(\mu)$. Note that systems (2.17)–(2.18) for different indices i are not coupled. Furthermore, since $\chi_i(\mu)$ and $R^{\perp}(\nu_i(\lambda), \lambda)W_\lambda \chi_i(\mu)$ are orthogonal, we have $\|\chi_i(\lambda)\| \geq \|\chi_i(\mu)\|$; and $\chi_i(\lambda)$ and $\chi_j(\lambda)$ are almost orthogonal for $i \neq j$. Assuming this system has a solution $(\nu_i(\lambda), \chi_i(\lambda))$, we see that, by Theorem 1.2 (d), with $P = P_{\mu i}$, $\chi_i(\lambda)$ are eigenfunctions of $H(\lambda)$ with the eigenvalues $\nu_i(\lambda)$ and (2.17) follows from the eigen-equation $H(\lambda)\chi_i(\lambda) = \nu_i(\lambda)\chi_i(\lambda)$.

For each i, we can reduce system (2.17)–(2.18) to the single equation for $\nu_i(\lambda)$, by treating $\chi_i(\lambda)$ in (2.17) as given by (2.18). This leads to the fixed point problem for the functions $\nu_i(\lambda)$:

$$\nu = F_i(\nu, \lambda),$$

where

$$F_i(\nu,\lambda) := \frac{\langle \chi_i(\nu,\lambda), H(\lambda)\chi_i(\nu,\lambda)\rangle}{\|\chi_i(\nu,\lambda)\|^2},$$
$$\chi_i(\nu,\lambda) := \chi_i(\mu) - R^\perp(\nu,\lambda) W_\lambda \chi_i(\mu).$$

Notice that since $P^\perp_{\mu i} H(\lambda) P^\perp_{\mu i}$ are self-adjoint, the resolvent $R^\perp(\nu,\lambda)$ and its derivatives in ν are uniformly bounded in a neighborhood of $(\nu_i(\mu),\mu)$, which does not contain branch points, except, possibly, for μ.

To be more specific, the resolvent $R^\perp(\nu,\lambda)$ is uniformly bounded in a neighborhood $\Omega_{\mu i}$ of $(\nu_i(\mu),\mu)$ whether μ is a branch point or not, as long as $\Omega_{\mu i}$ does not contain any other branch point than μ.

Hence $F_i(\nu,\lambda)$ is differentiable in ν and λ and $|\partial_\nu F_i(\nu,\lambda)| \to 0$, as $\lambda \to \mu$ (since $\partial_\nu \chi_i(\nu,\lambda) = R^\perp(\nu,\lambda)^2 W_\lambda \chi_i(\mu)$ and therefore

$$\|\partial_\nu \chi_i(\nu,\lambda)\| \le \|R^\perp(\nu,\lambda)^2 (H_0+\alpha)^{\frac{1}{2}}\| \|W_\lambda\|_{H_0,\alpha} \|(H_0+\alpha)^{\frac{1}{2}} \chi_i(\mu)\| \to 0$$

as $\lambda \to \mu$). Moreover, $\nu_i(\mu) - F_i(\nu_i(\mu),\mu) = 0$. Let $f_i(\nu,\lambda) := \nu - F_i(\nu,\lambda)$. Then, by the above, $f_i(\nu_i(\mu),\mu) = 0$ and $|\partial_\nu f_i(\nu,\lambda) - 1| \to 0$ as $\lambda \to \mu$. Thus, the implicit function theorem is applicable to the equation $f_i(\nu,\lambda) = 0$ and shows that there is a unique solution $\nu_i(\lambda)$ in a neighborhood of $(\nu_i(\mu),\mu)$ and that this solution is differentiable in λ.

Next, we have:

Lemma 2.3. *We have, for $\lambda \in \Omega'$,*

$$\nu_i'(\lambda) = \frac{\langle \chi_i(\lambda), U'(\lambda)\chi_i(\lambda)\rangle}{\|\chi_i(\lambda)\|^2}. \tag{2.19}$$

(Equation (2.19) *is closely related to the widely used Hellmann–Feynman theorem.)*

Proof. Let $\hat\chi_i(\lambda) := \frac{\chi_i(\lambda)}{\|\chi_i(\lambda)\|}$. Writing equation (2.17) as

$$\nu_i(\lambda) = \langle \hat\chi_i(\lambda), H(\lambda)\hat\chi_i(\lambda)\rangle$$

and differentiating this with respect to λ, we obtain

$$\nu_i'(\lambda) = \langle \hat\chi_i(\lambda), H'(\lambda)\hat\chi_i(\lambda)\rangle + \eta(\lambda),$$

where $\eta(\lambda) := \langle \hat\chi_i'(\lambda), H(\lambda)\hat\chi_i(\lambda)\rangle + \langle \hat\chi_i(\lambda), H(\lambda)\hat\chi_i'(\lambda)\rangle$. Now, moving $H(\lambda)$ in the last term to the l.h.s. and using that $H(\lambda)\hat\chi_i(\lambda) = \nu_i(\lambda)\hat\chi_i(\lambda)$, we find

$$\begin{aligned}\eta(\lambda) &= \nu_i(\lambda)[\langle \hat\chi_i'(\lambda), \hat\chi_i(\lambda)\rangle + \langle \hat\chi_i(\lambda), \hat\chi_i'(\lambda)\rangle] \\ &= \nu_i(\lambda)\partial_\lambda \langle \hat\chi_i(\lambda), \hat\chi_i(\lambda)\rangle = 0,\end{aligned}$$

which implies (2.19). ∎

To prove (2.16), we use formula (2.19) above and the normalization of $\hat\chi_i(\lambda)$ to estimate $\nu_i'(\lambda)$ as

$$|\nu_i'(\lambda)| \le \|U'(\lambda)\|. \tag{2.20}$$

To bound the right-hand side of (2.20), we use the analyticity of $U(\lambda)$ in Ω and estimate (2.3). Indeed, by the Cauchy integral formula, we have

$$\|U'(\lambda)\| \leq \frac{1}{R\gamma_0} \sup_{|\lambda'-\lambda|=R\gamma_0} \|U(\lambda')\|,$$

with $R \leq a - r$, so that $\{\lambda' \in \mathbb{C} : |\lambda' - \lambda| < R\gamma_0\} \subset \Omega$, for $\lambda \in \Omega'$. This, together with (2.3) and under the conditions of Lemma 2.1, gives the estimate

$$\|U'(\lambda)\| \leq \frac{k}{a-r} \frac{\Phi(W)}{\gamma_0} \tag{2.21}$$

for $\lambda \in \Omega'$. Combining (2.20) and (2.21), we arrive at (2.16).

Due to the definition of r after (2.15), we can rewrite estimate (2.13) as

$$|\nu_i(\lambda) - \lambda_0| \leq r\gamma_0.$$

By (2.15), this shows that $\nu_i(\lambda)$ maps the interval Ω' into itself which proves (iii).

Hence, under condition (1.4), the fixed point equations $\lambda = \nu_i(\lambda)$ have unique solutions on the interval Ω', proving Proposition 2.2. ∎

By Corollary 1.3, the eigenvalues of H in Ω' are in one-to-one correspondence with the solutions of the equations $\nu_i(\lambda) = \lambda$. By Proposition 2.2, these equations have unique solutions, say, λ_i. Then, estimate (2.13) implies inequality (1.6).

To obtain estimate (1.7), we recall from Theorem 1.2 that $Q(\lambda_j)\varphi_{0j} = \varphi_j$, where the operator $Q(\lambda)$ is given by

$$Q(\lambda) := \mathbf{1} - R^{\perp}(\lambda)P^{\perp}WP,$$

λ_j are the eigenvalues of H and φ_{0j} are eigenfunctions of H_0 corresponding to λ_0. This gives

$$\varphi_{0j} - \varphi_j = \varphi_{0j} - Q(\lambda_j)\varphi_{0j} = R^{\perp}(\lambda_j)P^{\perp}WP. \tag{2.22}$$

Now, as in the derivation of (2.9), using identity (2.8), estimates (2.5) and

$$\big\||H_\lambda|^{-\frac{1}{2}}\big\| \leq \frac{1}{\sqrt{(1-r)\gamma_0}}$$

and the estimate $\|K_\lambda\| \leq \frac{b}{1-a}$, see (2.6), we find, for $\lambda \in \Omega'$,

$$\|R^{\perp}(\lambda)H_0^{\frac{1}{2}}\| \leq \frac{k\lambda_*^{\frac{1}{2}}}{\gamma_0}, \tag{2.23}$$

noting that $r < a$. Using inequalities (2.23) and (2.10), we estimate the right-hand side of (2.22) as

$$\|R^{\perp}(\lambda_j)P^{\perp}WP\| \leq k\sqrt{\frac{\Phi(W)}{\gamma_0}}.$$

This, together with (2.22), gives (1.7). This proves Theorem 1.1. ∎

Remark 6. Differentiating (2.18) with respect to λ, setting $\lambda = \mu$, using $W_{\lambda=\mu} = 0$ and $W'_\lambda = U'(\lambda)$ and changing the notation μ to λ, we find the following formula for $\chi'_i(\lambda)$:

$$\chi'_i(\lambda) = -R^{\perp}(\nu_i(\lambda), \lambda)U'(\lambda)\chi_i(\lambda).$$

Finally, we prove relations (1.8) and (1.9) stated in the introduction.

Proof of equation (1.8). Equation (2.11) gives

$$H(\lambda) = (\lambda_0 + \langle W \rangle + \langle U(\lambda) \rangle)P,$$

where we use the notation $\langle A \rangle := \langle \varphi_0, A\varphi_0 \rangle$. Since P is a rank one projection, it follows that $H(\lambda)$ has only one eigenvalue and this eigenvalue is

$$\nu_1(\lambda) = \lambda_0 + \langle W \rangle + \langle U(\lambda) \rangle.$$

This formula and estimate (2.3) show that $\nu_1(\lambda)$ obeys the estimate

$$|\nu_1(\lambda) - \lambda_0 - \langle W \rangle| \le k\Phi(W).$$

By Proposition 2.2, the eigenvalue λ_1 of H satisfies the equation $\lambda_1 = \nu_1(\lambda_1)$. Then the estimate above implies inequality (1.8). ∎

Proof of equation (1.9). Let W be symmetric and denote by $\mu_{\min}$ the smallest eigenvalue of the matrix PWP. Then $U(\lambda) \le 0$ and (2.11) implies $H(\lambda) \le \lambda_0 P + PWP$. This, together with $\inf A \le \inf B$ for $A \le B$, gives the upper bound $\nu_1(\lambda) \le \lambda_0 + \mu_{\min}$ on the smallest eigenvalue-branch $\nu_1(\lambda)$ of $H(\lambda)$. For the lower bound, equations (2.3) and (2.11) imply the estimate

$$\lambda_0 + \mu_{\min} - k\Phi(W) \le \nu_1(\lambda).$$

The last two estimates and the equation $\nu_i(\lambda) = \lambda$ (see (1.12)) yield (1.9). ∎

3 Application: The ground state energy of the Helium-type ions

In this section, we will use the inequalities obtained above to estimate the ground state energy of the Helium-type ions, which is the simplest not completely solvable atomic quantum system. For simplicity, we assume that the nucleus is infinitely heavy, but we allow for a general nuclear charge ze, $z \ge 1$. Then the corresponding Schrödinger operator (describing two electrons of mass m and charge $-e$, and the nucleus of infinite mass and charge ze) is given by

$$H_{z,2} = \sum_{j=1}^{2} \left(-\frac{\hbar^2}{2m}\Delta_{x_j} - \frac{\kappa z e^2}{|x_j|} \right) + \frac{\kappa e^2}{|x_1 - x_2|}, \tag{3.1}$$

acting on the space $L^2_{\text{sym}}(\mathbb{R}^6)$ of L^2-functions symmetric (or antisymmetric) with respect to the permutation of x_1 and x_2.[1] Here $\kappa = \frac{1}{4\pi\varepsilon_0}$ is Coulomb's constant and ε_0 is the vacuum permittivity. For $z = 2$, $H_{z,2}$ describes the Helium atom, for $z = 1$, the negative ion of the Hydrogen and for $2 < z \leq 94$ (or ≤ 103, depending on what one counts as stable elements), Helium-type positive ion. (We can call (3.1) with $z > 103$ a *z-ion*.)

It is well known that $H_{z,2}$ has eigenvalues below its continuum. Variational techniques give excellent upper bounds on the eigenvalues of $H_{z,2}$, but good lower bounds are hard to come by. Thus, we formulate

Problem 3.1. Estimate the ground state energy of $H_{z,2}$.

The most difficult case is of $z = 1$, the negative ion of the hydrogen, and the problem simplifies as z increases.

Here we present fairly precise bounds on the ground state energy of $H_{z,2}$ implied by our actual estimates. However, the conditions under which these estimates are valid impose rather sever restrictions in z. We introduce the reference energy

$$E^{\text{ref}} := \frac{\kappa^2 e^4 m}{\hbar^2} = \alpha^2 mc^2$$

(twice the ground state energy of the hydrogen, or 2 Ry), where c is the speed of light in vacuum and

$$\alpha = \kappa \frac{e^2}{\hbar c}$$

is the fine structure constant, whose numerical value, approximately $\frac{1}{137}$. Let $E^{(z)}_{\#}$ stand for either $E^{\text{sym}}_{z,2}/E^{\text{ref}}$ or $E^{\text{as}}_{z,2}/E^{\text{ref}}$, the ground state energy of $H_{z,2}$ on either symmetric or anti-symmetric functions. We have:

Proposition 3.2. *Assume $z \geq 31.25$ for the symmetric space and $z \geq 170$ for the anti-symmetric one. Then the ground state energy of $H_{z,2}$ is bounded as*

$$-c^{\#}z^2 + w_1^{\#}z - \frac{10w_2^{\#}}{\gamma_0^{\#}} \leq E^{(z)}_{\#} \leq -c^{\#}z^2 + w_1^{\#}z, \tag{3.2}$$

where, for symmetric functions, $c = 1$, $\gamma_0 = \frac{3}{8}$ and $w_1 \approx 0.6$ and $w_2 \approx 0.27$, defined in equations (3.13) *and* (3.15)*, and, for anti-symmetric functions, $c^{\text{as}} = \frac{5}{8}$, $\gamma_0^{\text{as}} = \frac{5}{72}$ and $w_1^{\text{as}} \approx 0.20$ and $w_2^{\text{as}} \approx 0.01$, defined in* (3.12).

The approximate values of w_1, w_2, w_1^{as} and w_2^{as} are computed numerically in Appendix B. Here we report the stable digits of our computations.

[1]By the Pauli principle, the product of coordinate and spin wave functions should anti-symmetric with respect to permutation of the particle coordinates and spins. Hence, in the two particle case, after separation of spin variables, coordinate wave functions could be either antisymmetric or symmetric.

z	10	20	30	40	50
$-E_{\text{exact}}$ (from [16–18])	93.9	387.7	881.4	1575.2	2468.9
main part $-E_{z,2}^{\text{sym,lead}}$ in (3.2)	94	388	882	1576	2470
relative difference err_z	0.11 %	0.077 %	0.068 %	0.051 %	0.045 %
relative error term Δ_z in (3.2)	8.51 %	2.06 %	0.91 %	0.51 %	0.32 %

Table 1.1. Comparison of non-rigorous computations (see [8, 9, 16–18]) with the main part $-E_{z,2}^{\text{sym,lead}} := z^2 - w_1 z$ in equation (3.2), its relative contribution of the error estimation $\Delta_z := 10w_2/(-E_{z,2}^{\text{sym,lead}}\gamma_0)$ and relative difference $\text{err}_z := |E_{\text{exact}} - E_{z,2}^{\text{sym,lead}}|/(-E_{z,2}^{\text{sym,lead}})$.

The inequalities $z \geq 31.25$ and $z \geq 170$ come from condition (1.3), while estimates (3.2), from (1.8), with $b = 0.8$ and $a = 0.1$ (which give $k = 10$).

Table 1.1 compares the result for symmetric functions with computations in quantum chemistry. (We did not find results for the antisymmetric space.)

We observe that, except for $z = 50$, the results of [16–18] lie in the interval provided by the estimation (3.2).

For computations of the relativistic and QED contributions, see [10, 17, 19].

Now, we derive some consequences of estimates (3.2).

Let z_* be the smallest z for which the error bound in the symmetric case is less than or equal to the smallest explicit (subleading) term w_1. Inequality (3.2) shows that the latter bound is satisfied for z such that $w_1 z \geq \frac{10w_2}{\gamma_0}$, which shows that $z_* = \frac{10w_2}{\gamma_0 w_1}$ and consequently,

$$z_* \approx 12.$$

According to equation (3.2), the symmetric ground state energy is lower than the anti-symmetric one, if $-z^2 + w_1 z < -\frac{5}{8}z^2 + w_1^{\text{as}} z - 160\, w_2^{\text{as}}$. Using the values $w_1 \approx 0.6$, $w_2 \approx 0.3$, $w_1^{\text{as}} \approx 0.20$ and $w_2^{\text{as}} \approx 0.01$, we find that, based on (3.2), the ground state is symmetric and therefore its spin is 0 for

$$z \geq \frac{4}{3}\left(\frac{2}{5} + \sqrt{\frac{4}{25} + \frac{96}{5}}\right) = \frac{8}{3} \approx 2.6.$$

We conjecture that the symmetric ground state energy is lower than the anti-symmetric one for all z.

Finally, note that on symmetric functions, the eigenvalue of the unperturbed operator is simple and on anti-symmetric ones, has the degeneracy 4, see below.

Proof of Proposition 3.2. First, we rescale the Hamiltonian (3.1) as $x \to \frac{x}{\mu}$, with

$$\mu := \frac{z\kappa e^2 m}{\hbar^2} = \frac{z}{\text{the Bohr radius}},$$

to obtain

$$U_\mu^{-1} H_{z,2} U_\mu = \frac{z^2\kappa^2 e^4 m}{\hbar^2} H^{(z)},$$

where $U_\mu : \psi(x) \to \mu^3\psi(\mu x)$ and $H^{(z)}$ is the rescaled Hamiltonian given by

$$H^{(z)} = \sum_{j=1}^{2}\Big(-\frac{1}{2}\Delta_{x_j} - \frac{1}{|x_j|}\Big) + \frac{1/z}{|x_1 - x_2|}.$$

Thus, it suffices to estimate the ground state energy of $H^{(z)}$. We consider

$$W = \frac{1/z}{|x_1 - x_2|}$$

as the perturbation and $\sum_{j=1}^{2}(-\frac{1}{2}\Delta_{x_j} - \frac{1}{|x_j|})$ as the unperturbed operator.

First, we consider the rescaled Hamiltonian $H^{(z)}$ on the *symmetric* functions subspace. On symmetric functions, the ground state energy of $\sum_{j=1}^{2}(-\frac{1}{2}\Delta_{x_j} - \frac{1}{|x_j|})$ is -1 (see below), so we shift the operator $H^{(z)}$ by $1 + \beta$, for some $\beta > 0$, so that

$$H^{(z)} + 1 + \beta = H_0 + W,$$

with

$$H_0 := \sum_{j=1}^{2}\Big(-\frac{1}{2}\Delta_{x_j} - \frac{1}{|x_j|}\Big) + 1 + \beta \quad \text{and} \quad W = \frac{1/z}{|x_1 - x_2|}.$$

Now, the ground state energy of H_0 is β and we can use inequality (1.8) and Proposition 1.4 (c) to estimate the ground state energy of $H^{(z)} + 1 + \beta$.

By the HVZ theorem, the spectrum of H_0 on symmetric functions consists of the continuum $[e_1 + 1 + \beta, \infty)$ and the eigenvalues $e_m + e_n + 1 + \beta, m, n \geq 1$, where $e_n, n = 1, 2, \ldots$, denote the discrete eigenvalues of the Hydrogen Hamiltonian $-\frac{1}{2}\Delta_x - \frac{1}{|x|}$. The eigenvalues $e_n, n = 1, 2, \ldots$, are known explicitly:

$$e_n = -\frac{1}{2n^2}, \tag{3.3}$$

with the multiplicities of n^2 and the corresponding eigenfunctions are given by equation (A.1) below.

Then the ground state energy of H_0 is $\lambda_0 = 2e_1 + 1 + \beta = \beta$, as claimed, and the gap, $\gamma_0 = e_1 + e_2 - 2e_1 = e_2 - e_1 = \frac{3}{8}$.

Next, we show that condition (1.3) is satisfied for $z \geq 31.25$. We begin with the really rough estimate

$$\frac{1}{|x_1 - x_2|} \leq h_{x_1} + h_{x_2} + 10, \tag{3.4}$$

where $h_x := -\frac{1}{2}\Delta_x - \frac{1}{|x|}$. First, we use Hardy's inequality, $-\Delta \geq \frac{1}{4|x|^2}$, and the estimate $\frac{1}{4m|x|^2} - \frac{1}{|x|} \geq -m$ to obtain

$$-\frac{1}{m}\Delta - \frac{1}{|x|} \geq -m. \tag{3.5}$$

Next, passing to the relative and center-of-mass coordinates, $x - y$ and $\frac{1}{2}(x + y)$, we find

$$-\Delta_x - \Delta_y = -2\Delta_{x-y} - \frac{1}{2}\Delta_{\frac{1}{2}(x+y)} \geq -2\Delta_{x-y},$$

which, together with (3.5), yields

$$\frac{1}{|x_1 - x_2|} \le -\frac{1}{2}\Delta_{x_1 - x_2} + 2 \le -\frac{1}{4}(\Delta_{x_1} + \Delta_{x_2}) + 2.$$

Now, we use $-\frac{1}{2}\Delta_x = h_x + \frac{1}{|x|} \le h_x - \frac{1}{4}\Delta_x + 4$ to obtain

$$-\frac{1}{4}\Delta_x \le h_x + 4.$$

The last two inequalities yield (3.4). Equation (3.4), together with the relation

$$H_0 = h_{x_1} + h_{x_2} + 1 + \beta,$$

implies the estimate

$$\frac{1}{|x_1 - x_2|} \le H_0 + 9 - \beta. \tag{3.6}$$

Now, we check condition (1.3):

$$\|P^\perp W P^\perp\|_{H_0} \le \frac{b\gamma_0}{\lambda_*}.$$

Using the last relation, we estimate

$$\begin{aligned} z(H_0^\perp)^{-\frac{1}{2}} W (H_0^\perp)^{-\frac{1}{2}} &\le (H_0^\perp)^{-1}(H_0^\perp + 9 - \beta) \\ &= \mathbf{1} + (9 - \beta)(H_0^\perp)^{-1}. \end{aligned}$$

Recalling that $\lambda_0 = \beta$, we see that $H_0^\perp \ge \beta + \gamma_0$, so that the last inequality gives

$$0 \le (H_0^\perp)^{-\frac{1}{2}} W (H_0^\perp)^{-\frac{1}{2}} \le \frac{1}{z}\frac{9 + \gamma_0}{\beta + \gamma_0},$$

provided $\beta < 9$. This implies

$$\|P^\perp W P^\perp\|_{H_0} \le \frac{1}{z}\frac{9 + \gamma_0}{\beta + \gamma_0}. \tag{3.7}$$

Since $\lambda_* = \beta + \gamma_0$, condition (1.3) is satisfied, if $\frac{9+\gamma_0}{z} \le b\gamma_0$, which gives $z \ge \frac{9+\gamma_0}{b\gamma_0}$. Since $\gamma_0 = \frac{3}{8}$ and $b = 0.8$, this implies $z \ge 31.25$. We will need the following lemma, whose proof is given after the proof of the proposition.

Lemma 3.3. *Recall that P is the orthogonal projection onto the eigenspace of H_0 corresponding to the lowest eigenvalue β. We have*

$$\langle W \rangle = \|PWP\| = \frac{w_1}{z} \quad \text{and} \quad \Phi(W) \le \frac{w_2}{z^2 \gamma_0}, \tag{3.8}$$

with, recall, $\langle W \rangle := \langle \varphi_0, W\varphi_0 \rangle$, where φ_0 is the eigenfunction of H_0 corresponding to λ_0.

We check now the second condition of Theorem 1.1, (1.4). Recall the definition

$$r := \frac{\|PWP\| + k\Phi(W)}{\gamma_0}.$$

Then condition (1.4) can be written as $r < a$. By Lemma 3.3,

$$r \leq \frac{1}{\gamma_0}\Big(\frac{w_1}{z} + \frac{kw_2}{z^2\gamma_0}\Big).$$

Recall the condition $z \geq 31.25$ and $k = 10$, $w_1 \approx 0.6$ and $w_2 \approx 0.3$. Thus

$$\frac{kw_2}{z^2\gamma_0} \leq \frac{w_1}{z} \cdot \frac{kw_2}{31.25\gamma_0 w_1} \leq 0.5 \cdot \frac{w_1}{z}.$$

Then we find $r \leq \frac{1.5w_1}{z\gamma_0}$ and therefore condition (1.4) is satisfied if $a > \frac{1.5w_1}{z\gamma_0}$, or (for $a = 0.1$)

$$z > \frac{3}{2}\frac{8}{3}\frac{0.6}{a} = 24.$$

This is less restrictive than $z \geq 31.25$.

For the third condition (1.5), recalling that $\lambda_0 = \beta$, so that $\lambda_* = \gamma_0 + \beta$ ($= \lambda_1$, the second eigenvalue of H_0) and using both relations in (3.8), we obtain that condition (1.5) is satisfied if $\frac{kw_2}{z^2\gamma_0} < \frac{1}{2}(a\gamma_0 - \frac{w_1}{z})$, which is equivalent to

$$z > \frac{1}{2a\gamma_0}\Big(w_1 + \sqrt{w_1^2 + 8akw_2}\Big).$$

Using the values $\gamma_0 = \frac{3}{8}$, $w_1 \approx 0.6$ and $w_2 \approx 0.3$, and $b = 0.8$ and $a = 0.1$, giving $k = 10$, this shows that the latter inequality, and therefore (1.5), holds if $z > 30$, which is less restrictive than $z \geq 31.25$. Thus, all three conditions are satisfied for $z \geq 31.25$.

Next, we use (1.8) to estimate the ground state energy, $E_{\mathrm{sym}}^{(z)}$ of $H^{(z)}$. The first and second relations in (3.8), together with (1.8) and the fact that $U(\lambda) \leq 0$, gives the bounds

$$-\frac{kw_2}{\gamma_0 z^2} \leq E_{\mathrm{sym}}^{(z)} + 1 - \frac{w_1}{z} \leq 0,$$

which after rescaling gives (3.2).

Now, we consider the ground state of $H^{(z)}$ on *anti-symmetric* functions. In this case, the ground state energy of $\sum_{j=1}^{2}(-\frac{1}{2}\Delta_{x_j} - \frac{1}{|x_i|})$ is $e_1 + e_2 = -\frac{5}{8}$ of multiplicity 4, with thc ground statcs

$$\phi_{2\ell k}(x_1, x_2) := \frac{1}{\sqrt{2}}(\psi_{100}(x_1)\psi_{2\ell k}(x_2) - \psi_{2\ell k}(x_1)\psi_{100}(x_2)) \tag{3.9}$$

for $(\ell, k) = (0,0), (1,-1), (1,0), (1,1)$, where $\psi_{n\ell k}(x)$ for $\ell = 0, 1, 2, \ldots, n-1$, $k = -\ell, -\ell+1, \ldots, \ell$, $n = 1, 2, \ldots$ are the eigenfunctions of the Hydrogen-like Hamiltonian $-\frac{1}{2}\Delta_x - \frac{1}{|x|}$ corresponding to eigenvalues e_n, so that $\psi_{100}(x) = \psi_1(x)$, see Appendix A. Now, we define the unperturbed operator as

$$H_0 := \sum_{j=1}^{2}\Big(-\frac{1}{2}\Delta_{x_j} - \frac{1}{|x_j|}\Big) + \frac{5}{8} + \beta$$

to have β as the lowest eigenvalue of H_0.

For the gap, since the first two eigenvalues are

$$e_1 + e_2 + \frac{5}{8} + \beta = \beta \quad \text{and} \quad e_1 + e_3 + \frac{5}{8} + \beta \left(< e_2 + e_2 + \frac{5}{8} + \beta\right),$$

we have $\gamma_0^{\mathrm{as}} = e_1 + e_3 - (e_1 + e_2) = e_3 - e_2$ giving $\gamma_0^{\mathrm{as}} = \frac{5}{72}$.

For condition (1.3), note first that since now $H_0 = h_{x_1} + h_{x_2} + \frac{5}{8} + \beta$, equation (3.4) implies

$$\frac{1}{|x_1 - x_2|} \le H_0 + 9 + \frac{3}{8} - \beta,$$

instead of (3.6), which shows that (3.7) should be modified as

$$\|P^\perp W P^\perp\|_{H_0} \le \frac{1}{z}\frac{9 + \frac{3}{8} + \gamma_0^{\mathrm{as}}}{\beta + \gamma_0^{\mathrm{as}}}$$

in the anti-symmetric case. Since $\lambda_*^{\mathrm{as}} = \beta + \gamma_0^{\mathrm{as}}$, it follows that condition (1.3) is satisfied if $\frac{1}{z}(9 + \frac{3}{8} + \gamma_0^{\mathrm{as}}) \le b\gamma_0^{\mathrm{as}}$, which gives

$$z \ge \frac{9 + \frac{3}{8} + \gamma_0^{\mathrm{as}}}{b\gamma_0^{\mathrm{as}}}.$$

Since $\gamma_0^{\mathrm{as}} = \frac{5}{72}$ and $b = 0.8$, this implies $z \ge 170$ for the anti-symmetric case.

We skip the verification of conditions (1.4) and (1.5), which are simpler and similar to that of the symmetric case.

To estimate the ground state energy, $E_{\mathrm{as}}^{(z)}$, of $H^{(z)}$ on anti-symmetric functions, we use bound (1.9). First, we claim that $\Phi(W)$ defined in (1.2) satisfies (cf. Remark 2)

$$\Phi(W) \le \frac{\mu_{\max}}{\gamma_0^{\mathrm{as}}}, \tag{3.10}$$

where $\mu_{\max}$ is the largest eigenvalue of the positive, 4×4-matrix $PWP^\perp WP$. Indeed, recall that

$$\Phi(W) := \frac{\lambda_0\lambda_*\|P^\perp WP\|_{H_0}^2}{\gamma_0^{\mathrm{as}}}$$

with $\lambda_0 = \beta$ and observe that $\|PWP^\perp\|_{H_0} = \|A\|$, with $A := H_0^{-\frac{1}{2}} PWP^\perp H_0^{-\frac{1}{2}}$. Using now the relations

$$\begin{aligned} AA^* &= H_0^{-\frac{1}{2}} PWP^\perp H_0^{-1} WPH_0^{-\frac{1}{2}}, \\ H_0^{-\frac{1}{2}} P &= \lambda_0^{-\frac{1}{2}} P \quad (\text{since } H_0 P = \lambda_0 P), \\ H_0^{-1} P^\perp &\le \lambda_*^{-1} P^\perp, \end{aligned}$$

we find

$$AA^* \leq (\lambda_0\lambda_*)^{-1} PWP^\perp WP,$$

which, together with $\|A\| = \|A^*\| = \|AA^*\|^{\frac{1}{2}}$, gives

$$\|PWP^\perp\|_{H_0}^2 \leq (\lambda_0\lambda_*)^{-1} \|PWP^\perp WP\|. \tag{3.11}$$

Since $PWP^\perp WP \geq 0$, we have $\|PWP^\perp WP\| = \mu_{\max}$, which gives (3.10).

Hence, we have to compute the smallest eigenvalue $\mu_{\min}$ of the 4×4 matrix PWP and the largest eigenvalue, $\mu_{\max}$, of the positive, 4×4-matrix $PWP^\perp WP$, see (1.9).

By scaling, it is easy to see that the matrices PWP and $PWP^\perp WP$ are of the form $PWP = z^{-1}A$ and $PWP^\perp WP = z^{-2}B$, where A and B are z-independent 4×4 matrices. Hence

$$\mu_{\min} = \frac{w_1^{\mathrm{as}}}{z} \quad \text{and} \quad \mu_{\max} = \frac{w_2^{\mathrm{as}}}{z^2}, \tag{3.12}$$

for some positive z-independent constants w_1^{as} and w_2^{as}. These constants are computed in Appendix B. Now, we see that (1.9) and (3.10) give

$$\frac{w_1^{\mathrm{as}}}{z} - \frac{k w_2^{\mathrm{as}}}{\gamma_0^{\mathrm{as}} z^2} \leq E_{\mathrm{as}}^{(z)} + \frac{5}{8} \leq \frac{w_1^{\mathrm{as}}}{z},$$

which after rescaling and setting $k = 10$ gives (3.2) in the antisymmetric case. ∎

Proof of Lemma 3.3. The ground state energy, e_1, of $-\frac{1}{2}\Delta_x - \frac{1}{|x|}$ is non-degenerate, with the normalized ground state (known as the 1s wavefunction) given by (see, e.g., [7], or Wiki, with a_0 replaced by 1 due to the rescaling above)

$$\psi_1(x) = \frac{1}{\sqrt{\pi}} e^{-|x|}.$$

Hence, the ground state energy, 0, of H_0 is also non-degenerate, with the normalized ground state $\psi_1(x_1)\psi_1(x_2)$ (in the symmetric subspace).

First, we compute $\langle W \rangle$ and $\|PWP\|$. Let $(\psi \otimes \phi)(x, y) := \psi(x)\phi(y)$ and, for an operator A on $L^2(\mathbb{R}^6)$, we denote $\langle A \rangle := \langle \psi_1 \otimes \psi_1, A\psi_1 \otimes \psi_1 \rangle$. In the symmetric case, $PWP = \langle W \rangle P$, where

$$\begin{aligned} \langle W \rangle &:= \langle \psi_1 \otimes \psi_1, W\psi_1 \otimes \psi_1 \rangle \\ &= \frac{1}{z} \int_{\mathbb{R}^6} \frac{|\psi_1(x)\psi_1(y)|^2}{|x-y|} \, dx \, dy =: \frac{1}{z} w_1. \end{aligned} \tag{3.13}$$

Hence $\|PWP\| = \frac{w_1}{z}$. This gives the first relation in (3.8).

Now, using that $P^\perp = \mathbf{1} - P$, we compute

$$PWP^\perp WP = PW^2P - PWPWP = \frac{w_2}{z^2} P \tag{3.14}$$

and

$$\begin{aligned} w_2 &:= z^2(\langle W^2\rangle - \langle W\rangle^2) \\ &= \int_{\mathbb{R}^6} \frac{|\psi_1(x)\psi_1(y)|^2}{|x-y|^2}\,dx\,dy - \left(\int_{\mathbb{R}^6} \frac{|\psi_1(x)\psi_1(y)|^2}{|x-y|}\,dx\,dy\right)^2. \end{aligned} \tag{3.15}$$

By the Schwarz inequality and the normalization of ψ_1, we have $w_2 \geq 0$, which, together with (3.14), gives

$$\|PWP^{\perp}WP\| = \frac{w_2}{z^2}. \tag{3.16}$$

Equations (3.11) and (3.16) imply the second relation in (3.8). ■

A The eigenfunctions of the Hydrogen-like Hamiltonian

The eigenfunctions $\psi_{n\ell k}(x)$ are given by (see Wiki, "Hydrogen atom" with a_0 replaced by 1)

$$\psi_{n\ell k}(x) = \sqrt{\left(\frac{2z}{n}\right)^3 \frac{(n-\ell-1)!}{2n(n+\ell)!}}\, e^{-\frac{\rho}{2}} \rho^\ell L_{n-\ell-1}^{2\ell+1}(\rho) Y_\ell^k(\theta,\phi), \tag{A.1}$$

for $\ell = 0, 1, 2, \ldots, n-1$, $k = -\ell, -\ell+1, \ldots, \ell$, $n = 1, 2, \ldots$. Here (r, θ, ϕ) are the polar coordinates of x, $\rho = \frac{2zr}{n}$, $L_{n-\ell-1}^{2\ell+1}(\rho)$ is a generalized Laguerre polynomial of degree $n-\ell-1$, and $Y_\ell^k(\theta,\phi)$ is a spherical harmonic function of the degree ℓ and order k.[2]

B The numerical approximation of the constants w_1, w_2, w_1^{as} and w_2^{as}

Let us first focus on w_1 and w_2. From the definition

$$\begin{aligned} w_1 &:= \int_{\mathbb{R}^6} \frac{|\psi_1(x)\psi_1(y)|^2}{|x-y|}\,dy\,dx, \\ w_2 &:= \int_{\mathbb{R}^6} \frac{|\psi_1(x)\psi_1(y)|^2}{|x-y|^2}\,dy\,dx - w_1^2 \end{aligned}$$

[2]Quoting from Wikipedia (https://en.wikipedia.org/wiki/Hydrogen_atom, 30.10.2020): "...the generalized Laguerre polynomials are defined differently by different authors. The usage here is consistent with the definitions used by [7, p. 1136] and Wolfram Mathematica. In other places, the Laguerre polynomial includes a factor of $(n+\ell)!$. ...or the generalized Laguerre polynomial appearing in the hydrogen wave function is $L_{n+\ell}^{2\ell+1}(\rho)$ instead."

it becomes clear that we need to compute terms of the form

$$C_\alpha = \int_{\mathbb{R}^6} \frac{|\psi_1(x)\psi_1(y)|^2}{|x-y|^\alpha}\,dy\,dx = \int_{\mathbb{R}^3} |\psi_1(x)|^2 \int_{\mathbb{R}^3} \frac{|\psi_1(x-z)|^2}{|z|^\alpha}\,dz\,dx,$$

with $\alpha = 1, 2$. We thus introduce a numerical quadrature in order to approximate the values of these integrals. We do not aim within this work to obtain the most efficient implementation. We define a numerical quadrature grid in terms of spherical coordinates with integration points $x_{i,n}$ and weights $\omega_{i,n}$ given by

$$x_{i,n} = r_i s_n, \quad \omega_{i,n} = h r_i^3 \omega_n^{\text{leb}}$$

with

$$r_i = \exp(-\ln(R_{\max}) + (i-1)h), \quad i = 1, \ldots, N_r, \qquad h = \frac{2\ln(R_{\max})}{N_r - 1}$$

and where $\{s_n, \omega_n^{\text{leb}}\}$ denote Lebedev integration points on the unit sphere with N_{leb} points. Thus, N_r denotes the number of points along the r-coordinate, $R_{\max}$ the radius of the furthest points away from the origin.

We thus approximate C_α by

$$C_\alpha \approx \sum_{i=1}^{N_r} \sum_{n=1}^{N_{\text{leb}}} \omega_{i,n} |\psi_1(x_{i,n})|^2 \Phi_\alpha(x_{i,n})$$

with

$$\Phi_\alpha(x_{i,n}) = \sum_{j=1}^{N_r} \sum_{m=1}^{N_{\text{leb}}} \omega_{j,m} \frac{|\psi_1(x_{i,n} - x_{j,m})|^2}{|x_{j,m}|^\alpha}.$$

Then we approximate $w_1 \approx C_1$ and $W_2 \approx C_2 - C_1^2$.

We now shed our attention to the constants w_1^{as} and w_2^{as} and we first introduce the bi-electronic integrals, for $(\ell, k), (\ell', k') = (0,0), (1,-1), (1,0), (1,1)$, by

$$\mathsf{A}_{\ell,k}^{\ell',k'} = \int_{\mathbb{R}^3} \psi_{2\ell k}(x)\psi_{2\ell' k'}(x) \int_{\mathbb{R}^3} \frac{|\psi_1(y)|^2}{|x-y|}\,dy\,dx,$$

$$\mathsf{B}_{\ell,k}^{\ell',k'} = \int_{\mathbb{R}^3} \psi_1(x)\psi_{2\ell' k'}(x) \int_{\mathbb{R}^3} \frac{\psi_1(y)\psi_{2\ell k}(y)}{|x-y|}\,dy\,dx.$$

Then, using the definition (3.9) and symmetry properties, we derive the following expressions for the matrix elements:

$$\mathsf{M}_{\ell,k}^{\ell',k'} = z\,\langle \psi_{2\ell k} | W | \psi_{2\ell' k'} \rangle = \mathsf{A}_{\ell,k}^{\ell',k'} - \mathsf{B}_{\ell,k}^{\ell',k'}.$$

We then employ the same quadrature rule as above to approximate the matrix elements $\mathsf{A}_{\ell,k}^{\ell',k'}, \mathsf{B}_{\ell,k}^{\ell',k'}$ and compute the smallest eigenvalue of M as an approximation of w_1^{as}.

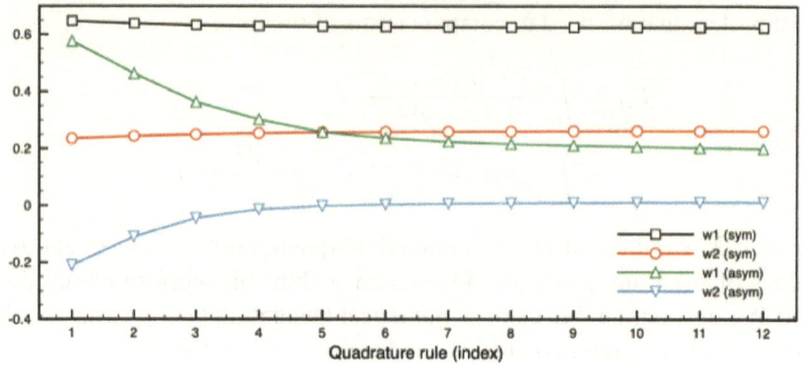

Figure 1. Value of the approximate coefficients w_1, w_2, w_1^{as} and w_2^{as} depending on the quadrature rule reported in Table 2.

Index	1	2	3	4	5	6	7	8	9	10	11	12
$R_{\max}$	16	18	20	22	24	26	28	30	32	34	34	34
N_r	10	12	14	16	18	20	22	24	26	28	30	32
N_{leb}	194	266	350	590	974	1454	2030	2702	3470	4334	5294	5810

Table 2. Parameters for the different quadrature rules.

Finally, the approximation of w_2^{as} involves the matrix

$$PWP^{\perp}WP = PW^2P - PWPWP,$$

whose elements are given by $\mathsf{Q} = \mathsf{N} - \mathsf{M}^2$ with

$$\mathsf{N}_{\ell,k}^{\ell',k'} = z^2 \langle \psi_{2\ell k} | W^2 | \psi_{2\ell' k'} \rangle = \mathsf{C}_{\ell,k}^{\ell',k'} - \mathsf{D}_{\ell,k}^{\ell',k'}$$

and

$$\mathsf{C}_{\ell,k}^{\ell',k'} = \int_{\mathbb{R}^3} \psi_{2\ell k}(x)\psi_{2\ell' k'}(x) \int_{\mathbb{R}^3} \frac{|\psi_1(y)|^2}{|x-y|^2}\,dy\,dx,$$

$$\mathsf{D}_{\ell,k}^{\ell',k'} = \int_{\mathbb{R}^3} \psi_1(x)\psi_{2\ell' k'}(x) \int_{\mathbb{R}^3} \frac{\psi_1(y)\psi_{2\ell k}(y)}{|x-y|^2}\,dy\,dx.$$

We then again, approximate the different integrals with the above introduced quadrature rule and find the maximal eigenvalue of Q in order to approximate w_2^{as}.

In Figure 1, we plot the approximate value of w_1, w_2, w_1^{as} and w_2^{as} for different values of the parameters $R_{\max}$, N_r and N_{leb} given in Table 2.

We obtain approximate values

$$w_1 \approx 0.625, \quad w_2 \approx 0.2628, \quad w_1^{\text{as}} \approx 0.3, \quad w_2^{\text{as}} \approx 0.01,$$

where the last reported digit is stable over the last three quadrature rules.

Acknowledgements. The second author is grateful to Volker Bach and Jürg Fröhlich for enjoyable collaboration. The authors are indebted to the anonymous referee for many useful suggestions and remarks.

Bibliography

[1] V. Bach, J. Fröhlich and I. M. Sigal, Quantum electrodynamics of confined nonrelativistic particles. *Adv. Math.* **137** (1998), 299–395

[2] H. L. Cycon, R. G. Froese, W. Kirsch and B. Simon, *Schrödinger operators with application to quantum mechanics and global geometry*. Study edn., Texts Monogr. Phys., Springer, Berlin, 1987

[3] G. Dusson, I. M. Sigal and B. Stamm, Analysis of the Feshbach–Schur method for the planewave discretizations of Schrödinger operators. Preprint 2020, arXiv:2008.10871

[4] S. J. Gustafson and I. M. Sigal, *Mathematical concepts of quantum mechanics*. Universitext, Springer, Cham, 2011

[5] P. D. Hislop and I. M. Sigal, *Introduction to spectral theory*. Appl. Math. Sci. 113, Springer, New York, 1996

[6] T. Kato, *Perturbation theory for linear operators*. 2nd edn., Springer, Berlin, 1976

[7] A. Messiah, *Quantum mechanics*. Dover, New York, 1999

[8] H. Nakashima and H. Nakatsuji, Solving the Schrödinger equation for helium atom and its isoelectronic ions with the free iterative complement interaction (ICI) method. *J. Chem. Phys.* **127** (2007), Article ID 224104

[9] H. Nakashima and H. Nakatsuji, Solving the electron-nuclear Schrödinger equation of helium atom and its isoelectronic ions with the free iterative-complement-interaction method. *J. Chem. Phys.* **128** (2008), Article ID 154107

[10] K. Pachucki, V. Patkos and V. A. Yerokhin, Testing fundamental interactions on the helium atom. *Phys. Rev. A* **95** (2017), Article ID 062510

[11] M. Reed and B. Simon, *Methods of modern mathematical physics. II. Fourier analysis, self-adjointness*. Academic Press, New York, 1975

[12] M. Reed and B. Simon, *Methods of modern mathematical physics. IV. Analysis of operators*. Academic Press, New York, 1978

[13] F. Rellich, *Perturbation theory of eigenvalue problems*. Assisted by J. Berkowitz. With a preface by Jacob T. Schwartz, Gordon and Breach Science Publishers, New York, 1969

[14] B. Simon, *A comprehensive course in analysis, Part 4: Operator theory*. American Mathematical Society, Providence, 2015

[15] B. Simon, Tosio Kato's work on non-relativistic quantum mechanics: Part 1 and Part 2. *Bull. Math. Sci.* **8** (2018), 121–232; (2019), Article ID 1950005

[16] A. V. Turbiner and J. C. Lopez Vieyra, Helium-like and Lithium-like ions: Ground state energy. Preprint 2017, arXiv:1707.07547v3

[17] A. V. Turbiner, J. C. Lopez Vieyra and H. Olivares-Pilón, Few-electron atomic ions in non-relativistic QED: The ground state. *Ann. Physics* **409** (2019), Article ID 167908

[18] A. V. Turbiner, J. C. Lopez Vieyra, J. C. del Valle and D. J. Nader, Ultra-compact accurate wave functions for He-like and Li-like iso-electronic sequences and variational calculus: I. Ground state. Int. J. Quantum Chem. (2020), Article ID e26586; arXiv:2007.11745

[19] V. A. Yerokhin and K. Pachucki, Theoretical energies of low-lying states of light helium-like ions. *Phys. Rev. A* **81** (2010), Article ID 022507

Adiabatic and non-adiabatic evolution of wave packets and applications to initial value representations

Clotilde Fermanian Kammerer, Caroline Lasser and Didier Robert

To Ari Laptev on the occasion of his 70th birthday

We review some recent results obtained for the time evolution of wave packets for systems of equations of pseudo-differential type, including Schrödinger ones, and discuss their application to the approximation of the associated unitary propagator. We start with scalar equations, propagation of coherent states, and applications to the Herman–Kluk approximation. Then we discuss the extension of these results to systems with eigenvalues of constant multiplicity or with smooth crossings.

1 Introduction

We consider semiclassical systems of the form

$$i\varepsilon\partial_t\psi^\varepsilon(t) = \widehat{H(t)}\psi^\varepsilon(t), \quad \psi^\varepsilon_{|t=t_0} = \psi^\varepsilon_0, \tag{1.1}$$

where (ψ^ε_0) is a bounded family in $L^2(\mathbb{R}^d, \mathbb{C}^N)$, $\|\psi^\varepsilon_0\|_{L^2(\mathbb{R}^d,\mathbb{C}^N)} = 1$ and $\widehat{H}(t)$ is the ε-Weyl quantization of a smooth Hermitian symbol $H(t,x,\xi)$ satisfying suitable growth assumptions. We are interested in approximate realizations of the unitary propagator associated with this equation relying on the use of continuous superpositions of Gaussian wave packets. Before explaining these ideas in more details (in particular the definition of the quantization that the reader will find below), let us first emphasize different types of systems that we have in mind.

The simplest case is the one where $H(t)$ is independent of (x,ξ), which implies that the operator $\widehat{H}(t)$ coincides with the matrix $H(t)$. In this case it is known that for $N \geqq 2$ new phenomena appear by comparison with the scalar case $N = 1$. When the eigenvalues of $H(t)$ are not crossing, the adiabatic theorem says that if ψ^ε_0 is an

Keywords: Time-dependent Schrödinger equation, eigenvalue crossing, wave packets, initial value representation

2020 Mathematics Subject Classification: Primary 35, 81; secondary 35C20, 35Q41, 81Q05, 81Q20

eigenfunction of $H(t_0)$, then at every time $\psi^\varepsilon(t)$ has an asymptotic expansion for $\varepsilon \to 0$, and the leading term is an eigenfunction of $H(t)$. The first complete proof was given by Friedrichs ([10] after first results by Born-Fock [3] and Kato [16]. In [10, Part II], Friedrichs considered a non-adiabatic situation where $H(t)$ is a 2×2 Hermitian matrix with two analytic eigenvalues $h_\pm(t)$ such that $h_+(t) \neq h_-(t)$ for $t \neq 0$ and

$$(h_+ - h_-)(0) = 0, \quad \frac{d}{dt}(h_+ - h_-)(0) \neq 0.$$

In this spirit, we here consider a toy model for space dependent Hamiltonians of the following form: We choose $d = 1$, $N = 2$, $\theta \in \mathbb{R}_+$, $k \in \mathbb{R}^*$, and

$$\widehat{H}_{k,\theta} = \frac{\varepsilon}{i}\frac{d}{dx}\mathbb{I}_{\mathbb{C}^2} + kx\begin{pmatrix} 0 & \mathrm{e}^{i\theta x} \\ \mathrm{e}^{-i\theta x} & 0 \end{pmatrix}. \tag{1.2}$$

Its semiclassical symbol

$$H_{k,\theta}(x,\xi) = \xi + kx\begin{pmatrix} 0 & \mathrm{e}^{i\theta x} \\ \mathrm{e}^{-i\theta x} & 0 \end{pmatrix}$$

has the eigenvalues and associated eigenvectors (the latter depending only in x)

$$h_\pm(x,\xi) = \xi \pm kx, \quad \vec{V}_\pm(x) = \frac{1}{\sqrt{2}}\begin{pmatrix} \mathrm{e}^{i\theta x} \\ \pm 1 \end{pmatrix}. \tag{1.3}$$

So for $k \neq 0$ we have a crossing at $x = 0$ that we call *smooth crossing* because there exists smooth eigenvalues and eigenprojectors. We shall use this toy-model all along this article. Notice that one can prove that the operator $\widehat{H}_{k,\theta}$ is essentially self-adjoint in $L^2(\mathbb{R}, \mathbb{C}^2)$ by using a commutator criterion with the standard harmonic oscillator (see [22, Appendix A]). As we shall see later, the solutions of the Schrödinger equation for $\widehat{H}_{k,\theta}$ can be computed by solving an ODE asymptotically as $\varepsilon \to 0$, see (5.4), and by using non-adiabatic asymptotic results from Friedrichs [10] and Hagedorn [12].

More generally, of major interest because of their applications to molecular dynamics, are the Schrödinger Hamiltonians with matrix-valued potential considered by Hagedorn in his monograph [13, Chapter 5]

$$\widehat{H}_S = -\frac{\varepsilon^2}{2}\Delta_x\mathbb{I}_{\mathbb{C}^N} + V(x), \quad V \in \mathcal{C}^\infty(\mathbb{R}^d, \mathbb{C}^{N\times N}), \tag{1.4}$$

or the models arising in solid state physics in the context of Bloch band decompositions as described in [27, Chapter 5]

$$\widehat{H}_A = A(-i\varepsilon\nabla_x) + W(x)\mathbb{I}_{\mathbb{C}^2}, \quad A \in \mathcal{C}^\infty(\mathbb{R}^d, \mathbb{C}^{N\times N}), \quad W \in \mathcal{C}^\infty(\mathbb{R}^d, \mathbb{C}). \tag{1.5}$$

The notation $\widehat{H}$ refers to the Weyl quantization that we shall use extensively in this article. For $a \in \mathcal{C}^\infty(\mathbb{R}^{2d}, \mathbb{C}^{N,N})$ with adequate control on the growth of derivatives, the operator $\widehat{a}$ is defined by its action on functions $f \in \mathcal{S}(\mathbb{R}^d, \mathbb{C}^N)$:

$$\mathrm{op}_\varepsilon^w(a) f(x) := \widehat{a} f(x) := (2\pi\varepsilon)^{-d} \int_{\mathbb{R}^d} a\left(\frac{x+y}{2}, \xi\right) e^{\frac{i}{\varepsilon}\xi\cdot(x-y)} f(y)\, dy\, d\xi.$$

It turns out that the propagator $\mathcal{U}_H^\varepsilon(t, t_0)$ associated with $\widehat{H}$ is well defined according to [22, Theorem 5.15] provided that the map

$$(t, z) \mapsto H(t, z)$$

is in $\mathcal{C}^\infty(\mathbb{R} \times \mathbb{R}^{2d}, \mathbb{C}^{N\times N})$ valued in the set of self-adjoint matrices and that it has subquadratic growth, i.e.

$$\forall \alpha \in \mathbb{N}^{2d},\ |\alpha| \geqslant 2,\ \exists C_\alpha > 0, \qquad \sup_{(t,z)\in\mathbb{R}\times\mathbb{R}^{2d}} \|\partial^\alpha H(t, z)\|_{\mathbb{C}^{N,N}} \leqslant C_\alpha. \tag{1.6}$$

Using the commutator methods of [22], we could extend main of our results to a more general setting. However, the assumptions in (1.6) are enough to guarantee the existence of solutions to equation (1.1) in $L^2(\mathbb{R}^d, \mathbb{C}^N)$. One of our objectives here is to describe different Gaussian-based approximations of the semigroup $\mathcal{U}_H^\varepsilon(t, t_0)$ in the limit $\varepsilon \to 0$.

Let g_z^ε denote the Gaussian wave packet centered in $z = (q, p) \in \mathbb{R}^{2d}$ with standard deviation $\sqrt{\varepsilon}$:

$$g_z^\varepsilon(x) = (\pi\varepsilon)^{-\frac{d}{4}} \exp\left(-\frac{1}{2\varepsilon}|x-q|^2 + \frac{i}{\varepsilon} p\cdot(x-q)\right). \tag{1.7}$$

The family of wave packets $(g_z^\varepsilon)_{z\in\mathbb{R}^{2d}}$ forms a continuous frame and provides for all square integrable functions $f \in L^2(\mathbb{R}^d)$ the reconstruction formula

$$f(x) = (2\pi\varepsilon)^{-d} \int_{z\in\mathbb{R}^{2d}} \langle g_z^\varepsilon, f\rangle g_z^\varepsilon(x)\, dz. \tag{1.8}$$

The leading idea is then to write the unitary propagation of general, square integrable initial data $\psi_0^\varepsilon \in L^2(\mathbb{R}^d)$ as

$$\mathcal{U}_H^\varepsilon(t, t_0)\psi_0^\varepsilon = (2\pi\varepsilon)^{-d} \int_{z\in\mathbb{R}^{2d}} \langle g_z^\varepsilon, \psi_0^\varepsilon\rangle\, \mathcal{U}_H^\varepsilon(t, t_0) g_z^\varepsilon\, dz, \tag{1.9}$$

and to take advantage of the specific properties of the propagation of Gaussian states to obtain an integral representation that allows in particular for an efficient numerical realization of the propagator.

Such a program has been completely accomplished in the scalar case ($N = 1$). It appeared first in the 1980s in theoretical chemistry [15, 17, 18] and has led to the so-called Herman–Kluk approximation. The mathematical proof of the convergence of this approximation is more recent [25, 26] and lead to numerical realization (see [20, 21]). Here, we revisit these results in Section 2 and Section 5, and discuss some extensions to the case of systems ($N \geqslant 1$), first for the gapped situation in Section 3 and then for smooth crossings in Section 4.

2 Propagation of Gaussian states and the Herman–Kluk approximation for scalar equations

In this section, we have $N = 1$ and equation (1.1) is associated with a scalar symbol $H(t) = h(t)\,\mathbb{I}_{\mathbb{C}}$ of subquadratic growth (1.6). In that case, the approximate propagation of Gaussian states is described by classical quantities, leading to a simple Herman–Kluk approximation. We set

$$J = \begin{pmatrix} 0 & \mathbb{I}_{\mathbb{R}^d} \\ -\mathbb{I}_{\mathbb{R}^d} & 0 \end{pmatrix}$$

and for $z_0 \in \mathbb{R}^{2d}$ we consider the classical Hamiltonian trajectory $z(t) = (q(t), p(t))$ defined by the ordinary differential equation

$$\dot{z}(t) = J\partial_z h(t, z(t)), \quad z(t_0) = z_0.$$

The associated flow map is then denoted by

$$\Phi_h^{t,t_0}(z_0) = z(t) = (q(t), p(t)), \tag{2.1}$$

and the blocks of its Jacobian matrix $F(t, t_0, z_0) = \partial_z \Phi_h^{t,t_0}(z_0)$ by

$$F(t, t_0, z_0) = \begin{pmatrix} A(t, t_0, z_0) & B(t, t_0, z_0) \\ C(t, t_0, z_0) & D(t, t_0, z_0) \end{pmatrix}.$$

We note that the Jacobian satisfies the linearized flow equation

$$\partial_t F(t, t_0, z_0) = J\, \mathrm{Hess}_z\, h(t, z(t)) F(t, t_0, z_0), \quad F(t_0, t_0, z_0) = \mathbb{I}_{\mathbb{R}^{2d}}. \tag{2.2}$$

Thus, F is smooth in t, t_0, z with any derivative in z bounded. We will also use the action integral

$$S(t, t_0, z_0) = \int_{t_0}^{t} (p(s)\cdot \dot{q}(s) - h(s, z(s)))\, ds. \tag{2.3}$$

Then the Herman–Kluk approximation (also called *frozen Gaussian approximation* in the literature) of the unitary propagator $\mathcal{U}_h^\varepsilon(t, t_0)$ writes as follows.

Theorem 2.1 ([25, 26]). *By assuming* (1.6), *the evolution through the scalar equation with* $N = 1$ *and* $H = h\mathbb{I}_{\mathbb{C}}$ *in* (1.1) *satisfies for every* $J \geqslant 0$,

$$\mathcal{U}_h^\varepsilon(t, t_0) = \mathcal{I}_h^{\varepsilon,J}(t, t_0) + O(\varepsilon^{J+1})$$

in the norm of bounded operators on $L^2(\mathbb{R}^d)$, *where the Herman–Kluk propagator is defined for all* $\psi \in L^2(\mathbb{R}^d)$,

$$\mathcal{I}_h^{\varepsilon,J}(t, t_0)\psi = (2\pi\varepsilon)^{-d} \int_{\mathbb{R}^{2d}} \langle g_z^\varepsilon, \psi\rangle u^{\varepsilon,J}(t, t_0, z)\, \mathrm{e}^{\frac{i}{\varepsilon}S(t,t_0,z)}\, g^\varepsilon_{\Phi_h^{t,t_0}(z)}\, dz, \tag{2.4}$$

with

$$u^{\varepsilon,J}(t,t_0,z) = \sum_{0\leqslant j\leqslant J} \varepsilon^j u_j(t,t_0,z),$$

where every u_j *is smooth in* t, t_0, z, *with any derivative in* z *bounded. The function* u_0 *is the Herman–Kluk prefactor*

$$u_0(t,t_0,z) = 2^{-\frac{d}{2}} \det^{\frac{1}{2}}\big(A(t,t_0,z) + D(t,t_0,z) + i(C(t,t_0,z) - B(t,t_0,z))\big), \quad (2.5)$$

which has the branch of the square root determined by continuity in time.

Let us remark that the Gaussian wave packets in (2.4) all have the same covariance matrix $\Gamma = \mathbb{I}_{\mathbb{C}^d}$, that is, in the terminology put forward by E. Heller [14], the Gaussians are frozen. The first statement of the form (2.4) at leading order ($J = 0$) is due to M. Herman and E. Kluk (1984) [15]. Then W. H. Miller (2002) [23] noticed that (2.4) can be deduced from the Van Vleck approximation of the propagator (hence related with a Feynman integral). In more recent work [25, 26], formula (2.4) was established using Fourier-integral operator techniques. Here we shall comment in more details on a proof based on the propagation of Gaussian wave packets, via a thawed Gaussian approximation (see Section 5). In particular, we shall see how the coefficients u_j can be computed.This proof has the advantage that it can be used in the case of systems of Schrödinger equations, which were not treated in the preceding references.

A Gaussian wave packet is a wave packet with profile function belonging to the class of Gaussian states with variance taken in the Siegel set $\mathfrak{S}^+(d)$ of $d \times d$ complex-valued symmetric matrices with positive imaginary part,

$$\mathfrak{S}^+(d) = \{\Gamma \in \mathbb{C}^{d\times d} : \Gamma = \Gamma^\tau,\ \Im\Gamma > 0\}.$$

With $\Gamma \in \mathfrak{S}^+(d)$ and $z \in \mathbb{R}$, we associate the Gaussian state

$$g_z^{\Gamma,\varepsilon}(x) = c_\Gamma(\pi\varepsilon)^{-\frac{d}{4}} \exp\Big(\frac{i}{\varepsilon} p\cdot(x-q) + \frac{i}{2\varepsilon}\Gamma(x-q)\cdot(x-q)\Big),$$

where $c_\Gamma = \det^{\frac{1}{4}}(\Im\Gamma)$ is a positive normalization constant in $L^2(\mathbb{R}^d)$. We note that the standardized Gaussian defined in (1.7) satisfies $g_z^\varepsilon = g_z^{i\mathbb{I},\varepsilon}$.

We shall use the notation $\mathrm{WP}_z^\varepsilon(\varphi)$ to denote the bounded family in $L^2(\mathbb{R}^d)$ associated with $\varphi \in \mathcal{S}(\mathbb{R}^d)$ and $z = (q,p) \in \mathbb{R}^{2d}$ by

$$\mathrm{WP}_z^\varepsilon(\varphi)(x) = \varepsilon^{-\frac{d}{4}} e^{\frac{i}{\varepsilon}p\cdot(x-q)}\, \varphi\Big(\frac{x-q}{\sqrt{\varepsilon}}\Big). \quad (2.6)$$

With wave packet notation, the above Gaussian states satisfy $g_z^{\Gamma,\varepsilon} = \mathrm{WP}_z^\varepsilon(g_0^{\Gamma,1})$.

Theorem 2.2 ([5, 24]). *By assuming* (1.6) *with* $N = 1$ *and* $H = h\mathbb{I}_{\mathbb{C}}$, *there exists a family of time-dependent Schwartz functions* $(\varphi_j(t))_{j\in\mathbb{N}}$ *such that for all* $N_0 \in \mathbb{N}$, *in* $L^2(\mathbb{R}^d)$,

$$\mathcal{U}_h^\varepsilon(t,t_0) g_{z_0}^{\Gamma_0,\varepsilon} = e^{\frac{i}{\varepsilon}S(t,t_0,z_0)} \left(g_{\Phi^{t,t_0}(z_0)}^{\Gamma(t,t_0,z_0),\varepsilon} + \sum_{j=1}^{N_0} \varepsilon^{\frac{j}{2}} \mathrm{WP}^\varepsilon_{\Phi^{t,t_0}(z_0)}(\varphi_j(t)) \right) + O(\varepsilon^{\frac{N_0+1}{2}}) \tag{2.7}$$

with

$$\Gamma(t,t_0,z_0) = (C(t,t_0,z_0) + D(t,t_0,z_0)\Gamma_0)(A(t,t_0,z_0) + B(t,t_0,z_0)\Gamma_0)^{-1} \tag{2.8}$$

and

$$c_{\Gamma(t,t_0,z_0)} = c_{\Gamma_0} \det{}^{-\frac{1}{2}}(A(t,t_0,z_0) + B(t,t_0,z_0)\Gamma_0),$$

where the complex square root is continuous in time.

Note that we have $\Gamma(t_0,t_0,z_0) = \Gamma_0$ in the statement above. Actually, explicit information is obtained on the profiles $(\varphi_j(t))_{j\in\mathbb{N}}$, in particular about $\varphi_1(t)$ (see [5, Section 4.1.2] or [8, Proposition 2.3]). As we shall see in Section 5, this result can be a starting point for proving the Herman–Kluk approximation of Theorem 2.1.

The result above also holds for general wave packets as in (2.6) with profiles that are not necessarily Gaussians (see [5, Section 4.1.2]). Moreover, also the norm can be generalized to

$$\|f\|_{\Sigma_k^\varepsilon} = \sup_{|\alpha|+|\beta|\leqslant k} \|x^\alpha(\varepsilon\partial_x)^\beta f\|_{L^2}, \quad k\in\mathbb{N}.$$

Example 2.3. Both Theorem 2.1 and 2.2 are exact when the Hamiltonian h is quadratic. For the Hamiltonians $h_\pm(z) = p \pm kq$ associated with the toy model (see (1.3)), the classical flow is linear

$$\Phi_\pm^{t,t_0}(z_0) =: (q_\pm(t), p_\pm(t)) = z_0 + (t-t_0)(1,\mp k), \quad z_0 = (q_0,p_0),$$

the width of the Gaussian is constant, $\Gamma_\pm(t,t_0,z_0) = \Gamma_0$, and the actions only depend on q_0 and are given by

$$S_\pm(t,t_0,q_0) = \mp k q_0(t-t_0) \mp \frac{k}{2}(t-t_0)^2.$$

3 Herman–Kluk formula in the adiabatic setting

We need adaptations for generalizing the scalar ideas to systems. The first one replaces the scalar Herman–Kluk prefactor $u_0(t,t_0,z_0)$ by a vector-valued one $\vec{U}(t,t_0,z_0)$ that is expanded in a basis of eigenvectors of $H(t,z)$, taken along the classical trajectory $\Phi_{h_j}^{t,t_0}(z)$ of the corresponding eigenvalues (denoted here by $h_j(t,z)$). This is done by parallel transport, and it is sufficient for an order ε approximation as long as the system is gapped.

3.1 Parallel transport

The following construction generalizes [4, Proposition 1.9], which was inspired by the work of G. Hagedorn, see [13, Proposition 3.1]. The details are given in [8]. In the sequel, we denote the complementary orthogonal projector by

$$\Pi^{\perp}(t,z) = \mathbb{I}_{\mathbb{C}^N} - \Pi(t,z)$$

and assume that

$$H(t,z) = h(t,z)\Pi(t,z) + h^{\perp}(t,z)\Pi^{\perp}(t,z) \tag{3.1}$$

with the second eigenvalue given by $h^{\perp}(t,z) = \mathrm{tr}(H(t,z)) - h(t,z)$. We introduce the auxiliary matrices

$$\begin{aligned}
\Omega(t,z) &= -\frac{1}{2}\big(h(t,z) - h^{\perp}(t,z)\big)\Pi(t,z)\{\Pi,\Pi\}(t,z)\Pi(t,z),\\
K(t,z) &= \Pi^{\perp}(t,z)\big(\partial_t\Pi(t,z) + \{h,\Pi\}(t,z)\big)\Pi(t,z),\\
\Theta(t,z) &= i\Omega(t,z) + i(K - K^*)(t,z),
\end{aligned}$$

that are smooth and satisfy some algebraic properties. In particular, Ω is skew-symmetric and Θ is self-adjoint: $\Omega = -\Omega^*$ and $\Theta = \Theta^*$.

Proposition 3.1 ([8]). *Let $H(t,z)$ be a smooth Hamiltonian that satisfies* (1.6) *and has a smooth spectral decomposition* (3.1). *We assume that both eigenvalues are of subquadratic growth as well. We consider $\vec{V}_0 \in \mathcal{C}_0^{\infty}(\mathbb{R}^{2d},\mathbb{C}^N)$ and $z_0 \in \mathbb{R}^{2d}$ such that there exists a neighborhood U of z_0 where for all $z \in U$*

$$\vec{V}_0(z) = \Pi(t_0,z)\vec{V}_0(z) \quad \textit{and} \quad \|\vec{V}_0(z)\|_{\mathbb{C}^N} = 1.$$

Then there exists a smooth normalized vector-valued function $\vec{V}(t,t_0)$ satisfying

$$\vec{V}(t,t_0,z) = \Pi(t,z)\vec{V}(t,t_0,z) \quad \textit{for all } z \in \Phi_h^{t,t_0}(U),$$

and such that for all $t \in \mathbb{R}$ and $z \in \Phi_h^{t,t_0}(U)$,

$$\partial_t\vec{V}(t,t_0,z) + \{h,\vec{V}\}(t,t_0,z) = -i\Theta(t,z)\vec{V}(t,t_0,z), \quad \vec{V}(t_0,t_0,z) = \vec{V}_0(z). \tag{3.2}$$

We note that the results of this Proposition are valid as long as smooth eigenprojectors and eigenvalues do exist. It does not require an explicit adiabatic situation and we will use that observation in Section 4. In the case of Schrödinger systems, one refers to (3.2) as parallel transport because the vectors $\partial_t\vec{V}(t) + \{h,\vec{V}\}(t)$ belong to the range of $\Pi^{\perp}(t)$ at any time of the evolution.

Example 3.2. For the toy model $H_{k,\theta}(x,\xi)$, the auxiliary matrices are

$$\Omega_{\pm} = 0, \quad K_{\pm} = \frac{i\theta}{4}\begin{pmatrix} 1 & \pm e^{i\theta x} \\ \mp e^{-i\theta x} & -1 \end{pmatrix}, \quad \Theta_{\pm} = \frac{\theta}{2}\begin{pmatrix} -1 & 0 \\ 0 & 1 \end{pmatrix}.$$

Initiating the parallel transport equation by the eigenvectors given in (1.3), we obtain

$$\vec{V}_{\pm}(t,t_0,q) = \frac{1}{\sqrt{2}}\begin{pmatrix} e^{i\frac{\theta}{2}(t-t_0)}\, e^{i\theta(q-t+t_0)} \\ \pm e^{-i\frac{\theta}{2}(t-t_0)} \end{pmatrix} = \frac{1}{\sqrt{2}}\begin{pmatrix} e^{-\frac{i}{2}\theta(t-t_0)}\, e^{i\theta q} \\ \pm e^{-\frac{i}{2}\theta(t-t_0)} \end{pmatrix}.$$

Note that they do not depend on p.

3.2 Herman–Kluk approximation in an adiabatic setting

In the context of the preceding section, and in presence of an eigenvalue gap,

$$\exists \delta > 0: \quad \operatorname{dist}(h(t,z), h^{\perp}(t,z)\}) \geq \delta \quad \text{for all } (t,z),$$

it is well known that one has *adiabatic decoupling*: If the initial data are scalar multiples of an eigenvector associated with the eigenvalue h, then the solution keeps this property to leading order in ε. For an approximation to higher order in ε, one has to consider perturbations of the eigenprojector, the so-called *superadiabatic projectors* (see [2, 27] and the new edition of [5] that should appear soon). Adiabatic theory implies a Herman–Kluk approximation of the propagator that we state next.

Theorem 3.3 ([8]). *In the situation of Proposition* 3.1, *we assume the existence of an eigenvalue gap for the eigenvalue h of H and consider initial data of the form*

$$\psi_0^\varepsilon = \widehat{\vec{V}_0} v_0^\varepsilon + O(\varepsilon) \quad \text{in } L^2(\mathbb{R}^d),$$

where $\vec{V}_0$ is a smooth eigenvector and $(v_0^\varepsilon)_{\varepsilon>0}$ a family of functions uniformly bounded in $L^2(\mathbb{R}^d,\mathbb{C})$. Then, in $L^2(\mathbb{R}^d)$,

$$\mathcal{U}_H^\varepsilon(t,t_0)\psi_0^\varepsilon = (2\pi\varepsilon)^{-d}\int_{\mathbb{R}^{2d}} \langle g_z^\varepsilon, v_0^\varepsilon\rangle \vec{U}(t,t_0,z)\, e^{\frac{i}{\varepsilon}S(t,t_0,z)}\, g^\varepsilon_{\Phi_h^{t,t_0}(z)}\, dz + O(\varepsilon)$$

with

$$\vec{U}(t,t_0,z) = u_0(t,t_0,z)\vec{V}(t,t_0,\Phi_h^{t,t_0}(z)),$$

where $u_0(t,t_0,z)$ is given by (2.5) *and the eigenvector $\vec{V}(t,t_0,z)$ by* (3.2).

We describe in Section 5 a proof of this result that crucially uses the approximate evolution of a Gaussian state, that is [24, Section 3],

$$\begin{aligned} &\mathcal{U}_H^\varepsilon(t,t_0)(\widehat{\vec{V}_0} g_z^{\Gamma,\varepsilon}) \\ &= e^{\frac{i}{\varepsilon}S(t,t_0,z)}\, \widehat{\vec{V}(t,t_0) g^{\Gamma(t,t_0,z),\varepsilon}_{\Phi_h^{t,t_0}(z)} \left(1 + \sqrt{\varepsilon}\vec{a}(t)\cdot\frac{(x-q(t))}{\sqrt{\varepsilon}}\right)} + O(\varepsilon) \end{aligned} \tag{3.3}$$

with $\vec{a}(t) = \vec{a}(t,t_0,z)$ a smooth and bounded map. Note that using superadiabatic projectors [24], one can push the asymptotics further and exhibit $O(\varepsilon)$ contributions that will have components on the other mode. Here again, a formula similar to (3.3) can be proved for general wave packets [5].

3.3 Algorithmic realization of the propagator

Numerical realizations of the Herman–Kluk approximation have been first developed in [19] and are still in practical use in theoretical chemistry [1,30]. Below, we follow the more recent account given in [20,21]. We stay with the assumptions of Theorem 3.3 and consider initial data associated with the (gapped) mode h.

Step 1: Initial sampling. Choose a set of numerical quadrature points

$$z_1 = (q_1, p_1), \dots, z_N = (q_N, p_N)$$

with associated weights $w_1, \dots, w_N > 0$, and evaluate the initial transform $\langle g_z^\varepsilon, v_0^\varepsilon \rangle$ in these points. This provides an approximation to the initial data,

$$\psi_0^\varepsilon(x) \sim (2\pi\varepsilon)^{-d} \sum_{1 \leqslant j \leqslant N} \langle g_{z_j}^\varepsilon, v_0^\varepsilon \rangle \vec{V}_0(z_j) g_{z_j}^\varepsilon(x) w_j,$$

which is of order ε in $L^2(\mathbb{R}^d)$, as long as the chosen quadrature rule is sufficiently accurate.

Step 2: Transport. For each of the points z_j, $j = 1, \dots, N$, compute

(1) the classical trajectories $z_j(t)$ defined by (2.1),

(2) the eigenvectors $\vec{V}(t, t_0, z_j(t))$ along the flow by use of equation (3.2),

(3) the Jacobian matrices $F(t, t_0, z_j)$ by solving equation (2.2),

(4) the action integrals $S(t, t_0, z_j)$ of equation (2.3),

(5) the Herman–Kluk prefactor $u_0(t, t_0, z_j)$ using the Jacobians and equation (2.5).

If the time-discretization is symplectic and sufficiently accurate, then the overall accuracy of order ε is not harmed, see [21, Theorem 2].

Step 3: Conclusion. At the end of these two steps, we are left with the Herman–Kluk quadrature formula

$$\psi^\varepsilon(t, x) \sim (2\pi\varepsilon)^{-d} \sum_{1 \leqslant j \leqslant N} \langle g_{z_j}^\varepsilon, v_0^\varepsilon \rangle \vec{U}(t, t_0, z)\, \mathrm{e}^{\frac{i}{\varepsilon} S(t, t_0, z_j)}\, g_{z_j(t)}^\varepsilon w_j.$$

Of course, the algorithm generalizes to an initial data which has several components on separated eigenspaces. In higher-dimensional applications, Monte-Carlo quadrature is often used. Then, the initial transform is written as

$$(2\pi\varepsilon)^{-d} \langle g_z^\varepsilon, v_0^\varepsilon \rangle\, dz = r_0^\varepsilon(z) \mu_0^\varepsilon(dz),$$

where

$$\mu_0^\varepsilon(dz) = \left(\int |\langle g_z^\varepsilon, v_0^\varepsilon \rangle|\, dz \right)^{-1} |\langle g_z^\varepsilon, v_0^\varepsilon \rangle|\, dz$$

is a probability measure and $r_0^\varepsilon(z)$ a complex-valued function of $L^1(\mathbb{R}^{2d}, d\mu_0^\varepsilon)$ according to the Radon–Nikodym Theorem. The quadrature nodes $z_1, \dots, z_N$ are then chosen independently and identically distributed according to the measure μ_0^ε, while the weights are all the same, $w_j = \frac{1}{N}$ for all j.

4 What about smooth crossings?

For simplicity we assume here that the Hermitian matrix $H(t,z)$ is 2×2 ($N = 2$) (the same results are also proved for two eigenvalues with arbitrary multiplicity [8]). We assume that it has smooth eigenvalues $h_1(t,z)$ and $h_2(t,z)$ and smooth eigenprojectors $\Pi_1(t,z)$ and $\Pi_2(t,z)$. To ensure control on the derivatives of the eigenprojectors, we suppose a non-crossing assumption at infinity: there exist $c_0, n_0, r_0 > 0$ such that

$$|h_1(t,z) - h_2(t,z)| \geqslant c_0 \langle z \rangle^{-n_0} \quad \text{for all } (t,z) \text{ with } |z| \geqslant r_0,$$

where we denote $\langle z \rangle = (1 + |z|^2)^{\frac{1}{2}}$. We assume the following form of the matrix:

Assumption (SC). There exist scalar functions $v, f \in \mathcal{C}^\infty(\mathbb{R}^{2d+1}, \mathbb{R})$ and a vector-valued function $u \in \mathcal{C}^\infty(\mathbb{R}^{2d+1}, \mathbb{R}^3)$ with $|u(t,z)| = 1$ for all (t,z) such that

$$H(t,z) = v(t,z)\mathrm{Id} + f(t,z) \begin{pmatrix} u_1(t,z) & u_2(t,z) + iu_3(t,z) \\ u_2(t,z) - iu_3(t,z) & -u_1(t,z) \end{pmatrix}.$$

Besides, the crossing is *generic* in Υ in the sense that

$$(\partial_t f + \{v, f\})(t^\flat, z^\flat) \neq 0 \quad \text{for all } (t^\flat, z^\flat) \in \Upsilon.$$

In that case, the crossing set Υ is a submanifold of codimension one, and all the classical trajectories that reach Υ are transverse to it.

Example 4.1. For the toy model (1.2), $u(x) = (0, \cos(\theta x), \sin(\theta x))$, $v(\xi) = \xi$ and $f(x) = kx$. Hence we have Assumption (SC): $\Upsilon = \{x = 0\}$ and $\{v, f\} = k \neq 0$.

In contrast to the previous adiabatic situation, initial data associated with one eigenspace generates a component on the other eigenspace, that is larger than the adiabatic $O(\varepsilon)$, namely $O(\sqrt{\varepsilon})$.

Starting at time t_0 with a Gaussian wave packet that is associated with the eigenvalue h_1 and localized far from the crossing set Υ, an approximation of the form (3.3) holds as long as the trajectory does not reach Υ. The apparition of a $\sqrt{\varepsilon}$ contribution on the other mode then occurs exactly on Υ. One can interpret this phenomenon in terms of *hops*: starting at time t_0 from some point $z_0 \notin \Upsilon$ for which the Hamiltonian trajectory $z_1(t,t_0) = \Phi_1^{t,t_0}(z_0)$ passes through the crossing at time $t = t^\flat(z_0)$ and point $z^\flat = z_1(t^\flat, t_0)$, we generate a new trajectory $\Phi_2^{t,t^\flat}(z^\flat)$ associated with the mode h_2. This results in the construction of a *hopping trajectory* that hops from one mode to the other one at time $t^\flat$.

Such an interpretation is crucial for describing the dynamics of systems with eigenvalue crossings. It has been introduced around the 1970s in the chemical literature for avoided and conical crossings of eigenvalues (see [28]) and has been widely used since then. The first mathematical results on the subject are more recent and analyze the propagation of Wigner functions through these singular crossings (see [6, 7]).

We now aim at a precise description of these new contributions in the case of smooth crossing, first for initial data that are Gaussian wave packets, then we lift this result to a Herman–Kluk formula for smooth crossings in the case of the toy model.

4.1 Wave packet propagation

Let us now give a precise statement for the propagation of wave packets through a smooth crossing. We start with initial data of the form

$$\psi_0^\varepsilon = \widehat{\vec{V}_0} v_0^\varepsilon, \quad v_0^\varepsilon = g_{z_0}^{\Gamma_0,\varepsilon}$$

with $H(t_0, z)\vec{V}_0(z) = h_1(t_0, z)\vec{V}_0(z)$ in a neighborhood of z_0. We use Proposition 3.1 to construct two families of time-dependent eigenvectors: $(\vec{V}_1(t, z))_{t\geqslant t_0}$ is associated with the eigenvalue $h_1(t, z)$ and initial data at time t_0 given by $\vec{V}_1(t_0, z) = \vec{V}_0(z)$, while $(\vec{V}_2(t, z))_{t\geqslant t^\flat}$ is constructed for $t \geqslant t^\flat$ (the crossing time introduced in the previous paragraph) for the eigenvalue $h_2(t, z)$ with initial data at time $t^\flat = t^\flat(z_0)$ satisfying

$$\vec{V}_2(t^\flat, z) = -\gamma(t^\flat, z)^{-1}\Pi_2(\partial_t \Pi_2 + \{v, \Pi_2\})\vec{V}_1(t^\flat, z)$$

with

$$\gamma(t^\flat, z) = \|(\partial_t \Pi_2 + \{v, \Pi_2\})\vec{V}_1(t^\flat, z)\|_{\mathbb{C}^N}.$$

We next introduce a family of transformations, which describes the non-adiabatic effects for a wave packet that passes the crossing. For parameters $(\mu, \alpha, \beta) \in \mathbb{R} \times \mathbb{R}^{2d}$ and $\varphi \in \mathcal{S}(\mathbb{R}^d)$, we set

$$\mathcal{T}_{\mu,\alpha,\beta}\varphi(y) = \int_{-\infty}^{+\infty} \mathrm{e}^{i(\mu-\frac{\alpha\cdot\beta}{2})s^2}\, \mathrm{e}^{is\beta\cdot y}\, \varphi(y - s\alpha)\, ds.$$

This operator maps $\mathcal{S}(\mathbb{R}^d)$ into itself if and only if $\mu \neq 0$. Moreover, for $\mu \neq 0$, it is a metaplectic transformation of the Hilbert space $L^2(\mathbb{R}^d)$, multiplied by a complex number. In particular, for any Gaussian function g^Γ, the function $\mathcal{T}_{\mu,\alpha,\beta}g^\Gamma$ is a Gaussian:

$$\mathcal{T}_{\mu,\alpha,\beta}\, g^\Gamma = c_{\mu,\alpha,\beta,\Gamma}\, g^{\Gamma_{\mu,\alpha,\beta,\Gamma}},$$

where $\Gamma_{\mu,\alpha,\beta,\Gamma} \in \mathfrak{S}^+(d)$ and $c_{\mu,\alpha,\beta,\Gamma} \in \mathbb{C}$ can be computed explicitly (see [8, Appendix E]).

Combining the parallel transport for the eigenvector and the metaplectic transformation for the non-adiabatic transitions, we obtain the following result.

Theorem 4.2 (Propagation through a smooth crossing). *Let Assumption* (SC) *on the Hamiltonian matrix $H(t)$ hold and assume that the crossing is generic. Assume that the initial data $(\psi_0^\varepsilon)_{\varepsilon>0}$ are wave packets as above. Let $T > 0$ be such that the* *interval* $[t_0, t^\flat]$ *is strictly included in the interval* $[t_0, t_0 + T]$*. Then, for all* $k \in \mathbb{N}$ *there exists a constant* $C > 0$ *such that for all* $t \in [t_0, t^\flat) \cup (t^\flat, t_0 + T]$ *and for all* $\varepsilon \leqslant |t - t^\flat|^{\frac{9}{2}}$,

$$\left\| \psi^\varepsilon(t) - \widehat{\vec{V}}_1(t) v_1^\varepsilon(t) - \sqrt{\varepsilon} \mathbf{1}_{t > t^\flat} \widehat{\vec{V}}_2(t) v_2^\varepsilon(t) \right\|_{L^2(\mathbb{R}^d)} \leqslant C \varepsilon^m,$$

with an exponent $m \geqslant \frac{5}{9}$*. The components of the approximate solution are*

$$v_1^\varepsilon(t) = \mathcal{U}^\varepsilon_{h_1}(t, t_0) g_{z_0}^{\Gamma_0, \varepsilon} \quad \text{and} \quad v_2^\varepsilon(t) = \mathcal{U}^\varepsilon_{h_2}(t, t^\flat) v_2^\varepsilon(t^\flat)$$

with

$$v_2^\varepsilon(t^\flat) = \gamma^\flat \, \mathrm{e}^{\frac{i S^\flat}{\varepsilon}} \, \mathrm{WP}^\varepsilon_{z^\flat}(\mathcal{T}^\flat \varphi_0(t^\flat)),$$

where $\varphi_0(t) = g_0^{\Gamma_1(t, t_0, z), 1}$ *is the leading-order profile of the coherent state* $v_1^\varepsilon(t)$ *given by Theorem 2.2, and*

$$\gamma^\flat = \gamma(t^\flat, z^\flat) = \| (\{v, \Pi_2\} + \partial_t \Pi_2) \vec{V}_1(t^\flat, z^\flat) \|_{\mathbb{C}^N}.$$

The transition operator $\mathcal{T}^\flat = \mathcal{T}_{\mu^\flat, \alpha^\flat, \beta^\flat}$ *is defined by the parameters*

$$\mu^\flat = \frac{1}{2} (\partial_t f + \{v, f\})(t^\flat, z^\flat) \quad \text{and} \quad (\alpha^\flat, \beta^\flat) = J d_z f(t^\flat, z^\flat).$$

The constant $C = C(T, k, z_0, \Gamma_0) > 0$ *is* ε*-independent but depends on the Hamiltonian* $H(t, z)$*, the final time* T*, and on the initial wave packet's center* z_0 *and width* Γ_0.

This theorem is an extension, to a more general setting, of results obtained in [13] for the Schrödinger equation (1.4). In [29] the authors have adapted Hagdorn's approach to smooth crossings of Bloch modes leading to a model problem of the form of equation (1.5). Our approach is different, much more explicit, and contains these two results as particular cases.

Note that by the transversality assumption we have $\mu^\flat \neq 0$, which guarantees that $\mathcal{T}^\flat \varphi_0(t^\flat)$ is Schwartz class. The coefficient $\gamma^\flat$ quantitatively describes the distortion of the eigenprojector $\Pi_1(t)$ during its evolution along the flow generated by $h_1(t)$. If the matrix $H(t, z)$ is diagonal (or diagonalizes in a fixed orthonormal basis that is (t, z)-independent), then $\gamma^\flat = 0$: the equations are decoupled (or can be decoupled), and one can then apply the result for a system of two independent equations with a scalar Hamiltonian and, of course, there is no interaction of order $\sqrt{\varepsilon}$ between the modes.

The previous theorem extends to more general wave packets as defined in (2.6) and also holds with respect to Σ^ε_k-norms for $k \in \mathbb{N}$ (see [8, Theorem 3.8]). As mentioned alongside the proof [8], the argument contains the germs for a full asymptotic expansion in powers of $\sqrt{\varepsilon}$ (with $\log \varepsilon$ corrections).

Example 4.3. Note that for the toy model $H_{k,\theta}$, we have at any point of $\Upsilon = \{x = 0\}$,

$$\mu^\flat = \frac{k}{2}, \quad \alpha^\flat = 0, \quad \beta^\flat = -k, \quad \gamma^\flat = \frac{|\theta|}{2},$$

and

$$\mathcal{T}^{\flat}\varphi(y) = \sqrt{\frac{2\pi}{ik}}\, \mathrm{e}^{\frac{k}{2i}y^2}\, \varphi(y) \quad \text{for all } \varphi \in \mathcal{S}(\mathbb{R}) \text{ and all } y \in \mathbb{R}.$$

Besides, the trajectories of the minus mode that reach Υ are those arising from points $z = (q, p)$ with $q < 0$. One then has $t^{\flat} = t_0 - q$, $p^{\flat} = p - kq$ and the trajectory on the plus mode issued from $z^{\flat} = (0, p^{\flat})$ is

$$\Phi_{+}^{t,t^{\flat}}(0, p^{\flat}) = (t - t^{\flat}, p^{\flat} - k(t - t^{\flat})) = (t - t_0 + q, p - 2kq - k(t - t_0)).$$

4.2 Towards a Herman–Kluk approximation

The preceding result implies that the leading order of the propagation is still driven by the modes in which the initial data had been taken and we have an Herman–Kluk formula similar to the one obtained in the adiabatic regime, however, with a remainder which is worse. One can conjecture that a more accurate Herman–Kluk approximation holds in a weaker sense (see [9]). We define the operator $\mathcal{I}^{\varepsilon}_{\mathrm{sc}}(t)$ by its actions on functions of the form

$$\psi_0^{\varepsilon} = \widehat{\vec{V}_0} v_0^{\varepsilon} + O(\sqrt{\varepsilon}) \text{ in } L^2(\mathbb{R}^d)$$

with $v_0^{\varepsilon} \in L^2(\mathbb{R}^d)$ as

$$\begin{aligned} \mathcal{I}^{\varepsilon}_{\mathrm{sc}}(t)\psi_0^{\varepsilon}(x) &= (2\pi\varepsilon)^{-d} \int_{\mathbb{R}^{2d}} \langle g_z^{\varepsilon}, v_0^{\varepsilon}\rangle \vec{U}_1(t, t_0, z)\, \mathrm{e}^{\frac{i}{\varepsilon}S_1(t,t_0,z)}\, g^{\varepsilon}_{\Phi_{h_1}^{t,t_0}(z)}\, dz \\ &\quad + \sqrt{\varepsilon}(2\pi\varepsilon)^{-d} \int_{\mathbb{R}^{2d}} \mathbf{1}_{t>t^{\flat}(z)} \langle g_z^{\varepsilon}, v_0^{\varepsilon}\rangle \vec{U}_2(t, t^{\flat}(z), z) \\ &\qquad \times \mathrm{e}^{\frac{i}{\varepsilon}S_1(t^{\flat}(z),t_0,z)+\frac{i}{\varepsilon}S_2(t,t^{\flat}(z),\mathfrak{z}^{\flat}(z))}\, g^{\varepsilon}_{\Phi_{h_2}^{t,t^{\flat}(z)}(\mathfrak{z}^{\flat})(z)}\, dz \end{aligned}$$

with

$$\mathfrak{z}^{\flat}(z) = \Phi_{h_1}^{t^{\flat}(z),t_0}(z)$$

and with some adequate formula for the prefactor $\vec{U}_2(t, t^{\flat}(z), z)$:

$$\vec{U}_2(t, t^{\flat}(z), z) = v_2(t, t^{\flat}(z), z)\vec{V}_2(t, t^{\flat}(z), \mathfrak{z}^{\flat}(z)),$$

the scalar prefactor $v_2(t, t^{\flat}(z), z)$ depending on classical quantities associated with the mode h_2 and taking into account the transfer coefficient $\gamma^{\flat}(z)$.

The conjecture is, that if Assumptions (SC) are satisfied, then, for all $\chi \in \mathcal{C}_0^{\infty}(\mathbb{R})$, in $L^2(\mathbb{R}^d)$, one has

$$\int_{\mathbb{R}} \chi(t)(\mathcal{I}^{\varepsilon}_{\mathrm{sc}}(t)\psi_0^{\varepsilon} - \mathcal{U}_H^{\varepsilon}(t, t_0)\psi_0^{\varepsilon})\, dt = o(\sqrt{\varepsilon}). \tag{4.1}$$

Estimates that are "averaged in time" have previously been obtained for systems (see for example [6, 7, 11]). They differ from pointwise estimates in the sense that they correspond to an observation on a time interval, that, of course can be assumed short, but not negligible.

An approximation as (4.1) can easily be proved for the toy-model (1.2) (see Section 5.5) below. The proof in the general case is a work in progress [9]. It involves more refined estimates than those of Theorem 3.3, that we present in the next section.

The authors believe that the technics used for treating the apparition of a new contribution when trajectories reach a hypersurface (the crossing one in this special case) will be useful for developing Herman–Kluk approximations for avoided crossings and conical ones, using the hopping trajectories of [7] and [6], respectively. However, we point out, that conical crossings pose the additional difficulty that Gaussians states do not remain Gaussian, even at leading order, as emphasized in [13].

5 A sketchy proof for Herman–Kluk approximations

We consider here the scalar and the adiabatic case and we discuss the proof of Theorems 2.1 and 3.3.

5.1 The proof strategy

As mentioned in the introduction, the underlying idea for constructing Gaussian based approximations for unitary propagators is to start form equation (1.9). We develop below a proof strategy based on this idea.

Step 1. For each $z \in \mathbb{R}^{2d}$, we build a thawed wave packet approximation $\psi^{\varepsilon}_{\mathrm{th},z}(t)$ to the Schrödinger system (1.1) with initial data

$$\psi^{\varepsilon}_{|t=t_0} = \psi^{\varepsilon}_0 = \begin{cases} g^{\varepsilon}_z & \text{for } N = 1, \\ \widehat{\vec{V}}_0 g^{\varepsilon}_z & \text{for } N > 1. \end{cases}$$

We prove that $\psi^{\varepsilon}_{\mathrm{th},z}(t)$ satisfies an evolution equation of the form

$$i\varepsilon\partial_t \psi^{\varepsilon}_{\mathrm{th},z}(t) = \widehat{H(t)}\psi^{\varepsilon}_{\mathrm{th},z}(t) + \varepsilon^2 R^{\varepsilon}_z(t), \quad \psi^{\varepsilon}_{\mathrm{th},z}(t_0) = \psi^{\varepsilon}_0 \tag{5.1}$$

with source term $R^{\varepsilon}_z(t)$.

Step 2. For the thawed Gaussian propagation of general initial data

$$\psi^{\varepsilon}_{|t=t_0} = \psi^{\varepsilon}_0 = \begin{cases} v^{\varepsilon}_0 & \text{for } N = 1, \\ \widehat{\vec{V}}_0 v^{\varepsilon}_0 & \text{for } N > 1, \end{cases}$$

with $v^{\varepsilon}_0 \in L^2(\mathbb{R}^d, \mathbb{C})$, we consider

$$\mathcal{I}^{\varepsilon}_{\mathrm{th}}(t, t_0)\psi^{\varepsilon}_0 = (2\pi\varepsilon)^{-d} \int_{z\in\mathbb{R}^{2d}} \langle g^{\varepsilon}_z, v^{\varepsilon}_0\rangle\, \psi^{\varepsilon}_{\mathrm{th},z}(t)\, dz.$$

The error $e^{\varepsilon}_{\mathrm{th}}(t) = \mathcal{U}^{\varepsilon}_H(t, t_0)\psi^{\varepsilon}_0 - \mathcal{I}^{\varepsilon}_{\mathrm{th}}(t, t_0)\psi^{\varepsilon}_0$ satisfies the evolution equation

$$i\varepsilon\partial_t e^{\varepsilon}_{\mathrm{th}}(t) = \widehat{H}(t)e^{\varepsilon}_{\mathrm{th}}(t) + \varepsilon^2 \Sigma^{\varepsilon}_{\mathrm{th}}(t), \quad e^{\varepsilon}_{\mathrm{th}}(t_0) = 0$$

with source term

$$\Sigma_{\mathrm{th}}^{\varepsilon}(t) = (2\pi\varepsilon)^{-d} \int_{\mathbb{R}^{2d}} \langle g_z^{\varepsilon}, \psi_0^{\varepsilon} \rangle R_z^{\varepsilon}(t)\, dz.$$

Since $\widehat{H}(t)$ is self-adjoint, the usual energy argument provides

$$\|e_{\mathrm{th}}^{\varepsilon}(t)\| \leqslant \varepsilon \int_{t_0}^{t} \|\Sigma_{\mathrm{th}}^{\varepsilon}(s)\| ds.$$

Step 3. For analyzing the source term $\Sigma_{\mathrm{th}}^{\varepsilon}(t)$, we consider the integral operator

$$\psi \mapsto (2\pi\varepsilon)^{-d} \int_{\mathbb{R}^{2d}} \langle g_z^{\varepsilon}, \psi \rangle R_z^{\varepsilon}(t)\, dz$$

and its Bargmann kernel

$$k_{\mathcal{B}}(t; X, Y) = (2\pi\varepsilon)^{-2d} \int_{\mathbb{R}^{2d}} \langle g_z^{\varepsilon}, g_Y^{\varepsilon} \rangle \langle g_X^{\varepsilon}, R_z^{\varepsilon}(t) \rangle\, dz, \quad X, Y \in \mathbb{R}^{2d}.$$

We aim at establishing constants $C_1(t), C_2(t) > 0$, that do not depend on the semiclassical parameter ε, such that

$$\sup_X \int_{\mathbb{R}^{2d}} |k_{\mathcal{B}}(t; X, Y)| dY \leqslant C_1(t), \quad \sup_Y \int_{\mathbb{R}^{2d}} |k_{\mathcal{B}}(t; X, Y)| dX \leqslant C_2(t), \tag{5.2}$$

since then, by the Schur test,

$$\|\Sigma_{\mathrm{th}}^{\varepsilon}(t)\| \leqslant \sqrt{C_1(t) C_2(t)}\, \|\psi_0^{\varepsilon}\|.$$

Step 4. For systems, that is, for $N > 1$, we use the additional observation that

$$\begin{aligned}&\mathcal{I}_{\mathrm{th}}^{\varepsilon}(t, t_0)\psi_0^{\varepsilon} \\ &\quad = (2\pi\varepsilon)^{-d} \int_{z \in \mathbb{R}^{2d}} \langle g_z^{\varepsilon}, v_0^{\varepsilon} \rangle \vec{V}(t, t_0, \Phi_h^{t,t_0}(z))\, \mathrm{e}^{\frac{i}{\varepsilon} S(t,t_0,z)}\, g_{\Phi_h^{t,t_0}(z)}^{\Gamma(t,t_0,z),\varepsilon}\, dz + O(\varepsilon).\end{aligned}$$

Step 5. We turn the thawed propagation in a frozen one, in proving that

$$\mathcal{I}_{\mathrm{th}}^{\varepsilon}(t, t_0) = \mathcal{I}_{\mathrm{fr}}^{\varepsilon}(t, t_0) + O(\varepsilon)$$

in the norm of bounded operators on $L^2(\mathbb{R}^d)$, where the frozen propagator is defined by the Herman–Kluk formula

$$\mathcal{I}_{\mathrm{fr}}^{\varepsilon}(t, t_0)\psi_0^{\varepsilon} = (2\pi\varepsilon)^{-d} \int_{z \in \mathbb{R}^{2d}} \langle g_z^{\varepsilon}, v_0^{\varepsilon} \rangle \vec{U}(t, t_0, z)\, \mathrm{e}^{\frac{i}{\varepsilon} S(t,t_0,z)}\, g_{\Phi_h^{t,t_0}(z)}^{\varepsilon}\, dz$$

with

$$\vec{U}(t, t_0, z) = \begin{cases} u_0(t, t_0, z) & \text{for } N = 1, \\ u_0(t, t_0, z)\vec{V}(t, t_0, \Phi_h^{t,t_0}(z)) & \text{for } N > 1. \end{cases}$$

Once the previous steps have been carried out, we have proven the basic $J = 0$ version of the scalar Herman–Kluk formula of Theorem 2.1 and its generalization to the adiabatic situation given in Theorem 3.3.

5.2 The thawed remainder

We first consider scalar wave packet propagation as described in Theorem 2.2 with an accuracy of order ε, that is, for $N_0 = 1$. The corresponding thawed Gaussian wave packet $\psi^\varepsilon_{\mathrm{th},z}(t)$ that is defined by the right hand side of (2.7) satisfies an evolution equation of the form (5.1) with a source term

$$R^\varepsilon_z(t) = \mathrm{e}^{\frac{i}{\varepsilon}S(t,t_0,z)}\ \mathrm{op}^w_\varepsilon(L_z(t,t_0))g^{\Gamma(t,t_0,z),\varepsilon}_{\Phi^{t,t_0}_h(z)},$$

where $w \mapsto L_z(t,t_0,w)$ is a smooth function, that is polynomially bounded. It depends on the remainder of Taylor expansions of $h(t,\cdot)$ around the point $\Phi^{t,t_0}_h(z)$, see [5, Section 4.3.1]. For adiabatic propagation by systems with eigenvalue gaps, as presented in Theorem 3.3, we work with

$$\psi^\varepsilon_{\mathrm{th},z}(t) = \mathrm{e}^{\frac{i}{\varepsilon}S(t,t_0,z)}\,\widehat{\vec{V}(t,t_0)}\left(1 + \sqrt{\varepsilon}\vec{a}(t,t_{,0},z)\cdot\frac{x-q(t,t_0,z)}{\sqrt{\varepsilon}}\right)g^{\Gamma(t,t_0,z),\varepsilon}_{\Phi^{t,t_0}_h(z)},$$

where the vector $\vec{a}(t,t_0,z;x)$ can be constructed explicitly. This wave packet satisfies an evolution equation of the form (5.1) with source term

$$R^\varepsilon_z(t) = \mathrm{e}^{\frac{i}{\varepsilon}S(t,t_0,z)}\ \mathrm{op}^w_\varepsilon(\vec{L}_z(t,t_0))g^{\Gamma(t,t_0,z),\varepsilon}_{\Phi^{t,t_0}_h(z)},$$

where the $w \mapsto \vec{L}_z(t,t_0,w)$ is now vector-valued and contains remainder terms of Taylor expansions of $h(t,\cdot)$ around the classical trajectory. We note that $\vec{L}_z(t,t_0,\cdot)$ has contributions both in the range of $\vec{V}(t,t_0,\cdot)$ but also in the orthogonal complement.

5.3 The Schur estimate

We now analyze the Bargmann kernel of the source term $\Sigma^\varepsilon_{\mathrm{th}}(t)$. Since

$$|\langle g^\varepsilon_z, g^\varepsilon_Y\rangle| = \mathrm{e}^{-\frac{|Y-z|^2}{4\varepsilon}},$$

we have

$$|k_{\mathcal{B}}(t;X,Y)| \leqslant (2\pi\varepsilon)^{-2d}\int_{\mathbb{R}^{2d}} \mathrm{e}^{-\frac{|Y-z|^2}{4\varepsilon}}\,|\langle g^\varepsilon_X, R^\varepsilon_z(t)\rangle|\,dz.$$

Hence, the crucial estimate that is required concerns the Bargmann transform of the remainder $R^\varepsilon_z(t)$. We write the remainder as

$$R^\varepsilon_z(t) = \mathrm{e}^{\frac{i}{\varepsilon}S(t,t_0,z)}\ \mathrm{op}^w_\varepsilon(L_z(t,t_0))g^{\Gamma_z,\varepsilon}_{\Phi_z},$$

where $w \mapsto L_z(t,t_0,w)$ is a smooth function on phase space, that grows at most polynomially, and $\Gamma_z = \Gamma(t,t_0,z)$, $\Phi_z = \Phi^{t,t_0}(z)$ are short-hand notations for the classical quantities defined in (2.1) and (2.8), respectively. We express the Bargmann transform as a phase space integral

$$\langle g^\varepsilon_X, R^\varepsilon_z(t)\rangle = \mathrm{e}^{\frac{i}{\varepsilon}S(t,t_0,z)}\int_{\mathbb{R}^{2d}} L_z(t,t_0,w)\mathrm{Wig}(g^\varepsilon_X, g^{\Gamma_z,\varepsilon}_{\Phi_z})(w)\,dw$$

with respect to the cross-Wigner function of two Gaussian wave packets with different centers and different width. One can prove (see [20, Lemma 5.20]) that

$$\begin{aligned}&\mathrm{Wig}(g_X^\varepsilon, g_{\Phi_z}^{\Gamma_z,\varepsilon})(w)\\&\quad = (\pi\varepsilon)^{-d}\,\gamma_{X,z}\,\exp\Big(\frac{i}{\varepsilon}J(X-\Phi_z)\cdot w + \frac{i}{2\varepsilon}G_z(t)(w-m_{X,z})\cdot(w-m_{X,z})\Big),\end{aligned}$$

where $m_{X,z} = \frac{1}{2}(X+\Phi_z)$ is the mean of the two centers, while $\gamma_{X,z}$ is a complex number with $|\gamma_{X,z}| \leqslant 1$ and $G_z(t) \in \mathfrak{S}^+(2d)$. Repeated integration by parts, see [20, Proposition 5.21], yields an upper bound

$$|\langle g_X^\varepsilon, R_z^\varepsilon(t)\rangle| \leqslant c_z(t)\Big\langle \frac{X-\Phi^{t,t_0}(z)}{\sqrt{\varepsilon}}\Big\rangle^{-(d+1)},$$

where the constant $c_z(t) > 0$ depends on bounds of the function $L_z(t,t_0,w)$ and is inversely proportional to the smallest eigenvalue of $\Im G_z(t)$. A combination of arguments given in the proofs of [20, Lemma 5.18] and [25, Lemma 3.2], reveals that the spectrum of $\Im G_z(t)$ is bounded away from zero uniformly in z, which implies the existence of constant $c(t) > 0$ such that

$$|\langle g_X^\varepsilon, R_z^\varepsilon(t)\rangle| \leqslant c(t)\Big\langle \frac{X-\Phi^{t,t_0}(z)}{\sqrt{\varepsilon}}\Big\rangle^{-(d+1)}.$$

This gives us enough decay to deduce the existence of constants $C_1(t), C_2(t) > 0$ such that the Schur estimate (5.2) holds.

5.4 Passing from thawed to frozen approximation

Here we present a slight variant of [25, Proposition 4.1] for passing from a thawed to a frozen Gaussian approximation by a linear deformation argument.

Proposition 5.1. *Let $\vec{U}(t,t_0,z)$ be a smooth function with values in $\mathbb{C}^N$, $N \geqslant 1$, whose derivatives are at most of polynomial growth. Then*

$$\begin{aligned}&(2\pi\varepsilon)^{-d}\int_{\mathbb{R}^{2d}} \langle g_z^\varepsilon,\psi\rangle \vec{U}(t,t_0,z)\,\mathrm{e}^{\frac{i}{\varepsilon}S(t,t_0,z)}\, g^{\Gamma(t,t_0,z),\varepsilon}_{\Phi_h^{t,t_0}(z)}\,dz\\&\quad = (2\pi\varepsilon)^{-d}\int_{\mathbb{R}^{2d}} \langle g_z^\varepsilon,\psi\rangle u_0(t,t_0,z)\,\vec{U}(t,t_0,z)\,\mathrm{e}^{\frac{i}{\varepsilon}S(t,t_0,z)}\, g^{\varepsilon}_{\Phi_h^{t,t_0}(z)}\,dz + O(\varepsilon)\end{aligned}$$

uniformly for all $\psi \in L^2(\mathbb{R}^d,\mathbb{C})$ with norm one.

Proof. For notational simplicity, we omit the time-dependance in $S = S(t,t_0,z)$, $\Phi = \Phi_h^{t,t_0}(z)$, $\Gamma = \Gamma(t,t_0,z)$, and $\vec{U} = \vec{U}(t,t_0,z)$. We consider both operators, the thawed and the frozen one, as special members of a class of linear operators of the form

$$\mathcal{I}\psi = (2\pi\varepsilon)^{-d}\int_{\mathbb{R}^{2d}} \langle g_z^\varepsilon,\psi\rangle (x-\Phi_q(z))^\alpha \vec{W}(z)\,\mathrm{e}^{\frac{i}{\varepsilon}S(z)}\, g^{\mathcal{G}(z),\varepsilon}_{\Phi(z)}\,dz,$$

that are defined by two smooth functions $\mathcal{G} : \mathbb{R}^{2d} \to \mathfrak{S}^+(d)$ and $\vec{W} : \mathbb{R}^{2d} \to \mathbb{C}^N$. The Siegel half-space $\mathfrak{S}^+(d)$ is invariant under inversion in the sense that any $G \in \mathfrak{S}^+(d)$ is invertible with $-G^{-1} \in \mathfrak{S}^+(d)$. We require that the smallest eigenvalue of $\Im(\mathcal{G}(z))$ and $\Im(-\mathcal{G}^{-1}(z))$ are bounded away from zero uniformly in z. The monomial powers $(x - \Phi_q(z))^\alpha$ with $\alpha \in \mathbb{N}_0^d$ are included for technical reasons, that will become clear soon. These operators are bounded on $L^2(\mathbb{R}^d)$ and satisfy

$$\|\mathcal{I}\| \leqslant C\varepsilon^{\lceil \frac{|\alpha|}{2} \rceil}, \tag{5.3}$$

where the constant $C > 0$ independent of ε, and $\lceil \frac{|\alpha|}{2} \rceil$ denotes the smallest integer larger or equal to $\frac{|\alpha|}{2}$, see [20, Proposition 5.12]. We linearly link the thawed matrix function $z \mapsto \Gamma(z)$ and the frozen $z \mapsto i\,\mathrm{Id}$ by setting

$$\mathcal{G}(z,s) = (1-s)\Gamma(z) + is\,\mathrm{Id} \in \mathfrak{S}^+(d), \quad s \in [0,1],$$

and consider the corresponding Gaussian function, that is only partially normalized,

$$\tilde{g}_{\Phi(z)}^{\mathcal{G}(z,s),\varepsilon}(x) = (\pi\varepsilon)^{-\frac{d}{4}}\, \mathrm{e}^{\frac{i}{\varepsilon}\Phi_p(z)\cdot(x-\Phi_q(z)) + \frac{i}{2\varepsilon}\mathcal{G}(z,s)(x-\Phi_q(z))\cdot(x-\Phi_q(z))}.$$

We now aim at constructing a smooth function $\vec{W}(z,s)$ with two properties.

(1) Firstly, we require that $\vec{W}(z,0) = \det^{-\frac{1}{2}}(A(z) + iB(z))\vec{U}(z)$, ensuring that the deformation value $s = 0$ corresponds to the thawed approximation.

(2) Secondly, we hope to achieve

$$\frac{\partial}{\partial s}(2\pi\varepsilon)^{-d} \int_{\mathbb{R}^{2d}} \langle g_z^\varepsilon, \psi\rangle \vec{W}(z,s)\, \mathrm{e}^{\frac{i}{\varepsilon}S(z)}\, \tilde{g}_{\Phi(z)}^{\mathcal{G}(z,s),\varepsilon}\, dz = O(\varepsilon),$$

uniformly for all $\psi \in L^2(\mathbb{R}^d)$ of norm one.

Once this construction has been carried out, we will verify that the deformation yields the frozen approximation for $s = 1$, that is, $\vec{W}(z,1) = u_0(z)\vec{U}(z)$. As a first step, we open the inner product involving the standard Gaussian g_z^ε and examine the multi-variate exponential function

$$\overline{g_z^\varepsilon(y)}\, \mathrm{e}^{\frac{i}{\varepsilon}S(z)}\, \tilde{g}_{\Phi(z)}^{\mathcal{G}(z,s),\varepsilon}.$$

Using the derivative properties of the action $S(z)$, one obtains the identity

$$\begin{aligned}
&(x - \Phi_q(z))\overline{g_z^\varepsilon(y)}\, \mathrm{e}^{\frac{i}{\varepsilon}S(z)}\, \tilde{g}_{\Phi(z)}^{\mathcal{G}(z,s),\varepsilon} \\
&\quad = \Big(\frac{\varepsilon}{i} M_{\mathcal{G}(z,s)}^{-T}(z)(i\partial_q + \partial_p) - f(x,z,s)\Big)\overline{g_z^\varepsilon(y)}\, \mathrm{e}^{\frac{i}{\varepsilon}S(z)}\, \tilde{g}_{\Phi(z)}^{\mathcal{G}(z,s),\varepsilon},
\end{aligned}$$

where

$$M_{\mathcal{G}(z,s)}(z) = -i\mathcal{G}(z,s)A(z) - \mathcal{G}(z,s)B(z) + iC(z) + D(z)$$

is an invertible complex $d \times d$ and

$$f(x,z,s) = M_{\mathcal{G}(z,s)}^{-T}(z)\Big(\frac{1}{2}((i\partial_{q_j} + \partial_{p_j})\mathcal{G}(z,s))(x - \Phi_q(z))\cdot(x - \Phi_q(z))\Big)_{j=1}^{d}$$

is a vector-valued function, that is quadratic in $x - \Phi_q(z)$. Since

$$\frac{i}{2\varepsilon}(x - \Phi_q(z)) \cdot f(x, z, s)$$

is cubic in $x - \Phi_q(z)$, we use (5.3) and recognize its contribution as a term of order ε. Thus, we have

$$\begin{aligned}(2\pi\varepsilon)^{-d} &\int_{\mathbb{R}^{2d}} \langle g_z^\varepsilon, \psi\rangle \vec{W}(z,s)\, e^{\frac{i}{\varepsilon}S(z)}\, \partial_s \tilde{g}_{\Phi(z)}^{\mathscr{G}(z,s),\varepsilon}\, dz \\ &= (2\pi\varepsilon)^{-d} \int_{\mathbb{R}^{2d}} \vec{W}(z,s) L(z,s)(x - \Phi_q(z)) \\ &\qquad\qquad \times (i\partial_q + \partial_p)\langle g_z^\varepsilon, \psi\rangle\, e^{\frac{i}{\varepsilon}S(z)}\, \tilde{g}_{\Phi(z)}^{\mathscr{G}(z,s),\varepsilon}\, dz + O(\varepsilon),\end{aligned}$$

with

$$L(z,s) = \frac{1}{2} M_{\mathscr{G}(z,s)}^{-1}(z)\partial_s \mathscr{G}(z,s).$$

We perform an integration by parts and arrive at

$$\begin{aligned}-(2\pi\varepsilon)^{-d} \int_{\mathbb{R}^{2d}} \sum_{k=1}^{d} (i\partial_{q_k} + \partial_{p_k})\big(\vec{W}(z,s)(L(z,s)(x - \Phi_q(z)))_k\big) \\ \times \langle g_z^\varepsilon, \psi\rangle\, e^{\frac{i}{\varepsilon}S(z)}\, \tilde{g}_{\Phi(z)}^{\mathscr{G}(z,s),\varepsilon}\, dz + O(\varepsilon).\end{aligned}$$

Computing the derivative, we obtain several terms that are linear in $x - \Phi_q(z)$, and thus of order ε. The contributions we have to keep are

$$\begin{aligned}-\vec{W}(z,s) &\sum_{k,\ell=1}^{d} L(z,s)_{k\ell}(i\partial_{q_k} + \partial_{p_k})(x - \Phi_q(z))_\ell \\ &= \vec{W}(z,s) \sum_{k,\ell=1}^{d} L(z,s)_{k\ell}(iA(z) + B(z))_{\ell k} \\ &= \vec{W}(z,s)\, \mathrm{tr}(L(z,s)(iA(z) + B(z))).\end{aligned}$$

We observe that

$$L(z,s)(iA(z) + B(z)) = -\frac{1}{2} M_{\mathscr{G}(z,s)}^{-1}(z)\partial_s M_{\mathscr{G}(z,s)}(z).$$

This suggests that $\vec{W}(z,s)$ should solve the differential equation

$$\partial_s \vec{W}(z,s) - \frac{1}{2}\,\mathrm{tr}\big(M_{\mathscr{G}(z,s)}^{-1}(z)\partial_s M_{\mathscr{G}(z,s)}(z)\big)\vec{W}(z,s) = 0,$$

that is,

$$\vec{W}(z,s) = 2^{-\frac{d}{2}} \det{}^{\frac{1}{2}}(M_{\mathscr{G}(z,s)}(z))\, \vec{U}(z),$$

by using Liouville's formula for the differentiation of determinants. Checking for the initial condition at $s = 0$, we observe that

$$\begin{aligned} M_{\mathscr{g}(z,0)}(z) &= -i\big(\Gamma(z)(A(z) - iB(z)) - (C(z) - iD(z))\big) \\ &= -i\big(\Gamma(z) - \overline{\Gamma(z)}\big)(A(z) - iB(z)) \\ &= 2\Im\Gamma(z)(A(z) - iB(z)) = 2\,(A(z) + iB(z))^{-T}, \end{aligned}$$

which implies $\vec{W}(z,0) = \det^{-\frac{1}{2}}(A(z) + iB(z))\vec{U}(z)$, indeed. ∎

5.5 Herman–Kluk approximation for the toy model

The Schrödinger equation associated with (1.2) turns out to be a transport equation. Then, by integrating along curves $s \mapsto (s, x + s)$, it reads

$$i\varepsilon \frac{d}{ds}\psi^\varepsilon(s, x+s) = k(x+s)V_\theta(x+s)\psi^\varepsilon(s, x+s) \tag{5.4}$$

with $V_\theta(x) = \left(\begin{smallmatrix} 0 & e^{i\theta x} \\ e^{-i\theta x} & 0 \end{smallmatrix}\right)$, and reduces to the system of ODEs

$$i\varepsilon \frac{d}{d\sigma}\eta^\varepsilon(\sigma) = k\sigma V_\theta(\sigma)\eta^\varepsilon(\sigma). \tag{5.5}$$

This problem was solved first in [10] and in a more general setting in [12] where an asymptotic expansion in power of $\varepsilon^{\frac{1}{2}}$ (with $\log\varepsilon$ corrections), at any order, is established for $\mathscr{R}^\varepsilon_{k,\theta}(\sigma, \sigma_0)$, the propagator (or resolvent matrix) of the linear differential equation (5.5); we shall use this result below. It is then possible to prove the Herman–Kluk approximation of the conjecture (4.1).

Proposition 5.2. *Consider $t_0 < 0$ and an initial data of the form*

$$\psi_0^\varepsilon(x) = \vec{V}_-(x)v_0^\varepsilon(x) + O(\varepsilon) \quad \text{in } L^2(\mathbb{R}^d),$$

where $\vec{V}_-(x)$ is the smooth eigenvector for the minus mode of (1.3). *Set*

$$\begin{aligned} \mathcal{J}^\varepsilon_{\mathrm{sc}}(v_0^\varepsilon) = (2\pi\varepsilon)^{-d} &\int_{z\in\mathbb{R}^{2d}} \langle v_0^\varepsilon, g_z^\varepsilon\rangle\, e^{\frac{i}{\varepsilon}S_-(t,t_0,q)}\, \vec{V}_-(t,t_0,q+t-t_0) g^\varepsilon_{\Phi_-^{t,t_0}(z)}\, dz \\ &+ \sqrt{\varepsilon}\kappa_-(2\pi\varepsilon)^{-d}\int_{z\in\mathbb{R}^{2d}} \mathbf{1}_{q>0}\mathbf{1}_{t>t^\flat(z)}\langle v_0^\varepsilon, g_z^\varepsilon\rangle \\ &\qquad\times e^{\frac{i}{\varepsilon}(S_-(t_0,t^\flat(z),q)+S_+(t,t^\flat(z),0))} \\ &\qquad\times \vec{V}_+(t,t^\flat(z),t-t^\flat(z)) g^{\varepsilon,1-ik}_{\Phi_+^{t,t^\flat(z)}(0,p^\flat(z))}\, dz, \end{aligned} \tag{5.6}$$

where

$$\kappa := \sqrt{\frac{2\pi}{ik}\frac{\theta}{2}}$$

and for $z = (q, p)$, $t^\flat(z) = t_0 - q$, *and* $p^\flat(z) = p - kq$ *have been computed in Example* 4.3. *Then, for all* $\chi \in L^1(\mathbb{R})$, *in* $L^2(\mathbb{R}^d)$,

$$\int_{\mathbb{R}} \chi(t)\big(\mathcal{I}^\varepsilon_{\mathrm{sc}}(v^\varepsilon_0) - \mathcal{U}^\varepsilon_{H_{k,\theta}}(t, t_0)\psi^\varepsilon_0(x)\big)\, dt = o(\sqrt{\varepsilon}).$$

The scalar Herman–Kluk prefactor is 1 because the width of the Gaussians wave packet is constant along the propagation (see Example 2.3). Note also (see Example 4.3)

$$\mathcal{T}^\flat g^1(y) = \mathrm{e}^{-\frac{i}{2\varepsilon}ky^2}\, g^1 = g^{1,(1-\frac{ik}{2})};$$

in the statement above, we have chosen not to froze the Gaussian after crossing time, which can be done as in Section 5.4. The value of the coefficient κ arises from Theorem 4.2 and Example 4.3.

The proof relies on the analysis of the propagator $\mathcal{R}^\varepsilon_{k,\theta}(\sigma, \sigma_0)$ as performed in [12, p. 280]. We denote by $\vec{Y}_\pm(\sigma, \sigma_0)$ the time-dependent eigenprovectors of $V_\theta(\sigma)$ for the eigenvalues $E_\pm(\sigma) = \pm k\sigma$, and with initial data $\vec{V}_\pm(\sigma_0)$ (see (1.3)). Using Remark 3.2, we obtain

$$\vec{Y}_\pm(\sigma, \sigma_0) = \vec{V}_\pm(\sigma, \sigma_0, \sigma) = \vec{V}_\pm(\sigma + r, \sigma_0 + r, \sigma) \quad \text{for all } r \in \mathbb{R}. \tag{5.7}$$

Then the solutions $\eta(\sigma)$ with initial data at time $\sigma_0 < 0$ of the form $\eta(\sigma_0) = \eta_0 \vec{V}_-(\sigma_0)$ satisfy:

(1) If $\sigma < 0$, then $\eta(\sigma) = \mathrm{e}^{\frac{i}{\varepsilon}k\int_{\sigma_0}^{\sigma}\tau\, d\tau}\, \vec{Y}_-(\sigma, \sigma_0)\eta_0 + O(\varepsilon)$.

(2) If $\sigma > 0$, then

$$\begin{aligned}\eta(\sigma) &= \mathrm{e}^{\frac{i}{\varepsilon}k\int_{\sigma_0}^{\sigma}\tau\, d\tau}\, \vec{Y}_-(\sigma, \sigma_0)\eta_0 \\ &\quad - \sqrt{\varepsilon}(1-i)\sqrt{\pi}\,(-k)^{-\frac{1}{2}}h(0)\,\mathrm{e}^{\frac{i}{\varepsilon}\lambda}\,\mathrm{e}^{-\frac{i}{\varepsilon}k\int_0^{\sigma}\tau\, d\tau}\,\vec{Y}_+(\sigma, 0)\eta_0 + O(\varepsilon)\end{aligned}$$

with

$$h(0) = \Big(\vec{Y}_+(\sigma, 0), \frac{d}{d\sigma}\vec{Y}_-(\sigma, 0)\Big)\Big|_{\sigma=0} = -\frac{i\theta}{2} \quad \text{and} \quad \lambda = k\int_{\sigma_0}^{0}\tau\, d\tau.$$

We observe that, with the notations of Example 4.3,

$$-(1-i)\sqrt{\pi}(-k)^{-\frac{1}{2}}h(0) = \sqrt{\frac{2\pi}{ik}}\frac{\theta}{2} = \sqrt{\frac{2\pi}{ik}}\gamma^\flat = \kappa.$$

Proof of Proposition 5.2 . By (5.4),

$$\psi^\varepsilon(t, x) := \mathcal{U}^\varepsilon_{H_{k,\theta}}(t, t_0)\psi^\varepsilon_0(x) = \mathcal{R}^\varepsilon_{k,\theta}(x, x - t + t_0)\psi^\varepsilon_0(x - t + t_0),$$

and, in view of Friedrichs' description as detailed in points (1) and (2) above, we deduce

$$\begin{aligned}\psi^\varepsilon(t, x) &= \mathrm{e}^{\frac{ik}{\varepsilon}x(t-t_0) - \frac{ik}{2\varepsilon}(t-t_0)^2}\,\vec{V}_-(t, t_0, x)v^\varepsilon_0(x - t + t_0) \\ &\quad + \sqrt{\varepsilon}\kappa\,\mathbf{1}_{0<x<t-t_0}\,\mathrm{e}^{\frac{ik}{\varepsilon}x(t-t_0) - \frac{ik}{2\varepsilon}(t-t_0)^2 - \frac{ik}{\varepsilon}x^2} \\ &\qquad \times v^\varepsilon_0(x - t + t_0)\vec{V}_+(x, 0, x) + o(\sqrt{\varepsilon}),\end{aligned}$$

where we have used

$$\vec{Y}_-(x, x-t+t_0) = \vec{V}_-(t, t_0, x), \quad \vec{Y}_+(x, 0) = \vec{V}_+(x, 0, x)$$

and

$$\lambda = -\frac{k}{2}(x-t+t_0)^2.$$

Using (1.8), we write

$$v_0^\varepsilon(x-t+t_0) = (2\pi\varepsilon)^{-d} \int_{z\in\mathbb{R}^{2d}} \langle v_0^\varepsilon, g_z^\varepsilon\rangle g_z^\varepsilon(x-t+t_0)\, dz$$

and we observe that, in view of $S_-(t, t_0, z) = kq(t-t_0) + \frac{k}{2}(t-t_0)^2$, we have

$$g_z^\varepsilon(x-t+t_0) = \mathrm{e}^{-\frac{ik}{\varepsilon}x(t-t_0)+\frac{ik}{2\varepsilon}(t-t_0)^2}\, \mathrm{e}^{\frac{i}{\varepsilon}S_-(t,t_0,z)}\, g^\varepsilon_{\Phi_-^{t,t_0}(z)}(x).$$

Similarly, using

$$S_+(t, t^\flat(z), q) = -\frac{k}{2}(t-t_0+q)^2, \quad S_-(t^\flat, t_0, q) = \frac{3k}{2}q^2$$

and

$$\Phi_+^{t,t^\flat(z)}(0, p^\flat(z)) = (t-t_0+q, p-2kq-k(t-t_0)),$$

we obtain

$$\begin{aligned}
g_z^\varepsilon(x-t+t_0) &= \mathrm{e}^{\frac{i}{\varepsilon}k(x-t+t_0)(2q+(t-t_0)}\, g^\varepsilon_{\Phi_+^{t,t^\flat(z)}(0,p^\flat(z))}(x) \\
&= \mathrm{e}^{\frac{i}{\varepsilon}S_-(t^\flat,t_0,q)+\frac{i}{\varepsilon}S_+(t,t^\flat,0)+\frac{k}{i\varepsilon}(x-q-t+t_0)^2}\, g^\varepsilon_{\Phi_+^{t,t^\flat(z)}(0,p^\flat(z))}(x) \\
&= \mathrm{e}^{\frac{i}{\varepsilon}S_-(t^\flat,t_0,q)+\frac{i}{\varepsilon}S_+(t,t^\flat,0)}\, \mathrm{WP}^\varepsilon_{\Phi_+^{t,t^\flat(z)}(0,p^\flat(z))}(\mathcal{T}^\flat g^1).
\end{aligned}$$

Putting these elements together, we are left with

$$\begin{aligned}
\psi^\varepsilon(t,x) = o(\sqrt{\varepsilon}) &+ (2\pi\varepsilon)^{-d} \int_{z\in\mathbb{R}^{2d}} \langle v_0^\varepsilon, g_z^\varepsilon\rangle\, \mathrm{e}^{\frac{i}{\varepsilon}S_-(t,z)}\, \vec{V}_-(t, t_0, x) g^\varepsilon_{\Phi_-^{t,t_0}(z)}\, dz \\
&+ \sqrt{\varepsilon}\kappa(2\pi\varepsilon)^{-d} \int_{z\in\mathbb{R}^{2d}} \mathbf{1}_{0<x<t-t_0} \langle v_0^\varepsilon, g_z^\varepsilon\rangle\, \mathrm{e}^{\frac{i}{\varepsilon}S_-(t^\flat,t_0,q)+\frac{i}{\varepsilon}S_+(t,t^\flat,0)} \\
&\qquad\qquad \times \vec{V}_+(x, 0, x) g^\varepsilon_{\Phi_+^{t,t^\flat}(z)}\, dz.
\end{aligned}$$

Using Taylor expansion and Lemma 5.3, we can transform the first part of the right-hand side of the preceding equation:

$$\begin{aligned}
&(2\pi\varepsilon)^{-d} \int_{z\in\mathbb{R}^{2d}} \langle v_0^\varepsilon, g_z^\varepsilon\rangle\, \mathrm{e}^{\frac{i}{\varepsilon}S_-(t,z)}\, \vec{V}_-(t, t_0, x) g^\varepsilon_{\Phi_-^{t,t_0}(z)}\, dz \\
&= (2\pi\varepsilon)^{-d} \int_{z\in\mathbb{R}^{2d}} \langle v_0^\varepsilon, g_z^\varepsilon\rangle\, \mathrm{e}^{\frac{i}{\varepsilon}S_-(t,z)}\, \vec{V}_-(t, t_0, q_-(t)) g^\varepsilon_{\Phi_-^{t,t_0}(z)}\, dz + O(\varepsilon)
\end{aligned}$$

in $L^2(\mathbb{R}^d)$, which allows to identify the first term of (5.6) since $q_-(t) = q + t - t_0$. In the second term, one can treat similarly the term $\vec{V}_+(x,0,x)$ that turns into (using $x = t - t_0 + q$ and (5.7))

$$\begin{aligned}\vec{V}_+(x,0,x) &= \vec{V}_+(t - t_0 + q, 0, t - t_0 + q)\\ &= \vec{V}_+(t, t_0 - q, t - t_0 + q)\\ &= \vec{V}_+(t, t^\flat(z), t - t^\flat(z)).\end{aligned}$$

It remains to turn the discontinuous function $x \mapsto \mathbf{1}_{0<x<t-t_0}$ into

$$\mathbf{1}_{0<q_+(t)<t-t_0} = \mathbf{1}_{q<0}\mathbf{1}_{t>t_0-q} = \mathbf{1}_{q<0}\mathbf{1}_{t>t^\flat(z)}.$$

One then takes advantage of the averaging in time to use the estimate of (5.3) despite the discontinuity. One regularizes the discontinuous function which will differ from its regularization on a set of small Lebesgue measure in the variable t. One can then use the preceding argument on the regularized term and gets rid of the correction ones by estimating them thanks to estimate (5.3) and using the smallness of integrals on χ on sets of small Lebesgue measures. ■

Acknowledgements. The first and the second author acknowledge support from the CNRS 80|Prime project *AlgDynQua*, the first author from the John-von-Neumann visiting professorship program of the Technische Universität München, and the second author from the Visiting Professorship program of the I-Site Future (Ecole des Ponts Paris-Tech).

Bibliography

[1] T. Begušić, M. Cordova and J. Vaniček, Single-Hessian thawed Gaussian approximation. *J. Chem. Phys.* (2019) **150**, Article ID 154117

[2] J.-M. Bily, *Propagation d'états cohérents et applications*. Ph.D. thesis, University of Nantes, Nantes, 2001

[3] M. Born and V. Fock, Beweis des Adiabatensatzes. *Z. Phys.* (1928), 165–180

[4] R. Carles and C. Fermanian Kammerer, A nonlinear adiabatic theorem for coherent states. *Nonlinearity* **24** (2011), 2143–2164

[5] M. Combescure and D. Robert, *Coherent states and applications in mathematical physics*. Theoret. Math. Phys., Springer, Dordrecht, 2012

[6] C. Fermanian Kammerer and C. Lasser, Propagation through generic level crossings: A surface hopping semigroup. *SIAM J. Math. Anal.* **40** (2008), 103–133

[7] C. Fermanian Kammerer and C. Lasser, An Egorov theorem for avoided crossings of eigenvalue surfaces. *Comm. Math. Phys.* **353** (2017), 1011–1057

[8] C. Fermanian Kammerer, C. Lasser and D. Robert, Propagation of wave packets for systems presenting codimension one crossings. preprint 2020, arXiv:2001.07484

[9] C. Fermanian Kammerer, C. Lasser and D. Robert, Herman–Kluk approximation for systems presenting codimension one crossings. Work in progress

[10] K. O. Friedrichs, *On the adiabatic theorem in quantum theory. Part I–II.* Courant Institute of Mathematical Sciences, New York, 1955

[11] P. Gérard, P. A. Markowich, N. J. Mauser and F. Poupaud, Homogenization limits and Wigner transforms. *Comm. Pure Appl. Math.* **50** (1997), 323–379

[12] G. A. Hagedorn, Adiabatic expansions near eigenvalue crossings. *Ann. Physics* **196** (1989), 278–295

[13] G. A. Hagedorn, Molecular propagation through electron energy level crossings. *Mem. Amer. Math. Soc.* **111** (1994), no. 536

[14] E. J. Heller, Frozen Gaussians: A very simple semiclassical approximation. *J. Chem. Phys.* **75** (1981), 2923–2931

[15] M. F. Herman and E. Kluk, A semiclassical justification for the use of non-spreading wavepackets in dynamics calculations. *Chem. Phys.* **91** (1984), no. 1, 27–34

[16] T. Kato, On the adiabatic theorem of quantum mechanics. *J. Phys. Soc. Japan* **5** (1950), 435–439

[17] K. Kay, Integral expressions for the semi-classical time-dependent propagator. *J. Chem. Phys.* **100** (1994), no. 6, 4377-4392

[18] K. Kay, The Herman–Kluk approximation: Derivation and semiclassical corrections. *Chem. Phys.* **322** (2006), 3–12

[19] E. Kluk, M. F. Herman and H. L. Davis, Comparison of the propagation of semiclassical frozen Gaussian wave functions with quantum propagation for a highly excited anharmonic oscillator. *J. Chem. Phys.* **84** (1986), 326–334

[20] C. Lasser and C. Lubich, Computing quantum dynamics in the semiclassical regime. *Acta Numer.* **29** (2020), 229–401

[21] C. Lasser and D. Sattlegger, Discretising the Herman–Kluk propagator. *Numer. Math.* **137** (2017), 119–157

[22] A. Maspero and D. Robert, On time dependent Schrödinger equations: Global well-posedness and growth of Sobolev norms. *J. Funct. Anal.* **273** (2017), 721–781

[23] W. H. Miller, An alternate derivation of the Herman–Kluk (coherent state) semiclassical initial value representation of the time evolution operator. *Mol. Phys.* 100 (2002), no. 4, 397–400

[24] D. Robert, Propagation of coherent states in quantum mechanics and applications. In *Partial differential equations and applications*, pp. 181–252, Sémin. Congr. 15, Société Mathématique de France, Paris, 2007

[25] D. Robert, On the Herman-Kluk semiclassical approximation. *Rev. Math. Phys.* **22** (2010), 1123–1145

[26] T. Swart and V. Rousse, A mathematical justification for the Herman–Kluk propagator. *Comm. Math. Phys.* **286** (2009), 725–750

[27] S. Teufel, *Adiabatic perturbation theory in quantum dynamics*. Lecture Notes in Math. 1821, Springer, Berlin, 2003

[28] J. C. Tully, Perspective: Nonadiabatic dynamics theory. *J. Chem. Phys.* **137** (2012), Article ID 22A301

[29] A. Watson and M. I. Weinstein, Wavepackets in inhomogeneous periodic media: Propagation through a one-dimensional band crossing. *Comm. Math. Phys.* **363** (2018), 655–698

[30] M. Werther, S. L. Choudhury and F. Großmann, Coherent state based solutions of the time-dependent Schrödinger equation: Hierarchy of approximations to the variational principle. *Int. Rev. Phys. Chem.* **40** (2020), 81–125

Length scales for BEC in the dilute Bose gas

Søren Fournais

To Ari Laptev on the occasion of his 70th birthday

We give a short proof of Bose Einstein Condensation of dilute Bose gases on length scales much longer than the Gross–Pitaevskii scale.

1 Introduction

In this article, I will review how recent techniques from [4, 8] can be used to give simple proofs of Bose–Einstein Condensation (BEC) on different length scales for the dilute Bose gas.

One of the outstanding open problems in mathematical physics is to prove BEC in the thermodynamic limit for a continuous, (weakly) interacting system.[1] Unfortunately, we will have nothing new to report on this question. However, the substantially easier problem of proving BEC on shorter length scales has seen marked progress in recent years [10] and [1–3]. The first paper cited gave the first proof of BEC on the so-called Gross–Pitaevskii length scale, which will be defined precisely below. On this length scale the energy gap between the ground state of the kinetic energy and the first excited state is of the same order of magnitude as the energy per particle in the interacting system. Therefore it is feasible, but still not easy, to prove condensation on this length scale.

The recent papers [4, 5, 8] are dedicated to the study of the energy asymptotics of the Bose gas. By adapting the methods of these papers, one obtains easy proofs of condensation on length scales substantially longer than the Gross–Pitaevskii scale (but far shorter than "thermodynamic length scales"). The reference [1] goes beyond the Gross–Pitaevskii length scale but with a very different method than ours and for a smaller range of lengths. The estimates of reference [10] can also be used to go beyond the Gross–Pitaevskii length scale, we will discuss this further below.

[1] See http://web.math.princeton.edu/~aizenman/OpenProblems_MathPhys/ for information on this and other challenging problems in Mathematical Physics.

Keywords: Many-body quantum mechanics, Dilute Bose gases, Bose–Einstein condensation, Bogolubov theory
2020 Mathematics Subject Classification: Primary 81V73; secondary 81V70

For background on the mathematics of the Bose gas we recommend the reviews [11, 15] and references therein.

We proceed to give the precise statement of the result in the present paper. We study the following Hamiltonian $H = H(N, L)$:

$$H = \sum_{j=1}^{N} -\Delta_j^{\mathrm{per}} + \sum_{j<k} v^{\mathrm{per}}(x_j - x_k).$$

Here Δ^{per} is the Laplace operator on $\Omega = (-\frac{L}{2}, \frac{L}{2})^3$ with periodic boundary conditions and $v^{\mathrm{per}} = \sum_{j\in\mathbb{Z}^3} v(x - Lj)$ is the periodic version of the potential v. In other words, we identify Ω with the torus $\mathbb{R}^3/L\mathbb{Z}^3$ and consider the Laplacian and the potential v as given on the torus.

The ground state energy $E = E(N, L)$ of H is defined to be

$$E(N, L) := \inf \operatorname{Spec} H(N, L).$$

The particle density in the box is $\rho := \frac{N}{L^3}$.

Define furthermore,

$$P_\Omega := L^{-3}|1\rangle\langle 1|, \quad Q_\Omega := \mathbb{1} - P_\Omega,$$

i.e. P_Ω is the orthogonal projection in $L^2(\Omega)$ onto the constant functions and Q_Ω is the orthogonal projection to the complement. Using these definitions, we define

$$n_0 := \sum_{j=1}^{N} P_{\Omega,j}, \quad n_+ := \sum_{j=1}^{N} Q_{\Omega,j} = N - n_0.$$

The definition is based on the idea that low energy eigenstates of the system should concentrate in the constant function. Thus, n_0 counts the number of particles "in the condensate" and n_+ the number of particles "excited out of the condensate".

That the ground state Ψ of $H(N, \Omega)$ exhibits Bose–Einstein condensation in the constant function means by definition that

$$\langle \Psi, n_0 \Psi \rangle \geq cN,$$

for some $c > 0$ depending only on the density and for all sufficiently large N.

We will study this problem for length scales L that depend on the density ρ in such a way that L diverges as ρ becomes small, i.e. as the system becomes dilute. We will prove that in this limit

$$\frac{\langle \Psi, n_0 \Psi \rangle}{N} \to 1, \quad \text{or equivalently} \quad \frac{\langle \Psi, n_+ \Psi \rangle}{N} \to 0. \tag{1.1}$$

We will work under the following assumption on the potential.

Assumption 1.1. The potential $v \neq 0$ is non-negative and spherically symmetric, i.e. $v(x) = v(|x|) \geq 0$, and of class $L^1(\mathbb{R}^3)$ with compact support. We fix $R > 0$ such that $\operatorname{supp} v \subset B(0, R)$.

Under Assumption 1.1 the potential v has a finite scattering length $a > 0$. For the convenience of the reader, we recall the definition of the scattering length in Appendix A. The measure of diluteness is the (dimensionless) quantity ρa^3, and we will study the limit $\rho a^3 \to 0$. The length scale L of the box is tied to the density ρ in such a way that

$$L = C_1(\rho a^3)^{-\delta}\frac{1}{\sqrt{\rho a}},$$

for some fixed constant $C_1 > 0$ and for $0 < \delta$. The quantity $(\rho a)^{-\frac{1}{2}}$ has the unit of a length and is the Gross–Pitaevskii length scale mentioned above. Therefore, we see that we consider L's much longer than the Gross–Pitaevskii length scale. For results about condensation on length scales shorter than $(\rho a)^{-\frac{1}{2}}$ see [15] and references therein.

Theorem 1.2. *Suppose v satisfies Assumption* 1.1. *Let $\epsilon \in [0, \frac{1}{2}]$ and $C_0 > 0$. Suppose that $\Psi \in \otimes_s^N L^2(\Omega)$ is a normalized state with*

$$\langle \Psi, H\Psi\rangle \leq 4\pi a\rho N + C_0 a\rho(\rho a^3)^{\frac{1}{2}-\epsilon}N. \tag{1.2}$$

Then

$$\frac{\langle \Psi, n_+\Psi\rangle}{N} \leq C\rho a L^2(\rho a^3)^{\frac{1}{2}-\epsilon}.$$

In particular, if $2\delta + \epsilon < \frac{1}{2}$, then we have complete condensation in the sense of (1.1).

Remark. The trial states of [7] can be used to obtain examples of states satisfying the upper bound (1.2) (with $\epsilon = 0$). More precise approximate ground states for the Bose gas were constructed in [16] leading to the upper bound in (1.4).

Remark. On a box of side-length L there is a gap in the spectrum of the kinetic energy above the ground state of order L^{-2}, more precisely the gap has size $(\frac{2\pi}{L})^2$. We therefore get the easy lower bound

$$\langle \Psi, H\Psi\rangle \geq \langle \Psi, \widetilde{H}\Psi\rangle + 2\pi^2 L^{-2}\langle \Psi, n_+\Psi\rangle,$$

where

$$\widetilde{H} := \sum_{j=1}^{N}\left(-\Delta_j^{\text{per}} - 2\pi^2 L^{-2} Q_{\Omega,j}\right) + \sum_{j<k} v^{\text{per}}(x_j - x_k).$$

Therefore, Theorem 1.2 follows upon proving the lower bound

$$\widetilde{H} \geq 4\pi a\rho N + C_0 a\rho(\rho a^3)^{\frac{1}{2}}N. \tag{1.3}$$

The purpose of this paper is to show how this lower bound follows relatively easily using a localization argument from [4, 5, 8] (see also [12, 13] for earlier versions of a similar argument), thereby proving condensation in the sense of Theorem 1.2. This proof is very general, in particular, it is uniform in the L^1-norm of the potential, so it allows for hard core potentials. Also, the assumption on compact support can be relaxed. We will not pursue these generalizations – they follow fairly easily from the considerations in this paper – see [4] for details.

Notice how this strategy depends strongly on the existence of a large enough gap in the kinetic energy. Therefore, it is clear that to prove BEC on much longer length scales other ideas will be needed. In particular, this technique does not work in the thermodynamic limit where one considers the limit $L \to \infty$ for fixed (small) density $\rho = \frac{N}{L^3}$.

Remark. For large boxes[2] we know that the energy per particle in the ground state has the form

$$E(N,L)/N = 4\pi a\rho\left(1 + \frac{128}{15\sqrt{\pi}}\sqrt{\rho a^3}\right) + \mathcal{O}(a\rho(\rho a^3)^{\frac{1}{2}+\eta}) \tag{1.4}$$

when ρa^3 is small and for some $\eta > 0$. Here the leading-order term was proved in [6] (upper bound), [14] (lower bound), and the correction term – called the Lee–Huang–Yang-term [9] – was rigorously established recently in [16] (upper bound) and [8] (lower bound).

From the discussion in the previous remark it is clear that one can improve Theorem 1.2 by going to the next order in the energy. More precisely, one can improve (1.3) by including the Lee–Huang–Yang term and calculating the energy per particle to precision $\mathcal{O}(a\rho(\rho a^3)^{\frac{1}{2}+\eta})$. This would allow us to establish BEC for states with sufficiently low energy on length scales a bit (depending on η) longer than $(\rho a^3)^{-\frac{1}{4}}\frac{1}{\sqrt{\rho a}}$. For reasons of clarity of exposition we refrain from carrying out this more precise analysis in this short paper: The gain η in (1.4) as proved in [8] is not very large, and the additional necessary analysis is somewhat more lengthy than what is possible here. However, it should be clear from this paper, how to adapt [8] to obtain the improved result.

Remark. In the literature (cf. [1–3]) the Hamiltonian is often written in another (equivalent) scaling, so it seems worthwhile to translate between the two. Consider the periodic box $[0,\lambda]^3$. Typically λ is chosen to be equal to unity, but we keep it here for dimensional reasons. Define $H^{\mathrm{GP}}(N)$ to be the operator

$$H^{\mathrm{GP}}(N) := \sum_{j=1}^{N} -\Delta_j^{\mathrm{per}} + \sum_{j<k} N^{2-2\kappa} v^{\mathrm{per}}(N^{1-\kappa}(y_j - y_k)),$$

where Δ^{per} is the Laplacian on the λ-torus. By the scaling $x = N^{1-\kappa}y$, we see that

$$H^{\mathrm{GP}}(N) \sim N^{2-2\kappa} H(N,L),$$

with $\sim$ denoting unitary equivalence and for $L = N^{1-\kappa}\lambda$. We get the density after scaling as $\rho = NL^{-3} = N^{3\kappa-2}\lambda^{-3}$, so in terms of the density we find

$$L = C(\lambda)(\rho a^3)^{\frac{-\kappa}{4-6\kappa}}\frac{1}{\sqrt{\rho a}}, \tag{1.5}$$

[2]More precisely, the statement is valid in the thermodynamic limit.

with $C(\lambda)$ being the constant $(\frac{a}{\lambda})^{\frac{1}{2}+\frac{3\kappa}{4-6\kappa}}$. Notice in passing that (1.5) also implies that $\kappa = \frac{2}{3}$ would correspond to the thermodynamic limit. Our results on BEC in the range $\delta \in (0, \frac{1}{4})$ when $\epsilon = 0$, therefore translate into BEC for the low energy states of the operator $H^{\mathrm{GP}}(N)$, when $\kappa \in (0, \frac{2}{5})$. In [1] a different method is used that among other results gives BEC for $\kappa \in (0, \frac{1}{43})$. The original proof of BEC in the Gross–Pitaevskii-limit in [10] is only stated for $\kappa = 0$, but it is clear from the given estimates (see in particular [11, equation (5.15)]) that the proof yields BEC for $\kappa < \frac{1}{10}$.

2 Energy in small boxes

We study a localized problem depending on a parameter ρ_μ. The following Section 3 will show how the proof of Theorem 1.2 reduces to this localized problem. In the end ρ_μ will be chosen to be the density $\rho = N/L^3$ in the original box, but we use a different notation to avoid confusion with the particle density of the localized problem. The analysis in the present section follows closely that of [8, Appendix B], but with some simplifications similar to [4]. The box is $\Lambda = (-\frac{\ell}{2}, \frac{\ell}{2})^3$ of size

$$\ell := K^{-1} \frac{1}{\sqrt{\rho_\mu a}}, \tag{2.1}$$

where $K > 1$ will be chosen sufficiently large at the end, in order for the kinetic energy gap in the box to dominate various error terms (see (2.16) below).

We choose and fix a localization function $\chi \in C^\infty(\mathbb{R}^3)$ with $\operatorname{supp} \chi \subset [-\frac{1}{2}, \frac{1}{2}]^3$. We choose χ even and satisfying

$$0 \le \chi, \quad \int \chi^2(x)\, dx = 1.$$

We will also use the notation

$$\chi_\Lambda(x) := \chi\Big(\frac{x}{\ell}\Big),$$

and

$$W(x) := \frac{v(x)}{\chi * \chi(\frac{x}{\ell})}. \tag{2.2}$$

Since v has support in $B(0, R)$, we see that W is well-defined if $\rho_\mu a^3$ is sufficiently small. Clearly W depends on ℓ and thus ρ_μ, but we will not reflect this in our notation.

Define furthermore, as operators on $L^2(\Lambda)$,

$$P := \ell^{-3} |1\rangle\langle 1|, \quad Q := \mathbb{1} - P,$$

i.e. P is the orthogonal projection in $L^2(\Lambda)$ onto the constant functions and Q is the projection to the orthogonal complement.

The localized Hamiltonian $\mathcal{H}_\Lambda$ acts on the symmetric Fock space $\mathcal{F}_s(L^2(\Lambda))$. It preserves particle number and is given as

$$(\mathcal{H}_\Lambda(\rho_\mu))_M := \sum_{i=1}^{M} \mathcal{T}^{(i)} - \rho_\mu \sum_{i=1}^{M} \int w_1(x_i, y)\, dy + \sum_{1\leq i<j\leq M} w(x_i, x_j), \tag{2.3}$$

on the M-particle sector. We will see in Section 3 how $\mathcal{H}_\Lambda(\rho_\mu)$ is related to the original $\widetilde{H}$ after localization through a sliding procedure. Here

$$\mathcal{T} := Q[\chi(-\Delta - s^{-2}\ell^{-2})_+\chi + b\ell^{-2}]Q,$$

with $s, b > 0$ fixed constants, and

$$\begin{aligned} w(x,y) &= \chi\Big(\frac{x}{\ell}\Big) W(x-y)\chi\Big(\frac{y}{\ell}\Big), \\ W_1(x) &= W(x)(1-\omega(x)) = \frac{g(x)}{\chi * \chi(\frac{x}{\ell})}, \\ w_1(x,y) &= w(x,y)(1-\omega(x-y)) = \chi\Big(\frac{x}{\ell}\Big) g(x-y)\chi\Big(\frac{y}{\ell}\Big), \end{aligned} \tag{2.4}$$

where ω and g are related to the scattering length and defined in (A.1). For later use, we also define

$$w_2(x,y) = w(x,y)(1-\omega^2(x-y)) = w_1(x,y)(1+\omega(x-y)). \tag{2.5}$$

Notice the identities

$$\begin{aligned} \iint w_1(x,y)\, dx\, dy &= 8\pi\ell^3 a, \\ \iint w_2(x,y)\, dx\, dy &= \ell^3\Big(8\pi a + \int g\omega(x)\, dx\Big). \end{aligned} \tag{2.6}$$

We will prove the following lower bound on $\mathcal{H}_\Lambda(\rho_\mu)$. Notice that this is a lower bound on the entire operator on Fock space, i.e. independent of particle number. This is useful because the term with ρ_μ plays the role of a chemical potential and therefore effectively determines a preferred number of particles in the box.

Theorem 2.1. *Suppose v satisfies Assumption* 1.1 *and let the localization function $\chi \in C_0^\infty((-\frac{1}{2}, \frac{1}{2})^3)$ be given. Suppose that K in* (2.1) *is chosen sufficiently large (depending only on χ, b, s and the radius R of* supp v*). Then there exists a constant $C_0 > 0$ such that for sufficiently small values of $\rho_\mu a^3$ we have*

$$\mathcal{H}_\Lambda(\rho_\mu) \geq -4\pi\rho_\mu^2 a\ell^3 - C_0\rho_\mu^2 a\ell^3(\rho_\mu a^3)^{\frac{1}{2}}. \tag{2.7}$$

Theorem 2.1 is actually the same as [4, Theorem 6.1]. For completeness, we give the main steps of the proof below.

Proof. We start by obtaining a control on the number of particles. Notice that $\mathcal{H}_\Lambda(\rho_\mu)$ preserves particle number, so it suffices to establish the lower bound stated in (2.7) for $\langle \Psi, (\mathcal{H}_\Lambda(\rho_\mu))_M \Psi\rangle$ for $\Psi \in \bigotimes_s^M L^2(\Lambda)$ for some arbitrary particle number M. Since, for $\rho_\mu a^3$ sufficiently small,

$$\rho_\mu \ell^3 = K^{-3}(\rho_\mu a^3)^{-\frac{1}{2}} \geq 1,$$

we can divide the M particles in ξ groups with the order of $\rho_\mu \ell^3$ particles in each, with the last group taking care of a possible remainder, i.e. given some $\Xi \geq 3$, we write

$$\{1, \dots, M\} = \bigcup_{j=1}^{\xi} S_j,$$

with

$$S_j \cap S_k = \emptyset \quad \text{for } j \neq k$$

and

$$\begin{aligned} &|S_\xi| \leq (\Xi + 1)\rho_\mu \ell^3, \\ &|S_j| \in [\Xi \rho_\mu \ell^3, (\Xi + 1)\rho_\mu \ell^3] \quad \text{for } j < \xi. \end{aligned} \tag{2.8}$$

Here the interval for $|S_j|$ is to allow for the fact that $\rho_\mu \ell^3$ might not be an integer.

Now we remove the positive potentials $w(x_i, x_j)$ from $(\mathcal{H}_\Lambda(\rho_\mu))_M$ except when i, j are in the same group S_k. i.e. discard the interaction between particles in different groups. In this way, we obtain the lower bound

$$\langle \Psi, (\mathcal{H}_\Lambda(\rho_\mu))_M \Psi\rangle \geq \sum_{j=1}^{\xi} \inf \operatorname{Spec}(\mathcal{H}_\Lambda(\rho_\mu))_{|S_j|}, \tag{2.9}$$

where the particle numbers $|S_j|$ satisfy (2.8). Therefore, from now on we can reduce the analysis to $(\mathcal{H}_\Lambda(\rho_\mu))_n$ with $n \leq (\Xi + 1)\rho_\mu \ell^3$.

Define the operators n_0, n_+ on $\mathcal{F}_s(L^2(\Lambda))$, given by

$$n_0 = \sum_i P_i,$$

$$n_+ = \sum_i Q_i.$$

An important step in the proof is the algebraic identity given in (2.10) below, which isolates a positive term $\mathcal{Q}_4^{\text{ren}}$ in the potential that can be discarded for a lower bound.

Lemma 2.2 (Potential energy decomposition). *We have*

$$\begin{aligned} &-\rho_\mu \sum_{i=1}^{n} \int w_1(x_i, y)\, dy + \frac{1}{2} \sum_{i \neq j} w(x_i, x_j) \\ &= \mathcal{Q}_0^{\text{ren}} + \mathcal{Q}_1^{\text{ren}} + \mathcal{Q}_2^{\text{ren}} + \mathcal{Q}_3^{\text{ren}} + \mathcal{Q}_4^{\text{ren}}, \end{aligned} \tag{2.10}$$

where

$$\mathcal{Q}_4^{\text{ren}} := \frac{1}{2}\sum_{i\neq j}\big[Q_iQ_j + (P_iP_j + P_iQ_j + Q_iP_j)\omega(x_i - x_j)\big]w(x_i,x_j) \\ \times\big[Q_jQ_i + \omega(x_i-x_j)(P_jP_i + P_jQ_i + Q_jP_i)\big],$$

$$\mathcal{Q}_3^{\text{ren}} := \sum_{i\neq j} P_iQ_jw_1(x_i,x_j)Q_jQ_i + \text{h.c.},$$

$$\mathcal{Q}_2^{\text{ren}} := \sum_{i\neq j} P_iQ_jw_2(x_i,x_j)P_jQ_i + \sum_{i\neq j} P_iQ_jw_2(x_i,x_j)Q_jP_i \\ - \rho_\mu\sum_{i=1}^{n} Q_i\int w_1(x_i,y)\,dy\,Q_i \\ - + \frac{1}{2}\sum_{i\neq j}(P_iP_jw_1(x_i,x_j)Q_jQ_i + \text{h.c.}),$$

$$\mathcal{Q}_1^{\text{ren}} := \sum_{i,j} P_jQ_iw_2(x_i,x_j)P_iP_j - \rho_\mu\sum_i Q_i\int w_1(x_i,y)\,dy\,P_i + \text{h.c.},$$

$$\mathcal{Q}_0^{\text{ren}} := \frac{1}{2}\sum_{i\neq j} P_iP_jw_2(x_i,x_j)P_jP_i - \rho_\mu\sum_i P_i\int w_1(x_i,y)\,dy\,P_i.$$

Proof. The identity in (2.10) is purely algebraic using that $P + Q = \mathbb{1}$, and the definitions (2.4) and (2.5). ∎

Remark. The identity in (2.10) is used in [8] to get the energy to Lee–Huang–Yang precision. If one is only aiming to prove Theorem 2.1 and is willing to let the error bound depend on $\int v$, then one can use the simpler

$$\widetilde{\mathcal{Q}}_4^{\text{ren}} := \frac{1}{2}\sum_{i\neq j}\big[Q_iQ_j + P_iP_j\omega(x_i-x_j)\big]w(x_i,x_j)\big[Q_jQ_i + \omega(x_i-x_j)P_jP_i\big].$$

This will then, of course, change the other terms in (2.10).

Notice that one of the effects of the identity in (2.10) is that all terms with three or fewer projections Q contain the potentials w_1 or w_2 instead of the original potential v. Integrals of w_1, w_2 are immediately controlled in terms of $a = \frac{1}{8\pi}\int g(x)\,dx$, which is possibly much smaller than $\frac{1}{8\pi}\int v(x)\,dx$. Therefore, using $\mathcal{Q}_4^{\text{ren}}$ instead of $\widetilde{\mathcal{Q}}_4^{\text{ren}}$ allows to get the uniformity of Theorem 2.1. Therefore, in this article, we follow [4, 8] and use the correct $\mathcal{Q}_4^{\text{ren}}$-term.

Using (2.10) we can simplify the potential energy significantly. We can use that $\mathcal{Q}_4^{\text{ren}}$ is a positive term to discard it. Also, we can use a Cauchy–Schwarz inequality on $\mathcal{Q}_3^{\text{ren}}$ to absorb part of $\mathcal{Q}_3^{\text{ren}}$ in the positive $\mathcal{Q}_4^{\text{ren}}$-term (before discarding it). This will, of course, introduce additional terms with two or fewer projections Q in them. Furthermore, we notice that for all functions F, $P_iF(x_i,y)P_i = \ell^{-3}P_i\int F(x,y)\,dx$, since P projects on the constant function. This can be used to simplify several terms. Finally, one can use a Cauchy–Schwarz inequality on the terms with one Q to estimate these in terms of $2Q$-terms and $0Q$-terms. This yields the following lemma, which is [8, Lemma B.2] (where the details of the proof can also be found).

Lemma 2.3 ([8, Lemma B.2]). *There is a constant $C > 0$ such that we have*

$$\begin{aligned} -\rho_\mu \sum_{i=1}^{n} \int w_1(x,y)\,dy + \frac{1}{2}\sum_{i\neq j} w(x_i,x_j) \\ \geq A_0 + A_2 - Ca(\rho_\mu + n_0\ell^{-3})n_+, \end{aligned} \tag{2.11}$$

where

$$\begin{aligned} A_0 = {} & \frac{n_0(n_0-1)}{2\ell^6} \iint w_2(x,y)\,dx\,dy \\ & - \Big(\rho_\mu \frac{n_0}{\ell^3} + \frac{1}{4}\Big(\rho_\mu - \frac{n_0-1}{\ell^3}\Big)^2\Big) \iint w_1(x,y)\,dx\,dy \end{aligned}$$

and

$$A_2 = \frac{1}{2}\sum_{i\neq j} P_i P_j w_1(x_i,x_j) Q_j Q_i + \text{h.c.}$$

Notice how, using the identities in (2.6) and $n_0 + n_+ = n$, we can get the estimate

$$\begin{aligned} & - \rho_\mu \sum_{i=1}^{n} \int w_1(x,y)\,dy + \frac{1}{2}\sum_{i\neq j} w(x_i,x_j) \\ & \geq A_2 + \frac{n^2}{2\ell^3}\Big(8\pi a + \int g\omega(x)\,dx\Big) - \Big(\rho_\mu \frac{n}{\ell^3} + \frac{1}{4}\Big(\rho_\mu - \frac{n}{\ell^3}\Big)^2\Big) 8\pi a \ell^3 \\ & \quad - Ca(\rho_\mu + n\ell^{-3})(n_+ + 1) \end{aligned} \tag{2.12}$$

from (2.11).

We combine the kinetic energy (without the gap) and the A_2-term from Lemma 2.3 in the following estimate. The analysis of this term uses second quantization.

Lemma 2.4. *On the n-particle sector, we have (with $\mathcal{T}$ from* (2.3) *and A_2 from Lemma* 2.3*)*

$$\begin{aligned} & \sum_{i=1}^{n} (\mathcal{T}^{(i)} - b\ell^{-2} Q_i) + A_2 \\ & \quad \geq - \frac{n(n+1)}{2\ell^3} \int g(x)\omega(x)\,dx - Ca\frac{n_0+1}{\ell^3} n_+ \\ & \qquad - Cna\ell^{-3}\Big(1 + \frac{a^2(n+1)^2}{\ell^2} + \frac{(n+1)R^2}{\ell^2}\Big). \end{aligned} \tag{2.13}$$

Proof. We define the following operators on $\mathcal{F}_s(L^2(\mathbb{R}^3))$:

$$b_k := \ell^{-\frac{3}{2}} a_0^\dagger a(Q\chi_\Lambda e^{-ikx}), \quad b_k^\dagger := \ell^{-\frac{3}{2}} a^\dagger(Q\chi_\Lambda e^{-ikx}) a_0,$$

where $a, a^\dagger$ are the usual creation/annihilation operators on $\mathcal{F}_s(L^2(\mathbb{R}^3))$, and

$$a_0 := \ell^{-\frac{3}{2}} a(\theta),$$

with $\theta(x) = \mathbb{1}_\Lambda(x)$. Notice for later use that, on the n-particle sector,

$$\begin{aligned}[b_k, b_k^\dagger] &= \ell^{-3}\big(a_0^\dagger a_0 \langle \chi_\Lambda e^{-ikx}, \chi_\Lambda e^{-ikx}\rangle - a^\dagger(Q\chi_\Lambda e^{-ikx}) a(Q\chi_\Lambda e^{-ikx})\big) \\ &\le n. \end{aligned} \tag{2.14}$$

A direct calculation gives

$$A_2 = \frac{1}{2}(2\pi)^{-3} \int \widehat{W_1}(p)\big(b_p^\dagger b_{-p}^\dagger + b_p b_{-p}\big)\, dp,$$

with W_1 from (2.4).

Furthermore, on the n-particle sector, we have, using that $n_0 \le n$, that

$$\sum_{i=1}^{n} (\mathcal{T}^{(i)} - b\ell^{-2} Q_i) \ge \frac{1}{2}(2\pi)^{-3} \int \frac{\ell^3}{n+1} \tau(p)\big(b_p^\dagger b_p + b_{-p}^\dagger b_{-p}\big)\, dp,$$

with $\tau(p) = (p^2 - s^{-2}\ell^{-2})_+$.

Finally, we add and subtract the term

$$\frac{1}{2}(2\pi)^{-3} \int 2\widehat{W_1}(0)\big(b_p^\dagger b_p + b_{-p}^\dagger b_{-p}\big)\, dp.$$

Notice that

$$\frac{1}{2}(2\pi)^{-3} \int 2\widehat{W_1}(0)\big(b_p^\dagger b_p + b_{-p}^\dagger b_{-p}\big)\, dp \le C a \frac{n_0 + 1}{\ell^3} n_+,$$

so the subtracted term is in agreement with (2.13).

We therefore estimate on the n-particle sector

$$\begin{aligned} &\sum_{i=1}^{n} (\mathcal{T}^{(i)} - b\ell^{-2} Q_i) + A_2 + \frac{1}{2}(2\pi)^{-3} \int 2\widehat{W_1}(0)\big(b_p^\dagger b_p + b_{-p}^\dagger b_{-p}\big)\, dp \\ &\quad \ge \frac{1}{2}(2\pi)^{-3} \int \mathcal{A}(p)\big(b_p^\dagger b_p + b_{-p}^\dagger b_{-p}\big) + \widehat{W_1}(p)\big(b_p^\dagger b_{-p}^\dagger + b_{-p} b_p\big)\, dp, \end{aligned}$$

where

$$\mathcal{A}(p) := \frac{\ell^3}{n+1} \tau(p) + 2\widehat{W_1}(0).$$

Upon calculating a square

$$(\alpha b_p^\dagger + \beta b_{-p})(\alpha b_p + \beta b_{-p}^\dagger) + (\alpha b_{-p}^\dagger + \beta b_p)(\alpha b_{-p} + \beta b_p^\dagger)$$

with appropriately adjusted coefficients α, β – and afterwards discarding the square as being positive – we get the estimate (see [8, Appendix A] for details)

$$\begin{aligned}
&\mathcal{A}(p)\big(b_p^\dagger b_p + b_{-p}^\dagger b_{-p}\big) + \widehat{W_1}(p)\big(b_p^\dagger b_{-p}^\dagger + b_{-p} b_p\big) \\
&\geq -\frac{1}{2}\Big(\mathcal{A}(p) - \sqrt{\mathcal{A}(p)^2 - \widehat{W_1}(p)^2}\Big)\big([b_p, b_p^\dagger] + [b_{-p}, b_{-p}^\dagger]\big) \\
&\geq -n\Big(\mathcal{A}(p) - \sqrt{\mathcal{A}(p)^2 - \widehat{W_1}(p)^2}\Big),
\end{aligned}$$

where the last inequality uses (2.14). Notice here how the added term $2\widehat{W_1}(0)$ ensures that $\mathcal{A}(p)^2 - \widehat{W_1}(p)^2 \geq 0$. Using again that

$$\frac{|\widehat{W_1}(p)|}{\mathcal{A}(p)} \leq \frac{1}{2}$$

to expand the square root, we find

$$\begin{aligned}
&\sum_{i=1}^n \big(\mathcal{T}^{(i)} - b\ell^{-2} Q_i\big) + A_2 + \frac{1}{2}(2\pi)^{-3} \int 2\widehat{W_1}(0)\big(b_p^\dagger b_p + b_{-p}^\dagger b_{-p}\big)\, dp \\
&\quad \geq -\frac{n}{2}(2\pi)^{-3} \int \frac{\widehat{W_1}(p)^2}{2\mathcal{A}(p)}\, dp - Cn \int \frac{\widehat{W_1}(0)^4}{2\mathcal{A}(p)^3}\, dp.
\end{aligned} \tag{2.15}$$

To estimate the last integral, we split in $\{|p|^2 \geq 2s^{-2}\ell^{-2}\} \cup \{|p|^2 \leq 2s^{-2}\ell^{-2}\}$. For the large values of $|p|$ we have

$$\mathcal{A}(p) \geq \frac{\ell^3}{2(n+1)} p^2.$$

Also, for all values of $|p|$, we have

$$\mathcal{A}(p) \geq 2\widehat{W}(0).$$

With this, it is easy to control the integral in p and get

$$n \int \frac{\widehat{W_1}(0)^4}{2\mathcal{A}(p)^3}\, dp \leq n \int \frac{\widehat{W_1}(0)^3}{2\mathcal{A}(p)^2}\, dp \leq Cn\bigg(\Big(\frac{n+1}{\ell^3}\Big)^2 a^3 \ell + a\ell^{-3}\bigg),$$

in agreement with the error bound in (2.13).

The first integral in (2.15) has to be calculated. We write it as

$$-\frac{n}{2}(2\pi)^{-3} \int \frac{\widehat{W_1}(p)^2}{2\mathcal{A}(p)}\, dp = \mathrm{I} + \mathrm{II},$$

with

$$\mathrm{I} := -\frac{n(n+1)}{2\ell^3}(2\pi)^{-3}\int \frac{\widehat{W_1}(p)^2}{2p^2}\,dp,$$

$$\mathrm{II} := -\frac{n}{2}(2\pi)^{-3}\int \frac{\widehat{W_1}(p)^2}{2\mathcal{A}(p)} - \frac{\widehat{W_1}(p)^2}{2\frac{\ell^3}{n+1}p^2}\,dp.$$

In II, we again estimate by splitting the integral as above. Notice how

$$\tau(p) = p^2 - s^{-2}\ell^{-2}$$

when $|p|^2 \geq 2s^{-2}\ell^{-2}$, so we easily get

$$\begin{aligned}|\mathrm{II}| &\leq Cna^2\Big(\int_{\{|p|^2\leq 2s^{-2}\ell^{-2}\}} a^{-1} + \frac{n+1}{\ell^3 p^2}\,dp \\ &\qquad + \int_{\{|p|^2\geq 2s^{-2}\ell^{-2}\}} \frac{\frac{\ell^3}{n+1}s^{-2}\ell^{-2} + 2\widehat{W_1}(0)}{(\frac{\ell^3}{n+1})^2 p^4}\,dp\Big) \\ &\leq Cn\Big(a\ell^{-3} + a^3\ell\Big(\frac{n+1}{\ell^3}\Big)^2\Big).\end{aligned}$$

This is also a bound that can be absorbed in the error term in (2.13).

Finally, we consider the integral I. Notice that

$$0 \leq W_1(x) \leq \Big(1 + C\Big(\frac{R}{\ell}\Big)^2\Big)g(x),$$

using (2.4) and a simple bound on the convolution (recall that $\operatorname{supp} v \subset B(0,R)$). Therefore,

$$\begin{aligned}(2\pi)^{-3}\int \frac{\widehat{W_1}(p)^2}{2p^2}\,dp &= \iint \frac{W_1(x)W_1(y)}{2|x-y|}\,dx\,dy \\ &\leq \Big(1 + C\Big(\frac{R}{\ell}\Big)^2\Big)(2\pi)^{-3}\int \frac{\widehat{g}(p)^2}{2p^2}\,dp \\ &= \Big(1 + C\Big(\frac{R}{\ell}\Big)^2\Big)\int g(x)\omega(x)\,dx,\end{aligned}$$

where we used (A.3) to get the last identity. Upon inserting this in the expression for I, this gives the main term in (2.13) as well as the error term depending on R^2.

This finishes the proof of Lemma 2.4. ■

Using Lemma 2.3 (in the form of (2.12)) and Lemma 2.4, we can now finish the proof of Theorem 2.1. Combining the terms, remembering that the "gap" of the kinetic energy was saved, we find on the n-particle subspace

$$\mathcal{H}_\Lambda(\rho_\mu))_n \geq E_{\mathrm{Main}} + E_{\mathrm{gap}} + E_{\mathrm{error}},$$

with

$$E_{\text{Main}} := \frac{n^2}{2\ell^3} 8\pi a - \left(\rho_\mu \frac{n}{\ell^3} + \frac{1}{4}\left(\rho_\mu - \frac{n}{\ell^3}\right)^2\right) 8\pi a \ell^3$$
$$= -4\pi a \rho_\mu^2 \ell^3 + 2\pi \frac{a}{\ell^3}(\rho_\mu \ell^3 - n)^2,$$

as well as

$$E_{\text{gap}} := \left(b\ell^{-2} - Ca\left(\frac{n+1}{\ell^3} + \rho_\mu\right)\right) n_+, \tag{2.16}$$

and

$$E_{\text{error}} := -Cna\ell^{-3}\left(1 + \frac{a^2(n+1)^2}{\ell^2} + \frac{(n+1)R^2}{\ell^2}\right) - Ca\rho_\mu.$$

When $n \leq (\Xi + 1)\rho_\mu \ell^3$, we find

$$E_{\text{error}} \geq -C\rho_\mu a\left(1 + \left(\frac{R}{a}\right)^2 \sqrt{\rho_\mu a^3}\right),$$

with C depending on Ξ and on K (from the definition (2.1) of ℓ).

Also for $n \leq (\Xi + 1)\rho_\mu \ell^3$, we may choose K large enough (depending on Ξ) to get

$$E_{\text{gap}} \geq \frac{b}{2\ell^2} n_+ \geq 0.$$

In conclusion, we get, for $n \leq (\Xi + 1)\rho_\mu \ell^3$,

$$\mathcal{H}_\Lambda(\rho_\mu))_n \geq -4\pi a \rho_\mu^2 \ell^3 + 2\pi \frac{a}{\ell^3}(\rho_\mu \ell^3 - n)^2 - C\rho_\mu a\left(1 + \left(\frac{R}{a}\right)^2 \sqrt{\rho_\mu a^3}\right). \tag{2.17}$$

In particular, for $n \in [\Xi \rho_\mu \ell^3, (\Xi + 1)\rho_\mu \ell^3]$, and using that $\Xi \geq 3$,

$$\mathcal{H}_\Lambda(\rho_\mu))_n \geq \rho_\mu a\left(4\pi \rho_\mu \ell^3 - C\left(1 + \left(\frac{R}{a}\right)^2 \sqrt{\rho_\mu a^3}\right)\right) \geq 0,$$

for $\rho_\mu a^3$ sufficiently small.

Upon inserting this as well as (2.17) (for the last group of particles) in (2.9), we get the bound (2.7).

This finishes the proof of Theorem 2.1. ∎

3 Localization to small boxes

To prove (1.3) we will localize to smaller boxes. In order not to have to worry about how the particles distribute themselves between the boxes it is convenient to reformulate our problem on Fock space. Consider, for given $\rho_\mu > 0$, the following operator $\mathcal{H}_{\rho_\mu}$

on the symmetric Fock space $\mathscr{F}_s(L^2(\Omega))$. The operator $\mathcal{H}_{\rho_\mu}$ commutes with particle number and satisfies, with $\mathcal{H}_{\rho_\mu,N}$ denoting the restriction of $\mathcal{H}_{\rho_\mu}$ to the N-particle subspace of $\mathscr{F}_s(L^2(\Omega))$,

$$\begin{aligned}\mathcal{H}_{\rho_\mu,N} &= \sum_{i=1}^N(-\Delta_i - 2\pi^2 L^{-2} Q_i) + \sum_{i<j} v^{\mathrm{per}}(x_i - x_j) - 8\pi a\rho_\mu N \\ &= \sum_{i=1}^N\Big(-\Delta_i - 2\pi^2 L^{-2} Q_i - \rho_\mu \int_{\mathbb{R}^3} g(x_i - y)\,dy\Big) + \sum_{i<j} v^{\mathrm{per}}(x_i - x_j),\end{aligned}$$

where g is defined in terms of the scattering solution (see (A.2)).

We will prove the following lower bound on $\mathcal{H}_{\rho_\mu}$.

Theorem 3.1. *Suppose that v satisfies Assumption 1.1. Then there exists $C_1 > 0$ such that*

$$\mathcal{H}_{\rho_\mu} \geq -4\pi\rho_\mu^2 aL^3\big(1 + C_1(\rho_\mu a^3)^{\frac{1}{2}}\big),$$

for all $\rho_\mu a^3$ sufficiently small.

Before proving Theorem 3.1 we show that (1.3) – and therefore our main result Theorem 1.2 – follows from it.

Proof of (1.3) *using Theorem* 3.1. To prove (1.3), let $\Psi \in \otimes_s^N L^2(\Omega) \subset \mathscr{F}_s(L^2(\Omega))$ be normalized. Then, using Theorem 3.1,

$$\begin{aligned}\langle\Psi, \widetilde{H}\Psi\rangle &\geq \langle\Psi, \mathcal{H}_{\rho_\mu}\Psi\rangle + 8\pi a\rho_\mu\rho L^3 \\ &\geq 4\pi a\rho^2 L^3 - 4\pi a(\rho - \rho_\mu)^2 L^3 - 4\pi\rho_\mu^2 aL^3 C_1(\rho_\mu a^3)^{\frac{1}{2}}.\end{aligned}$$

Estimate (1.3) follows upon choosing $\rho_\mu = \rho$. ■

Proof of Theorem 3.1. We will localize to boxes of size ℓ, with ℓ being as defined in equation (2.1). In particular, it is important that $R < \ell < \frac{1}{2}L$. It turns out that the Hamiltonians on the small boxes are all unitarily equivalent, so we really only have one box $\Lambda = [-\frac{\ell}{2}, \frac{\ell}{2}]^3$ and the Hamiltonian on this box can be identified with $\mathcal{H}_\Lambda(\rho_\mu)$ defined in (2.3). Therefore, Theorem 3.1 will be a consequence of Theorem 2.1. Below we give the details of this argument.

For given point $u \in \mathbb{R}^3$, define the sharp localization function θ_u to the box $\Lambda(u) := u + [-\frac{\ell}{2}, \frac{\ell}{2}]^3$, i.e.

$$\theta_u := \mathbb{1}_{\Lambda(u)}.$$

Recall that Ω is identified with the L-torus. Given θ_u, we define P_u, Q_u to be the maps $P_u, Q_u : L^2(\Omega) \to L^2(\Omega)$, defined by

$$P_u\varphi := \ell^{-3}\langle\theta_u, \varphi\rangle\theta_u, \quad Q_u\varphi := \theta_u\varphi - \ell^{-3}\langle\theta_u, \varphi\rangle\theta_u.$$

We also define, for $u \in \mathbb{R}^3$,

$$\chi_u(x) = \chi\Big(\frac{x-u}{\ell}\Big) = \chi_\Lambda(x-u).$$

Recall the definition of W from (2.2). We will also need the periodic versions

$$W^{\text{per}}(x) = \sum_{j \in \mathbb{Z}^3} W(x + Lj),$$

and similarly for χ_u^{per}.

Define the localized potentials

$$w_u(x, y) := \chi_u(x) W(x - y) \chi_u(y), \quad \text{so that } w(x, y) = w_{u=0}(x, y),$$

and

$$w_u^{\text{per}}(x, y) := \chi_u^{\text{per}}(x) W^{\text{per}}(x - y) \chi_u^{\text{per}}(y). \tag{3.1}$$

We also recall the functions W_1 and w_1 defined in (2.4). If we add a subscript u, we mean as above the translated versions $w_{1,u}(x, y) = w_1(x - \ell u, y - \ell u)$, and similarly we add a superscript "per" for the periodic versions as in (3.1).

We then get by a direct calculation:

Lemma 3.2. *For $2\ell < L$, we get*

$$-\rho_\mu \sum_{i=1}^{N} \int g(x_i - y)\, dy + \sum_{i<j} v^{\text{per}}(x_i - x_j) = \ell^{-3} \int_\Omega \mathcal{W}_{u,N}\, du,$$

with

$$\mathcal{W}_{u,N} := -\rho_\mu \sum_{i=1}^{N} \int_\Omega w_{1,u}^{\text{per}}(x_i, y)\, dy + \sum_{i<j} w_u^{\text{per}}(x_i, x_j).$$

Proof. The proof follows by direct calculation of the u-integral. ■

We estimate the kinetic energy $-\Delta^{\text{per}}$ in Ω below by an integral over kinetic energy operators in the boxes $\Lambda(u)$. The following theorem is essentially [4, Lemma 5.7].

Lemma 3.3 (Kinetic energy localization). *Assume that $2\ell < L$. There exists a constant $b > 0$ such that if s is small enough, then*

$$\ell^{-3} \int_\Omega \mathcal{T}_u\, du \leq -\Delta^{\text{per}} - 2\pi L^{-2} Q_L,$$

where Δ^{per} denotes the Laplacian on Ω with periodic boundary conditions, and where

$$\mathcal{T}_u := Q_u^* [\chi_u (-\Delta - s^{-2}\ell^{-2})_+ \chi_u + b\ell^{-2}] Q_u,$$

where Δ is the Laplacian on $\mathbb{R}^3$.

We leave the proof of Lemma 3.3 to the end of the section, and finish the proof of Theorem 3.1 first.

By combining Lemmas 3.3 and 3.2, we see that

$$\mathcal{H}_{\rho_\mu, N} \geq \ell^{-3} \int_\Omega (\mathcal{H}_{\Lambda(u)}(\rho_\mu))_N\, du, \tag{3.2}$$

with

$$(\mathcal{H}_{\Lambda(u)}(\rho_\mu))_N = \sum_{i=1}^{N} \mathcal{T}_u^{(i)} + \mathcal{W}_{u,N}.$$

Consider now the decomposition $L^2(\Omega) = L^2(\Lambda(u)) \oplus L^2(\Omega \setminus \Lambda(u))$. By using that $\mathcal{F}_s(L^2(\Omega))$ is unitarily equivalent to $\mathcal{F}_s(L^2(\Lambda(u))) \otimes \mathcal{F}_s(L^2(\Omega \setminus \Lambda(u)))$, it easy to see that

$$\inf \operatorname{Spec} \mathcal{H}_{\Lambda(u)}(\rho_\mu) = \inf \operatorname{Spec} \widetilde{\mathcal{H}}_{\Lambda(u)}(\rho_\mu),$$

where $\widetilde{\mathcal{H}}_{\Lambda(u)}(\rho_\mu)$ denotes the operator $\mathcal{H}_{\Lambda(u)}(\rho_\mu)$ considered on $\mathcal{F}_s(L^2(\Lambda(u)))$. Upon identifying $L^2(\Lambda(u)) \subset L^2(\Omega)$ with $L^2(\Lambda(u)) \subset L^2(\mathbb{R}^3)$, it is now clear that all the $\widetilde{\mathcal{H}}_{\Lambda(u)}(\rho_\mu)$ are pairwise unitarily equivalent – in particular, they are all unitarily equivalent to $\mathcal{H}_\Lambda(\rho_\mu)$ defined in (2.3). Therefore, by Theorem 2.1 we have

$$(\mathcal{H}_{\Lambda(u)}(\rho_\mu))_N \geq -4\pi\rho_\mu^2 a\ell^3 - C_0\rho_\mu^2 a\ell^3(\rho_\mu a^3)^{\frac{1}{2}},$$

for all N and all $u \in \Omega$. Upon inserting this bound in (3.2), we get the desired bound on $\mathcal{H}_{\rho_\mu}$. This finishes the proof of Theorem 3.1. ■

Proof of Lemma 3.3. Define, for $p \in \mathbb{R}^3$,

$$\tau(p) := (p^2 - s^{-2}\ell^{-2})_+.$$

An explicit calculation shows that for all $k, k' \in 2\pi L^{-1}\mathbb{Z}^3$,

$$L^{-3} \int_\Omega \langle e^{ikx}, \mathcal{T}_u e^{ik'x} \rangle \, du = \delta_{k,k'} \ell^3 F(k),$$

with

$$F(k) = \ell^{-3} \langle e^{ikx}, \mathcal{T}_{u=0} e^{ikx} \rangle.$$

So we have to show that, if b, s are small enough, then for all $k \in 2\pi L^{-1}\mathbb{Z}^3 \setminus \{0\}$,

$$F(k) \leq k^2 - 2\pi^2 L^{-2}. \tag{3.3}$$

It is easy to reduce by scaling to the case $\ell = 1$. We write, now with the convention that $\ell = 1$,

$$F(k) = F_1(k) + F_2(k),$$

where (using that χ is even to avoid complex conjugates)

$$F_1 = \mathrm{I} + \mathrm{II} + \mathrm{III}, \quad F_2 = b\big(1 - \widehat{\theta}(k)^2\big),$$

and

$$\begin{aligned}
\mathrm{I} &:= (2\pi)^{-3}(\tau * (\widehat{\chi_\Lambda}^2))(k), \\
\mathrm{II} &:= -2(2\pi)^{-3}[(\tau\widehat{\chi_\Lambda}) * \widehat{\chi_\Lambda}](k)\widehat{\theta}(k), \\
\mathrm{III} &:= (2\pi)^{-3}\widehat{\theta}(k)^2(\tau * (\widehat{\chi_\Lambda}^2))(0),
\end{aligned}$$

An explicit calculation of the Fourier transform of θ easily yields that

$$0 \le F_2(k) \le b \min\{\beta k^2, 1\}, \tag{3.4}$$

for suitable $\beta > 0$.

The estimate on F_1 is divided into two according to the size of $|k|$. For $|k| < \frac{1}{2}s^{-1}$ we prove that

$$0 \le F_1(k) \le Csk^2. \tag{3.5}$$

It is clear that $F_1(0) = 0$ and $\nabla_k F_1(0) = 0$ follows by parity. Therefore, by Taylor's Theorem it suffices to prove that $|\partial_{k_i}\partial_{k_j} F_1| \le Cs$. This is done by calculation on each of the three terms, using the decay and regularity of χ_Λ. For example, for the term III we have $|\partial_{k_i}\partial_{k_j}\widehat{\theta}(k)^2| \le C$, since θ is supported in the unit cube, and

$$(\tau * (\widehat{\chi_\Lambda}^2)(0) \le \int_{\{|p| \ge s^{-1}\}} p^2(\widehat{\chi_\Lambda}(p))^2\, dp \le Cs,$$

by the fast decay of $\widehat{\chi_\Lambda}$, which follows from its regularity. The restriction $|k| \le \frac{1}{2}s^{-1}$ is used in the estimate of I to ensure that $(p-k)^2 - s^{-2} \ge 0$ implies that

$$|p| \ge \frac{1}{2}s^{-1}$$

and therefore giving decay of $\widehat{\chi_\Lambda}$. We leave the remaining details to the reader.

Using (3.5) and (3.4), we get (3.3) for $|k| \le \frac{1}{2}s^{-1}$ as soon as $b\beta + Cs \le \frac{1}{2}$ (with C from (3.5) and β from (3.4)) and recalling that $k^2 \ge 4\pi^2 L^{-2}$ for all allowed k.

For $|k| \ge \frac{1}{2}s^{-1}$, we estimate the three terms in F_1 individually. Since χ is smooth, it is easy to see that

$$|\mathrm{II}| + |\mathrm{III}| \le C. \tag{3.6}$$

We estimate I with $\varepsilon \in (0,1)$, and using $[x+y]_+ \le [x]_+ + [y]_+$ and that χ is even, as

$$\begin{aligned}
\mathrm{I} &= (2\pi)^{-3}(\tau * (\widehat{\chi_\Lambda}^2)(k) \\
&= (2\pi)^{-3}\int ((p-k)^2 - s^{-2})\widehat{\chi_\Lambda}^2(p) + [s^{-2} - p^2 + 2kp - k^2]_+\widehat{\chi_\Lambda}^2(p)\, dp \\
&\le k^2 - s^{-2} + [s^{-2} - (1-\varepsilon)k^2]_+ + \varepsilon^{-1}(2\pi)^{-3}\int p^2\widehat{\chi_\Lambda}^2(p)\, dp.
\end{aligned}$$

Therefore, for $|k| \ge \frac{1}{2}s^{-1}$ we get

$$\mathrm{I} \le k^2 - (1-\varepsilon)\frac{1}{4}s^{-2} + \varepsilon^{-1}\int |\nabla\chi|^2\, dx.$$

Upon fixing $\varepsilon = \frac{1}{2}$ this becomes

$$\mathrm{I} \le k^2 - \frac{1}{8}s^{-2} + C, \tag{3.7}$$

for all $|k| \ge \frac{1}{2}s^{-1}$ and with C independent of s.

Upon adding (3.7), (3.4) and (3.6), see that (3.3) holds for $|k| \geq \frac{1}{2}s^{-1}$ for all fixed $b > 0$, as soon as s is small enough. ■

A Facts about the scattering solution

We briefly recall the definition of the scattering length and related quantities. For further details see [11, Appendix C].

We assume that v satisfies Assumption 1.1. The equation

$$(-\Delta + \frac{1}{2}v(x))(1 - \omega(x)) = 0, \quad \text{with } \omega \to 0 \text{ as } |x| \to \infty, \tag{A.1}$$

is called the scattering equation. The radial solution ω to this equation satisfies that $\omega(x) = \frac{a}{|x|}$ for some $a > 0$ and for x outside $\operatorname{supp} v$. This constant a is called the *scattering length* of the potential v. The function ω is radially symmetric and non-increasing with

$$0 \leq \omega(x) \leq 1.$$

We can rewrite (A.1) as

$$-\Delta\omega = \frac{1}{2}g, \quad \text{with } g := v(1 - \omega), \tag{A.2}$$

and get

$$a = (8\pi)^{-1} \int g(x)\, dx,$$

and that the Fourier transform satisfies

$$\widehat{\omega}(k) = \frac{\hat{g}(k)}{2k^2}. \tag{A.3}$$

Acknowledgements. This research was partly supported by the Charles Simonyi Endowment and by an EliteResearch Prize from the Danish Ministry of Higher Education and Science.

Bibliography

[1] A. Adhikari, C. Brennecke and B. Schlein, Bose–Einstein condensation beyond the Gross–Pitaevskii regime. *Ann. Henri Poincaré* **22** (2021), 1163–1233

[2] C. Boccato, C. Brennecke, S. Cenatiempo and B. Schlein, Complete Bose–Einstein condensation in the Gross–Pitaevskii regime. *Comm. Math. Phys.* **359** (2018), 975–1026

[3] C. Boccato, C. Brennecke, S. Cenatiempo and B. Schlein, Bogoliubov theory in the Gross–Pitaevskii limit. *Acta Math.* **222** (2019), 219–335

[4] B. Brietzke, S. Fournais and J. P. Solovej, A simple 2nd order lower bound to the energy of dilute Bose gases. *Comm. Math. Phys.* **376** (2020), 323–351

[5] B. Brietzke and J. P. Solovej, The second-order correction to the ground state energy of the dilute Bose gas. *Ann. Henri Poincaré* **21** (2020), 571–626

[6] F. J. Dyson, Ground-state energy of a hard-sphere gas. *Phys. Rev.* **106** (1957), 20–26

[7] L. Erdős, B. Schlein and H.-T. Yau, Ground-state energy of a low-density bose gas: A second-order upper bound. *Phys. Rev. A* **78** (2008), Article ID 053627

[8] S. Fournais and J. P. Solovej, The energy of dilute Bose gases. *Ann. of Math. (2)* **192** (2020), 893–976

[9] T. D. Lee, K. Huang and C. N. Yang, Eigenvalues and eigenfunctions of a Bose system of hard spheres and its low-temperature properties. *Phys. Rev. (2)* **106** (1957), 1135–1145

[10] E. H. Lieb and R. Seiringer, *Proof of Bose–Einstein condensation for dilute trapped gases. Phys. Rev. Lett.* **88** (2002), Article ID 170409-1–4

[11] E. H. Lieb, R. Seiringer, J. P. Solovej and J. Yngvason, *The mathematics of the Bose gas and its condensation*. Oberwolfach Semin. 34, Birkhäuser, Basel, 2005

[12] E. H. Lieb and J. P. Solovej, Ground state energy of the one-component charged Bose gas. *Comm. Math. Phys.* **217** (2001), 127–163

[13] E. H. Lieb and J. P. Solovej, Ground state energy of the two-component charged Bose gas. *Comm. Math. Phys.* **252** (2004), 485–534

[14] E. H. Lieb and J. Yngvason, Ground state energy of the low density Bose gas. In *Mathematical results in statistical mechanics (Marseilles, 1998)*, pp. 349–357, World Scientific Publishing, River Edge, 1999

[15] N. Rougerie, Scaling limits of bosonic ground states, from many-body to nonlinear Schrödinger. Preprint 2020, arXiv:2002.02678

[16] H.-T. Yau and J. Yin, The second order upper bound for the ground energy of a Bose gas. *J. Stat. Phys.* **136** (2009), 453–503

The periodic Lieb–Thirring inequality

Rupert L. Frank, David Gontier and Mathieu Lewin

To Ari Laptev on the occasion of his 70th birthday

We discuss the Lieb–Thirring inequality for periodic systems, which has the same optimal constant as the original inequality for finite systems. This allows us to formulate a new conjecture about the value of its best constant. To demonstrate the importance of periodic states, we prove that the 1D Lieb–Thirring inequality at the special exponent $\gamma = \frac{3}{2}$ admits a one-parameter family of periodic optimizers, interpolating between the one-bound state and the uniform potential. Finally, we provide numerical simulations in 2D which support our conjecture that optimizers could be periodic.

The Lieb–Thirring inequality [18, 19] plays an important role in mathematical physics. It has been used to give a short proof of the stability of matter [17, 18] and it is a fundamental tool for studying large fermionic systems. A famous problem is to determine its *optimal constant*. In this paper we discuss the extension of the Lieb–Thirring inequality to periodic systems, which has the same optimal constant as the original inequality. In light of our recent work [9], we then formulate a conjecture about the value of this best constant, different from the original Lieb–Thirring conjecture in [19]. In one space dimension we prove the optimality of some periodic potentials in the particular (completely integrable) case where the exponent equals $\gamma = \frac{3}{2}$. Finally, we provide numerical simulations in two space dimensions which show that periodic potentials give a better constant than all previously considered functions.

1 The periodic Lieb–Thirring inequality

Let us first recall the usual Lieb–Thirring inequality for finite systems. Let $d \geq 1$ and

$$\begin{cases} \gamma \geq \frac{1}{2} & \text{for } d = 1, \\ \gamma > 0 & \text{for } d = 2, \\ \gamma \geq 0 & \text{for } d \geq 3. \end{cases} \tag{1.1}$$

Keywords: Lieb–Thirring inequality, periodic Schrödinger operators
2020 Mathematics Subject Classification: 81Q10

The Lieb–Thirring inequality states that there is a universal constant $L_{\gamma,d} < \infty$ so that, for all $V \in L^{\gamma+d/2}(\mathbb{R}^d, \mathbb{R})$,

$$\sum_{n=1}^{\infty} |\lambda_n(-\Delta + V)|^\gamma \leq L_{\gamma,d} \int_{\mathbb{R}^d} V(x)_-^{\gamma+d/2} \, \mathrm{d}x. \tag{1.2}$$

Here $a_- = \max(0, -a)$ is the negative part and $\lambda_n(-\Delta + V)$ is the nth min-max level of $-\Delta + V$ in $L^2(\mathbb{R}^d)$, which equals the nth negative eigenvalue (counted with multiplicity) when it exists and 0 otherwise. Under our assumptions on V and γ, the operator $-\Delta + V$ has a bounded-below quadratic form and we always work with its Friedrichs extension. In the following $L_{\gamma,d}$ will always denote the *best (smallest) constant*. Inequality (1.2) was shown by Lieb and Thirring in [18, 19] when all the inequalities are strict in (1.1). The critical cases $\gamma = 0$ in $d \geq 3$ and $\gamma = \frac{1}{2}$ in $d = 1$ are respectively due to Cwikel, Lieb and Rozenblum [4, 16, 24] and Weidl [28]. We refer to the recent review [8] for an up-to-date account of what is known and what is open for the inequality (1.2).

Next, we turn to (infinite) periodic systems. Let $V \in L^{\gamma+d/2}_{\mathrm{loc}}(\mathbb{R}^d)$ be a periodic function, with γ as in (1.1). This means that there exists a discrete subgroup

$$\mathcal{L} = \left\{ \sum_{j=1}^{d} z_j v_j : z_j \in \mathbb{Z} \right\}$$

with $v_1, \ldots, v_d$ a basis of $\mathbb{R}^d$ such that $V(x + \ell) = V(x)$ for all $\ell \in \mathcal{L}$ and almost all $x \in \mathbb{R}^d$. We emphasize here that the lattice $\mathcal{L}$ is not fixed *a priori* but it is part of the definition of V. There are in fact several possible such lattices, for instance $k\mathcal{L}$ with $k \in \mathbb{N}$. Let $C = \{x \in \mathbb{R}^d : |x| = \min_{\ell \in \mathcal{L}} |x - \ell|\}$ be the Wigner–Seitz unit cell of $\mathcal{L}$. It is well known that the integral per unit volume converges to

$$\begin{aligned}\lim_{\Omega_n \nearrow \mathbb{R}^d} \frac{1}{|\Omega_n|} \int_{\Omega_n} V(x)_-^{\gamma+d/2} \, \mathrm{d}x &= \frac{1}{|C|} \int_C V(x)_-^{\gamma+d/2} \, \mathrm{d}x \\ &=: \fint V(x)_-^{\gamma+d/2} \, \mathrm{d}x \end{aligned} \tag{1.3}$$

under some natural conditions on the sequence of domains Ω_n. For instance we can just choose $\Omega_n = n\Omega$, where Ω is a fixed convex open nonempty set. Of course, $\fint V(x)_-^{\gamma+d/2} \, \mathrm{d}x$ in (1.3) does not depend on the chosen lattice $\mathcal{L}$ of periodicity for V.

The spectrum of $-\Delta + V$ is properly described using the Bloch–Floquet transform [23, Section XIII.16]. Let $\mathcal{L}^*$ be the dual lattice (such that $k \cdot \ell \in 2\pi\mathbb{Z}$ for all $k \in \mathcal{L}^*$ and $\ell \in \mathcal{L}$) and B its Wigner–Seitz cell (the Brillouin zone). For any quasimomentum $\xi \in B$, we denote by $H_\xi := |-i\nabla + \xi|^2 + V(x)$ the Schrödinger operator on $L^2(C)$ with periodic boundary conditions on ∂C. Its eigenvalues form a non-decreasing sequence $\varepsilon_n^V(\xi)$, each of which is Lipschitz in $\xi \in B$. The spectrum of the periodic Schrödinger operator is the union of the Bloch bands

$$\sigma(-\Delta + V) = \bigcup_{n \geq 1} \varepsilon_n^V(B).$$

Let us denote by $(-\Delta + V)_{|\Omega}$ the operator $-\Delta + V$ restricted to a domain Ω with Dirichlet boundary conditions at $\partial\Omega$. We then have the thermodynamic limits [6,22,23]

$$\begin{aligned}\lim_{n\to\infty} \frac{\sum_{j=1}^{\infty} \lambda_j\big((-\Delta+V)_{|\Omega_n}\big)_-^\gamma}{|\Omega_n|} &= \lim_{n\to\infty} \frac{\operatorname{Tr}\big(\mathbb{1}_{\Omega_n}(-\Delta+V)_-^\gamma \mathbb{1}_{\Omega_n}\big)}{|\Omega_n|} \\ &= \frac{1}{(2\pi)^d}\sum_{j\geq 1}\int_B \varepsilon_j^V(\xi)_-^\gamma \,\mathrm{d}\xi \\ &=: \underline{\operatorname{Tr}}(-\Delta+V)_-^\gamma. \end{aligned} \tag{1.4}$$

The trace per unit volume $\underline{\operatorname{Tr}}(-\Delta + V)_-^\gamma$ defined in (1.4) does not depend on the chosen $\mathscr{L}$, since it equals the first two limits. The following is a simple consequence of the usual Lieb–Thirring inequality but we have not found it stated explicitly anywhere.

Theorem 1.1 (Periodic Lieb–Thirring inequality). *Let $V \in L^{\gamma+d/2}_{\mathrm{loc}}(\mathbb{R}^d, \mathbb{R})$ be any periodic function with γ as in* (1.1). *Then we have*

$$\underline{\operatorname{Tr}}(-\Delta+V)_-^\gamma \leq L_{\gamma,d} \fint V(x)_-^{\gamma+d/2}\,\mathrm{d}x \tag{1.5}$$

with the same optimal constant $L_{\gamma,d}$ as in the original Lieb–Thirring inequality (1.2).

Proof. We quickly outline the argument. Let V be a periodic function as in the statement. From the variational principle and the Lieb–Thirring inequality (1.2), we have

$$\begin{aligned}\sum_{j\geq 1}\lambda_j\big((-\Delta+V)_{|\Omega}\big)_-^\gamma &\leq \sum_{j\geq 1}\lambda_j\left(-\Delta + V\mathbb{1}_{|\Omega}\right)_-^\gamma \\ &\leq L_{\gamma,d}\int_\Omega V(x)_-^{\gamma+d/2}\,\mathrm{d}x.\end{aligned}$$

Passing then to the limit $\Omega \nearrow \mathbb{R}^d$ using (1.3) and (1.4), we obtain (1.5). Let us now call $L^{\mathrm{per}}_{\gamma,d} \leq L_{\gamma,d}$ the best constant in (1.5) and show that it coincides with $L_{\gamma,d}$. Let $v \in C_c^\infty(\mathbb{R}^d, \mathbb{R})$ and define the $\ell\mathbb{Z}^d$-periodic function $V_\ell(x) = \sum_{z\in\mathbb{Z}^d} v(x+\ell z)$. The negative spectrum of the periodic operator $-\Delta + V_\ell$ is composed of very narrow bands about each eigenvalue $\lambda_n(-\Delta + v)$ (see [5]). Passing to the limit $\ell \to \infty$ provides the inequality

$$\sum_{j\geq 1}\lambda_j(-\Delta+v)_-^\gamma \leq L^{\mathrm{per}}_{\gamma,d}\int_{\mathbb{R}^d} v(x)_-^{\gamma+d/2}\,\mathrm{d}x.$$

Optimizing over v and using a simple density argument, we find that $L^{\mathrm{per}}_{\gamma,d} = L_{\gamma,d}$. ∎

We have considered the periodic case because such systems usually play an important role in statistical mechanics [3]. But a result similar to Theorem 1.1 holds for any potential V which has some kind of ergodicity allowing for the limits (1.3) and (1.4) to exist and be independent of Ω_n.

Next, we discuss our conjecture on $L_{\gamma,d}$. Let us define the best constant

$$L^{(N)}_{\gamma,d} := \sup_{V\in L^{\gamma+d/2}(\mathbb{R}^d)} \frac{\sum_{n=1}^{N} |\lambda_n(-\Delta+V)|^\gamma}{\int_{\mathbb{R}^d} V_-^{\gamma+d/2}}$$

for the inequality similar to (1.2) when only the N first eigenvalues are retained. The one-bound-state constant $L^{(1)}_{\gamma,d}$ can be expressed in terms of the constant for the Gagliardo–Nirenberg embedding $H^1(\mathbb{R}^d) \hookrightarrow L^p(\mathbb{R}^d)$ with $p = \frac{2\gamma+d}{\gamma+\frac{d}{2}-1}$ (see [8]). Another important number is the semiclassical constant

$$L^{\mathrm{sc}}_{\gamma,d} := (2\pi)^{-d} \int_{\mathbb{R}^d} (|p|^2-1)_-^\gamma \, \mathrm{d}p = \frac{\Gamma(\gamma+1)}{2^d \pi^{\frac{d}{2}} \Gamma(\gamma+\frac{d}{2}+1)}.$$

We have $L_{\gamma,d} \geq L^{\mathrm{sc}}_{\gamma,d}$ due to the semiclassical limit

$$\lim_{\hbar\to 0} \frac{\mathrm{Tr}\,(-\Delta+V(\hbar\cdot))_-^\gamma}{\int_{\mathbb{R}^d} V(\hbar x)_-^{\gamma+d/2}\,\mathrm{d}x} = \lim_{\hbar\to 0} \frac{\hbar^d\,\mathrm{Tr}\,(-\hbar^2\Delta+V)_-^\gamma}{\int_{\mathbb{R}^d} V(x)_-^{\gamma+d/2}\,\mathrm{d}x} = L^{\mathrm{sc}}_{\gamma,d}$$

for any $V \in C^0_c(\mathbb{R}^d,\mathbb{R})$ with $V_- \not\equiv 0$. A different way to understand $L^{\mathrm{sc}}_{\gamma,d}$ is to simply take a constant potential $V = -\mu < 0$ in the periodic inequality (1.5) and remark that

$$\fint V(x)_-^{\gamma+d/2}\,\mathrm{d}x = \mu_+^{\gamma+d/2},$$
$$\underline{\mathrm{Tr}}(-\Delta-\mu)_-^\gamma = (2\pi)^{-d}\int_{\mathbb{R}^d} (|p|^2-\mu)_-^\gamma\,\mathrm{d}p = \mu_+^{\gamma+d/2} L^{\mathrm{sc}}_{\gamma,d}.$$

The N-bound state constant $L^{(N)}_{\gamma,d}$ also has an interpretation in the setting of periodic systems. As we saw in the proof of Theorem 1.1, a potential $v \in L^{\gamma+d/2}(\mathbb{R}^d)$ with exactly N negative eigenvalues can be turned into a periodic potential

$$V_\ell(x) = \sum_{z\in\mathbb{Z}^d} v(x+\ell z)$$

of large periodicity ℓ with N negative Bloch bands converging to the eigenvalues of $-\Delta+v$ in the limit $\ell\to\infty$. We see that although the periodic model is perfect to describe the infinite uniform state, finite states can only be recovered asymptotically for an infinite period. For the usual Lieb–Thirring inequality, the reverse holds: finite bound-state potentials are well described but the semiclassical constant can only be obtained by a limiting procedure. Depending on what we think the optimal constant will be, it seems appropriate to study either (1.2) or (1.5).

The original Lieb–Thirring conjecture [19] stated that the optimal constant is given either by the one-bound state or by the semiclassical case: $L_{\gamma,d} \stackrel{?}{=} \max(L^{(1)}_{\gamma,d}, L^{\mathrm{sc}}_{\gamma,d})$. Rephrased for our periodic inequality (1.5), this would mean that either $V \equiv \mathrm{cnst}$ is optimal, or there is no optimizer since those would have an infinite period. In dimensions $d \leq 7$ the two curves $\gamma \mapsto (L^{(1)}_{\gamma,d}, L^{\mathrm{sc}}_{\gamma,d})$ cross at a unique critical exponent

$\gamma_{1\cap\mathrm{sc}}(d)$ equal to $\frac{3}{2}$ in 1D, and approximately equal to 1.165378 and 0.862689 in 2D and 3D respectively [9]. In dimensions $d \geq 8$ we always have $L^{(1)}_{\gamma,d} < L^{\mathrm{sc}}_{\gamma,d}$. It is known that $L_{\gamma,d} = L^{\mathrm{sc}}_{\gamma,d}$ for all $\gamma \geq \frac{3}{2}$ in all dimensions [2, 14, 19], and that $L_{1/2,1} = L^{(1)}_{1/2,1}$ (see [11]). But the exact value of $L_{\gamma,d}$ has not been found in all the other cases.

It is now understood that the situation ought to be more complicated than what was hoped in [19], except probably for $d = 1$. In [9] we proved that

$$L^{(1)}_{\gamma,d} < L^{(2)}_{\gamma,d} \leq L_{\gamma,d} \quad \text{for } \gamma > \max\left\{0, 2 - \frac{d}{2}\right\}$$

and

$$L^{(N)}_{\gamma,d} < L_{\gamma,d} \quad \text{for all } N \geq 1 \text{ when } \gamma \begin{cases} > \frac{3}{2} & \text{for } d = 1, \\ > 1 & \text{for } d = 2, \\ \geq 1 & \text{for } d \geq 3. \end{cases} \tag{1.6}$$

This showed that the one-bound state constant $L^{(1)}_{\gamma,d}$ cannot be optimal in regions where it was known that $L^{(1)}_{\gamma,d} > L^{\mathrm{sc}}_{\gamma,d}$, like $\frac{1}{2} < \gamma < 0.862689$ in 3D and $1 < \gamma < 1.165378$ in 2D. In fact, no finite-bound-state case can be optimal when γ satisfies (1.6). The result (1.6) will be generalized to all $\gamma > \max\{0, 2 - \frac{d}{2}\}$ in a forthcoming work.

In [9] we mentioned the possibility of a different scenario for the Lieb–Thirring optimal constant, which we would like to detail here. For $\gamma > 1$ it is possible to interpret the Lieb–Thirring problem as optimizing the state of a quantum system described by its one-particle density matrix Γ, in the presence of a local nonlinear attraction of the form $-\int_{\mathbb{R}^d} \Gamma(x,x)^p \,\mathrm{d}x$ and with a Tsallis-type entropy $\mathrm{Tr}(\Gamma^q)$ where $p = (\gamma + \frac{d}{2})'$ and $q = \gamma'$ are the corresponding dual exponents. This is explained later in Appendix A for completeness. In this physical interpretation, the property $L^{(N)}_{\gamma,d} < L_{\gamma,d}$ for all $N \geq 1$ means that the system is willing to form an infinite cluster of particles. Stable infinite systems are usually found in several possible phases depending on the value of the parameters, including fluids and solids [3]. In our situation a fluid corresponds to V being constant, in which case we obtain $L^{\mathrm{sc}}_{\gamma,d}$, as we have seen. The possibility of having a solid phase where V is periodic does not seem to have been considered before in the literature for the best Lieb–Thirring constant. More complicated phases are sometimes observed in statistical mechanics (for instance translation-invariance is rarely broken in 2D but rotation-invariance can be). Here we are in a mean-field setting where $V \equiv \mathrm{cnst}$ is the only translation-invariant state. This leads us to the following:

Conjecture 1.2 (Value of the Lieb–Thirring constant). For all $d \geq 1$ and γ satisfying (1.1), we have:

- *either* there exists $N \in \mathbb{N}$ and a potential $V \in L^{\gamma+d/2}(\mathbb{R}^d)$ with exactly N negative eigenvalues optimizing (1.2), so that $L_{\gamma,d} = L^{(N)}_{\gamma,d}$,
- *or* $L^{(N)}_{\gamma,d} < L_{\gamma,d}$ for all $N \in \mathbb{N}$ and there exists an optimal *periodic* potential $V \in L^{\gamma+d/2}_{\mathrm{loc}}(\mathbb{R}^d)$ optimizing (1.5). This potential can be constant (then $L_{\gamma,d} = L^{\mathrm{sc}}_{\gamma,d}$) or not.

The Lieb–Thirring inequalities (1.2) and (1.5) are invariant under the scaling $t^2V(tx)$ and optimizers are never unique. In the periodic case all periods are therefore possible by scaling. The inequalities are also invariant under space translations and the possibility of a (non-constant) periodic optimizer would be a breaking of this symmetry.

Our conjecture is consistent with what has been proved for $\gamma \geq \frac{3}{2}$ where the optimal potential is constant, and for $\gamma = \frac{1}{2}$ in dimension $d = 1$ where the one-bound-state case is best. A more precise conjecture would be that the system is in a fluid phase ($V \equiv$ cnst) for γ larger than some critical $\gamma_{\rm sc}(d)$, then goes to a solid phase when we decrease γ until it hits a point at which the period diverges and finite systems become better.

In dimension $d = 1$, numerics indicates that one should have $L_{\gamma,1} = L_{\gamma,1}^{(1)}$ for $\gamma \leq \frac{3}{2}$ and the fluid phase $L_{\gamma,1} = L_{\gamma,1}^{\rm sc}$ for $\gamma \geq \frac{3}{2}$, see [15] and Section 3 below. We will see in the next section that the solid phase actually occurs, but only at the special point $\gamma = \frac{3}{2}$. In some sense all the possible phase transitions seem to be compressed at the unique point $\gamma = \frac{3}{2}$. In dimension $d = 2$, the solid phase could be optimal in the region $1 < \gamma < \gamma_{\rm sc}(2)$ for some $\gamma_{\rm sc}(2) \in (1.165378, \frac{3}{2}]$, see Section 3. In dimension $d = 3$, it could occur for $\frac{1}{2} < \gamma < 1$, if we believe the Lieb–Thirring conjecture that the semiclassical constant becomes optimal at $\gamma_{\rm sc}(3) = 1$.

2 The one-dimensional integrable case $\gamma = \frac{3}{2}$

We provide here a new result for $\gamma = \frac{3}{2}$ in dimension $d = 1$. Using a link with the Korteweg–de Vries (KdV) equation, it was proved by Lieb and Thirring in [19] that

$$L_{3/2,1} = L_{3/2,1}^{(N)} = L_{3/2,1}^{\rm sc} = \frac{3}{16} \quad \text{for all } N \in \mathbb{N}. \tag{2.1}$$

In fact, $L_{3/2,1}^{(N)}$ is attained for every $N \in \mathbb{N}$ and thus the Lieb–Thirring inequality has infinitely many optimizers at $\gamma = \frac{3}{2}$, modulo space translations. We show that it also admits a continuous family of *periodic* optimizers, parametrized by their period.

Theorem 2.1 (Periodic optimizers in the integrable case). *Let $\gamma = \frac{3}{2}$ and $d = 1$. For all $0 < k < 1$, we have equality in* (1.5) *for the periodic Lamé potential*

$$V_k(x) = 2k^2 \operatorname{sn}(x|k)^2 - 1 - k^2$$

of minimal period $\ell = 2K(k) > 0$. Here $\operatorname{sn}(\cdot|k)$ is a Jacobi elliptic function with modulus k and $K(k)$ is the complete elliptic integral of the first kind with modulus k.

It is known that $V_k(x) \to -1$ uniformly as $k \to 0$ (resp. $\ell \to 0$) and that $V_k(x) \to -2(\cosh x)^{-2} =: V_1(x)$ locally uniformly as $k \to 1$ (resp. $\ell \to \infty$). See Figure 1.1. Since V_1 is the optimum for $L_{3/2,1}^{(1)}$, it follows that V_k interpolates continuously between the semiclassical and the one-particle regimes when we vary the periodicity ℓ. The potential V_k is a periodic traveling (cnoidal) wave for KdV [13] and it is in fact a periodic superposition of V_1 (see [27]). It is also very well known in the theory of

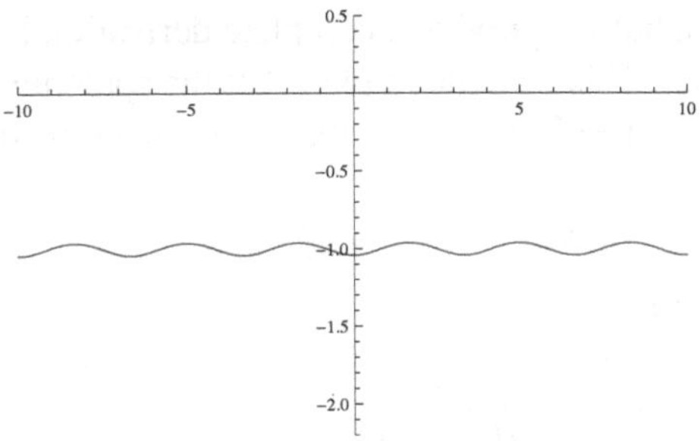

(a) $k = 0.2$ ($\ell = 3.32$)

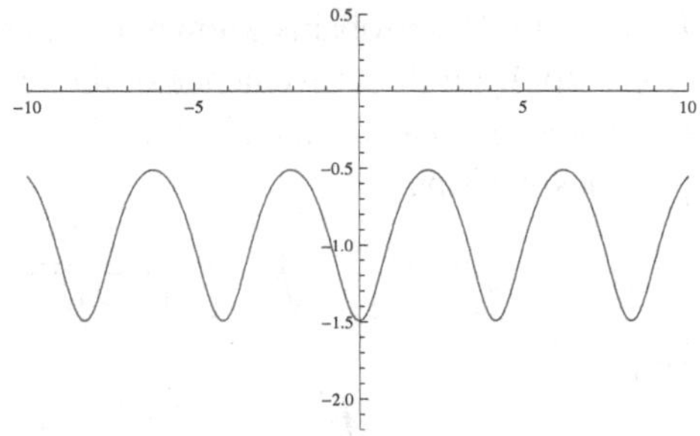

(b) $k = 0.7$ ($\ell = 4.15$)

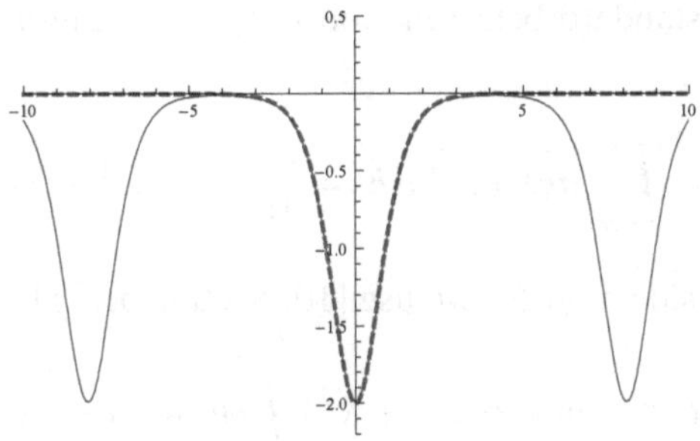

(c) $k = 0.995$ ($\ell = 8.08$)

Figure 1.1. Plot of V_k for different values of k and the period $\ell = 2K(k)$ (together with V_1 for the last).

one-dimensional periodic Schrödinger operators, since it produces a unique negative Bloch band and only one gap [12, 21, 26]. There are explicit families of periodic potentials with exactly K negative Bloch bands for any $K \geq 1$ (see [7]) but they will not be discussed here.

Proof. We fix $0 < k < 1$. It is well known that V_k is $2K(k)$ periodic [10]. Recall that the density of states $n(E)$ is defined by

$$\underline{\mathrm{Tr}}\, \mathbb{1}_{(-\infty,E]}(-\Delta + V_k) = \int_{-\infty}^{E} n(E')\, dE'.$$

Our proof is based on the following explicit formula:

$$n(E) = \frac{E + c}{2\pi\sqrt{(E+1)(E+k^2)E}}\big(-\mathbb{1}_{(-1,-k^2)}(E) + \mathbb{1}_{(0,+\infty)}(E)\big), \tag{2.2}$$

where[1]

$$c = \frac{k^2}{2K(k)} \int_{-K(k)}^{K(k)} \mathrm{sn}(x|k)^2\, dx.$$

[1] We do not need to compute c explicitly, although this could be done using [10, Section 5.134].

Formula (2.2) is probably known to experts, but we provide a complete derivation in Appendix B for the convenience of the reader. This formula implies that the spectrum has a unique isolated Bloch band: $\sigma(-\Delta + V_k) = [-1, -k^2] \cup [0, \infty)$, as mentioned previously. Using

$$\int_{k^2}^{1} \frac{t\,dt}{\sqrt{(1-t)(t-k^2)}} = \frac{\pi}{2}(1+k^2)$$

and

$$\int_{k^2}^{1} \frac{t^2\,dt}{\sqrt{(1-t)(t-k^2)}} = \frac{\pi}{8}(3+2k^2+3k^4)$$

(which follow from standard beta function integrals letting $t = (1-a)s + a$), we obtain

$$\underline{\mathrm{Tr}}(-\Delta + V)_{-}^{\frac{3}{2}} = \int_{-\infty}^{0} n(E)\, E_{-}^{\frac{3}{2}}\, dE = \frac{1}{16}(3+2k^2+3k^4) - \frac{c}{4}(1+k^2).$$

To compute the right side of (1.5), we use [10, Section 5.131]

$$\int du = \operatorname{sn} u \operatorname{cn} u \operatorname{dn} u + 2(1+k^2)\int \operatorname{sn}^2 u\, du - 3k^2 \int \operatorname{sn}^4 u\, du,$$

where we drop the parameter k from the notation for simplicity. Since

$$\operatorname{sn} 0 = \operatorname{sn} 2K = 0$$

by [10, Section 8.151], we infer that

$$\begin{aligned}\int_0^{2K} \operatorname{sn}^4 u\, du &= \frac{1}{3k^2}\Big(2(1+k^2)\int_0^{2K} \operatorname{sn}^2 u\, du - 2K\Big)\\ &= \frac{2K}{3k^2}\Big(2(1+k^2)\frac{c}{k^2} - 1\Big).\end{aligned}$$

Using $L_{3/2,1} = \frac{3}{16}$ as we have recalled in (2.1), this implies

$$\begin{aligned}\frac{1}{\ell}\int_{-\frac{\ell}{2}}^{\frac{\ell}{2}} V(x)_{-}^{2}\, dx &= \frac{1}{2K}\Big(4k^4 \int_0^{2K} \operatorname{sn}^4 u\, du\\ &\qquad - 4k^2(1+k^2)\int_0^{2K} \operatorname{sn}^2 u\, du + (1+k^2)^2 2K\Big)\\ &= \frac{4k^2}{3}\Big(2(1+k^2)\frac{c}{k^2} - 1\Big) - 4(1+k^2)c + (1+k^2)^2\\ &= (L_{3/2,1})^{-1}\Big(\frac{1}{16}(3+2k^2+3k^4) - \frac{c}{4}(1+k^2)\Big)\end{aligned}$$

and proves the assertion. ∎

3 Numerical simulations in 1D and 2D

We have computed a numerical approximation of the optimal periodic potential V in dimensions $d = 1$ and $d = 2$. In order to remove the scaling invariance, we fix a lattice $\mathscr{L}$ with unit cell C of volume $|C| = 1$ and Brillouin zone B, and work with the additional constraint that the norm of V is fixed. We also retain a fixed number K of Bloch bands. In other words, for $I > 0$ and $K \in \mathbb{N}$ we set

$$L_{\gamma,d,\mathscr{L}}(K,I) := \sup\Bigg\{\sum_{n=1}^{K} \frac{1}{|B|}\int_B \varepsilon_n^V(\xi)_-^\gamma \, d\xi : V \in L^{\gamma+\frac{d}{2}}(C), \\ \int_C V(x)_-^{\gamma+\frac{d}{2}} \, dx = I^{\gamma+\frac{d}{2}}\Bigg\}, \tag{3.1}$$

where $\varepsilon_n^V(\xi)$ is the nth eigenvalue of the operator $H_\xi = |-i\nabla + \xi|^2 + V$ with periodic boundary conditions on ∂C and quasimomentum ξ. The Lieb–Thirring constant equals

$$L_{\gamma,d} := \sup_{I>0,\, K\in\mathbb{N},\, \mathscr{L}} \frac{L_{\gamma,d,\mathscr{L}}(K,I)}{I^{\gamma+d/2}}.$$

After scaling, varying I is the same as changing the period of the potential.

We solve (3.1) with an iterative fixed point-type algorithm. At each iteration n, we compute the K first eigenvectors $u_{j,\xi}^{(n)}$ of $-\Delta + V^{(n)}$ with quasimomentum ξ, and set

$$\rho^{(n)}(x) := \frac{1}{|B|}\sum_{j=1}^{K}\int_B \varepsilon^{V^{(n)}}(\xi)_-^{\gamma-1} |u_{j,\xi}^{(n)}(x)|^2 \, d\xi,$$

where

$$V^{(n+1)}(x) = -a_n \rho^{(n)}(x)^{\frac{1}{\gamma+\frac{d}{2}-1}}, \tag{3.2}$$

with the constant $a_n > 0$ chosen so that $\|V^{(n+1)}\|_{L^{\gamma+d/2}(C)} = I$. The corresponding objective function in (3.1) can be seen to increase with the iterations. We stop the algorithm when $\|V^{(n+1)} - V^{(n)}\|_{L^{\gamma+d/2}(C)}$ is smaller than a prescribed small parameter. For the initial potential $V^{(0)}$ we use a periodic arrangement of Gaussians.

We represent a potential V by its values on a $(N_C)^d$ regular grid in C. Since the obtained potentials seem smooth, the Riemann sum converges fast to $\int_C V_-^{\gamma+d/2}$ as N_C gets large. The Brillouin zone integration is computed on a $(N_B)^d$ regular grid in B. When the operator $-\Delta + V$ has a gap above its Kth band (which was always the case in our computations), the Brillouin zone integration converges exponentially fast in N_B.

Results in one dimension for $K = 1$ Bloch band. Using Theorem 2.1 and the fact that $k \in (0,1) \mapsto \|\widetilde{V_k}\|_{L^2(0,1)}$ is increasing from π^2 to $+\infty$, where $\widetilde{V_k}$ is the 1-periodic rescaled version of V_k of Theorem 2.1, one can prove that for $\gamma = \frac{3}{2}$, the problem $L_{3/2,1,\mathbb{Z}}(1, I)$ admits as maximizer the constant potential $V = -I$ for

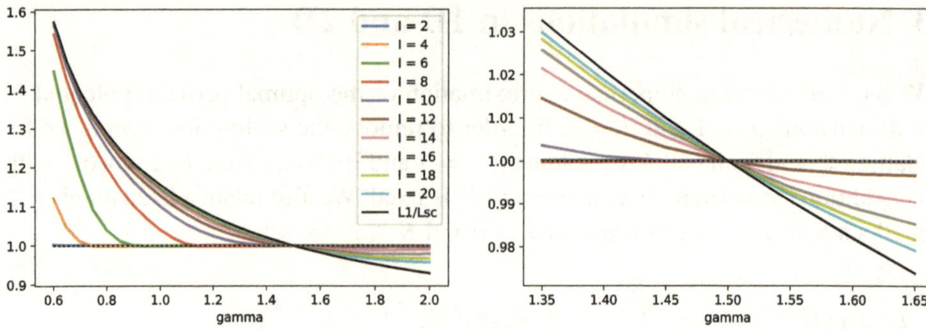

Figure 1.2. $L_{\gamma,1,\mathbb{Z}}(1,I)/L^{\rm sc}_{\gamma,1}$ for different values of $I>0$. The black curve is $L^{(1)}_{\gamma,1}/L^{\rm sc}_{\gamma,1}$.

$I\in(0,\pi^2]$, and some $\widetilde{V_k}$ for $I\in[\pi^2,\infty)$. For $I<\pi^2$, the corresponding Hamiltonian is gapless, while for $I>\pi^2$, the first band is isolated from the rest by a gap of size k^2.

In Figure 1.2, we provide numerical results for $L_{\gamma,1,\mathbb{Z}}(1,I)$ with $\gamma\in[0.6,2]$ and for different values of $I>0$. Each $I>0$ seems to give rise to a branch of periodic optimizers. When $I<\pi^2$ (that is, $I\in\{2,\dots,8\}$ in the picture), the branch coincides with the semiclassical constant in a neighborhood of $\gamma=\frac{3}{2}$. When $I>\pi^2$, the branch passes through the (rescaled) solution V_k at $\gamma=\frac{3}{2}$ and the corresponding potentials are never constant. All the curves cross at $\gamma=\frac{3}{2}$, as expected from Theorem 2.1. For $\gamma<\frac{3}{2}$, the curves are all below the one-bound-state constant $L^{(1)}_{\gamma,1}$, while for $\gamma>\frac{3}{2}$, they are all below the semiclassical constant $L^{\rm sc}_{\gamma,1}$. This is in agreement with the Lieb–Thirring conjecture in one dimension.

Results in two dimensions. In dimension $d=2$, we recall that the curves $L^{(1)}_{\gamma,2}$ and $L^{\rm sc}_{\gamma,2}$ cross at $\gamma_{1\cap\rm sc}(2)\simeq 1.165378$. In [9], we proved that the critical exponent at which the semiclassical constant becomes optimal satisfies $\gamma_{\rm sc}(2)>\gamma_{1\cap\rm sc}(2)$. We will here provide numerical evidence of the strict inequality, but also that $\gamma_{\rm sc}(2)$ could be quite close to $\gamma_{1\cap\rm sc}(2)$. As we will see, a very high precision is needed to be able to compare it with the semiclassical and one-bound state constants. We took $(N_C,N_B)=(40,30)$ and the computation of the optimal potential for one value of I and γ took approximately one hour on one processor. Several points were handled simultaneously using parallel computing.

In Figure 1.3, we display the curves

$$I\mapsto L_{\gamma,2,\mathcal{L}}(K,I)\quad \text{for } \gamma\in\{1.165300, 1.165400\},$$

that is, slightly below and above $\gamma_{1\cap\rm sc}(2)$. We considered three different lattices $\mathcal{L}$: triangular ($K=1$), square ($K=1$) and hexagonal ($K=2$). The scale in the y-axis is very fine and all quantities are computed to the order 10^{-7}. The curves are computed by solving (3.1) on a grid, while the horizontal black line $L^{(1)}_{\gamma,d}/L^{\rm sc}_{\gamma,d}$ is computed by finding the positive solution of the nonlinear Schrödinger equation using Runge–Kutta methods. The fact that the curves get very close when I increases strongly suggests that both numerical codes are valid with very high accuracy in this region.

At $\gamma = 1.165300 < \gamma_{1\cap \mathrm{sc}}(2)$, for each of the three lattices we can find a value of I so that the corresponding *periodic potential beats both the semiclassical and one-bound-state constants*. The triangular lattice provides the largest constant. When we increase γ, the three curves go down and end up touching the semiclassical constant slightly after $\gamma_{1\cap \mathrm{sc}}(2)$. The touching points are provided in Table 1.1. That the critical exponents are so close to $\gamma_{1\cap \mathrm{sc}}(2)$ could be a consequence of the exponentially small attraction between the particles [9]. For the second curve in Figure 1.3 we have $\gamma = 1.165400 > \gamma_{1\cap \mathrm{sc}}(2)$, and only the triangular lattice is above the semiclassical constant. In Figure 1.4 we display the periodic potentials at $\gamma = 1.165400$. We note that in all cases the obtained optimal potentials had exactly K negative bands.

If we double the period, which corresponds to taking $4K$ bands, we would obtain similar curves as in Figure 1.3 with a maximum located at I equal to about four times the values in Table 1.1. This maximum must be at least as high as the one we got for K bands. When we allow more and more bands, we in fact know from the proof of Theorem 1.1 that this local maximum converges to the optimal Lieb–Thirring constant, whatever lattice $\mathscr{L}$ we start with. In other words, the function $I \mapsto \max_{K\geq 1} L_{\gamma,d,\mathscr{L}}(K,I)$ in (3.1) is oscillating and converging to the best

	Triangular	Square	Hexagonal	$L^{(1)}_{\gamma,2}$
Critical γ	1.165417	1.165395	1.165390	1.165378
Corresponding I	28.7	33.1	77.2	–

Table 1.1. Approximate critical values of γ at which $\sup_I L_{\gamma,2,\mathscr{L}}(K,I) = L^{\mathrm{sc}}_{\gamma,2}$, for different lattices $\mathscr{L}$ (with the corresponding value of I), and for the constant $L^{(1)}_{\gamma,2}$.

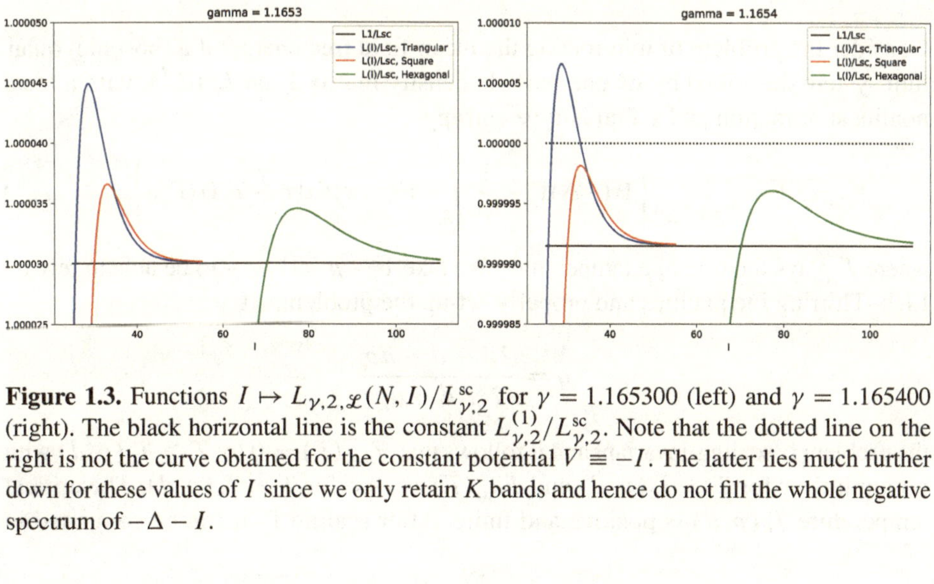

Figure 1.3. Functions $I \mapsto L_{\gamma,2,\mathscr{L}}(N,I)/L^{\mathrm{sc}}_{\gamma,2}$ for $\gamma = 1.165300$ (left) and $\gamma = 1.165400$ (right). The black horizontal line is the constant $L^{(1)}_{\gamma,2}/L^{\mathrm{sc}}_{\gamma,2}$. Note that the dotted line on the right is not the curve obtained for the constant potential $V \equiv -I$. The latter lies much further down for these values of I since we only retain K bands and hence do not fill the whole negative spectrum of $-\Delta - I$.

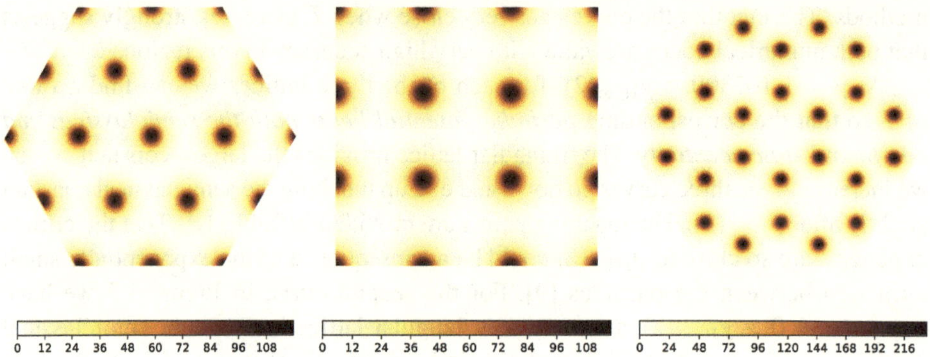

Figure 1.4. Absolute value of the optimal potential at $\gamma = 1.165400$ for the triangular (left), square (center) and hexagonal lattices (right), with the corresponding best I.

Lieb–Thirring constant $L_{\gamma,d}$ in the limit $I \to \infty$. In Figure 1.3 we only display the first bump of this function. Our Conjecture 1.2 would mean that for a certain lattice $\mathscr{L}$ these maxima are all the same. It is unfortunately hard to simulate a too large number of Bloch bands since more discretization points are then needed.

To conclude, *above* $\gamma_{1\cap \mathrm{sc}}(2)$ *we have found a periodic potential which beats the semiclassical constant and have thus shown that* $\gamma_{\mathrm{sc}}(2) \geq 1.165417$. Even *slightly below* $\gamma_{1\cap \mathrm{sc}}(2)$, *periodic potentials can do better than the one-bound state constant.* This shows that periodic potentials are important for the Lieb–Thirring inequality and gives evidence to our conjecture that they are optimizers.

A A minimization problem with entropy

Consider the problem of minimizing the mean-field free energy of a (bosonic) quantum system described by its one-particle density matrix Γ on $L^2(\mathbb{R}^d)$, with a local nonlinear attraction and a Tsallis-type entropy

$$F_{p,d}(T) = \inf_{\Gamma=\Gamma^*\geq 0}\left\{\mathrm{Tr}(-\Delta)\Gamma - \frac{1}{p}\int_{\mathbb{R}^d}\Gamma(x,x)^p\,\mathrm{d}x + T\,\mathrm{Tr}(\Gamma^q)\right\}, \tag{A.1}$$

where T plays the role of a temperature. We take $1 \leq p < 1 + \frac{2}{d}$ to be able to rely on Lieb–Thirring inequalities and properly set-up the problem. At

$$q = \frac{2p + d - dp}{2 + d - dp}$$

the problem is scaling-invariant and it follows that $F_{p,q}(T) = 0$ for $T \geq T_c(p,d)$ with no minimizer for $T > T_c(p,d)$ and $F_{p,q}(T) = -\infty$ for $T < T_c(p,d)$. The critical temperature $T_c(p,d)$ is positive and finite. After scaling Γ, it can be computed in terms of the best constant in the inequality

$$K_{p,d}\|\rho_\Gamma\|_{L^p(\mathbb{R}^d)}^{\frac{2p}{d(p-1)}} \le \left(\mathrm{Tr}(\Gamma^q)\right)^{\frac{p(2-d)+d}{dq(p-1)}} \mathrm{Tr}(-\Delta)\Gamma. \tag{A.2}$$

The inequality stays valid for $p = 1 + \frac{2}{d}$ and $q = +\infty$ with $(\mathrm{Tr}(\Gamma^q))^{\frac{1}{q}}$ replaced by the operator norm $\|\Gamma\|$ but in (A.1) this becomes a constraint $\|\Gamma\| \le 1$. For simplicity, we assume $p < 1 + \frac{2}{d}$. In [9, 20], the inequality in (A.2) was shown to be dual to the Lieb–Thirring inequality (1.2), when $p = (\gamma + \frac{d}{2})'$ and $q = \gamma'$, so that $K_{p,d}$ can be expressed in terms of $L_{\gamma,d}$. The periodic equivalent of Theorem 1.1 is

Theorem A.1 (Periodic dual Lieb–Thirring inequality). *Let $d \ge 1$ and $1 < p < 1 + \frac{2}{d}$. For every bounded periodic operator $\Gamma = \Gamma^* \ge 0$ we have*

$$K_{p,d}\left(\fint \rho_\Gamma^p\right)^{\frac{2}{d(p-1)}} \le \left(\underline{\mathrm{Tr}}(\Gamma^q)\right)^{\frac{p(2-d)+d}{dq(p-1)}} \underline{\mathrm{Tr}}(-\Delta)\Gamma, \tag{A.3}$$

with the same optimal constant $K_{p,d}$ as in (A.2).

Any periodic optimizer for (A.3), if it exists, solves the nonlinear equation

$$\Gamma = \left(-\Delta - a\rho_\Gamma^{\frac{1}{\gamma+\frac{d}{2}-1}}\right)_-^{\gamma-1}$$

with $p = (\gamma + \frac{d}{2} - 1)^{-1}$, $q = (\gamma - 1)^{-1}$ and some appropriate constant $a > 0$. This is the equation used in our fixed point algorithm (3.2).

B Computation of the density of states of V_k

Our goal in this appendix is to prove formula (2.2) for the density of states of V_k. For the convenience of the reader, we will provide here all the necessary tools from the theory of elliptic functions. There is a well developed theory of periodic Schrödinger operators in 1D with finitely many gaps, which is closely connected to integrable systems. The simplest case is for one-gap potential like V_k, where everything relies on elliptic functions. Recall that there are two theories of elliptic functions due to Weierstrass and Jacobi, respectively. We have stated our main theorem in terms of Jacobi elliptic functions, but it will be more convenient to compute the density of states in the set-up of Weierstrass elliptic functions. We will do this in Section B.2. In the first section we quickly review the theory of elliptic functions.

B.1 Weierstrass theory of elliptic functions

The whole discussion depends on two parameters $\omega_1, \omega_2 \in \mathbb{C} \setminus \{0\}$ with $\frac{\omega_2}{\omega_1} \notin \mathbb{R}$.[2] For our need, we focus on the case $\omega_1 \in \mathbb{R}$ (at the end, $\omega_1 = 2K(k)$ is the period),

[2]There are two conflicting notational conventions concerning the periods of elliptic functions. We follow here [1, Chapter 7] and denote by ω_1 and ω_2 the periods. Sometimes ω_1, ω_2 denote instead the *half-periods*.

and $\omega_2 \in i\mathbb{R}$. This choice simplifies some proofs, and we refer to [1,25] for a general discussion. Associated with these two numbers is a Weierstrass $\wp$-function

$$\wp(z) := \frac{1}{z^2} + \sum_{n\in\mathbb{Z}^2\setminus\{0\}} \left(\frac{1}{(z - n_1\omega_1 - n_2\omega_2)^2} - \frac{1}{(n_1\omega_1 + n_2\omega_2)^2} \right), \quad z \in \mathbb{C}.$$

This function is meromorphic with double poles on the grid $\{n_1\omega_1 + n_2\omega_2 : n \in \mathbb{Z}^2\}$, and periodic with periods ω_1 and ω_2. We always have $\wp(-z) = \wp(z)$, and since $\omega_1 \in \mathbb{R}$ and $\omega_2 \in i\mathbb{R}$, we also have

$$\wp\left(x - n_2\frac{\omega_2}{2}\right) \in \mathbb{R} \quad \text{and} \quad \wp\left(ix - n_1\frac{\omega_1}{2}\right) \in \mathbb{R} \quad \text{for all } x \in \mathbb{R}. \tag{B.1}$$

On the positively oriented rectangle

$$\mathcal{C} := \left[0, \frac{\omega_1}{2}\right] \cup \left[\frac{\omega_1}{2}, \frac{\omega_1 + \omega_2}{2}\right] \cup \left[\frac{\omega_1 + \omega_2}{2}, \frac{\omega_2}{2}\right] \cup \left[\frac{\omega_2}{2}, 0\right], \tag{B.2}$$

the function $\wp$ is real-valued, continuous except at 0, and satisfies $\wp(x) \to \infty$ and $\wp(ix) \to -\infty$ as $\mathbb{R} \ni x \to 0$. Since $\wp$ has order 2, for any $w \in \mathbb{C}$, the equation $\wp(z) = w$ has exactly two solutions in a fundamental cell. This implies first that $\wp$ is real-valued only on the grid (B.1), and that $\wp$ is strictly decreasing along the loop. In particular,

$$e_1 > e_2 > e_3 \quad \text{with } e_1 := \wp\left(\frac{\omega_1}{2}\right),\ e_2 := \wp\left(\frac{\omega_1 + \omega_2}{2}\right),\ e_3 := \wp\left(\frac{\omega_2}{2}\right).$$

By evenness and periodicity of $\wp$ we also have

$$\wp'\left(\frac{\omega_1}{2}\right) = \wp'\left(\frac{\omega_1 + \omega_2}{2}\right) = \wp'\left(\frac{\omega_2}{2}\right) = 0.$$

Since $\wp'$ is of order 3, these are the only roots of $\wp'$ in the fundamental cell. We set

$$g_2 = 60 \sum_{n\in\mathbb{Z}^2\setminus\{0\}} \frac{1}{(n_1\omega_1 + n_2\omega_2)^4} \quad \text{and} \quad g_3 = 140 \sum_{n\in\mathbb{Z}^2\setminus\{0\}} \frac{1}{(n_1\omega_1 + n_2\omega_2)^6}.$$

Comparing the expansions of $\wp'$ and $\wp$ near the poles, we obtain

$$\wp'(z)^2 = 4\wp(z)^3 - g_2\wp(z) - g_3 = 4(\wp(z) - e_1)(\wp(z) - e_2)(\wp(z) - e_3),$$

where we used in the last equality that $\wp'$ vanishes at $\frac{\omega_1}{2}$, $\frac{\omega_1+\omega_2}{2}$ and $\frac{\omega_2}{2}$. Together with the fact that $\wp$ is decreasing along the loop, this gives by separation of variables that

$$a - a_0 = i \int_{\wp(a_0)}^{\wp(a)} \frac{\mathrm{d}w}{\sqrt{4(e_1 - w)(e_2 - w)(e_3 - w)}} \quad \text{for all } a, a_0 \in (0, \tfrac{1}{2}\omega_2]. \tag{B.3}$$

We have similar formulas on the other parts of the loop $\mathcal{C}$, that we omit for brevity.

Before we turn to the link between the Weierstrass $\wp$ function and the Schrödinger equation, we need two more functions. The first one is the Weierstrass zeta function

$$\zeta(z) := \frac{1}{z} + \sum_{n\in\mathbb{Z}^2\setminus\{0\}} \left(\frac{1}{z - n_1\omega_1 - n_2\omega_2} + \frac{1}{n_1\omega_1 + n_2\omega_2} + \frac{z}{(n_1\omega_1 + n_2\omega_2)^2}\right)$$

for $z \in \mathbb{C}$. It has simple poles at $\{n_1\omega_1 + n_2\omega_2 : n \in \mathbb{Z}^2\}$ and satisfies

$$\zeta'(z) = -\wp(z). \tag{B.4}$$

We have $\zeta(-z) = -\zeta(z)$, $\zeta(z + \omega_1) = \zeta(z) + \eta_1$ and $\zeta(z + \omega_2) = \zeta(z) + \eta_2$, for

$$\eta_1 := 2\zeta\left(\frac{\omega_1}{2}\right) \quad \text{and} \quad \eta_2 := 2\zeta\left(\frac{\omega_2}{2}\right). \tag{B.5}$$

In our case where $\omega_1 \in \mathbb{R}$ and $\omega_2 \in i\mathbb{R}$, we get from (B.4) and the fact that $\wp$ is real-valued on the loop $\mathcal{C}$, that $\zeta(x) \in \mathbb{R}$ while $\zeta(ix) \in i\mathbb{R}$ for all $x \in \mathbb{R}$. This shows for instance that $\eta_1 \in \mathbb{R}$ and $\eta_2 \in i\mathbb{R}$. Moreover, one can show [1] that

$$\eta_1\omega_2 - \eta_2\omega_1 = 2\pi i. \tag{B.6}$$

As in (B.3), using (B.4) we see that, for all $a, a_0 \in (0, \frac{1}{2}\omega_2]$,

$$\zeta(a) - \zeta(a_0) = -i\int_{\wp(a_0)}^{\wp(a)} \frac{w\,\mathrm{d}w}{\sqrt{4(e_1 - w)(e_2 - w)(e_3 - w)}}. \tag{B.7}$$

The last function to be introduced is

$$\sigma(z) := z\prod_{n\in\mathbb{Z}^2\setminus\{0\}} \left(1 - \frac{z}{n_1\omega_1 + n_2\omega_2}\right) e^{\frac{z}{n_1\omega_1+n_2\omega_2} + \frac{1}{2}\frac{z^2}{(n_1\omega_1+n_2\omega_2)^2}}, \quad z \in \mathbb{C}.$$

This is an odd entire function which satisfies

$$\begin{aligned} \frac{\sigma'(z)}{\sigma(z)} &= \zeta(z), \\ \sigma(z + \omega_1) &= -\sigma(z)e^{\eta_1(z+\frac{\omega_1}{2})}, \\ \sigma(z + \omega_2) &= -\sigma(z)e^{\eta_2(z+\frac{\omega_2}{2})}. \end{aligned} \tag{B.8}$$

B.2 The Schrödinger equation

We let $\omega_1 \in (0, \infty)$ and $\omega_2 \in i(0, \infty)$ as before, and denote by $\wp$ the corresponding Weierstrass function. We consider the potential

$$W(x) := 2\wp\left(x + \frac{1}{2}\omega_2\right), \quad x \in \mathbb{R}.$$

It has (real) period ω_1, is real-valued and real analytic (since the line $\mathbb{R} + \omega_2/2$ does not hit the poles of $\wp$). Moreover, it is even and symmetric about $x = \frac{1}{2}\omega_1$. It is strictly increasing on $[0, \frac{1}{2}\omega_1]$ with $W(0) = 2e_3$ and $W(\frac{1}{2}\omega_1) = 2e_2$. We study the (periodic) Schrödinger equation in the form

$$\begin{cases} -\psi'' + W\psi = E\psi & \text{on } \mathbb{R}, \\ E = -\wp(a). \end{cases} \tag{B.9}$$

Here, $a \in \mathbb{C}$ is any parameter. Soon we will require $E \in \mathbb{R}$, in which case a must belong to the loop $\mathcal{C}$ defined in (B.2). As a runs positively through this loop, E goes from $-\infty$ to ∞. Let

$$\psi_\pm(x) := \frac{\sigma(x + \frac{1}{2}\omega_2 \pm a)}{\sigma(x + \frac{1}{2}\omega_2)} e^{\mp(x+\frac{1}{2}\omega_2)\zeta(a)}.$$

Lemma B.1. *The functions* $\psi_\pm$ *solve* (B.9). *If* a *belongs to the fundamental domain and* $a \notin \{0, \frac{1}{2}\omega_1, \frac{1}{2}\omega_2, \frac{1}{2}(\omega_1 + \omega_2)\}$, *then* ψ_+ *and* ψ_- *are linearly independent.*

We will see that the excluded values of a give the boundary of the spectrum.

Proof. We have, using (B.8),

$$\begin{aligned} \psi_\pm'(x) &= e^{\mp(x+\frac{1}{2}\omega_2)\zeta(a)} \bigg(\mp \zeta(a) \frac{\sigma(x + \frac{1}{2}\omega_2 \pm a)}{\sigma(x + \frac{1}{2}\omega_2)} + \frac{\sigma'(x + \frac{1}{2}\omega_2 \pm a)}{\sigma(x + \frac{1}{2}\omega_2)} \\ &\qquad - \frac{\sigma(x + \frac{1}{2}\omega_2 \pm a)\sigma'(x + \frac{1}{2}\omega_2)}{\sigma(x + \frac{1}{2}\omega_2)^2} \bigg) \\ &= \psi_\pm(x) \big(\mp\zeta(a) + \zeta(x + \tfrac{1}{2}\omega_2 \pm a) - \zeta(x + \tfrac{1}{2}\omega_2) \big) \end{aligned}$$

and this gives

$$\begin{aligned} \psi_\pm''(x) = \psi_\pm(x) \Big\{ &\big(\mp\zeta(a) + \zeta(x + \tfrac{1}{2}\omega_2 \pm a) - \zeta(x + \tfrac{1}{2}\omega_2) \big)^2 \\ &+ \big(\zeta'(x + \tfrac{1}{2}\omega_2 \pm a) - \zeta'(x + \tfrac{1}{2}\omega_2) \big) \Big\}. \end{aligned}$$

Thus, to prove the lemma, we need to show that the term in the curly bracket equals $2\wp(x + \frac{1}{2}\omega_2) + \wp(a)$. According to (B.4), this is the same as

$$\big(\mp\zeta(a) + \zeta(z \pm a) - \zeta(z) \big)^2 + \zeta'(z \pm a) + \zeta'(z) + \zeta'(a) = 0$$

for all $z = x + \frac{1}{2}\omega_2$. Using the oddness of ζ we rewrite the quantity under the square as $\mp\zeta(a) + \zeta(x \pm a) - \zeta(x) = \zeta(\mp a) + \zeta(x \pm a) + \zeta(-x)$ and note that the three numbers involved in the right side satisfy $(\mp a) + (z \pm a) + (-z) = 0$. Therefore, by [25, Exercise 5, Section 10.45] and the evenness of ζ, we have

$$\begin{aligned} \big(\zeta(\mp a) + \zeta(z \pm a) + \zeta(-z) \big)^2 &= -\big(\zeta'(\mp a) + \zeta'(z \pm a) + \zeta'(-z) \big) \\ &= -\big(\zeta'(a) + \zeta'(z \pm a) + \zeta'(z) \big), \end{aligned}$$

which proves the claimed identity. In order to show that the two functions are linearly independent, we compute their Wronskian, using the above formulas for $\psi'_\pm$:

$$\begin{aligned}\psi_+\psi'_- - \psi_-\psi'_+ &= \big(2\zeta(a) + \zeta(x + \tfrac{1}{2}\omega_2 - a) - \zeta(x + \tfrac{1}{2}\omega_2 + a)\big)\\ &\quad\times \frac{\sigma(x + \frac{1}{2}\omega_2 + a)\sigma(x + \frac{1}{2}\omega_2 - a)}{\sigma(x + \frac{1}{2}\omega_2)^2}.\end{aligned}$$

According to [25, equation (10.4.90)] we have

$$\frac{\sigma(x + \frac{1}{2}\omega_2 + a)\sigma(x + \frac{1}{2}\omega_2 - a)}{\sigma(x + \frac{1}{2}\omega_2)^2} = \big(\wp(a) - \wp(x + \tfrac{1}{2}\omega_2)\big)\sigma(a)^2$$

and according to oddness of ζ and [25, equation (10.4.91)], we have

$$\begin{aligned}2\zeta(a) &+ \zeta(x + \tfrac{1}{2}\omega_2 - a) - \zeta(x + \tfrac{1}{2}\omega_2 + a)\\ &= 2\zeta(a) - \zeta(a - (x + \tfrac{1}{2}\omega_2)) - \zeta(a + (\tfrac{1}{2}\omega_2))\\ &= -\frac{\wp'(a)}{\wp(a) - \wp(x + \frac{1}{2}\omega_2)}.\end{aligned}$$

Thus we find $\psi_+\psi'_- - \psi_-\psi'_+ = -\sigma(a)^2\wp'(a)$. Since σ does not vanish in the fundamental domain except at the origin (this follows from the product formula) and since $\wp'$ vanishes exactly at the excluded points, we see that the Wronskian is nonzero for a in the fundamental domain and different from the excluded values. ∎

The dispersion relation. From (B.8) we have $\psi_\pm(x + \omega_1) = e^{\mp\omega_1\zeta(a)\pm\eta_1 a}\psi_\pm(x)$. Setting

$$\xi_\pm := \mp i\Big(\zeta(a) - \frac{\eta_1}{\omega_1}a\Big) \quad \text{modulo } 2\pi, \tag{B.10}$$

we have $\psi_\pm(x + \omega_1) := e^{-i\omega_1\xi_\pm}\psi_\pm(x)$, so $\psi_\pm$ are Bloch–Floquet solutions for the energy $E = -\wp(a)$ whenever $\xi_\pm \in \mathbb{R}$. Since we want $E \in \mathbb{R}$, we need to take $a \in \mathcal{C}$. Together with (B.4), we can see that $\xi_\pm \in \mathbb{R}$ if and only if a belongs to the vertical segments $a \in (0, \frac{1}{2}\omega_2] \cup [\frac{1}{2}\omega_1, \frac{1}{2}(\omega_1 + \omega_2))$. This already proves that $-\partial_x^2 + W$ has spectrum $[-e_1, -e_2] \cup [-e_3, \infty)$, and in particular has a unique gap.

Formula (B.10) gives an implicit parametrization of the dispersion curves. We will now use the integral representations to rewrite $\xi_\pm$ in terms of E instead of a. First consider $a \in (0, \frac{1}{2}\omega_2]$, that is $E \in [-e_3, \infty)$. We set $a_0 := \frac{1}{2}\omega_2$, and get from (B.3)–(B.7) that

$$\zeta(a) - \frac{\eta_1}{\omega_1}a = \Big(\zeta(a_0) - \frac{\eta_1}{\omega_1}a_0\Big) - i\int_{\wp(a_0)}^{\wp(a)} \frac{(w + \frac{\eta_1}{\omega_1})\,\mathrm{d}w}{\sqrt{4(e_1 - w)(e_2 - w)(e_3 - w)}}.$$

By (B.5) and (B.6), the first term is $-i\frac{\pi}{\omega_1}$. This gives, for all $E \in [-e_3, \infty)$,

$$\xi_\pm(E) = \mp\int_{e_3}^{-E} \frac{(w + \frac{\eta_1}{\omega_1})\,\mathrm{d}w}{\sqrt{4(e_1 - w)(e_2 - w)(e_3 - w)}} \mp \frac{\pi}{\omega_1}.$$

For $a \in [\frac{1}{2}\omega_1, \frac{1}{2}(\omega_1 + \omega_2)]$, that is, $E \in [-e_1, -e_2]$, we choose $a_0 = \frac{1}{2}\omega_2$. This time, we have $\zeta(a_0) - \frac{\eta_1}{\omega_1}a_0 = 0$. Performing a similar reasoning, we obtain, for all $E \in (-e_1, -e_2)$,

$$\xi_\pm(E) = \int_{-E}^{e_1} \frac{(w + \frac{\eta_1}{\omega_1})\,\mathrm{d}w}{\sqrt{4(e_1 - w)(w - e_2)(w - e_3)}}.$$

The density of states. We have already shown that the spectrum of $-\partial_x^2 + W$ in $L^2(\mathbb{R})$ is $[-e_1, -e_2] \cup [-e_3, +\infty)$. We now compute the *integrated* density of states, which is equal to $N_W(E) = \pi^{-1}\xi_\pm(E)$ for $E \in \sigma(-\partial_x^2 + W)$, where the sign is chosen in such a way that N_W is increasing. Using the above formulas, we obtain

$$N_W(E) = \begin{cases} 0 & \text{if } E \in (-\infty, -e_1], \\ (2\pi)^{-1} \int_{-E}^{e_1} \frac{(w+\frac{\eta_1}{\omega_1})\,\mathrm{d}w}{\sqrt{(e_1-w)(w-e_2)(w-e_3)}} & \text{if } E \in [-e_1, -e_2], \\ \omega_1^{-1} & \text{if } E \in [-e_2, -e_3], \\ \omega_1^{-1} + (2\pi)^{-1} \int_{-E}^{e_3} \frac{(w+\frac{\eta_1}{\omega_1})\,\mathrm{d}w}{\sqrt{(e_1-w)(e_2-w)(e_3-w)}} & \text{if } E \in [-e_3, +\infty). \end{cases}$$

Taking a derivative gives the sought-after density of states

$$n_W(E) = N_W'(E) = \begin{cases} (2\pi)^{-1} \frac{-E+\frac{\eta_1}{\omega_1}}{\sqrt{(e_1+E)(-E-e_2)(-E-e_3)}} & \text{if } E \in (-e_1, -e_2), \\ (2\pi)^{-1} \frac{E-\frac{\eta_1}{\omega_1}}{\sqrt{(e_1+E)(e_2+E)(e_3+E)}} & \text{if } E \in (-e_3, +\infty), \\ 0 & \text{otherwise.} \end{cases} \tag{B.11}$$

Passing to Jacobi elliptic functions. We finally relate the potential W with the potential V_k of Theorem 2.1. From [10, Equation 8.169.1], we have

$$\wp(z) = e_3 + \frac{e_1 - e_3}{\mathrm{sn}(\sqrt{e_1 - e_3}z|k)^2} \quad \text{with } k = \sqrt{\frac{e_2 - e_3}{e_1 - e_3}}.$$

Let $0 < k < 1$, and choose

$$\omega_1 := 2K(k) \in \mathbb{R}^+ \quad \text{and} \quad \omega_2 := 2iK'(k) \in i\mathbb{R}^+.$$

Using $k\,\mathrm{sn}(u|k)\,\mathrm{sn}(u + iK'(k)|k) = 1$ (see [25, Equation 10.5.115]), we get

$$W(x) = 2\wp(x + \tfrac{1}{2}\omega_2) = 2e_3 + 2(e_1 - e_3)k^2\,\mathrm{sn}^2(\sqrt{e_1 - e_3}x|k)^2.$$

It remains to find the values of e_1, e_2 and e_3. First, since $x \mapsto \mathrm{sn}(x|k)^2$ has period $2K(k)$, while W has period ω_1, we deduce that $\sqrt{e_1 - e_3} = 1$. In addition, we always have

$$e_1 + e_2 + e_3 = 0,$$

because the coefficient in front of w^2 of the polynomial $4w^3 - g_2 w - g_3$ vanishes. So

$$\begin{cases} e_1 - e_3 = 1, \\ k^2(e_1 - e_3) = (e_2 - e_3), \\ e_1 + e_2 + e_3 = 0, \end{cases} \quad \text{which gives} \quad \begin{cases} e_1 = 1 - \frac{1}{3}(1+k^2), \\ e_2 = k^2 - \frac{1}{3}(1+k^2), \\ e_3 = -\frac{1}{3}(1+k^2). \end{cases}$$

Altogether, this proves that

$$W(x) = -\tfrac{2}{3}(1+k^2) + 2k^2 \operatorname{sn}^2(x|k)^2 = V_k(x) + \tfrac{1}{3}(1+k^2).$$

Since the operator $(-\partial_x^2 + W)$ has spectrum $[-e_1, -e_2] \cup [-e_3, \infty)$, the operator $(-\partial_x^2 + V_k)$ has spectrum $[-1, -k^2] \cup [0, \infty)$, as wanted. The claimed formula (2.2) for $n(E)$ easily follows from (B.11).

Acknowledgements. This project has received funding from the U.S. National Science Foundation (grant agreements DMS-1363432 and DMS-1954995 of Rupert L. Frank) and from the European Research Council (ERC) under the European Union's Horizon 2020 research and innovation programme (grant agreement MDFT 725528 of Mathieu Lewin).

Bibliography

[1] L. V. Ahlfors, *Complex analysis*. 3rd edn., McGraw–Hill, New York, 1978

[2] M. Aizenman and E. H. Lieb, On semiclassical bounds for eigenvalues of Schrödinger operators. *Phys. Lett. A* **66** (1978), 427–429

[3] X. Blanc and M. Lewin, The crystallization conjecture: A review. *EMS Surv. Math. Sci.* **2** (2015), 255–306

[4] M. Cwikel, Weak type estimates for singular values and the number of bound states of Schrödinger operators. *Ann. of Math. (2)* **106** (1977), 93–100

[5] F. Daumer, Équation de Schrödinger avec champ électrique périodique et champ magnétique constant dans l'approximation du tight binding. *Comm. Partial Differential Equations* **18** (1993), 1021–1041

[6] S. Doi, A. Iwatsuka and T. Mine, The uniqueness of the integrated density of states for the Schrödinger operators with magnetic fields. *Math. Z.* **237** (2001), 335–371

[7] B. A. Dubrovin and S. P. Novikov, Periodic and conditionally periodic analogs of the many-soliton solutions of the Korteweg–de Vries equation. *Ž. Èksper. Teoret. Fiz.* **67** (1974), 2131–2144

[8] R. L. Frank, The Lieb–Thirring inequality: Recent results and open problems. Preprint 2020, arXiv:2007.09326

[9] R. L. Frank, D. Gontier and M. Lewin, The nonlinear Schrödinger equation for orthonormal functions II. Application to Lieb–Thirring inequalities. *Comm. Math. Phys.*, to appear

[10] I. S. Gradshteyn and I. M. Ryzhik, *Table of integrals, series, and products*. 8th edn., Elsevier/Academic Press, Amsterdam, 2015

[11] D. Hundertmark, E. H. Lieb and L. E. Thomas, A sharp bound for an eigenvalue moment of the one-dimensional Schrödinger operator. *Adv. Theor. Math. Phys.* **2** (1998), 719–731

[12] E. L. Ince, Further investigations into the periodic Lamé functions. *Proc. Roy. Soc. Edinburgh* **60** (1940), 83–99

[13] D. J. Korteweg and G. de Vries, On the change of form of long waves advancing in a rectangular canal, and on a new type of long stationary waves. *Philos. Mag. (5)* **39** (1895), 422–443

[14] A. Laptev and T. Weidl, Sharp Lieb–Thirring inequalities in high dimensions. *Acta Math.* **184** (2000), 87–111

[15] A. Levitt, Best constants in Lieb–Thirring inequalities: A numerical investigation. *J. Spectr. Theory* **4** (2014), 153–175

[16] E. H. Lieb, Bounds on the eigenvalues of the Laplace and Schroedinger operators. *Bull. Amer. Math. Soc.* **82** (1976), 751–753

[17] E. H. Lieb and R. Seiringer, *The stability of matter in quantum mechanics*. Cambridge University Press, Cambridge, 2010

[18] E. H. Lieb and W. E. Thirring, *Bound on kinetic energy of fermions which proves stability of matter. Phys. Rev. Lett.* **35** (1975), 687–689

[19] E. H. Lieb and W. E. Thirring, Inequalities for the moments of the eigenvalues of the Schrödinger hamiltonian and their relation to Sobolev inequalities. In *Studies in mathematical physics*, pp. 269–303, Princeton University Press, Princeton, 1976

[20] P.-L. Lions and T. Paul, Sur les mesures de Wigner. *Rev. Mat. Iberoam.* **9** (1993), 553–618

[21] W. Magnus and S. Winkler, *Hill's equation*. Interscience Tracts Pure Appl. Math. 20, John Wiley & Sons, New York, 1966

[22] S. Nakamura, A remark on the Dirichlet–Neumann decoupling and the integrated density of states. *J. Funct. Anal.* **179** (2001), 136–152

[23] M. Reed and B. Simon, *Methods of modern mathematical physics. IV. Analysis of operators*. Academic Press, New York, 1978

[24] G. V. Rozenbljum, Distribution of the discrete spectrum of singular differential operators. *Dokl. Akad. Nauk SSSR* **202** (1972), 1012–1015

[25] B. Simon, *Basic complex analysis. A comprehensive course in analysis, Part 2A*. American Mathematical Society, Providence, 2015

[26] B. Sutherland, Some exact results for one-dimensional models of solids. *Phys. Rev. A* **8** (1973), 2514–2516

[27] M. Toda, Waves in nonlinear lattice. *Prog. Theor. Phys. Supp.* **45** (1970), 174–200

[28] T. Weidl, On the Lieb–Thirring constants $L_{\gamma,1}$ for $\gamma \geq 1/2$. *Comm. Math. Phys.* **178** (1996), 135–146

Semiclassical asymptotics for a class of singular Schrödinger operators

Rupert L. Frank and Simon Larson

To Ari Laptev on the occasion of his 70th birthday

Let $\Omega \subset \mathbb{R}^d$ be bounded with C^1 boundary. In this paper we consider Schrödinger operators $-\Delta + W$ on Ω with $W(x) \approx \operatorname{dist}(x, \partial\Omega)^{-2}$ as $\operatorname{dist}(x, \partial\Omega) \to 0$. Under weak assumptions on W we derive a two-term asymptotic formula for the sum of the eigenvalues of such operators.

1 Introduction

In this paper we consider semiclassical asymptotics for a class of Schrödinger operators on bounded sets $\Omega \subset \mathbb{R}^d$ with potentials that are singular at the boundary. Specifically, for a bounded open set $\Omega \subset \mathbb{R}^d$ with C^1 boundary we consider Schrödinger operators

$$-\Delta + W(x) \quad \text{with } W(x) \approx \operatorname{dist}(x, \Omega)^{-2} \text{ as } \operatorname{dist}(x, \partial\Omega) \to 0 \tag{1.1}$$

subject to Dirichlet boundary conditions. These operators have purely discrete spectrum and our main interest is towards the asymptotic behavior of their eigenvalues. Our main result is a two-term asymptotic formula for the sum of the eigenvalues.

Before we formulate our main result it is necessary to explain more precisely how (1.1) is to be interpreted. We shall assume that our potential decomposes as one part that is in $L^\infty_{\mathrm{loc}}(\Omega)$ and has the prescribed singular behavior at the boundary and a part that, in comparison, is well-behaved. To simplify the exposition we write

$$H_{\Omega,b,V}(h) = -h^2\Delta + h^2\Big(b^2(x) - \frac{1}{4}\Big)\frac{1}{\operatorname{dist}(x, \partial\Omega)^2} + h^2 V(x) - 1 \quad \text{for } h > 0.$$

Technically, the operator $H_{\Omega,b,V}(h)$ is defined through the quadratic form

$$\begin{aligned} u \mapsto \int_\Omega \Big(& h^2|\nabla u(x)|^2 + h^2\Big(b^2(x) - \frac{1}{4}\Big)\frac{|u(x)|^2}{\operatorname{dist}(x, \partial\Omega)^2} \\ & + h^2 V(x)|u(x)|^2 - |u(x)|^2\Big)\, dx \end{aligned} \tag{1.2}$$

Keywords: Schrödinger operator, semiclassical asymptotics
2020 Mathematics Subject Classification: 35P20

with form domain $\{u \in H_0^1(\Omega) : V_+ u^2 \in L^1(\Omega)\}$. Throughout, we shall assume that $V \in L^1(\Omega)$, $V_- \in L^{1+\frac{d}{2}}(\Omega)$, and that $b \in L^\infty(\Omega)$ is positive and satisfies

$$\lim_{r\to 0+} \int_{\partial\Omega} \Big[\sup_{y \in B_r(x)\cap\Omega} b(y) - \inf_{y\in B_r(x)\cap\Omega} b(y) \Big] \, d\mathcal{H}^{d-1}(x) = 0. \tag{1.3}$$

Here and in what follows, we define $x_\pm = \frac{|x|\pm x}{2}$ and note that with this convention both x_+ and x_- are non-negative. As a consequence of Hardy's inequality, the assumptions on V and b ensure that the quadratic form (1.2) is bounded from below and closed. Therefore, it generates a selfadjoint, bounded from below operator $H_{\Omega,b,V}(h)$ in $L^2(\Omega)$.

We emphasize that by positivity of b, we mean $\inf_\Omega b > 0$. This assumption can naturally be relaxed to require positivity only in a neighborhood of the boundary by adjusting V correspondingly. The regularity assumption (1.3) implies that $b|_{\partial\Omega}$ can be made sense of as an element of $L^\infty(\partial\Omega)$; indeed, by (1.3), b has a well-defined limit $\mathcal{H}^{d-1}$-almost everywhere on $\partial\Omega$, which is finite since $b \in L^\infty(\Omega)$. Our main result can now be stated as follows:

Theorem 1.1. *Let $\Omega \subset \mathbb{R}^d$ be open and bounded with C^1 boundary, $V \in L^1(\Omega)$ with $V_- \in L^{1+\frac{d}{2}}(\Omega)$, and let $b \in L^\infty(\Omega)$ be positive and satisfy* (1.3)*. Then, as $h \to 0^+$,*

$$\operatorname{Tr}(H_{\Omega,b,V}(h))_- = L_d h^{-d}|\Omega| - \frac{L_{d-1}}{2} h^{-d+1} \int_{\partial\Omega} b(x)\, d\mathcal{H}^{d-1}(x) + o(h^{-d+1}),$$

where $L_d = (4\pi)^{-\frac{d}{2}}\Gamma(2+\frac{d}{2})^{-1}$.

As a corollary of Theorem 1.1 we deduce:

Corollary 1.2. *Let $\Omega \subset \mathbb{R}^d$ be open and bounded with C^1 boundary. Then, with Δ_Ω denoting the Dirichlet Laplace operator in Ω, as $h \to 0^+$ and in the sense of measures*

$$h^{d+1} \frac{\mathbb{1}(-h^2\Delta_\Omega \le 1)(x,x)}{\operatorname{dist}(x,\partial\Omega)^2}\, dx \to \frac{L_{d-1}}{2} \mathcal{H}^{d-1}|_{\partial\Omega}.$$

Proof. The corollary follows from a standard Feynman–Hellmann argument (cf. [10]) and Theorem 1.1 applied with the potential $W(x) = \frac{tf(x)}{\operatorname{dist}(x,\partial\Omega)^2}$ for $f \in C(\overline{\Omega})$ and sending first h then t to zero. ∎

Spectral asymptotics for differential operators that degenerate at the boundary of the domain are not new. However, the results in the literature mainly concern cases where the operator degenerates at leading order and how this affects the first term in the asymptotics, see [1, 2] and references therein. While the class of operators considered here is significantly less singular, our interest is towards the effect of the degeneracy on the second term in the asymptotics.

In the special case of the Dirichlet Laplacian, i.e. $V \equiv 0$ and $b \equiv \frac{1}{2}$, Theorem 1.1 was proved in [5, 6]. The strategy of our proof follows closely that developed there, but several new obstacles need to be circumvented in the presence of the potential that is singular at the boundary. The idea is to localize the operator in balls whose size varies depending on the distance to the boundary and h. In a ball far from the boundary the influence of the boundary conditions and the potential both have a negligible effect and precise asymptotics can be obtained through standard methods. In a ball close to the boundary the regularity of the boundary allows to map the problem to a half-space where asymptotics are obtained by explicitly diagonalizing an effective operator. The main new ingredients needed here is to control how the straightening of the boundary affects the singular part of the potential and to understand how the potential enters in the resulting half-space problem.

The works [5, 6] for domains with C^1 boundaries were extended to the case of Lipschitz boundaries in [8], see also [7]. Since the (weak) Hardy constant can be smaller than $\frac{1}{4}$ for Lipschitz domains, it is not clear how to generalize the results of the present paper to this setting.

The plan for the paper is as follows. In Section 2 we recall a number of results concerning changes of variables mapping $\partial\Omega$ locally to a hyperplane. In particular, Lemma 2.2 describes how such a mapping affects the singular part of our potential. We also prove a local Hardy–Lieb–Thirring inequality, which will be crucial in controlling error terms appearing in our analysis, and which replaces the Lieb–Thirring inequality in [5] in the absence of a singular potential. In Section 3 we provide local asymptotics, both in the bulk of our domain and close to the boundary. Finally, in Section 4 we adapt the localization procedure developed in [5, 6, 8] to our current setting and use it to piece together the local asymptotics of Section 3, thus proving Theorem 1.1.

The letter C will denote a constant whose value can change at each occurrence.

2 Preliminaries

2.1 Straightening the boundary

Let $\mathbb{R}^d_+ = \{y \in \mathbb{R}^d : y_d > 0\}$. Let $B \subset \mathbb{R}^d$ be an open ball of radius ℓ centered at a point $x_0 \in \partial\Omega$. By rotating and translating we may assume that $x_0 = 0$ and that $\nu_0 = (0, \dots, 0, 1)$ is the inward pointing unit normal to $\partial\Omega$ at x_0. Since Ω is bounded with C^1 boundary, there is a non-decreasing modulus of continuity $\omega: \mathbb{R}_+ \to [0, 1]$ such that, if ℓ is small enough, there is a function $f: \mathbb{R}^{d-1} \to \mathbb{R}$ satisfying $|\nabla f(x')| \le \omega(|x'|)$ such that

$$\partial\Omega \cap B_{2\ell}(0) = \{(x', x_d) \in \mathbb{R}^{d-1} \times \mathbb{R} : x_d = f(x')\} \cap B_{2\ell}(0).$$

Note that, by the choice of coordinates, $f(0) = 0$ and $\nabla f(0) = 0$.

Set

$$\mathcal{X} = \{(x', x_d) \in \mathbb{R}^{d-1} \times \mathbb{R} : |x'| < 2\ell\}.$$

We define a diffeomorphism $\Phi: \mathcal{X} \to \mathbb{R}^d$ by $\Phi_j(x) = x_j$ for $j = 1, \dots, d-1$ and $\Phi_d(x) = x_d - f(x')$. Note that the Jacobian determinant of Φ equals 1 and that the inverse of Φ is well-defined on $\Phi(\mathcal{X}) = \mathcal{X}$. The inverse is given by $\Phi_j^{-1}(y) = y_j$ for $j = 1, \dots, d-1$ and $\Phi_d^{-1}(y) = y_d + f(y')$.

In the following lemma we gather some results whose proofs are standard and can be found, for instance, in [6, Section 4].

Lemma 2.1 (Straightening of the boundary). *Let B, Φ be as above and for $u\colon B \to \mathbb{R}$ set $\tilde{u} = u \circ \Phi^{-1}$. For $0 < \ell \le c(\omega)$ and with C depending only on d, we have:*

(1) *if $u \in L^1(B)$, then*

$$\int_B u(x)\,dx = \int_{\Phi(B)} \tilde{u}(y)\,dy.$$

(2) *if $u \in L^\infty(\partial\Omega \cap B)$, then*

$$\begin{aligned}&\left|\int_{\partial\Omega\cap B} u(x)\,d\mathcal{H}^{d-1}(x) - \int_{\partial\mathbb{R}^d_+\cap\Phi(B)} \tilde{u}(y)\,d\mathcal{H}^{d-1}(y)\right|\\ &\quad\le C\ell^{d-1}\omega(\ell)^2\|u\|_{L^\infty}.\end{aligned}$$

(3) *if $u \in H_0^1(\Omega \cap B)$, then $\tilde{u} \in H_0^1(\mathbb{R}^d_+ \cap \Phi(B))$ and*

$$\begin{aligned}&\left|\int_{\Omega\cap B} |\nabla u(x)|^2\,dx - \int_{\mathbb{R}^d_+\cap\Phi(B)} |\nabla\tilde{u}(y)|^2\,dy\right|\\ &\quad\le C\omega(\ell)\int_{\mathbb{R}^d_+\cap\Phi(B)} |\nabla\tilde{u}(y)|^2\,dy.\end{aligned}$$

(4) *if $u \in C_0^1(\mathbb{R}^d)$ is supported in $\overline{B}$, then, after extension by zero, $\tilde{u} \in C_0^1(\mathbb{R}^d)$ with* $\operatorname{supp}\tilde{u} \subseteq \overline{B_{2\ell}(0)}$ *and $\|\nabla\tilde{u}\|_{L^\infty} \le C\|\nabla u\|_{L^\infty}$.*

In addition to the properties in Lemma 2.1, we will need the following result, which enables us to control the change under Φ of the singular part of our potentials:

Lemma 2.2. *Let B, Φ be as above. There is a constant C depending only on d such that for any $x \in B \cap \Omega$,*

$$0 \le \frac{1}{\operatorname{dist}(x,\partial\Omega)^2} - \frac{1}{\operatorname{dist}(\Phi(x),\partial\mathbb{R}^d_+)^2} \le C\frac{\omega(2\ell)^2}{\operatorname{dist}(\Phi(x),\partial\mathbb{R}^d_+)^2}. \tag{2.1}$$

Proof. By the definition of f, $(x', f(x')) \in \partial\Omega$, thus

$$\operatorname{dist}(x,\partial\Omega) \le |x - (x', f(x'))| = |x_d - f(x')| = \operatorname{dist}(\Phi(x),\partial\mathbb{R}^d_+),$$

which implies the lower bound in (2.1).

To prove the upper bound, let $z = (z', f(z')) \in \partial\Omega$ be such that

$$\operatorname{dist}(x,\partial\Omega) = |x - z|.$$

Since $\partial\Omega$ is parametrized by f in the larger ball $B_{2\ell}(x_0)$, it is clear that such a point exists and that $z \in B_{2\ell}(x_0)$. The point z might not be uniquely determined but that will not play any role in what follows.

We begin by rewriting the expression we want to bound in terms of z:

$$
\begin{aligned}
\frac{1}{\operatorname{dist}(x,\partial\Omega)^2} &- \frac{1}{\operatorname{dist}(\Phi(x),\partial\mathbb{R}^d_+)^2} \\
&= \frac{1}{|x-z|^2} - \frac{1}{|x_d - f(x')|^2} \\
&= \frac{(f(x')-f(z'))(f(x')+f(z')-2x_d) - |x'-z'|^2}{|x-z|^2|x_d-f(x')|^2}.
\end{aligned}
$$

Since f is C^1 and by the definition of z, it holds that

$$
x = z + |x-z|\frac{(-\nabla f(z'),1)}{\sqrt{1+|\nabla f(z')|^2}}.
$$

Consequently,

$$
|x'-z'|^2 = |x-z|^2\frac{|\nabla f(z')|^2}{1+|\nabla f(z')|^2}, \quad |x_d - f(z')|^2 = \frac{|x-z|^2}{1+|\nabla f(z')|^2}. \tag{2.2}
$$

Note also that $f(x') \le f(z') \le x_d$. From the above identities one finds

$$
\begin{aligned}
\frac{1}{\operatorname{dist}(x,\partial\Omega)^2} &- \frac{1}{\operatorname{dist}(\Phi(x),\partial\mathbb{R}^d_+)^2} \\
&= \frac{1}{|x_d-f(x')|^2}\Bigg[\frac{|f(x')-f(z')|^2}{|x-z|^2} + 2\frac{|f(x')-f(z')|}{|x-z|\sqrt{1+|\nabla f(z')|^2}} \\
&\qquad\qquad - \frac{|\nabla f(z')|^2}{1+|\nabla f(z')|^2}\Bigg].
\end{aligned} \tag{2.3}
$$

By the fundamental theorem of calculus and (2.2),

$$
|f(x')-f(z')| = \left|(x'-z')\int_0^1 \nabla f(tx'+(1-t)z')\,dt\right| \le \omega(2\ell)^2|x-z|.
$$

Therefore,

$$
\frac{|f(x')-f(z')|^2}{|x-z|^2} + 2\frac{|f(x')-f(z')|}{|x-z|\sqrt{1+|\nabla f(z')|^2}} - \frac{|\nabla f(z')|^2}{1+|\nabla f(z')|^2} \le C\omega(2\ell)^2.
$$

Combined with (2.3) this completes the proof of Lemma 2.2. ■

2.2 A local Hardy–Lieb–Thirring inequality

The aim of this subsection is to prove a bound for localized traces of our operator. Before stating the result we recall the following Hardy inequality due to Davies [3] (obtained by combining his Theorems 2.3 and 2.4).

Lemma 2.3. *Let $\Omega \subset \mathbb{R}^d$ be open and bounded with C^1-boundary. Then for any $\varepsilon > 0$ there is a $c_H(\varepsilon, \Omega) \geq 0$ such that for all $u \in H_0^1(\Omega)$,*

$$\int_\Omega |\nabla u(x)|^2\,dx + \Big(\varepsilon - \frac{1}{4}\Big)\int_\Omega \frac{|u(x)|^2}{\operatorname{dist}(x,\partial\Omega)^2}\,dx \geq -c_H(\varepsilon,\Omega)\int_\Omega |u(x)|^2\,dx.$$

Remark. Lemma 2.3 can be proved in a direct manner by using a partition of unity and appealing to Lemmas 2.1 and 2.2. In particular, this allows one to quantify the best constant c_H in terms of the C^1-regularity of $\partial\Omega$. Indeed, such a proof yields the bound

$$c_H(\varepsilon,\Omega) \leq \frac{C}{\omega^{-1}(\varepsilon)^2}$$

for a constant C depending only on the dimension and ω^{-1} is the inverse of the C^1-modulus of continuity of $\partial\Omega$.

With Lemma 2.3 in hand we move on to the main result of this subsection. Specifically, the following local Hardy–Lieb–Thirring-type inequality for $H_{\Omega,b,V}$ (cf. [9]):

Lemma 2.4. *Let Ω, b, V be as in Theorem 1.1. Let $\phi \in C_0^1(\mathbb{R}^d)$ be supported in a ball $\overline{B}$ of radius ℓ and set $\underline{b} = \inf_{\Omega\cap B} b$. If $0 < h \leq K \min\{\ell, c_H(\frac{1}{2}\underline{b}^2, \Omega)^{-\frac{1}{2}}\}$, then*

$$\operatorname{Tr}(\phi H_{\Omega,b,V}(h)\phi)_- \leq C \min\{\underline{b}, 1\}^{-d}\ell^d h^{-d}\Big(1 + h^2\|V_-\|_{L^{1+\frac{d}{2}}(\Omega\cap B)}^{1+\frac{d}{2}}\Big),$$

where the constant C depends only on d, K, and $\|\phi\|_{L^\infty}$.

Proof. By assumption, $\underline{b} > 0$. By the variational principle and for any $\delta \in (0, \frac{1}{2}]$, we find

$$\begin{aligned}\phi H_{\Omega,b,V}(h)\phi \geq \phi\Big(&-h^2\delta\Delta - h^2V_-(x) - 1\\ &+ h^2(1-\delta)\Big(-\Delta + (1-\delta)^{-1}\Big(\underline{b}^2 - \frac{1}{4}\Big)\frac{1}{\operatorname{dist}(x,\partial\Omega)^2}\Big)\Big)\phi.\end{aligned}$$

Since $\delta \in (0, \frac{1}{2}]$, we have

$$(1-\delta)^{-1}\Big(\underline{b}^2 - \frac{1}{4}\Big) \geq (1+2\delta)\Big(\underline{b}^2 - \frac{1}{4}\Big) > \underline{b}^2 - \frac{\delta}{2} - \frac{1}{4}.$$

Thus, setting $\delta = \min\{\underline{b}^2, \frac{1}{2}\} \leq \frac{1}{2}$, Lemma 2.3 implies with $c_0 = c_H(\frac{1}{2}\underline{b}^2, \Omega)$ that

$$\phi H_{\Omega,b,V}(h)\phi \geq \phi(-h^2\delta\Delta - c_0h^2 - h^2V_-(x) - 1)\phi. \tag{2.4}$$

Consequently, for any $0 < \rho < 1$, the variational principle and (2.4) yields

$$\begin{aligned}\operatorname{Tr}(\phi H_{\Omega,b,V}(h)\phi)_- \leq{}& \operatorname{Tr}(\phi(-h^2\delta(1-\rho)\Delta - c_0h^2 - 1)\phi)_-\\ &+ \operatorname{Tr}(\phi(-h^2\delta\rho\Delta - h^2V_-)\phi)_-.\end{aligned}$$

Using the Berezin–Li–Yau inequality

$$\mathrm{Tr}(\phi(-h^2\delta(1-\rho)\Delta - c_0h^2 - 1)\phi)_- \le C(1+c_0h^2)^{1+\frac{d}{2}}(1-\rho)^{-\frac{d}{2}}\delta^{-\frac{d}{2}}h^{-d}\ell^d,$$

with $C > 0$ depending on d and $\|\phi\|_{L^\infty}$. For the remaining term the Lieb–Thirring inequality implies

$$\mathrm{Tr}(\phi(-h^2\delta\rho\Delta - h^2V_-)\phi)_- \le Ch^2\delta^{-\frac{d}{2}}\rho^{-\frac{d}{2}}\|V_-\|^{1+\frac{d}{2}}_{L^{1+\frac{d}{2}}(\Omega\cap B_\ell)},$$

for some constant $C > 0$ depending only on d. Gathering the estimates and setting $\rho = \frac{h^2}{2K^2\ell^2} < 1$ completes the proof. ■

3 Local asymptotics

3.1 Local asymptotics in the bulk

Lemma 3.1. *Let $\phi \in C_0^1(\mathbb{R}^d)$ be supported in a ball $\overline{B}$ of radius $\ell > 0$ and satisfy*

$$\|\nabla\phi\|_{L^\infty(\mathbb{R}^d)} \le M\ell^{-1}. \tag{3.1}$$

If $V \in L^1(B)$ is such that $V_- = V_0 + V_1$ with $0 \le V_0 \in L^\infty(B)$ and $V_1 \in L^{1+\frac{d}{2}}(B)$, then, for $0 < h \le K\min\{\ell, \|V_0\|_\infty^{-1/2}\}$,

$$\Big|\mathrm{Tr}(\phi(-h^2\Delta + h^2V - 1)\phi)_- - L_dh^{-d}\int_B \phi^2(x)\,dx\Big|$$
$$\le Ch^{-d+2}\Big[\ell^{d-2} + \ell^d\|V_0\|_{L^\infty(B)} + \ell^d\|V_1\|^{1+\frac{d}{2}}_{L^{1+\frac{d}{2}}(B)} + \|V_+\|_{L^1(B)}\Big],$$

where the constant C depends only on d, M, K.

Proof. Throughout the proof, we set $H_V = H_{\mathbb{R}^d,0,V} = -h^2\Delta + h^2V - 1$ in $L^2(\mathbb{R}^d)$.

To prove the lower bound, consider the operator γ with integral kernel

$$\gamma(x,y) = \frac{1}{(2\pi)^d}\chi(x)\int_{|\xi|<h^{-1}} e^{i\xi(x-y)}\,d\xi\,\chi(y),$$

where $\chi \in C_0^\infty(\mathbb{R}^d)$ with $0 \le \chi \le 1$ and $\chi \equiv 1$ on B. The operator γ is trace class and satisfies $0 \le \gamma \le \mathbf{1}$. Therefore, the variational principle implies that

$$\begin{aligned}
\mathrm{Tr}(\phi H_V\phi)_- &\ge \mathrm{Tr}(\phi H_{V_+}\phi)_- \\
&\ge -\mathrm{Tr}(\gamma\phi H_{V_+}\phi) \\
&= -\frac{1}{(2\pi)^d}\int_{|\xi|<h^{-1}} \big(h^2\|\nabla e^{i\xi\cdot}\phi\|^2_{L^2(\mathbb{R}^d)} \\
&\qquad\qquad + h^2\|V_+\phi^2\|_{L^1(\mathbb{R}^d)} - \|\phi\|^2_{L^2(\mathbb{R}^d)}\big)\,d\xi \\
&= L_dh^{-d}\int_B \phi^2(x)\,dx - Ch^{-d+2}\big(\|\nabla\phi\|^2_{L^2(\mathbb{R}^d)} + \|V_+\phi^2\|_{L^1(\mathbb{R}^d)}\big).
\end{aligned}$$

Since, by (3.1), $\|\phi\|_{L^\infty} \leq M$ and $\|\nabla\phi\|^2_{L^2(\mathbb{R}^d)} \leq C\ell^{d-2}$, this proves the lower bound.

It remains to prove the upper bound. For any $0 < \rho \leq \frac{1}{2}$,

$$\begin{aligned}\operatorname{Tr}(\phi H_V \phi)_- &\leq \operatorname{Tr}(\phi H_{V_-}\phi)_- \\ &\leq \operatorname{Tr}(\phi(-h^2(1-\rho)\Delta - h^2 V_0 - 1)\phi)_- + h^2 \operatorname{Tr}(\phi(-\rho\Delta - V_1)\phi)_-.\end{aligned}$$

To bound the second term, we apply the Lieb–Thirring inequality to conclude that

$$\begin{aligned}h^2 \operatorname{Tr}(\phi(-\rho\Delta - V_1)\phi)_- &\leq h^2 \operatorname{Tr}(\phi(-\rho\Delta - V_1 1_B)_-\phi) \\ &\leq Ch^2\rho^{-\frac{d}{2}} \int_B |V_1(x)|^{1+\frac{d}{2}}\,dx,\end{aligned}$$

where we again used $\|\phi\|_{L^\infty} \leq M$. Since $V_0 \in L^\infty(B)$, we can bound

$$\begin{aligned}\operatorname{Tr}(\phi(-h^2(1-\rho)\Delta - h^2 V_0 - 1)\phi)_- &\leq \operatorname{Tr}(\phi(-h^2(1-\rho)\Delta - h^2 \sup_B V_0 - 1)\phi)_- \\ &= (1 + h^2 \sup_B V_0) \operatorname{Tr}(\phi(-\tilde{h}^2\Delta - 1)\phi)_-\end{aligned}$$

with $\tilde{h} = h(1-\rho)^{\frac{1}{2}}(1 + h^2 \sup_B V_0)^{-\frac{1}{2}}$. By the Berezin–Li–Yau inequality,

$$\operatorname{Tr}(\phi(-\tilde{h}^2\Delta - 1)\phi)_- \leq L_d \tilde{h}^{-d} \int_B \phi^2(x)\,dx.$$

Combining the above we have arrived at

$$\begin{aligned}\operatorname{Tr}(\phi H_V \phi)_- &\leq L_d h^{-d} \int_B \phi^2(x)\,dx + Ch^2\rho^{-\frac{d}{2}} \int_B |V_1(x)|^{1+\frac{d}{2}}\,dx \\ &\quad + L_d h^{-d}\Big[(1-\rho)^{-\frac{d}{2}}\Big(1 + h^2 \sup_B V_0\Big)^{1+\frac{d}{2}} - 1\Big] \int_B \phi^2(x)\,dx \\ &\leq L_d h^{-d} \int_B \phi^2(x)\,dx + Ch^2\rho^{-\frac{d}{2}} \int_B |V_1(x)|^{1+\frac{d}{2}}\,dx \\ &\quad + Ch^{-d}\Big[\rho + h^2 \sup_B V_0\Big] \int_B \phi^2(x)\,dx,\end{aligned}$$

where C depend only on d, K, M. Setting $\rho = \frac{h^2}{2K^2\ell^2} \leq \frac{1}{2}$ and using $\int \phi^2 \leq C\ell^d$ completes the proof. ∎

3.2 Local asymptotics near the boundary

In this subsection we prove the following local asymptotic expansion close to the boundary.

Theorem 3.2. *Let Ω, b, V be as in Theorem* 1.1. *Let $\phi \in C_0^1(\mathbb{R}^d)$ be supported in a ball $\overline{B}$ of radius ℓ and satisfy*

$$\|\nabla\phi\|_{L^\infty(\mathbb{R}^d)} \leq M\ell^{-1}.$$

Assume that $\mathrm{dist}(B, \partial\Omega) \le 2\ell$, *and set* $\underline{b} = \inf_{B\cap\Omega} b$. *Then, for* $0 < \ell \le c(\Omega, \underline{b})$ *and* $0 < h \le K\ell$,

$$\begin{aligned}
&\Big| \mathrm{Tr}(\phi H_{\Omega,b,V}(h)\phi)_- - L_d h^{-d} \int_\Omega \phi^2(x)\,dx \\
&\qquad + \frac{L_{d-1}}{2} h^{-d+1} \int_{\partial\Omega} \phi^2(x) b(x)\, d\mathcal{H}^{d-1}(x) \Big| \\
&\le \ell^d h^{-d} o_{\ell\to 0^+}(1) \\
&\qquad + O(h^{-d+1}) \int_{\partial\Omega} \phi^2(x) \Big[\sup_{y\in B_{2\ell}(x)} b(y) - \inf_{y\in B_{2\ell}(x)} b(y) \Big]\, d\mathcal{H}^{d-1}(x) \\
&\qquad + O(h^{-d+2}) \Big(\ell^{d-2} \log\Big(\frac{\ell}{h}\Big) + \|V_+\|_{L^1(\Omega\cap B)} + \ell^d \|V_-\|^{1+\frac{d}{2}}_{L^{1+\frac{d}{2}}(\Omega\cap B)} \Big).
\end{aligned}$$

Moreover, the error terms and the implicit constants can be quantified in terms of the C^1*-regularity of* $\partial\Omega$ *and* $M, K, \|b\|_{L^\infty(\Omega\cap B)}, \underline{b}$.

The proof of Theorem 3.2 will be split into several lemmas, the first of which reduces our problem to the corresponding one in a half-space.

Lemma 3.3. *Let* Ω, b, V *be as in Theorem* 1.1. *Let* $\phi \in C_0^1(\mathbb{R}^d)$ *be supported in a ball* $\overline{B}$ *of radius* ℓ *such that* $\mathrm{dist}(B, \partial\Omega) \le 2\ell$, *and* $\inf_{B\cap\Omega} b = \underline{b} > 0$. *For* $0 < \ell \le c(\Omega, \underline{b})$ *and* $0 < h \le K\ell$ *with* $\tilde\phi = \phi \circ \Phi^{-1}$, $\tilde V = V \circ \Phi^{-1}$,

$$\begin{aligned}
&\mathrm{Tr}(\tilde\phi H_{\mathbb{R}^d_+,\overline{b},\tilde V}(h)\tilde\phi)_- - \ell^d h^{-d} o_{\ell\to 0^+}(1)\Big(1 + h^2 \|V_-\|^{1+\frac{d}{2}}_{L^{1+\frac{d}{2}}(\Omega\cap B)}\Big) \\
&\le \mathrm{Tr}(\phi H_{\Omega,b,V}(h)\phi)_- \\
&\le \mathrm{Tr}(\tilde\phi H_{\mathbb{R}^d_+,\underline{b},\tilde V}(h)\tilde\phi)_- + \ell^d h^{-d} o_{\ell\to 0^+}(1)\Big(1 + h^2 \|V_-\|^{1+\frac{d}{2}}_{L^{1+\frac{d}{2}}(\Omega\cap B)}\Big),
\end{aligned}$$

where $\overline{b} = \sup_{x\in B\cap\Omega} b(x)$. *Moreover, the error terms and the implicit constants can be quantified in terms of the* C^1*-regularity of* $\partial\Omega$ *and* $K, \underline{b}, \overline{b}, \|\phi\|_{L^\infty}$.

Proof. Provided ℓ is small enough there is a ball $B' \supset B$ with center on $\partial\Omega$ and radius 4ℓ that satisfies the assumptions in Section 2.1. Let Φ be the associated diffeomorphism. We split the proof into two parts, in the first part we prove the upper bound and in the second we prove the lower bound.

Part 1: Proof of the upper bound. By the variational principle

$$\mathrm{Tr}(\phi H_{\Omega,b,V}(h)\phi)_- \le \mathrm{Tr}(\phi H_{\Omega,\underline{b},V}(h)\phi)_-.$$

Moreover, by Lemma 2.1 there is $C_0 > 0$, depending only on d, such that

$$\begin{aligned}
\mathrm{Tr}(\phi H_{\Omega,\underline{b},V}(h)\phi)_- \le \mathrm{Tr}\Big(\tilde\phi\Big(&-h^2(1 - C_0\omega(4\ell))\Delta_{\mathbb{R}^d_+} \\
&+ h^2 \frac{\underline{b}^2 - \frac{1}{4}}{\mathrm{dist}(\Phi^{-1}(\cdot), \partial\Omega)^2} + h^2\tilde V - 1\Big)\tilde\phi\Big)_-.
\end{aligned}$$

We claim that

$$\begin{aligned}&\operatorname{Tr}\Big(\tilde\phi\Big(-h^2(1-C_0\omega(4\ell))\Delta_{\mathbb{R}^d_+}+h^2\frac{\underline{b}^2-\frac14}{\operatorname{dist}(\Phi^{-1}(\cdot),\partial\Omega)^2}+h^2\tilde V-1\Big)\tilde\phi\Big)_-\\&\le\operatorname{Tr}\Big(\tilde\phi\Big(-h^2(1-C_0\omega(4\ell))\Delta_{\mathbb{R}^d_+}+h^2\frac{\underline{b}^2-\frac14-C\omega(8\ell)^2}{\operatorname{dist}(\cdot,\partial\mathbb{R}^d_+)^2}+h^2\tilde V-1\Big)\tilde\phi\Big)_-\end{aligned}$$

for a constant C depending only on d. Indeed, if $\underline{b}\ge\frac12$ Lemma 2.2 and the variational principle implies

$$\begin{aligned}&\operatorname{Tr}\Big(\tilde\phi\Big(-h^2(1-C_0\omega(4\ell))\Delta_{\mathbb{R}^d_+}+h^2\frac{\underline{b}^2-\frac14}{\operatorname{dist}(\Phi^{-1}(\cdot),\partial\Omega)^2}+h^2\tilde V-1\Big)\tilde\phi\Big)_-\\&\le\operatorname{Tr}\Big(\tilde\phi\Big(-h^2(1-C_0\omega(4\ell))\Delta_{\mathbb{R}^d_+}+h^2\frac{\underline{b}^2-\frac14}{\operatorname{dist}(\cdot,\partial\mathbb{R}^d_+)^2}+h^2\tilde V-1\Big)\tilde\phi\Big)_-\\&\le\operatorname{Tr}\Big(\tilde\phi\Big(-h^2(1-C_0\omega(4\ell))\Delta_{\mathbb{R}^d_+}+h^2\frac{\underline{b}^2-\frac14-C\omega(8\ell)^2}{\operatorname{dist}(\cdot,\partial\mathbb{R}^d_+)^2}+h^2\tilde V-1\Big)\tilde\phi\Big)_-.\end{aligned}$$

Similarly, if $0<\underline{b}<\frac12$, Lemma 2.2 and the variational principle implies

$$\begin{aligned}&\operatorname{Tr}\Big(\tilde\phi\Big(-h^2(1-C_0\omega(4\ell))\Delta_{\mathbb{R}^d_+}+h^2\frac{\underline{b}^2-\frac14}{\operatorname{dist}(\Phi^{-1}(\cdot),\partial\Omega)^2}+h^2\tilde V-1\Big)\tilde\phi\Big)_-\\&\le\operatorname{Tr}\Big(\tilde\phi\Big(-h^2(1-C_0\omega(4\ell))\Delta_{\mathbb{R}^d_+}+h^2\frac{(\underline{b}^2-\frac14)(1+C\omega(8\ell)^2)}{\operatorname{dist}(\cdot,\partial\mathbb{R}^d_+)^2}+h^2\tilde V-1\Big)\tilde\phi\Big)_-\\&\le\operatorname{Tr}\Big(\tilde\phi\Big(-h^2(1-C_0\omega(4\ell))\Delta_{\mathbb{R}^d_+}+h^2\frac{\underline{b}^2-\frac14-C\omega(8\ell)^2}{\operatorname{dist}(\cdot,\partial\mathbb{R}^d_+)^2}+h^2\tilde V-1\Big)\tilde\phi\Big)_-.\end{aligned}$$

For any $2C_0\omega(4\ell)<\rho\le\frac12$ we estimate

$$\begin{aligned}&\operatorname{Tr}\Big(\tilde\phi\Big(-h^2(1-C_0\omega(4\ell))\Delta_{\mathbb{R}^d_+}+h^2\frac{\underline{b}^2-\frac14-C\omega(8\ell)^2}{\operatorname{dist}(\cdot,\partial\mathbb{R}^d_+)^2}+h^2\tilde V-1\Big)\tilde\phi\Big)_-\\&\le\operatorname{Tr}(\tilde\phi H_{\mathbb{R}^d_+,\underline{b},\tilde V}(h)\tilde\phi)_-\\&\qquad+\operatorname{Tr}\Big(\tilde\phi\Big(-h^2(\rho-C_0\omega(4\ell))\Delta_{\mathbb{R}^d_+}\\&\qquad\qquad+h^2\frac{\rho(\underline{b}^2-\frac14)-C\omega(8\ell)^2}{\operatorname{dist}(\cdot,\partial\mathbb{R}^d_+)^2}+h^2\rho\tilde V-\rho\Big)\tilde\phi\Big)_-.\end{aligned}$$

Provided

$$\begin{aligned}\frac{\rho(\underline{b}^2-\frac14)-C\omega(8\ell)^2}{\rho-C_0\omega(4\ell)}&=\Big(\underline{b}^2-\frac14\Big)\frac{1}{1-C_0\omega(4\ell)\rho^{-1}}-C\frac{\omega(8\ell)^2}{\rho-C_0\omega(4\ell)}\\&>-\frac14,\end{aligned}\tag{3.2}$$

we can apply the local Hardy–Lieb–Thirring inequality of Lemma 2.4 in $\mathbb{R}^d_+$ to bound

$$\begin{aligned}
&\operatorname{Tr}\Big(\tilde\phi\Big(-h^2(\rho - C_0\omega(4\ell))\Delta_{\mathbb{R}^d_+} + h^2\frac{\rho(\underline{b}^2-\frac14) - C\omega(8\ell)^2}{\operatorname{dist}(\,\cdot\,,\partial\mathbb{R}^d_+)^2} + h^2\rho\tilde V - \rho\Big)\tilde\phi\Big)_-\\
&\leq C\rho^{1+\frac d2}\ell^d h^{-d}(\rho - C_0\omega(4\ell))^{-\frac d2}\Big(1 + h^2\rho^{\frac d2}(\rho - C_0\omega(4\ell))^{-\frac d2}\|V_-\|^{1+\frac d2}_{L^{1+\frac d2}(\Omega\cap B)}\Big)\\
&\leq C\rho\ell^d h^{-d}\Big(1 + h^2\|V_-\|^{1+\frac d2}_{L^{1+\frac d2}(\Omega\cap B)}\Big).
\end{aligned}$$

Set $\rho = \sqrt{\omega(4\ell)} + \omega(8\ell)$. Then $\rho > 2C_0\omega(4\ell)$ and (3.2) are valid provided ℓ is small enough. Therefore, upon collecting the estimates above we arrive at the bound

$$\begin{aligned}
\operatorname{Tr}(\phi H_{\Omega,b,V}(h)\phi)_- &\leq \operatorname{Tr}(\tilde\phi H_{\mathbb{R}^d_+,\underline{b},\tilde V}(h)\tilde\phi)_-\\
&\quad + C\ell^d h^{-d}\Big(\sqrt{\omega(4\ell)} + \omega(8\ell)\Big)\Big(1 + h^2\|V_-\|^{1+\frac d2}_{L^{1+\frac d2}(\Omega\cap B)}\Big),
\end{aligned}$$

thus completing the proof of the upper bound.

Part 2: Proof of the lower bound. The proof of the lower bound proceeds as the upper bound but with the roles of Ω and $\mathbb{R}^d_+$ exchanged. By Lemma 2.1,

$$\begin{aligned}
\operatorname{Tr}(\tilde\phi H_{\mathbb{R}^d_+,\overline{b},\tilde V}(h)\tilde\phi)_- \leq \operatorname{Tr}\Big(\phi\Big(&-h^2(1 + C_0\omega(4\ell))^{-1}\Delta_\Omega\\
&+ h^2\frac{\overline{b}^2 - \frac14}{\operatorname{dist}(\Phi(\,\cdot\,),\partial\mathbb{R}^d_+)^2} + h^2V - 1\Big)\phi\Big)_-.
\end{aligned}$$

If ℓ is sufficiently small so that $C_0\omega(4\ell) \leq \frac12$, then

$$(1 + C_0\omega(4\ell))^{-1} \geq 1 - C_0\omega(4\ell) > 0,$$

and hence

$$\begin{aligned}
\operatorname{Tr}(\tilde\phi H_{\mathbb{R}^d_+,\overline{b},\tilde V}(h)\tilde\phi)_- \leq \operatorname{Tr}\Big(\phi\Big(&-h^2(1 - C_0\omega(4\ell))\Delta_\Omega\\
&+ h^2\frac{\overline{b}^2 - \frac14}{\operatorname{dist}(\Phi(\,\cdot\,),\partial\mathbb{R}^d_+)^2} + h^2V - 1\Big)\phi\Big)_-.
\end{aligned}$$

By splitting into cases depending on the sign of $\overline{b}^2 - \frac14$ as in the proof of the upper bound one finds

$$\begin{aligned}
&\operatorname{Tr}\Big(\phi\Big(-h^2(1 - C_0\omega(4\ell))\Delta_\Omega + h^2\frac{\overline{b}^2 - \frac14}{\operatorname{dist}(\Phi(\,\cdot\,),\partial\mathbb{R}^d_+)^2} + h^2V - 1\Big)\phi\Big)_-\\
&\quad\leq \operatorname{Tr}\Big(\phi\Big(-h^2(1 - C_0\omega(4\ell))\Delta_\Omega + h^2\frac{\overline{b}^2 - \frac14 - C\omega(8\ell)^2}{\operatorname{dist}(\,\cdot\,,\partial\Omega)^2} + h^2V - 1\Big)\phi\Big)_-
\end{aligned}$$

for a constant C depending on $d, \overline{b}$.

For any $2C_0\omega(4\ell) < \rho \le \frac{1}{2}$ we estimate

$$\begin{aligned}
&\operatorname{Tr}\Big(\phi\Big(-h^2(1-C_0\omega(4\ell))\Delta_\Omega + h^2\frac{\overline{b}^{-2}-\frac{1}{4}-C\omega(8\ell)^2}{\operatorname{dist}(\cdot,\partial\Omega)^2} + h^2V - 1\Big)\phi\Big)_- \\
&\le \operatorname{Tr}(\phi H_{\Omega,\overline{b},V}(h)\phi)_- + \operatorname{Tr}\Big(\phi\Big(-h^2(\rho - C_0\omega(4\ell))\Delta_\Omega \\
&\qquad + h^2\frac{\rho(\overline{b}^{-2}-\frac{1}{4})-C\omega(8\ell)^2}{\operatorname{dist}(\cdot,\partial\Omega)^2} + h^2\rho V - \rho\Big)\phi\Big)_- \\
&\le \operatorname{Tr}(\phi H_{\Omega,b,V}(h)\phi)_- + \operatorname{Tr}\Big(\phi\Big(-h^2(\rho - C_0\omega(4\ell))\Delta_\Omega \\
&\qquad + h^2\frac{\rho(\overline{b}^{-2}-\frac{1}{4})-C\omega(8\ell)^2}{\operatorname{dist}(\cdot,\partial\Omega)^2} + h^2\rho V - \rho\Big)\phi\Big)_-.
\end{aligned}$$

Provided the analogue of (3.2) with $\overline{b}$ instead of $\underline{b}$ holds we can apply the local Hardy–Lieb–Thirring inequality of Lemma 2.4 to bound

$$\begin{aligned}
&\operatorname{Tr}\Big(\phi\Big(-h^2(\rho - C_0\omega(4\ell))\Delta_\Omega + h^2\frac{\rho(\overline{b}^{-2}-\frac{1}{4})-C\omega(8\ell)^2}{\operatorname{dist}(\cdot,\partial\Omega)^2} + h^2\rho V - \rho\Big)\phi\Big)_- \\
&\le C\rho\ell^d h^{-d}\Big(1 + h^2\|V_-\|^{1+\frac{d}{2}}_{L^{1+\frac{d}{2}}(\Omega\cap B)}\Big).
\end{aligned}$$

Again we can set $\rho = \sqrt{\omega(4\ell)} + \omega(8\ell)$ and combine the above estimates to arrive at

$$\begin{aligned}
&\operatorname{Tr}(\tilde{\phi} H_{\mathbb{R}^d_+,\overline{b},\tilde{V}}(h)\tilde{\phi})_- \le \operatorname{Tr}(\phi H_{\Omega,b,V}(h)\phi)_- \\
&\qquad + C\ell^d h^{-d}\Big(\sqrt{\omega(4\ell)} + \omega(8\ell)\Big)\Big(1 + h^2\|V_-\|^{1+\frac{d}{2}}_{L^{1+\frac{d}{2}}(\Omega\cap B)}\Big).
\end{aligned}$$

This completes the proof of the lower bound and hence the proof of Lemma 3.3. ∎

The proof of Theorem 3.2 has been reduced to understanding the asymptotics of $\operatorname{Tr}(\phi H_{\mathbb{R}^d_+,b,V}(h)\phi)_-$ with $b(x) \equiv b > 0$.

Lemma 3.4. *Let Ω, V be as in Theorem 1.1. Let $\phi \in C_0^1(\mathbb{R}^d)$ be supported in a ball $\overline{B}$ of radius ℓ and satisfy*

$$\|\nabla\phi\|_{L^\infty} \le M\ell^{-1}. \tag{3.3}$$

With $b(x) \equiv b > 0$ we have, for $0 < h \le K\ell$,

$$\begin{aligned}
&\Big|\operatorname{Tr}(\phi H_{\mathbb{R}^d_+,b,V}(h)\phi)_- - L_d h^{-d}\int_{\mathbb{R}^d_+}\phi^2(y)\,dy \\
&\qquad + \frac{b\,L_{d-1}}{2}h^{-d+1}\int_{\partial\mathbb{R}^d_+}\phi^2(y)\,d\mathcal{H}^{d-1}(y)\Big| \\
&\le Ch^{-d+2}\Big(\ell^{d-2}\Big|\log\Big(\frac{\ell}{h}\Big)\Big| + \|V_+\|_{L^1(\mathbb{R}^d_+\cap B)} + \ell^d\|V_-\|^{1+\frac{d}{2}}_{L^{1+\frac{d}{2}}(\mathbb{R}^d_+\cap B)}\Big),
\end{aligned}$$

where C depends only on d, M, K, b and can be uniformly bounded for b in compact subsets of $[0, \infty)$.

Proof. Our proof proceeds by diagonalizing the operator $H_{\mathbb{R}^d_+,b,0}(h)$. For the general background on what follows, see [4, Chapter XIII].

For $f \in C^2(\mathbb{R}_+)$ define the differential expression

$$L_b f(x) = f''(x) - \left(b^2 - \frac{1}{4}\right)\frac{f(x)}{x^2}.$$

The operator $H_{\mathbb{R}^d_+,b,0}(h)$ can then be decomposed as

$$H_{\mathbb{R}^d_+,b,0}(h) = -h^2\Delta' - h^2 L_b,$$

where

$$\Delta' = \sum_{j=1}^{d-1} \frac{\partial^2}{\partial y_j^2}$$

and L_b acts in the y_d-coordinate.

For $b > 0, \mu \geq 0$ the ODE

$$-L_b u(x) = \mu u(x)$$

has two linearly independent solutions

$$\psi_{b,\mu}(x) = x^{\frac{1}{2}} J_b(x\sqrt{\mu})$$

and

$$\eta_{b,\mu}(x) = x^{\frac{1}{2}} Y_b(x\sqrt{\mu}).$$

If $b \geq \frac{1}{2}$, only ψ vanishes at $x = 0$ while for $b \in (0, \frac{1}{2})$ both solutions vanish, indeed

$$\psi \sim x^{\frac{1}{2}+b} \quad \text{and} \quad \eta \sim x^{\frac{1}{2}-b} \quad \text{as } x \to 0^+.$$

However, for any $b \neq \frac{1}{2}$ only the first solution $\psi_{b,\nu}$ is in H^1 around zero. In particular, our effective operator $H_{\mathbb{R}^d_+,b,0}(h)$ is diagonalized through a Fourier transform with respect to y' and a Hankel transform $\mathfrak{H}_b$ with respect to y_d. Recall that the Hankel transform $\mathfrak{H}_\alpha : L^2(\mathbb{R}_+) \to L^2(\mathbb{R}_+)$ is initially defined by

$$\mathfrak{H}_\alpha(g)(s) = \int_0^\infty g(t) J_\alpha(st)\sqrt{st}\, dt \quad \text{for } g \in L^1(\mathbb{R}_+)$$

and extended to $L^2(\mathbb{R}_+)$ in a similar manner as the Fourier transform. Moreover, $\mathfrak{H}_\alpha$ is unitary, is its own inverse $\mathfrak{H}_\alpha^2 = 1$. Moreover, for $G \in L^\infty(\mathbb{R}_+)$ with compact support and $f \in H_0^1(\mathbb{R}_+) \cap H^2(\mathbb{R}_+)$

$$\langle f, G(-L_b) f\rangle_{L^2(\mathbb{R}_+)} = \int_0^\infty G(s^2)|\mathfrak{H}_b(f)(s)|^2\, ds.$$

By a similar argument as in the proof of Lemma 2.4 the upper bound can be reduced to the case $V \equiv 0$. Indeed, for any $0 < \rho \leq \frac{1}{2}$,

$$\begin{aligned}
&\mathrm{Tr}(\phi H_{\mathbb{R}^d_+,b,V}(h)\phi)_- \\
&\leq \mathrm{Tr}(\phi H_{\mathbb{R}^d_+,b,0}(h(1-\rho))\phi)_- + \mathrm{Tr}\Big(\phi\Big(h^2\rho\Delta_{\mathbb{R}^d_+} + h^2\rho\frac{b^2-\frac{1}{4}}{\mathrm{dist}(\cdot,\partial\mathbb{R}^d_+)^2} - h^2V\Big)\phi\Big)_- \\
&\leq \mathrm{Tr}(\phi H_{\mathbb{R}^d_+,b,0}(h(1-\rho))\phi)_- + Ch^2\rho^{-\frac{d}{2}}\|V_-\|^{1+\frac{d}{2}}_{L^{1+\frac{d}{2}}(\mathbb{R}^d_+\cap B)}.
\end{aligned}$$

Set $\rho = \frac{h^2}{2K^2\ell^2}$ so that

$$h^2\rho^{-\frac{d}{2}} = O(\ell^d h^{-d+2}) \quad \text{and} \quad (h(1-\rho))^{-\beta} = h^{-\beta}(1 + O(\ell^{-2}h^2)).$$

The claimed upper bound now follows from the case $V \equiv 0$.

Using the inequality $\mathrm{Tr}(\phi H\phi)_- \leq \mathrm{Tr}(\phi H_-\phi)$, applying the Fourier transform with respect to y' and the Hankel transform in the y_d-direction yields

$$\begin{aligned}
&\mathrm{Tr}(\phi H_{\mathbb{R}^d_+,b,0}(h)\phi)_- \\
&\leq \mathrm{Tr}(\phi(H_{\mathbb{R}^d_+,b,0}(h))_-\phi) \\
&= \frac{1}{(2\pi)^{d-1}}\iint_{\mathbb{R}^d_+\times\mathbb{R}^d_+} \phi^2(y)(h^2|\xi|^2-1)_-\xi_d y_d J_b(\xi_d y_d)^2\,d\xi\,dy.
\end{aligned} \tag{3.4}$$

For the lower bound define the operator γ with integral kernel

$$\begin{aligned}
\gamma(x,y) = \frac{1}{(2\pi)^{d-1}}\chi(x)\int_{\mathbb{R}^d_+\cap B_{h^{-1}}(0)} &e^{i\xi'(x'-y')}\sqrt{\xi_d x_d}\,J_b(\xi_d x_d) \\
&\times \sqrt{\xi_d y_d}\,J_b(\xi_d y_d)\,d\xi\,\chi(y),
\end{aligned}$$

where $\chi \in C_0^\infty(\mathbb{R}^d)$ is such that $0 \leq \chi \leq 1$ and $\chi \equiv 1$ on $\operatorname{supp}\phi$. The operator γ is trace class, satisfies $0 \leq \gamma \leq \mathbf{1}$, and its range is contained in the domain of $H_{\mathbb{R}^d_+,b,V}$. Thus, by the variational principle,

$$\begin{aligned}
&-\mathrm{Tr}(\phi H_{\mathbb{R}^d_+,b,V}(h)\phi)_- \\
&\leq \mathrm{Tr}(\gamma\phi H_{\mathbb{R}^d_+,b,V_+}(h)\phi) \\
&= \frac{1}{(2\pi)^{d-1}}\iint_{\mathbb{R}^d_+\times\mathbb{R}^d_+}(h^2|\xi|^2-1)_-\phi^2(x)\xi_d x_d J_b(\xi_d x_d)^2\,d\xi\,dx \\
&\quad + h^{-d+2}\int_{\mathbb{R}^d_+}(V_+(x)\phi^2(x) + |\nabla\phi(x)|^2)\int_0^1(x_d t h^{-1})J_b(x_d t h^{-1})^2\,dt\,dx \\
&\leq \frac{1}{(2\pi)^{d-1}}\iint_{\mathbb{R}^d_+\times\mathbb{R}^d_+}(h^2|\xi|^2-1)_-\phi^2(x)\xi_d x_d J_b(\xi_d x_d)^2\,d\xi\,dx \\
&\quad + Ch^{-d+2}\int_{\mathbb{R}^d_+}(V_+(x)\phi^2(x) + |\nabla\phi(x)|^2)\,dx,
\end{aligned} \tag{3.5}$$

with constant C uniformly bounded for b in compact subsets of $[0,\infty)$, since we have $\|\sqrt{\cdot}\, J_b\|_{L^\infty(\mathbb{R}_+)} < \infty$ uniformly for b in compact subsets of $[0,\infty)$ (see [12, Chapter 7]). By (3.3) we can estimate

$$\|\phi\|_{L^\infty} \le M \quad \text{and} \quad \int_{\mathbb{R}^d_+} |\nabla\phi(x)|^2\,dx \le CM^2\ell^{d-2}.$$

What remains is to understand the common integral in (3.4) and (3.5). We begin by extracting the desired leading term:

$$\begin{aligned}
\frac{1}{(2\pi)^{d-1}} &\iint_{\mathbb{R}^d_+\times\mathbb{R}^d_+} \phi^2(y)(h^2|\xi|^2-1)_- \xi_d y_d J_b(\xi_d y_d)^2\,d\xi\,dy \\
&= L_d h^{-d}\int_{\mathbb{R}^d_+}\phi^2(y)\,dy \\
&\quad - L_{d-1}h^{-d+1}\int_0^\infty\int_{\mathbb{R}^{d-1}}\phi^2(y',ht)\,dy' \\
&\qquad \times \int_0^1 (1-\xi_d^2)^{\frac{d+1}{2}}\Big(\frac{1}{\pi}-\xi_d t J_b(\xi_d t)^2\Big)\,d\xi_d\,dt.
\end{aligned} \tag{3.6}$$

Define, for $b \ge 0$ and $t \ge 0$,

$$P_b(t) = \int_0^1 (1-\zeta^2)^{\frac{d+1}{2}}\Big(\frac{1}{\pi}-\zeta t J_b(\zeta t)^2\Big)\,d\zeta.$$

In Lemmas A.1 and A.2 we shall prove that

$$\int_0^\infty P_b(t)\,dt = \frac{b}{2} \quad \text{and} \quad P_b(t) = O(t^{-2}) \text{ as } t\to\infty, \tag{3.7}$$

with the implicit constant uniformly bounded for b in compact subsets of $[0,\infty)$.

Using (3.7), we can estimate

$$\begin{aligned}
\int_0^\infty &\int_{\mathbb{R}^{d-1}}\phi^2(y',ht)\,dy'\,P_b(t)\,dt \\
&= \int_0^{\frac{2\ell}{h}}\int_{\mathbb{R}^{d-1}}\phi^2(y',ht)\,dy'\,P_b(t)\,dt \\
&= \frac{b}{2}\int_{\mathbb{R}^{d-1}}\phi^2(y',0)\,dy' - \int_{\frac{2\ell}{h}}^\infty\int_{\mathbb{R}^{d-1}}\phi^2(y',0)\,dy'\,P_b(t)\,dt \\
&\qquad + 2\int_0^{\frac{2\ell}{h}} ht\int_{\mathbb{R}^{d-1}}\int_0^1 \phi(y',hts)\partial_{y_d}\phi(y',hts)\,ds\,dy'\,P_b(t)\,dt \\
&= \frac{b}{2}\int_{\mathbb{R}^{d-1}}\phi^2(y',0)\,dy' + O\Big(h\ell^{d-2}\Big|\log\Big(\frac{\ell}{h}\Big)\Big|\Big).
\end{aligned}$$

Combined with (3.6), (3.4), and (3.5) this completes the proof of Lemma 3.4. ∎

We are now ready to prove Theorem 3.2.

Proof of Theorem 3.2. By combining Lemma 3.3 and Lemma 3.4 the claimed estimate follows from

$$\int_{\partial\Omega} \phi^2(x)\Big[b(x) - \inf_{y\in\Omega\cap B} b(y)\Big]\, d\mathcal{H}^{d-1}(x) \\ \leq \int_{\partial\Omega} \phi^2(x)\Big[\sup_{y\in\Omega\cap B} b(y) - \inf_{y\in\Omega\cap B} b(y)\Big]\, d\mathcal{H}^{d-1}(x),$$

and the corresponding inequality for the sup and the fact that $\operatorname{supp}\phi \subseteq \overline{B} \subset \overline{B_{2\ell}(x)}$ for any $x \in \operatorname{supp}\phi$. ■

4 From local to global asymptotics

In this section we prove our main result by piecing together the local asymptotics obtained above. The key ingredient is the following construction of a continuum partition of unity due to Solovej and Spitzer [11].

Let

$$\ell(u) = \frac{1}{2}\max\{\operatorname{dist}(u, \Omega^c), 2\ell_0\}$$

with a small parameter $0 < \ell_0$ to be determined. Note that $0 < \ell \leq \max\{\frac{r_{in}(\Omega)}{2}, \ell_0\}$, where $r_{in}(\Omega)$ denotes the inradius, and, since $|\nabla \operatorname{dist}(u, \Omega^c)| = 1$ a.e., $\|\nabla\ell\|_{L^\infty} \leq \frac{1}{2}$. Note also that $\operatorname{dist}(B_{\ell(u)}, \Omega^c)) \leq 2\ell(u)$ if and only if $\operatorname{dist}(u, \partial\Omega) \leq 2\ell_0$, in which case $\ell(u) = \ell_0$. In particular, if $\operatorname{dist}(u, \Omega) > \ell_0$, then $B_{\ell(u)}(u) \cap \Omega = \emptyset$.

Fix a function $\phi \in C_0^\infty(\mathbb{R}^d)$ with $\operatorname{supp}\phi \subseteq \overline{B_1(0)}$ and $\|\phi\|_{L^2} = 1$. By [11, Theorem 22] (see also [8, Lemma 2.5]) the functions

$$\phi_u(x) = \phi\Big(\frac{x-u}{\ell(u)}\Big)\sqrt{1 + \nabla\ell(u)\cdot\frac{x-y}{\ell(u)}}, \quad x \in \mathbb{R}^d,\ u \in \mathbb{R}^d,$$

belong to $C_0^\infty(\mathbb{R}^d)$ with $\operatorname{supp}\phi_u \subseteq \overline{B_{\ell(u)}(u)}$, satisfy

$$\int_{\mathbb{R}^d} \phi_u(x)^2 \ell(u)^{-d}\, du = 1 \quad \text{for all } x \in \mathbb{R}^d \tag{4.1}$$

and, with a constant C depending only on d,

$$\|\phi_u\|_{L^\infty} \leq \sqrt{2}\,\|\phi\|_{L^\infty}, \quad \|\nabla\phi_u\|_{L^\infty} \leq C\ell(u)^{-1}\|\nabla\phi\|_{L^\infty} \quad \text{for all } u \in \mathbb{R}^d.$$

The application to our problem here is summarized in the following lemma.

Lemma 4.1. *Let Ω, b, V be as in Theorem 1.1 and define $\ell, \{\phi_u\}_{u\in\mathbb{R}^d}$ as above. Then, for $0 < \ell_0 \leq c(\Omega, b)$ and $0 < h \leq K\ell_0$,*

$$\Big|\operatorname{Tr}(H_{\Omega,b,V}(h))_- - \int_{\mathbb{R}^d} \operatorname{Tr}(\phi_u H_{\Omega,b,V}(h)\phi_u)_- \ell(u)^{-d}\, du\Big| \\ \leq Ch^{-d+2}\int_{\operatorname{dist}(u,\Omega)\leq\ell_0} \Big(1 + h^2\|V_-\|^{1+\frac{d}{2}}_{L^{1+\frac{d}{2}}(\Omega\cap B_{\ell(u)}(u))}\Big)\ell(u)^{-2}\, du,$$

where the constant C depends only on $\Omega, b, K, \|\phi\|_{L^\infty}$.

For the sake of brevity, we omit the proof of Lemma 4.1 and instead refer the reader to the proof of [8, Lemma 2.8]. Lemma 4.1 can be proved in the same manner but replacing the use of a local Berezin–Li–Yau inequality by an application of Lemma 2.4.

With the above results in hand we are ready to prove Theorem 1.1.

Proof of Theorem 1.1. Set $\ell_0 = \frac{h}{\varepsilon_0}$ with $0 < h \leq \frac{\varepsilon_0 r_{in}(\Omega)}{2}$ for a parameter $\varepsilon_0 \in (0, 1]$, which will eventually tend to zero.

We divide the set of $u \in \mathbb{R}^d$ such that $B_{\ell(u)}(u) \cap \Omega \neq \emptyset$ into two disjoint parts:

$$\Omega_* = \{u \in \mathbb{R}^d : 2\ell_0 < \delta_\Omega(u)\}, \quad \Omega^* = \{u \in \mathbb{R}^d : -\ell_0 < \delta_\Omega(u) \leq 2\ell_0\}, \quad (4.2)$$

where δ_Ω denotes the signed distance function to the boundary,

$$\delta_\Omega(y) = \operatorname{dist}(u, \Omega^c) - \operatorname{dist}(u, \Omega).$$

Note that for all $u \in \Omega^*$ we have $\ell(u) = \ell_0$.

By Lemma 4.1 we need to understand the integral with respect to u of the local traces $\operatorname{Tr}(\phi_u H_{\Omega,b,V}(h)\phi_u)_-$. Breaking the integral according to the partition (4.2), we have

$$\begin{aligned}
&\int_{\mathbb{R}^d} \operatorname{Tr}(\phi_u H_{\Omega,b,V}(h)\phi_u)_- \ell(u)^{-d}\, du \\
&\quad = \int_{\Omega_*} \operatorname{Tr}(\phi_u H_{\Omega,b,V}(h)\phi_u)_- \ell(u)^{-d}\, du + \int_{\Omega^*} \operatorname{Tr}(\phi_u H_{\Omega,b,V}(h)\phi_u)_- \ell_0^{-d}\, du.
\end{aligned}$$

For the first term Lemma 3.1 with

$$V_0(x) = \frac{(b(x)^2 - \frac{1}{4})_-}{\operatorname{dist}(x, \partial\Omega)^2}, \quad V_1 = V_-(x)$$

yields

$$\begin{aligned}
&\int_{\Omega_*} \operatorname{Tr}(\phi_u H_{\Omega,b,V}(h)\phi_u)_- \ell(u)^{-d}\, du \\
&\quad = L_d h^{-d} \int_{\Omega_*}\int_{\Omega} \phi_u^2(x)\ell(u)^{-d}\, dx\, du + O(h^{-d+2}) \int_{\Omega_*} \Big[\ell(u)^{-2}(1 + \|b\|_{L^\infty}^2) \\
&\qquad\qquad + \|V_-\|_{L^{1+\frac{d}{2}}(B_{\ell(u)}(u))}^{1+\frac{d}{2}} + \ell(u)^{-d}\|V_+\|_{L^1(B_{\ell(u)}(u))}\Big]\, du,
\end{aligned}$$

where we used

$$\|V_0\|_{L^\infty} \leq \frac{C}{(\operatorname{dist}(u, \partial\Omega) - \ell(u))^2} \leq C\ell(u)^{-2}$$

and

$$\frac{(b(x)^2 - \frac{1}{4})_+}{\operatorname{dist}(x, \partial\Omega)^2} \leq C\|b\|_{L^\infty}^2 \ell(u)^{-2}.$$

For the integral over the boundary region Ω^* Theorem 3.2, for ε_0, ℓ_0, h sufficiently small, implies

$$\begin{aligned}
&\int_{\Omega^*} \operatorname{Tr}(\phi_u H_{\Omega,b,V}(h)\phi_u)_- \ell_0^{-d}\, du \\
&= L_d h^{-d} \int_{\Omega^*}\int_{\Omega} \phi_u^2(x)\ell_0^{-d}\, dx\, du \\
&\quad - \frac{L_{d-1}}{2} h^{-d+1} \int_{\Omega^*}\int_{\partial\Omega} \phi_u^2(x) b(x) \ell_0^{-d}\, d\mathcal{H}^{d-1}(x)\, du \\
&\quad + O(h^{-d})|\Omega^*|(o_{\ell_0\to 0^+}(1) + \varepsilon_0^2|\log(\varepsilon_0)|) + h^{-d+1} o_{\ell_0\to 0^+}(1) \\
&\quad + O(h^{-d+2}) \int_{\Omega^*} \Big[\|V_-\|_{L^{1+\frac{d}{2}}(B_{\ell(u)}(u))}^{1+\frac{d}{2}} + \ell_0^{-d} \|V_+\|_{L^1(B_{\ell(u)}(u))}\Big]\, du.
\end{aligned}$$

Here we used the fact that b satisfies (1.3).

Combining the estimates for the contribution from the bulk and boundary region, using (4.1), and estimating the integrals of the norms of V_-, V_+, we find

$$\begin{aligned}
&\int_{\mathbb{R}^d} \operatorname{Tr}(\phi_u H_{\Omega,b,V}(h)\phi_u)_- \ell(u)^{-d}\, du \\
&= L_d h^{-d}|\Omega| - \frac{L_{d-1}}{2} h^{-d+1} \int_{\partial\Omega} b(x)\, d\mathcal{H}^{d-1}(x) \\
&\quad + O(h^{-d})|\Omega^*|(o_{\ell_0\to 0^+}(1) + \varepsilon_0^2|\log(\varepsilon_0)|) + h^{-d+1} o_{\ell_0\to 0^+}(1) \qquad (4.3) \\
&\quad + O(h^{-d+2})(1 + \|b\|_{L^\infty}^2) \int_{\Omega_*} \ell(u)^{-2}\, du \\
&\quad + O(h^{-d+2}) \Big[\|V_-\|_{L^{1+\frac{d}{2}}(\Omega)}^{1+\frac{d}{2}} + \|V_+\|_{L^1(\Omega)}\Big].
\end{aligned}$$

By [8, equations (4.6)–(4.8)], $\int_{\Omega_*} \ell(u)^{-2}\, du \le C\ell_0^{-1}$ and $|\Omega^*| \le C\ell_0$ with C depending only on Ω. Thus, by Lemma 4.1, (4.3), and since $\frac{h^2}{\ell(u)^2} \le \varepsilon_0^2$, we conclude that

$$\begin{aligned}
&h^{d-1}\Big|\operatorname{Tr}(H_{\Omega,b,V}(h))_- - L_d h^{-d}|\Omega| + \frac{L_{d-1}}{2} h^{-d+1} \int_{\partial\Omega} b(x)\, d\mathcal{H}^{d-1}(x)\Big| \\
&\le \varepsilon_0^{-1} o_{h/\varepsilon_0\to 0^+}(1) + O(\varepsilon_0|\log(\varepsilon_0)|) + o_{h/\varepsilon_0\to 0^+}(1) \\
&\qquad + O(\varepsilon_0)(1 + \|b\|_{L^\infty}^2) + O(h)\Big[\|V_-\|_{L^{1+\frac{d}{2}}(\Omega)}^{1+\frac{d}{2}} + \|V_+\|_{L^1(\Omega)}\Big].
\end{aligned}$$

Letting first h and then ε_0 tend to 0 completes the proof of Theorem 1.1. ∎

A Properties of P_ν

Our aim is to prove the following two lemmas.

Lemma A.1. *For $\nu \geq 0$ it holds that*

$$P_\nu(t) = \int_0^1 (1-\zeta^2)^{\frac{d+1}{2}} \Big(\frac{1}{\pi} - \zeta t J_\nu(\zeta t)^2\Big)\, d\zeta = O(t^{-2}) \quad \text{as } t \to \infty.$$

Moreover, the implicit constant is uniformly bounded for ν in compact subsets of $[0,\infty)$.

Lemma A.2. *For any $\nu \geq 0$ we have the identity*

$$\int_0^\infty P_\nu(t)\, dt = \int_0^\infty \int_0^1 (1-\zeta^2)^{\frac{d+1}{2}} \Big(\frac{1}{\pi} - \zeta t J_\nu(\zeta t)^2\Big)\, d\zeta\, dt = \frac{\nu}{2}.$$

We shall need the following asymptotic expansion for the Bessel function:

$$\begin{aligned} J_\nu(t) = \Big(\frac{2}{\pi t}\Big)^{\frac{1}{2}} \Big[&\cos\Big(t - \frac{\nu\pi}{2} - \frac{\pi}{4}\Big) \\ &- \frac{4\nu^2 - 1}{8t} \sin\Big(t - \frac{\nu\pi}{2} - \frac{\pi}{4}\Big) + O(t^{-2})\Big], \end{aligned} \tag{A.1}$$

where the implicit constant is uniformly bounded for ν in compact subsets of $[0,\infty)$ (see [12, Chapter 7]). We shall also make use of the following identity:

$$x J_\nu(x)^2 = \frac{d}{dx}\Big[\frac{x^2}{2} J_\nu(x)^2 + \frac{x^2}{2} J_{\nu+1}(x)^2 - \nu x J_\nu(x) J_{\nu+1}(x)\Big], \tag{A.2}$$

which is easily deduced from $J'_\nu(x) = \frac{1}{2}(J_{\nu-1}(x) - J_{\nu+1}(x))$ and the recursion formula $J_{\nu-1}(x) + J_{\nu+1}(x) = \frac{2\nu}{x} J_\nu(x)$.

Proof of Lemma A.1. By an integration by parts, (A.2), and since $|J_\nu(x)| \leq 1$,

$$\begin{aligned} P_\nu(t) = (d+1) \int_\delta^1 (1-\zeta^2)^{\frac{d-1}{2}} \Big[&\frac{\zeta^2}{\pi} - \frac{t\zeta^3}{2} J_\nu(\zeta t)^2 - \frac{t\zeta^3}{2} J_{\nu+1}(\zeta t)^2 \\ &+ \nu \zeta^2 J_\nu(\zeta t) J_{\nu+1}(\zeta t)\Big]\, d\zeta + O(t\delta^4 + \delta^3) \end{aligned}$$

for any $0 \leq \delta < 1$. Provided $\delta t \gtrsim 1$, (A.1) implies

$$\begin{aligned} &\frac{\zeta^2}{\pi} - \frac{t\zeta^3}{2} J_\nu(\zeta t)^2 - \frac{t\zeta^3}{2} J_{\nu+1}(\zeta t)^2 + \nu\zeta^2 J_\nu(\zeta t) J_{\nu+1}(\zeta t) \\ &\quad = \frac{\zeta}{2\pi t} \cos(2\zeta t - \pi\nu) + O(t^{-2}), \end{aligned}$$

with the implicit constant uniformly bounded for ν in compact subsets of $[0,\infty)$. Thus, we have arrived at

$$\begin{aligned} P_\nu(t) &= \frac{d+1}{2\pi t} \int_\delta^1 (1-\zeta^2)^{\frac{d-1}{2}} \zeta \cos(2\zeta t - \pi\nu)\, d\zeta + O(t^{-2} + t\delta^4 + \delta^3) \\ &= \frac{d+1}{2\pi t} \int_0^1 (1-\zeta^2)^{\frac{d-1}{2}} \zeta \cos(2\zeta t - \pi\nu)\, d\zeta + O(t^{-2}), \end{aligned}$$

where we chose $\delta = O(t^{-1})$. An integration by parts yields

$$\int_0^1 (1-\zeta^2)^{\frac{d-1}{2}} \zeta \cos(2\zeta t - \pi\nu)\, d\zeta$$
$$= \frac{1}{2t} \int_0^1 (1-\zeta^2)^{(d-3)/2} (d\zeta^2 - 1) \sin(2\zeta t - \pi\nu)\, d\zeta.$$

Since the integral on the right is bounded uniformly in ν, this completes the proof. ■

Proof of Lemma A.2. For any $T > 0$, by (A.2), Fubini's theorem, and a change of variables

$$\int_0^T P_\nu(t)\, dt = \int_0^1 (1-\zeta^2)^{\frac{d+1}{2}} \int_0^T \Big(\frac{1}{\pi} - \zeta t J_\nu(\zeta t)^2\Big)\, dt\, d\zeta$$
$$= \frac{T}{2}(P_\nu(T) + P_{\nu+1}(T)) + \nu \int_0^T \Big(1 - \frac{s^2}{T^2}\Big)^{\frac{d+1}{2}} J_\nu(s) J_{\nu+1}(s)\, ds.$$

By Lemma A.1 only the remaining integral contributes as $T \to \infty$. By [12, p. 406] and for $\nu > -1$, in the sense of an improper Riemann integral

$$\int_0^\infty J_\nu(s) J_{\nu+1}(s)\, ds = \frac{1}{2}.$$

The proof is completed by appealing to a simple Abelian theorem in Lemma A.3. ■

Lemma A.3. *If $f \in L^\infty(\mathbb{R}_+)$ and $\lim_{T\to\infty} \int_0^T f(t)\, dt = A$, then for all $\alpha > 0$,*

$$\lim_{T\to\infty} \int_0^T \Big(1 - \frac{t^2}{T^2}\Big)^\alpha f(t)\, dt = A.$$

Proof. By integration by parts and a change of variables,

$$\int_0^T \Big(1 - \frac{t^2}{T^2}\Big)^\alpha f(t)\, dt = \int_0^T \Big(-\frac{d}{dt}\Big(1 - \frac{t^2}{T^2}\Big)^\alpha\Big) \int_0^t f(s)\, ds\, dt$$
$$= 2\alpha \int_0^1 (1-\sigma^2)^{\alpha-1} \sigma \int_0^{\sigma T} f(s)\, ds\, d\sigma.$$

By our assumptions there is a $S_0 < \infty$ so that for $S \geq S_0$,

$$\Big|\int_0^S f(s)\, ds\Big| \leq |A| + 1.$$

Since f is bounded, we have

$$\Big|\int_0^S f(s)\, ds\Big| \leq S\|f\|_\infty.$$

Thus, for all σ, T,

$$\left|\int_0^{\sigma T} f(s)\,ds\right| \leq \max\{|A|+1, S_0\|f\|_\infty\}.$$

Since $\alpha > 0$, the function $\sigma \mapsto (1-\sigma^2)^{\alpha-1}\sigma$ is integrable and by dominated convergence,

$$\lim_{T\to\infty} 2\alpha \int_0^1 (1-\sigma^2)^{\alpha-1}\sigma \int_0^{\sigma T} f(s)\,ds d\sigma = 2\alpha A \int_0^1 (1-\sigma^2)^{\alpha-1}\sigma\,d\sigma = A.$$

This completes the proof of Lemma A.3. ∎

Acknowledgements. We are deeply grateful to Ari Laptev for sharing his fascination for spectral estimates and Hardy's inequality with us and we would like to dedicate this paper to him on the occasion of his 70th birthday.

U.S. National Science Foundation grants DMS-1363432 and DMS-1954995 (Rupert L. Frank) and Knut and Alice Wallenberg Foundation grant KAW 2018.0281 (Simon Larson) are acknowledged.

Bibliography

[1] M. Š. Birman and M. Z. Solomjak, Asymptotic properties of the spectrum of differential equations. *J. Soviet Math.* **12** (1979), no. 3, 247–283

[2] M. Š. Birman and M. Z. Solomjak, *Quantitative analysis in Sobolev imbedding theorems and applications to spectral theory*. Amer. Math. Soc. Transl. (2) 114, American Mathematical Society, Providence, 1980

[3] E. B. Davies, The Hardy constant. *Quart. J. Math. Oxford Ser. (2)* **46** (1995), 417–431

[4] N. Dunford and J. T. Schwartz, *Linear operators. Part II: Spectral theory. Self adjoint operators in Hilbert space*. With the assistance of William G. Bade and Robert G. Bartle, John Wiley & Sons, New York, 1963

[5] R. L. Frank and L. Geisinger, Two-term spectral asymptotics for the Dirichlet Laplacian on a bounded domain. In *Mathematical results in quantum physics*, pp. 138–147, World Sciientific Publishing, Hackensack, 2011

[6] R. L. Frank and L. Geisinger, Semi-classical analysis of the Laplace operator with Robin boundary conditions. *Bull. Math. Sci.* **2** (2012), 281–319

[7] R. L. Frank and S. Larson, On the error in the two-term Weyl formula for the Dirichlet Laplacian. *J. Math. Phys.* **61** (2020), Article ID 043504

[8] R. L. Frank and S. Larson, Two-term spectral asymptotics for the Dirichlet Laplacian in a Lipschitz domain. *J. Reine Angew. Math.* **766** (2020), 195–228

[9] R. L. Frank and M. Loss, Hardy–Sobolev–Maz'ya inequalities for arbitrary domains. *J. Math. Pures Appl. (9)* **97** (2012), 39–54

[10] E. H. Lieb and B. Simon, The Thomas–Fermi theory of atoms, molecules and solids. *Advances in Math.* **23** (1977), 22–116

[11] J. P. Solovej and W. L. Spitzer, A new coherent states approach to semiclassics which gives Scott's correction. *Comm. Math. Phys.* **241** (2003), 383–420

[12] G. N. Watson, *A treatise on the theory of Bessel functions*. Cambridge University Press, Cambridge, 1944

On the spectral properties of the Bloch–Torrey equation in infinite periodically perforated domains

Denis S. Grebenkov, Bernard Helffer and Nicolas Moutal

To Ari Laptev on the occasion of his 70th birthday.

We investigate spectral and asymptotic properties of the particular Schrödinger operator (also known as the Bloch–Torrey operator), $-\Delta + igx$, in infinite periodically perforated domains of $\mathbb{R}^d$. We consider Dirichlet realizations of this operator and formalize a numerical approach proposed in [17] for studying such operators. In particular, we discuss the existence of the spectrum of this operator and its asymptotic behavior as $g \to \infty$.

1 Introduction

The aim of this paper is to formalize on the mathematical side a numerical approach proposed in [17] for analyzing the Bloch–Torrey equation in infinite periodically perforated domains. More precisely, we consider the Dirichlet realization of the Bloch–Torrey operator

$$B := -\Delta + igx,$$

denoted respectively by $\mathcal{B}^D$, where Δ is the Laplace operator, x is one of the Cartesian coordinates, and g is a real non-zero parameter. The Neumann and Robin realizations, which are important in physical applications [7–9, 17, 19], could also be treated by the same techniques but the details of the proof are omitted in the present paper. The major novelty of this work (as compared to former studies in [2, 3, 10, 11]) is that we consider the Bloch–Torrey operator in a periodically perforated domain ($d \geq 2$),

$$\Omega = \mathbb{R}^d \setminus \left\{ \bigcup_{\gamma \in \mathbb{Z}^d} H_\gamma \right\}, \tag{1.1}$$

Keywords: Bloch–Torrey equation, Floquet theory, non-self-adjoint operators
2020 Mathematics Subject Classification: Primary 47B28; secondary 47B93

where

$$H_\gamma = \{\mathbf{x} \in \mathbb{R}^d : \mathbf{x} - \gamma \in H_0\} \tag{1.2}$$

and $H_0 \subset (-\frac{1}{2}, \frac{1}{2})^d$ is a domain with a smooth boundary.

A typical example is the case when $H_\gamma = B_2(\gamma, r)$, where $B_2(\gamma, r)$ is the disk of radius $r < \frac{1}{2}$ centered at γ. One of the major difficulties in the definition and study of such non-self-adjoint operators is that the potential igx is not periodic, unbounded and changing sign.

The Bloch–Torrey operator describes the diffusion-precession of spin-bearing particles in nuclear magnetic resonance experiments and helps in understanding the intricate relation between the geometric structure of a studied sample (domain) and the measured signal [7–9]. The spectral properties of this operator and its asymptotic behavior play thus a crucial role in this analysis.

The paper is organized as follows. In Section 2, we provide a rigorous definition of the considered realization of the Bloch–Torrey operator and describe its basic properties. In Section 3, we use the periodicity in y-direction via the Floquet theory to reduce the operator on the planar perforated domain to a family (indexed by a Floquet parameter) of operators on the infinite perforated cylinder. In particular, we formulate here two conjectures about their spectral properties. Section 4 analyzes the role of the pseudo-periodicity in x-direction, which is specific to the Bloch–Torrey operator. In Section 5, we formulate in an asymptotic regime the main results concerning the non-emptiness of the spectrum and the asymptotic properties. Finally, Section 6 concludes the paper and discusses the extensions in the proofs to more general settings.

2 The Bloch–Torrey operator in the perforated whole plane

We start from the Dirichlet realization of the Bloch–Torrey operator

$$B := -\Delta_{x,y} + igx$$

in $\Omega = \mathbb{R}^2 \setminus \bigcup_{\gamma \in \mathbb{Z}^2} H_\gamma$ as defined in (1.1)–(1.2) and where g is a non-zero real parameter (in the following, we set $g > 0$ for clarity). We introduce

$$\mathcal{H} := L^2(\Omega) \quad \text{and} \quad \mathcal{V} := \{u \in H_0^1(\Omega) : |x|^{\frac{1}{2}} u \in L^2(\Omega)\}$$

and we can now extend the operator initially defined on $C_0^\infty(\Omega)$ to get a closed operator on $\mathcal{H}$ by the following variant of the Friedrichs extension.

Proposition 2.1. *The Dirichlet realization $\mathcal{B}^D$ of B with domain*

$$D(\mathcal{B}^D) = \{u \in \mathcal{V} : \mathcal{B}^D u \in \mathcal{H}\}$$

is a closed accretive operator which generates a continuous semi-group on $L^2(\Omega)$.

Proof. We take

$$\mathcal{H} := L^2(\Omega) \quad \text{and} \quad \mathcal{V} := \{u \in H_0^1(\Omega) : |x|^{\frac{1}{2}} u \in L^2(\Omega)\}$$

and apply Theorem A.1 to the quadratic form

$$a(u,u) = \int_\Omega |\nabla u|^2 \, dx\, dy + i \int_\Omega x|u|^2 \, dx\, dy + C_0\|u\|^2,$$

where $\|u\|$ denotes the $L_2(\Omega)$ norm and $C_0 > 0$ is a large positive constant to be determined. For $\Phi_1 = \Phi_2$ we take the multiplication operator by $\frac{x}{\sqrt{1+x^2}}$. We simply then observe that

$$\Re a(u,u) = \int_\Omega |\nabla u|^2 \, dx\, dy + C_0\|u\|^2,$$

and then consider $\Im a(u, \Phi_1(u))$. By an easy computation, we can show the existence of $C > 0$ such that, for all $u \in \mathcal{V}$,

$$\Im a(u, \Phi_1(u)) \ge \int_\Omega x^2(1+x^2)^{-\frac{1}{2}} |u|^2 \, dx\, dy - C\,\|u\|^2_{H^1(\Omega)}$$

Hence the assumptions of the theorems are satisfied if we choose C_0 sufficiently large. What we call $\mathcal{B}^D$ is the operator $S - C_0$, where S is given by Theorem A.1. To get the semi-group property, we then apply the Hille–Yosida Theorem. ■

3 Floquet approach for y-periodic problems

The goal is now to analyze the spectrum of this operator and to possibly consider asymptotic problems in function of g.

3.1 Floquet decomposition

Let τ_2 be the translation by $(0, 1)$, i.e.

$$\tau_2 u(x,y) = u(x, y-1) \quad \text{for } u \in L^2(\Omega).$$

Observing that

$$\tau_2 \circ \mathcal{B}^D = \mathcal{B}^D \circ \tau_2,$$

we can, at least formally, apply a Floquet theorem in the y-variable and get the family of operators ($q \in \mathbb{R}$)

$$B_q := -\left(\frac{d}{dy} - iq\right)^2 - \frac{d^2}{dx^2} + igx$$

in

$$\Omega_0 = \left(\mathbb{R} \times \left(-\frac{1}{2}, \frac{1}{2}\right)\right) \setminus \left\{\bigcup_{n \in \mathbb{Z}} H_{(n,0)}\right\},$$

where we put the Dirichlet condition at the boundary of each $H_{(n,0)}$ and the periodicity condition on the remaining boundary

$$u\left(x, -\frac{1}{2}\right) = u\left(x, \frac{1}{2}\right), \quad \partial_y u\left(x, -\frac{1}{2}\right) = \partial_y u\left(x, \frac{1}{2}\right) \quad \text{for all } x \in \mathbb{R}.$$

Nevertheless, the best way is to consider it as an operator on the infinite perforated cylinder

$$\widehat{\Omega}_0 := (\mathbb{R} \times \mathbb{T}^1) \setminus \left\{\bigcup_{n \in \mathbb{Z}} H_{(n,0)}\right\},$$

where $\mathbb{T}^1 = \mathbb{R}/\mathbb{Z}$ with Dirichlet condition on $\partial\widehat{\Omega}_0$.

3.2 Spectral analysis

We now study the spectral properties of the Dirichlet realization $\mathcal{B}_q^D$ of B_q.

Proposition 3.1. *For any $q \in \mathbb{R}$, we can extend B_q as a closed operator $\mathcal{B}_q^D$ with the following properties:*

(1) *$\mathcal{B}_q^D$ is a closed accretive operator on $L^2(\Omega_0)$ and is the generator of a continuous semi-group.*

(2) *$\mathcal{B}_q^D$ has compact resolvent.*

Proof. The proof is the same as for Proposition 2.1 but this time we take

$$\mathcal{H}_0 := L^2(\Omega_0) \quad \text{and} \quad \mathcal{V}_0 := \{u \in H^1(\Omega_0) : |x|^{\frac{1}{2}} u \in L^2(\Omega_0) \text{ and (BC) holds}\},$$

where by (BC) we mean that u satisfies the Dirichlet condition at the boundary of the $H_{(0,n)}$ and that we have the periodicity condition $u(x, -\frac{1}{2}) = u(x, \frac{1}{2})$ for $x \in \mathbb{R}$. Actually, it is better to deal with $\widehat{\Omega}_0$ as mentioned above and to consider

$$\widehat{\mathcal{H}}_0 := L^2(\widehat{\Omega}_0) \quad \text{and} \quad \widehat{\mathcal{V}}_0 := \{u \in H_0^1(\widehat{\Omega}_0) : |x|^{\frac{1}{2}} u \in L^2(\widehat{\Omega}_0)\}.$$

The compact resolvent property is a consequence of the compact injection of $\mathcal{V}_0$ in $\mathcal{H}_0$.

Noting the inequality

$$\Re\langle \mathcal{B}_q^D u, u\rangle \geq 0 \quad \text{for all } u \in \mathcal{V}_0,$$

we can construct by the Hille–Yosida theorem the associated continuous semi-group and consider (as in [17])

$$\mathbb{R}_+ \ni t \mapsto \exp(-t\mathcal{B}_q^D). \qquad \blacksquare$$

3.3 Application of Floquet decomposition to spectral theory

The interest of the Floquet theory is to relate the spectrum of the initial operator to the union of the spectra of a family of simpler operators. It is based on the so-called Floquet decomposition whose definition is independent of the operator but is only related to the symmetry group property satisfied by the operator. The application to spectral theory is standard in the self-adjoint case (see for example in [18]) but we are not aware of a reference for the application in the non-self-adjoint case (see nevertheless [15, 16]). Therefore it is not completely clear that the proof easily goes on in the non-self-adjoint case and this is why we present in this subsection an approach covering this situation. Our main result is:

Proposition 3.2. *We have*

$$\overline{\bigcup_{q\in\mathbb{R}} \sigma(\mathcal{B}_q^D)} \subset \sigma(\mathcal{B}^D),$$

where, for $A \subset \mathbb{C}$, $\overline{A}$ denotes the closure of A.

Proof. We divide the proof into two parts.

Step 1. We first observe that $\mathcal{B}_q^D$ is unitarily equivalent to $\mathcal{B}_{q+2\pi}^D$. Hence it is enough to consider the problem for $q \in [0, 2\pi]$.

Step 2. This is a simple version of the Schnoll theorem relating the spectrum and the existence of generalized polynomially bounded eigenfunctions. Let $q \in \mathbb{R}$. We know from Proposition 3.1 that $\sigma(\mathcal{B}_q^D)$ is either empty or consists of eigenvalues with finite multiplicity. There is nothing to prove if $\sigma(\mathcal{B}_q^D)$ is empty. Let us assume now that it is not empty and let μ be an eigenvalue of $\mathcal{B}_q^D$ associated to a normalized eigenfunction $u_\mu(x, y)$. We extend u_μ as a periodic function in the whole domain Ω and consider a cutoff function χ which equals 1 on $(-\frac{1}{2}, \frac{1}{2})$ with support in $(-1, +1)$. We then consider

$$u_{\mu,n}(x, y) := e^{iyq}\chi\Big(\frac{y}{n}\Big)u_\mu(x, y).$$

We have, for $n \in \mathbb{N} \setminus \{0\}$,

$$\|u_{\mu,n}\|^2_{L^2(\Omega)} \geq \int_{\Omega\cap\{|y|<\frac{n}{2}\}} |u_\mu|^2\, dx\, dy = n \int_{\Omega\cap\{|y|<\frac{1}{2}\}} |u_\mu|^2\, dx\, dy = n \tag{3.1}$$

and

$$f_n := (\mathcal{B}^D - \mu)u_{\mu,n} = -\frac{2}{n}\chi'\Big(\frac{y}{n}\Big)\partial_y u_\mu(x, y) - \frac{1}{n^2}\chi''\Big(\frac{y}{n}\Big)u_\mu(x, y).$$

It is immediate to see that there exists a constant C such that

$$\|f_n\|_{L^2(\Omega)} \leq \frac{C}{\sqrt{n}}.$$

If μ was not in the spectrum of $\mathcal{B}_D$, we would have

$$u_{\mu,n} = (\mathcal{B}_D - \mu)^{-1} f_n = \mathcal{O}\left(\frac{1}{\sqrt{n}}\right),$$

in contradiction with (3.1). ∎

We know that in the self-adjoint case (i.e. the case of the Schrödinger operator with real periodic potential) the converse inclusion holds true. We notice in our situation two conjectures.

Conjecture 3.3. We have

$$\overline{\bigcup_{q\in\mathbb{R}} \sigma(\mathcal{B}_q^D)} = \sigma(\mathcal{B}^D).$$

We will see below that this conjecture is a consequence of the second one:

Conjecture 3.4. For $\mathcal{B}^D$, the following holds:

$$\overline{\rho(\mathcal{B}^D)} = \mathbb{C}, \tag{3.2}$$

where $\rho(\mathcal{B}^D)$ denotes the resolvent set of $\mathcal{B}^D$.

Remarks. We make the following observations.

(1) Property (3.2) would have been automatically satisfied if $\mathcal{B}^D$ was self-adjoint (because the spectrum would have been real).

(2) The proof of the second conjecture (which would imply the first one as can be seen from the proof below) does not seem at the moment easier as the first one.

(3) Validity of Conjecture 3.4 would probably be quite specific to the Bloch–Torrey operator and to the case $d = 2$. For instance, we suspect that the resolvent set of the operator $-\Delta + \cos x + i \cos y$ in $\mathbb{R}^2$ has not this property. We indeed suspect that the spectrum can be described by a Floquet theory as $\bigcup_{p,q}(\lambda_1(p) + \lambda_2(q))$, where $\lambda_1(p)$ is the Floquet eigenvalue of $(-\frac{d^2}{dx^2} + \cos x)$ and $\lambda_2(q)$ is the Floquet eigenvalue of $(-\frac{d^2}{dy^2} + i \cos y)$ which is not real-valued.

(4) It could be easier to show that the set

$$\Omega^R := \mathbb{C} \setminus \left(\overline{\bigcup_{q\in[0,2\pi]} \sigma(\mathcal{B}_q^D)}\right)$$

satisfies

$$\overline{\Omega^R} = \mathbb{C}.$$

The set $\bigcup_{q\in\mathbb{R}} \sigma(\mathcal{B}_q^D)$ seems indeed be a union of piecewise analytic curves in $\mathbb{C}$. We do not know if this property will lead to the proof of Conjecture 3.3. Again this question would be specific to the case $d = 2$.

(5) One could imagine that (for g large) some of these curves $[0, 2\pi] \ni q \mapsto \lambda_k(q)$ are simple closed curves in $\mathbb{C}$ implying the non-connectedness of Ω^R. This situation is of course excluded in the self-adjoint case.

We now prove a weaker version of Conjecture 3.3 for which we need more definitions. Observing that all our operators are accretive, we immediately see that

$$\mathbb{C}_- := \{z \in \mathbb{C} : \Re z < 0\} \subset \rho(\mathcal{B}^D) \cap \Omega^R.$$

We now denote by ω_D the connected component in $\Omega_D := \rho(\mathcal{B}^D)$ containing $\mathbb{C}_-$ and by ω_R the connected component in Ω^R containing $\mathbb{C}_-$. From Proposition 3.2 we know that

$$\omega_D \subset \omega_R.$$

We then prove:

Proposition 3.5. *There holds* $\omega_D = \omega_R$.

Proof. By Floquet theory, there is a unitary transform $\mathcal{U}$ such that

$$\mathcal{U}^{-1}(\mathcal{B}^D - \lambda)^{-1}\mathcal{U} = \int_0^{2\pi} (\mathcal{B}_q^D - \lambda)^{-1}\, dq. \tag{3.3}$$

It is clear that this formula holds if $\lambda \notin \sigma(\mathcal{B}^D)$ but we will show that, under assumption (3.2) this can be extended if the right-hand side is well defined.

Suppose indeed that $\lambda \notin \bigcup_{q\in[0,2\pi]} \sigma(\mathcal{B}_q^D)$. Hence $(\mathcal{B}_q^D - \lambda)^{-1}$ exists for any $q \in [0, 2\pi]$ and we have to verify (to give a meaning of the right-hand side in (3.3)) that we have a uniform control with respect to q. For this we observe that

$$\begin{aligned}(\mathcal{B}_q^D - \lambda) \circ (\mathcal{B}_{q_0}^D - \lambda)^{-1} = I &+ 2i(q - q_0)\partial_y(\mathcal{B}_{q_0}^D - \lambda)^{-1} \\ &+ (q^2 - q_0)^2(\mathcal{B}_{q_0}^D - \lambda)^{-1}.\end{aligned}$$

For $|q - q_0|$ small enough this implies

$$\begin{aligned}(\mathcal{B}_q^D - \lambda)^{-1} = (\mathcal{B}_{q_0}^D - \lambda)^{-1}\big(I &+ 2i(q - q_0)\partial_y(\mathcal{B}_{q_0}^D - \lambda)^{-1} \\ &+ (q^2 - q_0)^2(\mathcal{B}_{q_0}^D - \lambda)^{-1}\big)^{-1}.\end{aligned}$$

By compactness, we get the uniformity and define an holomorphic L^2-valued function $\widetilde{R}(\lambda)$ extending the resolvent $R(\lambda)$ to the open set Ω^R. Let us assume by contradiction that ω_D is strictly included in ω_R. Let us consider $\lambda_0 \in \omega_R \setminus \omega_D$. Using Lemma B.1 and the remark there, we can construct a sequence u_n of normalized functions such that $(\mathcal{B}^D - \lambda_0)u_n$ tends to zero. Now we consider the holomorphic functions $\omega_R \ni \lambda \mapsto v_n(\lambda) = \widetilde{R}(\lambda)(\mathcal{B}^D - \lambda)u_n$. On ω_D, one has $v_n(\lambda) = u_n$, therefore $v_n(\lambda)$ is constant and equal to u_n over ω_R. In particular, $\|v_n(\lambda)\| = 1$ for all integer n and all $\lambda \in \omega_R$. However, one has $\|v_n(\lambda_0)\| = \|\widetilde{R}(\lambda_0)(\mathcal{B}^D - \lambda_0)u_n\| \to 0$, hence we get a contradiction. ∎

Remark. Note that the proposition implies that $\partial\omega_D = \partial\omega_R$. One could hope that $\partial\omega_R = \partial\Omega^R$ which would correspond to the guess that all the other components of Ω^R are bounded by spectral curves (as mentioned above).

4 The Bloch–Torrey operator in a perforated cylinder continued

Here the analysis is quite specific of the Bloch–Torrey operator.

4.1 The pseudo-invariance in the x-variable

Let τ_1 be the translation by 1 in the x-variable, which is defined for $u \in L^2_{\text{loc}}(\overline{\Omega_0})$ by

$$\tau_1 u(x,y) = u(x-1,y).$$

Using this translation, one obtains that:

$$\text{If } \lambda \in \sigma(\mathcal{B}_q^D), \text{ then } \lambda + ig \in \sigma(\mathcal{B}_q^D).$$

This indeed results from the commutation relation

$$\tau_1 \circ \mathcal{B}_q^D = (\mathcal{B}_q^D - ig) \circ \tau_1. \tag{4.1}$$

Here it is important to note in the argument that we have used that

$$\tau_1 \Omega_0 = \Omega_0.$$

Remark. This commutation relation does not permit to use the standard Floquet theory (see [15,16,18,20]). We can try to recover this invariance by considering on the Hilbert space $\bigoplus_{n\in\mathbb{Z}} \mathcal{H}_n$ (with $\mathcal{H}_n$ isomorphic to $L^2(\widehat{\Omega}_0)$) the operator $\oplus_{n\in\mathbb{Z}}(\mathcal{B}_q^D + ign)$. If we denote by τ_0 the translation on $\ell^2(\mathbb{Z})$, we obtain the commutation of this operator with respect to $\tau_0^{-1}\tau_1$. One can then perform a Floquet decomposition in this context. This will give another way to obtain our next results but not more.

Using Proposition 3.1 (second assertion), we get:

Proposition 4.1. *For any $q \in \mathbb{R}$, there is a discrete sequence (finite, infinite or possibly empty) of eigenvalues $\lambda_k(q) \in \mathbb{R}_+ + ig(-\frac{1}{2}, \frac{1}{2}]$ such that the spectrum is given by*

$$\sigma(\mathcal{B}_q^D) = \bigcup_{k,n} (\lambda_k(q) + ign). \tag{4.2}$$

Moreover, if u is an eigenfunction of $\mathcal{B}_q^D$ for $\lambda_k(q)$, then $\tau_1^n u$ is an eigenfunction for

$$\lambda_{k,n}(q) := \lambda_k(q) + ign.$$

Nevertheless, it remains to analyze these sequences $\lambda_k(q)$ and to determine in particular if the spectrum is not empty (a question which is specific of the non-self-adjoint situation). We now consider the associate semi-group. We deduce from (4.1)

$$\tau_1 \circ \exp(-t\mathcal{B}_q^D) = \exp(itg)\exp(-t\mathcal{B}_q^D) \circ \tau_1. \tag{4.3}$$

Choosing

$$t_g = \frac{2\pi}{g},$$

we get

$$\tau_1 \circ \exp(-t_g\mathcal{B}_q^D) = \exp(-t_g\mathcal{B}_q^D) \circ \tau_1. \tag{4.4}$$

The idea is now to consider the spectrum of the operator

$$K_{g,q} := \exp(-t_g\mathcal{B}_q^D),$$

and to perform an adapted Floquet decomposition for this operator since we have recovered the commutation relation.

A natural question is then to ask if $K_{g,q}$ is a compact operator on $L^2(\widehat{\Omega}_0)$. This would be clear if $A := \mathcal{B}_q^D$ was self-adjoint or more generally sectorial. In this case, one can indeed prove that for any $T > 0$, $tA\exp(-tA)$ is bounded uniformly for $t \in (0, T)$ (see for example [6, Chapter 2] and references therein). Hence the operator $\exp(-tA) = (A+\lambda_0)^{-1}((A+\lambda_0)\exp(-tA))$ is compact as the composition of a compact operator and a bounded operator. But $\mathcal{B}_q^D$ is not sectorial (at least if the spectrum is not empty) due to (4.2).

On the other hand, the guess is that its spectrum is given as follows:

Conjecture 4.2. We have

$$\sigma(K_{g,q}) \setminus \{0\} = \bigcup_k \exp(-t_g\lambda_k(q)).$$

The first result, which may be non-intuitive, is:

Proposition 4.3. *If $K_{g,q}$ is a compact operator, then the spectrum of $\mathcal{B}_q^D$ is empty.*

Proof. We divide the proof into two parts.

Step 1. If λ_k is an eigenvalue of $\mathcal{B}_q^D$ and u_k is the corresponding normalized eigenfunction, then u_k is an eigenfunction of $\exp(-t_g\mathcal{B}_q^D)$ associated with $\lambda = \exp(-t_g\lambda_k)$. This proves at least that

$$\bigcup_k \exp(-t_g\lambda_k(q)) \subset \sigma(K_{g,q}) \setminus \{0\}. \tag{4.5}$$

Step 2. By (4.4) $\tau_1 u_k$ is also an eigenfunction. More generally, $u_{k,n} := \tau_1^n u_k$ is an eigenfunction for any $n \in \mathbb{Z}$. It remains to show that the vector space $\mathcal{U}$ formed by the $\tau_1^n u_k$ ($n \in \mathbb{Z}$) is not of finite dimension which will give a contradiction to the compactness. Suppose by contradiction that $\mathcal{U}$ is of finite dimension N_0, then there exists N_1 such that this space is generated by the $\tau_1^n u_k$ for $|n| \leq N_1$. Hence (uniform localization of a normalized element belonging to a finite subspace in $L^2(\Omega_0)$) for any $\varepsilon > 0$, there would exist $R > 0$ such that

$$\|\tau_1^n u\|_{L^2((-R,R)\times(-\frac{1}{2},\frac{1}{2})\cap\Omega_0)} \geq 1 - \varepsilon \quad \text{for all } n \in \mathbb{Z}.$$

But this cannot be true as $|n| \to +\infty$, since, for fixed R, $\|\tau_1^n u\|_{L^2((-R,R)\times(-\frac{1}{2},\frac{1}{2})\cap\Omega_0)}$ tends to 0. ■

Remark. In the self-adjoint case, we could have directly used that the eigenfunctions corresponding to different eigenvalues are orthogonal. This is not the case here. In the case $q = 0$, we can observe that if u is an eigenfunction of $\mathcal{B}_0^D$ associated with λ, then $\overline{u}$ is an eigenfunction of the realization of $-\Delta - ix$ for the eigenvalue $\overline{\lambda}$. Hence, we get

$$\langle u_{k,n}, \overline{u}_{k,n'} \rangle = 0 \quad \text{for } n \neq n'.$$

Remark. In the case without holes, the emptiness of the spectrum for the operators $-\frac{d^2}{dx^2} + igx - \frac{d^2}{dy^2}$ in $\mathbb{R}^2$ or $\mathbb{R} \times (-\frac{1}{2}, \frac{1}{2})$ is known for a long time (see [12] and the references therein). Nevertheless, our guess is that this is not always the case for our perforated situation (see for example [2] for the exterior problem).

4.2 Floquet decomposition in the x-direction

The authors of [17] analyze directly the spectrum of $K_{q,g}$. Observing as in [17] that $K_{q,g}$ commutes with the translation τ_1, we now consider a Floquet decomposition (relative to the translation in x). We recall from this decomposition that, if u is in $L^2(\Omega_0)$, then

$$u_p := \sum_n e^{-inp} \tau_1^n u \tag{4.6}$$

satisfies the p-Floquet condition

$$\tau_1 u_p = e^{ip} u_p,$$

is well defined in L^2_{loc} and satisfies

$$\int_{\Omega_0} |u(x, y)|^2 \, dx \, dy = \int_{(0,2\pi)} \|u_p(x, y)\|^2_{L^2(\Omega_0 \cap (-\frac{1}{2},\frac{1}{2})^2)} \, dp. \tag{4.7}$$

Conversely, if we have a measurable family (with respect to p) of functions u_p in $L^2(\Omega_0)$ satisfying the p-Floquet condition and such that

$$\int_{(0,2\pi)} \|u_p(x, y)\|^2_{L^2(\Omega_0 \cap (-\frac{1}{2},\frac{1}{2})^2)} \, dp < \infty,$$

then

$$u = \frac{1}{2\pi} \int_{(0,2\pi)} u_p(x, y) \, dp$$

is well defined in $L^2(\Omega_0)$ and satisfies (4.7). Now, if u is an eigenfunction associated with an eigenvalue λ of $\mathcal{B}_q^D$, then u_p satisfies

$$(\exp(-t_g \mathcal{B}_q^D))u_p = \exp(-t_g\lambda)\, u_p.$$

More generally, what one hopes from Floquet theory is that

$$\sigma(K_{q,g}) \setminus \{0\} = \bigcup_{p\in\mathbb{R}} (\sigma(K_{q,g,p}) \setminus \{0\}),$$

where $K_{q,g,p}$ is the restriction of $K_{q,g}$ to the functions satisfying the p-Floquet condition with respect to the translation τ_1.

The authors of [17] formally observe the following rather surprising property:

Proposition 4.4. *Suppose that u_p is a Floquet eigenfunction of $K_{g,q,p}$. Then, for any $t > 0$, the function*

$$u_{p,t} := \exp(-t\mathcal{B}_q^D)u_p$$

satisfies the Floquet condition with parameter $p - tg$ and

$$\exp(-t_g \mathcal{B}_q^D)u_{p,t} = \exp(-t_g\lambda)u_{p,t}.$$

Proof. By (4.3), we have

$$\tau_1 \underbrace{(\exp(-t\mathcal{B}_q^D))u_p}_{u_{p,t}} = \exp(itg)\exp(-t\mathcal{B}_q^D)\tau_1 u_p = \exp(-i(p-tg))u_{p,t},$$

as desired. ∎

As an application, this justifies at least formally to look numerically at the periodic problem associated with $\exp(-t_g\mathcal{B}_q^D)$ and to recover the spectrum of $\mathcal{B}_q$ by considering $-\log\frac{\mu}{t_g} + ig\mathbb{Z}$ (for $\mu \in \sigma(K_{q,g,0})$). Note that at this stage we have only proved (see (4.5)) that

$$\exp(-t_g\sigma(\mathcal{B}_q^D)) \subset \sigma(K_{q,g}).$$

From Floquet eigenfunctions of $K_{q,g}$ to eigenfunctions of $\mathcal{B}_q^D$. We get from formula (4.6) that for an eigenfunction u of $\mathcal{B}_q^D$,

$$u = \frac{1}{2\pi}\int_0^{2\pi} u_p\, dp. \tag{4.8}$$

This formula is standard due to the particular choice of the u_p through formula (4.6). Note for example that

$$\tau_1 u = \frac{1}{2\pi}\int_0^{2\pi} \tau_1 u_p\, dp = \frac{1}{2\pi}\int_0^{2\pi} e^{ip} u_p\, dp.$$

Hence the infinite-dimensional space $\mathrm{Span}(\tau_1^n u)$ is recovered by $\frac{1}{2\pi}\int_0^{2\pi} \beta_p u_p\, dp$ for some function β_p.

The next proposition is a kind of converse statement.

Proposition 4.5. *If μ is an eigenvalue of $K_{q,p=0}$ with corresponding eigenfunction u_0, then $-\log \frac{\mu}{t_g} + ig\mathbb{Z}$ belongs to the spectrum of $\mathcal{B}_q^D$ and, for each k, we can construct starting from u_0 an eigenfunction u_{λ_k} associated with $\lambda_k := -\log \frac{\mu}{t_g} + igk$.*

Proof. We divide the proof into two parts.

Step 1: The heuristics behind the proof. Heuristically, we will proceed in the following way. We start from an eigenfunction u_0 and associate with it

$$u_p = \exp\Big(-\frac{p}{g}(\mathcal{B}_q^D - \lambda_0)\Big)u_0. \tag{4.9}$$

Defining u by (4.8), we obtain formally:

$$\begin{aligned}(\mathcal{B}_q^D - \lambda_0)u &= \Big(\frac{1}{2\pi}\int_0^{2\pi}(\mathcal{B}_q^D - \lambda_0)\exp\Big(-\frac{p}{g}(\mathcal{B}_q^D - \lambda_0)\Big)\,dp\Big)u_0\\ &= \Big(\frac{g}{2\pi}\int_0^{2\pi}\Big(-\frac{d}{dp}\exp\Big(-\frac{p}{g}(\mathcal{B}_q^D - \lambda_0)\Big)\Big)\,dp\Big)u_0\\ &= \frac{g}{2\pi}\Big(I - \exp\Big(-\frac{2\pi}{g}(\mathcal{B}_q^D - \lambda_0)\Big)\Big)u_0\\ &= 0.\end{aligned} \tag{4.10}$$

We finally observe that u is not 0, thus we have indeed formally

$$\Sigma_n \tau_1^n u = u_0.$$

Step 2: Mathematical proof. In Step 1, we have been very formal, omitting in particular the questions relative to the domain for u. We now give a rigorous proof. Let us first state the following proposition.

Proposition 4.6. *For any $s \in \mathbb{R}$,*

$$\exp(-t\mathcal{B}_q^D)L^{2,s}(\Omega_0) \subset L^{2,s}(\Omega_0) \quad \text{for all } t \geq 0,$$

and there exists ω_s such that

$$\|\exp(-t\mathcal{B}_q^D)v\|_{L^{2,s}(\Omega_0)} \leq \exp(\omega_s t)\|v\|_{L^{2,s}(\Omega_0)} \quad \text{for all } v \in L^{2,s}(\Omega_0),$$

where $L^{2,s}$ is the weighted space

$$L^{2,s}(\Omega_0) = \{v \in L^2_{\mathrm{loc}}(\Omega_0) : \langle x\rangle^{\frac{s}{2}} v \in L^2(\Omega_0)\}$$

with $\langle x\rangle = \sqrt{1+x^2}$.

We can now give a mathematical sense to formulas (4.8)–(4.9). We observe that u_0 belongs to $L^{2,-s}(\Omega_0)$ for $s > \frac{1}{2}$. This implies by the proposition that u is well defined in particular in $L^{2,-s}(\Omega_0)$.

We now prove that u is actually in L^2. For this we use the Floquet theory. We observe that

$$u_p := \exp\Big(-\frac{p}{g}(\mathcal{B}_q^D - \lambda_0)\Big)u_0 \in L^{2,-s},$$

and that u_p satisfies the p-Floquet condition and the uniform bound

$$\|u_p\|_{L^2((0,1)^2\cap\Omega_0)} \le \check{C}\,\|u_0\|_{L^2((0,1)^2\cap\Omega_0)} \quad \text{for all } p \in [0, 2\pi],$$

for some constant $\check{C} > 0$. This implies, by the reverse Floquet decomposition formula (see around (4.7)), that u is indeed in $L^2(\Omega_0)$.

We now compute $\exp(-t(\mathcal{B}_q^D - \lambda_0))u$. It is clear using the periodicity of the map $\widehat{t} \mapsto \exp(-\widehat{t}(\mathcal{B}_q^D - \lambda_0))u_0$, that

$$\exp(-t(\mathcal{B}_q^D - \lambda_0))u = u \quad \text{in } L^{2,-s} \quad \text{for all } t \in \Big[0, \frac{2\pi}{g}\Big].$$

By semi-group theory, we immediately get that $u \in D(\mathcal{B}_q^D)$ and that $(\mathcal{B}_q^D - \lambda_0)u = 0$ as claimed in (4.10). This achieves the proof of Proposition 4.5 modulo the proof of Proposition 4.6. ∎

Proof of Proposition 4.6. To prove this proposition, we have to analyze the semi-group

$$t \mapsto \langle x\rangle^{\frac{s}{2}} \exp(-t\mathcal{B}_0^D)\langle x\rangle^{-\frac{s}{2}}$$

on $L^2(\Omega_0)$. Its associated infinitesimal generator is the unbounded closed operator

$$\mathcal{C}_s^D := \langle x\rangle^{\frac{s}{2}} \mathcal{B}_0^D \langle x\rangle^{-\frac{s}{2}}$$

on $L^2(\Omega_0)$. The associated differential operator is, with $\alpha_s := \langle x\rangle^{\frac{s}{2}}$,

$$-\frac{d^2}{dx^2} - \frac{d^2}{dy^2} + igx - 2\alpha_s'\alpha_s^{-1}\frac{d}{dx} - \alpha_s(\alpha_s^{-1})'',$$

which can be rewritten in the form

$$-\frac{d^2}{dx^2} - \frac{d^2}{dy^2} + igx + a_s(x)\frac{d}{dx} + b_s(x),$$

where a_s et b_s are bounded. The introduced perturbation does not change the form domain and using Hille–Yosida theorem and the results of Appendix A, we get the existence of ω_s such that

$$\|\langle x\rangle^{\frac{s}{2}} \exp(-t\mathcal{B}_0^D)\langle x\rangle^{-\frac{s}{2}}\|_{\mathcal{L}(L^2(\Omega_0))} \le \exp(\omega_s t) \quad \text{for all } t > 0.$$

This proves our proposition. ∎

Remark. We guess but have not proved that $K_{q,p=0}$ has only point spectrum, i.e. that its spectrum only consists of eigenvalues. Note that we could think of replacing in the above proof u_0 by a sequence of approximate eigenfunctions $u_0^{(n)}$. Formally (4.10) gives

$$(\mathcal{B}_0^D - \lambda_0)u^{(n)} = \frac{g}{2\pi}\Big(I - \exp\Big(-\frac{2\pi}{g}(\mathcal{B}_q^D - \lambda_0)\Big)\Big)u_0^{(n)}.$$

But on the right-hand side we have only a sequence of periodic functions which only tends to 0 on a fundamental domain. Hence, we do not have a Weyl sequence for $\mathcal{B}_0^D$ relative to λ_0. It remains some hope to proceed like in the proof of the Schnoll theorem by introducing cut-off functions.

5 Quasi-modes and non-emptiness of the spectrum

We would like to analyze the behavior of the operator as $g \to +\infty$ using the techniques of [2, 3, 13]. We take the semi-classical representation and look in the two-dimensional case at the asymptotic limit as $h \to 0$ of the spectrum of the semi-classical realization $\mathcal{A}_h^D$ of $-h^2\Delta + ix$.

5.1 Main results

The main result is:

Theorem 5.1. *Under Assumptions* (1.1)–(1.2) *and assuming that* H_0 *is a strictly convex set in* $\mathbb{R}^2$ *with boundary of positive curvature, we have*

$$\lim_{h\to 0} \frac{1}{h^{\frac{2}{3}}} \inf\{\Re\, \sigma(\mathcal{A}_h^D)\} = \frac{|a_1|}{2},$$

where $a_1 < 0$ *is the rightmost zero of the Airy function* Ai. *Moreover, for every* $\varepsilon > 0$, *there exist* $h_\varepsilon > 0$ *and* $C_\varepsilon > 0$ *such that*

$$\sup_{\substack{\gamma \le \frac{|a_1|}{2} \\ \nu \in \mathbb{R}}} \|(\mathcal{A}_h^D - (\gamma - \varepsilon)h^{\frac{2}{3}} - i\nu)^{-1}\| \le \frac{C_\varepsilon}{h^{\frac{2}{3}}} \quad \textit{for all } h \in (0, h_\varepsilon)$$

In particular, the spectrum of $\mathcal{A}_h^D$ *is not empty.*

If one now looks at the reduced problem on the cylinder $\mathbb{R} \times (-\frac{1}{2}, \frac{1}{2})$, the main result would be the same for $\mathcal{A}_h^{D,q}$ with the difference that we know that the spectrum is discrete.

Theorem 5.2. *Under the same assumptions as in Theorem* 5.1, *there exists* $\lambda(h, q)$ *such that*

$$\lim_{h\to 0} (\lambda(h,q) - ir)h^{-\frac{2}{3}} = |a_1| e^{-i\frac{\pi}{3}}$$

and such that

$$\lambda(h,q) + ik \in \sigma(\mathcal{A}_h^{D,q}) \quad \textit{for all } k \in \mathbb{Z}.$$

This result is essentially a reformulation of the results stated by Almog in [1] and in the case of the exterior problem in [2].

Remark. According to [2, 3, 13], similar results can be formulated for Neumann and Robin boundary conditions, by adapting the proofs from [2].

5.2 Proofs

Lower bound. The proof is identical to the exterior case considered in [2, Section 2.2]. The fact that there is an infinite number of holes instead one hole does not change the proof. The assumptions that $V(x, y) = x$ and the strict convexity of H_0 permits to verify all the assumptions appearing in this subsection. We recall the main lines with the simplifications that our potential V is simply $V(x, y) = x$.

By lower bound, we mean

$$\varliminf_{h\to 0} \frac{1}{h^{\frac{2}{3}}} \inf\{\Re\, \sigma(\mathcal{A}_h^D)\} \geq \frac{|a_1|}{2}. \tag{5.1}$$

We keep the notation of [3, Section 6] and [2]. For some $\frac{1}{3} < \rho < \frac{2}{3}$ and for every $h \in (0, h_0]$, we choose two sets of indices $\mathcal{J}_i(h)$, $\mathcal{J}_\partial(h)$, and a set of points

$$\{a_j(h) \in \Omega : j \in \mathcal{J}_i(h)\} \cup \{b_k(h) \in \partial\Omega : k \in \mathcal{J}_\partial(h)\} \tag{5.2a}$$

such that $B(a_j(h), h^\rho) \subset \Omega$,

$$\overline{\Omega} \subset \bigcup_{j\in\mathcal{J}_i(h)} B(a_j(h), h^\rho) \cup \bigcup_{k\in\mathcal{J}_\partial(h)} B(b_k(h), h^\rho), \tag{5.2b}$$

and such that the closed balls $\overline{B}(a_j(h), \frac{h^\rho}{2})$, $\overline{B}(b_k(h), \frac{h^\rho}{2})$ are all disjoint.

Now we construct in $\mathbb{R}^2$ two families of functions

$$(\chi_{j,h})_{j\in\mathcal{J}_i(h)} \quad \text{and} \quad (\zeta_{j,h})_{j\in\mathcal{J}_\partial(h)} \tag{5.2c}$$

such that, for every $x \in \overline{\Omega}$,

$$\sum_{j\in\mathcal{J}_i(h)} \chi_{j,h}(x)^2 + \sum_{k\in\mathcal{J}_\partial(h)} \zeta_{k,h}(x)^2 = 1, \tag{5.2d}$$

and such that

- $\operatorname{Supp} \chi_{j,h} \subset B(a_j(h), h^\rho)$ for $j \in \mathcal{J}_i(h)$,
- $\operatorname{Supp} \zeta_{j,h} \subset B(b_j(h), h^\rho)$ for $j \in \mathcal{J}_\partial(h)$,
- $\chi_{j,h} \equiv 1$ (respectively $\zeta_{j,h} \equiv 1$) on $\overline{B}(a_j(h), \frac{h^\rho}{2})$ (respectively $\overline{B}(b_j(h), \frac{h^\rho}{2})$).

We now define the approximate resolvent as in [2]

$$\mathcal{R}_h(z) = \sum_{j\in\mathcal{J}_i(h)} \chi_{j,h}(\mathcal{A}_h - z)^{-1}\chi_{j,h} + \sum_{j\in\mathcal{J}_\partial(h)} q_{j,h} R_{j,h}(z) q_{j,h},$$

where $R_{j,h}(z)$ is given by [2, equation (6.14)], and $q_{j,h} = \mathbf{1}_\Omega \zeta_{j,h}$. We write

$$\mathcal{R}_h(z) \circ (\mathcal{A}_h^D - z) = I + \mathcal{E}(h,z), \tag{5.3}$$

where

$$\mathcal{E}(h,z) = -h^2[\Delta, \chi_{j,h}](\mathcal{A}_h - z)^{-1}\chi_{j,h} + \sum_{j \in \mathcal{J}_\partial(h)} (\mathcal{A}_h - z) q_{j,h} R_{j,h}(z) q_{j,h}.$$

The control can be achieved as in [2]. We may thus conclude that for any $\varepsilon > 0$ there exists $C_\varepsilon > 0$ such that for sufficiently small h,

$$\sup_{\Re z \le h^{\frac{2}{3}}\left(\frac{|a_1|}{2} - \varepsilon\right)} \|\mathcal{E}(h,z)\| \le C\left(h^{2-2\rho-\frac{2}{3}} + h^{2\rho-\frac{2}{3}}\right).$$

Since for sufficiently small h, $(I + \mathcal{E}(h,z))$ becomes invertible, we can now use (5.3) to conclude that for any $\varepsilon > 0$ there exist $C_\varepsilon > 0$ and $h_\varepsilon > 0$ such that for any $h \in (0, h_\varepsilon]$,

$$\sup_{\Re z \le h^{\frac{2}{3}}\left(\frac{|a_1|}{2} - \varepsilon\right)} \|(\mathcal{A}_h^D - z)^{-1}\| \le \frac{C_\varepsilon}{h^{\frac{2}{3}}}.$$

This completes the proof of (5.1).

The proof of upper bounds. According to Proposition 3.2, it is enough to prove Theorem 5.2. Without loss of generality it is enough to present the proof for $q = 0$. Up to multiplication by e^{iyq}, the quasimode is the same and we can take it localized near the point $(r, 0)$ (see [3, formula (7.2)]). To prove that

$$\overline{\lim_{h\to 0}} \frac{1}{h^{\frac{2}{3}}} \inf\{\Re\, \sigma(\mathcal{A}_h^{D,q})\} \le \frac{|a_1|}{2}.$$

we use the same procedure presented in [3, Section 7].

6 Conclusion

The goal of the paper was to analyze mathematically the approach of [17] for the spectral analysis of the Bloch–Torrey equation in periodic perforated domains in $\mathbb{R}^d$.

We have proved that, at least in an asymptotic regime ($g \to +\infty$), the presence of holes creates indeed spectrum and that it was natural to detect these eigenvalues by analyzing an associate bounded operator on $L^2(\mathbb{T}^2)$ through an extension of the Floquet theory.

We have focused on the proofs for the Dirichlet realization $\mathcal{B}^D$ in planar domains. The precise definition of the realizations for other boundary conditions is treated in detail in the (1D)-case in [11] (see also [10] in (2D)) referring to the same variant of the Lax–Milgram theorem recalled in Appendix A. The semi-classical analysis of Section 5 relies on [2–5, 10, 13]. Note that [13] deals with problems without the restriction to the dimension. Nevertheless, as mentioned in the remarks of Section 3.3 some of the results and conjectures are specific to the dimension 2.

A Generalized Lax–Milgram theorem

In this Appendix we recall some results established in [4]. For other applications of these results we also mention [14] where magnetic Laplacians are considered.

We consider two Hilbert spaces $\mathcal{V}$ and $\mathcal{H}$ such that $\mathcal{V} \subset \mathcal{H}$, and that for some $C > 0$ and any $u \in \mathcal{V}$, we have

$$\|u\|_{\mathcal{H}} \leq C\,\|u\|_{\mathcal{V}}.$$

Suppose further that

$$\mathcal{V} \text{ is dense in } \mathcal{H}.$$

Consider a continuous sesquilinear form a defined on $\mathcal{V} \times \mathcal{V}$:

$$(u,v) \mapsto a(u,v).$$

Let

$$D(S) = \{u \in \mathcal{V} : v \mapsto a(u,v) \text{ is continuous on } \mathcal{V} \text{ in the norm of } \mathcal{H}\}, \tag{A.1}$$

and define the operator $S : D(S) \to \mathcal{H}$ by

$$a(u,v) = \langle Su, v\rangle_{\mathcal{H}} \quad \text{for all } u \in D(S) \text{ and all } v \in \mathcal{V}. \tag{A.2}$$

We have the following theorem:

Theorem A.1. *Assume that a is a continuous sesquilinear form satisfying for some* $\Phi_1, \Phi_2 \in \mathcal{L}(\mathcal{V})$,

$$\begin{aligned} |a(u,u)| + |a(u,\Phi_1(u))| &\geq \alpha \|u\|_{\mathcal{V}}^2 \quad \textit{for all } u \in \mathcal{V}, \\ |a(u,u)| + |a(\Phi_2(u),u)| &\geq \alpha \|u\|_{\mathcal{V}}^2 \quad \textit{for all } u \in \mathcal{V}. \end{aligned}$$

Assume further that Φ_1 *and* Φ_2 *extend into continuous linear maps in* $\mathcal{L}(\mathcal{H})$ *and let* S *be defined by* (A.1)–(A.2). *Then:*

(1) S *is bijective from* $D(S)$ *onto* $\mathcal{H}$ *and* $S^{-1} \in \mathcal{L}(\mathcal{H})$,

(2) $D(S)$ *is dense in both* $\mathcal{V}$ *and* $\mathcal{H}$,

(3) S *is closed,*

(4) *Let b denote the conjugate sesquilinear form of a, i.e.*

$$(u,v) \mapsto b(u,v) := \overline{a(v,u)}.$$

Let S_1 *denote the closed linear operator associated with b by the same construction. Then*

$$S^* = S_1 \quad \textit{and} \quad S_1^* = S.$$

Remark. We recall that the Hille–Yosida theorem (see [12, Theorem 13.22]) can be applied to the above defined operator S if we have

$$\Re\langle Su, u\rangle \geq -C\,\|u\|_{\mathcal{H}}^2 \quad \text{for all } u \in \mathcal{V}.$$

B Spectrum and Weyl's sequences

In this section, we adapt to the non-self-adjoint situation a characterization of the spectrum for a self-adjoint operator through Weyl's sequences.

Lemma B.1. *Let A be a closed operator in a Hilbert space H which has the property that, for any $\lambda \in \mathbb{C}$,*

$$(A-\lambda) \textit{ injective implies } (A^*-\overline{\lambda}) \textit{ injective.} \tag{B.1}$$

Under this assumption, the two assertions are equivalent:

(1) $\lambda \in \sigma(A)$.

(2) *There exists a sequence u_n in $D(A)$ such that $\|u_n\| = 1$ and $(A-\lambda)u_n \to 0$.*

Proof. The only difficulty is to prove that (1) implies (2). If there is no sequence $(u_n) \in D(A)$ such that (2) is satisfied, then there exists $c > 0$ such that

$$\|(A-\lambda)u\| \geq c\|u\| \quad \text{for all } u \in D(A).$$

From this we deduce that $(A-\lambda)$ is injective with closed range. Now we have

$$\operatorname{Range}(A-\lambda) = (\operatorname{Ker}(A^*-\overline{\lambda}))^{\perp}.$$

By (B.1) $(A^*-\overline{\lambda})$ is injective. Hence we get the surjectivity. ■

Remarks. We make the following observations.

- The property is evidently satisfied in the self-adjoint case because the spectrum is real.
- Property (B.1) is satisfied if H is a complex Hilbert space and if there is an antilinear involution Γ such that $\Gamma D(A) \subset D(A)$ and

$$\Gamma A = A^*\Gamma.$$

  In particular, it holds for our Bloch–Torrey operators by taking as Γ the complex conjugation.
- The lemma is not true without property (B.1). As suggested by the referee, one can indeed consider $H = L^2(0,1)$, $A = -i\frac{d}{dx}$, $D(A) = H_0^1(0,1)$. We have $\sigma(A) = \mathbb{C}$ and there is no a Weyl sequence for any $\lambda \in \mathbb{C}$. On the other hand $(A-\lambda)$ is injective for any λ and $(A^*-\overline{\lambda})$ is non-injective for any λ.

Acknowledgements. Denis S. Grebenkov acknowledges a partial financial support from the Alexander von Humboldt Foundation through a Bessel Research Award. The authors thank the anonymous referee for his remarks and suggestions, in particular when observing that the statement of Lemma B.1 that we initially gave in the submitted version, was wrong.

Bibliography

[1] Y. Almog, The stability of the normal state of superconductors in the presence of electric currents. *SIAM J. Math. Anal.* **40** (2008), 824–850

[2] Y. Almog, D. S. Grebenkov and B. Helffer, Spectral semi-classical analysis of a complex Schrödinger operator in exterior domains. *J. Math. Phys.* **59** (2018), Article ID 041501

[3] Y. Almog, D. S. Grebenkov and B. Helffer, On a Schrödinger operator with a purely imaginary potential in the semiclassical limit. *Comm. Partial Differential Equations* **44** (2019), 1542–1604

[4] Y. Almog and B. Helffer, On the spectrum of non-selfadjoint Schrödinger operators with compact resolvent. *Comm. Partial Differential Equations* **40** (2015), 1441–1466

[5] Y. Almog and R. Henry, Spectral analysis of a complex Schrödinger operator in the semiclassical limit. *SIAM J. Math. Anal.* **48** (2016), 2962–2993

[6] K.-J. Engel and R. Nagel, *One-parameter semigroups for linear evolution equations*. Grad. Texts in Math. 194, Springer, New York, 2000

[7] D. S. Grebenkov, NMR survey of reflected Brownian motion. *Rev. Mod. Phys.* **79** (2007), 1077–1137

[8] D. S. Grebenkov, Exploring diffusion across permeable barriers at high gradients. II. Localization regime. *J. Magn. Reson.* **248** (2014), 164–176

[9] D. S. Grebenkov, Diffusion MRI/NMR at high gradients: Challenges and perspectives. *Micro. Meso. Mater.* **269** (2017), 79–82

[10] D. S. Grebenkov and B. Helffer, On spectral properties of the Bloch–Torrey operator in two dimensions. *SIAM J. Math. Anal.* **50** (2018), 622–676

[11] D. S. Grebenkov, B. Helffer and R. Henry, The complex Airy operator on the line with a semipermeable barrier. *SIAM J. Math. Anal.* **49** (2017), 1844–1894

[12] B. Helffer, *Spectral theory and its applications*. Cambridge Stud. Adv. Math. 139, Cambridge University Press, Cambridge, 2013

[13] R. Henry, On the semi-classical analysis of Schrödinger operators with purely imaginary electric potentials in a bounded domain. Preprint 2014, arXiv:1405.6183

[14] D. Krejčiřík, N. Raymond, J. Royer and P. Siegl, Reduction of dimension as a consequence of norm-resolvent convergence and applications. *Mathematika* **64** (2018), 406–429

[15] P. Kuchment, *Floquet theory for partial differential equations*. Oper. Theory Adv. Appl. 60, Birkhäuser, Basel, 1993

[16] P. Kuchment, An overview of periodic elliptic operators. *Bull. Amer. Math. Soc. (N.S.)* **53** (2016), 343–414

[17] N. Moutal, A. Moutal and D. Grebenkov, Diffusion NMR in periodic media: Efficient computation and spectral properties. *J. Phys. A* **53** (2020), Article ID 325201

[18] M. Reed and B. Simon, *Methods of modern mathematical physics. I–IV*. Academic Press, New York, 1972–1978

[19] S. D. Stoller, W. Happer and F. J. Dyson, Transverse spin relaxation in inhomogeneous magnetic fields. *Phys. Rev. A* **44** (1991), 7459–7477

[20] C. H. Wilcox, Theory of Bloch waves. *J. Analyse Math.* **33** (1978), 146–167

Counting bound states with maximal Fourier multipliers

Dirk Hundertmark, Peer Kunstmann, Tobias Ried and Semjon Vugalter

To Ari who has been a source of inspiration and has set high standards for kindness

We report on a version of Cwikel's proof of the famous Cwikel–Lieb–Rozenblum (CLR) inequality which highlights the connection of the CLR inequality to maximal Fourier multipliers. This new approach enables us to get a constant at least ten times better than Cwikels in all dimensions. In dimensions $d \geq 5$ our results are better than all previously known ones.

1 Introduction

For the one-particle Schrödinger operator $P^2 + V$ with a real-valued potential V, a bound for the number of negative eigenvalues, including their multiplicities, with the right semi-classical behavior, goes back to Cwikel, Lieb, and Rozenblum [3, 15, 16, 18, 19], who give very different proofs for the bound

$$N(P^2 + V) \leq L_d \int_{\mathbb{R}^d} V_-(x)^{\frac{d}{2}} \,\mathrm{d}x \tag{1.1}$$

for the number of negative eigenvalues of a Schrödinger operator. This bound is a semi-classical bound since a simple scaling argument shows that the classical phase-space volume of the region of negative energy is given by

$$N^{\mathrm{cl}}(\eta^2 + V) = \iint_{\{\eta^2 + V(x) < 0\}} \frac{\mathrm{d}\eta \,\mathrm{d}x}{(2\pi)^d} = \frac{|B_1^d|}{(2\pi)^d} \int_{\mathbb{R}^d} V_-(x)^{\frac{d}{2}} \,\mathrm{d}x,$$

wherc $|B_1^d|$ is the volume of the unit ball in $\mathbb{R}^d$. The reason for a semi-classical bound of the form (1.1) is the uncertainty principle, which forces a quantum particle to occupy roughly a classical phase-space volume $(2\pi)^d$. Thus the classical phase-space volume where the classical Hamiltonian energy $H(\eta, x) = \eta^2 + V(x)$ is negative, should be

Keywords: CLR inequality, bound states, maximal Fourier multipliers
2020 Mathematics Subject Classification: Primary 35P15; secondary 35J10, 81Q10

able to control $N(P^2 + V)$. The CLR bound (1.1) shows that this is the case, even for very rough potentials, up to a factor $L_d = C_d (2\pi)^d |B_1^d|^{-1}$.

We will not go too much into the history of this famous bound. Let us mention that Rozenblum's paper [18] was an announcement of his result. Typically for the journal, it did not contain any proofs, full proofs were published in [1, 19]. Similarly, Lieb's paper [15] is an announcement of his result and the details of his proof were published later in [16, 21]. Rozenblum's approach was strongly motivated by the St. Petersburg school of mathematical physics around Birman and Solomyak, whose work had been virtually unnoticed in the west until the mid 1970s, see the "Added notes" in [20, p. 378]. Simon's work [20] motivated the work of Cwikel and Lieb.

While it is very difficult to get estimates, even on the order of magnitude, of the constant in the CLR bound from Rozenblum's proof, the proofs of Cwikel and Lieb give explicit, dimension dependent bounds, with a considerable growth in the dimension d. Lieb's bound on C_d grows like $\sqrt{\pi d}(1 + O(d^{-1}))$, see [21] or [17, Section 3.4] for an excellent discussion of Lieb's method and Remark 1 below for some explicit numbers, while Cwikel's bound grows much stronger.

However, any physically reasonable bound on C_d should not grow in the dimension, since semi-classical arguments tend to work better in high dimensions. The first dimension independent bound $C_d \leq 81$ was derived by extending Cwikel's original proof to operator-valued potentials in 2002 in [8]. This work extended an *induction in the dimension* argument[1] by Laptev and Weidl [13], who were the first to derive Lieb–Thirring bounds with the sharp classical Lieb–Thirring constant in all dimensions in certain cases. Although the upper bound from [8] does not depend on the dimension, it is certainly too large.

There have been several previous attempts to improve Lieb's result, for example due to Conlon [2], Li and Yau [14], Frank [5], and Weidl [23, 24], see also the reviews [6, 9, 12]. All these very different proofs shed a new light on the CLR bound, but failed to give better bounds on the involved constants than already achieved by Lieb more than 40 years ago.

By drastically simplifying and generalizing the ideas of Cwikel, in particular, connecting them with certain maximal Fourier multiplier bounds on $L^2(\mathbb{R}^d)$, our approach leads to a new class of variational problems, see Theorem 1.1 below, which allows us to improve Lieb's constants in dimensions $d \geq 5$. This is the first improvement on Lieb's constants since [15].

Theorem 1.1. *The number $N(P^2 + V)$ of negative energy bound states of $P^2 + V$ obeys the semi-classical bound*

$$N(P^2 + V) \leq C_d \frac{|B^d|}{(2\pi)^d} \int_{\mathbb{R}^d} V_-(x)^{\frac{d}{2}}\, \mathrm{d}x \tag{1.2}$$

[1] See also [11] for some indication of the *induction in dimension* trick.

with constant

$$C_d = \frac{d^{d+1}}{4(d-2)^{d-2}} M_d,$$

where

$$M_d = \inf\Big\{ \big(\|m_1\|_{L^2(\mathbb{R}_+,\frac{ds}{s})} \|m_2\|_{L^2(\mathbb{R}_+,\frac{ds}{s})}\big)^{d-2} \int_0^\infty (1 - t^{-1} m(t))^2 t^{1-d}\,dt \Big\}, \quad (1.3)$$

the infimum being taken over all $m_1, m_2 \in L^2(\mathbb{R}_+, \frac{ds}{s})$, and $m = m_1 * m_2$ denotes the convolution of m_1, m_2 on $(\mathbb{R}_+, \frac{ds}{s})$.

Table 1.1 compares the upper bounds on C_d, obtained with our method, with the best known ones so far.

d	Our results	Best known so far	Lower bound*
3	7.55151	6.86924	6.75000
4	6.32791	6.03398	5.33333
5	5.95405	5.96677	4.82253
6	5.77058	6.07489	4.55625
7	5.67647	6.24464	4.39229
8	5.63198	6.43921	4.28088
9	5.62080	6.64378	4.20028

Table 1.1. Comparison of the upper bounds on C_d (* for the constant achievable with our method).

Remarks. (1) Our bounds also hold for Schrödinger operators with operator-valued potentials *with the same constants*. The best known bound for a non-relativistic Schrödinger operator with operator-valued potentials was due to Frank, Lieb, and Seiringer [7] who showed $C_d \leq 10.332$ in this case.

(2) All bounds on C_d in the third column of Table 1.1 were obtained already in the original work of Lieb.[2] For dimensions $d = 3, \dots, 9$ our upper bounds are compared with the values of the lower bound (1.6) for the best constant achievable by our method in the last column of Table 1.1.

(3) The minimization problem for M_d in (1.3) is new and, as Theorem 1.1 shows, has considerable implications for the spectral theory of Schrödinger operators. For a more detailed discussion of upper and lower bounds on the variational problem (1.3), including numerical examples which yield the upper bounds from Table 1.1, see [10].

(4) For ultra-relativistic Schrödinger operators in three dimensions, we get the upper bound

$$N(|P| + V) \leq 5.77058 \int_{\mathbb{R}^3} V_-(x)^3\,dx$$

[2] The numbers are taken from Roepstorff's book [17, Table 3.1].

see [10], which improves the result of Daubechies [4], who gets

$$N(|P| + V) \le 6.08 \int_{\mathbb{R}^3} V_-(x)^3 \, dx.$$

We can also handle polyharmonic Schrödinger-type operators in [10] and derive a generalization of Cwikel's weak trace ideal bound.

(5) For an upper bound on C_d, a simple, but not optimal, choice in the variational problem (1.3) is $m_1(s) = \sqrt{2}s\mathbf{1}_{\{0<s\le 1\}}$ and $m_2(s) = \sqrt{2}s^{-1}\mathbf{1}_{\{s>1\}}$, in which case $\|m_1\|_{L^2(\mathbb{R}_+,\frac{ds}{s})} = \|m_2\|_{L^2(\mathbb{R}_+,\frac{ds}{s})} = 1$ and $m(t) = m_1 * m_2(t) = \min(t, t^{-1})$, so

$$\int_0^\infty (1 - t^{-1}m(t))^2 t^{1-d} \, dt = \int_1^\infty (1 - t^{-2})^2 t^{1-d} \, dt = \frac{8}{(d-2)d(d+2)}.$$

This gives the upper bound

$$C_d \le \frac{2\, d^d}{(d-2)^{d-1}(d+2)} \tag{1.4}$$

as a possible constant in the CLR inequality and yields $C_3 \le 10.8$, already an order of a magnitude smaller than Cwikel's bound. Moreover, combining this with "stripping-off-dimensions" ideas with the help of similar bounds for operator-valued potentials, one can get this bound also uniformly in the dimension. At the end of the introduction we will show that a lower bound on M_d is given by

$$M_d \ge \frac{2}{d(d-1)(d-2)}, \tag{1.5}$$

which yields a lower bound for the constant C_d achievable with our method,

$$C_d \ge \frac{d^d}{2(d-1)(d-2)^{d-1}} =: C_d^{\text{lower}}, \tag{1.6}$$

giving the last column in Table 1.1. In particular,

$$\frac{C_d}{C_d^{\text{lower}}} \le 4\frac{d-1}{d+2} < 4,$$

which shows that even our (very) easy upper bound (1.4) is less than a factor of 4 off the lower bound and, in particular, off the truth. In addition,

$$C_d^{\text{lower}} = \frac{d^2}{2(d-1)(d-2)}\left(1 + \frac{2}{d-2}\right)^{d-2} \xrightarrow{d\to\infty} \frac{e^2}{2} \ge 3.69452\,.$$

This comparison shows that there is not too much room to improve on the upper bounds we obtained, even if one finds the sharp value in the minimization problem for M_d in (1.3).

In the next section we explain the ideas behind our proof of Theorem 1.1, but before we do that, we prove the lower bound (1.5). This is based on the observation that by Cauchy–Schwarz, one has $\|m\|_\infty \le \|m_1\|_{L^2(\mathbb{R}_+,\frac{ds}{s})}\|m_2\|_{L^2(\mathbb{R}_+,\frac{ds}{s})}$ for $m = m_1 * m_2$. Thus,

$$\begin{aligned} M_d &\ge \inf_m \Big\{ \|m\|_\infty^{d-2} \int_0^\infty (1 - t^{-1}m(t))^2 t^{1-d}\,\mathrm{d}t \Big\} \\ &= \inf_{l>0} \inf_{\|m\|_\infty = l} \Big\{ l^{d-2} \int_0^\infty (1 - t^{-1}m(t))^2 t^{1-d}\,\mathrm{d}t \Big\}. \end{aligned}$$

To make the integral $\int_0^\infty (1 - t^{-1}m(t))^2 t^{1-d}\,\mathrm{d}t$ as small as possible under the constraint $\|m\|_\infty = l$, the optimal choice is $m(t) = m_l(t) = \min(t, l)$. Since

$$\int_0^\infty (1 - t^{-1}m_l(t))^2 t^{1-d}\,\mathrm{d}t = l^{2-d} \frac{2}{(d-2)(d-1)d},$$

we get

$$M_d \ge \inf_{l>0} \Big\{ l^{d-2} \int_0^\infty (1 - t^{-1}m_l(t))^2 t^{1-d}\,\mathrm{d}t \Big\} = \frac{2}{(d-2)(d-1)d}.$$

2 The splitting trick

The main idea in the proof of Theorem 1.1 is quite simple but powerful. We define $U := V_- \ge 0$. As quadratic forms we have that $P^2 + V \ge P^2 - U$. This and the Birman–Schwinger principle shows

$$N(P^2 + V) \le N(P^2 - U) = n(U^{\frac12}|P|^{-2}U^{\frac12}; 1),$$

where $n(A;\kappa)$ is the number of singular values $(s_j(A))_{j\in\mathbb{N}}$ greater than $\kappa > 0$ of a compact operator A.

We denote by $\mathcal{F}$ the Fourier transform and by $\mathcal{F}^{-1}$ its inverse, by M_h the operator of multiplication with a function h, and $A = A_f = M_f\mathcal{F}^{-1}M_{1/|\cdot|}$ for f a non-negative (measurable) function on $\mathbb{R}^d$. When $f(x) = U(x)^{\frac12}$, then

$$AA^* = U^{\frac12}|P|^{-2}U^{\frac12},$$

which has the same non-zero eigenvalues as A^*A. Thus

$$N(P^2 - U) = n(A_f; 1) \quad \text{with } f = U^{\frac12}.$$

In particular, the Chebyshev–Markov inequality gives

$$N(P^2 - U) = n(A_f; 1) \le \sum_j \frac{(s_j(A_f) - \mu)_+^2}{(1-\mu)^2}$$

for any $0 < \mu < 1$. Let us drop the dependence of A on f for the moment. We want to split $A = B + H$, where B is bounded and H is Hilbert–Schmidt. Note that Ky Fan's inequality for singular values [22, Theorem 1.7] yields

$$s_j(A) = s_j(B + H) \le \|B\| + s_j(H)$$

for all $j \in \mathbb{N}$. So if $\|B\| \le \mu < 1$ we get

$$N(P^2 - U) \le (1 - \mu)^{-2} \sum_{j \in \mathbb{N}} s_j(H)^2 = (1 - \mu)^{-2} \|H\|_{\mathrm{HS}}^2, \tag{2.1}$$

where $\|H\|_{\mathrm{HS}}$ denotes the Hilbert–Schmidt norm of the operator H.

In order to make the above argument work, one has to be able to split

$$A_f = B_f + H_f$$

in such a way that the Hilbert–Schmidt norm of H_f is easy to calculate and one has a good bound on the operator norm of B_f. It will turn out, see (2.4) below, that one has $\|H_f\|_{\mathrm{HS}}^2 = c \int_{R^d} f(x)^d \, \mathrm{d}x$, so the right-hand side of (2.1) has exactly the right (semi-classical) scaling in $f = U^{\frac{1}{2}}$.

This has two important consequences:

(1) In order to use (2.1), we must have an upper bound μ on the operator norm of B_f which is *independent* of f. Using a devil's advocate game, given $\varphi \in L^2$ one can choose $f \ge 0$ in such a way as to make $B_f \varphi$ pointwise as big as possible. That is, for given φ one can choose a measurable function $f_* \ge 0$ to have $B_{f_*}\varphi(x)$ as big as possible for almost all x. This leads naturally to an associated maximal operator, see (2.6) below. Cwikel constructed a decomposition of $A_f = B_f + H_f$ using a dyadic decomposition in the ranges of f and $\eta \mapsto \frac{1}{|\eta|}$. Collecting suitable terms for B_f, in the spirit of a discrete convolution, enabled him to get an operator bound on B_f, which is independent of f. We will do this in a simpler and more effective way. This allows us to get a constant which is much smaller, in general, than the original constant by Cwikel.

(2) Since H_{Lf} is a Hilbert–Schmidt operator for all $L > 0$, it follows that A_f is a compact operator. Indeed, we have

$$A_f = L^{-1} A_{Lf} = L^{-1} B_{Lf} + L^{-1} H_{Lf},$$

so that in operator norm

$$\|A_f - L^{-1} H_{Lf}\| \le L^{-1} \|B_{Lf}\| \le \frac{\mu}{L} \to 0 \quad \text{as } L \to \infty$$

because $\|B_{Lf}\| \le \mu$ independent of L. This shows that A_f is the norm limit of Hilbert–Schmidt operators, hence compact. Thus the Birman–Schwinger operator $U^{\frac{1}{2}} |P|^{-2} U^{\frac{1}{2}}$ is also compact.

Writing out the inverse Fourier transform, one sees that A_f has a kernel

$$A_f(x,\eta) = (2\pi)^{-\frac{d}{2}} e^{ix\cdot\eta} \frac{f(x)}{|\eta|},$$

that is,

$$A_f\varphi(x) = f(x)\mathcal{F}^{-1}\Big[\frac{1}{|\cdot|}\varphi\Big](x) = (2\pi)^{-\frac{d}{2}} \int_{\mathbb{R}^d} e^{ix\cdot\eta} \frac{f(x)}{|\eta|}\varphi(\eta)\,\mathrm{d}\eta,$$

at least for nice enough φ.

In order to write A_f as a sum of a bounded and a Hilbert–Schmidt operator, set $t = \frac{f(x)}{|\eta|}$, split $t = m(t) + t - m(t)$ for some bounded, measurable function $m : [0,\infty) \to \mathbb{R}$, and define $B_{f,m}$ and $H_{f,m}$ via their kernels

$$B_{f,m}(x,\eta) = (2\pi)^{-\frac{d}{2}} e^{ix\cdot\eta} m\Big(\frac{f(x)}{|\eta|}\Big), \tag{2.2}$$

$$H_{f,m}(x,\eta) = (2\pi)^{-\frac{d}{2}} e^{ix\cdot\eta} \Big(\frac{f(x)}{|\eta|} - m\Big(\frac{f(x)}{|\eta|}\Big)\Big). \tag{2.3}$$

It is then clear that $A_f = B_{f,m} + H_{f,m}$. It turns out that the Hilbert–Schmidt norm of $H_{f,m}$ is easy to calculate and it is not too hard to get an explicit bound on the operator norm of $B_{f,m}$ on L^2 under a suitable assumption on m.

Theorem 2.1. *The following statements hold.*

(1) *The Hilbert–Schmidt norm of $H_{f,m}$ is given by*

$$\|H_{f,m}\|_{\mathrm{HS}}^2 = \frac{d|B_1^d|}{(2\pi)^d} \int_0^\infty (1 - t^{-1}m(t))^2 t^{1-d}\,\mathrm{d}t \int_{\mathbb{R}^d} f(x)^d\,\mathrm{d}x. \tag{2.4}$$

(2) *If m is given by a convolution, that is,*

$$m(t) = m_1 * m_2(t) = \int_0^\infty m_1\Big(\frac{t}{s}\Big) m_2(s) \frac{\mathrm{d}s}{s},$$

then for all measurable non-negative functions f the operator $B_{f,m}$ is bounded on $L^2(\mathbb{R}^d)$ with

$$\|B_{f,m}\|_{L^2\to L^2} \le \Big(\int_0^\infty |m_1(s)|^2 \frac{\mathrm{d}s}{s}\Big)^{\frac{1}{2}} \Big(\int_0^\infty |m_2(s)|^2 \frac{\mathrm{d}s}{s}\Big)^{\frac{1}{2}}. \tag{2.5}$$

Remark. We stress that the bound on the operator norm of $B_{f,m}$ is *independent* of the choice of f, as required. This will turn out to be a natural consequence of the convolution structure of m.

Proof of Theorem 2.1. Since the operator $H_{f,m}$ has a kernel given by the right-hand side of equation (2.3), one computes its Hilbert–Schmidt norm using a simple scaling in the η integral as

$$\begin{aligned}\|H_{f,m}\|_{\mathrm{HS}}^2 &= \iint_{\mathbb{R}^d\times\mathbb{R}^d}\Big(\frac{f(x)}{|\eta|}-m\Big(\frac{f(x)}{|\eta|}\Big)\Big)^2\frac{\mathrm{d}x\,\mathrm{d}\eta}{(2\pi)^d}\\ &= \int_{\mathbb{R}^d} f(x)^d\,\mathrm{d}x\int_{\mathbb{R}^d}(|\eta|^{-1}-m(|\eta|^{-1}))^2\frac{\mathrm{d}\eta}{(2\pi)^d}.\end{aligned}$$

Going to spherical coordinates shows

$$\begin{aligned}\int_{\mathbb{R}^d}(|\eta|^{-1}-m(|\eta|^{-1}))^2\frac{\mathrm{d}\eta}{(2\pi)^d} &= \frac{|S^{d-1}|}{(2\pi)^d}\int_0^\infty(r^{-1}-m(r^{-1}))^2r^{d-1}\,\mathrm{d}r\\ &= \frac{d|B_1^d|}{(2\pi)^d}\int_0^\infty(1-t^{-1}m(t))^2t^{1-d}\,\mathrm{d}t,\end{aligned}$$

where $|S^{d-1}|$ is the surface area of the unit sphere in $\mathbb{R}^d$ and $|B_1^d|$ the volume of the unit ball in $\mathbb{R}^d$. This proves (2.4).

For the proof of the second half of Theorem 2.1 let

$$B_{t,m}\varphi := \mathcal{F}\Big[m\Big(\frac{t}{|\cdot|}\Big)\varphi\Big]$$

and note that for any measurable $f \geq 0$ we have

$$B_{f,m}\varphi(x) \leq B_m^*\varphi(x) := \sup_{t>0}|B_{t,m}\varphi(x)|, \tag{2.6}$$

where, strictly speaking, we should take the supremum in $t > 0$ over a dense subset of the positive reals, in order to ensure that the right-hand side above is measurable.[3] Note that by the convolution structure of m,

$$\begin{aligned}|B_{t,m}\varphi(x)| &= \Big|\int_0^\infty \mathcal{F}\Big[m_2\Big(\frac{s}{|\cdot|}\Big)\varphi\Big](x)m_1\Big(\frac{t}{s}\Big)\frac{\mathrm{d}s}{s}\Big|\\ &\leq \int_0^\infty\Big|\mathcal{F}\Big[m_2\Big(\frac{s}{|\cdot|}\Big)\varphi\Big](x)\Big|\Big|m_1\Big(\frac{t}{s}\Big)\Big|\frac{\mathrm{d}s}{s}\\ &\leq \Big(\int_0^\infty\Big|\mathcal{F}[m_2\Big(\frac{s}{|\cdot|}\Big)\varphi](x)\Big|^2\frac{\mathrm{d}s}{s}\Big)^{\frac12}\Big(\int_0^\infty\Big|m_1\Big(\frac{t}{s}\Big)\Big|^2\frac{\mathrm{d}s}{s}\Big)^{\frac12}\\ &= \Big(\int_0^\infty\Big|\mathcal{F}[m_2\Big(\frac{s}{|\cdot|}\Big)\varphi](x)\Big|^2\frac{\mathrm{d}s}{s}\Big)^{\frac12}\Big(\int_0^\infty|m_1(s)|^2\frac{\mathrm{d}s}{s}\Big)^{\frac12}\end{aligned} \tag{2.7}$$

since the measure $\mathrm{d}s/s$ on $(0,\infty)$ is invariant under scaling $s \mapsto ts$ and inversion $s \mapsto s^{-1}$.

[3] Since m is continuous, being a convolution of two L^2 functions, this will not make any difference.

As the right-hand side of (2.7) is *independent of* $t > 0$, we get

$$B_m^*\varphi(x)^2 \le \|m_1\|^2_{L^2(\mathbb{R}_+,\frac{ds}{s})} \int_0^\infty \left|\mathcal{F}\left[m_2\left(\frac{s}{|\cdot|}\right)\varphi\right](x)\right|^2 \frac{ds}{s},$$

thus

$$\begin{aligned}
\|B_m^*\varphi\|^2 &\le \|m_1\|^2_{L^2(\mathbb{R}_+,\frac{ds}{s})} \int_{\mathbb{R}^d}\int_0^\infty \left|\mathcal{F}\left[m_2\left(\frac{s}{|\cdot|}\right)\varphi\right](x)\right|^2 \frac{ds}{s}\,dx \\
&= \|m_1\|^2_{L^2(\mathbb{R}_+,\frac{ds}{s})} \int_0^\infty\int_{\mathbb{R}^d} \left|m_2\left(\frac{s}{|\eta|}\right)\varphi(\eta)\right|^2 d\eta\,\frac{ds}{s} \\
&= \|m_1\|^2_{L^2(\mathbb{R}_+,\frac{ds}{s})} \int_{\mathbb{R}^d}\int_0^\infty \left|m_2\left(\frac{s}{|\eta|}\right)\varphi(\eta)\right|^2 \frac{ds}{s}\,d\eta \\
&= \|m_1\|^2_{L^2(\mathbb{R}_+,\frac{ds}{s})}\|m_2\|^2_{L^2(\mathbb{R}_+,\frac{ds}{s})}\|\varphi\|^2
\end{aligned}$$

by interchanging the integrals and scaling out the factor $\frac{1}{|\eta|}$ in the last $\frac{ds}{s}$ integral as in (2.7). This proves the second part of Theorem 2.1. ∎

In the rest of this section we will discuss how Theorem 2.1 and the bound (2.1) lead to the Cwikel–Lieb–Rozenblum bound for a non-relativistic single-particle Schrödinger operator.

By scaling f by $\kappa > 0$ and using $\kappa A_f = A_{\kappa f} = B_{\kappa f,m} + H_{\kappa f,m}$, the argument leading to (2.1) yields

$$\begin{aligned}
N(P^2-U) &= n(A_{\kappa f};\kappa) \\
&\le (\kappa-\mu)^{-2}\,\|H_{\kappa f,m}\|^2_{\mathrm{HS}} \\
&= \frac{\kappa^d}{(\kappa-\mu)^2}\frac{d|B_1^d|}{(2\pi)^d}\int_0^\infty (1-t^{-1}m(t))^2 t^{1-d}\,dt \int_{\mathbb{R}^d} U(x)^{\frac{d}{2}}\,dx\,, \qquad (2.8)
\end{aligned}$$

as long as $\kappa > \mu \ge \|B_{\kappa f,m}\|$. It is important to note here that the last factor on the right-hand side of the above bound has the correct dependence on the potential U. That is, the factor in front of it, which depends on the upper bound μ on the operator norm of $B_{f,m}$, *has to be independent* of $f = \sqrt{U}$. The second part of Theorem 2.1 allows us to use

$$\mu = \|m_1\|_{L^2(\mathbb{R}_+,\frac{ds}{s})}\|m_2\|_{L^2(\mathbb{R}_+,\frac{ds}{s})}$$

as an upper bound for $\|B_{f,m}\|$, which is independent of f, so the same bound holds for $\|B_{\kappa f,m}\|$ for any $\kappa > 0$. We can now freely optimize in $\kappa > \mu$ in (2.8), to get

$$N(P^2-U) \le C_d \frac{|B_1^d|}{(2\pi)^d}\int_{\mathbb{R}^d} U(x)^{\frac{d}{2}}\,dx$$

with constant

$$C_d = \frac{d^{d+1}}{4(d-2)^{d-2}}\mu^{d-2}\int_0^\infty (1-t^{-1}m(t))^2 t^{1-d}\,dt\,.$$

This proves (1.2) and shows how bounds for maximal Fourier multipliers are at the heart of our proof of Theorem 1.1.

Acknowledgements. It is a pleasure to thank Michael Cwikel for many helpful comments and remarks on [10]. We gratefully acknowledge financial support by the Deutsche Forschungsgemeinschaft (DFG) – Project-ID 258734477 – SFB 1173. Dirk Hundertmark also thanks the Alfried Krupp von Bohlen und Halbach Foundation for financial support. We would also like to thank the Mathematisches Forschungsinstitut Oberwolfach (MFO) and the Centre International de Rencontres Mathématiques (CIRM) Luminy for their research in pairs programs, where part of this work was conceived.

Bibliography

[1] M. Š. Birman and M. Z. Solomjak, Quantitative analysis in Sobolev's imbedding theorems and applications to spectral theory. In *Tenth Mathematical School (Summer School, Kaciveli/Nalchik, 1972)*, pp. 5–189, Inst. Mat. Akad. Nauk Ukrain. SSR, Kiev, 1974 (in Russian); English transl. Amer. Math. Soc. Transl. Ser. 2 189, American Mathematical Society, Providence, 1980

[2] J. G. Conlon, A new proof of the Cwikel–Lieb–Rosenbljum bound. *Rocky Mountain J. Math.* **15** (1985), 117–122

[3] M. Cwikel, Weak type estimates for singular values and the number of bound states of Schrödinger operators. *Ann. of Math. (2)* **106** (1977), 93–100

[4] I. Daubechies, An uncertainty principle for fermions with generalized kinetic energy. *Comm. Math. Phys.* **90** (1983), 511–520

[5] R. L. Frank, Cwikel's theorem and the CLR inequality. *J. Spectr. Theory* **4** (2014), 1–21

[6] R. L. Frank, Eigenvalue bounds for the fractional Laplacian: a review. In *Recent developments in nonlocal theory*, pp. 210–235, De Gruyter, Berlin, 2018

[7] R. L. Frank, E. H. Lieb and R. Seiringer, Number of bound states of Schrödinger operators with matrix-valued potentials. *Lett. Math. Phys.* **82** (2007), 107–116

[8] D. Hundertmark, On the number of bound states for Schrödinger operators with operator-valued potentials. *Ark. Mat.* **40** (2002), 73–87

[9] D. Hundertmark, Some bound state problems in quantum mechanics. In *Spectral theory and mathematical physics: A Festschrift in honor of Barry Simon's 60th birthday*, pp. 463–496, Proc. Sympos. Pure Math. 76, American Mathematical Society, Providence, 2007

[10] D. Hundertmark, P. C. Kunstmann, T. Ried and S. Vugalter, Cwikel's bound reloaded. Preprint 2018, arXiv:1809.05069

[11] A. Laptev, Dirichlet and Neumann eigenvalue problems on domains in Euclidean spaces. *J. Funct. Anal.* **151** (1997), 531–545

[12] A. Laptev and T. Weidl, Recent results on Lieb–Thirring inequalities. In *Journées "Équations aux Dérivées Partielles" (La Chapelle sur Erdre, 2000)*, Exp. No. XX, 14, Université Nantes, Nantes, 2000

[13] A. Laptev and T. Weidl, Sharp Lieb–Thirring inequalities in high dimensions. *Acta Math.* **184** (2000), 87–111

[14] P. Li and S. T. Yau, On the Schrödinger equation and the eigenvalue problem. *Comm. Math. Phys.* **88** (1983), 309–318

[15] E. Lieb, Bounds on the eigenvalues of the Laplace and Schroedinger operators. *Bull. Amer. Math. Soc.* **82** (1976), 751–753

[16] E. H. Lieb, The number of bound states of one-body Schroedinger operators and the Weyl problem. In *Geometry of the Laplace operator (Proc. Sympos. Pure Math., Univ. Hawaii, Honolulu, Hawaii, 1979)*, pp. 241–252, Proc. Sympos. Pure Math. 36, American Mathematical Society, Providence, 1980

[17] G. Roepstorff, *Path integral approach to quantum physics. An introduction.* Texts Monogr. Phys., Springer, Berlin, 1994

[18] G. V. Rozenbljum, Distribution of the discrete spectrum of singular differential operators. *Dokl. Akad. Nauk SSSR* **202** (1972), 1012–1015 (in Russian); English transl. *Soviet Math. Dokl.* **13** (1972), 245–249

[19] G. V. Rozenbljum, Distribution of the discrete spectrum of singular differential operators. *Izv. Vysš. Učebn. Zaved. Matematika* (1976), no. 1(164), 75–86 (in Russian); English transl. *Soviet Math. (Iz. VUZ)* **20** (1976), 63–71

[20] B. Simon, Analysis with weak trace ideals and the number of bound states of Schrödinger operators. *Trans. Amer. Math. Soc.* **224** (1976), 367–380

[21] B. Simon, *Functional integration and quantum physics*. 2nd edn., AMS Chelsea Publishing, Providence, 2005

[22] B. Simon, *Trace ideals and their applications*. 2nd edn., Math. Surveys Monogr. 120, American Mathematical Society, Providence, 2005

[23] T. Weidl, Another look at Cwikel's inequality. In *Differential operators and spectral theory*, pp. 247–254, Amer. Math. Soc. Transl. Ser. 2 189, American Mathematical Society, Providence, 1999

[24] T. Weidl, Nonstandard Cwikel type estimates. In *Interpolation theory and applications*, pp. 337–357, Contemp. Math. 445, American Mathematical Society, Providence, 2007

Sharp dimension estimates of the attractor of the damped 2D Euler–Bardina equations

Alexei Ilyin and Sergey Zelik

To Ari Laptev on the occasion of his 70th birthday

We prove existence of the global attractor of the damped and driven 2D Euler–Bardina equations on the torus and give an explicit two-sided estimate of its dimension that is sharp as $\alpha \to 0^+$.

1 Introduction

The Navier–Stokes system remains over the last decades in the focus of the theory of infinite-dimensional dynamical systems (see, for example, [2, 12, 22, 29, 32] and the references therein). For a system defined on a bounded two-dimensional domain it was shown that the corresponding global attractor has finite fractal dimension. The idea to use the Lieb–Thirring inequalities [24] for orthonormal families played an essential role in deriving physically relevant upper bounds for the dimension. Furthermore, the upper bounds in case of the torus $\mathbb{T}^2$ are sharp up to a logarithmic correction [26].

Another model in incompressible hydrodynamics more recently studied from the point of view of attractors is the two-dimensional damped Euler system

$$\begin{cases} \partial_t u + (u, \nabla_x)u + \gamma u + \nabla_x p = g, \\ \operatorname{div} u = 0, \quad u(0) = u_0. \end{cases} \tag{1.1}$$

The linear damping term γu here makes the system dissipative and is important in various geophysical models [28]. The system is studied either on a 2D manifold (torus, sphere) or in a bounded 2d domain with stress-free boundary conditions. The natural phase space here is H^1 where it easy to prove the existence of a solution of class $L^\infty(0, T, H^1)$ and dissipativity. However, the solution in this class is not known to be unique.

Keywords: Damped Euler–Bardina equations, α models, attractors, dimension estimates
2020 Mathematics Subject Classification: 35B40, 35B41, 37L30, 35Q31

A weak attractor and a weak trajectory attractor for (1.1) were constructed in [14] and [9], respectively. It was then shown in [7] and [10] that these attractors are, in fact, compact in H^1 and the attraction holds in the norm of H^1.

Closely related to (1.1) is its Navier–Stokes perturbation

$$\begin{cases} \partial_t u + (u, \nabla_x)u + \nabla_x p + \gamma u = \nu\Delta_x u + g, & x \in \mathbb{T}^2, \\ \operatorname{div} u = 0, \quad u(0) = u_0, \end{cases} \tag{1.2}$$

which is studied in the vanishing viscosity limit $\nu \to 0$. It is proved in [7] that the attractors $\mathscr{A}_\nu$ of (1.2) tend in H^1 to the attractor $\mathscr{A}$ of (1.1), and, furthermore [16, 17], their fractal dimension satisfies an order-sharp (as $\nu \to 0^+$) two sided estimate

$$1.5 \cdot 10^{-6} \frac{\|\operatorname{curl} g\|_{L^2}^2}{\nu\gamma^3} \le \dim_F \mathscr{A}_\nu \le \frac{3\pi}{256} \frac{\|\operatorname{curl} g\|_{L^2}^2}{\nu\gamma^3}, \tag{1.3}$$

where the left-hand side estimate holds for a specially chosen Kolmogorov forcing, and in the right-hand side we used the recent estimate of the Lieb–Thirring constant on the torus [18]. This indicates (or at least suggests) that the problem of estimating the dimension of the attractor $\mathscr{A}$ of (1.1) is difficult and the attractor may well be infinite-dimensional.

In this work we use a different approximation of (1.1), namely, the so-called inviscid damped Euler–Bardina model (see [3, 4, 21] and the references therein)

$$\begin{cases} \partial_t u + (\bar{u}, \nabla_x)\bar{u} + \gamma u + \nabla_x p = g, \\ \operatorname{div} u = 0, \quad u(0) = u_0, \quad u = (1 - \alpha\Delta_x)\bar{u}, \end{cases} \tag{1.4}$$

on a 2D torus $\mathbb{T}^2 = (0, L)^2$ with the forcing $g \in H^1(\mathbb{T}^2)$ and $\alpha, \gamma > 0$. The system is studied in the phase space $H^{-1}(\mathbb{T}^2) \cap \{\operatorname{div} u = 0\} =: H^{-1}$. We also assume that $\int_{\mathbb{T}^2}(u, \bar{u}, g)\,dx = 0$. Here $\alpha = \alpha' L^2$ and $\alpha' > 0$ is a small dimensionless parameter, so that $\bar{u}$ is a smoothed (filtered) vector field.

The assumption that the mean value of g, u, $\bar{u}$ be zero involves (almost) no loss of generality, since otherwise, integrating (1.4) over $\mathbb{T}^2$ and taking into account $\operatorname{div} u = 0$, we obtain an equation for $\langle u\rangle$ (the mean value of u) $\partial_t\langle u\rangle + \gamma\langle u\rangle = \langle g\rangle$, whose solution tends exponentially to a constant vector $\frac{\langle g\rangle}{\gamma}$. Subtracting it from (1.4) we obtain a system of the type (1.4) with an exponentially small non-autonomous term.

Alternatively, equations (1.4) can be considered as a particular case of the so-called Kelvin–Voight regularization of damped Navier–Stokes equations. Indeed, rewriting it in terms of the variable $\bar{u}$, we arrive at the equations

$$\begin{cases} \partial_t\bar{u} - \alpha\Delta_x\partial_t\bar{u} + (\bar{u}, \nabla_x)\bar{u} + \gamma\alpha\bar{u} = \gamma\alpha\Delta_x\bar{u} + g, \\ \operatorname{div}\bar{u} = 0, \quad \bar{u}(0) = \bar{u}_0, \end{cases}$$

which are damped Kelvin–Voight Navier–Stokes equations with the specific choice of Ekman damping parameter $\bar{\gamma} := \gamma\alpha$ which coincides with the kinematic viscosity $\nu := \gamma\alpha$, see [4, 21] for more details. In the present paper we restrict our attention to 2D space periodic case since only in this case the sharp lower bounds for the attractor's dimension can be obtained. The explicit upper bounds for this dimension in bounded and unbounded 3D domains will be given in the forthcoming paper [15].

Let us describe the results of this paper. In Section 2 we prove that system (1.4) is dissipative in the phase space $H^{-1}(\mathbb{T}^2)$ and prove the existence of the global attractor. Since for $\alpha > 0$ the convective term is a bounded (compact) perturbation, the existence of the global attractor is essentially an ODE result.

As far as the attractor is concerned, the choice of the phase space can vary, however, in the next Section 3 the phase space $H^{-1}(\mathbb{T}^2)$ is most convenient for the estimates of the global Lyapunov exponents by means of the Constantin–Foias–Temam N-trace formula [2, 11, 32].

The following two-sided order-sharp estimate (as $\alpha \to 0^+$) of the dimension of the global attractor $\mathscr{A} = \mathscr{A}_\alpha$ of system (1.4) is the main result of this work

$$6.46 \cdot 10^{-7} \frac{\|\operatorname{curl} g\|_{L^2}^2}{\alpha\gamma^4} \le \dim_F \mathscr{A} \le \frac{1}{8\pi} \frac{\|\operatorname{curl} g\|_{L^2}^2}{\alpha\gamma^4}. \tag{1.5}$$

We observe that its looks somewhat similar to (1.3) with ν interchanged with α.

While the right-hand side estimate is universal, the lower bound here holds as in (1.3) for a family of Kolmogorov right-hand sides $g = g_s$ specially chosen in Section 4. For this purpose we use the instability analysis for the family of generalized Kolmogorov flows generated by the family of right-hand sides

$$g = g_s = \begin{cases} g_1 = \dfrac{1}{\sqrt{2\pi}} \gamma\lambda(s) \sin s x_2, \\ g_2 = 0, \end{cases}$$

where $s \in \mathbb{N}$. We show that the parameter $\lambda(s)$ can be chosen so that the unstable manifold of the stationary solution corresponding to g_s has dimension $O(s^2)$, while for $s := \alpha^{-\frac{1}{2}}$ we have

$$\|\operatorname{curl} g_s\|_{L^2} = O(1) \quad \text{as } \alpha \to 0.$$

Since the unstable manifold lies in the global attractor, it follows that we have a lower bound for the dimension of the attractor $\mathscr{A}_s$ corresponding to $g = g_s$, which can be also written in an explicit form given on the left-hand side in (1.5).

Finally, in the Appendix in Section 5 we prove the following spectral inequality that is an essential analytical ingredient in the proof of the upper bound for the dimension:

$$\|\rho\|_{L^2} \le \frac{1}{2\sqrt{\pi}} m^{-1} n^{\frac{1}{2}}, \quad \rho(x) = \sum_{j=1}^{n} |u_j(x)|^2,$$

where

$$u_j = (m^2 - \Delta_x)^{-\frac{1}{2}} \psi_j,$$

and $\{\psi_j\}_{j=1}^n$ is an orthonormal family in $L^2(\mathbb{T}^2)$ and each ψ_j has mean value zero. The inequality of this type is proved (among others) in [23] in the case of $\mathbb{R}^d$. The case of the torus requires a special treatment, since in this case the parameter m^2 can no longer be scaled out.

2 Global attractor

Theorem 2.1. *There exists a unique solution $u = u(t) \in H^{-1}$ of system (1.4), which defines the semigroup of solution operators acting in H^{-1}:*

$$S(t)u_0 = u(t), \quad S(0)u_0 = u_0.$$

Proof. Since for $u \in H^{-1}$ we have $\bar{u} \in H^1$, it follows that the nonlinear term $(\bar{u}, \nabla_x)\bar{u}$ is bounded and Lipschitz in H^{-1}. Therefore the existence of a local in time unique solution is straightforward. Furthermore, the solution is continuous with respect to the initial data $u_0 \in H^{-1}$. The fact that it exists for all $t > 0$ follows from a priori estimates derived below. ■

Vorticity equation. Taking curl from both sides of (1.4) we arrive at the vorticity equation for $\omega = \operatorname{curl} u$

$$\partial_t \omega + (\bar{u}, \nabla_x)\bar{\omega} + \gamma\omega = \operatorname{curl} g, \ \omega = (1 - \alpha\Delta_x)\bar{\omega}. \tag{2.1}$$

Multiplying this equation by $\bar{\omega}$ and integrating over x, we see that the nonlinear term vanishes, since $\operatorname{div} \bar{u} = 0$, and we obtain

$$\begin{aligned} &\frac{1}{2}\frac{d}{dt}\big(\|\bar{\omega}\|_{L^2}^2 + \alpha\|\nabla_x\bar{\omega}\|_{L^2}^2\big) + \gamma\big(\|\bar{\omega}\|_{L^2}^2 + \alpha\|\nabla_x\bar{\omega}\|_{L^2}^2\big) \\ &\quad = (\operatorname{curl} g, \bar{\omega}) = (g, \nabla_x^{\perp}\bar{\omega}) \le \|g\|_{L^2}\|\nabla_x\bar{\omega}\|_{L^2} \le \frac{1}{2\alpha\gamma}\|g\|_{L^2}^2 + \frac{\alpha\gamma}{2}\|\nabla_x\bar{\omega}\|_{L^2}^2. \end{aligned}$$

This gives by Gronwall's inequality the estimate

$$\|\bar{\omega}(t)\|_{L^2}^2 + \alpha\|\nabla_x\bar{\omega}(t)\|_{L^2}^2 \le \frac{1}{\alpha\gamma^2}\|g\|_{L^2}^2,$$

which holds for every trajectory $u(t)$ as $t \to \infty$.

Alternatively, without integration by parts of the curl on the right-hand side, we can write

$$\|\operatorname{curl} g\|_{L^2}\|\bar{\omega}\|_{L^2} \le \frac{1}{2\gamma}\|\operatorname{curl} g\|_{L^2}^2 + \frac{\gamma}{2}\|\bar{\omega}\|_{L^2}^2,$$

giving as $t \to \infty$

$$\|\bar{\omega}(t)\|_{L^2}^2 + \alpha\|\nabla_x\bar{\omega}(t)\|_{L^2}^2 \le \frac{1}{\gamma^2}\|\operatorname{curl} g\|_{L^2}^2,$$

so that as $t \to \infty$

$$\|\bar{\omega}(t)\|_{L^2}^2 + \alpha\|\nabla_x\bar{\omega}(t)\|_{L^2}^2 \le \frac{1}{\gamma^2}\min\left(\frac{\|g\|_{L^2}^2}{\alpha}, \|\operatorname{curl} g\|_{L^2}^2\right) =: R_0^2. \tag{2.2}$$

Proposition 2.2. *For any $R_1 > 0$ and any u_0 with $\|u_0\|_{H^{-1}} \le R_1$ there exists a time $T = T(R_1, R_0)$ such that for all $t \ge T(R_1, R_0)$ the solution $u(t) = S(t)u_0$ enters and never leaves the ball of radius $2R_0$ in H^{-1}:*

$$\|u(t)\|_{H^{-1}} \le 2R_0 \quad \textit{for } t \ge T(R_1, R_0).$$

In other words, this ball is an absorbing set for $S(t)$ in H^{-1}.

Proof. The claim of the proposition follows from (2.2) (where we may want to drop the term with α on the left-hand side) and the following equivalence of the norms

$$\|u\|_{H^{-1}} \sim \|\bar{u}\|_{H^1} = \|\nabla_x\bar{u}\|_{L^2} = \|\bar{\omega}\|_{L^2}. \qquad \blacksquare$$

Asymptotic compactness and global attractor. To prove the asymptotic compactness of the semigroup $S(t)$, we construct an attracting compact set in the absorbing ball in H^{-1}. For this purpose we decompose $S(t)$ as follows:

$$S(t) = \Sigma(T) + S_2(t), \quad S_2(t) = S(t) - \Sigma(t),$$

where $\Sigma(t)$ is exponentially contacting and $S_2(t)$ is uniformly compact. This decomposition (and its more elaborate variants) is very useful for dissipative hyperbolic problems [2, 8, 32] and the original idea goes back to [13].

Let $\Sigma(t)$ be the solution (semi)group of the linear equation

$$\partial_t v + \gamma v = 0, \quad \operatorname{div} v = 0, \quad v(0) = u_0,$$

and w (playing the role of $S_2(t)$) is the solution of

$$\partial_t w + \gamma w + \nabla_x p = G(t) := -(\bar{u}, \nabla_x)\bar{u} + g, \quad \operatorname{div} w = 0, \quad w(0) = 0,$$

where, of course, $u = v + w$. Obviously, $v(t) = e^{-\gamma t}u_0$ is exponentially decaying in H^{-1} (and in every H^s). The right-hand side $G(t)$ in the "linear" equation for w is bounded in $H^{-\delta}$ for every $\delta > 0$ uniformly in t. In fact, since $\bar{u}(t)$ is bounded in H^1, it follows from the Sobolev imbedding theorem that $\bar{u}(t) \subset L^p$ for $p < \infty$, and Hölder's inequality gives that $(\bar{u}, \nabla_x)\bar{u}$ is bounded in $L^{2-\varepsilon}$ for $\varepsilon > 0$, while, in turn, by duality $L^{2-\varepsilon} \subset H^{-\delta}$, $\delta = \delta(\varepsilon) = \frac{\varepsilon}{2-\varepsilon}$. In fact, again by the Sobolev imbedding theorem

$$H^\delta \subset L^{2/(1-\delta)} \iff L^{2-\varepsilon} = L^{2/(1+\delta)} = \left(L^{2/(1-\delta)}\right)^* \subset H^{-\delta}.$$

Taking into account that $w(0) = 0$, we see that the solution w is bounded in $H^{-\delta}$ uniformly with respect to t, and since the imbedding $H^{-\delta}(\mathbb{T}^2) \subset H^{-1}(\mathbb{T}^2)$ is compact, the asymptotic compactness of the solution semigroup $S(t)$ is established.

We recall the definition of the global attractor.

Definition 2.3. Let $S(t), t \geq 0$, be a semigroup in a Banach space H. The set $\mathscr{A} \subset H$ is a global attractor of the semigroup $S(t)$ if

(1) the set $\mathscr{A}$ is compact in H,

(2) it is strictly invariant: $S(t)\mathscr{A} = \mathscr{A}$,

(3) it attracts the images of bounded sets in H as $t \to \infty$, i.e., for every bounded set $B \subset H$ and every neighborhood $\mathcal{O}(\mathscr{A})$ of the set $\mathscr{A}$ in H there exists $T = T(B, \mathcal{O})$ such that for all $t \geq T$,

$$S(t)B \subset \mathcal{O}(\mathscr{A}).$$

The following general result (see, for instance, [2,8,32]) gives sufficient conditions for the existence of a global attractor.

Theorem 2.4. *Let $S(t)$ be a semigroup in a Banach space H. Suppose that*

(1) *$S(t)$ possesses a bounded absorbing ball $B \subset H$,*

(2) *for every fixed $t \geq 0$ the map $S(t) : B \to H$ is continuous,*

(3) *$S(t)$ is asymptotically compact.*

Then the semigroup $S(t)$ possesses a global attractor $\mathscr{A} \subset B$. Moreover, the attractor $\mathscr{A}$ has the following structure:

$$\mathscr{A} = \mathscr{K}\big|_{t=0}, \tag{2.3}$$

where $\mathscr{K} \subset L^\infty(\mathbb{R}, H)$ is the set of complete trajectories $u : \mathbb{R} \to H$ of semigroup $S(t)$ which are defined for all $t \in \mathbb{R}$ and bounded.

We are now prepared to state the main result of this section, whose proof directly follows from Theorem 2.4 since its assumptions were verified above.

Theorem 2.5. *The semigroup $S(t)$ corresponding to* (1.4) *possesses in the phase space H^{-1} a global attractor $\mathscr{A}$.*

3 Dimension estimate

Theorem 3.1. *The global attractor $\mathscr{A}$ corresponding to the regularized damped Euler system* (1.4) *has finite fractal dimension*

$$\dim_F \mathscr{A} \leq \frac{1}{8\pi} \frac{\|\operatorname{curl} g\|^2_{L^2}}{\alpha \gamma^4}. \tag{3.1}$$

Proof. Note that the differentiability of the solution semigroup $S(t) : H^{-1} \to H^{-1}$ is obvious here, so we only need to estimate the traces. The equation of variations for equation (1.4) reads

$$\begin{cases} \partial_t \theta = -\gamma\theta - (\bar{u}, \nabla_x)\bar{\theta} - (\bar{\theta}, \nabla_x)\bar{u} - \nabla_x p =: L_{u(t)}\theta, \\ \operatorname{div} \theta = 0, \quad \bar{\theta} = (1 - \alpha\Delta_x)^{-1}\theta. \end{cases}$$

We will estimate volume contraction factor in the space H^{-1} endowed with the norm

$$\begin{aligned}\|\theta\|^2_\alpha &:= (\theta,\bar\theta)_{L^2} = \|\bar\theta\|^2_{L^2} + \alpha\|\nabla_x\bar\theta\|^2_{L^2} \\ &= \|\bar\theta\|^2_{L^2} + \alpha\|\operatorname{curl}\bar\theta\|^2_{L^2} = \|(1-\alpha\Delta_x)^{-\frac12}\theta\|^2_{L^2},\end{aligned}$$

where for the vector Laplacian

$$-(u,\Delta_x u)_{L^2} = \|\nabla_x u\|^2_{L^2} = \|\operatorname{curl} u\|^2_{L^2} + \|\operatorname{div} u\|^2_{L^2},$$

and scalar product

$$(\theta,\xi)_\alpha = ((1-\alpha\Delta_x)^{-\frac12}\theta,(1-\alpha\Delta_x)^{-\frac12}\xi).$$

The numbers $q(n)$ (the sums of the first n global Lyapunov exponents) are estimated/defined as follows [32]:

$$q(n) \le (:=) \limsup_{t\to\infty}\ \sup_{u(t)\in\mathscr{A}}\ \sup_{\{\theta_j\}_{j=1}^n} \frac1t\int_0^t \sum_{j=1}^n (L_{u(\tau)}\theta_j,\theta_j)_\alpha\,d\tau,$$

where $\{\theta_j\}_{j=1}^n$ is an orthonormal family with respect to $(\cdot,\cdot)_\alpha$:

$$(\theta_i,\theta_j)_\alpha = \delta_{i\,j},\quad \operatorname{div}\theta_j = 0 \tag{3.2}$$

and $u(t)$ is an arbitrary trajectory on the attractor. Then

$$\begin{aligned}\sum_{j=1}^n (L_{u(t)}\theta_j,\theta_j)_\alpha &= \sum_{j=1}^n (L_{u(t)}\theta_j,\bar\theta_j) = -\sum_{j=1}^n \gamma\|\theta_j\|^2_\alpha - \sum_{j=1}^n ((\bar\theta_j,\nabla_x)\bar u,\bar\theta_j) \\ &= -\gamma n - \sum_{j=1}^n ((\bar\theta_j,\nabla_x)\bar u,\bar\theta_j).\end{aligned}$$

Next, $\theta_j\in H^{-1}$ are such that the family $\{(1-\alpha\Delta_x)^{-\frac12}\theta_j\}_{j=1}^n =: \{\psi_j\}_{j=1}^n$ is orthonormal in L^2. By definition,

$$\bar\theta_j = (1-\alpha\Delta_x)^{-1}\theta_j = (1-\alpha\Delta_x)^{-\frac12}\psi_j,$$

and therefore for the function

$$\rho(x) = \sum_{j=1}^n |\bar\theta_j(x)|^2$$

in view of (5.2) (with $p=2$) and (5.3) we have the estimate

$$\|\rho\|_{L^2} \le \mathrm{B}_2\frac{n^{\frac12}}{\sqrt\alpha},\quad \mathrm{B}_2 \le \frac{1}{2\sqrt\pi}. \tag{3.3}$$

Since $\operatorname{div} \bar{u} = 0$, we have pointwise

$$|(\bar{\theta}, \nabla_x)\bar{u} \cdot \bar{\theta}| \le \frac{1}{\sqrt{2}} |\bar{\theta}(x)|^2 |\nabla_x \bar{u}(x)|$$

and using (3.3) and (2.2), we obtain

$$\begin{aligned}\sum_{j=1}^{n} |((\bar{\theta}_j, \nabla_x)\bar{u}, \bar{\theta}_j)| &\le \frac{1}{\sqrt{2}} \int_{\mathbb{T}^2} \rho(x) |\nabla_x \bar{u}| \, dx \\ &\le \frac{1}{\sqrt{2}} \|\rho\|_{L^2} \|\nabla_x \bar{u}\|_{L^2} = \frac{1}{\sqrt{2}} \|\rho\|_{L^2} \|\bar{\omega}\|_{L^2} \\ &\le \frac{\mathrm{B}_2}{\sqrt{2}} \frac{n^{\frac{1}{2}}}{\sqrt{\alpha}} \frac{\|\operatorname{curl} g\|_{L^2}}{\gamma}.\end{aligned}$$

Since all our estimates are uniform with respect to time and the α-orthonormal family (3.2), it follows that

$$q(n) \le -\gamma n + \frac{\mathrm{B}_2}{\sqrt{2}} \frac{n^{\frac{1}{2}}}{\sqrt{\alpha}} \frac{\|\operatorname{curl} g\|_{L^2}}{\gamma}.$$

The number n^* such that for which $q(n^*) = 0$ and $q(n) < 0$ for $n > n^*$ is the upper bound both for the Hausdorff [2, 32] and the fractal [5, 6] dimension of the global attractor $\mathscr{A}$:

$$\dim_F \mathscr{A} \le \frac{\mathrm{B}_2^2}{2} \frac{\|\operatorname{curl} g\|_{L^2}^2}{\alpha \gamma^4} \le \frac{1}{8\pi} \frac{\|\operatorname{curl} g\|_{L^2}^2}{\alpha \gamma^4}. \qquad \blacksquare$$

4 A sharp lower bound

In this section we derive a sharp lower bound for the dimension of the global attractor $\mathscr{A}$ based on the generalized Kolmogorov flows [26, 27, 35].

We consider the vorticity equation (2.1) written in terms of ω only. If $\operatorname{curl} u = \omega$, then u is recovered as $u = -\nabla_x^{\perp}(\Delta_x^{-1}\omega)$, where $\nabla_x^{\perp} := (\partial_{x_2}, -\partial_{x_1})$, and (2.1) goes over to the equation

$$\partial_t \omega + J\big((\Delta_x - \alpha \Delta_x^2)^{-1}\omega, (1 - \alpha\Delta_x)^{-1}\omega\big) + \gamma\omega = \operatorname{curl} g, \tag{4.1}$$

where J is the Jacobian operator

$$J(a, b) = \nabla a \cdot \nabla^{\perp} b = \partial_{x_1} a \, \partial_{x_2} b - \partial_{x_2} a \, \partial_{x_1} b.$$

Next, let $g = g_s$ be a family of right-hand sides

$$g = g_s = \begin{cases} g_1 = \dfrac{1}{\sqrt{2}\pi} \gamma \lambda(s) \sin s x_2, \\ g_2 = 0, \end{cases} \tag{4.2}$$

where $s \in \mathbb{N}$, and $\lambda(s)$ is a parameter to be chosen later. Then

$$\operatorname{curl} g_s = -\frac{1}{\sqrt{2}\pi}\gamma\lambda(s)s\cos sx_2 \tag{4.3}$$

and

$$\|g_s\|_{L^2}^2 = \gamma^2\lambda(s)^2, \quad \|\operatorname{curl} g_s\|_{L^2}^2 = \gamma^2\lambda(s)^2 s^2. \tag{4.4}$$

Corresponding to the family (4.3) is the family of stationary solutions

$$\omega_s = -\frac{1}{\sqrt{2}\pi}\lambda(s)s\cos sx_2 \tag{4.5}$$

of equation (4.1). Since ω_s depends only on x_2, the nonlinear term vanishes:

$$J\big((\Delta_x - \alpha\Delta_x^2)^{-1}\omega_s, (1-\alpha\Delta_x)^{-1}\omega_s\big) \equiv 0.$$

We linearize (4.1) about the stationary solution (4.5) and consider the eigenvalue problem

$$\begin{aligned}\mathcal{L}_s\omega := {} & J\big((\Delta_x - \alpha\Delta_x^2)^{-1}\omega_s, (1-\alpha\Delta_x)^{-1}\omega\big) \\ & + J\big((\Delta_x - \alpha\Delta_x^2)^{-1}\omega, (1-\alpha\Delta_x)^{-1}\omega_s\big) + \gamma\omega = -\sigma\omega.\end{aligned} \tag{4.6}$$

The solutions with $\mathrm{Re}\sigma > 0$ will unstable eigenmodes.

We use the orthonormal basis of trigonometric functions, which are the eigenfunctions of the Laplacian,

$$\Big\{\frac{1}{\sqrt{2}\pi}\sin kx, \frac{1}{\sqrt{2}\pi}\cos kx\Big\}, \quad kx = k_1x_1 + k_2x_2,$$

where

$$k \in \mathbb{Z}_+^2 = \{k \in \mathbb{Z}_0^2,\, k_1 \ge 0,\, k_2 \ge 0\} \cup \{k \in \mathbb{Z}_0^2,\, k_1 \ge 1,\, k_2 \le 0\},$$

and write

$$\omega(x) = \frac{1}{\sqrt{2}\pi}\sum_{k\in\mathbb{Z}_+^2} a_k\cos kx + b_k\sin kx.$$

Substituting this into (4.6) and using the equality

$$J(a,b) = -J(b,a),$$

we obtain

$$\begin{aligned}&\frac{s\lambda(s)}{\sqrt{2}\pi(s^2+\alpha s^4)}\sum_{k\in\mathbb{Z}_+^2}\Big(\frac{k^2-s^2}{k^2+\alpha k^4}\Big)J(\cos sx_2, a_k\cos kx + b_k\sin kx) \\ &\quad + (\gamma+\sigma)\sum_{k\in\mathbb{Z}_+^2}(a_k\cos kx + b_k\sin kx) = 0.\end{aligned} \tag{4.7}$$

Next, we have the following two similar formulas:

$$
\begin{aligned}
&J(\cos sx_2, \cos(k_1x_1 + k_2x_2)) \\
&\quad = -k_1 s \sin sx_2 \sin(k_1x_1 + k_2x_2) \\
&\quad = \frac{k_1 s}{2}(\cos(k_1x_1 + (k_2+s)x_2) - \cos(k_1x_1 + (k_2-s)x_2), \\
&J(\cos sx_2, \sin(k_1x_1 + k_2x_2)) \\
&\quad = k_1 s \sin sx_2 \cos(k_1x_1 + k_2x_2) \\
&\quad = \frac{k_1 s}{2}(\sin(k_1x_1 + (k_2+s)x_2) - \sin(k_1x_1 + (k_2-s)x_2),
\end{aligned}
$$

which we substitute into (4.7) and collect the terms with $\cos(k_1x_1 + k_2x_2)$. We obtain the following equation for the coefficients a_{k_1,k_2} (the equation for b_{k_1,k_2} is exactly the same):

$$
\begin{aligned}
-\Lambda(s)k_1 &\left(\frac{k_1^2 + (k_2+s)^2 - s^2}{k_1^2 + (k_2+s)^2 + \alpha(k_1^2 + (k_2+s)^2)^2)}\right) a_{k_1\,k_2+s} \\
&+ \Lambda(s)k_1 \left(\frac{k_1^2 + (k_2-s)^2 - s^2}{k_1^2 + (k_2-s)^2 + \alpha(k_1^2 + (k_2-s)^2)^2)}\right) a_{k_1\,k_2-s} \\
&+ (\gamma + \sigma) a_{k_1\,k_2} = 0,
\end{aligned}
$$

where

$$
\Lambda = \Lambda(s) := \frac{s^2 \lambda(s)}{2\sqrt{2}\pi(s^2 + \alpha s^4)} = \frac{\lambda(s)}{2\sqrt{2}\pi(1 + \alpha s^2)}. \tag{4.8}
$$

We set here

$$
a_{k_1\,k_2}\left(\frac{k^2 - s^2}{k^2 + \alpha k^4}\right) =: c_{k_1\,k_2},
$$

and setting further

$$
k_1 = t, \quad k_2 = sn + r, \quad \text{and} \quad c_{t\,sn+r} = e_n
$$

for

$$
t = 1, 2, \ldots, \quad r \in \mathbb{Z}, \quad r_{\min} < r < r_{\max},
$$

where the numbers $r_{\min}, r_{\max}$ satisfy $r_{\max} - r_{\min} < s$ and will be specified below, we obtain for each t and r the following three-term recurrence relation:

$$
d_n e_n + e_{n-1} - e_{n+1} = 0, \quad n = 0, \pm 1, \pm 2, \ldots, \tag{4.9}
$$

where

$$
d_n = \frac{\left(t^2 + (sn+r)^2 + \alpha(t^2 + (sn+r)^2)^2\right)(\gamma + \sigma)}{\Lambda t(t^2 + (sn+r)^2 - s^2)}. \tag{4.10}
$$

We look for non-trivial decaying solutions $\{e_n\}$ of (4.9) and (4.10). Each non-trivial decaying solution with $\operatorname{Re}\sigma > 0$ produces an unstable eigenfunction ω of the eigenvalue problem (4.6).

Theorem 4.1. *Given an integer $s > 0$, let a fixed pair of integers t, r belong to a bounded region $A(\delta)$ defined by conditions*

$$t^2 + r^2 < \frac{s^2}{3}, \quad t^2 + (-s + r)^2 > s^2, \quad t^2 + (s + r)^2 > s^2, \quad t \geq \delta s, \tag{4.11}$$

where

$$r_{\min} < r < r_{\max}, \quad r_{\min} = -\frac{s}{6}, \, r_{\max} = \frac{s}{6}, \quad 0 < \delta < \frac{1}{\sqrt{3}}.$$

For any $\Lambda > 0$ there exists a unique real eigenvalue $\sigma = \sigma(\Lambda)$, which increases monotonically as $\Lambda \to \infty$ and satisfies the inequality

$$\Lambda \frac{21\sqrt{2}\delta^2 s}{55(1 + \alpha s^2)} - \gamma \leq \sigma(\Lambda) \leq \Lambda \frac{\sqrt{2}\delta^{-1} s}{(1 + \alpha s^2)} - \gamma. \tag{4.12}$$

The unique $\Lambda_0 = \Lambda_0(s)$ solving the equation

$$\sigma(\Lambda_0) = 0$$

satisfies the two-sided estimate

$$\frac{\gamma\delta}{\sqrt{2}} \frac{(1 + \alpha s^2)}{s} < \Lambda_0 < \frac{55\gamma\delta^{-2}}{21\sqrt{2}} \frac{1 + \alpha s^2}{s}. \tag{4.13}$$

Proof. We shall follow quite closely the argument in [19] and first observe that the following inequalities hold for any (t, r) satisfying (4.11):

$$s^2 \leq t^2 + (-s + r)^2 = \mathrm{dist}((0, s), (t, r))^2 \leq \mathrm{dist}((0, s), C)^2 = \frac{5}{3}s^2,$$
$$s^2 \leq t^2 + (s + r)^2 = \mathrm{dist}((0, -s), (t, r))^2 \leq \mathrm{dist}((0, -s), B)^2 = \frac{5}{3}s^2,$$

where $B = (\frac{\sqrt{11}s}{6}, \frac{s}{6})$ and $C = (\frac{\sqrt{11}s}{6}, -\frac{s}{6})$.

In view of (4.11) for any real σ satisfying $\sigma > -\gamma$ we have in (4.9) and (4.10)

$$d_0 < 0, \quad d_n > 0 \text{ for } n \neq 0 \quad \text{and} \quad \lim_{|n|\to\infty} d_n = \infty. \tag{4.14}$$

The main tool in the analysis of relation (4.9) are continued fractions and a variant of Pincherle's theorem (see [20, 26, 27, 35]) saying that under condition (4.14) recurrence relation (4.9) has a decaying solution $\{e_n\}$ with $\lim_{|n|\to\infty} e_n = 0$ if and only if

$$-d_0 = \cfrac{1}{d_{-1} + \cfrac{1}{d_{-2} + \cdots}} + \cfrac{1}{d_1 + \cfrac{1}{d_2 + \cdots}}.$$

Next, we set

$$f(\sigma) = -d_0 = \frac{(\gamma + \sigma)(t^2 + r^2 + \alpha(t^2 + r^2)^2)}{\Lambda t (s^2 - (t^2 + r^2))} \tag{4.15}$$

and

$$g(\sigma) = \cfrac{1}{d_{-1} + \cfrac{1}{d_{-2} + \cdots}} + \cfrac{1}{d_1 + \cfrac{1}{d_2 + \cdots}}. \tag{4.16}$$

It follows from (4.15) that

$$f(-\gamma) = 0, \qquad \text{and} \qquad f(\sigma) \to \infty \quad \text{as } \sigma \to \infty,$$

while (4.16) and (4.10) give that

$$g(\sigma) < \frac{1}{d_{-1}} + \frac{1}{d_1}, \quad g(\sigma) \to 0 \quad \text{as } \sigma \to \infty.$$

Hence, there exists a $\sigma > -\gamma$ such that

$$f(\sigma) = g(\sigma). \tag{4.17}$$

From elementary properties of continued fractions we deduce as in [26, 35] that $\sigma = \sigma(\Lambda)$ so obtained is unique and increases monotonically with Λ.

To establish (4.12), we deduce from (4.17) and (4.16) that

$$\cfrac{1}{d_{-1} + \cfrac{1}{d_{-2}}} + \cfrac{1}{d_1 + \cfrac{1}{d_2}} < f(\sigma) < \frac{1}{d_{-1}} + \frac{1}{d_1}. \tag{4.18}$$

Using (4.11), we see from (4.10) that

$$\begin{aligned}\frac{1}{d_{\pm 1}} &= \frac{\Lambda t}{\gamma + \sigma} \cdot \frac{t^2 + (s \pm r)^2 - s^2}{t^2 + (s \pm r)^2 + \alpha(t^2 + (s \pm r)^2)^2)} \\ &\le \frac{\Lambda t}{\gamma + \sigma} \cdot \frac{1}{1 + \alpha(t^2 + (s \pm r)^2)} \le \frac{\Lambda t}{\gamma + \sigma} \cdot \frac{1}{1 + \alpha s^2}\end{aligned}$$

and therefore from the right-hand side inequality in (4.18) it follows that

$$\begin{aligned}f(\sigma) &= \frac{(\gamma + \sigma)(t^2 + r^2 + \alpha(t^2 + r^2)^2)}{\Lambda t (s^2 - (t^2 + r^2))} \\ &< \frac{1}{d_{-1}} + \frac{1}{d_1} < \frac{2\Lambda t}{\gamma + \sigma} \cdot \frac{1}{1 + \alpha s^2},\end{aligned}$$

which gives the right-hand side inequality in (4.12):

$$\begin{aligned}(1 + \alpha s^2)(\gamma + \sigma)^2 &\le \frac{2\Lambda^2 t^2 (s^2 - (t^2 + r^2))}{t^2 + r^2 + \alpha(t^2 + r^2)^2} \\ &\le \frac{2\Lambda^2 t^2 s^2}{t^2 + \alpha t^4} = \frac{2\Lambda^2 s^2}{1 + \alpha t^2} \le \frac{2\Lambda^2 \delta^{-2} s^2}{1 + \alpha s^2}.\end{aligned}$$

From the left-hand side inequality in (4.18), where $d_{-1}, d_1, d_{-2}, d_2, f > 0$, we see that

$$fd_1 + \frac{f}{d_2} > 1 \quad \text{and} \quad fd_{-1} + \frac{f}{d_{-2}} > 1. \tag{4.19}$$

Next, it follows from (4.11) that $4sr \le \frac{2s^2}{3}$ and $|2s + r| \ge \frac{11s}{6}$. Therefore

$$\begin{aligned}\frac{f}{d_2} &= \frac{(t^2 + r^2 + \alpha(t^2 + r^2)^2)(t^2 + (2s + r)^2 - s^2)}{(s^2 - (t^2 + r^2))(t^2 + (2s + r)^2 + \alpha(t^2 + (2s + r)^2)^2)} \\ &< \frac{(s^2/3 + \alpha s^4/9)4s^2}{2s^2/3((11/6)^2 s^2 + \alpha(11/6)^4 s^4)} = \frac{2(1 + t/3)}{(11/6)^2 + t(11/6)^4}_{t=\alpha s^2} \\ &\le \frac{72}{121}.\end{aligned}$$

Along with (4.19) this implies that $fd_1 > \frac{49}{121}$, which for $r \ge 0$ gives that

$$\begin{aligned}\frac{49}{121} &< fd_1 \\ &= \frac{(\gamma + \sigma)^2}{\Lambda^2} \frac{(t^2 + r^2 + \alpha(t^2 + r^2)^2)(t^2 + (s + r)^2 + \alpha(t^2 + (s + r)^2)^2)}{t^2(s^2 - (t^2 + r^2))(t^2 + (s + r)^2 - s^2)} \\ &< \frac{(\gamma + \sigma)^2}{\Lambda^2} \frac{(s^2/3 + \alpha s^4/9)(5s^2/3 + \alpha s^4 25/9)}{t^2(2/3)s^2 t^2} \\ &< \frac{25}{18} \frac{(\gamma + \sigma)^2}{\Lambda^2} \frac{(1 + \alpha s^2)^2}{\delta^4 s^2},\end{aligned}$$

and proves the left-hand side inequality in (4.12). For $r < 0$ we use d_{-1} instead of d_1.

Finally, estimate (4.13) follows from (4.12) with $\sigma = 0$. ∎

This result has the following important implications for the attractors of the damped doubly regularized Euler equations (1.4). Namely, it says that estimate (3.1) is order-sharp in the limit as $\alpha \to 0^+$.

Corollary 4.2. *The parameter $\lambda(s)$ in the family* (4.2) *can be chosen so that the dimension of the corresponding global attractor $\mathscr{A} = \mathscr{A}_s$ satisfies*

$$\dim \mathscr{A} \ge c_1 \frac{\|\operatorname{curl} g\|_{L^2}^2}{\alpha \gamma^4}, \quad c_1 > 6.46 \cdot 10^{-7}.$$

Proof. Writing (4.13) in terms of $\lambda(s)$ (see (4.8)), we see that for

$$\lambda(s) = \frac{110\pi}{21} \gamma \delta^{-2} \frac{(1 + \alpha s^2)^2}{s}$$

each point in (t, r)-plane satisfying (4.11) produces an unstable (positive) eigenvalue σ of multiplicity two (the equation for the coefficients b_k is the same). Denoting by $d(s)$ the number of points of the integer lattice inside the region $A(\delta)$, we obviously have

$$d(s) := \#\{(t, r) \in D(s) = \mathbb{Z}^2 \cap A(\delta)\} \simeq a(\delta) \cdot s^2 \quad \text{as } s \to \infty,$$

where $a(\delta)\cdot s^2 = |A(\delta)|$ is the area of the region $A(\delta)$ (see [19, Figure 1]). Therefore the dimension of the unstable manifold around the stationary solution ω_s in (4.5) is at least $2a(\delta)\cdot s^2$. The solutions lying on this unstable manifold are bounded on the entire time axis $t\in\mathbb{R}$, since they tend to ω_s as $t\to-\infty$ and all solutions are bounded as $t\to\infty$ in view of (2.2). Therefore representation (2.3) implies that (see [2])

$$\dim A \ge 2d(s) \backsimeq 2a(\delta)\cdot s^2. \tag{4.20}$$

It remains to express this lower bound in terms of the physical parameters of the system. So far s was an arbitrary (large) parameter. We now set

$$s := \frac{1}{\sqrt{\alpha}},$$

and see from (4.4) that

$$\|\operatorname{curl} g_s\|^2_{L^2} = \gamma^2\lambda(s)^2 s^2 = \left(\frac{110\pi}{21}\right)^2 \gamma^4\delta^{-4}(1+\alpha s^2)^4 = \left(\frac{110\pi}{21}\right)^2\gamma^4\frac{2^4}{\delta^4}$$

and

$$\frac{\|\operatorname{curl} g_s\|^2_{L^2}}{\alpha\gamma^4} = \frac{1}{\alpha}\left(\frac{110\pi}{21}\right)^2\frac{2^4}{\delta^4}.$$

We finally obtain that (4.20) can be written in the form

$$\dim\mathscr{A} \ge \max_{0<\delta<\frac{1}{\sqrt{3}}} a(\delta)\delta^4\frac{1}{8}\left(\frac{21}{110\pi}\right)^2\frac{\|\operatorname{curl} g_s\|^2_{L^2}}{\alpha\gamma^4} = 6.46\cdot 10^{-7}\frac{\|\operatorname{curl} g_s\|^2_{L^2}}{\alpha\gamma^4},$$

where we optimized the result with respect to $\delta\in(0,3^{-\frac{1}{2}})$ ∎

5 Appendix

We first recall a result in [23], which we formulate for the torus $\mathbb{T}^2$.

Theorem 5.1. *Let a family $\{\psi_i\}_{i=1}^n$ be orthonormal in $L^2(\mathbb{T}^2)$ and let*

$$\int_{\mathbb{T}^2}\psi_i(x)\,dx = 0.$$

Then the function

$$\rho(x) = \sum_{i=1}^n |u_i(x)|^2,$$

where

$$u_i = (m^2-\Delta_x)^{-\frac{1}{2}}\psi_i,$$

satisfies for $1\le p<\infty$ and the constant $\mathrm{B}_p<\infty$ the estimate

$$\|\rho\|_{L^p} \le \mathrm{B}_p m^{-\frac{2}{p}} n^{\frac{1}{p}}. \tag{5.1}$$

In our notation this estimate reads as follows.

Corollary 5.2. *Let a family $\{\psi_i\}_{i=1}^n$ be orthonormal in $L^2(\mathbb{T}^2)$. Then the function*

$$\rho(x) = \sum_{i=1}^n |\bar\theta_i(x)|^2,$$

where

$$\bar\theta_i = (1-\alpha\Delta_x)^{-\frac12}\psi_i,$$

satisfies for $1 \le p < \infty$ the estimate

$$\|\rho\|_{L^p} \le \mathrm{B}_p \alpha^{\frac1p - 1} n^{\frac1p}. \tag{5.2}$$

Proof. We write

$$\bar\theta_i = (1-\alpha\Delta_x)^{-\frac12}\psi_i = \left(\frac{1}{\alpha} - \Delta_x\right)^{-\frac12}\frac{\psi_i}{\sqrt{\alpha}},$$

and (5.1) gives (5.2):

$$\left\|\alpha\sum_{i=1}^n |\bar\theta_i|^2\right\|_{L^p} \le \mathrm{B}_p \alpha^{\frac1p} n^{\frac1p},$$

ad desired. ∎

Next, we prove Theorem 5.1 for $p = 2$ and give the estimate of the constant.

Proposition 5.3. *Under the assumptions of Theorem* 5.1 *it holds for $p = 2$,*

$$\|\rho\|_{L^2} \le \mathrm{B}_2 m^{-1} n^{\frac12}, \quad \mathrm{B}_2 \le \frac{1}{2\sqrt{\pi}}. \tag{5.3}$$

Proof. We follow the idea in [23]. The main technical difference being that in the discrete case we cannot scale out m and therefore we have to estimate the corresponding series over $\mathbb{Z}_0^2$ for all $m > 0$. This can be done for all $1 \le p < \infty$, but each case requires a special treatment (at least in our proof).

For a non-negative function $V = V(x) \in L^\infty$ we set

$$H = V^{\frac12}(m^2 - \Delta_x)^{-\frac12}, \quad H^* = (m^2 - \Delta_x)^{-\frac12} V^{\frac12}.$$

We further set $K = H^*H$ and claim that

$$\operatorname{Tr} K^2 \le \frac{1}{4\pi}\frac{1}{m^2}\|V\|_{L^2}^2.$$

In fact,

$$\begin{aligned}\operatorname{Tr} K^2 &= \operatorname{Tr}\big((m^2-\Delta_x)^{-\frac12} V (m^2-\Delta_x)^{-\frac12}\big)^2\\ &\le \operatorname{Tr}\big((m^2-\Delta_x)^{-1} V^2 (m^2-\Delta_x)^{-1}\big)\\ &= \operatorname{Tr}\big(V^2(m^2-\Delta_x)^{-2}\big),\end{aligned}$$

where we used the Araki–Lieb–Thirring inequality for traces [1, 25]:

$$\operatorname{Tr}(BA^2B)^p \le \operatorname{Tr}(B^pA^{2p}B^p),$$

and the cyclicity property of the trace. Using the basis of orthonormal eigenfunctions of the Laplacian $(2\pi)^{-1}e^{ikx}$, $k \in \mathbb{Z}_0^2 = \mathbb{Z}^2 \setminus \{0\}$, in view of (5.5) below we find that

$$\begin{aligned}\operatorname{Tr} K^2 &\le \operatorname{Tr}\big(V^2(m^2-\Delta_x)^{-2}\big)\\ &= \frac{1}{4\pi^2}\sum_{k\in\mathbb{Z}_0^2}\frac{1}{(|k|^2+m^2)^2}\int_{\mathbb{T}^2}V^2(x)\,dx\\ &\le \frac{1}{4\pi}\frac{1}{m^2}\|V\|_{L^2}^2.\end{aligned} \tag{5.4}$$

We can now complete the proof as in [23]. We observe that

$$\begin{aligned}\int_{\mathbb{T}^2}\rho(x)V(x)\,dx &= \sum_{i=1}^n\int_{\mathbb{T}^2}|((m^2-\Delta_x)^{-\frac12}\psi_i)(x)|^2V(x)\,dx\\ &= \sum_{i=1}^n\|H\psi_i\|_{L^2}^2,\end{aligned}$$

and in view of orthonormality and the variational principle

$$\sum_{i=1}^n\|H\psi_i\|_{L^2}^2 = \sum_{i=1}^n(K\psi_i,\psi_i) \le \sum_{i=1}^n\lambda_i,$$

where λ_i are the eigenvalues of the self-adjoint compact operator K. This finally gives

$$\int_{\mathbb{T}^2}\rho(x)V(x)\,dx \le \sum_{i=1}^n\lambda_i \le n^{\frac12}\bigg(\sum_{i=1}^n\lambda_i^2\bigg)^{\frac12} \le n^{\frac12}(\operatorname{Tr}K^2)^{\frac12} \le \frac{n^{\frac12}m^{-1}}{2\sqrt{\pi}}\|V\|_{L^2}.$$

Setting $V(x) := \rho(x)$, we complete the proof of (5.3). ∎

Remark. In fact, we proved Proposition 5.3 in the scalar case, while θ_j in the proof of Theorem 3.1 are vector functions with mean value zero and $\operatorname{div}\theta_j = 0$. The result with the same constant and the same proof still holds in this case, since on the torus the Helmholtz–Leray projection commutes with the Laplacian and the orthonormal family of vector-valued eigenfunctions of the Stokes operator $v_k(x) = \frac{1}{2\pi}\frac{k^\perp}{|k|}e^{ikx}$, $k \in \mathbb{Z}_0^2$ satisfy $|v_k(x)| = \frac{1}{2\pi}$ as in the scalar case (see (5.4)).

Remark. Inequality (5.4) is nothing more than a special case of the Kato–Seiler–Simon inequality [30]

$$\|a(-i\nabla)b\|_{\mathfrak{S}_q} \le (2\pi)^{-\frac{d}{q}}\|a\|_{L^q}\|b\|_{L^q}$$

on the torus with $d = 2$, $q = 2$ and

$$a(k) = \frac{1}{|k|^2 + m^2}$$

(and with the same constant, since, as the next shows, $\|a\|_{l^2(\mathbb{Z}_0^2)} \le \frac{\sqrt{\pi}}{m}$).

Lemma 5.4. *For $m \ge 0$,*

$$F(m) := m^2 \sum_{k \in \mathbb{Z}_0^2} \frac{1}{(|k|^2 + m^2)^2} < \pi. \tag{5.5}$$

Proof. We assume that $m \ge 1$. We show below that (5.5) holds for $m \ge 1$, which proves the lemma, since $F'(m) > 0$ on $m \in (0, 1]$ and F is increasing on $m \in [0, 1]$.

We use the Poisson summation formula (see, e.g., [31])

$$\sum_{k \in \mathbb{Z}^d} f\left(\frac{k}{m}\right) = (2\pi)^{\frac{d}{2}} m^d \sum_{k \in \mathbb{Z}^d} \widehat{f}(2\pi k m),$$

where

$$\mathcal{F}(f)(\xi) = \widehat{f}(\xi) = (2\pi)^{-\frac{d}{2}} \int_{\mathbb{R}^d} f(x) e^{-i\xi x}\, dx.$$

For the function $f(x) = \frac{1}{(1+|x|^2)^{-2}}$, $x \in \mathbb{R}^2$, with $\int_{\mathbb{R}^2} f(x)\, dx = \pi$, this gives

$$F(m) = \frac{1}{m^2} \sum_{k \in \mathbb{Z}^2} f\left(\frac{k}{m}\right) - \frac{1}{m^2} f(0) = \pi - \frac{1}{m^2} + 2\pi \sum_{k \in \mathbb{Z}_0^2} \widehat{f}(2\pi m k). \tag{5.6}$$

Since f is radial, we have

$$\widehat{f}(\xi) = \int_0^\infty \frac{J_0(|\xi| r) r dr}{(1 + r^2)^2} = \frac{|\xi|}{2} K_1(|\xi|),$$

where K_1 is the modified Bessel function of the second kind, and where the second equality is [33, formula 13.51 (4)].

Therefore we have to show that

$$\sum_{k \in \mathbb{Z}_0^2} G(2\pi m|k|) < \frac{1}{m^2}, \quad G(x) = \pi x K_1(x).$$

Next, we use the estimate (see [34])

$$K_1(x) < \left(1 + \frac{1}{2x}\right) \sqrt{\frac{\pi}{2x}} e^{-x}, \quad x > 0,$$

which gives

$$G(2\pi m|k|) < \pi \left(\pi \sqrt{m|k|} + \frac{1}{4\sqrt{m|k|}}\right) e^{-2\pi m|k|}.$$

For the first term we use that

$$\sqrt{x}e^{-ax} \le \frac{1}{\sqrt{2ea}}$$

with $a = \frac{1}{2}\pi m$ and $x = |k|$ (and keep three quarters of the negative exponent), while for the second term we just replace $\frac{1}{\sqrt{m|k|}}$ by 1, since $m \ge 1$ and $k \ge 1$. This gives

$$G(2\pi m|k|) < \pi\left(\sqrt{\frac{\pi}{e}}e^{-\frac{3\pi m|k|}{2}} + \frac{1}{4}e^{-2\pi m|k|}\right).$$

Furthermore, we use that $|k| \ge \frac{1}{\sqrt{2}}(|k_1| + |k_2|)$ and, therefore,

$$G(2\pi m|k|) < \pi\left(\sqrt{\frac{\pi}{e}}e^{\frac{-3\pi m(|k_1|+|k_2|)}{2\sqrt{2}}} + \frac{1}{4}e^{-\sqrt{2}\pi m(|k_1|+|k_2|)}\right).$$

Thus, summing the geometric power series, we end up with

$$\begin{aligned} F(m) < \pi - \frac{1}{m^2} + \pi\sqrt{\frac{\pi}{e}}\left(\frac{4}{(e^{\frac{3\pi}{2\sqrt{2}}m} - 1)^2} + \frac{4}{e^{\frac{3\pi}{2\sqrt{2}}m} - 1}\right) \\ + \frac{\pi}{4}\left(\frac{4}{(e^{\sqrt{2}\pi m} - 1)^2} + \frac{4}{e^{\sqrt{2}\pi m} - 1}\right). \end{aligned}$$

Introducing the functions

$$\varphi(x) := \frac{x^2}{e^x - 1}, \quad \psi(x) := \frac{x}{e^x - 1},$$

we have to show that

$$\begin{aligned} \Psi(m) := 4\sqrt{\frac{\pi}{e}}\left(\frac{2\sqrt{2}}{3\pi}\right)^2\left(\varphi\left(\frac{3\pi}{2\sqrt{2}}m\right) + \psi\left(\frac{3\pi}{2\sqrt{2}}m\right)^2\right) \\ + \frac{1}{2\pi^2}\left(\varphi\left(\sqrt{2}\pi m\right) + \psi(\sqrt{2}\pi m)^2\right) - \frac{1}{\pi} < 0. \end{aligned}$$

Note that the function ψ is obviously decreasing for all $m \ge 0$, so all terms involving ψ are decreasing. The function $\varphi(x)$ has a global maximum at

$$x_0 = 1.5936\ldots,$$

and is decreasing when $x > x_0$. Since

$$m_1 := \frac{2\sqrt{2}}{3\pi}x_0 = 0.47824 < 1, \quad m_2 := \frac{1}{\sqrt{2}\pi}x_0 = 0.35868 < 1,$$

the function $\Psi(m)$ is decreasing for $m \ge 1$ and it is sufficient to verify the inequality for $m = 1$ only. Since

$$\Psi(1) = -0.141093\cdots < 0,$$

inequality (5.5) is proved. ∎

Remark. Since $f(x) = \frac{1}{(|x|^2+1)^2}$ is analytic, its Fourier transform decays exponentially and it follows from equation (5.6) that

$$F(m) = \pi - \frac{1}{m^2} + O(e^{-Cm}).$$

This immediately gives that inequality (5.5) holds on $[m_0, \infty)$ and we have to somehow specify m_0 and then use numerical calculations to verify (5.5) on a *finite* interval $(0, m_0)$ (precisely this is the case of a similar estimates in [18]). Lemma 5.4 is one of the few examples when this can be done purely analytically. The key points are of course the explicit formula for the Fourier transform and uniform founds for K_1.

Acknowledgements. The second author was supported by the EPSRC (United Kingdom) grant EP/P024920/1.

Bibliography

[1] H. Araki, On an inequality of Lieb and Thirring. *Lett. Math. Phys.* **19** (1990), 167–170

[2] A. V. Babin and M. I. Vishik, *Attractors of evolution equations*. Stud. Math. Appl.25, North-Holland, Amsterdam, 1992

[3] J. Bardina, J. Ferziger and W. Reynolds, *Improved subgrid scale models for large eddy simulation*. AIAA paper 80-1357, 1980

[4] Y. Cao, E. M. Lunasin and E. S. Titi, Global well-posedness of the three-dimensional viscous and inviscid simplified Bardina turbulence models. *Commun. Math. Sci.* **4** (2006), 823–848

[5] V. V. Chepyzhov and A. A. Ilyin, A note on the fractal dimension of attractors of dissipative dynamical systems. *Nonlinear Anal.* **44** (2001), 811–819

[6] V. V. Chepyzhov and A. A. Ilyin, On the fractal dimension of invariant sets; applications to Navier–Stokes equations. *Discrete Contin. Dyn. Syst.* **10** (2004), 117–135

[7] V. V. Chepyzhov, A. A. Ilyin and S. Zelik, Vanishing viscosity limit for global attractors for the damped Navier–Stokes system with stress free boundary conditions. *Phys. D* **376/377** (2018), 31–38

[8] V. V. Chepyzhov and M. I. Vishik, *Attractors for equations of mathematical physics*. Amer. Math. Soc. Colloq. Publ. 49, American Mathematical Society, Providence, RI, 2002

[9] V. V. Chepyzhov and M. I. Vishik, Trajectory attractors for dissipative 2D Euler and Navier–Stokes equations. *Russ. J. Math. Phys.* **15** (2008), 156–170

[10] V. V. Chepyzhov, M. I. Vishik and S. V. Zelik, Strong trajectory attractors for dissipative Euler equations. *J. Math. Pures Appl. (9)* **96** (2011), 395–407

[11] P. Constantin and C. Foias, Global Lyapunov exponents, Kaplan–Yorke formulas and the dimension of the attractors for 2D Navier–Stokes equations. *Comm. Pure Appl. Math.* **38** (1985), 1–27

[12] C. Foias, O. Manley, R. Rosa and R. Temam, *Navier–Stokes equations and turbulence*. Encyclopedia Math. Appl. 83, Cambridge University Press, Cambridge, 2001

[13] A. Haraux, Two remarks on hyperbolic dissipative problems. In *Nonlinear partial differential equations and their applications. Collège de France seminar, Vol. VII (Paris, 1983–1984)*, pp. 6, 161–179, Res. Notes in Math. 122, Pitman, Boston, 1985

[14] A. A. Ilyin, Euler equations with dissipation. *Mat. Sb.* **182** (1991), 1729–1739 (in Russian); English transl. *Math. USSR-Sb.* **74** (1993), 475–485

[15] A. A. Ilyin, A. G. Kostianko and S. V. Zelik, Finite-dimensional attractors for damped Euler–Bardina model in three dimensions. In preparation

[16] A. A. Ilyin, A. Miranville and E. S. Titi, Small viscosity sharp estimates for the global attractor of the 2-D damped-driven Navier–Stokes equations. *Commun. Math. Sci.* **2** (2004), 403–426

[17] A. A. Ilyin and A. A. Laptev, Lieb–Thirring inequalities on the torus. *Mat. Sb.* **207** (2016), 56–79 (in Russian); English transl. *Sb. Math.* **207** (2016), 1410–1434

[18] A. A. Ilyin, A. A. Laptev and S. Zelik, Lieb–Thirring constant on the sphere and on the torus. *J. Funct. Anal.* **279** (2020), Article ID 108784

[19] A. A. Ilyin and E. S. Titi, Attractors for the two-dimensional Navier–Stokes-α model: An α-dependence study. *J. Dynam. Differential Equations* **15** (2003), 751–778

[20] W. B. Jones and W. J. Thron, *Continued fractions. Analytic theory and applications*. Encyclopedia Math. Appl. 11, Addison-Wesley Publishing, Reading, 1980

[21] V. K. Kalantarov and E. S. Titi, Global attractors and determining modes for the 3D Navier–Stokes–Voight equations. *Chin. Ann. Math. Ser. B* **30** (2009), 697–714

[22] O. Ladyzhenskaya, *Attractors for semigroups and evolution equations*. Lezioni Lincee, Cambridge University Press, Cambridge, 1991

[23] E. Lieb, An L^p bound for the Riesz and Bessel potentials of orthonormal functions. *J. Funct. Anal.* **51** (1983), 159–165

[24] E. Lieb, On characteristic exponents in turbulence. *Comm. Math. Phys.* **92** (1984), 473–480

[25] E. Lieb and W. Thirring, Inequalities for the moments of the eigenvalues of the Schrödinger Hamiltonian and their relation to Sobolev inequalities. *Studies in Mathematical Physics. Essays in honor of Valentine Bargmann*, pp. 269–303, Princeton University Press, Princeton, 1976

[26] V. X. Liu, A sharp lower bound for the Hausdorff dimension of the global attractors of the 2D Navier–Stokes equations. *Comm. Math. Phys.* **158** (1993), 327–339

[27] L. D. Meshalkin and Ya. G. Sinai, Investigation of the stability of a stationary solution of a system of equations for the plane movement of an incompressible viscous liquid. *Prikl. Mat. Mekh.* **25** (1961), 1140–1143 (in Russian); English transl. *J. Appl. Math. Mech.* **25** (1961), 1700–1705

[28] J. Pedlosky, *Geophysical fluid dynamics*. Springer, New York, 1979

[29] G. R. Sell and Y. You, *Dynamics of evolutionary equations*. Appl. Math. Sci. 143, Springer, New York, 2002

[30] B. Simon, *Trace ideals and their applications*. 2nd edn., Math. Surveys Monogr. 120, American Mathematical Society, Providence, 2005

[31] E. M. Stein and G. Weiss, *Introduction to Fourier analysis on Euclidean spaces*. Princeton University Press, Princeton, 1971

[32] R. Temam, *Infinite-dimensional dynamical systems in mechanics and physics*. 2nd edn., Appl. Math. Sci. 68, Springer, New York, 1997

[33] G. N. Watson, *A treatise on the theory of Bessel functions*. Cambridge Math. Libr., Cambridge University Press, Cambridge, 1995

[34] Z.-H. Yang and Y.-M. Chu, On approximating the modified Bessel function of the second kind. *J. Inequal. Appl.* **41** (2017), Paper No. 41

[35] V. I. Yudovich, Example of the generation of a secondary stationary or periodic flow when there is loss of stability of the laminar flow of a viscous incompressible fluid. *Prikl. Mat. Mekh.* **29** (1965), 453–467 (in Russian); English transl. *J. Appl. Math. Mech.* **29** (1965), 587–603

Upper estimates for the electronic density in heavy atoms and molecules

Victor Ivrii

To Ari Laptev on the occasion of his 70th birthday

We derive an upper estimate for the electronic density $\rho_\Psi(x)$ in heavy atoms and molecules. While not sharp, on the distances $\gtrsim Z^{-1}$ from the nuclei it is still better than the known estimate CZ^3 (Z is the total charge of the nuclei, $Z \asymp N$ the total number of electrons).

1 Introduction

The present paper is a result of my rethinking of three rather old but still remarkable papers [5, 6, 13], which I discovered recently. The first of them derives the estimate electronic density $\rho_\Psi(x)$ from above via some integral also containing ρ_Ψ, the second one provides an estimate $\rho_\Psi(x) = O(Z^3)$, where Z is the total charge of nuclei and the third one derives the asymptotic of the averaged electronic density on the distances $O(Z^{-1})$ from the nuclei but its method works also on the larger distances.

Important role play arguments I borrowed from [1–3]. Later I became aware of [11, 14] and [4].

The purpose of this paper is to provide a better upper estimate for $\rho_\Psi(x)$ on distances larger than Z^{-1} from the nuclei.

Let us consider the following operator (quantum Hamiltonian):

$$\mathsf{H} = \mathsf{H}_N := \sum_{1\le j\le N} H_{V,x_j} + \sum_{1\le j<k\le N} |x_j - x_k|^{-1}$$

on

$$\mathfrak{H} = \bigwedge_{1\le n\le N} \mathscr{H}, \quad \mathscr{H} = \mathscr{L}^2(\mathbb{R}^3, \mathbb{C}^q)$$

with

$$H_V = -\Delta - V(x)$$

Keywords: Electronic density, Thomas–Fermi approximation
2020 Mathematics Subject Classification: Primary 35P20; secondary 81V70

describing N identical particles (electrons) in the external field with the scalar potential $-V$ and repelling each other according to the Coulomb law.

Here $x_j \in \mathbb{R}^3$ and $(x_1, \dots, x_N) \in \mathbb{R}^{3N}$. A potential $V(x)$ is assumed to be real-valued. Except when specifically mentioned we otherwise assume that

$$V(x) = \sum_{1\le m\le M} \frac{Z_m}{|x - \mathsf{y}_m|},$$

where $Z_m > 0$ and y_m are charges and locations of nuclei.

The mass is equal to $\frac{1}{2}$ and the Plank constant and the charge are equal to 1 here. We assume that $N \asymp Z = Z_1 + \dots + Z_M$.

Our purpose is to derive a pointwise upper estimate for the *electronic density*

$$\rho_\Psi(x) = N \int |\Psi(x, x_2, \dots, x_N)|^2 \, dx_2 \cdots dx_N.$$

Let

$$\ell(x) = \min_{1\le m\le M} |x - \mathsf{y}_m|$$

be the distance to the nearest nucleus. Our goal is to prove the following theorem:

Theorem 1.1. *Let*

$$\min_{1\le m<m'\le M} |\mathsf{y}_m - \mathsf{y}_{m'}| \ge Z^{-\frac{1}{3}+\sigma}$$

with $\sigma > 0$. *Then:*

(i) *For* $\ell(x) \le Z^{-\frac{1}{3}}$ *the following estimate holds:*

$$\rho_\Psi(x) \le C \begin{cases} Z^3 & \text{for } \ell(x) \le Z^{-\frac{8}{9}}, \\ Z^{\frac{19}{9}} \ell(x)^{-1} & \text{for } Z^{-\frac{8}{9}} \le \ell(x) \le Z^{-\frac{7}{9}}, \\ Z^{\frac{197}{90}} \ell(x)^{-\frac{9}{10}} & \text{for } Z^{-\frac{7}{9}} \le \ell(x) \le Z^{-\frac{1}{3}}. \end{cases}$$

(ii) *For* $\ell(x) \ge Z^{-\frac{1}{3}}$ *the following estimate holds:*

$$\rho_\Psi(x) \le C \begin{cases} Z^{\frac{17}{9}} \ell(x)^{-\frac{9}{5}} & \text{for } Z^{-\frac{1}{3}} \le \ell(x) \le Z^{-\frac{5}{18}}, \\ Z^{\frac{19}{9}} \ell(x)^{-1} & \text{for } \ell(x) \ge Z^{-\frac{5}{18}}. \end{cases}$$

(iii) *Furthermore, if* $(Z - N) \ge C_0 Z^{\frac{5}{6}}$, *then*

$$\rho_\Psi(x) \le Z^{\frac{19}{9}} \ell(x)^{-1} \quad \text{for } \ell(x) \ge C_0 (Z - N)_+^{-\frac{1}{3}}.$$

Remark 1. We would like to prove an estimate $\rho_\Psi(x) \le C\zeta(x)^3$, or to discover that it does not necessarily hold with

$$\zeta(x) = \begin{cases} Z^{\frac{1}{2}} \ell(x)^{-\frac{1}{2}} & \ell(x) \le Z^{-\frac{1}{3}}, \\ \ell(x)^{-2} & \ell(x) \ge Z^{-\frac{1}{3}}. \end{cases}$$

Plan of the paper. In Section 2 we prove a more subtle version of the main estimate of [5]. In Section 3 we provide upper estimates and asymptotics of ρ_Ψ integrated over small balls. In Section 4 we study the energy of electron-to-electron interaction (it involves a two-point correlation function) and in Section 5 we prove upper estimates for $\rho_\Psi(x)$.

2 Main intermediate inequality

We start from the main intermediate equality.

Proposition 2.1. *Let Ψ be an eigenfunction of H_N with an eigenvalue λ. Let $\phi(|x|)$ be a real-valued spherically symmetric function. Then*

$$\begin{aligned}\rho_\Psi(0) = (2\pi)^{-1}N \iint (\partial_r K(x))\Psi(x, x_2, \dots, x_N)\\ \times \Psi^*(x, x_2, \dots, x_N)\phi(|x|)\,dx\,dx_2\cdots dx_N \\ - (8\pi)^{-1}\int \rho_\Psi(x)\phi'''(|x|)\,dx,\end{aligned} \tag{2.1}$$

where

$$K(x) := -\mathsf{H}_N - (\partial_r^2 + 2r^{-1}\partial_r)$$

is an operator in the auxiliary space $\mathscr{H} := \bigotimes_{n=2,\dots,N}\mathscr{L}^2(\mathbb{R}^3, \mathbb{C}^q)\otimes\mathbb{C}^q$ with an inner product $\langle\cdot,\cdot\rangle$, $\phi'''(r) = \partial_r^3\phi(r)$ and $x = (r,\theta)\in\mathbb{R}^+\times\mathbb{S}^2$.

Proof. Let us consider Ψ as a function of $x\in\mathbb{R}^3$ with values in the auxiliary space $\mathscr{H}$, and let $u = r\Psi$, where (r,θ) are spherical coordinates in $\mathbb{R}^3$. Then similar to [5, (9)]

$$\begin{aligned}\langle\Psi(0),\Psi(0)\rangle = -(2\pi)^{-1}\int\langle\partial_r u, \partial_r^2 u\rangle\phi(r)r^{-2}\,dx\\ - (4\pi)^{-1}\int\langle\partial_r u, \partial_r u\rangle\phi'(r)r^{-2}\,dx\end{aligned}$$

and since $r^{-1}\partial_r^2 u = r\Delta_r\Psi := r(\partial_r^2 + 2r^{-1}\partial_r)\Psi$, the first term on the right is equal to

$$\begin{aligned}&-(2\pi)^{-1}\iint\langle\partial_r u, r\Delta_r\Psi\rangle\phi(r)\,dr\,d\theta\\ &= \pi^{-1}\iint\langle\partial_r u, (K+\lambda)u\rangle\phi(r)dr\,d\theta\\ &= -(2\pi)^{-1}\int\langle\Psi, K'\Psi\rangle\phi(r)\,dx - (2\pi)^{-1}\iint\langle\Psi,(K+\lambda)\Psi\rangle\phi'(r)\,dx\end{aligned} \tag{2.2}$$

because $\Delta_r\Psi = -2(K+\lambda)\Psi$, where $-K$ is the rest of the multiparticle Hamiltonian (including $-r^{-2}\Delta_\theta$) and we integrated by parts.

The first term in the latter formula is a corresponding term in [5], albeit truncated with ϕ, and we have two new terms

$$-(4\pi)^{-1}\iint \langle \partial_r(r\Psi), \partial_r(r\Psi)\rangle \phi'(r)\,dr\,d\theta - (2\pi)^{-1}\int \langle \Psi, (W+\lambda)\Psi\rangle \phi'(r)\,dx\,.$$

Integrating the first term by parts, we get

$$(4\pi)^{-1}\iint \langle r\Psi, \partial_r^2(r\Psi)\rangle \phi'(r)\,dr\,d\theta + (4\pi)^{-1}\iint \langle r\Psi, \partial_r(r\Psi)\rangle \phi''(r)\,dr\,d\theta,$$

where the first term cancels the second term in (2.2), while the second term integrates by parts one more time resulting in the last term in (2.1). ■

Applying (2.1) to our problem, and using skew-symmetry of Ψ, we get

$$\begin{aligned}\rho_\Psi(0) = (2\pi)^{-1}&\int \sum_m Z_m \frac{x\cdot(x-y_m)}{|x|\cdot|x-y_m|^3}\rho_\Psi(x)\phi(|x|)\,dx\\ &+ (2\pi)^{-1}N(N-1)\int \underbrace{\frac{x_1\cdot(x_2-x_1)}{|x_1|\cdot|x_2-x_1|^3}}\\ &\qquad\qquad \times|\Psi(x_1,x_2,\dots,x_N)|^2\,\underbrace{\phi(|x_1|)}\,dx_1\cdots dx_N\\ &- (2\pi)^{-1}N\int |x|^{-3}|\nabla_\theta\Psi(x,x_2,\dots,x_N)|^2\phi(|x|)\,dx\,dx_2\cdots dx_N\\ &- (8\pi)^{-1}\int \rho_\Psi(x)\phi'''(|x|)\,dx.\end{aligned}\tag{2.3}$$

Symmetrizing the second term with respect to x_1 and x_2 instead of the product of two indicated factors, we will get

$$\begin{aligned}&\frac{1}{4}(\phi(|x_1|)+\phi(|x_2|))\Bigl(\frac{x_1\cdot(x_2-x_1)}{|x_1|\cdot|x_2-x_1|^3} - \frac{x_2\cdot(x_2-x_1)}{|x_2|\cdot|x_2-x_1|^3}\Bigr)\\ &\qquad+\frac{1}{4}(\phi(|x_1|)-\phi(|x_2|))\Bigl(\frac{x_1\cdot(x_2-x_1)}{|x_1|\cdot|x_2-x_1|^3} + \frac{x_2\cdot(x_2-x_1)}{|x_2|\cdot|x_2-x_1|^3}\Bigr)\end{aligned}$$

with the big parenthesis on the first line equal to

$$-\frac{|x_1|+|x_2|}{|x_1-x_2|^3}\Bigl(1-\frac{x_1\cdot x_2}{|x_1|\cdot|x_2|}\Bigr)$$

and the big parenthesis on the second line equal to

$$-\frac{|x_1|-|x_2|}{|x_1-x_2|^3}\Bigl(1+\frac{x_1\cdot x_2}{|x_1|\cdot|x_2|}\Bigr).$$

One can see that the former is negative and the latter, multiplied by $(\phi(|x_1|)-\phi(|x_2|))$, is non-negative if ϕ is a non-decreasing function. Let us shift the origin to point x and observe that the first term in (2.3) is equal to

$$(2\pi)^{-1}\int \sum_m Z_m \frac{(x-\mathsf{x})\cdot(x-\mathsf{y}_m)}{|x-\mathsf{x}|\cdot|x-\mathsf{y}_m|^3}\rho_\Psi(x)\phi(|x-\mathsf{x}|)\,dx.$$

Consider first case $\phi = 1$. Then we get

$$\rho_\Psi(\mathsf{x}) \le (2\pi)^{-1}\int \sum_m Z_m|x-\mathsf{y}_m|^{-2}\rho_\Psi(x)\,dx. \tag{2.4}$$

Indeed, the second term on the right-hand expression of (2.3) is non-positive due to above analysis, so is the third term, and the fourth term vanishes while the first term does not exceed the right-hand expression

Applying Proposition 3.1 below, we arrive at the following estimate:

$$\rho_\Psi(x) \le CZ^3. \tag{2.5}$$

In the general case we have the following:

Proposition 2.2. *In the framework of Proposition* 2.1,

$$\begin{aligned}\rho_\Psi(\mathsf{x}) \le{}& (2\pi)^{-1}\int \sum_m Z_m \frac{(x-\mathsf{x})\cdot(x-\mathsf{y}_m)}{|x-\mathsf{x}|\cdot|x-\mathsf{y}_m|^3}\rho_\Psi(x)\phi(|x-\mathsf{x}|)\,dx\\ &+ Ct^{-1}\iint_{B(\mathsf{x},t)\times B(\mathsf{x},t)} |x-y|^{-1}\rho^{(2)}_\Psi(x,y)\,dx\,dy\\ &+ C\iint_{B(\mathsf{x},t)\times(\mathbb{R}^3\setminus B(\mathsf{x},t))} |y-\mathsf{x}|^{-2}\rho^{(2)}_\Psi(x,y)\,dx\,dy,\end{aligned} \tag{2.6}$$

where

$$\rho^{(2)}_\Psi(x,y) := N(N-1)\int |\Psi(x,y,x_3,\dots,x_N)|^2\,dx_3\cdots dx_N$$

is a two-point correlation function.

Recall that

$$\int \rho^{(2)}_\Psi(x,y)\,dy = (N-1)\rho_\Psi(x).$$

Remark 2. We make the following observations.

(i) Inequality (2.4) for $M = 1$ and $\mathsf{x} = \mathsf{y}_1$ is the main result of [5]. Our main achievement so far is an introduction of the truncation ϕ. However it brings three new terms in the right-hand expression of the estimate.

(ii) Estimate (2.5) (with a specified albeit not sharp constant) was proven in [13] for $x = \mathsf{y}_m$.

(iii) This estimate definitely has the correct magnitude as $|x-\mathsf{y}_m| \lesssim Z^{-1}$, $Z_m \asymp Z$.

3 Estimates of the averaged electronic density

We will need the following [8, estimate (3.3)]:

$$\int U\rho_\Psi\,dx \le \operatorname{Tr}(H^-_{W+\nu}) - \operatorname{Tr}(H^-_{W+U+\nu}) + CZ^{\frac{5}{3}-\delta} \tag{3.1}$$

with $\delta = \delta(\sigma)$, $\delta > 0$ for $\sigma > 0$ and $\delta = 0$ for $\sigma = 0$.

First, we use this estimate in the very rough form:

Proposition 3.1. *The following estimate holds:*

$$\int |x - \mathsf{y}_m|^{-2}\rho_\Psi(x)\,dx \le CZ^2. \tag{3.2}$$

Proof. Let $\psi(x)$, $\psi_0(x)$ be cut-off functions, $\psi(x) = 0$ in $\{x : \ell(x) \le b\}$, $\psi_0(x) = 0$ in $\{x : \ell(x) \ge 2b\}$, $\psi + \psi_0 = 1$, $b = Z^{-1}$. Then

$$\operatorname{Tr}(H^-_{W+U+\nu}) = \operatorname{Tr}(H^-_{W+U+\nu}\psi_0) + \operatorname{Tr}(H^-_{W+U+\nu}\psi). \tag{3.3}$$

Using the semiclassical methods of [7, Section 25.4] in the simplest form, we conclude that for $U = \epsilon|x - \mathsf{y}_m|^{-2}$ the second term on the right (with an opposite sign) could be replaced by its Weyl approximation

$$-\frac{2}{5}\kappa\int (W + U + \nu)_+^{\frac{5}{2}}\psi(x)\,dx \tag{3.4}$$

with an error not exceeding CZ^2, where here and below $\kappa = \frac{q}{6\pi^2}$. The same is true for $U = 0$. One can see easily that the difference between expression (3.4) and the same expression for $U = 0$ does not exceed

$$C\int \bigl[(W + \nu)_+^{\frac{3}{2}}U + U^{\frac{5}{2}}\bigr]\,dx,$$

which does not exceed CZ^2.

Consider the first term in the right-hand expression of (3.3). Using variational methods of [7], Section 9.1, we can reduce it to the analysis of the same operator in $\mathcal{X}^0 := \{x : \ell(x) \le 4b\}$ with the Dirichlet boundary conditions on ∂X. Observing that eigenvalue counting function for such operator is $O(1 + \lambda^{\frac{3}{2}}Z^{-3})$ (for ϵ sufficiently small), we conclude that the first term in (3.3) also does not exceed CZ^2. Estimate (3.2) has been proven. ∎

Let us return to (3.1) and consider $U = \zeta^2\phi_t(x;\mathsf{x})$, where x is a fixed point with

$$\ell(\mathsf{x}) := \min_m |\mathsf{x} - \mathsf{y}_m| \ge Z^{-1}, \tag{3.5}$$

$$\zeta(\mathsf{x}) := \max\bigl(Z^{\frac{1}{2}}\ell(\mathsf{x})^{-\frac{1}{2}}, \ell(\mathsf{x})^{-2}\bigr) \tag{3.6}$$

and $\phi_t(x;\mathsf{x}) = \phi_0(t^{-1}|x - \mathsf{x}|)$, $\phi \in \mathscr{C}_0^\infty([-1,1])$, $0 \le \phi \le 1$. We assume that

$$\zeta^{-1} \le t \le \frac{\ell}{2} \tag{3.7}$$

with $\ell = \ell(\mathsf{x})$, $\zeta = \zeta(\mathsf{x})$, where the last inequality allows us to apply semiclassical methods. Consider with $0 \le \varsigma \le 1$

$$\mathrm{Tr}(H^-_{W+\nu}) - \mathrm{Tr}(H^-_{W+\varsigma U+\nu}) = \int_0^\varsigma \mathrm{Tr}\big(U\big[\theta(-H_{W+sU+\nu}) - \theta(-H_{W+\nu})\big]\big)\,ds$$

and apply semiclassical methods to the right-hand expression. Then we get

$$\begin{aligned}&\mathrm{Tr}(H^-_{W+\nu}) - \mathrm{Tr}(H^-_{W+\varsigma U+\nu})\\ &\quad= \kappa \int\int_0^\varsigma \big(U\big[(W+sU+\nu)_+^{\frac{3}{2}} - (W+\nu)_+^{\frac{3}{2}}\big]\big)\,dx\,ds + O(\varsigma\zeta^4 t^2)\\ &\quad= \frac{2}{5}\kappa \int \big((W+\varsigma U+\nu)_+^{\frac{5}{2}} - (W+\nu)_+^{\frac{5}{2}}\big)\,dx + O(\varsigma\zeta^4 t^2).\end{aligned}$$

Indeed, the factor U is $O(\zeta^2)$ and therefore the semiclassical error is $O(\zeta^4 t^2)$ since the effective semiclassical parameter is $h = (\zeta t)^{-1}$. Observe that the principal part in the right-hand expression does is $O(\varsigma\zeta^5 t^3)$. Then after division by $\varsigma\zeta^2$, (3.1) becomes

$$\int \phi_t(x;\mathsf{x})\rho_\Psi(x)\,dx \le \int \phi_t(x;\mathsf{x})\rho(x)\,dx + C\big(\zeta^2 t^2 + \varsigma\zeta^3 t^3 + \varsigma^{-1} Z^{\frac{5}{3}-\delta}\big). \quad (3.8)$$

Replacing $\phi_t(x;\mathsf{x})$ by $-\phi_t(x;\mathsf{x})$ in this inequality and minimizing by $\varsigma \in (0,1]$, we arrive at the first statement of the following proposition:

Proposition 3.2. *The following statements hold.*

(i) *Under assumptions* (3.5)–(3.7)

$$\Big|\int (\rho_\Psi(x) - \rho(x))\phi_t(x;\mathsf{x})\,dx\Big| \le C\big(\zeta^2 t^2 + \zeta^{\frac{1}{2}} t^{\frac{3}{2}} Z^{\frac{5}{6}-\frac{\delta}{2}} + \zeta^{-2} Z^{\frac{5}{3}-\delta}\big). \quad (3.9)$$

(ii) *Further,*

$$\Big|\int \rho_\Psi(x)\phi_t(x;\mathsf{x})\,dx\Big| \le C\big(\zeta^3 t^3 + t^{\frac{6}{5}} Z^{1-\frac{3\delta}{5}}\big).$$

(iii) *Furthermore, if $N < Z$, then*

$$\Big|\int \rho_\Psi(x)\phi_t(x;\mathsf{x})\,dx\Big| \le C t^{\frac{6}{5}} Z^{1-\frac{3\delta}{5}} \quad \textit{for } \ell(\mathsf{x}) \ge C_0 (Z-N)_+^{-\frac{1}{3}}.$$

To prove the second statement, we consider $\varsigma > 0$ (without restriction $\varsigma \le 1$); then instead of (3.8) we have

$$\int \phi_t(x;\mathsf{x})\rho_\Psi(x)\,dx \le C\big(\zeta^3 t^3 + \varsigma^{\frac{3}{2}}\zeta^3 t^3 + \varsigma^{-1}\zeta^{-2} Z^{\frac{5}{3}-\delta}\big) \quad (3.10)$$

and we optimize it by $\varsigma > 0$.

The third statement follows from the same arguments and the fact that $\rho^{\mathsf{TF}}(x) = 0$ for $\ell(x) \ge C_0(Z-N)_+^{-\frac{1}{3}}$ and therefore (3.10) holds without the first term in the right-hand expression.

4 Estimates of the correlation function

We will need the following [7, Proposition 25.5.1] (first proven in [12]):

Proposition 4.1. *Let $\theta \in \mathscr{C}^\infty(\mathbb{R}^3)$ such that*

$$0 \le \theta \le 1.$$

Let $\chi \in \mathscr{C}^\infty(\mathbb{R}^6)$ and

$$\begin{aligned} \mathcal{J} &= \Bigl| \int \bigl(\rho^{(2)}_\Psi(x,y) - \rho(y)\rho_\Psi(x)\bigr)\theta(x)\chi(x,y)\,dx\,dy \Bigr| \\ &\le C\sup_x \|\nabla_y \chi\|_{\mathscr{L}^2(\mathbb{R}^3_y)}\bigl((Q+\varepsilon^{-1}N+T)^{\frac12}\Theta + P^{\frac12}\Theta^{\frac12}\bigr) \\ &\qquad + C\varepsilon N\|\nabla_y\chi\|_{\mathscr{L}^\infty}\Theta \end{aligned} \tag{4.1}$$

with

$$Q = \mathsf{D}(\rho_\Psi - \rho^{\mathsf{TF}}, \rho_\Psi - \rho^{\mathsf{TF}}), \quad \Theta = \Theta_\Psi := \int \theta(x)\rho_\Psi(x)\,dx,$$

$$T = \sup_{\operatorname{supp}(\theta)} W, \qquad P = \int |\nabla\theta^{\frac12}|^2\rho_\Psi\,dx,$$

respectively, and arbitrary $\varepsilon \le Z^{-\frac23}$.

We cannot apply it directly to estimate the second to the last term in (2.6) because of singularities. Let us consider

$$\iint_{\mathbb{R}^3\times\mathbb{R}^3} |x-y|^{-1}\rho^{(2)}_\Psi(x,y)\,dx\,dy.$$

Let us make an ℓ-admissible partition of unity ϕ_ι in with ℓ-admissible $\phi_\iota^{\frac12}$. We set $\ell(x) = Z^{-1}$ if $|x - \mathsf{y}_m| \le Z^{-1}$. Let us consider first

$$\int |x-y|^{-1}\rho^{(2)}_\Psi(x,y)\phi_\iota(x)\phi_\varkappa(y)\,dx\,dy$$

in the case of ϕ_ι and ϕ_κ having disjoint supports. Without any loss of the generality we can consider $\ell_x \le \ell_y$, where subscripts x, y are referring to supports of ϕ_ι, ϕ_κ, respectively.

Let $\theta(x) = \phi_\iota(x)$ and

$$\chi(x,y) = \bar\chi(x,y) := |x-y|^{-1}\bar\phi_\iota(x)\phi_\kappa(y),$$

where $\bar\phi_\iota$ are ℓ-admissible and equal 1 in the ℓ-vicinity of $\operatorname{supp}(\phi_\iota)$. Then $T = \zeta_x$ and for $\ell_x \le Z^{-\frac{5}{21}}$ in virtue of Proposition 3.2[1]

$$\Theta_\Psi \asymp \zeta_x^3\ell_x^3, \quad P \asymp \zeta_x^3\ell_x \tag{4.2}$$

[1]Indeed, $\zeta^3\ell^3 \ge Z^{\frac53-\delta}\zeta^{-2}$ if and only if $\ell \le Z^{-\frac{5}{21}+\frac{\delta}{7}}$.

and

$$\|\nabla_y \chi\|_{\mathscr{L}^2(\mathbb{R}^3_y))} \asymp d_{x,y}^{-1}\ell_y^{\frac{1}{2}}, \quad \|\nabla \chi\|_{\mathscr{L}^\infty} \asymp d_{x,y}^{-1}\ell_y^{-1}, \tag{4.3}$$

where $d_{x,y} \geq \ell_y$ is the distance between the supports of ϕ_ι and $\phi_\varkappa$. Then the right-hand expression of (4.1) is

$$C\zeta_x^3\ell_x^3\big(\ell_y^{-\frac{1}{2}}(Z^{\frac{5}{6}} + \zeta_x) + \varepsilon^{-\frac{1}{2}}\ell_y^{-\frac{1}{2}}Z^{\frac{1}{2}} + \varepsilon Z\ell_y^{-2} + \ell_y^{-\frac{1}{2}}\ell_x^{-1}\big)$$

and minimizing by $\varepsilon \leq Z^{-\frac{2}{3}}$, we get

$$C\zeta_x^3\ell_x^3\big(\ell_y^{-\frac{1}{2}}(Z^{\frac{5}{6}-\delta} + \zeta_x) + Z^{\frac{2}{3}}\ell_y^{-1} + \ell_y^{-\frac{1}{2}}Z^{\frac{5}{6}} + \ell_y^{-\frac{1}{2}}\ell_x^{-1}\big).$$

Observe that all powers of ℓ_y are negative. Therefore summation over all elements of the y-partition results in the same expression albeit with ℓ_y replaced by $\ell_x = \ell$:

$$C\zeta^3\ell^2\big(\ell^{\frac{1}{2}}(Z^{\frac{5}{6}} + \zeta) + Z^{\frac{2}{3}} + \ell^{-\frac{1}{2}} + \ell^{\frac{1}{2}}Z^{\frac{5}{6}}\big).$$

For $\ell \leq Z^{-\frac{1}{3}}$ we have $\zeta = Z^{\frac{1}{2}}\ell^{-\frac{1}{2}}$ and all powers are positive with the exception of one term, where the power is 0, and for $\ell \geq Z^{-\frac{1}{3}}$ we have $\zeta = \ell^{-2}$ and all powers are negative. Therefore summation over all elements of x-partition results in the same expression albeit with $\ell = Z^{-\frac{1}{3}}$, $\zeta = Z^{\frac{2}{3}}$, with the exception of one term which gains a logarithmic factor. We get CZ^2. Then

$$\Big|\sum_{\iota,\varkappa}\iint \big(\rho_\Psi^{(2)}(x,y) - \rho(y)\rho_\Psi(x)\big)\phi_\iota(x)\phi_\kappa(y), dx\,dy\Big| \leq CZ^2 \tag{4.4}$$

with summation over indicated pairs of elements of the partition (disjoint, with $\ell_x \leq \min(Z^{-\frac{5}{21}+\frac{\delta}{7}}, \ell_y)$).

Let us prove the following:

Claim 4.2. *Estimate* (4.4) *also holds with* $\rho_\Psi(x)$ *replaced by* $\rho(x)$ *and therefore it holds for a sum over pairs of elements with* $\min(\ell_x, \ell_y) \leq \ell^* = Z^{-\frac{5}{21}+\frac{\delta}{7}}$.

Indeed, in virtue of the proof of Proposition 3.2 (before minimizing by ς) the error

$$\Big|\iint \big(\rho_\Psi(x) - \rho(x)\big)\rho(y)\phi_\iota(x)\phi_\kappa(y)\,dx\,dy\Big| \tag{4.5}$$

on each pair of elements does not exceed $C\zeta_y^3\ell_y^2$ with all powers of ℓ_y positive for $\ell_y \leq Z^{-\frac{1}{3}}$ and negative for $\ell_y \geq Z^{-\frac{1}{3}}$. Then summation with respect to y-partition (recall, that $\ell_y \geq \ell_x$) results in

$$C\big(\zeta_x^2\ell_x^2 + \varsigma\zeta_x^3\ell_x^3 + \varsigma^{-1}\zeta_x^{-2}Z^{\frac{5}{3}-\delta}\big) \times \begin{cases} Z^{\frac{4}{3}} & \text{for } \ell_x \leq Z^{-\frac{1}{3}},\\ \zeta_x^3\ell_x^2 & \text{for } \ell_x \geq Z^{-\frac{1}{3}},\end{cases}$$

with the first line corresponding to $\ell_y = Z^{-\frac{1}{3}}$, $\zeta_y = Z^{\frac{2}{3}}$ and the second line corresponding to $\ell_y = \ell_x$, $\zeta_y = \zeta_x$.

Powers of ℓ_x are positive for $\ell_x \le Z^{-\frac{1}{3}}$ and negative for $\ell_x \ge Z^{-\frac{1}{3}}$, and summation with respect to x-partition results in the value as $\ell_x = Z^{-\frac{1}{3}}$, $\zeta_x = Z^{\frac{2}{3}}$, which is

$$CZ^2 + C\varsigma Z^{\frac{7}{3}} + C\varsigma^{-1} Z^{\frac{5}{3}-\delta}.$$

Minimizing by $\varsigma = Z^{-\frac{1}{3}}$, we conclude that the sum of expressions (4.5) over required pairs does not exceed CZ^2, which in turn implies (4.4).

Consider now the case when the supports of elements are not disjoint. Then we take

$$\chi(x,y) = \bar{\chi}(x,y)\eta\Big(\frac{|x-y|}{s}\Big) = |x-y|^{-1}\bar{\phi}_\iota(x)\phi_\kappa(y)\eta\Big(\frac{|x-y|}{s}\Big)$$

with $\eta(t)$ smooth function, equal 0 at $(0,\frac{1}{2})$ and 1 at $(1,\infty)$; $s \le Z^{-\frac{1}{3}}$ will be selected later.[2] Then while (4.2) is preserved, (4.3) should be replaced by

$$\|\nabla_y \chi\|_{\mathscr{L}^2(\mathbb{R}^3_y))} \asymp s^{-\frac{1}{2}}, \quad \|\nabla_y \chi\|_{\mathscr{L}^\infty} \asymp s^{-2}.$$

The right-hand expression (4.1) is

$$Cs^{-\frac{1}{2}}\zeta^3\ell^3\big((Z^{\frac{5}{6}} + \zeta) + \varepsilon^{-\frac{1}{2}}Z^{\frac{1}{2}} + \varepsilon s^{-\frac{3}{2}}Z + \ell^{-1}\big),$$

and minimizing by $\varepsilon \le Z^{-\frac{2}{3}}$, we get

$$\begin{aligned}\Big|\int \phi_\iota(x)\chi(x,y)\big(\rho^{(2)}_\Psi(x,y) - \rho_\Psi(x)\rho(y)\big)\,dx\,dy\Big|\\ \le Cs^{-\frac{1}{2}}\zeta^3\ell^3\big(Z^{\frac{5}{6}} + \zeta + s^{-\frac{1}{2}}Z^{\frac{2}{3}} + \ell^{-1}\big).\end{aligned} \tag{4.6}$$

Note that summation of estimate (4.6) over a partition returns its value as $\ell = Z^{-\frac{1}{3}}$, namely, $Cs^{-1}Z^{\frac{5}{3}}$..

Consider for t with $s \le t \le \ell$ the zone $\{(x,y) : |x-y| \asymp y\}$ and make there a t-admissible subpartition with respect to x, y (this means that the radii of the supports of subelements ϕ_* do not exceed t, and $|D^\alpha \phi_*| \le c_\alpha t^{-|\alpha|}$). Then the contribution of each pair of subelements to

$$\Big|\int \phi_\iota(x)\chi(x,y)\big(\rho_\Psi(x) - \rho(x)\big)\rho(y)\,dx\,dy\Big|$$

does not exceed

$$C\big(\zeta^2 t^2 + \varsigma\zeta^3 t^3 + \varsigma^{-1}\zeta^{-2}Z^{\frac{5}{3}-\delta}\big)\zeta^3 t^2$$

and since there are $\asymp \ell^3 t^{-3}$ of such pairs, we get

$$C\big(\zeta^2 t^2 + \varsigma\zeta^3 t^3 + \varsigma^{-1}\zeta^{-2}Z^{\frac{5}{3}-\delta}\big)\zeta^3\ell^3 t^{-1}.$$

[2] Since in this case $\ell_x = \ell_y$ and $\zeta_x = \zeta_y$, we skip subscripts.

Then summation over t with $s \le t \le \ell$ returns

$$C\big(\zeta^2\ell + \varsigma\zeta^3\ell^2 + \varsigma^{-1}s^{-1}\zeta^{-2}Z^{\frac{5}{3}-\delta}\big)\zeta^3\ell^3$$

and summation over the partition returns its value as $\ell = Z^{-\frac{1}{3}}$, namely

$$C\big(Z^2 + \varsigma Z^{\frac{7}{3}} + \varsigma^{-1}s^{-1}Z^{\frac{4}{3}-\delta}\big).$$

Minimizing by $\varsigma = (sZ)^{-\frac{1}{2}}$, we get $C(Z^2 + s^{-\frac{1}{2}}Z^{\frac{11}{6}-\delta})$.

On the other hand,

$$\iint \phi_\iota(x)\big(\bar{\chi}(x,y) - \chi(x,y)\big)\rho(x)\rho(y)\,dx\,dy \asymp s^2\zeta^6\ell^3,$$

and summation over $\ell \ge Z^{-\frac{1}{3}}$ returns its value at $Z^{-\frac{1}{3}}$, which is Cs^2Z^3, but summation over $\ell \le Z^{-\frac{1}{3}}$ returns $s^2Z^3\log Z$. To remedy this, we replace for $\ell \le Z^{-\frac{1}{3}}$ constant s by $s_x = s(\ell_x Z^{\frac{1}{3}})^{\delta'}$ with small $\delta' > 0$. It will not affect our previous estimates.

Consider the sum of these three right-hand expressions

$$C\big(Z^2 + s^{-1}Z^{\frac{5}{3}} + s^{-\frac{1}{2}}Z^{\frac{11}{6}-\delta} + s^2Z^3\big)$$

and minimize it by s; we get $CZ^{\frac{19}{9}}$ achieved as $s = Z^{-\frac{4}{9}}$.

Since we want $s \le \ell$, we finally set

$$s_x = \begin{cases} Z^{-\frac{4}{9}} & \text{for } \ell_x \ge Z^{-\frac{1}{3}}, \\ \min(Z^{-\frac{4}{9}}(\ell_x Z^{\frac{1}{3}})^{\delta'}, \ell_x) & \text{for } \ell_x \le Z^{-\frac{1}{3}}. \end{cases}$$

Observe that

$$\iint_{\{x:\ell_x \ge Z^{-\frac{5}{21}}\}} |x-y|^{-1}\rho(x)\rho(y) \asymp Z^{4\frac{1}{2}1}$$

and we arrive at

$$\iint_{\Omega} |x-y|^{-1}\big(\rho^{(2)}_{\Psi}(x,y) - \rho(x)\rho(y)\big)\,dx\,dy \le CZ^{\frac{19}{9}}$$

and

$$\iint_{\mathbb{R}^3_x\times\mathbb{R}^3_y\setminus\Omega} |x-y|^{-1}\rho(x)\rho(y)\,dx\,dy \le CZ^{\frac{19}{9}}$$

with

$$\Omega = \big\{(x,y) : \ell_x \le Z^{-\frac{5}{21}},\ |x-y| \ge s_y\big\}.$$

Therefore

$$\int_{\Omega} |x-y|^{-1}\rho^{(2)}_{\Psi}(x,y)\,dx\,dy \ge \int_{\mathbb{R}^6} |x-y|^{-1}\rho(x)\rho(y)\,dx\,dy - CZ^{\frac{19}{9}}. \quad (4.7)$$

However, we know that (see, e.g., [7, Section 25.2])

$$E_N \geq \operatorname{Tr}\bigl((H_W - \nu)^-\bigr) - \mathsf{D}(\rho_\Psi, \rho) + \frac{1}{2}\int |x-y|^{-1}\rho_\Psi^{(2)}(x,y)\,dx\,dy$$

and

$$E_N \leq \operatorname{Tr}\bigl((H_W - \nu)^-\bigr) - \frac{1}{2}\mathsf{D}(\rho, \rho) + CZ^{\frac{5}{3}}$$

with $\rho = \rho^{\mathsf{TF}}$, $W = W^{\mathsf{TF}}$ and $\mathsf{D}(f,g) := \iint |x-y|^{-1} f(x)g(x)\,dx$. Then

$$\int |x-y|^{-1}\rho_\Psi^{(2)}(x,y)\,dx\,dy \leq 2\mathsf{D}(\rho_\Psi - \rho, \rho) + \mathsf{D}(\rho,\rho) + CZ^{\frac{5}{3}}$$

and from

$$|\mathsf{D}(\rho_\Psi - \rho, \rho)| \leq \mathsf{D}(\rho_\Psi - \rho, \rho_\Psi - \rho)^{\frac{1}{2}}\mathsf{D}(\rho,\rho)^{\frac{1}{2}} \leq CZ^{\frac{5}{6}} \times Z^{\frac{7}{6}} = CZ^2$$

we conclude that

$$\int |x-y|^{-1}\rho_\Psi^{(2)}(x,y)\,dx\,dy \leq \mathsf{D}(\rho,\rho) + CZ^2.$$

Combining with (4.7), we conclude that

$$\int_{\mathcal{Z}} |x-y|^{-1}\rho_\Psi^{(2)}(x,y)\,dx\,dy \leq CZ^{\frac{19}{9}} \tag{4.8}$$

for $\mathcal{Z} = \mathbb{R}^3_x \times \mathbb{R}^3_y \setminus \Omega$.

5 Proof of Theorem 1.1

In the last two terms

$$Ct^{-1}\iint_{B(\mathsf{x},t)\times B(\mathsf{x},t)} |x-y|^{-1}\rho_\Psi^{(2)}(x,y)\,dx\,dy \\ + C\iint_{B(\mathsf{x},t)\times(\mathbb{R}^3\setminus B(\mathsf{x},t))} |y-\mathsf{x}|^{-2}\rho_\Psi^{(2)}(x,y)\,dx\,dy,$$

in (2.6) we replace $\rho_\Psi^{(2)}(x,y)$ by $\rho(x)\rho(y)$ and get

$$Ct^{-1}\iint_{B(\mathsf{x},t)\times B(\mathsf{x},t)} |x-y|^{-1}\rho(x)\rho(y)\,dx\,dy \\ + C\iint_{B(\mathsf{x},t)\times(\mathbb{R}^3\setminus B(\mathsf{x},t))} |\mathsf{x}-y|^{-2}\rho(x)\rho(y)\,dx\,dy \tag{5.1}$$

and the first term does not exceed $C\zeta^6 t^4$, while the second term does not exceed

$$C\zeta^3 t^3 \int_{\mathbb{R}^3\setminus B(\mathsf{x},t)} |\mathsf{x}-y|^{-2}\rho(y)\,dy. \tag{5.2}$$

The largest error comes from the first term when the integral is taken over the domain $B(\mathsf{x},t)\times B(\mathsf{x},t)\cap \mathcal{Z}$ and in virtue of estimate (4.8) it does not exceed $Ct^{-1}Z^{\frac{19}{9}}$, all other errors are lesser (to prove it we need just to repeat arguments of the previous subsection).

Observe that for $\ell := \ell_{\mathsf{x}} \le Z^{-\frac{1}{3}}$ the largest contribution to the integral in (5.2) comes from the layer $\{y : \ell_y \asymp \ell_{\mathsf{x}}\}$ and it is of magnitude $\zeta^3\ell_x$. On the other hand, for $\ell_{\mathsf{x}} \ge Z^{-\frac{1}{3}}$ the largest contribution to the integral in (5.1) comes from the layer $\{y : \ell_y \asymp Z^{-\frac{1}{3}}\}$ and it is of magnitude $Z\ell^{-2}$; the first term in (5.1) is smaller.

Therefore we estimate the two last terms in (2.6) by

$$Ct^{-1}Z^{\frac{19}{9}} + C\begin{cases} Z^3\ell^{-2}t^3 & \text{for } \ell \le Z^{-\frac{1}{3}},\\ Z\ell^{-8}t^3 & \text{for } \ell \ge Z^{-\frac{1}{3}}.\end{cases}$$

Consider the second term in (2.6):

$$(2\pi)^{-1}\int \sum_m Z_m \frac{(x-\mathsf{x})\cdot(x-\mathsf{y}_m)}{|x-\mathsf{x}|\cdot|x-\mathsf{y}_m|^3}\rho_\Psi(x)\phi(|x-\mathsf{x}|)\,dx.$$

We replace in the integral in the right-hand expression $\rho_\Psi(x)$ by $\rho(x)$ and get

$$(2\pi)^{-1}\sum_m Z_m \int \frac{(x-\mathsf{x})\cdot(x-\mathsf{y}_m)}{|x-\mathsf{x}|\cdot|x-\mathsf{y}_m|^3}\rho(x)\phi(|x-\mathsf{x}|)\,dx \tag{5.3}$$

with an error

$$(2\pi)^{-1}\sum_m Z_m \int \frac{(x-\mathsf{x})\cdot(x-\mathsf{y}_m)}{|x-\mathsf{x}|\cdot|x-\mathsf{y}_m|^3}(\rho_\Psi(x)-\rho(x))\phi(|x-\mathsf{x}|)\,dx \tag{5.4}$$

and one can see easily that (5.3) does not exceed $CZ\zeta^3\ell^{-3}t^4$.[3]

To estimate (5.4), we make a partition in $B(\mathsf{x},t)$ with subelements supported in the layers $\{x : |x-\mathsf{x}| \asymp t'\}$ with $p < t' \le t$ and in $B(\mathsf{x},p)$ with $\zeta^{-1} \le p \le t$. According to (3.9) the contribution of each layer does not exceed $CZ\ell^{-2}(\zeta^2t'^2 + \zeta^{-2}Z^{\frac{5}{3}-\delta})$ and summation over the layers returns its value as $t'=t$, with $\zeta^{-2}Z^{\frac{5}{3}-\delta}$ acquiring logarithmic factor which we compensate by decreasing δ:

$$CZ\ell^{-2}(\zeta^2t^2 + \zeta^{-2}Z^{\frac{5}{3}-\delta}).$$

Meanwhile, contribution of the ball $B(\mathsf{x},p)$ into the integral (5.4) does not exceed $CZ\ell^{-2}\|\rho_\Psi-\rho\|_{\mathscr{L}^1(B(\mathsf{x},p))}$ and to estimate it we use [9, Theorem 1.1] with $a=\ell$

[3] Indeed, it suffices to take a half-sum of the integrand in (5.3) with its value at symmetric about x point, because both $|x-\mathsf{y}_m|$ and $\rho(x)$ satisfy $|\nabla f| \le Cf\ell^{-1}$.

and $\mu = p^3\ell^{-3}$:

$$
\begin{aligned}
&\|\rho_\Psi - \rho^{\mathsf{TF}}\|_{\mathscr{L}^1(B(\mathsf{x},s))}\\
&\quad\le C\begin{cases} p^2\ell^{-\frac{2}{3}}Z^{\frac{11}{9}-\frac{\delta}{3}} + p^3Z\ell^{-2} & \text{as } p\ge Z^{-\frac{5}{18}-\delta}\ell^{\frac{5}{6}},\, Z^{-1}\le \ell\le Z^{-\frac{1}{3}},\\ p^2\ell^{-\frac{8}{3}}Z^{\frac{5}{9}-\frac{\delta}{3}} & \text{as } p\ge Z^{\frac{5}{9}-\delta}\ell^{\frac{10}{3}},\ \ Z^{-\frac{1}{3}}\le\ell\le Z^{-\frac{5}{21}}.\end{cases}
\end{aligned}
\tag{5.5}
$$

Therefore (2.6) implies

$$
\begin{aligned}
\rho_\Psi(\mathsf{x}) &\le C\big(Z\zeta^3\ell^{-3}t^4 + Z^{\frac{19}{9}}t^{-1} + Z\zeta^2\ell^{-2}t^2\big)\\
&\qquad + C\big(\zeta^3 + Z^{\frac{8}{3}-\delta}\zeta^{-2}\ell^{-2} + Z\ell^{-2}\|\rho_\Psi - \rho\|_{\mathscr{L}^1(B(\mathsf{x},p))}\big),
\end{aligned}
\tag{5.6}
$$

where only first line depends on t. One can see easily that the third term in the first line does not exceed the sum of two first terms. Further, the second term there is larger than $Z^{\frac{19}{9}}\ell^{-1}$ which is larger than CZ^3 as $\ell\le Z^{-\frac{8}{9}}$ and since we already have the estimate (2.5), we should consider only $\ell\ge Z^{-\frac{8}{9}}$. Furthermore, $Z^{\frac{19}{9}}\ell^{-1}\ge\zeta^3$.

Finally, optimizing the remaining two terms in the first line of (5.6) by t with $\zeta^{-1}\le t\le\ell$, we get

$$
\begin{aligned}
\rho_\Psi(\mathsf{x}) &\le C\big(Z^{\frac{17}{9}}\zeta^{\frac{3}{5}}\ell^{-\frac{3}{5}} + Z\zeta^{-1}\ell^{-3} + Z^{\frac{19}{9}}\ell^{-1}\big)\\
&\qquad + C\big(Z^{\frac{8}{3}-\delta}\zeta^{-2}\ell^{-2} + Z\ell^{-2}\|\rho_\Psi - \rho\|_{\mathscr{L}^1(B(\mathsf{x},p))}\big).
\end{aligned}
\tag{5.7}
$$

Let us compare the terms there.

(i) Let $Z^{-\frac{8}{9}}\le\ell\le Z^{-\frac{1}{3}}$. One can see easily that the first line is defined by the third term $Z^{\frac{19}{9}}\ell^{-1}$ for $Z^{-\frac{8}{9}}\le\ell\le Z^{-\frac{7}{9}}$ and by the first term, which is $Z^{\frac{197}{90}}\ell^{-\frac{9}{10}}$, for $Z^{-\frac{7}{9}}\le\ell\le Z^{-\frac{1}{3}}$.

One can see easily that the first term in the second line of (5.7) is smaller than $Z\ell^{-1}$. Using the first case in (5.5) with $p=\max(\zeta^{-1},\ell^{\frac{5}{6}}Z^{-\frac{5}{18}})$ and $\delta=0$, we see that the second term in the second line is smaller than the first line as well. Thus we arrive at Theorem 1.1, statement (i).

(ii) Let $\ell\ge Z^{-\frac{1}{3}}$. Then one can see easily that the first line of (5.7) is defined by the first term, which is $Z^{\frac{17}{9}}\ell^{-\frac{9}{5}}$ for $Z^{-\frac{1}{3}}\le\ell\le Z^{-\frac{5}{18}}$ and by $Z\ell^{-1}$ for $\ell\ge Z^{-\frac{5}{18}}$.

Consider the second line and impose condition $\ell\le Z^{-\frac{2}{9}}$. Then the first line dominates the first term here. Using the second case in (5.5) with $p=\max(\zeta^{-1},\ell^{\frac{10}{3}}Z^{\frac{5}{9}})$ and $\delta=0$, we see that the first line dominates the last term in the second line as well.

Further, let $\ell\ge Z^{-\frac{2}{9}}$. Recall that the second line (except $C\zeta^3$) was a result of the estimate of the second term in (2.6), which, however, could be estimated by

$$
CZ\ell^{-2}\int_{B(\mathsf{x},\ell(\mathsf{x}))}\rho_\Psi(x)\,dx. \tag{5.8}
$$

It is well known that $\int\rho_\Psi\le CZ$ and therefore (5.8) does not exceed $CZ^2\ell^{-2}$ which covers $\ell\ge Z^{-\frac{1}{9}}$.

Furthermore, in the remaining range $Z^{-\frac{2}{9}} \le \ell \le Z^{-\frac{1}{9}}$ we can use statement (ii) of Proposition 3.2 to show that the first term in the second line does not exceed $Z^{\frac{19}{9}}\ell^{-1}$ while the second term there is estimated again by the second case in (5.5). Thus we arrive at Theorem 1.1, statement (ii).

(iii) Finally, using Proposition 3.2 (iii), we prove Theorem 1.1, statement (iii).

Bibliography

[1] V. Bach, Error bound for the Hartree-Fock energy of atoms and molecules. *Comm. Math. Phys.* **147** (1992), 527–548

[2] D. M. Fradkin and F. Calogero, Integral expressions for arbitrary phase shift differences. *Nuclear Phys.* **75** (1966), 475–480

[3] G. M. Graf and J. P Solovej, A correlation estimate with applications to quantum systems with Coulomb interactions. *Rev. Math. Phys.* **6** (1994), 977–997. Reprinted in *The state of matter. A volume dedicated to E. H. Lieb*. Adv. Ser. Math. Phys. 20, pp. 142–166, World Scientific Publishing, Singapore, 1994

[4] O. J. Heilmann and E. H. Lieb, The electron density near the nucleus of a large atom. *Phys. Rev. A* **52** (1995), no 5, 3628–3643

[5] M. Hoffmann-Ostenhof, T. Hoffmann-Ostenhof and W. Thirring, Simple bounds to the atomic one-electron density at the nucleus and to expectation values of one-electron operators. *J. Phys. B* **11** (1978), L571–L575

[6] A. Iantchenko, E. H. Lieb and H. Siedentop, Proof of a conjecture about atomic and molecular cores related to Scott's correction. *J. Reine Angew. Math.* **472** (1996), 177–195

[7] V. Ivrii, *Microlocal analysis, sharp spectral asymptotics and applications. Volume V: Applications to quantum theory and miscellaneous problems*. Springer, New York, 2019

[8] V. Ivrii, Strong Scott conjecture. Preprint 2020, arXiv:1908.05478

[9] V. Ivrii, Thomas–Fermi approximation to electronic density. Preprint 2020, arXiv: 1911.03510v4

[10] E. H. Lieb and W. E. Thirring, Inequalities for the moments of the eigenvalues of the Schrödinger Hamiltonian and their relation to Sobolev inequalities, in *Studies in mathematical physics*, pp. 269–303, Princeton University Press, Princeton, 1976

[11] N. Rakowsky and H. Siedentop, An upper bound on the ionic ground state density at the nucleus. *Rep. Math. Phys.* **36** (1995), 91–98

[12] M. B. Ruskai and J. P. Solovej, Asymptotic neutrality of polyatomic molecules. In *Schrödinger operators (Aarhus, 1991)*, pp. 153–174, Lecture Notes in Phys. 403, Springer, Berlin, 1992

[13] H. Siedentop, An upper bound for the atomic ground state density at the nucleus. *Lett. Math. Phys.* **32** (1994), 221–229

[14] H. Siedentop, A proof of the strong Scott conjecture. In *Quasiclassical methods (Minneapolis, 1995)*, pp. 145–154, IMA Vol. Math. Appl. 95, Springer, New York, 1997

Non-linear Schrödinger equation in a uniform magnetic field

Thomas F. Kieffer and Michael Loss

To Ari who has influenced our research in many and fruitful ways

The aim of this paper is to study, in dimensions 2 and 3, the pure-power nonlinear Schrödinger equation with an external uniform magnetic field included. In particular, we derive a general criteria on the initial data and the power of the non-linearity so that the corresponding solution blows up in finite time, and we show that the time for blow up to occur decreases as the strength of the magnetic field increases. In addition, we also discuss some observations about Strichartz estimates in two dimensions for the Mehler kernel, as well as similar blow-up results for the non-linear Pauli equation.

1 Introduction

The importance of Virial Identities in the theory of partial differential equations can hardly be overstated. They can be used for proving absence of bound states in Schrödinger operators (see [14]) and the well known Pohozaev identity is a Virial Identity in disguise. What is maybe more surprising is that a Virial Identity is an entry point to scattering theory and has lead to what is now called "Mourre Estimates". A very readable introduction can be found in [4]. Another application, and this is the topic of this paper, is the study of blow-up of solutions for the non-linear Schrödinger equation

$$\begin{cases} i\partial_t \psi = -\Delta\psi + \mu|\psi|^{p-1}\psi, \\ \psi(0)(\mathbf{x}) = \psi_0(\mathbf{x}), \end{cases} \tag{1.1}$$

due to Glassey [9]. This system has a conserved energy

$$E[\psi(t)] = \|\nabla\psi(t)\|^2_{L^2(\mathbb{R}^d)} + \frac{2\mu}{p+1}\|\psi(t)\|^{p+1}_{L^{p+1}(\mathbb{R}^d)}, \tag{1.2}$$

and one computes with Glassey

$$\frac{1}{4}\frac{\mathrm{d}^2}{\mathrm{d}t^2}\langle\psi, |\mathbf{x}|^2\psi\rangle_{L^2(\mathbb{R}^d)} = 2\|\nabla\psi\|^2_{L^2(\mathbb{R}^d)} + \mu d\frac{p-1}{p+1}\|\psi\|^{p+1}_{L^{p+1}(\mathbb{R}^d)}.$$

Keywords: Non-linear magnetic Schrödinger equation, blowup, magnetic Strichartz inequalities
2020 Mathematics Subject Classification: Primary 35B44; secondary 35K10

Thus, when $p \geq 1 + \frac{4}{d}$ and $\mu < 0$, then

$$\frac{1}{4}\frac{\mathrm{d}^2}{\mathrm{d}t^2}\langle \psi, |\mathbf{x}|^2 \psi \rangle_{L^2(\mathbb{R}^d)} \leq 2E[\psi_0].$$

If the energy is strictly negative, then the function $\langle \psi, |\mathbf{x}|^2 \psi \rangle_{L^2(\mathbb{R}^d)}(t)$ is strictly concave and must hit zero at some time T provided that the solution were to exist up to that point in time. Thus, the solution ceases to exist after a finite time.

The initial value problem (1.1) has been studied extensively by many authors; see, for example, [16, Chapter 3] or [2] for detailed reviews. Local well-posedness in $H^1(\mathbb{R}^d)$ of (1.1) for $1 < p < 1 + \frac{4}{d-2}$ essentially follows from a fixed point argument utilizing the Strichartz estimates for the Schrödinger unitary group $e^{it\Delta}$. Such estimates read

$$\|e^{it\Delta} f\|_{L_t^q L^r(\mathbb{R}\times\mathbb{R}^d)} \leq C \|f\|_{L^2(\mathbb{R}^d)} \tag{1.3}$$

for all $f \in L^2(\mathbb{R}^d)$ and all *Schrödinger admissible* exponents (q, r) satisfying

$$q \in [2, \infty], \quad \frac{2}{q} = d\left(\frac{1}{2} - \frac{1}{r}\right), \quad (q, r, d) \neq (2, \infty, 2).$$

Here $C > 0$ is a constant depending on the numbers (q, r, d). Global well-posedness for $p < 1 + \frac{4}{d}$ follows along familiar lines using the energy conservation and the Gagliardo–Nirenberg inequalities.

The aim of this paper is to generalize the aforementioned blow-up result to the pure-power NLS equation with a uniform magnetic field included, referred to as the constant field non-linear magnetic Schrödinger (NLMS) equation. In particular, we derive a general criteria on the initial data $\psi_0 \in H^1_{\mathbf{A}}$ so that finite time blow-up occurs when $p \geq 1 + \frac{4}{d}$. It is easy to see by a scaling argument that the length of the time interval of local existence is inversely proportional to the strength of the magnetic field (see Theorem 2.5). Since Strichartz estimates are at the heart of some of the estimates for the non-linear Schrödinger equation, we add some simple and, we think, interesting remarks about Strichartz estimates in two dimensions for the Mehler kernel (see Theorem 2.2). We end the paper with a few simple remarks concerning blow-up results for the non-linear Pauli equation (see Theorems 3.1 and 3.3).

A resolution of the problem of finite time blow-up for the constant field NLMS equation was already claimed by G. Ribeiro in 1990 [10] in three dimensions, and subsequently generalized by A. Garcia in 2012 [8]. However, we believe that our approach brings a new perspective to this problem as we observe that a crucial feature is the existence of an additional conserved quantity related to the angular momentum (see (2.11)). By using this additional conserved quantity, a more exact, as compared to the results of [8, 10], virial identity for the second time derivative of the expectation value of $|\mathbf{x}|^2$ is derived (see (2.12)). This new identity yields a different sufficient condition on the initial data which guarantees finite time blow-up, and may also be solved exactly in two dimensions with critical power for the non-linearity (see (2.13)).

The paper is organized as follows. In Section 2 we introduce the NLMS equation, discussing local/global well-posedness and Strichartz estimates in Section 2.1, and study finite time blow-up in Section 2.2 and Section 2.3. We conclude with Section 3 which generalizes the results concerning the NLMS equation to the so-called non-linear Pauli equation.

2 Non-linear magnetic Schrödinger equation

The Cauchy problem for the NLMS equation in $d \geq 2$ space dimensions[1] reads

$$\begin{cases} i\partial_t \psi = (\mathbf{p} + \mathbf{A})^2 \psi + \mu |\psi|^{p-1} \psi, \\ \psi(0, \mathbf{x}) = \psi_0(\mathbf{x}), \end{cases} \tag{2.1}$$

where $\mathbf{A}$ is the magnetic vector potential. We assume that the particle has a negative charge. We typically consider (2.1) as an initial value problem in the Hilbert space

$$H^1_{\mathbf{A}}(\mathbb{R}^d; \mathbb{C}) = \{f \in L^2(\mathbb{R}^d; \mathbb{C}) : (\mathbf{p} + \mathbf{A}) f \in L^2(\mathbb{R}^d; \mathbb{C}^d)\},$$

which is equipped with the norm $\|(\mathbf{p} + \mathbf{A}) f\|_{L^2(\mathbb{R}^d)}$, and we will always assume the operator $(\mathbf{p} + \mathbf{A})^2$ is essentially self-adjoint on $L^2(\mathbb{R}^d; \mathbb{C})$ with domain

$$H^2_{\mathbf{A}}(\mathbb{R}^d; \mathbb{C}) = \{f \in H^1_{\mathbf{A}}(\mathbb{R}^d; \mathbb{C}) : (\mathbf{p} + \mathbf{A})^2 f \in L^2(\mathbb{R}^d; \mathbb{C})\}.$$

If $\mathbf{A} \in L^4(\mathbb{R}^d; \mathbb{R}^d)$ and $\operatorname{div} \mathbf{A} \in L^2(\mathbb{R}^d)$, then it follows from the Leinfelder–Simader theorem [13] that $(\mathbf{p} + \mathbf{A})^2$ is essentially self-adjoint on $C_c^\infty(\mathbb{R}^d)$.

The total energy associated with the NLMS equation (2.1) is[2]

$$E_{\mathrm{S}}[\psi, \mathbf{A}](t) = T_{\mathrm{S}}[\psi(t), \mathbf{A}] + \frac{2\mu}{p+1} \|\psi(t)\|^{p+1}_{L^{p+1}(\mathbb{R}^d)}, \tag{2.2}$$

where

$$T_{\mathrm{S}}[\psi, \mathbf{A}] = \|(\mathbf{p} + \mathbf{A})\psi\|^2_{L^2(\mathbb{R}^d)}.$$

Note that for $\mathbf{A} = 0$, $E_{\mathrm{S}}[\psi, \mathbf{A}]$ is given by (1.2). Often the magnetic vector potential $\mathbf{A}$ is understood, and thus we will usually suppress the $\mathbf{A}$-dependence of E_{S}, simply writing $E_{\mathrm{S}}[\psi]$, and likewise for T_{S}. It is straightforward to check via differentiation that, at least formally, $\|\psi(t)\|^2_{L^2(\mathbb{R}^d)}$ and the total energy (2.2) are conserved along the flow generated by (2.1).

---

[1]We only discuss (2.1) in dimension $d \geq 2$ because in dimension $d = 1$ it is always possible to pass from $(\mathbf{p} + \mathbf{A})^2$ to the free Hamiltonian $-\Delta$ via a gauge transformation.

[2]The subscript "S" is placed on certain quantities to distinguish them from similar quantities that come up in the discussion of the non-linear Pauli equation in Section 3.

Remark. It is common to a priori "fix a gauge" for the vector potential $\mathbf{A}$. When a gauge is chosen, we will always pick the *symmetric gauge* $\mathbf{A} = \frac{B}{2}\mathbf{x}_\perp$, where

$$\mathbf{x}_\perp = \begin{cases} (-x_2, x_1), & d = 2, \\ (-x_2, x_1, 0), & d = 3. \end{cases}$$

However, unless otherwise specified, we will generally not fix the gauge for reasons that will become clear later.

2.1 Strichartz estimates and well-posedness

Local well-posedness for the NLMS equation (2.1) with initial data in $H^1_{\mathbf{A}}(\mathbb{R}^3)$ first appeared in [3]. There the authors consider more general non-linearities and external potentials, and also study the orbital stability of the ground state associated with (2.1). Local well-posedness results for uniform magnetic fields in dimensions $d \geq 2$, and for more general magnetic fields, may be found in [8, Theorem 2.2], and are proved using the Strichartz estimates of [5,6]. For the sake of completeness, we state the local well-posedness result of [3] for the pure-power non-linearity case and no external potential.

Theorem 2.1 (Cazenave and Esteban, 1988 [3]). *Let $d = 3$, $\mu \in \mathbb{R}$, $p \in (1,5)$, and $\mathbf{A} = \frac{B}{2}\mathbf{x}_\perp$. For all $\psi_0 \in H^1_{\mathbf{A}}$ we have the following.*

(1) *There exists a unique maximal solution $\psi \in C([0,T_*), H^1_{\mathbf{A}}) \cap C^1([0,T_*), H^{-1}_{\mathbf{A}})$ of* (2.1). *If $T_* < \infty$, then $\|(\mathbf{p}+\mathbf{A})\psi(t)\|_{L^2(\mathbb{R}^d)} \to \infty$ as $t \uparrow T_*$.*

(2) *The mapping $\psi_0 \mapsto T_*(\psi_0)$ is lower semi-continuous and, if $t \in [0, T_*(\psi_0))$ and $(\phi_n)_{n\geq 1} \subset H^1_{\mathbf{A}}$ converges to ψ_0 as $n \to \infty$, in $H^1_{\mathbf{A}}$, then the corresponding sequence of solutions $(\psi_n)_{n\geq 1}$ to problem* (2.1) *verify $\psi_n \to \psi$ as $n \to \infty$, in $C([0,t], H^1_{\mathbf{A}})$.*

(3) *If $\psi_0 \in H^2_{\mathbf{A}}$, then $\psi \in C([0,T_*), H^2_{\mathbf{A}}) \cap C^1([0,T_*), L^2)$.*

(4) $\|\psi(t)\|_{L^2(\mathbb{R}^d)} = \|\psi_0\|_{L^2(\mathbb{R}^d)}$ *and* $E_{\mathrm{S}}[\psi(t), \mathbf{A}] = E_{\mathrm{S}}[\psi_0, \mathbf{A}]$.

Remark. Though Theorem 2.1 is proved in the symmetric gauge, by applying a gauge transformation we can obtain a similar theorem in other gauges. Therefore, the restriction to the symmetric gauge is one of convenience, not necessity.

As will be discussed below, Theorem 2.1 also applies in two dimensions. Global existence of $H^1_{\mathbf{A}}(\mathbb{R}^d)$-solutions to (2.1) for $p < 1 + \frac{4}{d}$ and $\mu < 0$ follows from the diamagnetic inequality (see [15] and also [12]) combined with the same estimates that were applied to (1.1) to obtained global existence there.[3] Indeed, the diamagnetic

[3]For the defocussing case $\mu > 0$, global existence follows for any $1 < p < 1 + \frac{4}{d-2}$ and $\psi_0 \in H^1_{\mathbf{A}}(\mathbb{R}^d)$. This follows from $\|(\mathbf{p}+\mathbf{A})\psi(t)\|^2_{L^2(\mathbb{R}^d)} \leq E_{\mathrm{S}}[\psi_0, \mathbf{A}]$, i.e., the conservation of energy.

inequality reads

$$|\nabla|\psi|(\mathbf{x})| \leq |(\mathbf{p}+\mathbf{A})\psi(\mathbf{x})| \quad \text{for a.e. } \mathbf{x} \in \mathbb{R}^d. \tag{2.3}$$

From this, the Gagliardo–Nirenberg inequality, and the conservation charge and energy for (2.1), we have the following bound on the kinetic energy:

$$\|(\mathbf{p}+\mathbf{A})\psi\|^2_{L^2(\mathbb{R}^d)} \leq |E_{\mathrm{S}}[\psi_0]| + \frac{2C_{p+1}}{p+1}\|(\mathbf{p}+\mathbf{A})\psi\|^{\frac{1}{2}d(p-1)}_{L^2(\mathbb{R}^d)}.$$

Hence, if $p < 1 + \frac{4}{d}$, then a uniform bound on $\|(\mathbf{p}+\mathbf{A})\psi\|_{L^2(\mathbb{R}^d)}$ follows. According to the blow-up alternative of Theorem 2.1, we have global well-posedness of the Cauchy problem (2.1) for $1 < p < 1 + \frac{4}{d}$.

The proof of Theorem 2.1 relies on homogeneous and non-homogeneous Strichartz estimates for the unitary time evolution $U_{\mathrm{S}}(t) = \exp\{-it(\mathbf{p}+\mathbf{A})^2\}$ when $\mathbf{A} = \frac{B}{2}\mathbf{x}_\perp$. To discuss these estimates in detail, we first write

$$U_{\mathrm{S}}(t) = \begin{cases} M(t), & d = 2, \\ e^{it\partial_3^2} M(t), & d = 3, \end{cases} \tag{2.4}$$

where $M(t)$ is the operator given by

$$M(t) = \exp\Big\{-it\Big[\Big(p_1 - \frac{B}{2}x_2\Big) + \Big(p_2 + \frac{B}{2}x_1\Big)\Big]\Big\}. \tag{2.5}$$

It is possible to write the integral kernel of the operator $M(t)$ explicitly; see, for example, [1] for a derivation. Known as the *Mehler kernel*, and denoted by the same symbol $M(t) : \mathbb{R}^4 \to \mathbb{C}$, it reads

$$M(\mathbf{x},\mathbf{y},t) = \frac{B}{4\pi \sin(Bt)} \exp\Big\{\frac{B}{4i}\big(\cot(Bt)|\mathbf{x}-\mathbf{y}|^2 - 2\mathbf{x}\wedge\mathbf{y}\big)\Big\}, \tag{2.6}$$

where $\mathbf{x}\wedge\mathbf{y} = x_1y_2 - x_2y_1$. Using the representation (2.6), one has the following L^p-estimate on the unitary time evolution for $d = 3$:

$$\|U_{\mathrm{S}}(t)\psi_0\|_{L^p(\mathbb{R}^3)} \leq \bigg(\frac{|B|}{\sqrt{|t|}|\sin(Bt)|}\bigg)^{1-\frac{2}{p}} \|\psi_0\|_{L^{p'}(\mathbb{R}^3)} \tag{2.7}$$

for all $\psi_0 \in L^{p'}(\mathbb{R}^3)$. We note that the time decay in (2.7) is due to the free motion in the third direction. By using (2.7), it is then possible to show

$$\|U_{\mathrm{S}}\psi_0\|_{L^q_t L^r_x([0,T]\times\mathbb{R}^3)} \leq C_1\|\psi_0\|_{L^2(\mathbb{R}^3)} \tag{2.8}$$

and

$$\bigg\|\int_0^t U_{\mathrm{S}}(t-\tau)\psi(\tau)\,\mathrm{d}\tau\bigg\|_{L^q_t L^r_x([0,T]\times\mathbb{R}^3)} \leq C_2\|\psi\|_{L^{\tilde{q}'}_t L^{\tilde{r}'}_x([0,T]\times\mathbb{R}^3)} \tag{2.9}$$

for all Schrödinger admissible (see (1.3)) (q,r) and $(\tilde{q},\tilde{r})$ with $d = 3$, and where $C_1 > 0$ depends only on r and T, and $C_2 > 0$ depends only on r, $\tilde{r}$, and T.

The authors in [3] prove Theorem 2.1 for $d = 3$ using estimates (2.8) and (2.9). However, using the following theorem regarding Strichartz estimates for $U_S(t) \equiv M(t)$ for the two-dimensional case, we may easily prove Theorem 2.1 in the $d = 2$ case as well. Note that the magnetic evolution is periodic with period $\frac{\pi}{B}$ which is essentially the Larmor period. Thus, there is no decay in time and one has to consider the evolution for $0 \le t \le \frac{\pi}{B}$. The following may be somewhat surprising.

Theorem 2.2. *For the unitary propagator* $U_S(t) \equiv M(t)$ *given by* (2.5)–(2.6) *we have the identity*

$$\|M(t)\psi_0\|_{L_t^q L_x^r((0,\frac{\pi}{B})\times\mathbb{R}^2)} = (4\pi)^{1-\frac{4}{q}} \|e^{it\Delta}\psi_0\|_{L_t^q L_x^r(\mathbb{R}\times\mathbb{R}^2)} \tag{2.10}$$

for all $\psi_0 \in L^2(\mathbb{R}^2)$ *and all Schrödinger admissible exponents* (q, r).

Proof. Observe that we may write the Mehler kernel (2.6) as

$$M(\mathbf{x}, \mathbf{y}) = \frac{B}{4\pi \sin(Bt)} \exp\left\{\frac{B}{4i} \cot(Bt)\big(|\mathbf{x}|^2 + |\mathbf{y}|^2\big)\right\} \exp\left\{i\,\frac{B\mathbf{y} \cdot R(Bt)\mathbf{x}}{2\sin(Bt)}\right\},$$

where $R(\theta)$ is the usual 2×2 rotation matrix given by

$$R(\theta) = \begin{pmatrix} \cos(\theta) & -\sin(\theta) \\ \sin(\theta) & \cos(\theta) \end{pmatrix}.$$

The unitary propagator $U_S(t) \equiv M(t)$ acting on $\psi_0 : \mathbb{R}^2 \to \mathbb{C}$ may be written

$$(M(t)\psi_0)(\mathbf{x}) = \frac{B \exp\{\frac{B}{4i} \cot(Bt)|\mathbf{x}|^2\}}{4\pi \sin(Bt)} \int_{\mathbb{R}^2} \exp\left\{i\,\frac{B\mathbf{y} \cdot R(Bt)\mathbf{x}}{2\sin(Bt)}\right\} g(\mathbf{y}, t)\, d\mathbf{y},$$

where

$$g(\mathbf{y}, t) = \exp\left\{\frac{B}{4i} \cot(Bt)|\mathbf{y}|^2\right\} \psi_0(\mathbf{y}).$$

Therefore,

$$(M(t)\psi_0)(\mathbf{x}) = \frac{B \exp\{\frac{B}{4i} \cot(Bt)|\mathbf{x}|^2\}}{4\pi \sin(Bt)} \mathcal{F} g\left(-\frac{BR(Bt)\mathbf{x}}{4\pi \sin(Bt)}, t\right),$$

where $\mathcal{F}$ is the Fourier transform defined by

$$(\mathcal{F} f)(\mathbf{k}) = \int_{\mathbb{R}^d} e^{-2\pi i \mathbf{k}\cdot\mathbf{x}} f(\mathbf{x})\, d\mathbf{x}.$$

Letting $r \ge 2$, we may now compute

$$\begin{aligned} \int_{\mathbb{R}^2} |(M(t)\psi_0)(\mathbf{x})|^r\, d\mathbf{x} &= \left|\frac{B}{4\pi \sin(Bt)}\right|^r \int_{\mathbb{R}^2} \left|\mathcal{F} g\left(\frac{B}{4\pi \sin(Bt)}\mathbf{x}, t\right)\right|^r d\mathbf{x} \\ &= \left|\frac{B}{4\pi \sin(Bt)}\right|^{r-2} \int_{\mathbb{R}^2} |\mathcal{F} g(\mathbf{x}, t)|^r\, d\mathbf{x}. \end{aligned}$$

If we raise this to the $\frac{q}{r}$-power, integrate over t from 0 to $\frac{\pi}{B}$, and use the substitution $s = B\cot(Bt)$, we find

$$\int_0^{\frac{\pi}{B}} \left(\int_{\mathbb{R}^2} |(M(t)\psi_0)(\mathbf{x})|^r \,\mathrm{d}\mathbf{x} \right)^{\frac{q}{r}} \mathrm{d}t$$
$$= \frac{1}{(4\pi)^2} \int_{-\infty}^{\infty} \left| \frac{B}{4\pi \sin(Bt)} \right|^{\frac{q}{r}(r-2)-2} \left(\int_{\mathbb{R}^2} \left| \mathcal{F}\left(e^{-\frac{is|\cdot|^2}{4}} \psi_0\right)(\mathbf{x}) \right|^r \mathrm{d}\mathbf{x} \right)^{\frac{q}{r}} \mathrm{d}s.$$

If (q, r) are Schrödinger admissible in dimension 2, we have

$$\frac{q}{r}(r-2) - 2 = 0.$$

This last observation together with the previous calculation yields (2.10). ∎

Theorem 2.2 states in essence that the Strichartz estimates for the Mehler kernel (2.6) are the same as those for the free Schrödinger evolution. In particular, we bring attention to an interesting corollary to Theorem 2.2.

Corollary 2.3. *For* $(q, r) = (4, 4)$, *identity* (2.10) *implies the sharp constant for the Strichartz estimate for the Mehler kernel* (2.6) *is the same as for the free Schrödinger evolution, namely* $\frac{1}{\sqrt{2}}$ *(see* [11]*).*

We refer the reader to [11] for a derivation of the sharp constant for the free Schrödinger Strichartz estimate in the $(q, r, d) = (4, 4, 2)$ case. The proof of the previous corollary follows directly from formula (2.10). Lastly, we mention that a simple formula like (2.10) does not seem possible in 3D.

2.2 Finite time blow-up for uniform magnetic fields

We introduce the function space

$$\Sigma_{\mathrm{S}} = \left\{ f \in H^1_{\mathbf{A}}(\mathbb{R}^d; \mathbb{C}) : \int_{\mathbb{R}^d} |\mathbf{x}|^2 |f(\mathbf{x})|^2 \,\mathrm{d}\mathbf{x} < \infty \right\},$$

and define $F_{\mathrm{S}} : \Sigma_{\mathrm{S}} \to \mathbb{R}$ by

$$F_{\mathrm{S}}[\psi, \mathbf{A}] = E_{\mathrm{S}}[\psi, \mathbf{A}] - B\,\mathrm{Re}\langle \mathbf{x}_\perp \psi, (\mathbf{p} + \mathbf{A})\psi \rangle_{L^2(\mathbb{R}^d)} + \frac{B^2}{2} \|\rho\psi\|^2_{L^2(\mathbb{R}^d)}, \tag{2.11}$$

where $\rho = \sqrt{x_1^2 + x_2^2}$.[4] The functional $F_{\mathrm{S}}[\psi, \mathbf{A}]$ defined on Σ_{S} will play a key role in providing a sufficient condition for blow-up of solutions to (2.1) when $p \geq 1 + \frac{4}{d}$ and $\mu < 0$. Since the vector potential $\mathbf{A}$ will be fixed we will usually suppress the $\mathbf{A}$-dependence of F_{S} and simply write $F_{\mathrm{S}}[\psi]$.

---

[4] We will frequently switch between the polar/cylindrical coordinates (ρ, ϑ) and (ρ, ϑ, x_3) and the Cartesian coordinates (x_1, x_2) and (x_1, x_2, x_3) for $\mathbb{R}^2$ and $\mathbb{R}^3$, respectively. Here $\rho = \sqrt{x_1^2 + x_2^2}$ and $\tan\vartheta = \frac{x_2}{x_1}$.

Observe that the functional F_S is gauge invariant, which is the main reason why we did not choose to fix any particular gauge from the start. Indeed, if we select the symmetric gauge $\mathbf{A} = \frac{B}{2}\mathbf{x}_\perp$, then the term

$$-B\,\mathrm{Re}\langle \mathbf{x}_\perp \psi, (\mathbf{p}+\mathbf{A})\psi\rangle_{L^2(\mathbb{R}^d)} + \frac{B^2}{2}\|\rho\psi\|^2_{L^2(\mathbb{R}^d)}$$

in (2.11) simply equals $-B\langle \psi, L_3\psi\rangle_{L^2(\mathbb{R}^d)}$, where

$$L_3 \equiv \mathbf{x}_\perp \cdot \mathbf{p} = -i\partial_\vartheta = -i(-x_2\partial_1 + x_1\partial_2)$$

is the x_3-component of the angular momentum. Since we are considering a uniform magnetic field along the x_3-axis, it is reasonable to assert the x_3-component of the angular momentum is preserved. Indeed, at least formally,

$$\frac{\mathrm{d}}{\mathrm{d}t}\langle \psi, L_3\psi\rangle_{L^2(\mathbb{R}^d)} = \int_{\mathbb{R}^d} |\psi|^2\partial_\vartheta|\psi|^{p-1} = \frac{p-1}{p+1}\int_{\mathbb{R}^d}\partial_\vartheta|\psi|^{p+1} = 0.$$

The next lemma makes this precise, showing that, in any gauge, F_S is conserved under the time evolution of (2.1).

Lemma 2.4. *Let $d \in \{2,3\}$ and $\mathbf{A} \in L^2_{\mathrm{loc}}(\mathbb{R}^d;\mathbb{R}^d)$ generate a uniform magnetic field $\mathbf{B} = (0,0,B)$. Let $\psi_0 \in H^1_{\mathbf{A}}(\mathbb{R}^d)$ and $\psi \in C([0,T_*), H^1_{\mathbf{A}}) \cap C^1([0,T_*), H^{-1}_{\mathbf{A}})$ denote the corresponding solution to* (2.1). *Then $F_S[\psi(t)] = F_S[\psi_0]$.*

The proof of Lemma 2.4 is given at the end of Section 2.3. Using that $F_S[\psi]$ is conserved we have the following theorem concerning the second time derivative of the expectation value of $|\mathbf{x}|^2$.

Theorem 2.5. *Let $d \in \{2,3\}$, let $1 < p < 1 + \frac{4}{d-2}$, let $\mathbf{A} \in L^2_{\mathrm{loc}}(\mathbb{R}^d;\mathbb{R}^d)$ generate a uniform magnetic field $\mathbf{B} = (0,0,B)$, and let $\psi_0 \in \Sigma_S$. Let $\psi \in C([0,T_*);\Sigma_S) \cap C^1([0,T_*); H^{-1}_{\mathbf{A}})$ be the corresponding maximal solution to the initial value problem* (2.1). *Then the function $g(t) = \frac{1}{4}\|\mathbf{x}\psi(t)\|^2_{L^2(\mathbb{R}^d)}$ satisfies the virial identity*

$$\ddot{g}(t) = 2F_S[\psi_0] + \mu d\frac{p-(1+\frac{4}{d})}{p+1}\|\psi(t)\|^{p+1}_{L^{p+1}(\mathbb{R}^d)} - B^2\|\rho\psi(t)\|^2_{L^2(\mathbb{R}^d)}. \quad (2.12)$$

In particular, if $d = 2$ and $p = 3$, then (2.12) *becomes a second-order equation for g that can be solved exactly:*

$$g(t) = \frac{F_S[\psi_0]}{2B^2} + \left(g_0 - \frac{F_S[\psi_0]}{2B^2}\right)\cos(2Bt) + \frac{\dot{g}_0}{2B}\sin(2Bt), \quad (2.13)$$

where $g_0 = \frac{1}{4}\|\rho\psi_0\|^2_{L^2(\mathbb{R}^2)}$ and $\dot{g}_0 = \mathrm{Re}\langle \mathbf{x}\psi_0, (\mathbf{p}+\mathbf{A})\psi_0\rangle_{L^2(\mathbb{R}^2)}$.

Remark. It is not obvious that, for $\psi_0 \in \Sigma_S$, the corresponding solution ψ is in the space $C([0,T_*);\Sigma_S) \cap C^1([0,T_*); H^{-1}_{\mathbf{A}})$. This fact is shown in [10] in the course of proving of Theorem 1.2 there.

The proof of Theorem 2.5 is reserved for Section 2.3. Following a similar reasoning as discussed in Section 1 for (1.1), observe that if $\mu < 0$ and $p \geq 1 + \frac{4}{d}$, then $\ddot{g}(t) \leq 2F_{\mathrm{S}}[\psi_0]$. So if $F_{\mathrm{S}}[\psi_0] < 0$, then $\ddot{g}(t) < 0$ for all times $t \in [0, T_*)$. If $T_* = +\infty$, then g would necessarily hit 0 at some $t_* > 0$, implying (by the uncertainty principle) that $\|(\mathbf{p} + \mathbf{A})\psi(t_*)\|_{L^2(\mathbb{R}^d)} = +\infty$. This contradicts the blow-up alternative of Theorem 2.1.

Corollary 2.6. *Suppose $\mu < 0$, $1 + \frac{4}{d} \leq p < 1 + \frac{4}{d-2}$, $\psi_0 \in \Sigma_{\mathrm{S}}$, and $F_{\mathrm{S}}[\psi_0] < 0$. Then the corresponding solution to (2.1) blows up in finite time. If $F_{\mathrm{S}}[\psi_0] = 0$, then blow-up occurs when $\dot{g}(0) = \mathrm{Re}\langle \mathbf{x}\psi_0, (\mathbf{p} + \mathbf{A})\psi_0\rangle_{L^2(\mathbb{R}^d)} < 0$.*

Note that, because of the oscillations, (2.13) implies the time for blow-up to occur decreases as the magnetic field strength increases.

Consider the explicit solution (2.13) for $d = 2$ and $p = 3$ from Theorem 2.5. The condition for $g(t) < 0$ at some time $t \in (0, T_*)$ is given by

$$\left(g_0 - \frac{F_{\mathrm{S}}[\psi_0]}{2B^2}\right)^2 + \left(\frac{\dot{g}_0}{2B}\right)^2 > \left(\frac{F_{\mathrm{S}}[\psi_0]}{2B^2}\right)^2.$$

After some algebra this simplifies to

$$F_{\mathrm{S}}[\psi_0] g_0 < B^2\left(g_0^2 + \frac{\dot{g}_0^2}{4B^2}\right). \tag{2.14}$$

Observe that (2.14) is much weaker than demanding $F_{\mathrm{S}}[\psi_0] < 0$ for blow-up to occur. Consider the special case $\dot{g}_0 = 0$. Using the definition (2.11) of $F_{\mathrm{S}}[\psi_0]$ and choosing the symmetric gauge $\mathbf{A} = \frac{B}{2}\mathbf{x}_\perp$, (2.14) then reduces to $E_{\mathrm{S}}[\psi_0, \mathbf{0}] < 0$. This is clearly only satisfied in the focusing case $\mu < 0$.

In the general case $\dot{g}_0 \neq 0$, (2.14) reduces to

$$F_{\mathrm{S}}[\psi_0] - \frac{B^2}{4}\|\rho\psi_0\|^2_{L^2(\mathbb{R}^2)} < \frac{\mathrm{Re}\langle \mathbf{x}\psi_0, (\mathbf{p} + \mathbf{A})\psi_0\rangle^2_{L^2(\mathbb{R}^2)}}{\|\rho\psi_0\|^2_{L^2(\mathbb{R}^2)}}. \tag{2.15}$$

By choosing again the symmetric gauge $\mathbf{A} = \frac{B}{2}\mathbf{x}_\perp$ and using the definition (2.2) of E_{S}, the expression (2.15) further reduces to

$$E_{\mathrm{S}}[\psi_0, \mathbf{0}] < \frac{\mathrm{Re}\langle \mathbf{x}\psi_0, \mathbf{p}\psi_0\rangle^2_{L^2(\mathbb{R}^2)}}{\|\rho\psi_0\|^2_{L^2(\mathbb{R}^2)}}. \tag{2.16}$$

Since $\mathrm{Re}\langle \mathbf{x}\psi_0, \mathbf{p}\psi_0\rangle_{L^2(\mathbb{R}^2)} \leq \|\rho\psi_0\|_{L^2(\mathbb{R}^2)}\|\nabla\psi_0\|_{L^2(\mathbb{R}^2)}$, we see that inequality (2.16) is only satisfied if $\mu < 0$. We note these observations are a consistency check with the earlier observation that the defocusing $H^1_{\mathbf{A}}(\mathbb{R}^2)$-subcritical NLMS equation is globally well-posed.

On a final note for this section, the sufficient condition for blow-up, namely $F_{\mathrm{S}}[\psi_0] < 0$, is significantly different than the one found in [8, 10]. There, the authors claim, based on a virial identity argument, that $E_S[\psi_0, \mathbf{A}] < 0$ is sufficient for blow-up to occur. For this reason we spend the next several paragraphs closely analyzing the relationship between $F_S[\psi_0]$ and $E_S[\psi_0]$. Choose the symmetric gauge $\mathbf{A} = \frac{B}{2}\mathbf{x}_\perp$, $B > 0$, and fix $\psi_0 \in \Sigma_S$. By expanding the kinetic energy $T_S[\psi_0, \mathbf{A}] = \|(\mathbf{p} + \mathbf{A})\psi_0\|^2_{L^2(\mathbb{R}^d)}$, we have that

$$F_S[\psi_0, \mathbf{A}] = E_0 + \frac{B^2}{4}\|\rho\psi_0\|^2_{L^2(\mathbb{R}^d)} \tag{2.17}$$

and

$$E_S[\psi_0, \mathbf{A}] = E_0 - B\langle L_3\rangle_0 + \frac{B^2}{4}\|\rho\psi_0\|^2_{L^2(\mathbb{R}^d)}, \tag{2.18}$$

where

$$E_0 \equiv E_S[\psi_0, \mathbf{0}] = \|\nabla\psi_0\|^2_{L^2(\mathbb{R}^d)} - \frac{2\mu}{p+1}\|\psi_0\|^{p+1}_{L^{p+1}(\mathbb{R}^d)}$$

and

$$\langle L_3\rangle_0 \equiv \langle L_3\psi_0, \psi_0\rangle_{L^2(\mathbb{R}^d)}.$$

Clearly, for $B = 0$, $F_S[\psi_0, \mathbf{0}] = E_0$. From expressions (2.17) and (2.18) we can immediately conclude

$$\begin{cases} F_S[\psi_0, \mathbf{A}] < E_S[\psi_0, \mathbf{A}] & \text{when } \langle L_3\rangle_0 < 0, \\ F_S[\psi_0, \mathbf{A}] > E_S[\psi_0, \mathbf{A}] & \text{when } \langle L_3\rangle_0 > 0, \\ F_S[\psi_0, \mathbf{A}] = E_S[\psi_0, \mathbf{A}] & \text{when } \langle L_3\rangle_0 = 0. \end{cases}$$

The previous inequalities suggest that, when $\langle L_3\rangle_0 < 0$, it is possible to derive a general criterion on the magnetic field strength $B > 0$ and the initial data $\psi_0 \in \Sigma_S$ that gives $E_S[\psi_0, \mathbf{A}] > 0$ and $F_S[\psi_0, \mathbf{A}] < 0$, and vice versa when $\langle L_3\rangle_0 > 0$. We first note that it is clear from (2.17) that we must assume $E_0 < 0$ for $F_S[\psi_0, \mathbf{A}] < 0$ to be possible. Assuming $E_0 < 0$, from (2.17) we see that if

$$B^2 < \frac{4|E_0|}{\|\rho\psi_0\|^2_{L^2(\mathbb{R}^d)}},$$

then $F_S[\psi_0, \mathbf{A}] < 0$ when $E_S[\psi_0, \mathbf{0}] < 0$. Furthermore, assuming $\langle L_3\rangle_0 < 0$, from (2.18) we observe that to have $E_S[\psi_0, \mathbf{A}] > 0$ we must have

$$B|\langle L_3\rangle_0| + \frac{B^2}{4}\|\rho\psi_0\|^2_{L^2(\mathbb{R}^d)} > |E_0|.$$

Using this estimate to ensure $E_S[\psi_0, \mathbf{A}] > 0$, we choose $B > |\frac{E_0}{\langle L_3\rangle_0}|$. Therefore, for $E_S[\psi_0, \mathbf{A}] > 0$ and $F_S[\psi_0, \mathbf{A}] < 0$, we may choose $B > 0$ such that

$$\left|\frac{E_0}{\langle L_3\rangle_0}\right| < B < \frac{2\sqrt{|E_0|}}{\|\rho\psi_0\|_{L^2(\mathbb{R}^d)}}.$$

Thus, for such a $B > 0$ to exist, it will be necessary to have

$$\sqrt{|E_0|}\|\rho\psi_0\|_{L^2(\mathbb{R}^d)} < 2|\langle L_3\rangle_0|.$$

Such an inequality can certainly be satisfied. Indeed, take $d = 2$, $p = 5$, and

$$\psi_0(\rho, \vartheta) = u(\rho)e^{-i\vartheta}$$

as an L_3-eigenstate with eigenvalue -1 with

$$u(\rho) = \frac{800\rho}{\sqrt{\pi}} e^{-400\rho^2}.$$

Then, for this state,

$$E_0 = 1600\left(1 - \frac{800}{81\pi^2}\right) \quad \text{and} \quad \frac{|\langle L_3 \rangle_0|^2}{\|\rho\psi_0\|^2_{L^2(\mathbb{R}^d)}} = 800\pi,$$

and for any $2 < B < 106$ will produce a positive $E_S[\psi_0, \mathbf{A}]$, but negative $F_S[\psi_0, \mathbf{A}]$.

2.3 Virial identity and proof of main result

The virial identities in this section for the NLMS equation (2.1) are already present in the literature. The linear case in any space dimension $d \geq 2$ is covered in [7, Theorem 1.2], while the non-linear generalization can be found in [8, Theorem 3.1]. We rederive these identities for completeness, as well as express them in a form that will be useful for the proof of Theorem 2.5. We treat the case of any dimension $d \geq 2$ and a general, time-independent, external magnetic field (i.e., not necessarily a uniform field). The vector potential $\mathbf{A} : \mathbb{R}^d \to \mathbb{R}^d$ generates the matrix-valued magnetic field $\mathbf{B} : \mathbb{R}^d \to M_{n\times n}(\mathbb{R})$ with components $B_{ij} = \partial_j A_i - \partial_i A_j$. We record this first virial identity as the following lemma.

Lemma 2.7 ([7, 8]). *Let*

$$g(t) = \frac{1}{4}\|\mathbf{x}\psi(t)\|^2_{L^2(\mathbb{R}^d)}.$$

Then, for any solution ψ to the NLMS equation (2.1) *with initial data $\psi_0 \in H^2_{\mathbf{A}}(\mathbb{R}^d; \mathbb{C})$, the following virial identity holds:*

$$\ddot{g} = 2T_S[\psi, \mathbf{A}] + \mu d \frac{p-1}{p+1}\|\psi\|^{p+1}_{L^{p+1}(\mathbb{R}^d)} - 2\,\mathrm{Re}\langle \mathbf{B}\mathbf{x}\psi, (\mathbf{p} + \mathbf{A})\psi\rangle_{L^2(\mathbb{R}^d)}. \quad (2.19)$$

Proof. As known from Theorem 2.1, the corresponding solution $\psi \in C([0, T_*), H^2_{\mathbf{A}}) \cap C^1([0, T_*), L^2)$ and therefore all the following computations are justified. Consider the function $f(t) = \langle \psi(t), G(\mathbf{x})\psi(t)\rangle$, where $G : \mathbb{R}^d \to \mathbb{R}$ is a differentiable, radial multiplier to be specified later. We denote $H_S = \boldsymbol{\pi}^2 + \mu|\psi|^{p-1}$, where $\boldsymbol{\pi} = \mathbf{p} + \mathbf{A}$. Using the identity $[A^2, B] = A[A, B] + [A, B]A$ and taking the time derivative of f, we find

$$\begin{aligned} \dot{f} &= \langle i[\boldsymbol{\pi}^2, G]\psi, \psi\rangle_{L^2(\mathbb{R}^d)} \\ &= \langle i(\boldsymbol{\pi} \cdot [\mathbf{p}, G] + [\mathbf{p}, G] \cdot \boldsymbol{\pi})\psi, \psi\rangle_{L^2(\mathbb{R}^d)} \\ &= 2\,\mathrm{Re}\langle \nabla G \cdot \boldsymbol{\pi}\psi, \psi\rangle_{L^2(\mathbb{R}^d)}. \end{aligned}$$

Choosing $G(\mathbf{x}) = |\mathbf{x}|^2$, we arrive at

$$\dot{g} = \mathrm{Re}\langle \mathbf{x} \cdot \boldsymbol{\pi}\psi, \psi\rangle_{L^2(\mathbb{R}^d)} = \langle (D + \mathbf{x} \cdot \mathbf{A})\psi, \psi\rangle_{L^2(\mathbb{R}^d)},$$

where $D = \frac{1}{2}(\mathbf{x} \cdot \mathbf{p} + \mathbf{p} \cdot \mathbf{x})$ is the dilation operator.

For the second time derivative we find

$$\ddot{g} = \langle i[H_{\mathrm{S}}, (D + \mathbf{x} \cdot \mathbf{A})]\psi, \psi\rangle_{L^2(\mathbb{R}^d)}. \tag{2.20}$$

Recall that

$$\frac{\mathrm{d}}{\mathrm{d}\theta} e^{-i\theta D} H_{\mathrm{S}} e^{i\theta D}\Big|_{\theta=0} = i[H_{\mathrm{S}}, D],$$

and that $(e^{i\theta D} f)(\mathbf{x}) = e^{\frac{d\theta}{2}} f(e^{\theta}\mathbf{x})$. Our task is to compute $e^{-i\theta D} H_{\mathrm{S}} e^{i\theta D}$. This is straightforward. For example, for a suitable f, we have

$$e^{-i\theta D} \boldsymbol{\pi} e^{i\theta D} f(\mathbf{x}) = ((e^{\theta}\mathbf{p} + \mathbf{A}(e^{-\theta}\cdot)) f)(\mathbf{x}),$$

which implies

$$e^{-i\theta D} \boldsymbol{\pi}^2 e^{i\theta D} = e^{2\theta} \boldsymbol{\pi}_\theta^2,$$

where $\mathbf{A}_\theta(\mathbf{x}) = e^{-\theta}\mathbf{A}(e^{-\theta}\mathbf{x})$ and $\boldsymbol{\pi}_\theta = \mathbf{p} + \mathbf{A}_\theta$. Similar computations yield

$$e^{-i\theta D} H_{\mathrm{S}} e^{i\theta D} = e^{2\theta} \boldsymbol{\pi}_\theta^2 + \mu|\psi(e^{-\theta}\cdot)|^{p-1}.$$

Differentiating the previous expression with respect to θ and evaluating it at $\theta = 0$, we find

$$i[H_{\mathrm{S}}, D] = 2\boldsymbol{\pi}^2 - \mu(\mathbf{x} \cdot \nabla)|\psi|^{p-1} + \frac{\mathrm{d}}{\mathrm{d}\theta}\boldsymbol{\pi}_\theta^2\Big|_{\theta=0}. \tag{2.21}$$

To complete the computation of equation (2.20), we must workout the commutator $[H_{\mathrm{S}}, \mathbf{x} \cdot \mathbf{A}]$, which reduces to $[\boldsymbol{\pi}^2, \mathbf{x} \cdot \mathbf{A}]$. One finds

$$[\boldsymbol{\pi}^2, \mathbf{x} \cdot \mathbf{A}] = \boldsymbol{\pi} \cdot [\mathbf{p}, \mathbf{x} \cdot \mathbf{A}] + [\mathbf{p}, \mathbf{x} \cdot \mathbf{A}] \cdot \boldsymbol{\pi}$$

and

$$[\nabla, \mathbf{x} \cdot \mathbf{A}] = \nabla(\mathbf{x} \cdot \mathbf{A}) = \mathbf{A} + (\mathbf{x} \cdot \nabla)\mathbf{A} - \sum_{ij} x_j B_{ij} \mathbf{e}_i.$$

Since

$$\mathbf{A}(\mathbf{x}) + (\mathbf{x} \cdot \nabla)\mathbf{A}(\mathbf{x}) = -\frac{\mathrm{d}}{\mathrm{d}\theta}\Big|_{\theta=0} \mathbf{A}_\theta(\mathbf{x}),$$

we conclude that

$$i[H_{\mathrm{S}}, \mathbf{x} \cdot \mathbf{A}] = -\frac{\mathrm{d}}{\mathrm{d}\theta}\Big|_{\theta=0} \boldsymbol{\pi}_\theta^2 - (\boldsymbol{\pi} \cdot \mathbf{B}\mathbf{x} + \mathbf{B}\mathbf{x} \cdot \boldsymbol{\pi}). \tag{2.22}$$

Combining (2.20), (2.21), and (2.22), we conclude (2.19). ■

Proof of Lemma 2.4 *and Theorem* 2.5. We specialize to the case $d = 3$, as $d = 2$ is similar. We will consider $\psi_0 \in H^2_{\mathbf{A}}(\mathbb{R}^3) \cap \Sigma_{\mathrm{S}}$, as the general case of $\psi_0 \in \Sigma_{\mathrm{S}}$ will follow from the continuous dependence portion of Theorem 2.1. Again, we denote $\boldsymbol{\pi} = \mathbf{p} + \mathbf{A}$.

Since the energy $E_S[\psi, \mathbf{A}]$ is conserved, the proof of Lemma 2.4 boils down to showing

$$-B \operatorname{Re}\langle \mathbf{x}_\perp \psi, \boldsymbol{\pi}\psi\rangle_{L^2(\mathbb{R}^3)} + \frac{B^2}{2}\|\rho\psi\|^2_{L^2(\mathbb{R}^3)}$$

is conserved. We start by computing the time derivative of $\langle \psi, \mathbf{x}_\perp \cdot \boldsymbol{\pi}\psi\rangle_{L^2(\mathbb{R}^3)}$. We first note

$$\begin{aligned}\frac{\mathrm{d}}{\mathrm{d}t}\langle \psi, \mathbf{x}_\perp \cdot \boldsymbol{\pi}\psi\rangle_{L^2(\mathbb{R}^3)} &= \langle i[\boldsymbol{\pi}^2, \mathbf{x}_\perp \cdot \boldsymbol{\pi}]\psi, \psi\rangle_{L^2(\mathbb{R}^3)} \\ &\quad + \mu\langle i(L_3|\psi|^{p-1})\psi, \psi\rangle_{L^2(\mathbb{R}^3)},\end{aligned} \tag{2.23}$$

where we recall that $L_3 \equiv \mathbf{x}_\perp \cdot \mathbf{p} = -i\partial_\vartheta$. The second commutator on the right-hand side of (2.23) is straightforward to compute:

$$\langle i(L_3|\psi|^{p-1})\psi, \psi\rangle_{L^2(\mathbb{R}^3)} = \frac{p-1}{p+1}\int_{\mathbb{R}^3} \partial_\vartheta |\psi|^{p+1} = 0.$$

To compute the first commutator on the right-hand side of (2.23), we note

$$[\boldsymbol{\pi}^2, \mathbf{x}_\perp \cdot \boldsymbol{\pi}] = \sum_{j,k=1}^{3} \big(\pi_j[\pi_j, (\mathbf{x}_\perp)_k\pi_k] + [\pi_j, (\mathbf{x}_\perp)_k\pi_k]\pi_j\big).$$

Noting that $[\pi_3, \pi_k] = -i(\partial_3 A_k - \partial_k A_3) = 0$ in a uniform magnetic field directed along the x_3-axis, the above sum reduces to

$$\begin{aligned}[\boldsymbol{\pi}^2, \mathbf{x}_\perp \cdot \boldsymbol{\pi}] &= \sum_{j,k=1}^{2} \big(\pi_j[\pi_j, (\mathbf{x}_\perp)_k\pi_k] + [\pi_j, (\mathbf{x}_\perp)_k\pi_k]\pi_j\big) \\ &= \pi_1[\pi_1, x_1\pi_2] + [\pi_1, x_1\pi_2]\pi_1 - [\pi_2, x_2\pi_1]\pi_2 - \pi_2[\pi_2, x_2\pi_1].\end{aligned} \tag{2.24}$$

Since $\partial_1 A_2 - \partial_2 A_1 = B$, for the first commutator in (2.24) we find

$$i[\pi_1, x_1\pi_2] = \pi_2 + ix_1[\pi_1, \pi_2] = \pi_2 + Bx_1,$$

and for the second commutator in (2.24) we find

$$i[\pi_2, x_2\pi_1] = \pi_1 + ix_2[\pi_2, \pi_1] = \pi_1 - Bx_2.$$

Plugging the previous two commutators back into (2.24), we conclude

$$\begin{aligned}i[\boldsymbol{\pi}^2, \mathbf{x}_\perp \cdot \boldsymbol{\pi}] &= B[\pi_1 x_1 + x_1\pi_1 + \pi_2 x_2 + x_2\pi_2] \\ &= -2iB + 2B(x_1, x_2, 0) \cdot (\mathbf{p} + \mathbf{A}).\end{aligned}$$

Therefore, (2.23) becomes

$$\frac{\mathrm{d}}{\mathrm{d}t}\operatorname{Re}\langle \mathbf{x}_\perp \psi, \boldsymbol{\pi}\psi\rangle_{L^2(\mathbb{R}^3)} = 2B\operatorname{Re}\langle (x_1, x_2, 0)\cdot(\mathbf{p}+\mathbf{A})\psi, \psi\rangle_{L^2(\mathbb{R}^3)}. \tag{2.25}$$

It is easily verified that the right-hand side of (2.25) is proportional to the time derivative of $\|\rho\psi\|^2_{L^2(\mathbb{R}^3)}$, where $\rho^2 = x_1^2 + x_2^2$. That is, we have the desired identity

$$\frac{\mathrm{d}}{\mathrm{d}t}\operatorname{Re}\langle \mathbf{x}_\perp \psi, \boldsymbol{\pi}\psi\rangle_{L^2(\mathbb{R}^3)} = \frac{B}{2}\frac{\mathrm{d}}{\mathrm{d}t}\|\rho\psi\|^2_{L^2(\mathbb{R}^3)}. \tag{2.26}$$

To prove Theorem 2.5, we note that, in dimensions $d \in \{2,3\}$, with $\mathbf{B} = (0,0,B)$ a uniform field, (2.19) becomes

$$\ddot{g} = 2T_{\mathrm{S}}[\psi, \mathbf{A}] + \mu d \frac{p-1}{p+1}\|\psi\|^{p+1}_{L^{p+1}(\mathbb{R}^d)} - 2\operatorname{Re}\langle B\mathbf{x}_\perp\psi, \boldsymbol{\pi}\psi\rangle_{L^2(\mathbb{R}^d)}. \tag{2.27}$$

Thus, finishing the proof of Theorem 2.5 simply amounts to rewriting identity (2.27) and using (2.26). ∎

3 The non-linear Pauli equation

In this final section, we consider generalizing earlier results on the NLMS equation to the non-linear Pauli (NLP) equation. In space dimensions $d \in \{2,3\}$, the NLP equation[5] reads

$$\begin{cases} i\partial_t\psi = [\boldsymbol{\sigma}\cdot(\mathbf{p}+\mathbf{A})]^2\psi + \mu|\psi|^{p-1}\psi, \\ \psi(0,\mathbf{x}) = \psi_0(\mathbf{x}), \end{cases} \tag{3.1}$$

where $\psi : \mathbb{R}^d \to \mathbb{C}^2$, and

$$\boldsymbol{\sigma} = \begin{cases} (\sigma_1,\sigma_2,\sigma_3), & d = 3, \\ (\sigma_1,\sigma_2), & d = 2, \end{cases}$$

is the vector of Pauli matrices, which are 2×2 Hermitian matrices assumed to satisfy the commutation relations $[\sigma^j,\sigma^k] = 2i\epsilon_{jk\ell}\sigma^\ell$ and anticommutation relations $\{\sigma^j,\sigma^k\} = 2\delta_{jk}I$. A typical representation is

$$\sigma_1 = \begin{pmatrix} 0 & 1 \\ 1 & 0 \end{pmatrix}, \quad \sigma_2 = \begin{pmatrix} 0 & -i \\ i & 0 \end{pmatrix}, \quad \sigma_3 = \begin{pmatrix} 1 & 0 \\ 0 & -1 \end{pmatrix}.$$

We typically consider (3.1) as an initial value problem in the space $H^1_{\mathbf{A}}(\mathbb{R}^d;\mathbb{C}^2)$, which is the obvious generalization of the space $H^1_{\mathbf{A}}(\mathbb{R}^d;\mathbb{C})$ discussed in Section 2. Again, we will always assume $\mathbf{A} \in L^2_{\mathrm{loc}}(\mathbb{R}^d;\mathbb{R}^d)$ is such that $[\boldsymbol{\sigma}\cdot(\mathbf{p}+\mathbf{A})]^2$ is an essentially self-adjoint operator on $L^2(\mathbb{R}^d;\mathbb{C}^2)$ with domain $H^2_{\mathbf{A}}(\mathbb{R}^d;\mathbb{C}^2)$. The total energy of (3.1) is

$$E_{\mathrm{P}}[\psi,\mathbf{A}](t) = T_{\mathrm{P}}[\psi,\mathbf{A}] + \frac{2\mu}{p+1}\|\psi(t)\|^{p+1}_{L^{p+1}(\mathbb{R}^d;\mathbb{C}^2)}, \tag{3.2}$$

[5]Note that for $d = 2$, the NLP equation is not equivalent to the NLMS equation via a gauge transformation. This is a consequence of the non-linearity “mixing” the components of ψ.

respectively, where

$$T_{\mathrm{P}}[\psi, \mathbf{A}] = \|\boldsymbol{\sigma} \cdot (\mathbf{p} + \mathbf{A})\psi\|^2_{L^2(\mathbb{R}^d;\mathbb{C}^2)}$$

is the total Pauli kinetic energy. Again, we usually suppress the $\mathbf{A}$-dependence of E_{P} and T_{P}. At least formally, the L^2-norm $\|\psi(t)\|^2_{L^2(\mathbb{R}^d;\mathbb{C}^2)}$ and the total energy (3.2) are conversed along the flow generated by (3.1).

Consider the case of a uniform magnetic field $\mathbf{B} = (0, 0, B)$, $B \in \mathbb{R}\backslash\{0\}$. Using the algebraic properties of the Pauli matrices, we note that

$$[\boldsymbol{\sigma} \cdot (\mathbf{p} + \mathbf{A})]^2 = (\mathbf{p} + \mathbf{A})^2 + B\sigma_3.$$

As a consequence, we have that the unitary time evolution for the Pauli operator

$$U_{\mathrm{P}}(t) = \exp\{-it[\boldsymbol{\sigma} \cdot (\mathbf{p} + \mathbf{A})]^2\}$$

is equal to

$$U_{\mathrm{P}}(t) = e^{-iBt\sigma_3} U_{\mathrm{S}}(t),$$

where $U_{\mathrm{S}}(t)$ is given by (2.4). Hence, estimates (2.7), (2.8), and (2.9) continue to hold with $U_{\mathrm{S}}(t)$ replaced with $U_{\mathrm{P}}(t)$. Likewise, Theorem 2.2 easily generalizes to $U_{\mathrm{P}}(t)$ in $d = 2$ dimensions. Therefore, using the same proof that was used for Theorem 2.1, we have the following theorem.

Theorem 3.1. *Let $d \in \{2, 3\}$, $\mu \in \mathbb{R}$, $1 < p < 1 + \frac{4}{d-2}$, and $\mathbf{A} = \frac{B}{2}\mathbf{x}_\perp$. For all $\psi_0 \in H^1_{\mathbf{A}}(\mathbb{R}^d; \mathbb{C}^2)$ we have the following.*

(1) *There exists a unique maximal solution $\psi \in C([0, T_*), H^1_{\mathbf{A}}) \cap C^1([0, T_*), H^{-1}_{\mathbf{A}})$ of (3.1). If $T_* < \infty$, then $\|(\mathbf{p} + \mathbf{A})\psi(t)\|_{L^2(\mathbb{R}^d)} \to \infty$ at $t \uparrow T_*$.*

(2) *The mapping $\psi_0 \mapsto T_*(\psi_0)$ is lower semi-continuous and, if $t \in [0, T_*(\psi_0))$ and $(\phi_n)_{n\geq 1} \subset H^1_{\mathbf{A}}$ converges to ψ_0 as $n \to \infty$, in $H^1_{\mathbf{A}}$, then the corresponding sequence of solutions $(\psi_n)_{n\geq 1}$ to problem (3.1) verify $\psi_n \to \psi$ as $n \to \infty$, in $C([0, t], H^1_{\mathbf{A}})$.*

(3) *If $\psi_0 \in H^2_{\mathbf{A}}$, then $\psi \in C([0, T_*), H^2_{\mathbf{A}}) \cap C^1([0, T_*), L^2)$.*

(4) *$\|\psi(t)\|_{L^2(\mathbb{R}^d)} = \|\psi_0\|_{L^2(\mathbb{R}^d)}$ and $E_{\mathrm{P}}[\psi(t), \mathbf{A}] = E_{\mathrm{P}}[\psi_0, \mathbf{A}]$.*

In general, the diamagnetic inequality (2.3) no longer holds when $\mathbf{p} + \mathbf{A}$ is replaced by $\boldsymbol{\sigma} \cdot (\mathbf{p} + \mathbf{A})$. However, in three dimensions, as a result of the estimate

$$\langle \boldsymbol{\sigma} \cdot \mathbf{B}\psi, \psi\rangle_{L^2(\mathbb{R}^3;\mathbb{C}^2)} = B \int_{\mathbb{R}^3} (|\psi_1|^2 - |\psi_2|^2) \leq B\|\psi\|^2_{L^2(\mathbb{R}^3;\mathbb{C}^2)},$$

we may still obtain a uniform bound on $\|(\mathbf{p} + \mathbf{A})\psi\|_{L^2(\mathbb{R}^3;\mathbb{C}^2)}$ when $p < \frac{7}{3}$ in a similar manner as the magnetic Schrödinger case discussed in Section 2.1. A similar estimate in two dimensions shows the same conclusion holds, but now with $p < 3$. Therefore, by Theorem 3.1 we conclude global well-posedness of the Cauchy problem (3.1) in the range $1 < p < 1 + \frac{4}{d}$, $d \in \{2, 3\}$, with a uniform magnetic field.

Our blow-up result for (3.1) is very similar to that for the NLMS equation (2.1). To state the result, we introduce the function space

$$\Sigma_{\mathrm{P}} := \left\{ f \in H^1_{\mathbf{A}}(\mathbb{R}^d;\mathbb{C}^2) : \int_{\mathbb{R}^d} |\mathbf{x}|^2 |f(\mathbf{x})|^2 \,\mathrm{d}\mathbf{x} < \infty \right\},$$

and define $F_{\mathrm{P}} : \Sigma_{\mathrm{P}} \to \mathbb{R}$ by

$$F_{\mathrm{P}}[\psi] = E_{\mathrm{P}}[\psi] - B\,\mathrm{Re}\langle \boldsymbol{\sigma}\cdot\mathbf{x}_\perp\psi, \boldsymbol{\sigma}\cdot(\mathbf{p}+\mathbf{A})\psi\rangle_{L^2(\mathbb{R}^d;\mathbb{C}^2)} + \frac{B^2}{2}\|\rho\psi\|^2_{L^2(\mathbb{R}^d;\mathbb{C}^2)}.$$

Similar to F_{S}, the next lemma shows that F_{P} is conserved under the time evolution of (3.1).

Lemma 3.2. *Let $d \in \{2,3\}$. Let $\mathbf{A} \in L^2_{\mathrm{loc}}(\mathbb{R}^d;\mathbb{R}^d)$ generate a uniform magnetic field $\mathbf{B} = (0,0,B)$. Let $\psi_0 \in H^1_{\mathbf{A}}(\mathbb{R}^d;\mathbb{C}^2)$ and let $\psi \in C([0,T_*),H^1_{\mathbf{A}}) \cap C^1([0,T_*),H^{-1}_{\mathbf{A}})$ denote the corresponding solution to* (2.1). *Then $F_{\mathrm{P}}[\psi(t)] = F_{\mathrm{P}}[\psi_0]$.*

The proof of Lemma 3.2 is almost identical to that of Lemma 2.4 and is reserved for the end of Section 2.3. Using that $F_{\mathrm{P}}[\psi]$ is conserved, we have the following theorem concerning the second time derivative of the expectation value of $|\mathbf{x}|^2$.

Theorem 3.3. *Let $d \in \{2,3\}$, let $1 < p < 1 + \frac{4}{d-2}$, let $\mathbf{A} \in L^2_{\mathrm{loc}}(\mathbb{R}^d;\mathbb{R}^d)$ generate a uniform magnetic field $\mathbf{B} = (0,0,B)$, and let $\psi_0 \in \Sigma_{\mathrm{P}}$. Let $T_* \in (0,\infty]$ be the time so that $\psi \in C([0,T_*);\Sigma_{\mathrm{P}}) \cap C^1([0,T_*);H^{-1}_{\mathbf{A}})$ is the corresponding maximal solution to the Cauchy problem* (3.1). *Then the function $g(t) = \frac{1}{4}\|\mathbf{x}\psi(t)\|^2_{L^2(\mathbb{R}^d;\mathbb{C}^2)}$ satisfies the virial identity*

$$\ddot{g}(t) = 2F_{\mathrm{P}}[\psi_0] + \mu d\,\frac{p-(1+\frac{4}{d})}{p+1}\|\psi(t)\|^{p+1}_{L^{p+1}(\mathbb{R}^d)} - B^2\|\rho\psi(t)\|^2_{L^2(\mathbb{R}^d;\mathbb{C}^2)}. \tag{3.3}$$

In particular, if $d = 2$ and $p = 3$, then (3.3) *becomes a second-order equation for g that can be solved exactly:*

$$g(t) = \frac{F_{\mathrm{P}}[\psi_0]}{2B^2} + \left(g_0 - \frac{F_{\mathrm{P}}[\psi_0]}{2B^2}\right)\cos(2Bt) + \frac{\dot{g}_0}{2B}\sin(2Bt),$$

where $g_0 = \frac{1}{4}\|\rho\psi_0\|^2_{L^2(\mathbb{R}^2;\mathbb{C}^2)}$ and $\dot{g}_0 = \mathrm{Re}\langle \mathbf{x}\psi_0, (\mathbf{p}+\mathbf{A})\psi_0\rangle_{L^2(\mathbb{R}^2;\mathbb{C}^2)}$.

Corollary 3.4. *Suppose $\mu < 0$, $1 + \frac{4}{d} \le p < 1 + \frac{4}{d-2}$, $\psi_0 \in \Sigma_{\mathrm{P}}$, and $F_{\mathrm{P}}[\psi_0] < 0$. Then the corresponding solution to* (2.1) *blows up in finite time. If $F_{\mathrm{P}}[\psi_0] = 0$, then blow-up occurs when $\dot{g}(0) = \mathrm{Re}\langle \mathbf{x}\psi_0, (\mathbf{p}+\mathbf{A})\psi_0\rangle_{L^2(\mathbb{R}^d)} < 0$.*

As with the NLMS equation, both Theorem 3.3 and its corollary are proved by deriving a virial identity for the second time derivative of $\langle \psi, |\mathbf{x}|^2\psi\rangle_{L^2(\mathbb{R}^d;\mathbb{C}^2)}$. As before, we first treat the case of a general, time-independent, external magnetic field (not necessarily a uniform field). We record the virial identity for the NLP equation as the following lemma.

Lemma 3.5. *Let*

$$g(t) = \frac{1}{4}\|\mathbf{x}\psi(t)\|^2_{L^2(\mathbb{R}^d;\mathbb{C}^2)}.$$

Then, for any solution ψ to the NLP equation (3.1) *with initial data $\psi_0 \in H^2_{\mathbf{A}}(\mathbb{R}^d;\mathbb{C}^2)$, the following virial identities holds: In dimension $d = 3$,*

$$\begin{aligned}\ddot{g} &= 2T_{\mathrm{P}}[\psi] + 3\mu\frac{p-1}{p+1}\|\psi\|^{p+1}_{L^{p+1}(\mathbb{R}^2;\mathbb{C}^2)} \\ &\quad + 2\,\mathrm{Re}\langle\boldsymbol{\sigma}\cdot(\mathbf{x}\wedge\mathbf{B})\psi, \boldsymbol{\sigma}\cdot(\mathbf{p}+\mathbf{A})\psi\rangle_{L^2(\mathbb{R}^2;\mathbb{C}^2)},\end{aligned} \tag{3.4}$$

and, assuming **B** *is aligned with the x_3-axis, in dimension $d = 2$,*

$$\begin{aligned}\ddot{g} &= 2T_{\mathrm{P}}[\psi] + 2\mu\frac{p-1}{p+1}\|\psi\|^{p+1}_{L^{p+1}(\mathbb{R}^2;\mathbb{C}^2)} \\ &\quad - 2\,\mathrm{Re}\langle B\boldsymbol{\sigma}\cdot\mathbf{x}_\perp\psi, \boldsymbol{\sigma}\cdot(\mathbf{p}+\mathbf{A})\psi\rangle_{L^2(\mathbb{R}^2;\mathbb{C}^2)}.\end{aligned} \tag{3.5}$$

Proof. We denote

$$H_{\mathrm{P}} = [\boldsymbol{\sigma}\cdot\boldsymbol{\pi}]^2 + \mu|\psi|^{p-1},$$

where $\boldsymbol{\pi} = \mathbf{p} + \mathbf{A}$. For simplicity we focus on the $d = 3$ case, as the $d = 2$ case will be nearly identical. Computing first time derivative of g is essentially the same as the first time derivative of g in the proof of Lemma 2.7. For the second time derivative we find

$$\ddot{g} = \langle i[H_{\mathrm{P}}, (D + \mathbf{x}\cdot\mathbf{A})]\psi, \psi\rangle_{L^2(\mathbb{R}^3;\mathbb{C}^2)},$$

where

$$H_{\mathrm{P}} = [\boldsymbol{\sigma}\cdot\boldsymbol{\pi}]^2 + \mu|\psi|^{p-1}\psi.$$

We worked out the commutator $[H_{\mathrm{P}}, D]$ in the same way as in the proof of Lemma 2.7. We find

$$i[H_{\mathrm{P}}, D] = 2[\boldsymbol{\sigma}\cdot\boldsymbol{\pi}]^2 - \mu(\mathbf{x}\cdot\nabla)|\psi|^{p-1} + \frac{\mathrm{d}}{\mathrm{d}\theta}[\boldsymbol{\sigma}\cdot\boldsymbol{\pi}_\theta]^2\Big|_{\theta=0},$$

where $\boldsymbol{\pi}_\theta = \mathbf{p} + \mathbf{A}_\theta$.

Likewise, for the computation of

$$\begin{aligned}\langle i[H_{\mathrm{P}}, \mathbf{x}\cdot\mathbf{A}]\psi, \psi\rangle_{L^2(\mathbb{R}^3;\mathbb{C}^2)} &= \langle i[[\boldsymbol{\sigma}\cdot\boldsymbol{\pi}]^2, \mathbf{x}\cdot\mathbf{A}]\psi, \psi\rangle_{L^2(\mathbb{R}^3;\mathbb{C}^2)} \\ &= \langle i[\boldsymbol{\pi}^2, \mathbf{x}\cdot\mathbf{A}]\psi, \psi\rangle_{L^2(\mathbb{R}^3;\mathbb{C}^2)}\end{aligned}$$

we refer to the proof of Lemma 2.7. In total we arrive at

$$\begin{aligned}g &= 2T_{\mathrm{P}}[\psi, \mathbf{A}] + 3\mu\frac{p-1}{p+1}\|\psi\|^{p+1}_{L^{p+1}(\mathbb{R}^d)} \\ &\quad + \frac{\mathrm{d}}{\mathrm{d}\theta}\langle\boldsymbol{\sigma}\cdot\mathbf{B}_\theta\psi, \psi\rangle_{L^2(\mathbb{R}^d)} + 2\,\mathrm{Re}\langle\mathbf{x}\wedge\mathbf{B}\psi, \boldsymbol{\pi}\psi\rangle_{L^2(\mathbb{R}^d)}.\end{aligned} \tag{3.6}$$

where

$$\mathbf{B}_\theta(\mathbf{x}) = e^{-2\theta}\mathbf{B}(e^{-\theta}\mathbf{x}).$$

We may simplify the expression (3.6) by observing the following calculation:

$$\begin{aligned}\int_{\mathbb{R}^3} (\mathbf{x} \wedge \mathbf{B}) \cdot \operatorname{curl} \langle \psi, \boldsymbol{\sigma} \psi \rangle_{\mathbb{C}^2} \, \mathrm{d}\mathbf{x} &= \int_{\mathbb{R}^3} \operatorname{curl} (\mathbf{x} \wedge \mathbf{B}) \cdot \langle \psi, \boldsymbol{\sigma} \psi \rangle_{\mathbb{C}^2} \, \mathrm{d}\mathbf{x} \\ &= \int_{\mathbb{R}^3} (-2\mathbf{B} - (\mathbf{x} \cdot \nabla)\mathbf{B}) \cdot \langle \psi, \boldsymbol{\sigma} \psi \rangle_{L^2(\mathbb{R}^3;\mathbb{C}^2)} \, \mathrm{d}\mathbf{x} \\ &= \frac{\mathrm{d}}{\mathrm{d}\theta}\bigg|_{\theta=0} \langle \boldsymbol{\sigma} \cdot \mathbf{B}_\theta \psi, \psi \rangle_{L^2(\mathbb{R}^3;\mathbb{C}^2)}.\end{aligned}$$

Combining the previous observation with (3.6), we arrive at (3.4). ■

Proof of Lemma 3.2 *and Theorem* 3.3. The proof of Lemma 3.2 and, hence, Theorem 3.3 is nearly identical to the proof for the NLMS equation case. In particular, we have the identity

$$\frac{\mathrm{d}}{\mathrm{d}t} \operatorname{Re} \langle \boldsymbol{\sigma} \cdot \mathbf{x}_\perp \psi, \boldsymbol{\sigma} \cdot (\mathbf{p} + \mathbf{A}) \psi \rangle_{L^2(\mathbb{R}^d)} = \frac{B}{2} \frac{\mathrm{d}}{\mathrm{d}t} \| \rho \psi \|^2_{L^2(\mathbb{R}^d)},$$

which upon integration in time together with (3.4)–(3.5) yields the desired result. ■

Acknowledgements. The authors are grateful to Jan-Philip Solovej for many discussions and for the hospitality at the University of Copenhagen. They would like to thank the anonymous referee for helpful remarks. This work was partially funded by NSF grant DMS-1856645.

Bibliography

[1] J. Avron, I. Herbst and B. Simon, Schrödinger operators with magnetic fields. I. General interactions. *Duke Math. J.* **45** (1978), 847–883

[2] T. Cazenave, *An introduction to nonlinear Schrödinger equations*. Textos Mét. Matem. 26, Universidade Federal do Rio de Janeiro, Rio de Janeiro, 1989

[3] T. Cazenave and M. J. Esteban, On the stability of stationary states for nonlinear Schrödinger equations with an external magnetic field. *Mat. Apl. Comput.* **7** (1988), 155–168

[4] H. L. Cycon, R. G. Froese, W. Kirsch and B. Simon, *Schrödinger operators with application to quantum mechanics and global geometry*. Study edn., Texts Monogr. Phys., Springer, Berlin, 1987

[5] P. D'Ancona and L. Fanelli, Strichartz and smoothing estimates of dispersive equations with magnetic potentials. *Comm. Partial Differential Equations* **33** (2008), 1082–1112

[6] P. D'Ancona, L. Fanelli, L. Vega and N. Visciglia, Endpoint Strichartz estimates for the magnetic Schrödinger equation. *J. Funct. Anal.* **258** (2010), 3227–3240

[7] L. Fanelli and L. Vega, Magnetic virial identities, weak dispersion and Strichartz inequalities. *Math. Ann.* **344** (2009), 249–278

[8] A. Garcia, Magnetic virial identities and applications to blow-up for Schrödinger and wave equations. *J. Phys. A* **45** (2012), Article ID 015202

[9] R. T. Glassey, On the blowing up of solutions to the Cauchy problem for nonlinear Schrödinger equations. *J. Math. Phys.* **18** (1977), 1794–1797

[10] J. M. Gonçalves Ribeiro, Finite time blow-up for some nonlinear Schrödinger equations with an external magnetic field. *Nonlinear Anal.* **16** (1991), 941–948

[11] D. Hundertmark and V. Zharnitsky, On sharp Strichartz inequalities in low dimensions. *Int. Math. Res. Not.* **2006** (2006), Article ID 34080

[12] H. Leinfelder, Gauge invariance of Schrödinger operators and related spectral properties. *J. Operator Theory* **9** (1983), 163–179

[13] H. Leinfelder and C. G. Simader, Schrödinger operators with singular magnetic vector potentials. *Math. Z.* **176** (1981), 1–19

[14] M. Reed and B. Simon, *Methods of modern mathematical physics. IV. Analysis of operators*. Academic Press, New York, 1978

[15] B. Simon, Maximal and minimal Schrödinger forms. *J. Operator Theory* **1** (1979), 37–47

[16] T. Tao, *Nonlinear dispersive equations*. CBMS Reg. Conf. Ser. Math. 106, American Mathematical Society, Providence, 2006

Large $|k|$ behavior of d-bar problems for domains with a smooth boundary

Christian Klein, Johannes Sjöstrand and Nikola Stoilov

To Ari Laptev on the occasion of his 70th birthday

In this work we study the large $|k|$ behavior of complex geometric optics solutions to a system of d-bar equations for a potential being the characteristic function of a strictly convex set with smooth boundary, by using almost holomorphic functions. This is an extension of our previous work where we consider sets with real-analytic boundary.

1 Introduction

This note is concerned with solutions to the Dirac system

$$\begin{cases} \bar{\partial}\phi_1 = \dfrac{1}{2} q e^{\bar{k}\bar{z}-kz}\phi_2, \\ \partial\phi_2 = \sigma\dfrac{1}{2}\bar{q} e^{kz-\bar{k}\bar{z}}\phi_1, \quad \sigma = \pm 1, \end{cases} \tag{1.1}$$

subject to the asymptotic conditions

$$\lim_{|z|\to\infty} \phi_1 = 1, \qquad \lim_{|z|\to\infty} \phi_2 = 0, \tag{1.2}$$

where $q = q(x, y)$ is a complex-valued field, the *spectral parameter* $k \in \mathbb{C}$ is independent of $z = x + \mathrm{i}y$, and

$$\partial := \frac{1}{2}\left(\frac{\partial}{\partial x} - \mathrm{i}\frac{\partial}{\partial y}\right) \quad \text{and} \quad \bar{\partial} := \frac{1}{2}\left(\frac{\partial}{\partial x} + \mathrm{i}\frac{\partial}{\partial y}\right).$$

The functions $\phi_i(z; k)$, $i = 1, 2$, depend on z and k, where it is understood that they need not be holomorphic in either variable.

This system appears in the scattering theory of the integrable Davey–Stewartson (DS) II equation (see [3])

$$\begin{cases} iq_t + (q_{xx} - q_{yy}) + 2\sigma(\Phi + |q|^2)q = 0, \\ \Phi_{xx} + \Phi_{yy} + 2(|q|^2)_{xx} = 0, \end{cases} \tag{1.3}$$

Keywords: d-bar problems, Dirac systems, semiclassical limit, smooth boundaries
2020 Mathematics Subject Classification: Primary 81Q20; secondary 30E20

a two-dimensional nonlinear Schrödinger equation; here the parameter $\sigma = 1$ in the *defocussing* case, and $\sigma = -1$ in the *focusing* case. DS systems appear in the modulational regime of many dispersive equations as for instance the water wave systems. Note that they are in general not integrable unless the parameters are as in (1.3).

The scattering data for the integrable DS (1.3) are given in terms of the reflection coefficient $R = R(k)$,

$$\overline{R} = \frac{2\sigma}{\pi} \int_{\mathbb{C}} \mathrm{e}^{kz-\bar{k}\bar{z}} \overline{q}(z)\phi_1(z;k)\, \mathrm{d}^2 z.$$

The time dependence of the reflection coefficient is trivial,

$$R(k;t) = R(k,0)e^{4it\Re(k^2)}.$$

The inverse scattering transform for DS II is then given by (1.1) and (1.2) after replacing q by R and vice versa, the derivatives with respect to z by the corresponding derivatives with respect to k, and asymptotic conditions for $k \to \infty$ instead of $z \to \infty$, see [4].

System (1.1) appears also in the context of Calderón's problem [2] in electrical impedance tomography (EIT). This denotes the reconstruction of the conductivity in a given domain $\Omega \in \mathbb{R}^2$ from measurements of the electrical current through its boundary, induced by an applied voltage, i.e., from the Dirichlet-to-Neumann map, see [9, 11]. The basic idea of EIT is to construct *complex geometric optics* (CGO) solutions to the conductivity equation in some domain Ω,

$$\nabla \cdot (\rho(x,y)\nabla u(x,y)) = 0, \quad (x,y) \in \Omega,$$

i.e., solutions with an exponential behavior at infinity. In [1], this was done in form of system (1.1) by putting $q = -\frac{1}{2}\partial \ln \rho$ for conductivities $\rho \in C^1(\Omega)$. The reconstruction of the conductivity from the Dirichlet-to-Neumann map is also achieved via a d-bar problem, see [11]. For more applications of system (1.1) and references, see [6].

In [6], we have addressed the large $|k|$ behavior of the solutions to (1.1) and (1.2). It was shown that the k dependence for large $|k|$ strongly depends on the regularity of the potential q in (1.1). Of particular interest is the case of potentials being the characteristic function of a compact domain with strictly convex boundary. It was shown in [6] that in this case $\phi_2 = \mathcal{O}(|k|^{-\frac{1}{2}})$ in polynomially weighted L^2-spaces. The leading-order contribution to ϕ_2 is equal to $\overline{f(z,k)}/(2\pi)$, where f is given by the integral

$$f(z,k) = \int_{\Omega} \frac{1}{z-w} e^{\overline{kw}-kw} L(\mathrm{d}w) = \iint_{\Omega} \frac{e^{\overline{kw}-kw}}{z-w} \frac{\mathrm{d}\overline{w} \wedge \mathrm{d}w}{2i} \tag{1.4}$$

for $z, k \in \mathbb{C}$, $|k| \gg 1$. Here

$$L(\mathrm{d}w) = \mathrm{dRe}(w) \wedge \mathrm{dIm}(w) = \frac{\mathrm{d}\overline{w} \wedge \mathrm{d}w}{2i}$$

is the Lebesgue measure. Note that the second term on the right of (1.4) is simply the solid Cauchy transform of $qe^{\overline{k\bar{z}}-kz}$, i.e., a standard representation of $\overline{\partial}^{-1}$. The large $|k|$ asymptotics of f was computed in [6] when $\partial\Omega$ is analytic.

In this note we extend the results of [6, Section 5] to the case of strictly convex domains $\Omega \Subset \mathbb{C}$ with a boundary that is smooth but not necessarily analytic. In doing so, we follow the complex analysis arguments in [6], but make an essential use of almost holomorphic extensions, introduced by Hörmander [5] and others. See [5] with the attached commentary and additional bibliography by one of us. Almost holomorphic extensions play an important role in the global theory of Fourier integral operators with complex phase [8]. Complex phase arguments were also used by Laptev, Safarov and Vassiliev [7] in their approach to Fourier integral operators with real phase and their global representations as superpositions of Gaussian type operators, and at one point the authors evoke almost analytic (i.e., almost holomorphic) extensions.

2 Main result

As in [6, Section 5.2] we use Stokes' formula to get

$$f(z,k) = \frac{1}{2i\overline{k}} \int_{\partial\Omega} \frac{1}{z-w} e^{\overline{kw}-kw}\, dw + \frac{\pi}{\overline{k}} e^{\overline{kz}-kz} 1_\Omega(z). \tag{2.1}$$

We write

$$kw - \overline{kw} = \frac{i}{h}\Re(w\overline{\omega}) \quad \text{with } h = \frac{1}{|k|},\ \omega = \frac{2\overline{k}}{|k|}.$$

Consider the function

$$iu_0(w) = kw - \overline{kw} \quad \text{on } \partial\Omega$$

(suppressing sometimes the large parameter from the notation). In [6] we use the analyticity of $\partial\Omega$ to introduce a holomorphic extension $iu(w)$ of $iu_0(w)$ from $\partial\Omega$ to $\mathrm{neigh}(\partial\Omega,\mathbb{C})$.

When $\partial\Omega$ is only smooth, we take instead an almost holomorphic extension $u \in C^\infty(\mathrm{neigh}(\partial\Omega,\mathbb{C}))$ so that

$$u|_{\partial\Omega} = u_0, \quad \partial_{\overline{w}} u(w) = \mathcal{O}(\mathrm{dist}(w,\partial\Omega)^\infty), \tag{2.2}$$

see [8, 10]. We recall that u is unique up to infinite order: If $\tilde{u}$ is another analytic holomorphic extension of u_0, then

$$u - \tilde{u} = \mathcal{O}(\mathrm{dist}(\,\cdot\,,\partial\Omega)^\infty) \quad \text{in } \mathrm{neigh}(\partial\Omega,\mathbb{C}).$$

As in [6, equation (5.13)] we parametrize points w in a neighborhood of $\partial\Omega$ by

$$w = \gamma(t) + is\dot{\gamma}(t), \quad t \in \mathbb{R}/L\mathbb{Z},\ s \in \mathrm{neigh}(0,\mathbb{R}),$$

where $L > 0$ is the length of $\partial\Omega$. Here $t \to \gamma(t)$ is a parametrization of $\partial\Omega$ with positive orientation and with $|\dot{\gamma}(t)| = 1$ such that γ can be viewed as an L-periodic function. As in [6, Section 5], we can view u as a function $u = u(t, s)$, $t \in \mathbb{R}/L\mathbb{Z}$, $s \in \mathrm{neigh}(0, \mathbb{R})$, satisfying

$$\partial_t u + \mathrm{i}\Big(1 + \mathrm{i}s\frac{\ddot{\gamma}}{\dot{\gamma}}\Big)\partial_s u = \mathcal{O}(s^\infty),$$

and with $u(t, s) = u_0(t) + s u_1(t) + s^2 u_2(t) + \mathcal{O}(s^3)$, we get

$$\partial_t u_0(t) = \frac{|k|}{2}(\dot{\gamma}(t)\overline{\omega} + \overline{\dot{\gamma}}(t)\omega) = |k|\langle\dot{\gamma}(t), \omega\rangle_{\mathbb{R}^2}, \tag{2.3}$$

$$\partial_t^2 u_0(t) = \frac{|k|}{2}(\ddot{\gamma}(t)\overline{\omega} + \overline{\ddot{\gamma}}(t)\omega) = |k|\langle\ddot{\gamma}(t), \omega\rangle_{\mathbb{R}^2}. \tag{2.4}$$

From (2.3) we see that $\gamma(t)$ is a critical point of $u_0(t)$ if and only if ω (which is non-vanishing) is normal to $\partial\Omega$ at $\gamma(t)$. From $\langle\dot{\gamma}(t), \dot{\gamma}(t)\rangle = 1$ we know that

$$\langle\dot{\gamma}(t), \ddot{\gamma}(t)\rangle = 0,$$

and hence $\ddot{\gamma}(t)$ is normal to $\partial\Omega$ everywhere. Thus at a critical point of u_0 we have $\ddot{\gamma}(t) \in \mathbb{R}\omega$. It follows from (2.4) that a critical point is non-degenerate precisely when $\ddot{\gamma}(t) \neq 0$, i.e., when $\partial\Omega$ has non-vanishing curvature there. Such a point is

- a local maximum if $\ddot{\gamma}(t) = c\omega$, $c < 0$, and
- a local minimum if $\ddot{\gamma}(t) = c\omega$, $c > 0$.

Now recall the assumption that

$$\Omega \text{ is strictly convex.}$$

Then at every point in $\partial\Omega$, $\ddot{\gamma}(t)$ is non-vanishing and of the form $c(t)\nu(t)$, where $c(t) > 0$ and $\nu(t) = \mathrm{i}\dot{\gamma}(t)$ is the interior unit normal (recall that γ is positively oriented).

For a fixed $k \neq 0$, we can decompose

$$\partial\Omega = \{w_-(k)\} \cup \Gamma_+ \cup \{w_+(k)\} \cup \Gamma_-,$$

ordered in the positive direction when starting and ending at $w_-(k)$. Here

- $w_-(k)$ is the south pole, where $\nu = c\omega$ for some $c < 0$. Equivalently this is the global maximum point of u_0.
- Γ_+ is the open boundary segment connecting $w_-(k)$ to $w_+(k)$ in the positive direction.
- $w_+(k)$ is the north pole, where $\nu = c\omega$ for some $c > 0$. Equivalently this is the global minimum point of u_0.
- Γ_- is the open boundary segment connecting $w_+(k)$ to $w_-(k)$ in the positive direction.

Notice that

$$\Gamma_\pm = \{\gamma(t) \in \partial\Omega : \mp\partial_t u_0(\gamma(t)) > 0\}.$$

On $\partial\Omega$ we have

$$e^{\overline{kw}-kw} = e^{-iu_0(w)}, \quad u_0(w) = |k|\Re(w\overline{\omega}), \ w = \gamma(t).$$

Formula (2.1) reads

$$f(z,k) = \frac{1}{2i\overline{k}} \int_{\partial\Omega} \frac{1}{z-w} e^{-iu(w,k)}\, dw + \frac{\pi}{\overline{k}} e^{-i|k|\Re(z\overline{\omega})} 1_\Omega(z), \tag{2.5}$$

where $u(\cdot\,,k)$ is an almost holomorphic extension of $u_0(w) = u_0(w,k)$ to a neighborhood of $\partial\Omega$.

We introduce the same closed contour Γ as in [6, Section 5.3]. Recall that Γ is a deformation of $\partial\Omega$ obtained by pushing Γ_+ inward and Γ_- outward. This is obtained by means of a smooth vector field υ, defined near $\partial\Omega$, transversal to $\partial\Omega$ and pointing outward, so that

$$\Gamma = \{\exp(t\upsilon(w))|_{t=\tau(w)} : w \in \partial\Omega\}$$

when $\tau(w)$ is a suitable function such that $\sup|\tau| \ll 1$, $\tau < 0$ on Γ_+, $\tau > 0$ on Γ_-. Near w_+ we can choose a real Morse coordinate μ on Γ, centered at w_+ such that as a function of μ,

$$u = u(w_+(k),k) + \frac{|k|\mu^2}{2}. \tag{2.6}$$

Let μ also denote an almost holomorphic extension. Then in a complex neighborhood of w_+, we have u as a function of μ

$$u = u(w_+(k),k) + \frac{|k|\mu^2}{2} + |k|\mathcal{O}((\Im\mu)^\infty). \tag{2.7}$$

Near w_+ we choose Γ to coincide with $\exp(-\frac{i\pi}{4})\mathbb{R}$ in the complex μ-coordinate. Thus along this part of Γ, we have

$$-i\big(u(\cdot\,,k) - u(w_+(k),k)\big) = -\frac{|k||\mu|^2}{2} + |k|\mathcal{O}(|\mu|^\infty).$$

Near w_- we have a similar construction. Along Γ we have

$$|e^{-iu(w,k)}| \le e^{-\frac{|k|\operatorname{dist}(w,\{w_+(k),w_-(k)\})^2}{C}}.$$

Let $\Omega_+ \subset \overline{\Omega}$ and $\Omega_- \subset \mathbb{C}\setminus\Omega$ be the points swept over by the deformation of Γ_+ and Γ_-, respectively, i.e.,

$$\Omega_+ = \{\exp t\upsilon(w) : w \in \Gamma_+,\ \tau(w) \le t \le 0\},$$
$$\Omega_- = \{\exp t\upsilon(w) : w \in \Gamma_-,\ 0 \le t \le \tau(w)\},$$

and notice that

$$\Gamma = \partial((\Omega \setminus \Omega_+) \cup \Omega_-).$$

Assume for simplicity that $z \notin \partial\Omega_+ \cup \partial\Omega_-$ and put

$$F(z) = F_\Gamma(z) = \int_\Gamma \frac{1}{z-w} e^{-iu(w,k)}\, \mathrm{d}w.$$

When Γ is real-analytic, we can choose u holomorphic and using the residue theorem we got in [6]

$$\begin{aligned} f(z,k) &= \frac{1}{2i\overline{k}} \int_{\partial\Omega} \frac{1}{z-w} e^{-iu(w,k)}\, \mathrm{d}w + \frac{\pi}{\overline{k}} e^{-i|k|\Re(z\overline{\omega})} 1_\Omega(z) \\ &= \frac{1}{2i\overline{k}} F(z) + \frac{\pi}{\overline{k}} \big(e^{-iu(z,k)}(1_{\Omega_-}(z) - 1_{\Omega_+}(z)) + e^{-i|k|\Re(z\overline{\omega})} 1_\Omega(z)\big). \quad (2.8) \end{aligned}$$

In the smooth case we get a correction term that we compute with the help of Stokes' theorem. Let $V \Subset \mathbb{C}$ be open with a smooth positively oriented boundary. If g is a distribution on V, smooth near ∂V and with a smooth extension to a neighborhood of ∂V, then Stokes' formula gives

$$\frac{1}{2\mathrm{i}} \int_{\partial V} g(w)\, \mathrm{d}w = \int_V \partial_{\overline{w}} g \frac{\mathrm{d}\overline{w} \wedge \mathrm{d}w}{2\mathrm{i}} = \int_V \partial_{\overline{w}} g L(\mathrm{d}w).$$

More generally, this holds if $g = g_1 + g_2$, where g_1 is a distribution with compact support contained in V and $g_2 \in H^1(V)$; the standard Sobolev space of order 1. With $g(w) = \frac{1}{z-w} e^{-\mathrm{i}u(w,k)}$ and δ_z denoting the delta mass at the point z, we get

$$\partial_{\overline{w}} g = -\pi \delta_z(w) e^{-\mathrm{i}u(z,k)} - \frac{\mathrm{i}\partial_{\overline{w}} u}{z-w} e^{-iu(w,k)},$$

hence

$$\frac{1}{2\mathrm{i}} \int_{\partial V} \frac{1}{z-w} e^{-\mathrm{i}u(w,k)\,\mathrm{d}w} = -\pi e^{-iu(z,k)} 1_V(z) - \mathrm{i} \int_V \frac{\partial_{\overline{w}} u(w,k) e^{-\mathrm{i}u(w,k)}}{z-w} L(\mathrm{d}w).$$

We apply this with V equal to Ω_+ and Ω_- and get

$$\begin{aligned} &\frac{1}{2\mathrm{i}} \int_{\partial\Omega} \frac{1}{z-w} e^{-\mathrm{i}u(w,k)}\, \mathrm{d}w - \frac{1}{2\mathrm{i}} \int_\Gamma \frac{1}{z-w} e^{-\mathrm{i}u(w,k)}\, \mathrm{d}w \\ &\quad = -\pi e^{-iu(z,k)} 1_{\Omega_+}(z) + \pi e^{-\mathrm{i}u(z,k)} 1_{\Omega_-}(z) \\ &\qquad - \mathrm{i} \int \frac{\partial_{\overline{w}} u(w,k)}{z-w} e^{-\mathrm{i}u(w,k)} (1_{\Omega_+}(w) - 1_{\Omega_-}(w)) L(\mathrm{d}w). \end{aligned}$$

The last term gives rise to a new term in (2.8) and we get

$$\begin{aligned} f(z,k) = \frac{1}{2i\overline{k}} F(z) &+ \frac{\pi}{\overline{k}} \big(e^{-iu(z,k)}(1_{\Omega_-}(z) - 1_{\Omega_+}(z)) + e^{-i|k|\Re(z\overline{\omega})} 1_\Omega(z)\big) \\ &+ \frac{i}{\overline{k}} \int \frac{\partial_{\overline{w}} u(w,k)}{z-w} e^{-\mathrm{i}u(w,k)} \big(1_{\Omega_-}(w) - 1_{\Omega_+}(w)\big) L(\mathrm{d}w). \quad (2.9) \end{aligned}$$

In order to estimate the last term, we notice that

$$\big|e^{-\mathrm{i}u(w,k)}\big| \le e^{-\frac{|k|\mathrm{dist}(w,\partial\Omega)}{C}}, \quad w \in (\Omega_+ \cup \Omega_-) \setminus \mathrm{neigh}(\{w_+, w_-\}). \quad (2.10)$$

In a neighborhood of w_+, we use the Morse coordinates in (2.7) and notice that with $t = \Re(\mu)$, $s = \Im(\mu)$ we have in

$$\mathrm{neigh}(w_+) \cap \Omega_- : 0 < t < \frac{1}{\mathcal{O}(1)}, \quad -t < s < 0,$$

and in

$$\mathrm{neigh}(w_+) \cap \Omega_+ : -\frac{1}{\mathcal{O}(1)} < t < 0, \quad 0 < s < -t.$$

In both sets, $-\Im(u) = -ts + \mathcal{O}(s^\infty) \geq s^2 - \mathcal{O}(s^\infty)$, so

$$\left|e^{-iu(w,k)}\right| \leq e^{-|k|(s^2+\mathcal{O}(s^\infty))} \quad \text{in neigh}(w_+) \cap (\Omega_- \cup \Omega_+). \tag{2.11}$$

The same estimate holds near w_- in the corresponding Morse coordinate. Combining (2.10), (2.11) and (2.2), we get

$$\left|(\bar{\partial}_w u)e^{iu(w,k)}\right| = \mathcal{O}(|k|^{-\infty}) \quad \text{uniformly in } \Omega_+ \cup \Omega_-.$$

We conclude that the last term in (2.9) is $\mathcal{O}(\langle z\rangle^{-1}|k|^{-\infty})$.

The analysis of F_Γ goes through without any changes as in [6] : When

$$\mathrm{dist}(z, \{w_+(k), w_-(k)\}) \geq \frac{1}{\mathcal{O}(1)},$$

then F_Γ can be expanded by the method of stationary phase – steepest descent and has the asymptotic behavior:

$$\begin{aligned}\sqrt{2\pi}\Bigg(&\frac{1}{z-w_+(k)} e^{-\frac{iu(w_+(k),k)-i\pi}{4}} \frac{\dot{\gamma}(t_+(k))}{|\partial_t^2 u(t_+(k))|^{\frac{1}{2}}} \\ &+ \frac{1}{z-w_-(k)} e^{-\frac{iu(w_-(k),k)+i\pi}{4}} \frac{\dot{\gamma}(t_-(k))}{|\partial_t^2 u(t_-(k))|^{\frac{1}{2}}}\Bigg) + \mathcal{O}(\langle z\rangle^{-1}k^{-\frac{3}{2}}).\end{aligned} \tag{2.12}$$

Here γ is an arc length parametrization of $\partial\Omega$, $w_\pm(k) = \gamma(t_\pm(k))$ and we identify $u(w)$ with $u(\gamma(t))$ whenever convenient. The leading term is $\mathcal{O}(k^{-\frac{1}{2}}\langle z\rangle^{-1})$.

The detailed description of the asymptotics of f when $k \to \infty$ was given in [6, Theorem 5.2] which we repeat here for the convenience of the reader:

Theorem 5.2 ([6]). *Let $\Omega \Subset \mathbb{C}$ be strictly convex with real analytic boundary and let $f(z,k)$, $z, k \in \mathbb{C}$, $|k| \geq 1$ be the function appearing in* (1.4). *Let*

$$iu_0(w) = kw - \overline{kw} = i|k|\Re(z\overline{\omega})$$

and let u be a holomorphic extension of $u_{0|\partial\Omega}$ to a neighborhood of $\partial\Omega$.

Assuming for simplicity that $z \notin \partial\Omega$, we have (2.5):

$$f(z,k) = \frac{1}{2i\bar{k}} \int_{\partial\Omega} \frac{1}{z-w} e^{-iu(w,k)}\, dw + (\pi/\bar{k})e^{-i\Re(z\overline{\omega})} 1_\Omega(z).$$

Let $w_+(k), w_-(k) \in \partial\Omega$ be the points of minimum and maximum of the (real-valued) function u_0, also characterized by $\omega \in \mathbb{R}_\pm \nu(w_\pm(k))$, where $\nu(w)$ denotes the interior normal of $\partial\Omega$ at w. Choose a parametrization $\mathbb{R}/|\partial\Omega|\mathbb{Z} \ni t \mapsto \gamma(t) \in \partial\Omega$ with positive (anti-clock-wise) orientation and $|\dot\gamma(t)| = 1$, so that $\nu(\gamma(t)) = \ddot\gamma(t)$.

Let $\Gamma_+ \subset \partial\Omega$ be the open oriented boundary segment from $w_-(k)$ to $w_+(k)$ and let Γ_- be the similar one from $w_+(k)$ to $w_-(k)$. Let Γ be a deformation of the oriented boundary $\partial\Omega$, and recall from [6] *that Γ is obtained by pushing Γ_+ inward and Γ_- outward, keeping $w_\pm(k)$ fixed and so that Γ coincides near $w_+(k)$ with the (image of the) oriented line $e^{-\frac{i\pi}{4}}\mathbb{R}$ in the Morse μ-coordinates (as in* (2.6)*) and similarly near $w_-(k)$. Let $\Omega_\pm \Subset \mathbb{C}$ be the closed sets swept over, when deforming $\partial\Omega$ to Γ. Let* Lead (2.12) = Lead$_+$ (2.12) + Lead$_-$ (2.12) *denote the leading term of* (2.12) *with the natural decomposition into contributions from $w_+(k)$ and $w_-(k)$. Let*

$$d(z,k) = |k| \min(1, |z - w_+(k)|^2, |z - w_-(k)|^2).$$

We first consider the case when $d(z,k) \geq 1$. Then we have

$$\begin{aligned} f(z,k) &+ \mathcal{O}\Big(\frac{d(z,k)^{-\frac{3}{2}}}{|k|\langle z\rangle}\Big) \\ &= \frac{\text{Lead}\,(2.12)}{2i\overline{k}} + \frac{\pi}{\overline{k}}\big(e^{-i\Re(z,\overline{\omega})}1_\Omega(z) - e^{-iu(z,k)}1_{\Omega_+}(z) + e^{-iu(z,k)}1_{\Omega_-}(z)\big). \end{aligned}$$

To cover the remaining case, it suffices to consider the cases when $|z - w_+(k)| \ll 1$ and $|z - w_-(k)| \ll 1$. Both cases are similar and we formulate the result only when $|z - w_+(k)| \ll 1$:

$$\begin{aligned} f(z,k) &+ \frac{\mathcal{O}(1)}{|k|^{\frac{3}{2}}(1 + d^{\frac{1}{2}})} \\ &= \frac{F^\sigma_{N,2}(z) + \text{Lead}_-\,(2.12)}{2i\overline{k}} \\ &\quad + \frac{\pi}{\overline{k}}\big(e^{-i\Re(z,\overline{\omega})}1_\Omega(z) - e^{-iu(z,k)}1_{\Omega_+}(z) + e^{-iu(z,k)}1_{\Omega_-}(z)\big). \end{aligned}$$

Here, $\sigma = r$ when z is inside Γ, and $\sigma = \ell$ when z is outside. The special function $F^\sigma_{N,2}(z)$ was introduced in [6] *and we have*

$$\frac{F^\sigma_{N,2}}{\overline{k}} = \frac{\mathcal{O}(1)}{|k|(1 + d^{\frac{1}{2}})}.$$

With the results of this section, we can generalize [6, Theorem 5.2] as follows.

Theorem 2.1. *The detailed description of the asymptotics of f when $k \to \infty$ in* [6, Theorem 5.2] *remains valid if we replace the assumption that $\partial\Omega$ is real-analytic, by the weaker assumption that $\partial\Omega$ is smooth.*

The results in [6, Section 5.5] about the leading approximations of ϕ_1, ϕ_2 in certain polynomially weighted L^2-norms remain valid after the same weakening of the boundary regularity assumption for Ω.

Acknowledgements. This work is partially supported by the ANR-FWF project ANuI – ANR-17-CE40-0035, the isite BFC project NAANoD, the EIPHI Graduate School (contract ANR-17-EURE-0002) and by the European Union Horizon 2020 research and innovation program under the Marie Sklodowska-Curie RISE 2017 grant agreement no. 778010 IPaDEGAN.

Bibliography

[1] R. M. Brown and G. A. Uhlmann, Uniqueness in the inverse conductivity problem for nonsmooth conductivities in two dimensions. *Comm. Partial Differential Equations* **22** (1997), 1009–1027

[2] A.-P. Calderón, On an inverse boundary value problem. In *Seminar on Numerical Analysis and its Applications to Continuum Physics (Rio de Janeiro, 1980)*, pp. 65–73, Soc. Brasil. Mat., Rio de Janeiro, 1980

[3] A. S. Fokas, Inverse scattering of first-order systems in the plane related to nonlinear multidimensional equations. *Phys. Rev. Lett.* **51** (1983), 3–6

[4] A. S. Fokas and M. J. Ablowitz, On the inverse scattering transform of multidimensional nonlinear equations related to first-order systems in the plane. *J. Math. Phys.* **25** (1984), 2494–2505

[5] L. Hörmander, *Fourier integral operators*. Lectures at the Nordic Summer School of Mathematics, Tjörn, 1969, https://portal.research.lu.se/portal/en/publications/fourier-integral-operators(7730e675-f8df-4464-9cfe-2faeb4619441).html

[6] C. Klein, J. Sjöstrand and N. M. Stoilov, On the large $|k|$ behavior of complex geometric optics solutions to d-bar problems. Preprint 2020, arXiv:2009.06909

[7] A. Laptev, Y. Safarov and D. Vassiliev, On global representation of Lagrangian distributions and solutions of hyperbolic equations. *Comm. Pure Appl. Math.* **47** (1994), 1411–1456

[8] A. Melin and J. Sjöstrand, Fourier integral operators with complex-valued phase functions. In *Fourier integral operators and partial differential equations (Colloq. Internat., Univ. Nice, Nice, 1974)*, pp. 120–223. Lecture Notes in Math. 459, Springer, Berlin, 1975

[9] J. L. Mueller and S. Siltanen, *Linear and nonlinear inverse problems with practical applications*. Computat. Sci. Eng. 10, Society for Industrial and Applied Mathematics (SIAM), Philadelphia, 2012

[10] J. Sjöstrand and M. Zworski, Complex scaling and the distribution of scattering poles. *J. Amer. Math. Soc.* **4** (1991), 729–769

[11] G. Uhlmann, Electrical impedance tomography and Calderón's problem. *Inverse Problems* **25** (2009), Article ID 123011

Heat kernel estimates for two-dimensional relativistic Hamiltonians with magnetic field

Hynek Kovařík

To Ari Laptev on the occasion of his 70th birthday

We study the long time behavior of the heat kernel generated by relativistic Hamiltonians with radial magnetic field. The main result of the paper shows how the magnetic field accelerates the pointwise decay of the heat kernel. The case of an Aharonov–Bohm magnetic field is discussed as well.

1 Introduction

Consider a two-dimensional magnetic Laplacian formally given by

$$H = P^2, \quad P = i\nabla + A$$

in $L^2(\mathbb{R}^2)$, where $A : \mathbb{R}^2 \to \mathbb{R}^2$ is a vector potential generating a magnetic field $B : \mathbb{R}^2 \to \mathbb{R}$ through the relation $\operatorname{rot} A = B$. Recall that if $A \in L^2_{\mathrm{loc}}(\mathbb{R}^2)$, then P^2 is the unique self-adjoint operator associated with the closed quadratic form

$$Q[u] = \|(i\nabla + A)u\|_2^2, \quad u \in D(Q),$$

with the form domain

$$D(Q) = \{u \in L^2(\mathbb{R}^2) : Pu \in L^2(\mathbb{R}^2)\}.$$

The main object of our interest in this paper is the integral kernel of the semigroup generated by the relativistic Hamiltonian

$$\mathscr{H} = \mathscr{H}(A, m) = \sqrt{P^2 + m^2} - m,$$

where $m \geq 0$ is the mass of the particle. In particular, we are interested in the long time behavior of $e^{-t\mathscr{H}(A,m)}(x, y)$ and in its dependence on the magnetic field. Note

Keywords: Relativistic Hamiltonian, magnetic field, heat kernel
2020 Mathematics Subject Classification: Primary 47D08; secondary 35P05

that for a massless particle in the absence of magnetic field, we have

$$e^{-t\mathscr{H}(0,0)}(x,y) = e^{-t\sqrt{-\Delta}}(x,y) = \frac{t}{2\pi(t^2+|x-y|^2)^{\frac{3}{2}}}, \quad x,y\in\mathbb{R}^2,\, t>0, \tag{1.1}$$

see e.g. [12, Section 7.11]. Hence

$$e^{-t\mathscr{H}(0,0)}(x,y) \le \frac{1}{2\pi t^2}, \quad t>0,$$

uniformly in x and y. By the diamagnetic inequality this upper bound can be extended to $\mathscr{H}(A,0)$;

$$\left|e^{-t\mathscr{H}(A,0)}(x,y)\right| \le \frac{1}{2\pi t^2} \quad t>0, \tag{1.2}$$

see equation (2.6) below.

However, Laptev and Weidl proved in [11], under mild regularity and decay assumptions on B, that the operator H satisfies a Hardy-type inequality

$$\int_{\mathbb{R}^2} |(i\nabla + A)u|^2 \ge \int_{\mathbb{R}^2} w|u|^2 \quad \text{for all } u\in D(Q),$$

where $w \gneqq 0$ is a weight functions such that $w(x) = \mathcal{O}(|x|^{-2})$ as $|x|\to\infty$. Hence the presence of a magnetic field removes the singularity of the Green of $-\Delta$ at zero energy. This suggest that it should be possible to improve the decay rate in t of the upper bound (1.2), probably at a cost of spatial weights. One of the results of this paper, Theorem 2.1, confirms this heuristic expectation for radial magnetic fields with compact support.

In the massive case when $m>0$ the semigroup generated by $\mathscr{H}(A,m)$ exhibits different behavior for $t\to 0$ and for $t\to\infty$. However, for large times one observes again a faster time decay of the associated heat kernel with respect to the heat kernel generated by the non-magnetic operator, see Theorem 2.2. In Section 3 we obtain analogous results for the Aharonov–Bohm-type magnetic fields.

Remark. It should be noted that the path integral methods developed in [6] could possibly provide a tool for alternative proofs or even more general results; in particular, for non-radial magnetic fields.

2 Radial magnetic field

2.1 Preliminaries

In this section we will always assume that $B\in L^1(\mathbb{R}^2)$. Let

$$\alpha = \frac{1}{2\pi}\int_{\mathbb{R}^2} B$$

be the total (normalized) magnetic flux, and let

$$\kappa = \min_{m\in\mathbb{Z}} |m+\alpha| \in [0, \tfrac{1}{2}]$$

be the distance between α and the set of integers. Recall also that for any $A \in L^2_{\mathrm{loc}}(\mathbb{R}^2)$ the semigroup e^{-tH} satisfies the diamagnetic inequality

$$\left|e^{-tH}(x,y)\right| \le \frac{1}{4\pi t} e^{-\frac{|x-y|^2}{4t}} \quad \text{a.e.} x, y \in \mathbb{R}^2,\ t > 0, \tag{2.1}$$

see [2,5,7,9,13].

2.2 The case $m = 0$

Theorem 2.1. *Assume that B is radial, continuous and compactly supported. Then*

$$e^{-t\mathscr{H}} : L^p(\mathbb{R}^2) \to L^q(\mathbb{R}^2), \quad 1 \le p \le q \le \infty, \tag{2.2}$$

and its kernel satisfies

$$\left|e^{-t\mathscr{H}(A,0)}(x,y)\right| \lesssim (1+|x|)^{\beta}(1+|y|)^{\beta} t^{-2-2\beta} \qquad \text{if } \kappa > 0, \tag{2.3}$$

$$\left|e^{-t\mathscr{H}(A,0)}(x,y)\right| \lesssim \log(2+|x|)^{\theta}\log(2+|y|)^{\theta} t^{-2}[\log(2+t)]^{-2\theta} \quad \text{if } \kappa = 0 \tag{2.4}$$

for any $\beta \in [0,\kappa]$ and any $\theta \in [0,1]$, respectively.

Proof. We use the Poincaré gauge for the vector potential A;

$$A(x) = \frac{(-x_2, x_1)}{|x|^2}\int_0^{|x|} B(r) r\, dr, \quad x \in \mathbb{R}^2.$$

Then $\operatorname{rot} A = B$ and $A \in C^1(\mathbb{R}^2) \cap L^\infty(\mathbb{R}^2)$. In view of [3, Theorem 6.1] the semigroup $e^{-tH} : L^p(\mathbb{R}^2) \to L^q(\mathbb{R}^2)$, $1 \le p \le q \le \infty$, is thus strongly continuous in $t > 0$, and its integral kernel $e^{-tH}(x,y)$ is jointly continuous in (x,y,t). On the other hand, inequality (2.1) shows that e^{-tH} is a contraction on $L^p(\mathbb{R}^2)$ for any $p \in [1,\infty]$. Hence by [8, Theorems 3.9.7 and 4.3.1] and [8, Example 3.9.16] we have

$$e^{-t\mathscr{H}(A,0)} = e^{-t\sqrt{H}} = \frac{t}{\sqrt{4\pi}}\int_0^\infty s^{-\frac{3}{2}} e^{-\frac{t^2}{4s}} e^{-sH}\, ds, \tag{2.5}$$

with $e^{-t\mathscr{H}(A,0)}$ being a strongly continuous contraction semigroup on $L^p(\mathbb{R}^2)$, where $p \in [1,\infty]$. Moreover, the above mentioned properties of $e^{-tH}(x,y)$ in combination with (2.5) imply that $e^{-t\mathscr{H}(A,0)}$ admits an integral kernel $e^{-t\mathscr{H}(A,0)}(x,y)$ which is jointly continuous in (x,y,t). Note also that

$$\begin{aligned}\int_0^\infty s^{-\frac{5}{2}} e^{-\frac{t^2}{4s}}\, ds &= t^{-3}\int_0^\infty s^{-\frac{5}{2}} e^{-\frac{1}{4s}}\, ds \\ &= 8t^{-3}\int_0^\infty \sqrt{r} e^{-r}\, dr = 8t^{-3}\Gamma\left(\frac{3}{2}\right) = 4t^{-3}\sqrt{\pi}.\end{aligned}$$

This in combination with (2.1) and (2.5) yields

$$\left|e^{-t\mathscr{H}(A,0)}(x,y)\right| \leq \frac{t}{(4\pi)^{\frac{3}{2}}} \int_0^\infty s^{-\frac{5}{2}} e^{-\frac{t^2}{4s}}\, ds = \frac{1}{2\pi t^2}. \tag{2.6}$$

Hence $e^{-t\mathscr{H}(A,0)} : L^1(\mathbb{R}^2) \to L^\infty(\mathbb{R}^2)$, and (2.2) follows.

Let us now assume that $\kappa > 0$. Then, by [10, Theorem 3.3.],

$$\left|e^{-tH}(x,y)\right| \leq C_0(1+|x|)^\kappa (1+|y|)^\kappa t^{-1-\kappa} \quad x,y \in \mathbb{R}^2,\ t>0, \tag{2.7}$$

holds for some C_0. Hence by (2.5)

$$\begin{aligned}|e^{-t\mathscr{H}(A,0)}(x,y)| &\lesssim (1+|x|)^\kappa (1+|y|)^\kappa t \int_0^\infty s^{-\frac{5}{2}-\kappa} e^{-\frac{t^2}{4s}}\, ds \\ &\lesssim (1+|x|)^\kappa (1+|y|)^\kappa t^{-2-2\kappa}.\end{aligned}$$

In view of (2.6) this proves (2.3). In order to prove (2.4) we note that for $\kappa = 0$,

$$e^{-tH}(x,x) \lesssim \log(2+|x|)\log(2+|y|)\, t^{-1}(\log(2+t))^{-2}, \tag{2.8}$$

see [10, Theorems 3.1 and 3.2]. Thus, proceeding as above we get

$$\begin{aligned}|e^{-t\mathscr{H}(A,0)}(x,y)| &\lesssim \log(2+|x|)\log(2+|y|) t \int_0^\infty s^{-\frac{5}{2}}(\log(2+s))^{-2} e^{-\frac{t^2}{4s}}\, ds \\ &= \log(2+|x|)\log(2+|y|) t^{-2} \int_0^\infty r^{-\frac{3}{2}}(\log(2+rt^2))^{-2} e^{-\frac{1}{4r}}\, dr \\ &\leq \log(2+|x|)\log(2+|y|)\Big(t^{-2}\int_0^{t^{-1}} r^{-\frac{3}{2}} e^{-\frac{1}{4r}}\, dr \\ &\qquad + (\log(2+t))^{-2}\int_{t^{-1}}^\infty r^{-\frac{3}{2}} e^{-\frac{1}{4r}}\, dr\Big) \\ &\lesssim \log(2+|x|)\log(2+|y|) t^{-2}\big[\log(2+t)\big]^{-2},\end{aligned} \tag{2.9}$$

where we have used the fact that

$$(1+t^n)\int_0^{t^{-1}} r^{-\frac{3}{2}} e^{-\frac{1}{4r}}\, dr = \mathcal{O}(1) \quad \text{for all } n \geq 1.$$

Inequality (2.4) thus follows from (2.6) and (2.9). ■

Remark. The faster time decay of $e^{-t\mathscr{H}(A,0)}(x,y)$ with respect to $e^{-t\mathscr{H}(0,0)}(x,y)$ is compensated by the spacial grow of x and y, as expected.

2.3 The case $m > 0$

For particles with positive mass we have the following theorem.

Theorem 2.2. *Let $m > 0$. Under the assumptions of Theorem 2.1 we have*

$$|e^{-t\mathscr{H}(A,m)}(x,y)| \lesssim (1+|x|)^{\beta}(1+|y|)^{\beta} t^{-1-\beta} \qquad \text{if } \kappa > 0,$$
$$|e^{-t\mathscr{H}(A,m)}(x,y)| \lesssim \log(2+|x|)^{\theta} \log(2+|y|)^{\theta} t^{-1} \big[\log(2+t)\big]^{-2\theta} \qquad \text{if } \kappa = 0$$

for $t \geq 1$, and any $\beta \in [0,\kappa]$ and any $\theta \in [0,1]$, respectively. Moreover, if $t \leq 1$, then

$$|e^{-t\mathscr{H}(A,m)}(x,y)| \lesssim t^{-2}. \tag{2.10}$$

The proof of Theorem 2.2 is based on the following technical result.

Lemma 2.3. *For any $a > 0$ there exist constants $C_1(a)$ and $C_2(a)$ such that*

$$\int_0^\infty r^a e^{-t(r-\frac{m}{2r})^2}\, dr \leq C_1(a) m^{\frac{a}{2}} t^{-\frac{1}{2}} + C_2(a) t^{-\frac{1+a}{2}} \tag{2.11}$$

holds for all $t > 0$.

Proof. Note that the function $s \mapsto s + \sqrt{s^2+2m}$ is an increasing bijection which maps $\mathbb{R}$ onto $(0,\infty)$. Hence we will apply the substitution

$$r = \frac{s + \sqrt{s^2+2m}}{2},$$

and split the integration in (2.11) into two parts as follows:

$$\begin{aligned}
\int_0^{\sqrt{\frac{m}{2}}} r^a e^{-t(r-\frac{m}{2r})^2}\, dr &= 2^{-1-a} \int_{-\infty}^0 \big(s+\sqrt{s^2+2m}\big)^a \Big(1+\frac{s}{\sqrt{s^2+2m}}\Big) e^{-ts^2}\, ds \\
&\leq 2^{-1-a} \int_{-\infty}^0 \big(s+\sqrt{s^2+2m}\big)^a e^{-ts^2}\, ds \\
&= 2^{-1-a} \int_{-\infty}^0 \Big(\frac{2m}{\sqrt{s^2+2m}-s}\Big)^a e^{-ts^2}\, ds \\
&\leq 2^{-1-a} (2m)^{\frac{a}{2}} \int_{-\infty}^0 e^{-ts^2}\, ds = 2^{-1-a} (2m)^{\frac{a}{2}} \frac{\sqrt{\pi}}{2} t^{-\frac{1}{2}}
\end{aligned}$$

and

$$\begin{aligned}
\int_{\sqrt{\frac{m}{2}}}^\infty r^a e^{-t(r-\frac{m}{2r})^2}\, dr &= 2^{-1-a} \int_0^\infty \big(s+\sqrt{s^2+2m}\big)^a \Big(1+\frac{s}{\sqrt{s^2+2m}}\Big) e^{-ts^2}\, ds \\
&\leq 2^{-a} \int_0^\infty \big(s+\sqrt{s^2+2m}\big)^a e^{-ts^2}\, ds \\
&\lesssim \int_0^\infty \big(2s+\sqrt{2m}\big)^a e^{-ts^2}\, ds \lesssim \int_0^\infty \big(s^a + m^{\frac{a}{2}}\big) e^{-ts^2}\, ds \\
&\lesssim t^{-\frac{1+a}{2}} + m^{\frac{a}{2}} t^{-\frac{1}{2}}.
\end{aligned}$$

The lemma is proved. ∎

Proof of Theorem 2.2. By (2.5),

$$e^{-t\mathscr{H}(A,m)}(x,y) = e^{mt}\frac{t}{\sqrt{4\pi}}\int_0^\infty s^{-\frac{3}{2}}e^{-\frac{t^2}{4s}-m^2 s}e^{-sH}(x,y)\,ds$$
$$= \frac{t}{\sqrt{4\pi}}\int_0^\infty s^{-\frac{3}{2}}e^{-(\frac{t}{2\sqrt{s}}-m\sqrt{s})^2}e^{-sH}(x,y)\,ds.$$

Hence using (2.1) and the substitution

$$r = \frac{\sqrt{t}}{2\sqrt{s}}, \tag{2.12}$$

we obtain from Lemma 2.3

$$\begin{aligned}\left|e^{-t\mathscr{H}(A,m)}(x,y)\right| &\lesssim t\int_0^\infty s^{-\frac{5}{2}}e^{-(\frac{t}{2\sqrt{s}}-m\sqrt{s})^2}\,ds\\ &\lesssim t^{-\frac{1}{2}}\int_0^\infty r^2 e^{-t(r-\frac{m}{2r})^2}\,dr\\ &\lesssim t^{-1}+t^{-2}.\end{aligned} \tag{2.13}$$

This proves (2.10). Similarly, if $\kappa > 0$ and $t \geq 1$, then using equations (2.7), (2.12) and Lemma 2.3, we get

$$\begin{aligned}\left|e^{-t\mathscr{H}(A,m)}(x,y)\right| &\lesssim (1+|x|)^\kappa(1+|y|)^\kappa\frac{t}{\sqrt{4\pi}}\int_0^\infty s^{-1-\kappa}e^{-(\frac{t}{2\sqrt{s}}-m\sqrt{s})^2}\frac{ds}{s^{\frac{3}{2}}}\\ &= \frac{2^{3+2\kappa}}{\sqrt{\pi}}(1+|x|)^\kappa(1+|y|)^\kappa t^{-\frac{1}{2}-\kappa}\int_0^\infty r^{2+2\kappa}e^{-t(r-\frac{m}{2r})^2}\,dr\\ &\lesssim (1+|x|)^\kappa(1+|y|)^\kappa t^{-1-\kappa}.\end{aligned}$$

The upper bound (2.3) then follows from Lemma 2.3 and (2.13). In the same way we deduce from (2.8) that if $t \geq 1$ and $\kappa = 0$, then

$$\begin{aligned}&\left|e^{-t\mathscr{H}(A,m)}(x,y)\right|\\ &\quad\lesssim \frac{\log(2+|x|)\log(2+|y|)}{\sqrt{t}}\int_0^\infty r^2\Big[\log\Big(2+\frac{t}{4r^2}\Big)\Big]^{-2}e^{-t(r-\frac{m}{2r})^2}\,dr\\ &\quad\lesssim \log(2+|x|)\log(2+|y|)t^{-\frac{1}{2}}\Big(\int_0^{t^{\frac{1}{4}}} r^2\Big[\log\Big(2+\frac{\sqrt{t}}{4}\Big)\Big]^{-2}e^{-t(r-\frac{m}{2r})^2}\,dr\\ &\qquad\qquad + \int_{t^{\frac{1}{4}}}^\infty r^2 e^{-t(r-\frac{m}{2r})^2}\,dr\Big)\\ &\quad\lesssim \log(2+|x|)\log(2+|y|)t^{-1}\big[\log(2+t)\big]^{-2}.\end{aligned}$$

To complete the proof, it suffices to use (2.13) once more. ∎

Remark. Note that in the massive case, contrary to the case $m = 0$, the semigroup $e^{-t\mathscr{H}(A,m)}(x, y)$ exhibits different behavior for long respectively short times. Indeed, by (2.13) we have

$$\|e^{-t\mathscr{H}(A,m)}\|_{L^1 \to L^\infty} = \mathcal{O}(t^{-2}), \quad t \to 0,$$
$$\|e^{-t\mathscr{H}(A,m)}\|_{L^1 \to L^\infty} = \mathcal{O}(t^{-1}), \quad t \to \infty.$$

This is caused by the different behavior of the symbol $\sqrt{P^2 + m^2} - m$ for $P \to 0$ and for $P \to \infty$, respectively. Analogous effect occurs when $A = 0$, see [12, Section 7.11].

3 Aharonov–Bohm-type magnetic fields

In this section we consider the Aharonov–Bohm magnetic field in $\mathbb{R}^2$. The latter is generated by the vector potential A whose radial and azimuthal components (in the polar coordinates) are given by

$$A(r, \theta) = (a_1(r, \theta), a_2(r, \theta)), \quad a_1 = 0, \ a_2(r) = \left(0, \frac{\alpha}{r}\right),$$

where α is the magnetic. Note that $A \notin L^2_{\mathrm{loc}}(\mathbb{R}^2)$. We will therefore proceed in a different way than in the previous section and define the semigroup $e^{-t\sqrt{H_\alpha}}$ with the help of the partial wave decomposition. We limit ourselves to the analysis of the massless case, $m = 0$. The main result of section is stated in Theorem 3.2.

We define the Hamiltonian H_α as the Friedrichs extension of $(-i\nabla + A)^2$ on $C_0^\infty(\mathbb{R}^2 \setminus \{0\})$. In other words, H_α is the self-adjoint operator in $L^2(\mathbb{R}^2)$ generated by the closure, in $L^2(\mathbb{R}^2)$, of the quadratic form

$$Q_\alpha[u] = \int_0^{2\pi} \int_0^\infty (|\partial_r u|^2 + r^{-2}|(-i\partial_\theta + \alpha)u|^2) r \, dr \, d\theta$$

for $u \in C_0^\infty((0, \infty) \times [0, 2\pi))$.

3.1 Partial wave decomposition

Given a function $f : \mathbb{R}^2 \to \mathbb{R}$ we will often use the polar coordinate representation

$$f(x, y) = f(r, r', \theta, \theta') \iff x = r(\cos\theta, \sin\theta), y = r'(\cos\theta', \sin\theta').$$

By expanding a given function $u \in L^2(\mathbb{R}_+ \times (0, 2\pi))$ into a Fourier series with respect to the basis $\{e^{im\theta}\}_{m \in \mathbb{Z}}$ of $L^2((0, 2\pi))$, we obtain a direct sum decomposition

$$L^2(\mathbb{R}^2) = \sum_{m \in \mathbb{Z}} \oplus \mathscr{L}_m,$$

where

$$\mathcal{L}_m = \left\{ g \in L^2(\mathbb{R}^2) : g(x) = f(r)e^{im\theta} \text{ a.e.}, \int_0^\infty |f(r)|^2 r\, dr < \infty \right\}.$$

Since the vector potential A is radial, the operator H_α can be decomposed accordingly to the direct sum

$$H_\alpha = \sum_{m\in\mathbb{Z}} \oplus (h_m \otimes \mathrm{id})\Pi_m,$$

where h_m are operators generated by the closures, in $L^2(\mathbb{R}_+, r\,dr)$, of the quadratic forms

$$q_m[f] = \int_0^\infty \left(|f'|^2 + \frac{(\alpha+m)^2}{r^2} |f|^2 \right) r\, dr$$

defined initially on $C_0^\infty(0,\infty)$, and $\Pi_m : L^2(\mathbb{R}^2) \to \mathcal{L}_m$ is the projector acting as

$$(\Pi_m u)(r,\theta) = \frac{1}{2\pi} \int_0^{2\pi} e^{im(\theta-\theta')} u(r,\theta')\, d\theta'.$$

Consider now the operator

$$L_m = U h_m U^{-1} \quad \text{in } L^2(\mathbb{R}_+, dr), \tag{3.1}$$

where $U : L^2(\mathbb{R}_+, r\,dr) \to L^2(\mathbb{R}_+, dr)$ is the unitary mapping acting as

$$(Uf)(r) = r^{\frac{1}{2}} f(r).$$

Note that L_m is subject to Dirichlet boundary condition at zero and that it coincides with the Friedrichs extension of the differential operator

$$-\frac{d^2}{dr^2} + \frac{(m+\alpha)^2 - \frac{1}{4}}{r^2}$$

defined on $C_0^\infty(\mathbb{R}_+)$. From [10, Section 5] we know that

$$W_m L_m W_m^{-1} \varphi(p) = p\varphi(p), \quad \varphi \in W_m(D(L_m)), \tag{3.2}$$

where the mappings

$$W_m, W_m^{-1} : L^2(\mathbb{R}_+) \to L^2(\mathbb{R}_+)$$

given by

$$(W_m u)(p) = \int_0^\infty u(r)\sqrt{r}\, J_{|m+\alpha|}(r\sqrt{p})\, dr,$$

$$(W_m^{-1}\varphi)(r) = \frac{1}{2}\int_0^\infty \varphi(p)\sqrt{r}\, J_{|m+\alpha|}(r\sqrt{p})\, dp$$

extend to unitary operators on $L^2(\mathbb{R}_+)$.

3.2 The semigroup $e^{-t\sqrt{H_\alpha}}$

By the spectral theorem,

$$e^{-t\sqrt{H_\alpha}} = \sum_{m\in\mathbb{Z}} \oplus \big(e^{-t\sqrt{h_m}} \otimes \mathrm{id}\big)\Pi_m. \tag{3.3}$$

Denote by $p_m(r,r',t)$ the integral kernel of $e^{-t\sqrt{h_m}}$ in $L^2(\mathbb{R}_+, r\,dr)$. From (3.1) it follows that

$$p_m(r,r',t) = \frac{1}{\sqrt{rr'}} e^{-t\sqrt{L_m}}(r,r').$$

On the other hand, in view of (3.2) we get

$$\begin{aligned}\big(e^{-t\sqrt{L_m}} g\big)(r) &= \big(W_m^{-1} e^{-t\sqrt{p}} W_m g\big)(r)\\ &= \frac{1}{2}\int_0^\infty \sqrt{rr'} \int_0^\infty e^{-t\sqrt{p}} J_{|m+\alpha|}(r\sqrt{p}) J_{|m+\alpha|}(r'\sqrt{p})\, dp\, g(r')\, dr'.\end{aligned}$$

Hence

$$\begin{aligned}p_m(r,r',t) &= \frac{1}{2}\int_0^\infty e^{-t\sqrt{p}} J_{|m+\alpha|}(r\sqrt{p}) J_{|m+\alpha|}(r'\sqrt{p})\, dp\\ &= \int_0^\infty e^{-tp} J_{|m+\alpha|}(rp) J_{|m+\alpha|}(r'p)\, p\, dp.\end{aligned}$$

In order to simplify the notation we will use in the sequel the shorthand

$$z := \frac{r^2}{t^2}. \tag{3.4}$$

From [4, equation 4.14.(16)] we then get the explicit expression for p_m on the diagonal

$$\begin{aligned}p_m(r,r,t) &= \int_0^\infty e^{-tp} J^2_{|m+\alpha|}(rp)\, p\, dp \qquad (3.5)\\ &= \frac{4^\nu}{\pi}(2\nu+1)\, t^{-2}\Big(\frac{r}{t}\Big)^{2\nu} \frac{\Gamma^2(\nu+\frac{1}{2})}{\Gamma(2\nu+1)} F\Big(\nu+\frac{1}{2}, \nu+\frac{3}{2}, 2\nu+1; -4z\Big),\end{aligned}$$

where

$$\nu = |m+\alpha| \tag{3.6}$$

and $F(a,b,c;w)$ denotes the Gauss hypergeometric series, see for example [1, equation 15.1.1]. Using its integral representation

$$F(a,b,c;w) = \frac{\Gamma(c)}{\Gamma(b)\Gamma(c-b)} \int_0^1 s^{b-1}(1-s)^{c-b-1}(1-sw)^{-a}\, ds,$$

where $\operatorname{Re} c > \operatorname{Re} b > 0$, see [1, equation 15.3.1], in combination with the transformation formula [1, equation 15.3.5]

$$F(a,b,c;w) = (1-w)^{-b} F\Big(b, c-a, c; \frac{w}{w-1}\Big),$$

we find that

$$
\begin{aligned}
& p_m(r,r,t) \\
&= \frac{2\nu+1}{\pi t^2}(4z)^\nu \frac{\Gamma^2(\nu+\frac12)}{\Gamma(2\nu+1)}(1+4z)^{-\nu-\frac32} F\Big(\nu+\frac32, \nu+\frac12, 2\nu+1; \frac{4z}{4z+1}\Big) \\
&= \frac{2\nu+1}{\pi t^2}(4z)^\nu (1+4z)^{-\nu-\frac32}\int_0^1 s^{\nu-\frac12}(1-s)^{\nu-\frac12}\Big(1-\frac{4zs}{4z+1}\Big)^{-\nu-\frac32} ds \\
&= \frac{2\nu+1}{\pi t^2}(4z)^\nu \int_0^1 s^{\nu-\frac12}(1-s)^{\nu-\frac12}(1+4z(1-s))^{-\nu-\frac32}\, ds \\
&= \frac{2\nu+1}{\pi t^2}(4z)^\nu \int_0^1 s^{\nu-\frac12}(1-s)^{\nu-\frac12}(1+4zs)^{-\nu-\frac32}\, ds. \qquad (3.7)
\end{aligned}
$$

We have:

Lemma 3.1. *There exists $\varepsilon_0 > 0$ such that for every $\varepsilon \in (0, \varepsilon_0)$ there is $C_\varepsilon > 0$ for which the upper bound*

$$
(1+r)^{-\frac32-\varepsilon}(1+r')^{-\frac32-\varepsilon} p_m(r,r',t) \le C_\varepsilon \frac{t^{-2-2\kappa}}{(|m+\alpha|+1)^{1+\varepsilon}}
$$

holds for all $m \in \mathbb{Z}$ and all $t \ge 1$.

Proof. Put

$$
\varepsilon_0 := \min\Big\{|m+\alpha| - \frac32 : m \in \mathbb{Z} \wedge |m+\alpha| > \frac32\Big\}. \qquad (3.8)
$$

Clearly we have $\varepsilon_0 > 0$. Let $\varepsilon < \varepsilon_0$ and put $\rho = \frac32 + \varepsilon$. Keeping in mind the notation (3.6), we will distinguish two cases depending on the value of ν.

Assume first that $\nu > \frac32$. In view of (3.4) and (3.7),

$$
\begin{aligned}
\frac{p_m(r,r,t)}{(1+r^2)^\rho} &= \frac{4\nu+2}{\pi t^3}\frac{r(4z)^{\nu-\frac12}}{(1+r^2)^\rho}\int_0^1 s^{\nu-\frac12}(1-s)^{\nu-\frac12}(1+4zs)^{-\nu-\frac32}\, ds \\
&\lesssim t^{-3}\nu(4z)^{\nu-\rho}\int_0^1 s^{\nu-\frac12}(1-s)^{\nu-\frac12}(1+4zs)^{-\nu-\frac32}\, ds \\
&= t^{-3}\nu(4z)^{\nu-\rho}\int_0^1 s^{\nu-\frac12}(1+4zs)^{\rho-\nu}(1-s)^{\nu-\frac12}(1+4zs)^{-\rho-\frac32}\, ds,
\end{aligned}
$$

where we have used the fact that $z \le r^2$ by assumption. From (3.8) it follows that $\rho < \nu$. Hence

$$
\begin{aligned}
(1+r^2)^{-\rho} p_m(r,r,t) &\lesssim t^{-3}\nu(4z)^{\nu-\rho}\int_0^1 s^{\nu-\frac12}(4zs)^{\rho-\nu}(1-s)^{\nu-\frac12}\, ds \\
&= t^{-3}\nu\int_0^1 s^{\rho-\frac12}(1-s)^{\nu-\frac12}\, ds = t^{-3}\nu B\Big(\rho+\frac12, \nu+\frac12\Big) \\
&= t^{-3}\nu\frac{\Gamma(\rho+\frac12)\Gamma(\nu+\frac12)}{\Gamma(\nu+\rho+1)},
\end{aligned}
$$

where $B(\cdot,\cdot)$ denotes the Euler beta function. Moreover, by the Stirling formula, see e.g. [1, equation 6.1.37], we have

$$\frac{\nu\Gamma(\nu+\frac{1}{2})}{\Gamma(\nu+\rho+1)} \sim \nu^{-\rho+\frac{1}{2}} = \nu^{-1-\varepsilon}, \quad \nu \to \infty.$$

Therefore there exists a constant C_1 such that

$$(1+r^2)^{-\rho} p_m(r,r,t) \le C_1\, t^{-3} \nu^{-1-\varepsilon} \quad \text{for all } \nu > \frac{3}{2}. \tag{3.9}$$

Now let $0 \le \nu \le \frac{3}{2}$. In this case we have $\nu \le \rho$ and (3.7) thus implies that

$$(1+r^2)^{-\rho} p_m(r,r,t) \le \frac{32}{\pi t^2}(1+r)^{-2\rho}\frac{r^{2\nu}}{t^{2\nu}} \le \frac{32}{\pi} t^{-2-2\kappa},$$

since $\nu \ge \kappa$ by definition. This together with (3.6) and (3.9) gives

$$(1+r^2)^{-\rho} p_m(r,r,t) \le C_2 t^{-2-2\kappa}(|m+\alpha|+1)^{-1-\varepsilon} \quad \text{for all } t \ge 1$$

holds for all ν and some C_2. To complete the proof, it suffices to use the semigroup property of $e^{-t\sqrt{h_m}}$, which implies that

$$p_m(r,r',t) \le \sqrt{p_m(r,r,t)\, p_m(r',r',t)},$$

as desired. ∎

As a consequence of Lemma 3.1 and equations (3.3), (3.5) we obtain:

Theorem 3.2. *Let ε_0 be given by* (3.8). *Then for every $\varepsilon \in (0,\varepsilon_0)$ there exists a constant K_ε such that*

$$\|(1+|x|)^{-\frac{3}{2}-\varepsilon} e^{-t\sqrt{H_\alpha}} (1+|x|)^{-\frac{3}{2}-\varepsilon}\|_{L^1(\mathbb{R}^2)\to L^\infty(\mathbb{R}^2)} \le K_\varepsilon\, t^{-2-2\kappa}.$$

Moreover,

$$e^{-t\sqrt{H_\alpha}}(x,x) = \frac{t^{-2}}{2\pi^2} \sum_{m\in\mathbb{Z}} 4^\nu (2\nu+1)\Big(\frac{|x|}{t}\Big)^{2\nu} \frac{\Gamma^2(\nu+\frac{1}{2})}{\Gamma(2\nu+1)} \times F\Big(\nu+\frac{1}{2}, \nu+\frac{3}{2}, 2\nu+1; -\frac{4|x|^2}{t^2}\Big),$$

where $\nu = |m+\alpha|$.

Remark. The decay rate $t^{-2-2\kappa}$ in Theorem 3.2 is sharp. This follows from (3.3) and the asymptotic behavior of $p_m(r,r,t)$ as $t\to\infty$. Indeed, let $k\in\mathbb{Z}$ be such that $\kappa = |k+\alpha|$. Equation (3.7) then implies that

$$\lim_{t\to\infty} t^{2+2\kappa} p_k(r,r,t) = \frac{2\kappa+1}{\pi}(2r)^{2\kappa} B\Big(\kappa+\frac{1}{2}, \kappa+\frac{1}{2}\Big).$$

It should be also noted that if $\alpha \in \mathbb{Z}$, in which case $\kappa = 0$, then the operator H_α is unitarily equivalent to the Laplacian in $L^2(\mathbb{R}^2)$ which satisfies

$$\|e^{-t\sqrt{-\Delta}}\|_{L^1 \to L^\infty} = \frac{1}{2\pi t^2}$$

for all $t > 0$, see (1.1).

Acknowledgements. The author would like to thank Gabriele Grillo for useful discussions.

Bibliography

[1] M. Abramowitz and I. Stegun, *Handbook of mathematical functions with formulas, graphs, and mathematical tables*. 10th edn., John Wiley & Sons, New York, 1972

[2] J. Avron, I. Herbst and B. Simon, Schrödinger operators with magnetic fields. I. General interactions. *Duke Math. J.* **45** (1978), 847–883

[3] K. Broderix, D. Hundertmark and H. Leschke, Continuity properties of Schrödinger semigroups with magnetic fields. *Rev. Math. Phys.* **12** (2000), 181–225

[4] A. Erdélyi, W. Magnus, F. Oberhettinger and F. G. Tricomi, *Tables of integral transforms. Vol. 2*. McGraw–Hill, New York, 1954

[5] H. Hess, R. Schrader and D. A. Uhlenbrock, Domination of semigroups and generalization of Kato's inequality. *Duke Math. J.* **44** (1977), 893–904

[6] F. Hiroshima, T. Ichinose and J. Lőrinczi, Path integral representation for Schrödinger operators with Bernstein functions of the Laplacian. *Rev. Math. Phys.* **24** (2012), Article ID 1250013

[7] D. Hundertmark and B. Simon, A diamagnetic inequality for semigroup differences. *J. Reine Angew. Math.* **571** (2004), 107–130

[8] N. Jacob, *Pseudo-differential Operators & Markov Processes*. Imperial College Press, London, 2001

[9] T. Kato, Remarks on Schrödinger operators with vector potentials. *Integral Equations Operator Theory* **1** (1978), 103–113

[10] H. Kovařík, Heat kernels of two-dimensional magnetic Schrödinger and Pauli operators. *Calc. Var. Partial Differential Equations* **44** (2012), 351–374

[11] A. Laptev and T. Weidl, Hardy inequalities for magnetic Dirichlet forms. In *Mathematical results in quantum mechanics (Prague, 1998)*, pp. 299–305, Oper. Theory Adv. Appl. 108, Birkhäuser, Basel, 1999

[12] E. H. Lieb and M. Loss, *Analysis*. 2nd edn., Grad. Stud. Math. 14, American Mathematical Society, Providence, 2001

[13] B. Simon, Schrödinger operators with singular magnetic vector potentials. *Math. Z.* **131** (1973), 361–370

A version of Watson lemma for Laplace integrals and some applications

Stanislas Kupin and Sergey Naboko†

To Ari Laptev, a mathematician and a personality

Let $f : \mathbb{R}_+ \to \mathbb{C}$ be a bounded measurable function. Suppose that $f(t) \to 0$ at logarithmic (or k-logarithmic) rate as $t \to 0+$. We consider the Laplace integral of the function f, i.e.,

$$I_n = \int_0^\infty f(t)e^{-nt}\, dt$$

and obtain its asymptotics for $n \to +\infty$, which is a version of the classical Watson's lemma for the integral. Actually, the result is proved for a larger class of "slowly oscillating" functions satisfying some mild regularity conditions.

1 Introduction

One cannot over-estimate the rôle of asymptotic analysis in modern mathematics. We quote a short presentation of the topic from the book by Bleistein and Handelsman [1, p. vii]: "Asymptotic analysis is that branch of mathematics devoted to the study of the behavior of functions in particular limits of interest. [...] Although the subject matter might appear narrow at first glance, in actuality its scope is quite large, and it is particularly relevant to applied mathematics. Indeed, the solutions to a large class of applied problems can, by means of integral transforms, be represented by definite integrals. Exact numerical values are often difficult to obtain from such representations, in which event one must resort to the method of approximation."

A particular aspect of the described activity is the study of asymptotic behavior of Laplace-type integrals, and it is also the theme of the present article. Probably, one of the most classical results in this direction is the so-called Watson's lemma from 1918,

†Sergey Naboko passed away on December 24, 2020.

Keywords: Asymptotic behavior of a Laplace integral, generalized Watson lemma, special class of functions "slowly decaying to zero" at the origin
2020 Mathematics Subject Classification: Primary 47B35; secondary 30H20, 42C10

see [8]. We give its simplest formulation in the "real" case (i.e., the asymptotic behavior of the integral is studied for the real values of the parameter).

Before going to the formulation of the lemma, we recall some basic notions of asymptotic analysis. Let $f, g : I := [a,b] \to \mathbb{C}$ be two functions, and $t_0 \in (a,b)$, $-\infty \le a < b \le +\infty$. First, we write $f(t) = O(g(t))$ as $t \to t_0$ if and only if

$$|f(t)| \le C|g(t)|, \quad t \to t_0.$$

Second, $f(t) = o(g(t))$ for $t \to t_0$ if and only if

$$\lim_{t\to t_0} \frac{|f(t)|}{|g(t)|} = 0.$$

Of course, the above definitions are rewritten correspondingly when $t_0 = a, b$ (i.e., for $t_0 = a$, we take $t \to a + 0$ instead of $t \to t_0$, etc.). Third, let $t_0 = a$ and $\{a_m\}_{m\in\mathbb{Z}_+}$, $a_m \ge 0$, be a (strictly) increasing sequence. We say that f admits an asymptotic expansion

$$f(t) \sim \sum_{m=0}^{\infty} c_m (t - t_0)^{a_m}, \quad c_m \in \mathbb{C}, \tag{1.1}$$

on the right neighborhood of t_0 (i.e., $t \to t_0 + 0$) if and only if for any natural N,

$$f(t) = \sum_{m=0}^{N} c_m (t - t_0)^{a_m} + o((t - t_0)^{a_N}).$$

The natural rewriting of (1.1) for $t_0 = b$ or $t_0 \in (a,b)$ is also obvious. Notice that the above definition makes a perfect sense even if the series $\sum_{m=0}^{\infty} c_m (t - t_0)^{a_m}$ is divergent.

These definitions at hand, let $f : \mathbb{R}_+ \to \mathbb{C}$ be a locally integrable function satisfying two conditions:

(1) there is an $a > 0$ such that $|f(t)| = O(e^{at})$, where $t \to +\infty$,

(2) we have

$$f(t) \sim \sum_{m=0}^{\infty} c_m t^{a_m}, \quad t \to 0+, \tag{1.2}$$

where $(a_m)_{m\in\mathbb{Z}_+}, a_m \ge 0$, is a (strictly) increasing sequence.

The original version of the below theorem is in [8], see also Bleistein and Handelsman [1, Theorem in Section 4.1], Simon [7, Section 15.2] for a modern presentation.

Theorem 1.1 (Watson's lemma). *Let the function f satisfy conditions* (1)–(2). *Consider the integral*

$$I_n = \int_0^{\infty} e^{-nt} f(t)\, dt. \tag{1.3}$$

Then

$$I_n \sim \sum_{m=0}^{\infty} \frac{c_m \Gamma(a_m + 1)}{n^{a_m+1}}, \quad n \to +\infty,$$

where $\Gamma(\cdot)$ is the Euler gamma-function.

This result is widely used for obtaining explicit asymptotics of Laplace integrals. It is quite natural that it has a number of different versions and generalizations. For instance, the asymptotics of integral (1.3) in certain angular regions of complex plane are given in Copson [3], Simon [7] and Wong [9]; this is the so-called "complex case" of the lemma. Bleistein and Handelsman [1, Sections 8.2 and 8.3] and Simon [7, Theorem 15.2.2] contain also the generalization of the result to several dimensions. In [1, Chapters 4–5], Watson's lemma is viewed as a rather particular case of more general results on Mellin transforms in the complex plane.

The point is that all widely known versions of Watson's lemma deal with the case when the function f admits a power asymptotic expansion on the (right) neighborhood of point $t_0 = 0$, see (1.2). The powers appearing in this expansion do not need to be necessarily natural, see [7, Theorem 15.2.7] and [9, Theorems in Section I.5].

We are interested in a substantially different situation, where the function f admits an expansion of powers of logarithmic-type functions looking as

$$f(t) \sim \sum_{m=0}^{\infty} c_m \left(\frac{1}{\log_{*k}(\frac{1}{t})} \right)^{a_m}, \quad t \to 0+, \tag{1.4}$$

see (3.1) for the definition of the function $\log_{*k}$. Clearly, the terms of this asymptotic expansion decay much slower than the terms of the "classical" expansion (1.2).

For instance, the article by Erdélyi [5] studies, among other cases, the behavior of Laplace integral of a function f of the form (1.4) admitting an asymptotic expansion in terms of the form $(\log(\frac{1}{|t|}))^{-a_m}$, $a_m > 0$, in a neighborhood of the origin. In this paper, $k = 1$ and this is barely the only reference we are aware of regarding the Laplace integrals for functions f of the form similar to (1.4).

In particular, there are no results of this kind in the standard literature, see [1–4,7,9], nor there seems to be a straightforward way to derive them from known claims. This is the *first* reason for writing this short note for us. The *second* reason is an interesting application of the obtained results to the study of compact Toeplitz operators on Bergman spaces and banded (Jacobi) matrices on Hilbert spaces, see Koita, Kupin, Naboko and Touré [6] in this connection. Hence, the problem is to understand if there is a counterpart of Watson's results for functions f admitting expansions similar to(1.4) at $t_0 = 0$.

We present a large class of functions satisfying rather mild conditions for which a counterpart of Theorem 1.1 holds. Observe that it is much larger than the class treated in [5], see assumptions (A1)–(A3) below. In a quite interesting manner, the terms of asymptotic expansion of these functions can decay to zero at $t_0 = 0$ very slowly. For instance, the above asymptotic expansions (1.4) in terms of $\log_{*k}$ are admissible from this point of view.

We need one more definition now. Suppose that $f, g : I \times J := [a,b] \times [c,d] \to \mathbb{R}$ are (positive) functions depending on two variables $(x,t) \in I \times J$. We write

$$f(x,\cdot) = \diamond(g(x,\cdot)) = \diamond_t(g(x,\cdot))$$

if and only if there is a constant C with the property

$$|f(x,t)| \le C|g(x,t)|$$

for all $(x,t) \in I \times J$. Saying this a bit differently, $f(x,\cdot)$ is an O-function of $g(x,\cdot)$ *uniformly in* t. The variable of uniform dependence is sometimes indicated as a subindex of the symbol $\diamond$ (i.e., $\diamond_t$ in the present situation).

Let $\varphi : \mathbb{R}_+ \to \mathbb{C}$ be a measurable function. Suppose *it is positive in a right neighborhood of* 0 and it meets the following **assumptions**:

(A1) $\varphi \in L^\infty(\mathbb{R}_+)$, and $\lim_{x\to 0+} \varphi(x) = 0$,

(A2) there are $\delta_0 = \delta_0(\varphi) > 0$ and $\epsilon_0 = \epsilon_0(\varphi) > 0$ such that for any $0 \le \epsilon < \epsilon_0$,

$$\varphi(x^{1+\epsilon}) = \varphi(x)(1 + \diamond(\epsilon)) = \varphi(x)(1 + \diamond_x(\epsilon)), \tag{1.5}$$

where $0 \le x < \delta_0$,

(A3) there holds

$$\lim_{x\to 0+} \frac{\log \varphi(x)}{\log x} = 0. \tag{1.6}$$

We continue with several remarks related to theses conditions. Without loss of generality, we can assume that $\phi(x) \ge 0$ for $0 \le x < \delta_0$.

Remark. We make the following observations.

(1) The function φ is continuous for $0 \le x < \delta_0$ by (A2).

(2) Suppose that there is an x', $0 \le x' < \delta_0$, such that $\varphi(x') = 0$. Then, by assumption (A2) once again, φ is identically zero in $[0, x']$.

The last point of this remark says that we can assume for our purposes that $\varphi(x) > 0$, $0 < x < \delta_0$.

Remark. A simple change of variables in (1.5) shows that assumption (A2) holds for negative ϵ, $-\epsilon_0 < \epsilon \le 0$, as well.

Remark. A rather simple scaling argument gives that assumptions (A1) and (A2) yield assumption (A3). More precisely, assuming that $|\diamond_x(\epsilon)| \le C_0\epsilon$ in (1.5), we have

$$C\left(\log\left(\frac{1}{x}\right)\right)^{-2C_0} \le \varphi(x)$$

in a right neighborhood of the point $x_0 = 0$, see Lemma 2.1 in this connection.

Roughly speaking, assumption (A2) above says that the function φ is "slowly oscillating" when its argument changes in power scale. Condition (A3) claims that $\varphi(x)$ tends to 0 when $x \to 0+$ slower than any power x^α, $\alpha > 0$. We prefer to keep assumption (A3) "as it is" for the transparency of presentation.

We are interested in the asymptotic behavior of the following Laplace integral:

$$I_n := \int_0^\infty e^{-ny}\varphi(y)\,dy. \tag{1.7}$$

Theorem 1.2. *Let φ be a function satisfying above conditions* (A1)–(A3). *We have*

$$I_n = \frac{1}{n}\varphi\left(\frac{1}{n}\right)(1 + o(1)), \quad n \to +\infty.$$

One can notice that hypothesis (A1) on φ yields that the integral

$$\int_{\delta_0}^{\infty} e^{-ny}\varphi(y)\,dy$$

goes to zero at exponential rate, and Lemma 2.1 shows that it does not contribute to the leading term of asymptotics of I_n we are interested in. Moreover, assumption (A1) on the function φ can be weakened. Namely, we can replace it with the following relation:

(A1') the function φ is locally bounded; for a fixed $a > 0$ we have $\varphi(x) = O(e^{ax})$ as $x \to +\infty$ and $\lim_{x\to 0+}\varphi(x) = 0$ exists.

One can prove a counterpart of Theorem 1.1 in the scale of powers of φ obeying conditions (A1)–(A3).

Theorem 1.3 (a version of Watson's lemma). *Let φ be as in assumptions* (A1)–(A3) *and let $f : \mathbb{R}_+ \to \mathbb{C}$ be a locally integrable function having the properties:*

(1) *there is an $a > 0$ such that $|f(t)| = O(e^{at})$ for $t \to +\infty$,*

(2) *we have*

$$f(t) \sim \sum_{m=0}^{\infty} c_m \varphi(t)^{a_m}, \quad t \to 0+,$$

where $\{a_m\}_{m\in\mathbb{Z}_+}, a_m \geq 0$, is a (strictly) increasing sequence.

Then

$$I_n = \int_0^{\infty} e^{-nt} f(t)\,dt \sim \sum_{m=0}^{\infty} \frac{c_m}{n}\varphi\left(\frac{1}{n}\right)^{a_m}, \quad n \to +\infty.$$

The passing from Theorem 1.2 to Theorem 1.3 is quite elementary and follows closely the proof of [1, Theorem in Section 4.1]. It uses the crucial fact that assumption (A2) is invariant with respect to the transformation $\varphi \mapsto \varphi^{\gamma}, \gamma > 0$.

The meaningful constants are numbered as C_1, C_2, etc. Inessential constants are denoted by c, C and change from one relation to another.

2 Proofs of the results

Lemma 2.1. *Assume that the function φ satisfies condition* (A2). *Let $|\diamondsuit_x(\epsilon)| \leq C_0\epsilon$ for some $C_0 > 0$, see* (1.5). *Then we have*

$$C\left(\log\left(\frac{1}{x}\right)\right)^{-2C_0} \leq \varphi(x) \tag{2.1}$$

in a right neighborhood of the origin. Above,

$$C := \alpha\left((1+\epsilon)\log\left(\frac{1}{\delta_0}\right)\right)^{2C_0},$$

where α is defined in (2.2) *below.*

Proof. For a $x \in (0,1]$, we can rewrite property (A2) "backwards" (see (1.5)), i.e.,

$$\varphi(x) = \varphi(x^{\frac{1}{1+\epsilon}})(1 + \Diamond(\epsilon)).$$

Suppose without loss of generality that $C_0\epsilon \le \frac{1}{2}$ and $\delta_0 < 1$. Set

$$\alpha := \inf_{x\in[\delta_0,1]} \varphi(x) > 0 \tag{2.2}$$

by our assumptions on the function φ. Now, we take an arbitrary $x \in (0, \delta_0]$, and define a sequence

$$x_0 := x, \quad x_n = x^{\frac{1}{(1+\epsilon)^n}}.$$

By construction, $x_{n-k} = x_n^{(1+\epsilon)^k}$, $k = 0, \dots, n$. It is also clear that $\{x_n\}_n$ is strictly increasing, and $\lim_{n\to+\infty} x_n = 1$.

Pick n so that $x_{n-1} \le \delta_0 < x_n$. That is,

$$x^{\frac{1}{(1+\epsilon)^{n-1}}} \le \delta_0 < x^{\frac{1}{(1+\epsilon)^n}} \tag{2.3}$$

or

$$n = \left\lfloor \frac{1}{\log(1+\epsilon)} \log\frac{\log x}{\log \delta_0} \right\rfloor + 1, \tag{2.4}$$

where $\lfloor\cdot\rfloor$ is the integer part of a real number. On the other hand, we see by (A2) that

$$\begin{aligned}\varphi(x) = \varphi(x_0) &\ge \varphi(x_1)(1 - C_0\epsilon) \ge \varphi(x_2)(1 - C_0\epsilon)^2 \\ &\ge \dots \ge \varphi(x_n)(1 - C_0\epsilon)^n \ge \alpha(1 - C_0\epsilon)^n,\end{aligned}$$

and, taking into account (2.3) and (2.4),

$$\begin{aligned}(1 - C_0\epsilon)^n &\ge \frac{1}{(1+2C_0\epsilon)^n} \ge \frac{1}{(1+\epsilon)^{2C_0 n}} \\ &\ge \left(\frac{(1+\epsilon)\log\delta_0}{\log x}\right)^{2C_0} = \left(\frac{(1+\epsilon)\log(\frac{1}{\delta_0})}{\log(\frac{1}{x})}\right)^{2C_0}.\end{aligned}$$

The lemma follows. ∎

Notice that with a little more effort one can improve (2.1) up to

$$C\left(\log\left(\frac{1}{x}\right)\right)^{-C_0} \le \varphi(x)$$

in an appropriate right neighborhood of the origin. The latter bound is sharp and cannot be improved.

Proof of Theorem 1.2. As follows from the discussion preceding the formulation of the Theorem, see Remarks following assumptions (A1)–(A3), we suppose without loss of generality that $\varphi : \mathbb{R}_+ \to \mathbb{C}$ *is continuous and strictly positive on the interval* $(0, \delta_0)$.

Consider integral (1.7). We start making the change of variable $y = \frac{t}{n}$. We take two parameters $0 < \eta \ll 1$ and $0 < \epsilon < \epsilon_0$ and fix them. We shall have $\epsilon = \epsilon(\eta)$ and their choice will be made precise later, see (2.12). The integral I_n is divided in three pieces as

$$\begin{aligned} I_n &= \frac{1}{n}\int_0^\infty e^{-t}\varphi\Big(\frac{t}{n}\Big)\,dt \\ &= \frac{1}{n}\bigg\{\int_0^{n^{-\epsilon}} e^{-t}\varphi\Big(\frac{t}{n}\Big)\,dt + \int_{n^{-\epsilon}}^{n^{\epsilon}} e^{-t}\varphi\Big(\frac{t}{n}\Big)\,dt + \int_{n^{\epsilon}}^{\infty} e^{-t}\varphi\Big(\frac{t}{n}\Big)\,dt\bigg\} \\ &=: \frac{1}{n}\{J_{1n} + J_{2n} + J_{3n}\}. \end{aligned}$$

Condition (A3) (see (1.6)) implies that

$$\frac{\log\varphi(x)}{\log x} = c(x), \qquad \lim_{x\to 0+} c(x) = 0, \tag{2.5}$$

or $\varphi(x) = x^{c(x)}$ for $x \to 0+$. Consequently, for a $0 < \epsilon < \epsilon_0$ there is a $\delta = \delta(\epsilon) > 0$ such that $0 \le c(x) < \epsilon$ for $x \in [0, \delta)$. It is convenient to choose $0 < \delta < \delta_0$, see (1.5). In particular, there is an $N_0 = N_0(\epsilon)$ large enough so that for $n \ge N_0$ we have $\frac{1}{n^{1-\epsilon}} < \delta$. In particular, $\frac{1}{n} < \frac{1}{n^{1-\epsilon}} < \delta$, and so $0 \le c(\frac{1}{n}) < \epsilon$. Hence we see that

$$\varphi\Big(\frac{1}{n}\Big) = \Big(\frac{1}{n}\Big)^{c(\frac{1}{n})} \ge \Big(\frac{1}{n}\Big)^{\epsilon}. \tag{2.6}$$

For the integral J_{3n} and $n \ge N_0$, we obtain

$$\begin{aligned} |J_{3n}| &= \bigg|\int_{n^{\epsilon}}^{\infty} e^{-t}\varphi\Big(\frac{t}{n}\Big)\,dt\bigg| \\ &\le \|\varphi\|_{L^\infty(\mathbb{R}_+)}\int_{n^{\epsilon}}^{\infty} e^{-t}\,dt \\ &= \|\varphi\|_{L^\infty(\mathbb{R}_+)}e^{-n^{\epsilon}} \le \|\varphi\|_{L^\infty(\mathbb{R}_+)}\frac{1}{\frac{n^{2\epsilon}}{2!}} \\ &= 2\|\varphi\|_{L^\infty(\mathbb{R}_+)}\varphi\Big(\frac{1}{n}\Big)^2 = \varphi\Big(\frac{1}{n}\Big)\cdot R_{3n}, \end{aligned} \tag{2.7}$$

where $R_{3n} := 2\|\varphi\|_{L^\infty(\mathbb{R}_+)}\varphi(\frac{1}{n}) \to 0$ as $n \to +\infty$. Pick $N_3 = N_3(\varphi, \eta)$ so that we get for $n \ge N_3$,

$$R_{3n} = 2\|\varphi\|_{L^\infty(\mathbb{R}_+)}\varphi\Big(\frac{1}{n}\Big) < \frac{\eta}{6}. \tag{2.8}$$

Set $N_3' = N_3'(\varphi, \epsilon, \eta) := \max\{N_0, N_3\}$.

Again, we suppose $n \geq N_0$ defined above, so that relation (2.6) holds. We have for the bound on J_{1n},

$$\begin{aligned} |J_{1n}| &= \left| \int_0^{n^{-\epsilon}} e^{-t} \varphi\left(\frac{t}{n}\right) dt \right| \leq \frac{1}{n^{\epsilon}} \cdot \sup_{\tau \in (0, n^{-(1+\epsilon)})} |\varphi(\tau)| \\ &\leq \frac{1}{n^{\epsilon}} \cdot \sup_{\tau \in (0, \frac{1}{n})} |\varphi(\tau)| \leq \varphi\left(\frac{1}{n}\right) \cdot R_{1n}, \end{aligned} \tag{2.9}$$

where

$$R_{1n} := \sup_{\tau \in (0, \frac{1}{n})} |\varphi(\tau)| \to 0 \quad \text{as } n \to +\infty$$

by property (1) of the function φ. We pick $N_1 = N_1(\varphi, \eta)$ so that $R_{1n} < \frac{\eta}{6}$ for $n \geq N_1$. Set $N_1' = N_1'(\varphi, \epsilon, \eta) := \max\{N_0, N_1\}$.

The considerations pertaining to the bound on J_{2n} are slightly more involved. Indeed, for $t \in [n^{-\epsilon}, n^{\epsilon}]$, and $x = \frac{t}{n}$, we have $x \in [n^{-1-\epsilon}, n^{-1+\epsilon}]$. That is,

$$x = \frac{1}{n}\left(\frac{1}{n}\right)^{\gamma},$$

where $\gamma = \gamma_x \in [-\epsilon, \epsilon]$. For $x = \frac{t}{n}$ and $n \geq N_0$, we infer

$$x = \frac{t}{n} = \left(\frac{1}{n}\right)^{1+\gamma} \leq n^{-(1-\epsilon)} < \delta < \delta_0,$$

so that $x = \frac{t}{n} \in [0, \delta)$, $\delta < \delta_0$. Here, we used the definitions of N_0 and δ, see two lines above relation (2.6). Condition (A2) on φ (see (1.5)) implies that

$$\begin{aligned} \varphi(x) &= \varphi\left(\frac{t}{n}\right) = \varphi\left(\frac{1}{n}\right)(1 + \diamond_t(\gamma)) = \varphi\left(\frac{1}{n}\right)(1 + \diamond_t(\epsilon)) \\ &= \varphi\left(\frac{1}{n}\right)(1 + \psi_{\epsilon}(t)), \end{aligned} \tag{2.10}$$

and $|\psi_{\epsilon}(t)| \leq C_1 \epsilon$ for $t \in [0, n^{\epsilon})$. Here, the constant C_1 depends on δ_0, but not on t. We continue as

$$\begin{aligned} \left|J_{2n} - \varphi\left(\frac{1}{n}\right)\right| &= \left| \int_{n^{-\epsilon}}^{n^{\epsilon}} e^{-t} \varphi\left(\frac{t}{n}\right) dt - \varphi\left(\frac{1}{n}\right) \right| \\ &= \left| \int_{n^{-\epsilon}}^{n^{\epsilon}} e^{-t} \varphi\left(\frac{1}{n}\right) [1 + \psi_{\epsilon}(t)]\, dt - \varphi\left(\frac{1}{n}\right) \right| \\ &\leq \left| \varphi\left(\frac{1}{n}\right) \left(\int_{n^{-\epsilon}}^{n^{\epsilon}} e^{-t}\, dt - 1 \right) \right| + \left| \varphi\left(\frac{1}{n}\right) \int_{n^{-\epsilon}}^{n^{\epsilon}} e^{-t} \psi_{\epsilon}(t)\, dt \right| \\ &\leq \varphi\left(\frac{1}{n}\right) \left\{ |e^{-n^{-\epsilon}} - e^{-n^{\epsilon}} - 1| + C_1 \epsilon \right\} \\ &\leq \varphi\left(\frac{1}{n}\right) \cdot R_{2n}, \end{aligned} \tag{2.11}$$

where

$$R_{2n} := \{|e^{-n^{-\epsilon}} - 1| + e^{-n^{\epsilon}} + C_1\epsilon\}.$$

For a given $\eta > 0$, we fix $0 < \epsilon < \epsilon_0$ so that

$$C_1\epsilon < \frac{\eta}{6}. \tag{2.12}$$

Choose also $N_2 = N_2(\varphi, \epsilon)$ with the property $R_{2n} < \eta/3$ for $n \geq N_2$. As before, set $N_2' = N_2'(\varphi, \epsilon, \eta) = \max\{N_0, N_2\}$.

Putting together computations (2.7)–(2.11), we obtain

$$\left|I_n - \frac{1}{n}\varphi\left(\frac{1}{n}\right)\right| \leq \frac{1}{n}\varphi\left(\frac{1}{n}\right)(R_{1n} + R_{2n} + R_{3n}).$$

Now, let an $\eta > 0$ be given. Take $\epsilon > 0$ defined by (2.12) and $\delta = \delta(\epsilon)$ defined in the line following (2.5). Considering $n \geq M := \max\{N_1', N_2', N_3'\}$, see the lines below relations (2.8), (2.9), (2.12), we see

$$(R_{1n} + R_{2n} + R_{3n}) < \eta.$$

The theorem is proved. ∎

3 Some corollaries

Take a natural $k > 0$. Define the function $\exp_{*k} : \mathbb{R} \to \mathbb{R}_+$ as

$$\exp_{*k} x = \underbrace{\exp \ldots \exp}_{k} x, \quad x \in \mathbb{R},$$

and

$$\log_{*k} x := \begin{cases} \underbrace{\log \ldots \log}_{k} x, & x > \exp_{*k} 1, \\ 0, & 0 < x \leq \exp_{*k} 1. \end{cases} \tag{3.1}$$

It is plain that $\log_{*k}(\exp_{*k}(x)) = x$, $x \in \mathbb{R}$, and $\exp_{*k}(\log_{*k}(x)) = x$, $x > \exp_{*k} 1$. For brevity, we call these functions k-exp-function and k-log-function, respectively.

Lemma 3.1. *Set $l_k := \exp_{*k} 1$. The k-log-function satisfies assumption* (A2) *on the half-axis $L_k := [l_k, +\infty)$, see* (1.5).

Proof. The proof is simple, but we sketch it briefly for the completeness of the presentation. Indeed, the claim is clear for $k = 1$. For $k = 2$ and $x > l_2$, we have

$$\begin{aligned}\log_{*2}(x^{1+\epsilon}) &= \log((1+\epsilon)\log x) = \log\log x + \log(1+\epsilon) \\ &= \log\log(x)\left(1 + \frac{\log(1+\epsilon)}{\log\log(x)}\right) = \log_{*2}(x)\left(1 + \frac{\log(1+\epsilon)}{\log_{*2}(x)}\right),\end{aligned}$$

where we use that the function $(\log_{*2} x)^{-1}$ is uniformly bounded by one on L_2 and $\log(1+\epsilon) \leq \epsilon$.

We also write down the computation for $k = 3$. Indeed, for $x \in L_3$

$$\begin{aligned}\log_{*3}(x^{1+\epsilon}) &= \log_{*2}((1+\epsilon)\log x)) = \log(\log\log x + \log(1+\epsilon))\\ &= \log\left(\log\log x\left[1 + \frac{\log(1+\epsilon)}{\log\log x}\right]\right)\\ &= \log_{*3}x\left(1 + \frac{\log[1 + \frac{\log(1+\epsilon)}{\log_{*2}x}]}{\log_{*3}x}\right) =: \log_{*3}x(1 + R_{*3,x}(\epsilon)).\end{aligned}$$

Once again, the function $(\log_{*3}x)^{-1}$ is uniformly bounded by one and $R_{*3,x} = \Diamond_x(\epsilon)$ for $x \in L_3$.

For the general $k \in \mathbb{N}$, we obtain by induction

$$\log_{*k}(x^{1+\epsilon}) = \log_{*k}x\,(1 + R_{*k,x}(\epsilon)),$$

where

$$R_{*k,x}(\epsilon) = \frac{\log\left(1 + \frac{\log\left(1+\cdots+\left[\frac{1+\log(1+\epsilon)}{\log_{*2}x}\right]\right)}{\log_{*k-1}x}\right)}{\log_{*k}x},$$

and, as before, the function $R_{*k,x} = \Diamond_x(\epsilon)$ satisfies condition (A2), see (1.5). ∎

For a natural $k > 0$ and a $\gamma > 0$, it is convenient to set

$$\varphi_{k,\gamma}(t) := \begin{cases}\left(\frac{1}{\log_{*k}(\frac{1}{t})}\right)^{\gamma}, & 0 < t \le \frac{1}{l_k},\\ 0, & t > \frac{1}{l_k}.\end{cases} \tag{3.2}$$

Observe that the function $\varphi_{k,\gamma}$ obeys relations (A1)–(A3). Indeed, $\varphi_{k,\gamma}$ satisfies trivially conditions (A1) and (A3). For any $\epsilon > 0$, the function $\varphi(x) = \log_{*k}(x)$ satisfies relation (1.5) on L_k (i.e., on the left neighborhood of $x_0 = +\infty$) by Lemma 3.1. It is bounded from above on L_k. This yields that $\varphi^{-\gamma}$, $\gamma > 0$, satisfies relation (1.5) as well. For $x \in L_k := [l_k, +\infty)$, we put $t := \frac{1}{x} \in (0, \frac{1}{l_k})$ and we obtain that $\varphi_{k,\gamma}$, see equation (3.2) above, obeys (A2) on $(0, \frac{1}{l_k})$ (i.e., on the right neighborhood of $t_0 = 0$).

The following corollaries follow at once from Theorem 1.2.

Corollary 3.2. *For a $\gamma > 0$, we have*

$$\begin{aligned}\int_0^\infty e^{-nt}\varphi_{k,\gamma}(t)\,dt &= \int_0^{\frac{1}{l_k}} e^{-nt}\log_{*k}\left(\frac{1}{t}\right)^{-\gamma} dt\\ &= \frac{1}{n}(\log_{*k}(n))^{-\gamma}(1+o(1)).\end{aligned} \tag{3.3}$$

Corollary 3.3. *For a $\gamma > 0$, we have*

$$\int_0^1 r^n \frac{1}{(1+\log(\frac{1}{1-r}))^\gamma}\,dr = \frac{1}{n}\frac{1}{(\log n)^\gamma}(1+o(1)),$$

$$\int_0^1 r^n \frac{1}{(1+\log_{*k}(\frac{1}{1-r}))^\gamma}\,dr = \frac{1}{n}\frac{1}{(\log_{*k} n)^\gamma}(1+o(1)).$$

Proof. We give the proof of the first claim of the corollary, the second one is completely analogous. The obvious point is to reduce the first relation to (3.3). Let $\delta > 0$ be small enough. Make the change of variable $r = e^{-t}$, $t \in (0, +\infty)$,

$$\int_0^1 r^n \frac{1}{(1+\log(\frac{1}{1-r}))^\gamma}\, dr = \int_0^\infty e^{-(n+1)t} \frac{1}{(1+\log(\frac{1}{1-e^{-t}}))^\gamma}\, dt$$
$$= \int_0^\delta \dots + \int_\delta^\infty \dots .$$

Clearly, the second integral goes to zero as fast as $O(e^{-n\delta})$ for $n \to +\infty$, since the function

$$f(t) = \frac{1}{(1+\log(\frac{1}{1-e^{-t}}))^\gamma}$$

is uniformly bounded on $\mathbb{R}_+$. For the first integral, an easy computation gives

$$\left| \int_0^\delta e^{-(n+1)t} \frac{1}{(1+\log(\frac{1}{1-e^{-t}}))^\gamma}\, dt - \int_0^\delta e^{-(n+1)t} \frac{1}{(1+\log(\frac{1}{t}))^\gamma}\, dt \right|$$
$$\le \frac{C}{n} \frac{1}{(\log n)^{\gamma+1}} = o\Big(\frac{1}{n(\log n)^\gamma}\Big),$$

where $C = C(\delta)$. Consequently, it remains to compute the asymptotics of the integral

$$\int_0^\delta e^{-(n+1)t} \frac{1}{(1+\log(\frac{1}{t}))^\gamma}\, dt. \tag{3.4}$$

As in Corollary 3.2, the function

$$\psi(t) = \frac{1}{(1+\log(\frac{1}{t}))^\gamma}$$

follows assumptions (A1)–(A3) (compare it to the function $\varphi_{1,\gamma}$ defined in (3.2)). Hence the asymptotics of (3.4) follows by Theorem 1.2, and the proof is completed. ■

Corollary 3.4. *Let $g : [0, 1] \to \mathbb{C}$ be a measurable bounded function admitting the following limit $g(1) := \lim_{r\to 1-} g(r) \neq 0$. For a $\gamma > 0$, we have*

$$\int_0^1 r^n \frac{g(r)}{(1+\log(\frac{1}{1-r}))^\gamma}\, dr = \frac{g(1)}{n(\log n)^\gamma}(1+o(1)),$$
$$\int_0^1 r^n \frac{g(r)}{(1+\log_{*k}(\frac{1}{1-r}))^\gamma}\, dr = \frac{g(1)}{n(\log_{*k} n)^\gamma}(1+o(1)).$$

Corollary 3.4 follows at once from Corollary 3.3, since the function

$$\frac{g(r)}{(1+\log_{*k}(\frac{1}{1-r}))^\gamma}$$

behaves like $g(1)(1+\log_{*k}(\frac{1}{1-r}))^{-\gamma}$ for $r \to 1-0$.

As a concluding remark, we mention that Corollaries 3.3 and 3.4 are used in [6] to compute the spectral asymptotics for special compact Toeplitz operators with non-radial symbols on Bergman space. These results on Toeplitz operators are then applied to the study of banded (Jacobi) matrices, see [6, Lemmas 2.1 and 5.1].

Acknowledgements. The authors are grateful to Leonid Golinskii for helpful discussions. They also thank the anonymous referee for remarks which led to improving the presentation of the paper. This work has been partially supported by the project ANR-18-CE40-0035.

Sergey Naboko was supported by the grant RScF-20-11-20032 and a Knut and Alice Wallenberg Foundation grant. This work was done in October and November 2019, during Sergey Naboko's visit to the University of Bordeaux, the hospitality of which is acknowledged.

Bibliography

[1] N. Bleistein and R. A. Handelsman, *Asymptotic expansions of integrals*. 2nd edn., Dover Publications, New York, 1986

[2] N. G. de Bruijn, *Asymptotic methods in analysis*. 3rd edn., Dover Publications, New York, 1981

[3] E. T. Copson, *Asymptotic expansions*. Reprint of the 1965 original, Cambridge Tracts in Math. 55, Cambridge University Press, Cambridge, 2004

[4] A. Erdélyi, *Asymptotic expansions*. Dover Publications, New York, 1956

[5] A. Erdélyi, General asymptotic expansions of Laplace integrals. *Arch. Ration. Mech. Anal.* **7** (1961), 1–20

[6] M. Koita, S. Kupin, S. Naboko and B. Touré, On spectral properties of compact Toeplitz operators on Bergman space with logarithmically decaying symbol and applications to banded matrices. Preprint 2020, arXiv:2006.02586

[7] B. Simon, *Advanced complex analysis. A comprehensive course in analysis. Part 2B*. American Mathematical Society, Providence, 2015

[8] G. N. Watson, The Harmonic Functions Associated with the Parabolic Cylinder. *Proc. Lond. Math. Soc. (2)* **17** (1918), 116–148

[9] R. Wong, *Asymptotic approximations of integrals*. Class. Appl. Math. 34, Society for Industrial and Applied Mathematics (SIAM), Philadelphia, 2001

Wehrl-type coherent state entropy inequalities for SU(1, 1) and its $AX + B$ subgroup

Elliott H. Lieb and Jan Philip Solovej

Dedicated to our friend and colleague Ari Laptev on his 70th birthday.

We discuss the Wehrl-type entropy inequality conjecture for the group SU(1, 1) and for its subgroup $AX + B$ (or affine group), their representations on $L^2(\mathbb{R}_+)$, and their coherent states. For $AX + B$ the Wehrl-type conjecture for L^p-norms of these coherent states (also known as the Renyi entropies) is proved in the case that p is an even integer. We also show how the general $AX + B$ case reduces to an unsolved problem about analytic functions on the upper half-plane and the unit disk.

1 Introduction

The starting point for this work,[1] historically, was Wehrl's definition of semiclassical entropy and his conjecture about its minimum value [27]. Given a density matrix ρ on $L^2(\mathbb{R})$ (a positive operator whose trace is 1) we define its classical probability density (or Husimi function) as follows:

$$\rho^{\mathrm{cl}}(p,q) := \langle p,q|\rho|p,q\rangle.$$

Here $|p,q\rangle$ is the (Schrödinger, Klauder, Glauber) coherent state, which is a normalized function in $L^2(\mathbb{R})$ parametrized by $p, q \in \mathbb{R}^2$, which is the classical phase space for a particle in one-dimension. It is given (with $\hbar = 1$) by

$$|p,q\rangle(x) = \pi^{-\frac{1}{4}} \exp[-(x-q)^2 + ipx].$$

We use the usual Dirac notation in which $|\cdots\rangle$ is a vector and $\langle\cdots|\cdots\rangle$ is the inner product conjugate linear in the first variable and linear in the second.

[1]This paper is a polished version of a paper on arxiv [23]. It is not the final version, however, and should be regarded as work in progress.

Keywords: Coherent states, affine group, $AX + B$ group
2020 Mathematics Subject Classification: 81R05, 81R30, 81S10, 22E46

The von Neumann entropy of any quantum state with density matrix ρ is

$$S^{\mathrm{Q}}(\rho) := -\operatorname{Tr} \rho \log \rho$$

and the classical entropy of any continuous probability density $\tilde{\rho}(p,q)$ is

$$S^{\mathrm{cl}}(\tilde{\rho}) := -\int \tilde{\rho} \log \tilde{\rho} \, \mathrm{d}p \, \mathrm{d}q.$$

The S^{Q} entropy is non-negative, while $S^{\mathrm{cl}}(\tilde{\rho}) \geq 0$ for any $\tilde{\rho}$ that is pointwise ≤ 1, as is the case for ρ^{cl}.

For the reader's convenience we recall some basic facts of, and the interest in, the Wehrl entropy. A classical probability does not always lead to a positive entropy and it often leads to an entropy that is $-\infty$. The quantum von Neumann entropy is always non-negative. Wehrl's contribution was to derive a classical probability distribution from a quantum state, which has several very desirable features. One is that it is always non-negative. Another is that it is monotone, meaning that the Wehrl entropy of a quantum state of a tensor product is always greater than or equal to the entropy of each subsystem obtained by taking partial traces. This monotonicity holds for marginals of classical probabilities, but not always for the quantum von Neumann entropy (the entropy of the universe may be smaller that the entropy of a peanut). In short the Wehrl entropy combines some of the desirable properties of the classical and quantum entropies.

The minimum possible von Neumann entropy is zero and occurs when ρ is any pure state, while that of the classical $S^{\mathrm{cl}}(\rho^{\mathrm{cl}})$ is strictly positive; Wehrl's conjecture was that the minimum is 1 and occurs when

$$\rho = |p,q\rangle\langle p,q|.$$

That is, when ρ is a pure state projector onto any coherent state. This conjecture was proved by one of us [19]. The uniqueness of this choice of ρ was shown by Carlen [12]. Recently De Palma [15] determined the minimal Wehrl entropy when the von Neumann entropy of ρ is fixed to some positive value.

The Wehrl conjecture was generalized in [19] to the theorem that the operator norm $\|\rho^{\mathrm{cl}}\|_p$, $p > 1$, was maximized for the same choice of ρ. This is equivalent to saying that the classical Renyi entropy of ρ^{cl} is minimized for this same choice of ρ. Later, in our joint paper [21], the same upper bound result was shown to hold for the integral of *any* convex function of ρ^{cl}, not just $x \to x^p$. Note that minus a convex function is a concave function so maximizing integrals of convex functions is the same as minimizing integrals of concave functions.

We achieved the generalization to all convex functions in [21] by considering a generalization of the map from quantum states to the classical Husimi functions. Instead of only considering maps from quantum states to classical states we may consider maps between quantum states. In 1991 in [20] we defined a general formalism for defining such maps between different unitary representations of the same group. In [20] we named these maps *quantum coherent operators*. They are, in fact, examples of quantum channels, i.e., completely positive trace preserving maps. Moreover, they are covariant in the sense that applying a unitary transformation to the state prior to the map is the same as applying it after the map. Later such covariant quantum channels were studied in [1, 2, 8, 9, 18].

The possibility of using coherent states to associate a classical density to a quantum state ρ has a group-theoretic interpretation, which we will explain below. In the case of the classical coherent states discussed above this relates to the Heisenberg group and the coherent states are heighest weight vectors in the unitary irreducible representation. This suggests that we can look at other groups, their representations, and associated coherent states (heighest weight vectors) and ask whether the analog of the Wehrl conjecture holds. The first question in this direction was raised in [19] for the group SU(2) where the representations are labeled by the quantum spin $J = 0, \frac{1}{2}, 1, \frac{3}{2}, \dots$. The spin cases $J = \frac{1}{2}, \frac{3}{2}$ were proved by Schupp in [25] and in full generality by us in [21].

The obvious next case is SU(N) for general N which has many kinds of representations. We showed the Wehrl hypothesis for all the symmetric representations in [22]. Here we turn our attention to another group of some physical but also mathematical interest which is the group SU(1, 1) and its subgroup the $AX + B$ group or affine group. These groups, like the Heisenberg group, are not compact and have only infinite-dimensional unitary representations. The affine group is not even unimodular which means that the left and right invariant Haar measures are different. The group SU(1, 1) is not simply connected. The purely mathematical interest in the affine group is in signal processing (see [13]). The coherent states for the affine group are like continuous families of wavelets.

The $AX + B$ (or affine) group is $G = \mathbb{R}_+ \times \mathbb{R}$ with the composition rule

$$(a, b) \cdot (a', b') = (aa', ab' + b).$$

This comes from thinking of the group acting on the real line as

$$x \mapsto ax + b,$$

hence the name $ax + b$ group. It is *not* unimodular and the *left* Haar measure on the group is given by

$$a^{-2}\, \mathrm{d}a\, \mathrm{d}b$$

(for reference, the right Haar measure is $a^{-1}\, \mathrm{d}a\, \mathrm{d}b$.)

The group SU(1, 1) is the group of 2×2 complex matrices of determinant one, that leave the form $|z_1|^2 - |z_2|^2$ invariant, i.e., all matrices of the form

$$\begin{pmatrix} \alpha & \beta \\ \overline{\beta} & \overline{\alpha} \end{pmatrix}, \tag{1.1}$$

where $|\alpha|^2 - |\beta|^2 = 1$. It is easily seen that we may consider the affine $AX + B$ group as a subgroup of SU(1, 1) through the injective group homomorphism

$$(a, b) \mapsto (\alpha, \beta) = \frac{1}{2\sqrt{a}}(a + 1 + ib, a - 1 + ib).$$

Both the $AX + B$ group and SU(1, 1) act transitively on the open complex unit disk by

$$z \mapsto \frac{\alpha z + \overline{\beta}}{\beta z + \overline{\alpha}}.$$

This is equivalent to a transitive action on the upper half-plane. For the $AX + B$ group the action on the upper half-plane is particularly simple as it becomes

$$z \mapsto az + b.$$

The coherent states for the two groups will naturally be parametrized by points on the unit disk or equivalently by points on the upper half-plane.

In Section 2 we introduce the unitary representations of the affine group and its coherent states and state the Wehrl entropy conjecture in this case. In Section 3 we generalize the conjecture to a larger class of functions, not just the entropy and we give an elementary proof in special cases. In Section 4 we formulate the generalized Wehrl inequality as a statement about analytic functions on the upper half-plane or the unit disk. In Section 5 we consider SU(1, 1) and a discrete series of unitary representations and its corresponding coherent states. We formulate the generalized Wehrl conjecture for SU(1, 1) and show that it is a consequence of the generalized Wehrl conjecture for the affine $AX + B$ group. In Section 6 we define for SU(1, 1) certain covariant quantum channels that map density matrices on one representation space to density matrices on another. These quantum channels are analogous to those SU(2) and SU(N) channels we introduced in [21, 22]. We conjecture that the majorization results from [21, 22] hold also in the case of SU(1, 1). We finally prove that this majorization conjecture implies our generalized Wehrl inequality by taking an appropriate semiclassical limit.

2 Unitary representations and coherent states for the $AX + B$ group

We begin with the subgroup of SU(1, 1) since that is easier than the full SU(1, 1) group and where we have the most results.

The group has two irreducible faithful unitary representations [3, 17] which may be realized on $L^2(\mathbb{R}_+, dk)$ either by

$$[U(a,b)f](k) = \exp(-2\pi i bk) a^{\frac{1}{2}} f(ak)$$

or with $e^{-2\pi i bk}$ replaced by $e^{+2\pi i bk}$. For the representation above the coherent states we shall consider are given in terms of fiducial vectors

$$\eta_\alpha(k) = C(\alpha) k^\alpha \exp(-k),$$

where the parameter α is positive and which we will henceforth keep fixed. These functions are identified as affine coherent states (extremal weight vectors) for the representation of the affine group in [14,26]. The constant $C(\alpha)$ is chosen such that

$$\int_0^\infty \frac{|\eta_\alpha(k)|^2}{k}\, \mathrm{d}k = 1,$$

i.e.,

$$C(\alpha) = 2^\alpha\, \Gamma(2\alpha)^{-\frac{1}{2}}.$$

Then

$$\int_{-\infty}^{\infty}\int_0^\infty U(a,b)|\eta_\alpha\rangle\langle\eta_\alpha|U(a,b)^* a^{-2}\, \mathrm{d}a\, \mathrm{d}b = I. \tag{2.1}$$

For any function $f \in L^2(\mathbb{R}_+)$ we may introduce the *coherent state transform*

$$h_f(a,b) = \langle U(a,b)\eta_\alpha|f\rangle = C(\alpha)a^{\alpha+\frac{1}{2}}\int_0^\infty \exp(-ka + 2\pi i bk)k^\alpha f(k)\, \mathrm{d}k. \tag{2.2}$$

As in the case of the classical coherent states, we then have

$$\int_{-\infty}^{\infty}\int_0^\infty |h_f(a,b)|^2 a^{-2}\, \mathrm{d}a\, \mathrm{d}b = \int_0^\infty |f(k)|^2\, \mathrm{d}k,$$

and if f is normalized in L^2, we may consider $|h_f(a,b)|^2$ as a probability density relative to the Haar measure $a^{-2}\, \mathrm{d}a\, \mathrm{d}b$. Observe that in contrast to the classical case we do not have $|h_f(a,b)|^2 \le 1$, but rather

$$|h_f(a,b)|^2 \le \int |\eta_\alpha(k)|^2\, \mathrm{d}k = \alpha.$$

This is a consequence of the fact that (2.1) requires $\int_0^\infty \frac{|\eta_\alpha(k)|^2}{k}\, \mathrm{d}k = 1$, which is different from the L^2-normalization. This difference is due to the group not being uni-modular. The *Wehrl entropy*

$$S^{\mathrm{W}}(f) = -\int |h_f(a,b)|^2 \ln(|h_f(a,b)|^2) a^{-2}\, \mathrm{d}a\, \mathrm{d}b$$

is, therefore, *not necessarily non-negative*, but by the definition, and by the normalization of h_f, it is bounded below by $-\ln\alpha$. The natural generalization of Wehrl's conjecture in this case is that the entropy $S^{\mathrm{W}}(f)$ is minimal for $f = \alpha^{-\frac{1}{2}}\eta_\alpha$, in which case it is $1 + (2\alpha)^{-1} - \ln\alpha$.

In the following section we state a more general conjecture and prove it in special cases. In the last section of the paper we show that our conjecture and theorem are equivalent to L^p estimates for analytic functions in the complex upper half-plane.

3 Generalized conjecture for the $AX + B$ group and a partial result

We make the following more general conjecture.

Conjecture 3.1 (Wehrl-type conjecture for the $AX + B$ group). Let $\alpha > 0$ be fixed. If $G : [0, \alpha] \to \mathbb{R}$ is a convex function, then

$$\int_{-\infty}^{\infty} \int_{0}^{\infty} G(|h_f(a,b)|^2) a^{-2}\, da\, db$$

is maximized among all normalized $f \in L^2(\mathbb{R}_+)$ if and only if (up to a phase) f has the form $f = \alpha^{-\frac{1}{2}} U(a,b)\eta_\alpha$ for some $a > 0$ and $b \in \mathbb{R}$.

Remark. The maximal value above may be infinite. If $G(t) = t^s$ for $s \geq 1$ the maximal value of the integral is conjectured to be

$$\frac{2\alpha^s}{(2\alpha + 1)s - 1}.$$

The analog of Wehrl's original entropy conjecture follows from this conjecture by taking minus a derivative at $s = 1$ as in [19] and gives the minimal entropy

$$S_{\min}^W = 1 + (2\alpha)^{-1} - \ln \alpha,$$

which agrees with the lower bound $-\ln \alpha$ mentioned above.

The conjecture was investigated by J. Bandyopadhyay in [4]. She proved the conjecture for $G(t) = t^s$ with $s = 1 + (2\alpha + 1)^{-1}$ as part of her study of coherent states for representations of SU(1, 1) in the same setting as in our Section 5.

The following theorem, which has an elementary proof, yields the conjecture for integer s. As we explain in the next section this result also follows from an application of [11, Theorem 3.1] and from [5] in their works on holomorphic functions. An alternative proof with a few generalizations of the result below was given recently in [6]. In [5] a conjecture is formulated that is equivalent to the above for $G(t) = t^s$ for all $s \geq 1$.

Main Theorem 3.2. *The statement of Conjecture* 3.1 *holds for the special cases of* $G(t) = t^s$, *for* s *being a positive integer.*

Proof. We want to prove that

$$\int_{-\infty}^{\infty} \int_{0}^{\infty} |h_f(a,b)|^{2s} a^{-2}\, da\, db$$

is maximized if and only if

$$f(k) = A k^{\alpha} \exp(-Bk)$$

with $A, B \in \mathbb{C}$ such that $f \in L^2(\mathbb{R}_+)$ is normalized. The proof will rely only on a Schwarz inequality and existence and uniqueness will follow from the corresponding uniqueness of optimizers for Schwarz inequalities.

If s is a positive integer, we can write

$$h_f(a,b)^s = C(\alpha)^s a^{s(\alpha+\frac{1}{2})} \int_0^\infty \cdots \int_0^\infty \exp[(-a+2\pi i b)(k_1+\cdots+k_s)] \\ \times (k_1\cdots k_s)^\alpha f(k_1)\cdots f(k_s)\,\mathrm{d}k_1\cdots \mathrm{d}k_s.$$

Hence doing the b integration gives

$$\int_{-\infty}^\infty \int_0^\infty |h_f(a,b)|^{2s} a^{-2}\,\mathrm{d}a\,\mathrm{d}b \\ = C(\alpha)^{2s}\Gamma(s(2\alpha+1)-1)\int_0^\infty \cdots \int_0^\infty [2(k_1+\cdots k_s)]^{-s(2\alpha+1)+1} \\ \times k_1^\alpha f(k_1)\cdots k_s^\alpha f(k_s) k_{s+1}^\alpha \overline{f(k_{s+1})}\cdots k_{2s}^\alpha \overline{f(k_{2s})}\,\mathrm{d}k_1\cdots \mathrm{d}k_{2s-1}.$$

where $k_{2s} = k_1+\cdots+k_s-(k_{s+1}+\cdots+k_{2s-1})$. We now do the a integration and arrive at

$$\int_{-\infty}^\infty \int_0^\infty |h_f(a,b)|^{2s} a^{-2}\,\mathrm{d}a\,\mathrm{d}b \\ = C(\alpha)^{2s}\Gamma(s(2\alpha+1)-1)\int_0^\infty \cdots \int_0^\infty [2(k_1+\cdots k_s)]^{-s(2\alpha+1)+1} \\ \times k_1^\alpha f(k_1)\cdots k_s^\alpha f(k_s) k_{s+1}^\alpha \overline{f(k_{s+1})}\cdots k_{2s}^\alpha \overline{f(k_{2s})}\,\mathrm{d}k_1\cdots \mathrm{d}k_{2s-1}.$$

We now change variables to

$$r = k_1+\cdots+k_s, \qquad u_j = \frac{k_j}{r}, \quad v_j = \frac{k_{s+j}}{r}, \quad j = 1,\dots,s-1.$$

The Jacobian determinant for this change of variables is easily found to be

$$\left|\det\left[\frac{\partial(k_1,\dots,k_{2s-1})}{\partial(r,u_1,\dots,u_{2s-1},v_1,\dots,v_{2s-1})}\right]\right| = r^{2s-2}.$$

We arrive at

$$\int_{-\infty}^\infty \int_0^\infty |h_f(a,b)|^{2s} a^{-2}\,\mathrm{d}a\,\mathrm{d}b \\ = C(\alpha)^{2s}\Gamma(s(2\alpha+1)-1)\int_0^\infty \underset{u_1+\cdots+u_{s-1}\le 1}{\int\cdots\int}\ \underset{v_1+\cdots+v_{s-1}\le 1}{\int\cdots\int} 2^{-s(2\alpha+1)+1} r^{s-1} \\ \times [u_1\cdots u_{s-1}(1-u_1-\cdots-u_{s-1})]^\alpha [v_1\cdots v_{s-1}(1-v_1-\cdots-v_{s-1})]^\alpha \\ \times f(u_1 r)\cdots f(u_{s-1}r) f(r(1-u_1-\cdots-u_{s-1})) \\ \times \overline{f(v_1 r)}\cdots \overline{f(v_{s-1}r) f(r(1-v_1-\cdots-v_{s-1}))} \\ \times \mathrm{d}u_1\cdots \mathrm{d}u_{s-1}\,\mathrm{d}v_1\cdots \mathrm{d}v_{s-1}\,\mathrm{d}r.$$

Let us apply the Cauchy–Schwarz inequality for each fixed r to conclude that

$$
\begin{aligned}
&\int_{-\infty}^{\infty}\int_0^{\infty} |h_f(a,b)|^{2s} a^{-2}\,da\,db \\
&\quad\leq 2^{-s(2\alpha+1)+1} C(\alpha)^{2s}\Gamma(s(2\alpha+1)-1) \\
&\qquad\times\Bigg[\underset{u_1+\cdots+u_{s-1}\leq 1}{\int\cdots\int} [u_1\cdots u_{s-1}(1-u_1-\cdots-u_{s-1})]^{2\alpha}\,du_1\cdots du_{s-1}\Bigg] \\
&\qquad\times\int_0^{\infty}\Bigg[\underset{u_1+\cdots+u_{s-1}\leq 1}{\int\cdots\int} |f(u_1 r)\cdots f(u_{s-1}r) f(r(1-u_1-\cdots-u_{s-1}))|^2 \\
&\qquad\qquad\qquad\qquad \times du_1\cdots du_{s-1}\Bigg] r^{s-1}\,dr \\
&\quad= 2^{-s(2\alpha+1)+1} C(\alpha)^{2s}\Gamma(s(2\alpha+1)-1)\left(\int_0^{\infty}|f(r)|^2\,dr\right)^s \\
&\qquad\times\Bigg[\underset{u_1+\cdots+u_{s-1}\leq 1}{\int\cdots\int} [u_1\cdots u_{s-1}(1-u_1-\cdots-u_{s-1})]^{2\alpha}\,du_1\cdots du_{s-1}\Bigg].
\end{aligned}
$$

If f is normalized, this is a number depending on α and s. The important observation is that the upper bound is achieved if and only if there is a function $K(r)$ depending on r such that

$$
f(u_1)\cdots f(u_{s-1}) f(r-u_1-\cdots-u_{s-1}) = K(r)[u_1\cdots u_{s-1}(r-u_1-\cdots-u_{s-1})]^{\alpha}
$$

for almost all $0\leq u_1,\ldots,u_{s-1},r$ satisfying $u_1+\cdots+u_{s-1}\leq r$. If we define

$$
g(u) = u^{-\alpha} f(u)
$$

and introduce the variable $u_s = r-(u_1+\cdots+u_{s-1})$, we may rewrite this is as

$$
g(u_1)\cdots g(u_s) = K(u_1+\cdots+u_s)
$$

for almost all $0\leq u_1,\ldots,u_s$. It is not difficult to show that if a locally integrable function g satisfies this, it must be smooth and it follows easily that

$$
g(u) = A\exp(-Bu)
$$

for complex numbers A, B which is exactly what we wanted to prove.

The maximal value can be found by a straightforward computation. ■

4 An analytic formulation

Using the Bergman–Paley–Wiener Theorem in [16], we can rephrase our conjecture and theorem in terms of analytic functions on the complex upper half-plane

$\mathbb{C}_+ = \{z \in \mathbb{C} : \Im z > 0\}$. Introducing the weighted Bergman space,

$$\mathcal{A}^2_\beta(\mathbb{C}_+) = \{F \in L^2(\mathbb{C}_+, (\Im z)^\beta d^2 z) : F \text{ analytic}\}$$

for $\beta > -1$. The Bergman–Paley–Wiener Theorem in this context says that there is a unitary map

$$L^2(\mathbb{R}_+) \ni f \mapsto F \in \mathcal{A}^2_{2\alpha-1}(\mathbb{C}_+),$$

given by

$$F(z) = \frac{2^\alpha}{\sqrt{2\pi\Gamma(2\alpha)}} \int_0^\infty e^{ikz} k^\alpha f(k)\, \mathrm{d}k.$$

If we write $z = 2\pi b + ia$, we recognize the coherent state transform in (2.2) to be

$$h_f(a, b) = \sqrt{2\pi}(\Im z)^{\alpha+\frac{1}{2}} F(z).$$

The action of the affine group on $\mathbb{R}$ may be extended to the upper half-plane $\mathbb{C}_+$ by $z \mapsto az + 2\pi b$. The representation of the $AX + B$ group on the Bergman space is then

$$\begin{aligned}(U(a,b)^* F)(z) &= \frac{2^\alpha}{\sqrt{2\pi\Gamma(2\alpha)}} \int_0^\infty e^{ikz} k^\alpha (U(a,b)^* f)(k)\, \mathrm{d}k \\ &= a^{\alpha+\frac{1}{2}} F(az + 2\pi b).\end{aligned}$$

In the analytic representation our conjecture states that if $G : [0, \frac{\alpha}{2\pi}] \to \mathbb{R}$ is convex, then

$$\int_{\mathbb{C}_+} G(|F(z)|^2 (\Im z)^{2\alpha+1}) |\Im z|^{-2}\, \mathrm{d}^2 z \tag{4.1}$$

is maximized among normalized functions F in $\mathcal{A}^2_{2\alpha-1}(\mathbb{C}_+)$ if and only if F is proportional to $(z - z_0)^{-(2\alpha+1)}$ for some z_0 in the lower half-plane.

We may also formulate the conjecture for analytic functions on the unit disk $\mathbb{D}$. Here it states that if $G : [0, \frac{\alpha}{2\pi}] \to \mathbb{R}$ is convex, then

$$\int_{\mathbb{D}} G(|F(z)|^2 (1 - |z|^2)^{2\alpha+1})(1 - |z|^2)^{-2}\, \mathrm{d}^2 z \tag{4.2}$$

is maximized among all analytic functions F on the unit disk with

$$\int_{\mathbb{D}} |F(z)|^2 (1 - |z|^2)^{2\alpha-1}\, \mathrm{d}^2 z = 1$$

if and only if $F(z)$ is proportional to $(1 - \zeta z)^{-(2\alpha+1)}$ for some ζ inside the unit disk.

Our theorem from the previous section is therefore equivalent to the following result.

Theorem 4.1. *If $G(t) = t^s$, with s a positive integer, then the integrals* (4.1) *and* (4.2) *satisfy the maximization properties stated above.*

5 Some unitary SU(1, 1) representations and their coherent states

We may represent SU(1, 1) unitarily on the bosonic Fock space over two modes

$$\mathcal{F}(\mathbb{C}^2) = \bigoplus_{N=0}^{\infty} \mathcal{H}_N, \quad \mathcal{H}_N = \bigoplus_{\mathrm{Sym}}^{N} \mathbb{C}^2, \; \mathcal{H}_0 = \mathbb{C}.$$

On this space we have the action of the creation and annihilation operators a_1, a_2 and $a_1^\dagger, a_2\dagger$ annihilation and creation particles in the two modes represented by the standard basis in $\mathbb{C}^2$. The unitary representation of SU(1, 1) on $\mathcal{F}(\mathbb{C}^2)$ is such that the element (1.1) in SU(1, 1) acts as the unitary that transforms the creation and annihilation operators according to

$$\begin{pmatrix} a_1 \\ a_2^\dagger \end{pmatrix} \to \begin{pmatrix} \alpha & \beta \\ \overline{\beta} & \overline{\alpha} \end{pmatrix} \begin{pmatrix} a_1 \\ a_2^\dagger \end{pmatrix} = \begin{pmatrix} a_1 + \beta a_2^\dagger \\ \overline{\beta} a_1 + \overline{a}_2^\dagger \end{pmatrix},$$

i.e., a Bogolubov transformation. It is clear that this defines a unitary representation of SU(1, 1). It is however not irreducible on $\mathcal{F}(\mathbb{C}^2)$. We see that the operator

$$\widehat{K} = a_1^\dagger a_1 - a_2^\dagger a_2$$

is left invariant by the action of the Bogolubov transformation. Hence we may write

$$\mathcal{F}(\mathbb{C}^2) = \bigoplus_{K=-\infty}^{\infty} \mathcal{D}_K,$$

where $\mathcal{D}_K$ is the subspace where $\widehat{K}$ is equal to K. Each of the spaces $\mathcal{D}_K$ are invariant and irreducible for the action of SU(1, 1) moreover the representation on $\mathcal{D}_K$ is the same as the representation $\mathcal{D}_{-K}$ as this corresponds just to the interchanging a_1, a_2. We will thus focus on $\mathcal{D}_K$ for $K \geq 0$.

The Lie algebra of SU(1, 1) is generated by the three elements $\widehat{K}_0$, $\widehat{K}_\pm$ with the commutation relations (see [24])

$$[\widehat{K}_0, \widehat{K}_\pm] = \pm \widehat{K}_\pm, \quad [\widehat{K}_-, \widehat{K}_+] = 2\widehat{K}_0.$$

On the Fock space they are represented as

$$\widehat{K}_0 = \frac{1}{2}(a_1^\dagger a_1 + a_2^\dagger a_2 + 1), \quad \widehat{K}_+ = a_1^\dagger a_2^\dagger, \quad \widehat{K}_- = a_1 a_2.$$

We see that an orthonormal basis for $\mathcal{D}_K$,$K \geq 0$ is given by

$$|n\rangle_K = \left(\frac{n!(n+K)!}{K!}\right)^{\frac{1}{2}} K_+^n |0\rangle_K,$$

where $|0\rangle_K$ is defined (up to a phase) by

$$a_1^\dagger a_1|0\rangle_K = K|0\rangle_K, \quad a_2|0\rangle_K = 0.$$

This easily shows that $\mathcal{D}_K$ define irreducible representation spaces for SU(1, 1).

The spaces $\mathcal{D}_K$, $K > 0$, correspond to one of the discrete series of the representations for SU(1, 1). There is another discrete series of representations. As for the affine $AX + B$ group the other discrete series is obtained by first mapping the matrix in (1.1) to the matrix with complex conjugate entries.

In each of the representation spaces $\mathcal{D}_K$, $K \geq 0$ we define the family of coherent states as the unitary transforms of the minimal weight vector $|0\rangle_K$. They can be parametrized by point on the unit disk $\mathbb{D}$ in $\mathbb{C}$.

Definition 5.1 (Coherent states for SU(1, 1)). For each $z \in \mathbb{D}$ we define the coherent state $|z\rangle_K$ by the action of

$$(1 - |z|^2)^{-\frac{1}{2}} \begin{pmatrix} 1 & z \\ \bar{z} & 1 \end{pmatrix}$$

on $|0\rangle_K$.

The coherent state $|z\rangle_K \in \mathcal{D}_K$ can be described up to a phase by the property

$$(za_1^\dagger + a_2)|z\rangle_K = 0.$$

From this equation we deduce the explicit form

$$|z\rangle_K = (1 - |z|^2)^{\frac{K+1}{2}} \sum_{j=0}^{\infty} \sqrt{\binom{j+K}{K}}\, z^j |j\rangle_K.$$

We see that

$$\frac{K}{\pi} \int_{\mathbb{D}} |z\rangle_K \langle z|(1 - |z|^2)^{-2}\, \mathrm{d}^2 z = I_{\mathcal{D}_K}.$$

Using these coherent states, we can calculate the coherent state transform of $|\zeta\rangle_K$

$${}_K\langle z|\zeta\rangle_K = (1 - |z|^2)^{\frac{K+1}{2}} (1 - |\zeta|^2)^{\frac{K+1}{2}} (1 - \bar{z}\zeta)^{-K-1} m$$

and we see that these indeed agree with the maximizers we conjectured in Section 4 if $K = 2\alpha$.

Conjecture 5.2 (Generalized Wehrl for SU(1, 1)). For any unit vector $\psi \in \mathcal{D}_K$ and any convex function $G : [0, 1] \to \mathbb{R}$ we have that

$$\int_{\mathbb{D}} G(|{}_K\langle z|\psi\rangle_K|^2)(1 - |z|^2)^{-2}\, \mathrm{d}^2 z$$

is maximized if and only if $\psi = |\zeta\rangle_K$ (up to an overall phase) for some $\zeta \in \mathbb{D}$.

This conjecture is equivalent to the $AX + B$ conjecture for half-integer α.

6 The SU(1, 1) quantum channels

If K and L are non-negative integers, we may identify the representation space $\mathcal{D}_{K+L}$ as a subspace of $\mathcal{D}_K \otimes \mathcal{D}_L$. Indeed, we can map $|0\rangle_{K+L}$ to $|0\rangle_K \otimes |0\rangle_L$ and then lift to a representation by

$$K_+^n|0\rangle_{K+L} \mapsto (K_+ \otimes I_{\mathcal{D}_L} + I_{\mathcal{D}_K} \otimes K_+)^n|0\rangle_K \otimes |0\rangle_L.$$

Let us denote by P the projection from $\mathcal{D}_K \otimes \mathcal{D}_L$ onto $\mathcal{D}_{K+L}$ considered as a subspace of $\mathcal{D}_K \otimes \mathcal{D}_L$.

We then have a covariant quantum channel (completely positive trace preserving map)

$$\Phi^L(\rho) = cP\rho \otimes I_L P$$

(for an appropriate c) from density matrices on $\mathcal{D}_K$ to density matrices on $\mathcal{D}_{K+L}$. We conjecture the following:

Conjecture 6.1 (Majorization of Φ^L). For all density matrices ρ on $\mathcal{D}_K$ the eigenvalues of $\Phi^L(\rho)$ will be majorized by the eigenvalues of $\Phi^L(|z\rangle_K\langle z|)$.

Theorem 6.2. *Conjecture* 6.1 *implies Conjecture* 5.2 *except for the uniqueness of the maximizers.*

We shall not give the proof of this theorem here, but it goes along the same lines as the proof of [21, Theorem 2.1].

Acknowledgements. Many thanks to Eric Carlen, Rupert Frank and Antti Haimi for valuable suggestions about references and for suggestions on the expansion of the initial version of this paper from $AX + B$ to SU(1, 1). Thanks also to John Klauder who helped us understand the $AX + B$ group. This work was supported by the Villum Centre of Excellence for the Mathematics of Quantum Theory (QMATH) and the ERC Advanced grant 321029.

Bibliography

[1] M. Al Nuwairan, Potential examples for non-additivity of the minimal output entropy. Preprint 2013, arXiv:1312.2200

[2] M. Al Nuwairan, The extreme points of SU(2)-irreducibly covariant channels. *Internat. J. Math.* **25** (2014), Article ID 1450048

[3] E. W. Aslaksen and J. R. Klauder, Unitary representations of the affine group. *J. Math. Phys.* **9** (1968), 206–211

[4] J. Bandyopadhyay, Optimal concentration for SU(1, 1) coherent state transforms and an analogue of the Lieb–Wehrl conjecture for SU(1, 1). *Comm. Math. Phys.* **285** (2009), 1065–1086

[5] F. Bayart, O. F. Brevig, A. Haimi, J. Ortega-Cerdà and K.-M. Perfekt, Contractive inequalities for Bergman spaces and multiplicative Hankel forms. *Trans. Amer. Math. Soc.* **371** (2019), 681–707

[6] D. Békollè, J. Gonessa and B. F Sehba, About a conjecture of Lieb–Solovej. Preprint 2020, arXiv:2010.14809

[7] B. G. Bodmann, A lower bound for the Wehrl entropy of quantum spin with sharp high-spin asymptotics. *Comm. Math. Phys.* **250** (2004), 287–300

[8] M. Brannan and B. Collins, Highly entangled, non-random subspaces of tensor products from quantum groups. *Comm. Math. Phys.* **358** (2018), 1007–1025

[9] M. Brannan, B. Collins, H. H. Lee and S.-G. Youn, Temperley–Lieb quantum channels. *Comm. Math. Phys.* **376** (2020), 795–839

[10] O. F. Brevig, J. Ortega-Cerdà, K. Seip and J. Zhao, Contractive inequalities for Hardy spaces. *Funct. Approx. Comment. Math.* **59** (2018), 41–56

[11] J. Burbea, Sharp inequalities for holomorphic functions. *Illinois J. Math.* **31** (1987), 248–264

[12] E. A. Carlen, Some integral identities and inequalities for entire functions and their application to the coherent state transform. *J. Funct. Anal.* **97** (1991), 231–249

[13] I. Daubechies, Orthonormal bases of coherent states: The canonical case and the $AX + B$ group. In *Coherent states, past, present and future*, pp. 103–107, World Scientific Publishing, Singapore, 1994

[14] I. Daubechies, J. R. Klauder and T. Paul, Wiener measures for path integrals with affine kinematic variables. *J. Math. Phys.* **28** (1987), 85–102

[15] G. De Palma, The Wehrl entropy has Gaussian optimizers. *Lett. Math. Phys.* **108** (2018), 97–116

[16] P. Duren, E. A. Gallardo-Gutiérrez and A. Montes-Rodríguez, A Paley–Wiener theorem for Bergman spaces with application to invariant subspaces. *Bull. Lond. Math. Soc.* **39** (2007), 459–466

[17] I. Gelfand and M. Neumark, Unitary representations of the group of linear transformations of the straight line. *C. R. (Doklady) Acad. Sci URSS (N.S.)* **55** (1947), 567–570

[18] A. S. Holevo, Additivity Conjecture and Covariant Channels. *Int. J. Quantum Inf.* **3** (2005), 41–48

[19] E. H. Lieb, Proof of an entropy conjecture of Wehrl. *Comm. Math. Phys.* **62** (1978), 35–41

[20] E. H. Lieb and J. P. Solovej, Quantum coherent operators: A generalization of coherent states. *Lett. Math. Phys.* **22** (1991), 145–154

[21] E. H. Lieb and J. P. Solovej, Proof of an entropy conjecture for Bloch coherent spin states and its generalizations. *Acta Math.* **212** (2014), 379–398

[22] E. H. Lieb and J. P. Solovej, Proof of the Wehrl-type entropy conjecture for symmetric $SU(N)$ coherent states. *Comm. Math. Phys.* **348** (2016), 567–578

[23] E. H. Lieb and J. P. Solovej, A Wehrl-type entropy inequality for the affine $AX + B$ group, Preprint 2019, arXiv:1906.00223v1

[24] A. Perelomov, *Generalized coherent states and their applications*. Texts Monogr. Phys., Springer, Berlin, 1986

[25] P. Schupp, On Lieb's conjecture for the Wehrl entropy of Bloch coherent states. *Comm. Math. Phys.* **207** (1999), 481–493

[26] G. Watson and J. R. Klauder, Generalized affine coherent states: A natural framework for the quantization of metric-like variables. *J. Math. Phys.* **41** (2000), 8072–8082

[27] A. Wehrl, General properties of entropy. *Rev. Modern Phys.* **50** (1978), 221–260

Blow-ups for the Horn–Kapranov parametrization of the classical discriminant

Evgeny Mikhalkin, Vitaly Stepanenko and Avgust Tsikh

Dedicated to the 70th birthday of our colleague and friend Ari Laptev

The Horn–Kapranov parametrizations describe the singular sets of hypergeometric functions in several variables. These parametrizations are inverses of logarithmic Gauss maps for A-discriminants. In this paper we demonstrate that, despite the multivalued nature of the indicated parametrizations, their blow-ups properties are the same as for single-valued meromorphic mappings. As an application, a new proof of factorization identities for the classical discriminant is given.

1 Introduction

Given a meromorphic mapping $f : X \to Y$ of complex analytic sets (spaces), it is reasonable to treat it as an analytic subset of $X \times Y$ that is the closure $\overline{G}_f$ of the graph of f. Fibers in $\overline{G}_f$ over uncertainty points can be regarded as result of "blow-up" or "shrinkage" procedures [13]. The most transparent scheme for such procedures can be seen for mappings that are inverses of the logarithmic Gauss map [9]. In the modern theory of hypergeometric functions [8], these inverses are known as Horn–Kapranov parametrizations [2, 14, 15].

Our goal in this paper is to study the Horn–Kapranov parametrization and to give a new proof of factorization identities for truncations of the classical discriminant. Such identities were obtained in [7, Chapter 10], using a complicated technique of the theory of A-determinants. In our approach, instead of meromorphic mappings we consider multivalued mappings with abelian monodromy groups.

The classical discriminant relates to a polynomial in one variable

$$f(y) = a_0 + a_1 y + \cdots + a_n y^n. \tag{1.1}$$

Keywords: Discriminant, Horn–Kapranov parametrization, Newton polytope, logarithmic Gauss map, blow-up

2020 Mathematics Subject Classification: Primary 32H04; secondary 32S25

It is an irreducible polynomial

$$\Delta_n = \Delta_n(a_0, a_1, \dots, a_n)$$

with integer coefficients that vanishes if and only if f has multiple roots. Discriminants play an important role in mathematics, which is demonstrated in the fundamental monographs [7,16]. The interest in the topic of discriminant identities considered in this article is inspired by studying the universal algebraic function [1,6,11,12,16] and by research in singularity theory [17], algebraic (tropical) geometry and mathematical physics [3,5].

It is well known [4,7] that the Newton polytope $\mathcal{N}(\Delta_n) \subset \mathbb{R}^{n+1}$ of the indicated discriminant is combinatorially equivalent to an $(n-1)$-dimensional cube. Since such a cube has 2^{n-1} vertices, it is natural to encode these vertices by all possible subsets of the set $\{1, \dots, n-1\}$. The polytope $\mathcal{N}(\Delta_n)$ has $n-1$ hyperfaces $\{h_k^0\}$ lying in the coordinate hyperplanes $\{t_k = 0\}$, $k = 1, \dots, n-1$ (it is assumed that the coordinates $t_0, t_1, \dots, t_n$ are chosen in the ambient space $\mathbb{R}^{n+1}$). Each hyperface h_k^0 has 2^{n-2} vertices, which are encoded by subsets $I \subset \{1, \dots, n-1\}$ that do not contain k. Let h_k denote the face opposite to h_k^0, whose vertices are encoded by subsets I containing k (an explicit equation for a hyperplane containing k is given by formula (2.7) in Section 2).

Our main objects of interest in this paper are truncations of the discriminant Δ_n to the faces of its Newton polytope. Recall that the *truncation* of a polynomial Δ to a face h of its Newton polytope $\mathcal{N}(\Delta)$ is the sum of all monomials from Δ whose exponents belong to h. This truncation is denoted by $\Delta|_h$.

The truncations $\Delta_n|_{h_k^0}$ to the coordinate hyperfaces h_k^0 are irreducible polynomials. In the recent paper [10] we wrote a sketch of proof of the following formula:

$$\Delta_n\big|_{h_k} = a_k^2\, \Delta_k(a_0, a_1, \dots, a_k)\Delta_{n-k}(a_k, a_{k+1}, \dots, a_n), \tag{1.2}$$

where Δ_k, Δ_{n-k} *are the discriminants of polynomials*

$$a_0 + a_1 y + \dots + a_k y^k, \quad a_k + a_{k+1} y + \dots + a_n y^{n-k}$$

of degrees k and $n-k$, respectively.

Note that for $k = 1$ the first discriminant on the right-hand side of (1.2) is identically equal to 1: $\Delta_1(a_0, a_1) \equiv 1$. The same is true for the second factor $\Delta_1(a_{n-1}, a_n)$ when $k = n-1$. Thus, the truncations of Δ_n to the hyperfaces h_1 and h_{n-1} coincide, up to the monomials a_1^2 and a_{n-1}^2, respectively, with the discriminants Δ_{n-1} of polynomials of degree $n-1$.

As an example consider the case $n = 3$ when f is a cubic polynomial

$$f = a_0 + a_1 y + a_2 y^2 + a_3 y^3$$

and the discriminant Δ_3 is equal to

$$\Delta_3(a_0, a_1, a_2, a_3) = -27a_0^2 a_3^2 - 4a_1^3 a_3 - 4a_0 a_2^3 + 18a_0 a_1 a_2 a_3 + a_1^2 a_2^2.$$

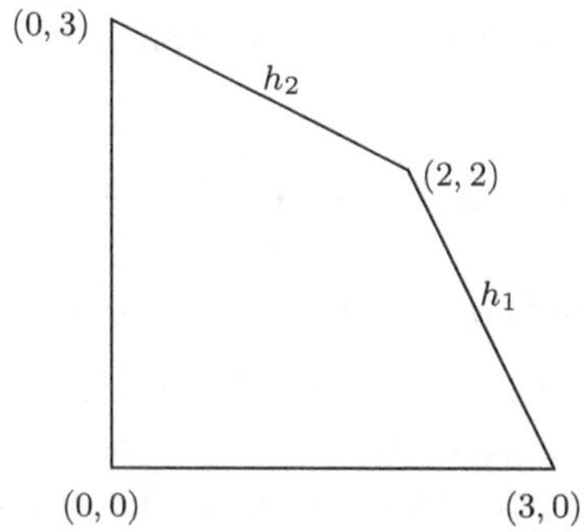

Figure 1.1. The Newton polytope for $\mathcal{N}(\Delta_3^{\mathrm{red}})$.

For a better understanding the geometry of the Newton polytope $\mathcal{N}(\Delta_3) \subset \mathbb{R}^4$ it is convenient to consider its projection onto the subspace $\mathbb{R}^2$ of variables t_1, t_2. For this it is enough to pass from the polynomial f to its reduction f^{red} setting $a_0 = a_3 = 1$. The corresponding reduced discriminant is the following:

$$\Delta_3^{\mathrm{red}}(a_1, a_2) = -27 - 4a_1^3 - 4a_2^3 + 18a_1a_2 + a_1^2a_2^2.$$

For its Newton polytope see Figure 1.1.

The truncations to h_1 and h_2 of the reduced discriminant are

$$a_1^2a_2^2 - 4a_1^3 = a_1^2 \cdot 1 \cdot (a_2^2 - 4a_1) = a_1^2 \cdot \Delta_1(1, a_1) \cdot \Delta_2(a_1, a_2, 1),$$
$$a_1^2a_2^2 - 4a_2^3 = a_2^2 \cdot (a_1^2 - 4a_2) \cdot 1 = a_2^2 \cdot \Delta_2(1, a_1, a_2)\Delta_1(a_2, 1),$$

respectively. For the original discriminant these expressions become as

$$\Delta_3\big|_{h_1} = a_1^2a_2^2 - 4a_1^3a_3 = a_1^2 \cdot \Delta_1(a_0, a_1) \cdot \Delta_2(a_1, a_2, a_3),$$
$$\Delta_3\big|_{h_2} = a_1^2a_2^2 - 4a_0a_2^3 = a_2^2 \cdot \Delta_2(a_0, a_1, a_2)\Delta_1(a_2, a_3),$$

which corresponds to (1.2).

In the present paper we prove more general formulas concerning the truncations of Δ_n to faces

$$h_K := h_{k_1} \cap \cdots \cap h_{k_p}$$

of codimension p, obtained by intersection of p non-coordinate hyperfaces. The multi-index $K = \{k_1, \ldots, k_p\}$ determines a partition of the tuple $\{0, 1, \ldots, n\}$ into $p + 1$ subtuples (segments)

$$K_i = \{k_i, k_i + 1, \ldots, k_{i+1}\}, \quad i = 0, 1, \ldots, p,$$

assuming that $k_0 = 0$ and $k_{p+1} = n$. Let $l_i := k_{i+1} - k_i$ denote the length of K_i and

$$f_{K_i} := a_{k_i} + a_{k_i+1}y + \cdots + a_{k_{i+1}}y^{l_i}.$$

Theorem 1.1. *In previous notations, the truncation of Δ_n to h_K admits the following factorization:*

$$\Delta_n\big|_{h_K} = a_K^2 \prod_{i=0}^{p} \Delta_{l_i}(f_{K_i}), \tag{1.3}$$

where $a_K^2 = a_{k_1}^2 \dots a_{k_p}^2$, and Δ_{l_i} are the discriminants of polynomials of degrees l_i.

In order to prove this theorem, we start with a similar assertion for the extremal part of Δ_n (Lemma 2.1). Thus we conclude that the Newton polytopes of the polynomials in both sides of (1.3) coincide. Then we prove that the zero set of the truncation $a_K^{-2}\Delta_n|_{h_K}$ contains the union of zero sets of discriminants $\Delta_{l_i}(f_{K_i})$ (Lemma 4.2). To this end, we use the Horn–Kapranov parametrizations of the zero sets of these discriminants. The final conclusion is based on general facts of the intersection theory.

2 Extremal discriminant and factorization of its truncations

Given a discriminant Δ_n, the sum of its monomials with exponents ranging over the vertex set of the polytope $\mathcal{N}(\Delta_n)$ we denote by $\widehat{\Delta}_n$ and call it the *extremal discriminant.*

Lemma 2.1. *The truncation of $\widehat{\Delta}_n$ to the face h_K admits the factorization*

$$\widehat{\Delta}_n\big|_{h_K} = a_K^2 \prod_{i=0}^{p} \widehat{\Delta}_{l_i}(f_{K_i}), \tag{2.1}$$

where $a_K^2 = a_{k_1}^2 \dots a_{k_p}^2$, and Δ_{l_i} are the discriminants of polynomials of degrees l_i.

For the proof of this statement we use the formulas for monomials of $\widehat{\Delta}_n$, i.e., for extremal monomials of Δ_n.These formulas were derived by Gelfand, Kapranov and Zelevinsky (see [7, Chapter 12, Theorems 2.2 and 2.3]) with the help of complicated algebraic technique of the theory of A-determinants. Note that another proof of this theorem was given by Batyrev [4] using only the classical Newton method for finding branches of an algebraic function. This assertion about monomials is as follows.

Claim. *The Newton polytope of the discriminant of polynomial* (1.1) *is combinatorially equivalent to an $(n-1)$-dimensional cube; it contains 2^{n-1} vertices, which are in bijective correspondence with all possible subsets $I \subset \{1, 2, \dots, n-1\}$. The vertex* $\mathrm{v}(I)$ *corresponding to a subset $I = \{i_1 < i_2 < \dots < i_s\}$ has the coordinates*

$$\begin{aligned} &\mathrm{v}_0 = i_1 - 1, \quad \mathrm{v}_n = n - i_s - 1, && \\ &\mathrm{v}_{i_\nu} = i_{\nu+1} - i_{\nu-1} && \text{for } i_\nu \in I, \\ &\mathrm{v}_i = 0 && \text{for } i \notin I \cup \{0, n\}. \end{aligned} \tag{2.2}$$

Let $l_\nu = i_{\nu+1} - i_\nu$, $0 \le \nu \le s$, $i_0 = 0$, $i_{s+1} = n$. *Then the monomial* $a^{\mathrm{v}(I)}$ *is involved in* Δ *with the coefficient*

$$C_{\mathrm{v}(I)} = C(I) = \prod_{\nu=0}^{s} (-1)^{\frac{l_\nu(l_\nu-1)}{2}} l_\nu^{l_\nu}. \tag{2.3}$$

Since the discriminant is known to be bihomogeneous, the polytope $\mathcal{N}(\Delta_n)$ lies in a plane of $\mathbb{R}^{n+1}$ of codimension 2 defined by two equations

$$\sum_{j=0}^{n} t_j = 2(n-1), \quad \sum_{j=1}^{n} j t_j = n(n-1).$$

Lemma 2.2. *Within the indicated plane the Newton polytope* $\mathcal{N}(\Delta_n)$ *is defined by* $2(n-1)$ *inequalities*

$$t_k \ge 0, \qquad k = 1, \dots, n-1, \tag{2.4}$$

$$\sum_{j=1}^{k} (n-k) j t_j + \sum_{j=k+1}^{n-1} k(n-j) t_j \ge nk(n-k), \quad k = 1, \dots, n-1. \tag{2.5}$$

Remark 1. Note that these inequalities involve only coordinates $t_1, \dots, t_{n-1}$, so (2.4) and (2.5) determine the projection of $\mathcal{N}(\Delta_n)$ onto the subspace $\mathbb{R}^{n-1}$ of these coordinates. It follows that the system of inequalities (2.4)–(2.5) simultaneously defines the Newton polytope of the polynomial $\Delta_n(1, a_1, \dots, a_{n-1}, 1)$, i.e., the discriminant of the reduced equation

$$f^{\mathrm{red}} = 1 + a_1 y + \dots + a_{n-1} y^{n-1} + y^n = 0. \tag{2.6}$$

Proof of Lemma 2.2. For each of $k \in \{1, \dots, n-1\}$ consider an $(n-1)$-dimensional vector

$$\mu^{(k)} = ((n-k) \cdot 1, \dots, (n-k) \cdot k, k \cdot (n-k-1), \dots, k \cdot 1),$$

composed by the coefficients of linear functions participating in (2.5). In accordance with Remark 1 we may identity $\mu^{(k)}$ with its embedding $(0, \mu^k, 0) \in (\mathbb{R}^{n+1})^*$. The projection (on the subspace $\{t_0 = t_n = 0\} \subset \mathbb{R}^{n+1}$) of the vertex $\mathrm{v}(I) \in \mathcal{N}(\Delta_n)$ belongs to the plane

$$F_k := \{t \in \mathbb{R}^{n-1} : \langle \mu^{(k)}, t \rangle = nk(n-k)\} \tag{2.7}$$

if and only if the subset $I = \{i_1 < \dots, i_s\}$ contains k. Let us prove this.

If $k \in I$, say $k = i_\tau$ for some $\tau \in \{1, \dots, s\}$, then by (2.2) one has

$$\langle \mu^{(k)}, \mathrm{v}(I) \rangle = \sum_{l=1}^{\tau} (n-k) i_l (i_{l+1} - i_{l-1}) + \sum_{l=\tau+1}^{s} k(n-i_l)(i_{l+1} - i_{l-1}).$$

Here the first sum is equal $(n-k)i_\tau i_{\tau+1}$ (recall that $i_0 = 0$), and the second sum is equal to

$$k(n^2 + i_\tau i_{\tau+1} - n(i_\tau + i_{\tau+1})).$$

Since $i_\tau = k$, we get

$$\langle \mu^{(k)}, v(I)\rangle = nk(n-k),$$

i.e., equality in a not strict inequality (2.5).

Consider now the case when $k \notin I = \{i_1 < \cdots < i_s\}$. Assume that $i_p < k < i_{p+1}$, and consider the set

$$I' = \{i_1 < \cdots < i_p < k < i_{p+1} < \cdots < i_s\}.$$

The vectors $\mathrm{v}(I)$, $\mathrm{v}(I')$ differ only in coordinates with indices (numbers) i_p, k, i_{p+1}:

$$\begin{aligned} \mathrm{v}_{i_p} &= i_{p+1} - i_{p-1}, & \mathrm{v}'_{i_p} &= k - i_{p-1}, \\ \mathrm{v}_k &= 0, & \mathrm{v}'_k &= i_{p+1} - i_p, \\ \mathrm{v}_{i_{p+1}} &= i_{p+2} - i_p, & \mathrm{v}'_{i_{p+1}} &= i_{p+2} - k. \end{aligned}$$

So the corresponding scalar products are equal

$$\begin{aligned} \langle \mu^{(k)}, \mathrm{v}(I)\rangle &= \cdots + (n-i_p)i_p(i_{p+1} - i_{p-1}) + 0 \\ &\quad + i_{p+1}(n - i_{p+1} - 1)(i_{p+2} - i_p) + \cdots, \\ \langle \mu^{(k)}, \mathrm{v}(I')\rangle &= \cdots + (n-i_p)i_p(k - i_{p-1}) + (n-k)k(i_{p+1} - i_p) \\ &\quad + i_{p+1}(n - i_{p+1} - 1)(i_{p+2} - k) + \cdots \end{aligned}$$

and a difference of these values is equal to

$$-i_p^2(i_{p+1} - k) - k^2(i_{p+1} - i_p) - (i_{p+1} + 1)i_{p+1}(k - i_p),$$

i.e., is negative. Note that $k \in I'$, therefore, taking into account the proved above, one has

$$\langle \mu^{(k)}, \mathrm{v}(I') = nk(n-k)\rangle.$$

As a result, in the case when $k \notin I$ we arrived the strict inequality.

$$\langle \mu^{(k)}, \mathrm{v}(I) < nk(n-k)\rangle.$$

Lemma 2.2 is proved. ∎

Proof of Lemma 2.1. According to formulas (2.2) and (2.3), the extremal discriminant $\widehat{\Delta}_{l_j}$ participating in the right-hand side of the equation (2.1) admits the following representations.

(A) For $j = 0$:

$$\widehat{\Delta}_{l_0}(a_0, a_1, \ldots, a_{k_1}) = \sum_{\{I^{(0)}\}} C(I^{(0)}) a^{\mathrm{v}(I^{(0)})},$$

where

$$I^{(0)} = \{i_1^{(0)}, \dots, i_{s_0}^{(0)}\}$$

runs through all possible subsets of $\{1, 2, \dots, k_1 - 1\}$ and

$$a^{\nu(I^{(0)})} = a_0^{i_1^{(0)}-1} a_{i_1^{(0)}}^{i_2^{(0)}} a_{i_2^{(0)}}^{i_3^{(0)}-i_1^{(0)}} \dots a_{i_{s_0}^{(0)}}^{k_1-i_{s_0-1}^{(0)}} a_{k_1}^{k_1-i_{s_0}^{(0)}-1}.$$

(B) For $1 \le j \le p-1$:

$$\widehat{\Delta}_{l_j}(a_{k_j}, a_{k_j+1}, \dots, a_{k_{j+1}}) = \sum_{\{I^{(j)}\}} C(I^{(j)}) a^{\nu(I^{(j)})},$$

where

$$I^{(j)} = \{i_1^{(j)}, \dots, i_{s_j}^{(j)}\}$$

runs through all possible subsets of $\{k_j + 1, \dots, k_{j+1} - 1\}$ and

$$a^{\nu(I^{(j)})} = a_{k_j}^{i_1^{(j)}-1} a_{i_1^{(j)}}^{i_2^{(j)}} a_{i_2^{(j)}}^{i_3^{(j)}-i_1^{(j)}} \dots a_{i_{s_j}^{(j)}}^{k_{j+1}-i_{s_j-1}^{(j)}} a_{k_{j+1}}^{k_{j+1}-i_{s_j}^{(j)}-1}.$$

(C) For $j = p$:

$$\widehat{\Delta}_{l_p}(a_{k_p}, a_{k_p+1}, \dots, a_n) = \sum_{\{I_p\}} C(I^{(p)}) a^{\nu(I^{(p)})},$$

where

$$I^{(p)} = \{i_1^{(p)}, \dots, i_{s_p}^{(p)}\}$$

runs through all possible subsets of $k_p + 1, \dots, n-1$ and

$$a^{\nu(I_p)} = a_{k_p}^{i_1^{(p)}-1} a_{i_1^{(p)}}^{i_2^{(p)}} \dots a_{i_s^{(p)}}^{n-i_{s_p-1}^{(p)}} a_n^{n-i_{s_p}^{(p)}-1}.$$

Each tuple of $p+1$ sets $I^{(0)}, \dots, I^{(p)}$ defines in $\{1, \dots, n-1\}$ a subset $\widetilde{I}$ of the type

$$\{i_1^{(0)} < \dots < i_{s_0}^{(0)} < k_1 < i_1^{(1)} + k_1 < \dots < i_{s_1}^{(1)} + k_1 < k_2 < \dots < k_p < \dots < i_{s_p}^{(p)} + k_p\},$$

which contains $k_1, \dots, k_p$. And, vice versa, each such subset $\widetilde{I}$ defines the tuple $I^{(0)}, \dots, I^{(p)}$ of subsets, described in (A), (B), (C). Since in the list of subsets

$$\begin{gathered}
\{i_1^{(0)}, i_2^{(0)}, \dots, i_{s_0}^{(0)}, k_1\}, \\
\{k_1, i_1^{(1)} + k_1, i_2^{(1)} + k_1, \dots, k_2\}, \\
\vdots \\
\{k_p, i_1^{(p)} + k_p, \dots, i_{s_p}^{(p)} + k_p\}
\end{gathered}$$

any pair of adjacent subsets intersect in a single element, we can conclude by (2.3) that

$$C(I^{(0)}) \cdots C(I^{(p)}) = C(\widetilde{I}).$$

It proves Lemma 2.1. ∎

3 Parametrizations of zero sets of discriminants

It suffices to prove the assertion of the theorem for the discriminant

$$\Delta_n^{\mathrm{red}}(x_1, \ldots, x_{n-1}) = \Delta_n(1, x_1, \ldots, x_{n-1}, 1)$$

which corresponds to the reduced polynomial (2.6). In this case the identity in (1.3) becomes

$$\Delta_n^{\mathrm{red}}(x)\big|_{h_K} = a_K^2 \Delta_{l_0}(1, x_1, \ldots, x_{k_1}) \left[\prod_{i=1}^{p-1} \Delta_{l_i}(f_{K_i}) \right] \Delta_{l_p}(x_{k_p}, \ldots, x_{n-1}, 1), \quad (3.1)$$

where the first and the last discriminants Δ_{l_0} and Δ_{l_p} correspond to the "half-reduced" polynomial, obtained from f_{K_0} and f_{K_p} by fixing $a_0 = 1$ and $a_n = 1$ in f_{K_0} and f_{K_p}, respectively.

Let us comment formula (3.1) by the reduced discriminant of order 4:

$$\begin{aligned}\Delta_4^{\mathrm{red}} = {} & 256 - 192x_1x_3 - 128x_2^2 + 144x_2x_3^2 - 4x_1^2x_2^3 - 4x_2^3x_3^2 - 4x_1^3x_3^3 \\ & + 144x_1^2x_2 - 6x_1^2x_3^2 + 16x_2^4 - 27x_1^4 - 27x_3^4 - 80x_1x_2^2x_3 \\ & + 18x_1x_2x_3^3 + 18x_1^3x_2x_3 + x_1^2x_2^2x_3^2.\end{aligned}$$

For its Newton polytope see Figure 1.2. Recall that the hyperfaces h_k are opposite to the coordinate planes $t_k = 0$. The truncation of this discriminant to the hyperface h_2 is equal to

$$\begin{aligned}\Delta_4^{\mathrm{red}}\big|_{h_2} &= 16x_2^4 + x_1^2x_2^2x_3^2 - 4x_2^2x_3^2 - 4x_1^2x_2^3 \\ &= x_2^2(x_1^2 - 4x_2)(x_3^2 - 4x_2) \\ &= x_2^2\Delta_2(1, x_1, x_2)\Delta_2(x_2, x_3, 1).\end{aligned}$$

The truncation to the edge $h_1 \cap h_2$ (the segment between points $(2, 2, 2)$ and $(2, 3, 0)$) is equal to

$$\begin{aligned}x_1^2x_2^2x_3^2 - 4x_1^2x_2^3 &= x_1^2x_2^2 \cdot 1 \cdot 1 \cdot (x_3^2 - 4x_2) \\ &= x_1^2x_2^2\Delta_1(1, x_1)\Delta_1(x_1, x_2)\Delta_2(x_2, x_3, 1).\end{aligned}$$

Our proof of the theorem is based on Horn–Kapranov parametrization for the reduced discriminant set

$$\nabla_n^{\mathrm{red}} := \{x \in \mathbb{C}^{n-1} : \Delta_n^{\mathrm{red}}(x) = 0\},$$

which was obtained in [14] with the use of the Silvester determinant for Δ_n and elementary properties of the logarithmic Gauss map for ∇_n^{red}. This parametrization is n-valued and it is given by

$$x_j = -\frac{ns_j}{\langle \alpha, s\rangle}\left(\frac{\langle \alpha, s\rangle}{\langle \beta, s\rangle}\right)^{\frac{j}{n}}, \quad j = 1, \ldots, n-1, \quad (3.2)$$

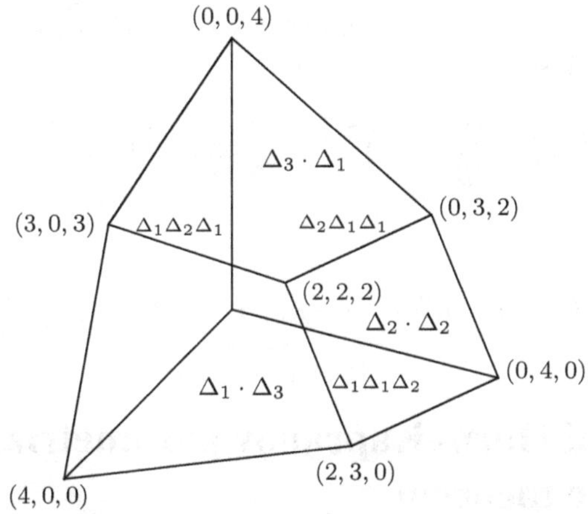

Figure 1.2. Truncated factorization for Δ_4^{red}.

where β and α are integer vectors $(1, 2, \dots, n-1)$ and $(n-1, n-2, \dots, 1)$, respectively, and $s = (s_1 : \dots : s_{n-1})$ is a parameter from $\mathbb{CP}^{n-2} \setminus \{s : \langle \alpha, s\rangle\langle \beta, s\rangle = 0\}$.

In view of bihomogeneity the complete and reduced discriminants $\Delta_n(a)$ and $\Delta_n^{\mathrm{red}}(x)$ are related by a substitution

$$x_j(a) = \frac{a_j}{a_0}\left(\frac{a_0}{a_n}\right)^{\frac{j}{n}}, \quad j = 1, \dots, n-1.$$

Therefore, in view of (3.2), the complete discriminant set $\nabla_n = \{\Delta_n(a) = 0\}$ admits the parametrization

$$\begin{cases} a_0 = s_0, \\ a_j = -\dfrac{ns_j}{\langle \alpha, s\rangle} s_0 \left(\dfrac{s_n}{s_0}\dfrac{\langle \alpha, s\rangle}{\langle \beta, s\rangle}\right)^{\frac{j}{n}}, \quad j = 1, \dots, n-1, \\ a_n = s_n, \end{cases} \tag{3.3}$$

where $(s_0, s_n) \in (\mathbb{C} \setminus 0)^2$. Setting $n = k$ and $a_0 = s_0 = 1$, we conclude that the discriminant set $\Delta_k(1, x_1, \dots, x_k) = 0$ of the "semi-reduced" polynomial

$$1 + x_1 y + \dots + x_k y^k$$

has parametrization of the form

$$x_j = -\frac{ks_j}{\langle \alpha', s'\rangle}\left(\frac{s_k\langle \alpha', s'\rangle}{\langle \beta', s'\rangle}\right)^{\frac{j}{k}}, \quad j = 1, \dots, k-1, \qquad x_k = s_k, \tag{3.4}$$

where

$$\beta' = (1, 2, \dots, k-1), \quad \alpha' = (k-1, k-2, \dots, 1), \quad s' = (s_1 : \dots : s_{k-1}).$$

For the parametrization of the discriminant set $\Delta_{n-k}(x_k, x_{k+1}, \dots, x_{n-1}, 1) = 0$, a similar formula holds:

$$x_j = -\frac{(n-k)s_j}{\langle \alpha'', s'' \rangle} s_k \Big(\frac{1}{s_k} \frac{\langle \alpha'', s'' \rangle}{\langle \beta'', s'' \rangle} \Big)^{\frac{j-k}{n-k}}, \quad j = k+1, \dots, n-1, \qquad x_k = s_k,$$

where

$$\beta'' = (1, \dots, n-k-1), \quad \alpha'' = (n-k-1, \dots, 1), \quad s'' = (s_{k+1} : \dots : s_{n-1}).$$

4 Blow-ups of Horn–Kapranov parametrizations and proof of the theorem

Firstly we describe the relationship between the reduced discriminant and its truncation to the face $h_K = h_{k_1} \cap \dots \cap h_{k_p}$. Let $x = (x_1, \dots, x_{n-1})$ and $\Delta_n^{\mathrm{red}}(x)$ is a reduced discriminant. In view of (2.7) the normal space to the face h_K is generated by vectors $\mu^{(k_1)}, \dots, \mu^{(k_p)}$. Denote this normal space by $F_K^{\perp}$.

Let μ be the $p \times (n-1)$-matrix composed by the rows $\mu^{(k_j)}$ and we denote by $\mu_{(1)}, \dots, \mu_{(n-1)}$ the columns of μ.

Consider the sublattice

$$M := \mathbb{Z}^{n-1} \cap F_k^{\perp} \subset \mathbb{Z}^{n-1}$$

of rank p, and choose in M a basis $\omega^{(1)}, \dots, \omega^{(p)}$ with non-negative coordinates. Let A be an integer $p \times p$-matrix such that $A\omega = \mu$, where ω is a $p \times (n-1)$-matrix composed by the rows $\omega^{(1)}, \dots, \omega^{(p)}$.

In the complex algebraic torus $(\mathbb{C} \setminus 0)^{n-1}$ with coordinates x consider a complex p-dimensional torus $x = \tau^{\omega}$, $\tau \in (\mathbb{C} \setminus 0)^p$, where the matrix monomial τ^{ω} means the usual monomial map

$$x_j = \tau^{\omega_{(j)}} = \tau_1^{\omega_{(j)}^{(1)}} \dots \tau_p^{\omega_{(j)}^{(p)}}, \quad j = 1, \dots, n-1,$$

where $\omega_{(j)}$ are the columns of ω. Define the function

$$H_K(\tau; x) := \tau_1^{d_1} \dots \tau_p^{d_p} \cdot \Delta_n^{\mathrm{red}} \Big(\frac{x_1}{\tau^{\omega_{(1)}}}, \dots, \frac{x_{n-1}}{\tau^{\omega_{(n-1)}}} \Big), \tag{4.1}$$

where d_i is the weighted degree of Δ_n^{red} respective the weight $\omega_{(i)}$, i.e., the maximum among the scalar products $\langle \lambda, \omega^{(i)} \rangle$ when λ runs over all exponents of monomials from $\Delta_n^{\mathrm{red}}(x)$. It is clear that all monomials $c_\lambda x^\lambda$ from $\Delta_n^{\mathrm{red}}(x)$, for which

$$\langle \lambda, \omega^{(i)} \rangle = d_i, \quad i = 1, \dots, p,$$

compose the truncation $\Delta_n^{\mathrm{red}}|_{h_K}$.

Lemma 4.1. *When $\tau \to 0$, the function $H_K(\tau; x)$ converges to the truncation of the discriminant to h_K:*

$$H_K(\tau; x) \xrightarrow{\tau \to 0} \Delta_n^{\mathrm{red}}(x)\big|_{h_K}. \tag{4.2}$$

To prove Theorem 1.1, we analyze parametrization (3.2) of reduced discriminant set. At the points of the union of hyperplanes

$$\{\langle \alpha, s\rangle = 0\} \cup \{\langle \beta, s\rangle = 0\} \subset \mathbb{CP}^{n-2},$$

where $s_j \neq 0$, parametrization (3.2) does not take finite value. However at uncertainly points, where some coordinate functions s_j vanish simultaneously with at least one of the forms $\langle \alpha, s\rangle$, $\langle \beta, s\rangle$, this parametrization gives limit positions for the discriminant set. According to the theory of correspondences [13, Section 2B] these limit positions are interpreted as blow-ups in the space $\mathbb{CP}_s^{n-2} \times \mathbb{C}_x^{n-1}$, which contains the graph of the map (3.2).

Lemma 4.2. *In equality* (3.1) *the zero set of the truncation $\Delta_n^{\mathrm{red}}|_{h_K}$ on the left-hand side contains the zero of the product on the right-hand side.*

Proof. To prove Lemma 4.2, we examen the zero sets

$$Z_\tau = \{x \in \mathbb{C}^{n-1} : H_K(\tau; x) = 0\}, \quad \tau \neq 0$$

of the function (4.1). According to (3.2), these sets admit the parametrization

$$x_j = -\tau^{\omega(j)} \frac{n s_j}{\langle \alpha, s\rangle} \left(\frac{\langle \alpha, s\rangle}{\langle \beta, s\rangle} \right)^{\frac{j}{n}}, \quad j = 1, \dots, n-1. \tag{4.3}$$

We have to prove that the zero set of the truncation to h_K for the reduced n-discriminant contains the zero sets of two extreme (on the right-hand side of (3.1)) semi-reduced discriminants and of each complete discriminant Δ_{l_i}, $1 \le i \le p-1$. Situations for extreme discriminants Δ_{l_0} and Δ_{l_p} are of the same type, so we consider only the semi-reduced discriminant Δ_{l_0} and an arbitrary complete discriminant Δ_{l_i} (they are not present, if $p = 1$). Therefore, for the normal vectors $\mu^{(k_1)}, \dots, \mu^{(k_p)}$, we need only their coordinates indicated by tuples

$$K_0 = \{1, \dots, k_1\} \quad \text{and} \quad K_i = \{k_i, k_i + 1, \dots, k_{i+1}\}, \quad 1 \le i \le p-1.$$

Recall that the columns of the matrix μ we denote by $\mu_{(j)}$. Due to Lemma 2.2, the two blocks of the matrix μ, corresponding to K_0 and K_i, are as follows.

	$\mu_{(1)}$	$\dots$	$\mu_{(k_1)}$	$\mu_{(k_i)}$	$\mu_{(k_i+1)}$	$\dots$	$\mu_{(k_{i+1})}$
$\mu^{(k_1)}$	$(n-k_1)\cdot 1$	$\dots$	$(n-k_1)k_1$	$k_1(n-k_i)$	$k_1(n-k_i-1)$	$\dots$	$k_1(n-k_{i+1})$
$\vdots$	$\vdots$	$\vdots$	$\vdots$	$\vdots$	$\vdots$	$\vdots$	$\vdots$
$\mu^{(k_i)}$	$(n-k_i)\cdot 1$	$\dots$	$(n-k_i)k_1$	$(n-k_i)k_i$	$k_i(n-k_i-1)$	$\dots$	$k_i(n-k_{i+1})$
$\vdots$	$\vdots$	$\vdots$	$\vdots$	$\vdots$	$\vdots$	$\vdots$	$\vdots$
$\mu^{(k_p)}$	$(n-k_p)\cdot 1$	$\dots$	$(n-k_p)k_1$	$(n-k_p)k_i$	$(n-k_p)(k_i+1)$	$\dots$	$(n-k_p)k_{i+1}$.

Here the first k_1 columns $\mu_{(j)}$, $j \in K_0$ are related by

$$\mu_{(j)} = \frac{j}{k_1}\mu_{(k_1)}, \quad j = 1, \dots, k_1 - 1.$$

The columns indicated by $j \in K_i$ are related by

$$\mu_{(k_i+t)} = \left(1 - \frac{t}{l_i}\right)\mu_{(k_i)} + \frac{t}{l_i}\mu_{(k_{i+1})}, \quad t = 1, \dots, l_i - 1,$$

where $l_i = k_{i+1} - k_i$. Since the matrix $\omega = A^{-1}\mu$ obtained from μ by linear transform, the mentioned relations are saved for the columns of the matrix ω:

$$\omega_{(j)} = \frac{j}{k_1}\omega_{(k_1)}, \qquad j = 1, \dots, k_1 - 1. \tag{4.4}$$

$$\omega_{(k_i+t)} = \left(1 - \frac{t}{l_i}\right)\omega_{(k_i)} + \frac{t}{l_i}\omega_{(k_{i+1})}, \quad t = 1, \dots, l_i - 1. \tag{4.5}$$

In view of (4.3) and (4.4) one has

$$x_j = (-1)^{1-\frac{j}{k_1}} \frac{n^{1-\frac{j}{k_1}} s_j}{\langle \alpha, s \rangle} \left(\frac{x_{k_1}\langle \alpha, s \rangle}{s_{k_1}}\right)^{\frac{j}{k_1}}, \quad j = 1, \dots, k_1 - 1. \tag{4.6}$$

Let us define in the projective space with homogeneous coordinates $(s_1 : \cdots : s_{n-1})$ the plane σ' by equations

$$\langle \beta, s \rangle = 0, \quad s_{k_1+1} = \cdots = s_{n-1} = 0.$$

The computations give

$$s_{k_1}\big|_{\sigma'} = -\frac{1}{k_1}\langle \beta', s' \rangle, \quad \langle \alpha, s \rangle\big|_{\sigma'} = \frac{n}{k_1}\langle \alpha', s' \rangle,$$

where

$$s' = (s_1 : \cdots : s_{k_1-1}), \quad \beta' = (1, \dots, k_1 - 1), \quad \alpha' = (k_1 - 1, \dots, 1).$$

Substituting the resulting restrictions $s_{k_1}|_{\sigma'}$ and $\langle \alpha, s \rangle|_{\sigma'}$ into formula (4.6) yields the following expressions for the restrictions x_j:

$$x_j\big|_{\sigma'} = -\frac{k_1 s_j}{\langle \alpha', s' \rangle} \left(\frac{s_{k_1}\langle \alpha', s' \rangle}{\langle \beta', s' \rangle}\right)^{\frac{j}{k_1}}, \quad j = 1, \dots, k_1 - 1, \qquad x_{k_1} = s_{k_1}.$$

Thus, in view of (3.4), we conclude that, under the projection onto the subspace of coordinates $x_1, \dots, x_{k_1}$ the limit (as $\tau \to 0$) set of the left-hand side of (4.2) is mapped into the discriminant set $\{\Delta_{l_0}(1, x_1, \dots, x_{k_1}) = 0\}$.

Now we have to study the contribution by limited positions of the parametrization (4.3), connected by relations (4.5). Due to these relations we exclude in (4.3) the parameters $\tau = (\tau_1, \dots, \tau_p)$, considering the ratios

$$\frac{x_{k_i+t}}{x_{k_i}^{1-\frac{t}{l_i}} x_{k_{i+1}}^{\frac{t}{l_i}}}, \quad t = 1, \dots, l_i - 1.$$

Computations give

$$\begin{aligned}\frac{x_{k_i+t}}{x_{k_i}^{1-\frac{t}{l_i}} x_{k_{i+1}}^{\frac{t}{l_i}}} &= \frac{n s_{k_i+t}}{\langle\alpha, s\rangle}\left(\frac{\langle\alpha, s\rangle}{\langle\beta, s\rangle}\right)^{\frac{k_i+t}{n}}\left[\frac{\langle\alpha, s\rangle}{n s_{k_i}}\left(\frac{\langle\beta, s\rangle}{\langle\alpha, s\rangle}\right)^{\frac{k_i}{n}}\right]^{1-\frac{t}{l_i}} \\ &\quad\times\left[\frac{\langle\alpha, s\rangle}{n s_{k_{i+1}}}\left(\frac{\langle\beta, s\rangle}{\langle\alpha, s\rangle}\right)^{\frac{k_{i+1}}{n}}\right]^{\frac{t}{l_i}} \\ &= \frac{s_{k_i+t}}{s_{k_i}^{1-\frac{t}{l_i}} s_{k_{i+1}}^{\frac{t}{l_i}}} = \frac{s_{k_i+t}}{s_{k_i}}\left(\frac{s_{k_i}}{s_{k_i+t}}\right)^{\frac{t}{l_i}},\end{aligned}$$

and therefore

$$x_{k_i+t} = x_{k_i}\frac{s_{k_i+t}}{s_{k_i}}\left(\frac{x_{k_{i+1}}}{x_{k_i}}\frac{s_{k_i}}{s_{k_{i+1}}}\right)^{\frac{t}{l_i}}.$$

Consider the restriction of x_{k_i+t} on the plane

$$\sigma'' = \{\langle\alpha, s\rangle = \langle\beta, s\rangle = 0 : s_j = 0 \text{ for all } j \notin K_i\}.$$

For that we solve the system of equations

$$\begin{cases} k_i s_{k_i} + (k_i + 1)s_{k_i+1} + \dots + k_{i+1}s_{k_{i+1}} = 0, \\ (n - k_i)s_{k_i} + (n - k_i - 1)s_{k_i+1} + \dots + (n - k_{i+1})s_{k_{i+1}} = 0 \end{cases}$$

respective the unknown s_{k_i} and $s_{k_{i+1}}$. By the first equation we get

$$s_{k_i} = -\frac{1}{k_i}\big((k_i + 1)s_{k_i+1} + (k_i + 2)s_{k_i+2} + \dots + (k_{i+1} - 1)s_{k_{i+1}-1} + k_{i+1}s_{k_{i+1}}\big).$$

After substitution of this expression into second equation we arrive at the equation

$$-\frac{n}{k_i}\big(1 \cdot s_{k_i+1} + 2 \cdot s_{k_i+2} + \dots + (l_i - 1)s_{k_i+l_i-1} + l_i s_{k_{i+1}}\big) = 0$$

(recall that $l_i = k_{i+1} - k_i$). It gives the expressions

$$\begin{aligned} s_{k_{i+1}} &= -\frac{1}{l_i}\big(1 \cdot s_{k_i+1} + 2 \cdot s_{k_i+2} + \dots + (l_i - 1)s_{k_{i+1}-1}\big), \\ s_{k_i} &= -\frac{1}{l_i}\big((l_i - 1)s_{k_i+1} + (l_i - 2)s_{k_i+2} + \dots + 1 \cdot s_{k_{i+1}}\big). \end{aligned}$$

Finally, for $t = 1, \dots, l_i - 1$ we get

$$x_{k_i+t} = -\frac{l_i s_{k_i+t}}{\langle \alpha'', s'' \rangle} s_{k_i} \left(\frac{s_{k_{i+1}}}{s_{k_i}} \frac{\langle \alpha'', s'' \rangle}{\langle \beta'', s'' \rangle} \right)^{\frac{t}{l_i}}, \quad x_{k_i} = s_{k_i}, \quad x_{k_{i+1}} = s_{k_{i+1}},$$

where

$$s'' = (s_{k_i}, \dots, s_{k_{i+1}}), \quad \beta'' = (1, 2, \dots, l_i - 1), \quad \alpha'' = (l_i - 1, \dots, 2, 1).$$

By (3.3) under the projection onto subspace of coordinates $x_{k_i}, x_{k_i+1}, \dots, x_{k_{i+1}}$ the limit zero set of the left-hand side of formula (3.1) is mapped into the discriminant set $\{\Delta_{l_i}(f_{K_i}) = 0\}$.

Noting that $\Delta_n^{\mathrm{red}}|_{h_K}$ is divisible by x_K^2, we finish the proof Lemma 4.2. ∎

Equality (3.1) is proved as follows. By Lemma 2.1, the Newton polytopes of the polynomials on the left and right-hand sides of (3.1) coincide. In this case, intersection theory and Lemma 4.2 imply that these polynomials have identical zero sets. However, by Lemma 2.1, their extremal parts coincide, so the polynomials coincide as well.

Acknowledgements. A. Tsikh was supported by the Russian Science Foundation No. 20-11-20117. The authors wish to thank V. Batyrev, A. Esterov and V. Vassiliev for very helpful discussions.

Bibliography

[1] I. A. Antipova, E. N. Mikhalkin and A. K. Tsikh, Rational expressions for multiple roots of algebraic equations. *Sb. Math.* **209** (2018), 1419–1444

[2] I. A. Antipova and A. K. Tsikh, The discriminant locus of a system of n Laurent polynomials in n variables. *Izv. Math.* **76** (2012), 881–906

[3] V. Batyrev, Truncated discriminants and toric moduli spaces. In progress

[4] V. Batyrev, Winter School lectures in Arizona. 2004, http://swc.math.arizona.edu/aws/2004/04BatyrevNotes.pdf

[5] A. Dickenstein, E. M. Feichtner and B. Sturmfels, Tropical discriminants. *J. Amer. Math. Soc.* **20** (2007), 1111–1133

[6] A. Esterov, Galois theory for general systems of polynomial equations. *Compos. Math.* **155** (2019), 229–245

[7] I. M. Gelfand, M. M. Kapranov and A. V. Zelevinsky, *Discriminants, resultants, and multidimensional determinants*. Math. Theory Appl., Birkhäuser Boston, Boston, 1994

[8] I. M. Gelfand, A. V. Zelevinsky and M. M. Kapranov, Hypergeometric functions and toric varieties. *Funct. Anal. Appl.* **23** (1989), 94–106

[9] M. M. Kapranov, A characterization of A-discriminantal hypersurfaces in terms of the logarithmic Gauss map. *Math. Ann.* **290** (1991), 277–285

[10] E. N. Mikhalkin, V. A. Stepanenko and A. K. Tsikh, Geometry of factorization identities for discriminants. *Dokl. Math.* **102** (2020), no. 1, 279–282

[11] E. N. Mikhalkin and A. K. Tsikh, On the structure of the classical discriminant. *J. Sib. Fed. Univ. Ser. Math. Phys.* **8** (2015), no. 4, 426–436

[12] E. N. Mikhalkin and A. K. Tsikh, Singular strata of cuspidal type for the classical discriminant. *Sb. Math.* **206** (2015), 282–310

[13] D. Mumford, *Algebraic geometry. I.* Springer, Berlin, 1976

[14] M. Passare and A. Tsikh, Algebraic equations and hypergeometric series. In *The legacy of Niels Henrik Abel*, pp. 653–672, Springer, Berlin, 2004

[15] T. M. Sadykov and A. K. Tsikh, *Hypergeometric and algebraic functions in several variables* (in Russian). Nauka, Moscow, 2014

[16] V. A. Vassiliev, *Topology of discriminants complements* (in Russian). Moscow, 1997

[17] V. A. Vassiliev, Stable cohomology of spaces of non-resultant systems of homogeneous polynomials in $\mathbb{R}^n$. *Dokl. Math.* **98** (2018), 330–333

Eigenvalue estimates and asymptotics for weighted pseudodifferential operators with singular measures in the critical case

Grigori Rozenblum and Eugene Shargorodsky

To Ari Laptev, with best wishes, on the occasion of his 70th birthday

We consider self-adjoint operators of the form $\mathbf{T}_{P,\mathfrak{A}} = \mathfrak{A}^* P \mathfrak{A}$ in a domain $\Omega \subset \mathbb{R}^{\mathbf{N}}$, where $\mathfrak{A}$ is an order $-l = -\frac{\mathbf{N}}{2}$ pseudodifferential operator in Ω and P is a signed Borel measure with compact support in Ω. Measure P may contain singular component. For a wide class of measures we establish eigenvalue estimates for operator $\mathbf{T}_{P,\mathfrak{A}}$. In case of measure P being absolutely continuous with respect to the Hausdorff measure on a Lipschitz surface of an arbitrary dimension, we find the eigenvalue asymptotics. The order of eigenvalue estimates and asymptotics does not depend on dimensional characteristics of the measure, in particular, on the dimension of the surface supporting the measure.

1 Introduction

We study the eigenvalue distribution for self-adjoint compact operators of the type $\mathfrak{A}^* P \mathfrak{A}$, were $\mathfrak{A}$ is a pseudodifferential operator of negative order $-l$ in a domain $\Omega \subset \mathbb{R}^{\mathbf{N}}$ and P is a signed measure in Ω. A number of spectral problems can be reduced to this one, the most important one, probably, being $\lambda(-\Delta)^l u = Pu$, closely related to the Schrödinger operator. In particular, if P is an absolutely continuous measure, the spectral asymptotics for such operators has been justified under rather mild conditions imposed on P, see [3]. The case of a singular measure on Ω is not that well studied. In 1951, M. G. Krein discovered that for the "singular string" described by the equation $-\lambda u'' = Pu$ with the Dirichlet boundary conditions at the endpoints of an interval and with P being a Borel measure, the leading term in the

Keywords: Eigenvalue estimates, eigenvalue asymptotics, pseudodifferential operators, singular measures
2020 Mathematics Subject Classification: Primary 47A75; secondary 47G30, 58C40, 58J50

asymptotics of the eigenvalues is determined by the absolutely continuous part P_{ac} of P only, while the singular part P_{sing} makes a weaker contribution. Further on, in papers by Birman, Solomyak, and Borzov, see, e.g., [3], this property of singular measures was proved for a wide class of "high-order" spectral problems, in particular for $\lambda(-\Delta)^l u = Pu$ in a domain $\Omega \subset \mathbb{R}^{\mathbf{N}}$, provided $2l > \mathbf{N}$. For "low-order problems", $2l < \mathbf{N}$, the influence of the singular part of the measure is different. The known cases concern P_{sing} concentrated on a surface of codimension 1, and here, in the opposite, it makes a leading-order contribution to the spectral asymptotics, see [6]. The intermediate, "critical" case $2l = \mathbf{N}$ has been studied even less (it is the common wisdom that for many questions in spectral theory this case is the hardest one). Until very recently, there were very few results here. In dimension $\mathbf{N} = 2$, if a measure P is concentrated at the boundary of Ω (which is equivalent to the Steklov problem) or on a smooth curve inside Ω, the eigenvalue asymptotics has the same order as for the regular Dirichlet problem, namely, $\lambda_k \sim Ck^{-1}$, see also general results in [1,6,10].

Recently, the interest to the critical case has revived, due to some new applications, see, e.g., [11,16]. So, a class of singular measures was considered in [8]. For a singular measure $P = V\mu$ in $\mathbb{R}^2$, where μ is Ahlfors α-regular, $\alpha \in (0,2)$, and V belongs to a certain Orlicz class, an estimate for the eigenvalues of the problem $-\lambda\Delta u = Pu$ was obtained, $|\lambda_k| \le C(V,\mu)k^{-1}$. Thus, unlike other cases, the order in the eigenvalue *estimate* does not depend on the (Hausdorff) dimension of the support of the measure. The question of sharpness of these estimates was not touched upon. In another field, the critical case was involved in studies related to noncommutative integration, see [11].

In this paper, we consider a class of singular measures in the critical case and prove order-sharp estimates and asymptotic formulae for eigenvalues for the corresponding spectral problems. Namely, the measure $P = P_{sing}$ is supposed to be supported on a compact Lipschitz surface Σ of codimension $\mathfrak{d}$ in $\mathbb{R}^{\mathbf{N}}$ and absolutely continuous with respect to the surface measure μ_Σ induced by the embedding of Σ into $\mathbb{R}^{\mathbf{N}}$, $P = V\mu_\Sigma$. We find that the eigenvalues λ_k have asymptotics of order k^{-1}, so the order of asymptotics does not depend on the dimension or codimension of the surface. We consider the case of compactly supported measures only and therefore do not touch upon effects related with infinity, which, as it is known, may influence the eigenvalue behavior drastically, even for a nice measure, see, e.g., [2]. If P is concentrated on several surfaces, of different dimensions, their contributions to the eigenvalue asymptotics add up. The same happens if the measure has both an absolutely continuous and a singular part.

In the usual way, the estimates and asymptotics of eigenvalues of the weighted problem, by means of the Birman–Schwinger principle, produce similar results for the number of negative eigenvalues of the Schrödiger-like operator as the coupling constant grows. This relation was explored in [8].

Some of the results of the paper were presented in the short note [13] without proofs.

2 Setting and main results

Since the pathbreaking papers by Birman and Solomyak, it is known that it is often useful to reduce a spectral problem for a differential equation to an eigenvalue problem for a compact operator. We consider a class of compact operators encompassing the above problems, as well as their pseudodifferential versions, as particular cases. Let $\mathfrak{A}$ be an operator in a bounded domain $\Omega \subset \mathbb{R}^{\mathbf{N}}$ such that the localization of $\mathfrak{A}$ to a proper subdomain $\Omega' \subset \Omega$ is a pseudodifferential operator of order $-l = -\frac{\mathbf{N}}{2}$, up to a smoothing additive term. The operators $(-\Delta)^{-\frac{l}{2}}$ and $(-\Delta+1)^{-\frac{l}{2}}$ in $\mathbb{R}^d$, restricted to Ω, or $(-\Delta_{\mathcal{D}})^{-\frac{l}{2}}$, where $-\Delta_{\mathcal{D}}$ is the Dirichlet Laplacian in Ω, are typical examples of such $\mathfrak{A}$. By localization we mean multiplication of $\mathfrak{A}$ on both sides by a smooth cut-off function $\varsigma \in C_0^\infty(\Omega)$, $\varsigma|_{\Omega'} = 1$. Our results do not depend on the particular localization chosen, see Section 3; it is convenient to assume that the above multiplication has already been performed and the cut-offs are incorporated in $\mathfrak{A}$.

Let Σ be a compact Lipschitz surface in Ω', of dimension d, $1 \le d \le \mathbf{N}-1$, and, correspondingly, of codimension $\mathfrak{d} = \mathbf{N} - d$. Thus, locally, in appropriate coordinates, $X = (\mathbf{x}; \mathbf{y}) := (x_1, \dots, x_d; y_{d+1}, \dots, y_{\mathfrak{d}})$, the corresponding piece of the surface Σ is described by an equation $\mathbf{y} = \varphi(\mathbf{x})$, where φ is a Lipschitz vector-function. For brevity, we describe our constructions in a single co-ordinate neighborhood; the resulting formulae are glued together in a standard manner. The embedding $\Sigma \subset \mathbb{R}^{\mathbf{N}}$ generates the surface measure μ_Σ on Σ,

$$d\mu_\Sigma = \sigma(\mathbf{x})d\mathbf{x}, \quad \sigma(\mathbf{x}) = [\det(\mathbf{1} + (\nabla\varphi)^*(\nabla\varphi))]^{\frac{1}{2}} d\mathbf{x},$$

which, in turn, generates a singular measure on Ω, supported in Σ, which we also denote by μ_Σ, as long as this does not cause confusion.

Let μ be a Radon measure on Ω. We denote by $\mathbf{M}$ its support, the smallest closed set of full measure. The Orlicz space $L^\Psi(\mathbf{M}, \mu)$, $\Psi(t) = (1+t)\log(1+t) - t$, consists of μ-measurable functions V on $\mathbf{M}$, satisfying $\int_{\mathbf{M}} \Psi(|V(X)|)\, d\mu(X) < \infty$. For a μ-measurable subset $E \subset \Omega$, we define the norm

$$\|V\|_E^{(\mathrm{av},\Psi,\mu)} := \sup\left\{\left|\int_{E\cap\mathbf{M}} Vg\, d\mu\right| : \int_{E\cap\mathbf{M}} \Phi(|g|)\, d\mu \le \mu(E\cap\mathbf{M})\right\}$$

if $\mu(E\cap\mathbf{M}) > 0$, and $\|V\|_E^{(\mathrm{av},\Psi,\mu)} = 0$ otherwise; here Φ is the Orlicz complementary function to Ψ, $\Phi(t) = e^t - 1 - t$. Such *averaged* norms, first introduced by Solomyak in [17], have played an important role in the study of the eigenvalue distribution in the critical case. For a real-valued function $V \in L_1(\mathbf{M}, \mu)$, we consider the quadratic form

$$\mathbf{F}_V[u] = \mathbf{F}_V[u; \mathfrak{A}; \mu] := \int_{\mathbf{M}} V(X)|(\mathfrak{A}u)(X)|^2\, d\mu(X), \quad u \in L_2(\Omega). \tag{2.1}$$

We will see later that, defined initially on continuous functions, this quadratic form is bounded in $L_2(\Omega)$, as soon as V belongs to $L^\Psi(\mathbf{M}, \mu)$, and can be extended by continuity to the whole of $L_2(\Omega)$; in this way it defines a bounded selfadjoint operator. This operator is denoted by $\mathbf{T}(V, \mu) = \mathbf{T}(V, \mu, \mathfrak{A})$.

The case of principal interest for us is μ being the measure μ_Σ and $\mathbf{M}$ being the surface Σ. Here we will use the notation $L^\Psi = L^\Psi(\Sigma)$ for the Orlicz space, $\|V\|_E^{(\mathrm{av},\Psi)}$ for the averaged norm and $\mathbf{T}(V) \equiv \mathbf{T}(V, \Sigma) \equiv \mathbf{T}(V, \Sigma, \mathfrak{A})$ for the operator defined by the form (2.1) with $\mu = \mu_\Sigma$.

For a compact self-adjoint operator $\mathbf{T}$ in a Hilbert space, we denote by $\lambda_k^\pm(\mathbf{T})$ the positive (negative) eigenvalues of $\mathbf{T}$ in the non-increasing order of their absolute values, repeated according to their multiplicities. By $n_\pm(\lambda, \mathbf{T})$ we denote the counting function of $\lambda_k^\pm(\mathbf{T})$. The notation $n(\lambda, \mathbf{T})$ is used for the counting function of singular numbers of the (not necessarily self-adjoint) operator $\mathbf{T}$. When the operator is associated with a quadratic form $\mathbf{F}$, the notation $n_\pm(\lambda, \mathbf{F})$, etc., is sometimes used.

Our first main result is the following eigenvalue estimate.

Theorem 2.1. *Let Σ be a compact Lipschitz surface of dimension $d < \mathbf{N}$ in $\Omega' \subset \mathbb{R}^\mathbf{N}$ and $V \in L^\Psi(\Sigma)$. Then, for the operator $\mathbf{T} = \mathbf{T}(V, \Sigma, \mathfrak{A})$, the estimate*

$$n(\lambda, \mathbf{T}) \le C \|V\|_\Sigma^{(\mathrm{av},\Psi)} \lambda^{-1} \tag{2.2}$$

holds with a constant C depending on the surface Σ and the operator $\mathfrak{A}$ but independent of the function V.

Theorem 2.1 extends to the case of singular measures supported on surfaces in $\mathbb{R}^\mathbf{N}$ the estimates obtained by Solomyak in [17] for domains (cubes) in $\mathbb{R}^\mathbf{N}$ for an even $\mathbf{N}$ and in [18] for an odd $\mathbf{N}$. In both cases, the operator $\mathfrak{A}_0 = (1-\Delta)^{-\frac{\mathbf{N}}{4}}$ played the role of $\mathfrak{A}$. The passage to a more general $\mathfrak{A}$ is easy and is carried out at the end of Section 5.3.

Theorem 2.1 follows from a spectral estimate in a more general setting extending the considerations in [8].

Definition 2.2. Let μ be a positive Radon measure on $\mathbb{R}^\mathbf{N}$. We say that it is α-Ahlfors regular (an α-AR-measure), $\alpha \in (0, \mathbf{N}]$, if there exist positive constants c_0 and c_1 such that

$$c_0 r^\alpha \le \mu(B(X,r)) \le c_1 r^\alpha \tag{2.3}$$

for all $0 < r \le \operatorname{diam}(\operatorname{supp}\mu)$ and all $X \in \operatorname{supp}\mu$, where $B(X,r)$ is the ball of radius r centered at X and the constants c_0 and c_1 are independent of the balls.

If μ is an α-AR-measure, then it is equivalent to the α-dimensional Hausdorff measure on its support (see, e.g., [5, Lemma 1.2]). The measure μ_Σ for a compact Lipschitz surface of dimension d in $\mathbb{R}^\mathbf{N}$ is, obviously, d-AR.

Theorem 2.3. *Let μ be a compactly supported α-AR measure, $0 < \alpha \le \mathbf{N}$, and $V \in L^\Psi(\mu)$. Then, for the operator $\mathbf{T}(V, \mu, \mathfrak{A})$, the estimate*

$$n_\pm(\lambda, \mathbf{T}(V, \mu, \mathfrak{A})) \le C(\alpha, \mu, \mathfrak{A}) \|V\|_\mathbf{M}^{(\mathrm{av},\Psi)} \lambda^{-1}$$

holds with a constant $C(\alpha, \mu, \mathfrak{A})$ depending on the domain Ω, the measure μ, and the operator $\mathfrak{A}$ but not on the function V.

To formulate the result on the eigenvalue asymptotics, we need more notation. According to the Rademacher Theorem, the function φ is differentiable μ_Σ-almost everywhere. At such "regular" points X_0, the tangent d-dimensional plane $T_{X_0}\Sigma$ and the normal $\mathfrak{d}$-dimensional plane $N_{X_0}\Sigma$ exist. The principal symbol $a_{-l}(X, \Xi)$ of the operator $\mathfrak{A}$ can be expressed in a neighborhood of such a point in the coordinates $(\mathbf{x}, \mathbf{y}; \xi, \eta)$, with $\mathbf{x} \in T_{X_0}\Sigma$, $\mathbf{y} \in N_{X_0}\Sigma$ and the corresponding co-variables ξ, η; we denote it again $a_{-l}(X_0; \xi, \eta)$. In this notation, we set

$$r_{-d}(X_0, \xi) = (2\pi)^{-\mathfrak{d}} \int_{N_{X_0}\Sigma} |a_{-l}(X_0; \xi, \eta)|^2 \, d\eta, \quad \xi \in T^*_{X_0}\Sigma.$$

This function is defined almost everywhere on $T^*\Sigma$ and is order $-d$ homogeneous in ξ. Now we formulate our result on eigenvalue asymptotics.

Theorem 2.4. *Under the conditions of Theorem* 2.1, *the counting function for the eigenvalues of the operator* $\mathbf{T}$ *has the following asymptotics:*

$$n_\pm(\lambda, \mathbf{T}(V, \Sigma)) \sim \lambda^{-1} C_\pm(V, \Sigma, \mathfrak{A}) \tag{2.4}$$

with

$$C_\pm = d^{-1}(2\pi)^{-d} \int_{S^*\Sigma} V_\pm(X) r_{-d}(X, \xi) \, d\mu_\Sigma(X) \, d\xi,$$

where the integration is performed over the cosphere bundle of Σ *and* $V_\pm(X)$ *denotes the positive, respectively, negative part of the function* V.

If several Lipschitz surfaces, of possibly different dimensions, are present and the measure P has a possibly nontrivial absolutely continuous part, the above eigenvalue asymptotics still holds, with the coefficient being the sum of the coefficients calculated for all components of the measure.

An important particular case of our general considerations concerns $\mathfrak{A}$ being an appropriate negative power of the Laplace operator with some boundary conditions. More precisely, $\mathfrak{A}_0 = (-\Delta)^{-\frac{\mathbf{N}}{4}} + S$, where S is an operator smoothing inside Ω (typically, a singular Green operator.) In this case, the eigenvalue problem for the operator $\mathbf{T}$ can be reduced, up to negligible terms (which we disregard throughout this section), to the weighted polyharmonic eigenvalue problem

$$\lambda(-\Delta)^l f = Pf$$

understood in the distributional sense:

$$\lambda \int_\Omega f(-\Delta)^l h \, dX = \int_\Omega f h P, \quad h \in \mathcal{D}(\Omega), \tag{2.5}$$

or, for $P = V\mu_\Sigma$,

$$\lambda \int_\Omega f(-\Delta)^l h \, dX = \int_\Sigma f h V \, d\mu_\Sigma, \quad h \in \mathcal{D}(\Omega), \tag{2.6}$$

with some boundary conditions understood, again, in the distributional sense. If the geometry of Σ is sufficiently "nice", then the spectral problem (2.5) or (2.6) can be expressed more explicitly. For example, if $\mathbf{N} = 2$, $l = 1$, $P = V\mu_\Sigma$, and Σ is a Lipschitz curve *inside* Ω, we arrive (see, e.g., [1]) at the *transmission (conjugation)* problem

$$\begin{cases} -\Delta f = 0 & \text{outside } \Sigma, \\ \lambda[f_n](X) = V(X)f(X) & \text{on } \Sigma, \end{cases} \qquad f \in H^1(\Omega), \tag{2.7}$$

where $[f_n]$ is the jump of the normal derivative f_n at Σ. Note that if Σ *is* the boundary of Ω, we obtain a Steklov-type problem, associated with the Neumann-to-Dirichlet operator,

$$\begin{cases} -\Delta f = 0 & \text{for } X \in \Omega, \\ \lambda f_n(X) = V(X)f(X) & \text{on } \Sigma. \end{cases}$$

This case is *not* covered by the reasoning in this paper as the surface Σ is *not* contained in Ω. We stress here that the question on the eigenvalue asymptotics for the Steklov problem with Lipschitz boundary is still open, even in the 2-dimensional case, while for the transmission problem with a nice weight on a Lipschitz surface the eigenvalue asymptotics was established in [15].

Let now, still for $\mathbf{N} = 2$, $\Sigma \subset \Omega$, the measure P have both absolutely continuous and singular parts, $P = V_0\, dX + V_1\mu_\Sigma$ for a Lipschitz curve $\Sigma \subset \Omega$, with $V_0 \in L^\Psi(\Omega)$ and $V_1 \in L^\Psi(\Sigma)$. The eigenvalue problem (2.5) takes the form

$$\begin{cases} -\lambda\Delta f(X) = V_0(X)f(X) & \text{for } X \in \Omega \setminus \Sigma, \\ \lambda[f_n](X) = V_1(X)f(X) & \text{on } \Sigma. \end{cases}$$

So, here the spectral parameter is present both in the differential equation and the transmission condition. Our results show that they both contribute to the leading term in the eigenvalue asymptotics. *Boundary* problems of this type have been considered by Kozhevnikov, see [9].

Let us now pass to the case $\mathbf{N} = 4$, $l = 2$. Here we have the following choice for the dimension d of Σ: $d = 1, 2$, or 3. For $d = 3$, $\mathfrak{d} = 1$, we arrive again at a transmission problem similar to (2.7):

$$\begin{cases} \Delta^2 f(X) = 0 & \text{for } X \in \Omega \setminus \Sigma, \\ \lambda[(-\Delta f)_n](X) = V(X)f(X) & \text{on } \Sigma, \end{cases} \qquad f \in H^2(\Omega). \tag{2.8}$$

For $d = 2$, $\mathfrak{d} = 2$, let $X = (\mathbf{x}, \mathbf{y}) \in \Omega \subset \mathbb{R}^4$, $\mathbf{x}, \mathbf{y} \in \mathbb{R}^2$, let $\Sigma = \{X : \mathbf{y} = 0\} \cap \Omega$ be (a piece of) the 2-dimensional plane in $\mathbb{R}^4$, and let $V = V(\mathbf{x})$. Then, after some simple calculations, we arrive at the following problem:

$$\begin{cases} \Delta^2 f(X) = 0 & \text{for } X \in \Omega \setminus \Sigma, \\ \lambda \lim\limits_{\delta \to 0} \displaystyle\int_{|\mathbf{y}|=\delta} (-\Delta f)_n(\mathbf{x}, \mathbf{y})\, ds(\mathbf{y}) = V(\mathbf{x})f(\mathbf{x}, 0), \end{cases} \tag{2.9}$$

where, for each fixed $\mathbf{x}$, the integral of the normal derivative of the Laplacian of f is taken over the δ-circle in the $\mathbf{y}$-plane. Finally, for $d = 1$, when the manifold Σ is the line $\mathbf{y} = 0$ in the coordinates $X = (\mathbf{x}, \mathbf{y})$, $\mathbf{x} \in \mathbb{R}^1$, $\mathbf{y} \in \mathbb{R}^3$ in $\mathbb{R}^4$, the resulting problem is

$$\begin{cases} \Delta^2 f(X) = 0 & \text{for } X \in \Omega \setminus \Sigma, \\ \lambda \lim_{\delta \to 0} \int_{|\mathbf{y}|=\delta} (-\Delta f)_n(\mathbf{x}, \mathbf{y})\, ds(\mathbf{y}) = V(\mathbf{x}) f(\mathbf{x}, 0), \end{cases} \tag{2.10}$$

where, for each fixed $\mathbf{x}$, the integration is performed over the 2D sphere $|\mathbf{y}| = \delta$. One can interpret the conditions on the surface Σ in (2.9) and (2.10) as multi-dimensional versions of the transmission conditions in (2.7) and (2.8). Our results show that all these spectral problems have the same order of the eigenvalue asymptotics.

If the measure P has a nonzero absolutely continuous part with density V_0, one should replace in (2.8)–(2.10) the equation $\Delta^2 f(X) = 0$ with

$$\lambda \Delta^2 f(X) = V_0(X) f(X),$$

so, again, we arrive at spectral problems containing the spectral parameter both in the equation and in the transmission conditions.

As we already mentioned, if the support of the singular measure consists of several disjoint surfaces, of possibly different dimensions, their contributions to the eigenvalue estimates have the same order and the coefficients in the asymptotics add up.

Corollary 2.5. *Let a measure P with a compact support in $\Omega \subset \mathbb{R}^{\mathbf{N}}$ have the absolutely continuous part $P_{\mathrm{ac}} = V_0(X)dX$, where $V_0 \in L^{\Psi}(\Omega, dX)$, and the singular part $P_{\mathrm{sing}} = \sum P_j$, where $P_j = V_j \mu_{\Sigma_j}$, Σ_j are disjoint Lipschitz surfaces of dimension $d_j < \mathbf{N}$, μ_{Σ_j} are the measures induced by the embeddings of Σ_j into $\mathbb{R}^{\mathbf{N}}$, and $V_j \in L^{\Psi}(\Sigma_j, \mu_{\Sigma_j})$. Then, for the operator in $L_2(\Omega)$ defined by the quadratic form $\mathbf{F}_P[u] = \int_\Omega |(\mathfrak{A}u)(X)|^2 P$, the following asymptotic formula holds:*

$$n_\pm(\lambda, \mathbf{F}) \sim \lambda^{-1} C(P, \mathfrak{A}), \quad \lambda \to 0; \tag{2.11}$$

here $C(P, \mathfrak{A})$ is the sum of the asymptotic coefficients in (2.4) *corresponding to all surfaces Σ_j, plus the term coming from P_{ac}, $C(P, \mathfrak{A}) = C(P_{\mathrm{ac}}, \mathfrak{A}) + \sum C(V_j, \Sigma_j)$, where*

$$C(P_{\mathrm{ac}}, \mathfrak{A}) = \mathbf{N}^{-1}(2\pi)^{-\mathbf{N}} \int_{S^*\Omega} V_{0,\pm}(X) |a_{-l}(X, \Xi)|^2\, dX\, d\Xi.$$

The disjointness conditions above can be considerably relaxed, see Section 7.

3 Some reductions

This section contains some technical observations that are used further on in the paper to reduce general eigenvalue estimates and asymptotics to more convenient setting.

Similar arguments for spectral problems for differential operators are a well-known part of mathematical folklore. They have been used systematically, starting with the papers by Birman and Solomyak, for the past 50 years. They are usually proved in a line or two. Our pseudodifferential versions require some additional, somewhat technical, reasoning, which nevertheless follows the classical pattern, and the results are quite natural. Readers familiar with the classical versions and not interested in these details can skip this section without detriment to understanding the rest of the paper. For brevity, we take a surface of dimension $d = \mathbf{N}$ to mean a domain in $\mathbb{R}^{\mathbf{N}}$. For an operator $\mathbf{T}$, we denote by $\mathbf{D}^{\pm}(\mathbf{T})$ the quantity

$$\mathbf{D}^{\pm}(\mathbf{T}) := \limsup_{\lambda\to 0} \lambda n_{\pm}(\lambda, \mathbf{T}) = \limsup_{k\to\infty} \pm k\lambda_k^{\pm}(\mathbf{T}).$$

Observation 1 (Localization 1). *It is sufficient to prove the eigenvalue estimate for a surface being described by only one coordinate neighborhood. Indeed, if Σ is split into a finite number J of disjoint surfaces Σ_j, $j = 1, \dots, J$, of possibly different dimensions, and $P_j = V_j\mu_{\Sigma_j}$ is the restriction of P to Σ_j, then*

$$\mathbf{D}^{\pm}(\mathbf{T}(V, \Sigma, \mathfrak{A})) \le J \sum_j \mathbf{D}^{\pm}(\mathbf{T}(V_j, \Sigma_j, \mathfrak{A})).$$

This property follows from the Ky Fan inequality for the sum of operators,

$$\mathbf{T}(V, \Sigma, \mathfrak{A}) = \sum_j \mathbf{T}(V_j, \Sigma_j, \mathfrak{A}).$$

Observation 2 (Lower-order terms). *Let $\mathfrak{B}$ be a pseudodifferential operator of order $\beta < -l$. Then $\mathbf{D}^{\pm}(\mathbf{T}(V, \Sigma, \mathfrak{B})) = 0$. This follows from the (already mentioned) result in* [3] *that a finite singular measure gives a lower-order contribution to the eigenvalue distribution for higher-order problems.*

Observation 3 (Perturbations by lower-order terms). *Let $\mathfrak{B}$ be as above. Then*

$$\mathbf{D}^{\pm}(\mathbf{T}(V, \Sigma, \mathfrak{A} + \mathfrak{B})) = \mathbf{D}^{\pm}(\mathbf{T}(V, \Sigma, \mathfrak{A})).$$

This follows from the inequality

$$(1-\epsilon)|a|^2 + \left(1 - \frac{1}{\epsilon}\right)|b|^2 \le |a+b|^2 \le (1+\epsilon)|a|^2 + \left(1 + \frac{1}{\epsilon}\right)|b|^2$$

with $a = (\mathfrak{A}u)(X)$, $b = (\mathfrak{B}u)(X)$ and the previous observation.

Observation 4 (Localization 2). *Let Ω_1 be an open subset in Ω such that $\Sigma \cap \overline{\Omega_1} = \varnothing$, and let $V \in L^{\Psi}(\Sigma)$. Then for the eigenvalues of the operator $\mathbf{T}(V, \Omega_1)$ on $L_2(\Omega_1)$ defined by the form $\mathbf{F}_{V,\Omega_1}[u] = \int_\Sigma V(X)|(\mathfrak{A}u)(X)|^2\, d\mu$, $u \in L_2(\Omega_1)$, one has the following estimate:*

$$n_{\pm}(\lambda, \mathbf{T}(V, \Omega_1)) = o(\lambda^{-1}) \quad \text{as } \lambda \to 0.$$

Proof. Let $\chi \in C_0^\infty$ be a smooth function equal to 0 in a neighborhood of $\overline{\Omega_1}$ and to 1 in a neighborhood of Σ. Then

$$\mathbf{F}_{V,\Omega_1}[u] = \int_\Sigma V(X)|(\chi(X)\mathfrak{A}u)(X)|^2\, d\mu = \int_\Sigma V(X)|([\chi,\mathfrak{A}]u)(X)|^2\, d\mu.$$

The commutator $[\chi, \mathfrak{A}]$ is a pseudodifferential operator of order $-l-1$, and therefore, by Observation 2, the eigenvalues of the corresponding operator $\mathbf{T}(V, \Omega_1)$ decay faster than k^{-1}. ■

It follows, in particular, that the eigenvalue counting function gets a lower-order perturbation if we perturb the operator $\mathfrak{A}$ outside a neighborhood of the surface Σ. In particular, this gives us freedom in choosing cut-off functions away from Σ or adding operators smoothing away from the boundary – the possibility already mentioned. It is possible, again, at the expense of a lower-order perturbation to the eigenvalue counting function, to increase or decrease the domain Ω – as long as Σ remains being inside the domain.

The next, more complicated, statement is used in the study of eigenvalue asymptotics. It says that if two measures have supports separated by a positive distance, then, up to a lower-order term, the counting functions behave additively with respect to the measures. This includes the important case when absolutely continuous measures are present. If both measures are absolutely continuous, this is a classical fact.

Lemma 3.1 (Localization 3). *Let $P = P_1 + P_2$, where $P_j = V_j\mu_{\Sigma_j}$ is a measure supported on a compact Lipschitz surface Σ_j of dimension $d_j \in [1, \mathbf{N}]$, $j = 1, 2$ (the cases $d_1 = \mathbf{N}$ and $d_2 = \mathbf{N}$ correspond to Σ_1, respectively Σ_2, being domains in $\Omega \subset \mathbb{R}^\mathbf{N}$ and the measures being absolutely continuous with respect to the Lebesgue measure). Suppose that* $\operatorname{dist}(\operatorname{supp} P_1, \operatorname{supp} P_2) > 0$. *Then*

$$n_\pm(\lambda, \mathbf{T}(P_1 + P_2)) = n_\pm(\lambda, \mathbf{T}(P_1)) + n_\pm(\lambda, \mathbf{T}(P_2)) + o(\lambda^{-1}) \quad \text{as } \lambda \to 0. \tag{3.1}$$

Proof. Consider two disjoint open sets $\Omega_1, \Omega_2 \subset \Omega$ with the property that $\overline{\Sigma_j} \subset \Omega_j$ and $\overline{\Omega_1} \cup \overline{\Omega_2} \supset \Omega$. Note that every function $u \in L_2(\Omega)$ splits into the (orthogonal) sum $u = u_1 \oplus u_2$, $u_j \in L_2(\Omega_j)$. The quadratic form of the operator $\mathbf{T}(P_1 + P_2)$ splits as follows:

$$\begin{aligned}
\mathbf{F}(P_1 + P_2)[u] &:= (\mathbf{T}(P_1 + P_2)u, u)_{L_2(\Omega)} \\
&= \int_\Omega V_1(X)|\mathfrak{A}(u_1)(X) + \mathfrak{A}(u_2)(X)|^2\, d\mu_{\Sigma_1} \\
&\quad + \int_\Omega V_2(X)|\mathfrak{A}(u_1)(X) + \mathfrak{A}(u_2)(X)|^2\, d\mu_{\Sigma_2} \\
&= \mathbf{F}_1(P_1)[u_1] + \mathbf{F}_2(P_2)[u_2] + \mathbf{F}_R[u_1, u_2] \\
&:= \int_\Omega V_1(X)|\mathfrak{A}(u_1)(X)|^2\, d\mu_{\Sigma_1} + \int_\Omega V_2(X)|\mathfrak{A}(u_2)(X)|^2\, d\mu_{\Sigma_2} \\
&\quad + \mathbf{F}_R[u_1, u_2],
\end{aligned} \tag{3.2}$$

where the remainder term $\mathbf{F}_R[u_1, u_2]$ is a form in functions $u_j \in L_2(\Omega_j)$ with the following property: if a term in $\mathbf{F}_R$ contains V_j, then it necessarily contains u_{3-j}, so it always contains a measure and a function with disjoint supports. If such a term has the form $\int_\Omega V_1|\mathfrak{A}u_2|^2\, d\mu_{\Sigma_1}$, the corresponding operator $\mathbf{T}$ satisfies $n_\pm(\lambda, \mathbf{T}) = o(\lambda^{-1})$ by Observation 4. If, on the other hand, such a term has the form

$$\int_\Omega V_1(\mathfrak{A}u_1)\overline{(\mathfrak{A}u_2)}\, d\mu_{\Sigma_1},$$

then by the Schwarz inequality,

$$\left|\int_\Omega V_1(\mathfrak{A}u_1)\overline{(\mathfrak{A}u_2)}\, d\mu_{\Sigma_1}\right| \le \left(\int_\Omega |V_1||\mathfrak{A}u_1|^2\, d\mu_{\Sigma_1}\right)^{\frac{1}{2}}\left(\int_\Omega |V_1||\mathfrak{A}u_2|^2\, d\mu_{\Sigma_1}\right)^{\frac{1}{2}},$$

and the last factor, again, provides the required o estimate for the eigenvalues. Now we observe that the forms $\mathbf{F}_1(P_1)[u_1]$ and $\mathbf{F}_2(P_2)[u_2]$ in (3.2) act on orthogonal subspaces. Then the spectrum of the sum of the corresponding operators $\mathbf{T}_1$ and $\mathbf{T}_2$ equals the union of the spectra of the summands, and hence

$$n_\pm(\lambda, \mathbf{T}_1 + \mathbf{T}_2) = n_\pm(\lambda, \mathbf{T}_1) + n_\pm(\lambda, \mathbf{T}_2).$$

Since the term $\mathbf{F}_R$ in (3.2) makes a weaker contribution, we have

$$n_\pm(\lambda, \mathbf{T}(P_1 + P_2)) = n_\pm(\lambda, \mathbf{T}_1) + n_\pm(\lambda, \mathbf{T}_2) + o(\lambda^{-1}). \tag{3.3}$$

Now consider the operator $\mathbf{T}(P_1)$. It has the quadratic form

$$\mathbf{F}(P_1)[u] = \int V_1(X)|\mathfrak{A}(u_1 \oplus u_2)|(X)\, d\mu_{\Sigma_1}.$$

Similarly to (3.2), we represent it as

$$\mathbf{F}(P_1)[u] = \int_\Omega V_1(X)|\mathfrak{A}(u_1)(X)|^2\, d\mu_{\Sigma_1} + \mathbf{F}_{R_1}[u_1, u_2],$$

with $\mathbf{F}_{R_1}$ having the same structure as $\mathbf{F}_R$ in (3.2). Again, the form $\mathbf{F}_{R_1}$ generates an operator with eigenvalues decaying faster than k^{-1}, and we obtain

$$n_\pm(\lambda, \mathbf{T}(P_1)) = n_\pm(\lambda, \mathbf{T}_1) + o(\lambda^{-1}). \tag{3.4}$$

In the same way,

$$n_\pm(\lambda, \mathbf{T}(P_2)) = n_\pm(\lambda, \mathbf{T}_2) + o(\lambda^{-1}). \tag{3.5}$$

Finally, we substitute (3.4) and (3.5) into (3.3) to obtain (3.1). ∎

The following corollary of Lemma 3.1 allows one to separate the positive and the negative parts of the function V when studying the distribution of the positive and the negative eigenvalues of $\mathbf{T}(V, \Sigma, \mathfrak{A})$.

Corollary 3.2. *Let Σ be a Lipschitz surface and $V \in L^{\Psi}(\Sigma)$. Let $\Sigma_{\pm}$ be relatively open subsets of Σ such that $V_{\pm} \geq 0$ in $\Sigma_{\pm}$, $V = 0$ in $\Sigma \setminus (\Sigma_+ \cup \Sigma_-)$, and* $\operatorname{dist}(\overline{\Sigma_+}, \overline{\Sigma_-}) > 0$. *Then*

$$n_{\pm}(\lambda, \mathbf{T}(V, \Sigma)) = n_{+}(\lambda, \mathbf{T}(V_{\pm}, \Sigma)) + o(\lambda^{-1}) \quad \text{as } \lambda \to 0.$$

In other words, up to a lower-order remainder, the behavior of the positive, respectively negative, eigenvalues of the operator $\mathbf{T}(V)$ is determined by the positive, respectively negative, part of the density V. To prove this property, we can use (3.3), taking as P_1 the restriction of the measure P to the set Σ_+, and as P_2 its restriction to Σ_-, and recall that $n_-(\lambda, \mathbf{T}(V, \Sigma_+)) = n_+(\lambda, \mathbf{T}(V, \Sigma_-)) = 0$.

4 Geometry considerations

Our proof of Theorems 2.1 and 2.3 relies upon certain geometric observations that might be of an independent interest.

Let $\mathcal{A} \subset \mathbb{R}^{\mathbf{N}}$ be a k-dimensional affine subspace, $0 \leq k < \mathbf{N}$, $\mathcal{A} = a + \mathcal{L}$, where $a \in \mathbb{R}^{\mathbf{N}}$ and $\mathcal{L}$ is a k-dimensional linear subspace in $\mathbb{R}^{\mathbf{N}}$. In the case $k = 0$, $\mathcal{L}$ is a 0-dimensional linear subspace, i.e., $\mathcal{L} = \{0\}$, and $\mathcal{A}$ is just the singleton $\{a\}$. The polar plane of $\mathcal{A}$ is the $(\mathbf{N} - k)$-dimensional linear subspace of $\mathbb{R}^{\mathbf{N}}$,

$$\mathcal{A}^{\perp} := \mathcal{L}^{\perp} = \{Y \in \mathbb{R}^{\mathbf{N}} : (X, Y) = 0 \text{ for all } X \in \mathcal{L}\},$$

where $(\cdot, \cdot)$ denotes the standard inner product in $\mathbb{R}^{\mathbf{N}}$. We will say that $\mathcal{A}$ is orthogonal to a vector $b \in \mathbb{R}^{\mathbf{N}} \setminus \{0\}$ if $b \in \mathcal{A}^{\perp}$. For $k = \mathbf{N} - 1$, $\mathcal{A}^{\perp}$ is the 1-dimensional linear subspace spanned by b. For a linear subspace $\mathcal{M}$ in $\mathbb{R}^{\mathbf{N}}$, we denote by $\mathcal{M}_{\mathbb{S}}$ the trace of $\mathcal{M}$ on $\mathbb{S}^{\mathbf{N}-1}$, $\mathcal{M}_{\mathbb{S}} := \mathcal{M} \cap \mathbb{S}^{\mathbf{N}-1}$, where $\mathbb{S}^{\mathbf{N}-1}$ is the unit sphere in $\mathbb{R}^{\mathbf{N}}$, $\mathbb{S}^{\mathbf{N}-1} := \{X \in \mathbb{R}^{\mathbf{N}} : |X| = 1\}$.

Lemma 4.1. *Let $\mathfrak{W}$ be an at most countable family of proper linear subspaces of $\mathbb{R}^{\mathbf{N}}$. Then there exists an orthonormal basis $\mathbf{e}_1, \dots, \mathbf{e}_{\mathbf{N}}$ of $\mathbb{R}^{\mathbf{N}}$ such that*

$$\mathbf{e}_j \notin \bigcup_{\mathcal{M} \in \mathfrak{W}} \mathcal{M}, \quad j = 1, \dots, \mathbf{N}.$$

Proof. The proof is by induction on $\mathbf{N}$. There is nothing to prove if $\mathbf{N} = 1$ since the only proper linear subspace of $\mathbb{R}$ is $\mathcal{M} = \{0\}$. Suppose that the statement is true for $\mathbf{N} = \mathbf{N}_0$ and let us prove it for $\mathbf{N} = \mathbf{N}_0 + 1$. Let $\mathfrak{W}'$ be the (possibly empty) subset of $\mathfrak{W}$ consisting of all $\mathbf{N}_0$-dimensional $\mathcal{M} \in \mathfrak{W}$. For every such $\mathcal{M}$, $\mathcal{M}^{\perp}$ is 1-dimensional and $\mathcal{M}^{\perp}_{\mathbb{S}}$ consists of two points. Since $\bigcup_{\mathcal{M} \in \mathfrak{W}} \mathcal{M}_{\mathbb{S}}$ is an at most countable union of spheres of dimension at most $\mathbf{N}_0 - 1$, the set

$$\Theta := \Big(\bigcup_{\mathcal{M} \in \mathfrak{W}'} \mathcal{M}^{\perp}_{\mathbb{S}}\Big) \cup \Big(\bigcup_{\mathcal{M} \in \mathfrak{W}} \mathcal{M}_{\mathbb{S}}\Big)$$

has $\mathbf{N}_0$-dimensional Lebesgue measure equal to 0. So, $\Theta \neq \mathbb{S}^{\mathbf{N}_0}$. Take any vector $\mathbf{e}_{\mathbf{N}_0+1} \in \mathbb{S}^{\mathbf{N}_0} \setminus \Theta$ and let $\mathcal{M}_0$ be the $\mathbf{N}_0$-dimensional linear subspace of $\mathbb{R}^{\mathbf{N}_0+1}$ orthogonal to $\mathbf{e}_{\mathbf{N}_0+1}$. Since

$$\mathbf{e}_{\mathbf{N}_0+1} \notin \bigcup_{\mathcal{M} \in \mathfrak{W}'} \mathcal{M}_{\mathbb{S}}^{\perp},$$

it follows that $\mathcal{M}_0$ does not coincide with any element of $\mathfrak{W}'$, and hence the dimension of $\mathcal{M}_0 \cap \mathcal{M}$ is at most $\mathbf{N}_0 - 1$ for every $\mathcal{M} \in \mathfrak{W}$. Then, by the inductive assumption, there exists an orthonormal basis $\mathbf{e}_1, \dots, \mathbf{e}_{\mathbf{N}_0}$ of $\mathcal{M}_0$ such that

$$\mathbf{e}_j \notin \bigcup_{\mathcal{M} \in \mathfrak{W}} (\mathcal{M}_0 \cap \mathcal{M}) = \mathcal{M}_0 \cap \Big(\bigcup_{\mathcal{M} \in \mathfrak{W}} \mathcal{M} \Big), \quad j = 1, \dots, \mathbf{N}_0.$$

It is clear that $\mathbf{e}_1, \dots, \mathbf{e}_{\mathbf{N}_0}, \mathbf{e}_{\mathbf{N}_0+1}$ is an orthonormal basis of $\mathbb{R}^{\mathbf{N}_0+1}$ and

$$\mathbf{e}_j \notin \bigcup_{\mathcal{M} \in \mathfrak{W}} \mathcal{M}, \quad j = 1, \dots, \mathbf{N}_0 + 1.$$

■

Lemma 4.2. *Let μ be a σ-finite Borel measure on $\mathbb{R}^{\mathbf{N}}$ such that $\mu(\mathcal{A}) = 0$ for every $(k-1)$-dimensional affine subspace $\mathcal{A} \subset \mathbb{R}^{\mathbf{N}}$ for some $k \in [1, \mathbf{N}-1]$. Then the set $\mathfrak{X}_k$ of k-dimensional affine subspaces $\mathcal{E}$ of $\mathbb{R}^{\mathbf{N}}$ such that $\mu(\mathcal{E}) > 0$ is at most countable.*

Proof. The proof is similar to that of [8, Lemma 2.13]. It is sufficient to prove the lemma for finite measures as the general case then follows easily from the assumption that μ is σ-finite. Suppose that $\mathfrak{X}_k$ is uncountable. Then there exists a $\delta > 0$ such that the set

$$\mathfrak{X}_{k,\delta} := \{\mathcal{E} \in \mathfrak{X}_k : \mu(\mathcal{E}) > \delta\}$$

is infinite. Otherwise,

$$\mathfrak{X}_k = \bigcup_{m \in \mathbb{N}} \mathfrak{X}_{k,\frac{1}{m}}$$

would have been at most countable. Now take distinct $\mathcal{E}_1, \dots, \mathcal{E}_j, \dots \in \mathfrak{X}_{k,\delta}$. We have $\mu(\mathcal{E}_j) > \delta$ for all $j \in \mathbb{N}$.

Since $\mathcal{E}_j \cap \mathcal{E}_{j'}$, $j \neq j'$, is an affine subspace of dimension at most $k-1$, it follows that

$$\mu\Big(\bigcup_{j \neq j'} (\mathcal{E}_j \cap \mathcal{E}_{j'}) \Big) = 0.$$

Let

$$\tilde{\mathcal{E}}_j := \mathcal{E}_j \setminus \bigcup_{j' \neq j} (\mathcal{E}_{j'} \cap \mathcal{E}_j).$$

Then $\tilde{\mathcal{E}}_j \cap \tilde{\mathcal{E}}_{j'} = \varnothing$, $j \neq j'$, and $\mu(\tilde{\mathcal{E}}_j) = \mu(\mathcal{E}_j)$. So, since μ is finite, we have

$$\sum_{j \in \mathbb{N}} \mu(\tilde{\mathcal{E}}_j) = \mu\Big(\bigcup_{j \in \mathbb{N}} \tilde{\mathcal{E}}_j \Big) < \infty.$$

On the other hand, $\mu(\tilde{\mathcal{E}}_j) = \mu(\mathcal{E}_j) > \delta$ implies that $\sum_{l \in \mathbb{N}} \mu_k(\tilde{\mathcal{E}}_j) \geq \sum_{j \in \mathbb{N}} \delta = \infty$. This contradiction means that $\mathfrak{X}_k$ is at most countable. ■

We arrive at our main geometric statement, considerably more general than what we actually need here.

Theorem 4.3. *Let μ be a σ-finite Borel measure on $\mathbb{R}^{\mathbf{N}}$ without point masses. Then there exists an orthonormal basis $\mathbf{e}_1, \dots, \mathbf{e}_{\mathbf{N}}$ of $\mathbb{R}^{\mathbf{N}}$ such that $\mu(\mathcal{E}) = 0$ for every $(\mathbf{N}-1)$-dimensional affine subspace $\mathcal{E}$ of $\mathbb{R}^{\mathbf{N}}$ orthogonal to an element of this basis.*

Proof. Let us first explain the idea of the proof. By Lemma 4.2 for $k = 1$, there is an at most countable set of 1-dimensional affine subspaces whose μ-measure is positive. Subtracting the portion of measure μ living on all these subspaces and applying Lemma 4.2 to the resulting measure, this time with $k = 2$, we conclude that at most countably many 2-dimensional affine subspaces are charged positively. We subtract the part of our measure living on these subspaces, and so on. This procedure is repeated in all dimensions, after which it turns out that the remaining measure is zero on *all* affine subspaces in $\mathbb{R}^{\mathbf{N}}$, and the proof is completed by applying Lemma 4.1.

Now the formal proof. Denote $\mu = \mu_1$. Lemma 4.2 with $k = 1$ implies that the set $\mathfrak{X}_1$ of all 1-dimensional affine subspaces $\mathcal{E}$ of $\mathbb{R}^{\mathbf{N}}$ such that $\mu_1(\mathcal{E}) > 0$ is at most countable. We introduce $\mathcal{Z}_1 := \bigcup_{\mathcal{E}\in\mathfrak{X}_1} \mathcal{E}$, $\mu_1^0(E) := \mu_1(E \cap \mathcal{Z}_1)$ for every Borel set $E \subseteq \mathbb{R}^{\mathbf{N}}$, and $\mu_2 := \mu_1 - \mu_1^0$.

Measure μ_2 annuls every 1-dimensional affine subspace $\mathcal{E}$. Indeed,

$$\mu_2(\mathcal{E}) \begin{cases} \le \mu_1(\mathcal{E}) = 0 & \text{if } \mathcal{E} \notin \mathfrak{X}_1, \\ = \mu_1(\mathcal{E}) - \mu_1^0(\mathcal{E}) = \mu_1(\mathcal{E}) - \mu_1(\mathcal{E}) = 0 & \text{if } \mathcal{E} \in \mathfrak{X}_1. \end{cases}$$

Therefore we can apply Lemma 4.2 again, this time with $k = 2$ and $\mu = \mu_2$.

Repeating this procedure, we obtain at the k-th step a σ-finite measure μ_k on $\mathbb{R}^{\mathbf{N}}$ such that $\mu_k(\mathcal{A}) = 0$ for every $(k-1)$-dimensional affine subspace $\mathcal{A}$ of $\mathbb{R}^{\mathbf{N}}$ and the set $\mathfrak{X}_k$ of all k-dimensional affine subspaces $\mathcal{E}$ of $\mathbb{R}^{\mathbf{N}}$ with $\mu_k(\mathcal{E}) > 0$ is at most countable. Similarly to the above, we set $\mathcal{Z}_k := \bigcup_{\mathcal{E}\in\mathfrak{X}_k} \mathcal{E}$, $\mu_k^0(E) := \mu_k(E \cap \mathcal{Z}_k)$ for every Borel set $E \subseteq \mathbb{R}^{\mathbf{N}}$, and $\mu_{k+1} := \mu_k - \mu_k^0$. Then, similarly to the above, $\mu_{k+1}(\mathcal{E}) = 0$ for every k-dimensional affine subspace $\mathcal{E}$.

For every $\mathcal{E} \in \mathfrak{X}_k$, $\mathcal{E}^\perp$ is an $(\mathbf{N}-k)$-dimensional linear subspace of $\mathbb{R}^{\mathbf{N}}$, where $1 \le k \le \mathbf{N}-1$. By Lemma 4.1, there exists an orthonormal basis $\mathbf{e}_1, \dots, \mathbf{e}_{\mathbf{N}}$ of $\mathbb{R}^{\mathbf{N}}$ such that

$$\mathbf{e}_j \notin \Xi := \bigcup_{k=1}^{\mathbf{N}-1} \bigcup_{\mathcal{E}\in\mathfrak{X}_k} \mathcal{E}^\perp, \quad j = 1, \dots, \mathbf{N}. \tag{4.1}$$

Take any $j = 1, \dots, \mathbf{N}$. Let $\mathcal{E}_0$ be an $(\mathbf{N}-1)$-dimensional affine subspace orthogonal to $\mathbf{e}_j$. Then, for any $k = 1, \dots, \mathbf{N}-1$,

$$\mathcal{E} \in \mathfrak{X}_k \implies \mathcal{E} \not\subseteq \mathcal{E}_0. \tag{4.2}$$

Indeed, if $\mathcal{E} \subseteq \mathcal{E}_0$, then

$$\mathbf{e}_j \in \mathcal{E}_0^\perp \subseteq \mathcal{E}^\perp \subseteq \Xi,$$

which contradicts (4.1).

Since $\mathcal{E}_0 \notin \mathfrak{X}_{\mathrm{N}-1}$, we have $\mu_{\mathrm{N}-1}(\mathcal{E}_0) = 0$. For any $\mathcal{E} \in \mathfrak{X}_k$, (4.2) implies that $\mathcal{E}_0 \cap \mathcal{E}$ is an affine subspace of dimension at most $k-1$. Then $\mu_k(\mathcal{E}_0 \cap \mathcal{E}) = 0$ and

$$\mu_k^0(\mathcal{E}_0) = \mu_k(\mathcal{E}_0 \cap \mathcal{Z}_k) \leq \sum_{\mathcal{E} \in \mathfrak{X}_k} \mu_k(\mathcal{E}_0 \cap \mathcal{E}) = 0.$$

Hence

$$\begin{aligned} 0 &= \mu_{\mathrm{N}-1}(\mathcal{E}_0) = \mu_{\mathrm{N}-2}(\mathcal{E}_0) - \mu_{\mathrm{N}-2}^0(\mathcal{E}_0) = \mu_{\mathrm{N}-2}(\mathcal{E}_0) = \mu_{\mathrm{N}-3}(\mathcal{E}_0) \\ &= \cdots = \mu_1(\mathcal{E}_0) = \mu(\mathcal{E}_0). \end{aligned}$$

∎

5 Estimates

The method of deriving eigenvalue estimates for nonsmooth spectral problems developed by Birman and Solomyak in the late 1960s has remained since then the most efficient approach to such singular problems. The method is based on constructing piecewise polynomial approximations of functions in Sobolev spaces. We follow the presentation in [17], with necessary modifications caused by the presence of singular measures.

5.1 A homogeneous Hölder inequality

We recall that for a bounded domain $G \subseteq \mathbb{R}^{\mathrm{N}}$, the Sobolev norm and the homogeneous Sobolev semi-norm are defined by

$$\|u\|^2_{H^s(G)} := \sum_{|\nu|_1 \leq s} \|\partial^\nu u\|^2_{L_2(G)},$$
$$\|u\|^2_{H^s_{\mathrm{hom}}(G)} := \sum_{|\nu|_1 = s} \|\partial^\nu u\|^2_{L_2(G)}$$

for an integer s, and by

$$\|u\|^2_{H^s(G)} := \|u\|^2_{H^m(G)} + \|u\|^2_{H^s_{\mathrm{hom}}(G)},$$
$$\|u\|^2_{H^s_{\mathrm{hom}}(G)} := \sum_{|\nu|_1 = m} \int_G \int_G \frac{|(\partial^\nu u)(X) - (\partial^\nu u)(Y)|^2}{|X - Y|_2^{\mathrm{N}+1}} \, dX \, dY$$

for a half-integer $s = m + \frac{1}{2}$. (Here $|\cdot|_q$ denotes the standard norm in ℓ^q.) The space $\overset{\circ}{H}{}^s(G)$, $s > \frac{1}{2}$, is defined as the closure of $C_0^\infty(G)$ in H^s norm which is equivalent on the space $\overset{\circ}{H}{}^s(G)$ to the homogeneous Sobolev norm. The semi-norm $\|\cdot\|^2_{H^s_{\mathrm{hom}}(G)}$ possesses the homogeneity property

$$\|u(R\,\cdot)\|_{H^s_{\mathrm{hom}}(G)} = R^{s-\frac{\mathrm{N}}{2}} \|u\|_{H^s_{\mathrm{hom}}(RG)}$$

for all $R > 0$. Our case of interest is $s = l = \frac{\mathbf{N}}{2}$; here the homogeneous semi-norm is invariant under dilations. Although this semi-norm is not, in general, invariant under rotations, it is easy to see that under rotations, this semi-norm transforms to an equivalent one, and there exist positive constants $m_{\mathbf{N}}, M_{\mathbf{N}}$ depending only on $\mathbf{N}$ and such that, for all $u \in H^l(G)$,

$$m_{\mathbf{N}} \|u(\mathbf{Z}(\cdot))\|_{H^l_{\text{hom}}(G)} \le \|u\|_{H^l_{\text{hom}}(\mathbf{Z}(G))} \le M_{\mathbf{N}} \|u(\mathbf{Z}(\cdot))\|_{H^l_{\text{hom}}(G)}$$

for every affine transformation $\mathbf{Z}$ of the form

$$X = \mathbf{Z}(Y) = RUY + X_0$$

with $X_0 \in \mathbb{R}^{\mathbf{N}}$, $R > 0$, and an orthogonal matrix $U \in \mathrm{O}(\mathbf{N})$.

Recall that $\Psi(t) = (1+t)\ln(1+t) - t$ and $\Phi(t) = e^t - 1 - t$ are complementary Orlicz functions.

Lemma 5.1 (cf. [12, Corollary 11.8 (2)]). *Let $G \subset \mathbb{R}^{\mathbf{N}}$ be a bounded domain with Lipschitz boundary. If a positive Borel measure μ on $\overline{G}$ satisfies, for some $\alpha > 0$,*

$$\mu(B(X,r)) \le Kr^\alpha \quad \textit{for all } X \in \overline{G} \textit{ and all } r \in \left(0, \frac{1}{2}\right),$$

then the inequality

$$\|w^2\|_{\overline{G}}^{\Phi,\mu} \le A_1 \|w\|^2_{H^l(G)} \quad \textit{for all } w \in H^l(G) \cap C(\overline{G}) \tag{5.1}$$

holds with a constant $A_1 = A_1(G, \alpha, K)$.

Proof. The proof of the lemma relies on [12, Theorem 11.8] and is almost identical to that of [8, Lemma 5.2]. ∎

Inequality (5.1) implies that the embedding $H^l(G) \cap C(\overline{G})$ into the exponential Orlicz space $L^{\Phi,\mu}$, defined initially on continuous functions, extends by continuity to the whole $H^l(G)$. This continuation will be assumed already performed further on. The boundedness of the trace operator $\Gamma : H^l(G) \to L_2(\mathbf{M}, \mu)$, $f \mapsto f|_{\mathbf{M}}$, used at least twice here, follows from the embedding $L^{\Phi,\mu}(\mathbf{M}) \subset L_1(\mathbf{M}, \mu)$ (recall that $\mathbf{M}$ is the support of the measure μ.)

We will also need the following version of the Poincaré inequality that can be traced back to Sobolev (a very detailed proof can be found in [18, Theorem 2.4] for G being a cube). For every bounded set $G \subset \mathbb{R}^{\mathbf{N}}$ with Lipschitz boundary and every $s > 0$, there exists a constant $C_s(G) > 0$ such that

$$\|u\|_{H^s(G)} \le C_s(G) \|u\|_{H^s_{\text{hom}}(G)} \tag{5.2}$$

for all $u \in H^s(G)$ orthogonal in $L_2(G)$ to every polynomial of degree strictly less than s. We will denote the optimal constant in (5.2) with $s = \frac{\mathbf{N}}{2}$ by $C(G) := C_{\frac{\mathbf{N}}{2}}(G)$.

Our eigenvalue estimates rely on the following inequality.

Lemma 5.2. *Let $Q \subset \mathbb{R}^{\mathbf{N}}$ be a cube and $\mu = \mu_\Sigma$. Then*

$$\int_{\overline{Q}} V(X)|f(X)|^2\, d\mu(X) \leq A_2 \|V\|_{\overline{Q}}^{(\mathrm{av},\Psi,\mu)} \|f\|^2_{H^l_{\mathrm{hom}}(Q)} \tag{5.3}$$

for all $f \in H^l(Q)$ orthogonal in $L_2(Q)$ to every polynomial of degree strictly less than l, with constant A_2 depending only on $\mathbf{N}$, the constants c_1, c_2, and the exponent α in (2.3).

Proof. If $Q = Q_1$ is a unit cube, then this is an immediate generalization of the proof of [8, Lemma 5.3]. For a cube of an arbitrary size, one should make a dilation (compression) in (5.3) of Q to Q_1. Under this transformation, the homogeneous norm in (5.3) is invariant. As for the norm $\|V\|_{\overline{Q}}^{(\mathrm{av},\Psi,\mu)}$, it is not invariant under dilations, but is "almost" invariant in the sense that the dilated (compressed) norm is majorized by the initial one, with constant depending on c_0, c_1 and α in (2.3), independently of the dilation coefficient. Estimate (5.3) is proved in [8, Lemma 5.4], and the proof does not depend on dimension and uses nothing but the Ahlfors regularity of μ, so it applies to our case without changes. ∎

5.2 Coverings and piecewise polynomial approximations

The construction to follow is an adaptation of the method developed by Solomyak in [17]. It stems from the approach initiated by Birman, Solomyak, with a contribution by Rozenblum, in the late 1960s – early 1970s, see the nice exposition in [4]. Recently, this construction was used again for obtaining eigenvalue estimates for the weighted (poly-)harmonic operator in the critical case, see [8, 18]. We present the main idea and the structure of the proof first, and then fill in the required details.

Theorem 5.3. *Let μ be a Borel measure in $G \subset \mathbb{R}^{\mathbf{N}}$ satisfying estimate* (2.3) *with $\alpha > 0$ in a bounded domain $G \subset \mathbb{R}^{\mathbf{N}}$ and let $V \in L^\Psi(\mathbf{M}, \mu)$. Denote by $\mathring{\mathbf{T}}(V, \mu)$ the operator defined in the space $\mathring{H}^l(G)$ by the quadratic form $\int V(X)|f(X)|^2\, d\mu$, $f \in \mathring{H}^l(G)$, with the norm $\|f\|_{H^l_{\mathrm{hom}}(G)}$ in the latter space. Then*

$$n_\pm(\lambda, \mathring{\mathbf{T}}(V, \mu)) \leq C\lambda^{-1} \|V\|_G^{(\mathrm{av},\Psi,\mu)} \tag{5.4}$$

with a constant C independent of V.

Proof. For $\alpha = \mathbf{N}$ and μ being the Lebesgue measure, this result has been proved in [17] for even $\mathbf{N}$ and in [18] for odd $\mathbf{N}$. So, let $\alpha < \mathbf{N}$. It suffices to consider the case $V \geq 0$. We use one of the possible formulations of the variational principle for compact self-adjoint operators. If $\mathbf{T} \geq 0$ is such an operator on a Hilbert space $\mathcal{H}$, with the quadratic form $\mathbf{t}[f]$, then for the eigenvalue counting function $n(\lambda, \mathbf{T})$,

$$n(\lambda, \mathbf{T}) = \min \operatorname{codim}\{\mathcal{Y} \subset \mathcal{H} : \mathbf{t}[f] \leq \lambda \|f\|^2, f \in \mathcal{Y}\}. \tag{5.5}$$

Here the codimension $\operatorname{codim} \mathcal{Y}$ is the number of linearly independent functionals that have $\mathcal{Y}$ as their common null space. Equality (5.5) hints at how to prove eigenvalue estimates. Let us, for some $\lambda > 0$, construct *some* subspace $\mathcal{Y} = \mathcal{Y}(\lambda)$ on which the inequality in (5.5) holds. Then the quantity $n(\lambda, \mathbf{T})$ is not greater than the codimension of the subspace we constructed, $n(\lambda, \mathbf{T}) \le \operatorname{codim} \mathcal{Y}$. The more efficient we are in constructing $\mathcal{Y}$, the sharper the estimate is.

Let $\mathcal{H}$ be the Sobolev space $\mathring{H}^l(G)$. In constructing the subspace $\mathcal{Y}$, we take some finite covering $\Upsilon = \Upsilon(\lambda)$ of G by cubes. With each cube $Q \in \Upsilon$, we associate a set of functionals, the $L_2(Q)$-scalar products with polynomials of degree less than l. Thus, with each cube, $\mathbf{d}(\mathbf{N}, l)$, the dimension of the space of polynomials of interest, functionals are associated, altogether $|\Upsilon|\mathbf{d}(\mathbf{N}, l)$ of them, so, the common null space $\mathcal{Y} = \mathcal{Y}[\Upsilon]$ of these functionals has just this codimension (or less). Let $\Lambda(\Upsilon)$ be defined as $\sup_{f \in \mathcal{Y}[\Upsilon]} \frac{\mathbf{t}[f]}{\|f\|^2}$. By the variational principle (5.5),

$$n(\Lambda(\Upsilon), \mathbf{T}) \le |\Upsilon|\mathbf{d}(\mathbf{N}, l). \tag{5.6}$$

So, given $\lambda > 0$, we need to construct a covering $\Upsilon = \Upsilon(\lambda)$ such that $\Lambda(\Upsilon) \le \lambda$, and at the same time, $|\Upsilon|$ should be under control, and its value will produce the required estimate via (5.6).

In our case, the quadratic form defining our operator is

$$\mathbf{t}[f] = \int_G V(X)|f(X)|^2 \, d\mu$$

with $V \ge 0$ (we identify measure μ with its natural extension by zero to the whole of G). The same integral over a cube Q will be denoted by $\mathbf{t}_Q[f]$. If $f \in \mathcal{Y}[\Upsilon]$, then on each cube Q of the covering Υ, the restriction f_Q of f to Q is orthogonal to polynomials of degree less than l. Suppose that for *such* functions f_Q an estimate of the form

$$\mathbf{t}_Q[f] = \mathbf{t}[f_Q] \le \mathbf{J}(Q)\|f\|^2_{H^l_{\mathrm{hom}}(Q)}, \quad f \in \mathcal{Y}[\Upsilon], \tag{5.7}$$

holds, with some function of cubes $\mathbf{J}(Q)$. We need further the function $\mathbf{J}$ to be *upper semi-additive*; this means that if Q_ι is a family of disjoint cubes, all of them inside a cube Q^0, then

$$\sum \mathbf{J}(Q_\iota) \le \mathbf{J}(Q^0). \tag{5.8}$$

We sum (5.7) over all cubes in the covering Υ to obtain

$$\begin{aligned} \mathbf{t}[f] &\le \sum_{Q \in \Upsilon} \mathbf{t}_Q[f] = \sum_{Q \in \Upsilon} \mathbf{t}[f_Q] \le \sum_{Q \in \Upsilon} \mathbf{J}(Q)\|f\|^2_{H^l_{\mathrm{hom}}(Q)} \\ &\le \max_{Q \in \Upsilon} \mathbf{J}(Q) \sum_{Q \in \Upsilon} \|f\|^2_{H^l_{\mathrm{hom}}(Q)}. \end{aligned} \tag{5.9}$$

Now, suppose that the covering Υ has a controlled finite multiplicity, i.e., every point in $\mathbf{M}$ is covered by no more than $\kappa = \kappa(\mathbf{N})$ different cubes in Υ. Then the sum on the right-hand side in (5.9) can be majorized by $\kappa\|f\|^2_{H^l_{\mathrm{hom}}(G)}$, which gives us

$$\mathbf{t}_Q[f] \le \kappa \max_{Q \in \Upsilon} \mathbf{J}(Q)\|f\|^2_{H^l_{\mathrm{hom}}(G)}.$$

We finally arrive at the estimate

$$\Lambda(\Upsilon) \le \kappa \max_{Q\in\Upsilon} \mathbf{J}(Q). \tag{5.10}$$

So, we reach our aim, the estimate (5.4), as soon as we construct the covering Υ such that

$$\kappa \max_{Q\in\Upsilon} \mathbf{J}(Q) < \lambda \quad \text{and} \quad |\Upsilon|\mathbf{d}(\mathbf{N}, l) \le C\lambda^{-1}\|V\|_G^{(\mathrm{av},\Psi,\mu)}.$$

We set

$$\mathbf{J}(Q) = A_2\|V\|_Q^{(\mathrm{av},\Psi,\mu)}$$

with A_2 being the constant in (5.3). By Lemma 5.2, the required inequality (5.7) is satisfied for this particular choice of $\mathbf{J}$. Next, this function $\mathbf{J}$ is upper semi-additive in the sense of (5.8). This property of the averaged Orlicz norm with respect to the Lebesgue measure was established in [17, Lemma 3] and then extended to AR-measures, see [8, Lemma 2.8].

Now we construct the covering $\Upsilon(\lambda)$. First, by our Theorem 4.3, there exists a cube Q_0 such that for every cube in $\mathbb{R}^{\mathbf{N}}$ with edges parallel to the ones of Q_0 (we call such cubes parallel to Q_0), its faces have zero μ-measure. We fix such a cube Q_0 and in the future *all* cubes under our consideration will be parallel to Q_0. Consider a "large" cube $\mathbf{Q}$ such that the cube concentric with $\mathbf{Q}$ and with three times shorter edges still contains G inside. We replace in the eigenvalue problem the domain G by this larger cube $\mathbf{Q}$. By the usual variational principle, any eigenvalue estimate in $\mathbf{Q}$ implies automatically the same estimate for the initial problem in G.

For any point $X \in G$, we consider the family of cubes $Q_X(t)$, $0 < t < \infty$, of size t centered at X. By our choice of Q_0, the function $\mu(Q_X(t))$ is continuous. Moreover, the function $\rho_X(t) = \mathbf{J}(Q_X(t))$ is a continuous function of t and it tends to zero as $t \to 0$. The proof of this, rather elementary, fact is presented in [8] for the case $\mathbf{N} = 2, 0 < \alpha < 2$, and a very detailed proof for $\alpha = \mathbf{N}$ is included in [18]. Both proofs carry over to our case automatically, without any modifications, since they are dimension-independent and it is only the continuity of $\mu(Q_X(t))$ that is used there.

The function $\rho_X(t)$ stabilizes for large t to ρ_∞ when $Q_X(t) \supset \mathbf{Q}$. With some constant $\mathbf{k}$, to be determined later, and $\lambda < \mathbf{k}\rho_\infty$, we find by continuity a value of $t = t(X)$ such that $\rho_X(t(X)) = \lambda\mathbf{k}^{-1}$. The set of all cubes $\{Q_X(t(X)) : X \in G\}$ forms a covering of $\mathbf{M}$, and by the Besicovitch Covering Lemma (see, e.g., [7, Chapter 1, Theorem 1.1]) one can find a finite sub-covering $\Upsilon = \Upsilon(\lambda)$ of finite multiplicity $\kappa = \kappa(\mathbf{N})$; moreover, this covering can be split into finitely many families, $\Upsilon_{\mathbf{n}}, \mathbf{n} \le \mathbf{n}_0$, with $\mathbf{n}_0$ depending only on the dimension $\mathbf{N}$, so that the cubes in each family are disjoint. This will be the covering we are looking for.

In order to estimate the quantity of cubes in Υ, we use the fact that the function $\mathbf{J}$ is upper semi-additive. Therefore, for each of the families $\Upsilon_{\mathbf{n}}$,

$$|\Upsilon_{\mathbf{n}}|\mathbf{k}^{-1}\lambda = \sum_{Q\in\Upsilon_{\mathbf{n}}} \mathbf{J}(Q) \le \mathbf{J}(\mathbf{Q}) = A_2\|V\|_G^{(\mathrm{av},\Psi,\mu)}, \quad \mathbf{n} = 1, \dots, \mathbf{n}_0,$$

and hence

$$|\Upsilon| = \sum |\Upsilon_{\mathbf{n}}| \le \mathbf{n}_0 \lambda^{-1} \mathbf{k} A_2 \|V\|_G^{(\mathrm{av},\Psi,\mu)}.$$

Now we can choose the constant $\mathbf{k}$ and arrive at the required estimate. Indeed, by (5.6) and (5.10),

$$n(\kappa\lambda\mathbf{k}^{-1}, \mathbf{T}) \le \lambda^{-1} A_2 \mathbf{n}_0 \mathbf{k} \mathbf{d}(\mathbf{N}, l) \|V\|_G^{(\mathrm{av},\Psi,\mu)}.$$

Taking $\mathbf{k} = \kappa$, we obtain estimate (5.4). ■

The approach, just presented, was called "piecewise polynomial approximation" by its authors. It measures how fast a function in the Sobolev space can be approximated in the weighted L_2 norm by functions that are polynomial on cubes in the covering Υ.

5.3 Eigenvalue estimates

We can conclude the *proof* of Theorem 2.3. Fix a bounded domain $G = \Omega' \subset \Omega$ with smooth boundary, $\mathbf{M} \subset G$. We consider first the case when $\mathfrak{A} = \mathfrak{A}_0 = (-\Delta_{\mathcal{D}} + 1)^{-\frac{l}{2}}$, where $(-\Delta_{\mathcal{D}})$ is the Laplace operator in G with the Dirichlet boundary conditions. Having proved the eigenvalue estimate for this case, we will then use a simple argument to justify the required estimate for a general operator $\mathfrak{A}$.

Consider the quadratic form $\mathbf{F}_V[u, \mu, \mathfrak{A}_0]$, $u \in L_2(G)$. Denote $f = \mathfrak{A}_0 u$, so that

$$u = (-\Delta_{\mathcal{D}} + 1)^{\frac{l}{2}} f, \quad f \in \mathring{H}^l(G).$$

The quadratic form $\mathbf{F}_V[u, \mu, \mathfrak{A}_0]$ is thus transformed to

$$\mathbf{F}_V^{(l)}[f] = \int_{\mathbf{M}} V|f|^2 \, d\mu, \quad f \in \mathring{H}^l(G).$$

The operator $\mathring{\mathbf{T}}(V, \mu)$ defined by the quadratic form $\mathbf{F}_V^{(l)}$ on the Hilbert space $\mathring{H}^l(G)$ is exactly the operator considered in Theorem 5.3. Due to our substitution,

$$u = (-\Delta_{\mathcal{D}} + 1)^{\frac{l}{2}} f,$$

the operator $\mathring{\mathbf{T}}(V, \mu)$ on $\mathring{H}^l(G)$ is similar to the operator $\mathbf{T}(V, \Sigma, \mathfrak{A}_0)$ on $L_2(G)$, therefore they have the same spectrum, and the required estimate (2.2) for this case follows. Note that the choice of $G = \Omega' \subset \Omega$ is arbitrary and may influence only the constant in (2.2).

Now we pass to the general case, again for $V \ge 0$. As explained in Sections 2 and 3, we may assume that the pseudodifferential operator $\mathfrak{A}$ contains cut-offs to Ω'. We denote by Γ the operator of restriction of functions in $H^l(\Omega')$ to $\mathbf{M}$; this is a bounded operator $\Gamma : H^l(\Omega) \to L_2(\mathbf{M}, \mu)$ by Lemma 5.1. We set $|V| = W^2$, $W \ge 0$. Then the quadratic form $\mathbf{F}_{|V|,\mathfrak{A}}[u]$ can be represented as follows:

$$\mathbf{F}_{V,\mathfrak{A}}[u] = \langle W\Gamma\mathfrak{A}u, W\Gamma\mathfrak{A}u \rangle_{L_2(\mathbf{M},\mu)} = \langle (W\Gamma\mathfrak{A})^*(W\Gamma\mathfrak{A})u, u \rangle_{L_2(\Omega')}.$$

Therefore,

$$\mathbf{T}(|V|,\mu,\mathfrak{A}) = \mathfrak{A}^*\Gamma^*|V|\Gamma\mathfrak{A}.$$

Thus, compared with the operator $\mathbf{T}(|V|,\mu,\mathfrak{A}_0)$, we have

$$\mathbf{T}(|V|,\mu,\mathfrak{A}) = (\mathfrak{A}_0^{-1}\mathfrak{A})^*\mathbf{T}(|V|,\mu,\mathfrak{A}_0)(\mathfrak{A}_0^{-1}\mathfrak{A}).$$

Since $\mathfrak{A}$ is a pseudodifferential operator of order $-l$, it follows that $\mathfrak{A}_0^{-1}\mathfrak{A}$ is a bounded operator on $L_2(\Omega')$, together with $(\mathfrak{A}_0^{-1}\mathfrak{A})^*$. Hence the eigenvalue estimate for $\mathbf{T}(|V|,\mu,\mathfrak{A}_0)$, already justified, is preserved after multiplication by bounded operators.

Proof of Theorem 2.1. It follows immediately from Theorem 2.3 since the measure μ_Σ on a compact Lipschitz surface Σ of dimension d satisfies (2.3) with $\alpha = d$. ■

6 Approximation of the weight

To perform the last reduction, we use systematically the Asymptotic Perturbation Lemma by Birman and Solomyak (see, e.g., [3, Lemma 1.5] or [14, Lemma 6.1]). By this fundamental lemma, if for an operator $\mathbf{T}$, there exists a family of approximating operators $\mathbf{T}_\varepsilon$, $0<\varepsilon<\varepsilon_0$, such that for their eigenvalues the asymptotic formula $n_\pm(\lambda,\mathbf{T}_\varepsilon)\sim A_{\pm,\varepsilon}\lambda^{-q}$ is known and for the difference $\mathbf{T}'_\varepsilon = \mathbf{T}-\mathbf{T}_\varepsilon$ the eigenvalue estimate $\limsup \lambda^q n(\lambda,\mathbf{T}'_\varepsilon)\le\varepsilon$ is proved, then the asymptotic formula $n_\pm(\lambda,\mathbf{T})\sim A_\pm\lambda^{-q}$ holds with coefficients $A_\pm = \lim A_{\pm,\varepsilon}$. In our case, the estimates will be provided by Theorem 2.1. So, let $V\in L^\Psi(\Sigma)$. We approximate V by sufficiently regular functions.

Lemma 6.1. *For any* $\varepsilon>0$, *there exists* $V_\varepsilon\in C_0^\infty(\Omega)$ *such that* $\|V-V_\varepsilon\|_\Sigma^{(\mathrm{av},\Psi)}<\varepsilon$.

Proof. For $d=\mathbf{N}$ this is a well-known statement about the density of smooth functions. So, let $d<\mathbf{N}$. As usual, we can assume that the surface Σ is covered by one local chart, $\Sigma:\mathbf{y}=\varphi(\mathbf{x})$, $\mathbf{x}\in\mathcal{D}\subset\mathbb{R}^d$. First, we truncate the function V at some level, i.e., we set

$$V'_\varepsilon(X) = \begin{cases} V(X) & \text{if } |V(X)|\le N,\\ N\operatorname{sign}V(X) & \text{otherwise.}\end{cases}$$

The function V'_ε is bounded and, since $V\in L^\Psi(\Sigma)$, the cut-off level N can be chosen so that

$$\|V-V'_\varepsilon\|_\Sigma^{(\mathrm{av},\Psi)}<\frac{\varepsilon}{3}.$$

Next, we extend the function V'_ε defined on Σ to $\mathcal{D}\times\mathbb{R}^{\mathfrak{d}}$ by setting

$$V''_\varepsilon(\mathbf{x},\mathbf{y}) = V'_\varepsilon(\mathbf{x},\varphi(\mathbf{x})),\quad \mathbf{x}\in\mathcal{D}.$$

On Σ, we still have

$$\|V-V''_\varepsilon\|_\Sigma^{(\mathrm{av},\Psi)}<\frac{\varepsilon}{3}.$$

The resulting function V_ε'' belongs to $L_2(\mathcal{D} \times (-r, r))$ for any $r > 0$. Now consider the convolution of V_ε'' with some mollifier $\omega(\mathbf{x})$. We obtain a smooth function V_ε''' depending, again, on $\mathbf{x}$ only. The mollifier ω can be chosen so that

$$\|V_\varepsilon'' - V_\varepsilon'''\|_\Sigma^{(\mathrm{av},\Psi)} \leq C\|V_\varepsilon''(\cdot,\mathbf{y}) - V_\varepsilon'''(\cdot,\mathbf{y})\|_{L_2(\mathcal{D})} < \frac{\varepsilon}{3}.$$

Finally, multiplying V_ε''' by a smooth compactly supported cut-off function χ, which equals 1 near Σ, we obtain the required approximation V_ε. ■

We need approximating functions to satisfy an additional property (cf. Corollary 3.2).

Lemma 6.2. *The function V_ε can be approximated by a function $\tilde{V}_\varepsilon \in C_0^\infty(\Omega)$ such that $\|V_\varepsilon - \tilde{V}_\varepsilon\|_\Sigma^{(\mathrm{av},\Psi)} < \varepsilon$ and there exist open sets $\Omega_\pm \subset \Omega$ such that $\pm\tilde{V}_\varepsilon(X) \geq 0$ for $X \in \Omega_\pm$, $\tilde{V}_\varepsilon(X) = 0$ for $X \in \Omega \setminus (\Omega_+ \cup \Omega_-)$, and, finally,* $\operatorname{dist}(\Omega_+, \Omega_-) > 0$.

Proof. Let the cylinder $\mathcal{D} \times [-N, N]$ contain the support of V_ε. Consider the closed set $\Omega_0 = \{X \in \overline{\mathcal{D}} \times [-N, N] : V_\varepsilon(X) = 0\}$. Since V_ε is uniformly continuous on the compact set $\overline{\mathcal{D}} \times [-N, N]$, for any $\delta_1 > 0$ there exists $\delta_2 > 0$ such that $|V_\varepsilon(X)| < \delta_1$ for X in the δ_2-neighborhood of Ω_0. Take a function $\chi_\varepsilon \in C_0^\infty(\Omega)$ with the property that $\chi_\varepsilon(X) \in [0, 1]$, $\chi_\varepsilon(X) = 0$ in the $\frac{\delta_2}{2}$-neighborhood of Ω_0 and $\chi_\varepsilon(X) = 1$ outside the δ_2-neighborhood of Ω_0. Set $\tilde{V}_\varepsilon = \chi_\varepsilon V_\varepsilon$. Then the sets $\Omega_\pm = \{X : \pm\tilde{V}_\varepsilon(X) > 0\}$ satisfy the conditions of the lemma. Indeed, if $|X_+ - X_-| < \frac{\delta_2}{2}$ and $\pm\tilde{V}_\varepsilon(X_\pm) > 0$, then, by the continuity of V_ε, there must exist a point $X_0 \in \Omega_0$ on the straight interval connecting X_+ and X_- such that at least one of the distances $|X_\pm - X_0|$ is smaller than $\frac{\delta_2}{2}$, and this contradicts the construction of $\Omega_\pm$. Moreover, $|V_\varepsilon - \tilde{V}_\varepsilon| \leq \delta_1$ everywhere in Ω, and, by choosing δ_1 sufficiently small, we obtain the required approximation property. ■

7 The asymptotic formula

We present the proof of Theorem 2.4 for the case when there is just one surface Σ.

Proof. By the Asymptotic Perturbation Lemma, together with our Lemmas 6.1, 6.2 and Theorem 2.1, it is sufficient to prove the eigenvalue asymptotics formula for the function $\tilde{V}_\varepsilon$, $\varepsilon > 0$, as above. We will denote it simply by V in what follows.

By Corollary 3.2, to study the asymptotics of positive (negative) eigenvalues of $\mathbf{T}(V)$, it is sufficient to consider the operator with a nonnegative (nonpositive) function V.

So, with a smooth $V \geq 0$, we perform a reduction to the integral operator. We have already made the first step. We set $V = W^2$ with $W \in C_0^\infty$, $W \geq 0$, and denote by Γ the operator of restriction of functions in $H^l(\Omega)$ to the surface Σ,

$$\Gamma : H^l(\Omega) \to L_2(\Sigma, \mu_\Sigma).$$

The form (2.1) can now be written as

$$\mathbf{F}_V[u] = \langle (W\Gamma\mathfrak{A})u, (W\Gamma\mathfrak{A})u\rangle_{L_2(\Sigma,\mu_\Sigma)} = \langle (W\Gamma\mathfrak{A})^*(W\Gamma\mathfrak{A})u, u\rangle_{L_2(\Omega)},$$

with $(W\Gamma\mathfrak{A})$ considered as acting from $L_2(\Omega)$ to $L_2(\Sigma, \mu_\Sigma)$.

It follows that

$$\mathbf{T}(V) = (W\Gamma\mathfrak{A})^*(W\Gamma\mathfrak{A}).$$

The nonzero eigenvalues of the operator $\mathbf{T}$ in $L_2(\Omega)$ coincide with the nonzero eigenvalues of the operator

$$\mathbf{L} = (W\Gamma\mathfrak{A})(W\Gamma\mathfrak{A})^* = W[\Gamma(\mathfrak{A}\mathfrak{A}^*)\Gamma^*]W$$

in $L_2(\Sigma, \mu_\Sigma)$. Here $\mathfrak{A}\mathfrak{A}^*$ is, up to a smoothing term, a nonnegative pseudodifferential operator of order $-2l = -\mathbf{N}$, with the principal symbol $|a_{-l}(X, \Xi)|^2$, thus it is a weakly polar integral operator. The corresponding Schwartz kernel $\mathcal{L}(x, y)$ has the leading singularity $A(X)\log|x - y| + \vartheta(X, X - Y)$, where $\vartheta(X, X - Y)$ is a function positively homogeneous in $X - Y$ of order zero and smooth in the first variable. Therefore, the operator $\mathbf{L}$ is an integral operator on Σ,

$$(\mathbf{L}v)(X) = \int_\Sigma W(X)\mathcal{L}(X, Y)W(Y)v(Y)\,d\mu_\Sigma(Y), \tag{7.1}$$

see, e.g., [19, Chapter 2, especially Proposition 2.6]. The function W, defined initially on Σ, is, in fact, the trace on Σ of a smooth function $\tilde{W}(X) = (\tilde{V}(X))^{\frac{1}{2}}$ defined on $\mathbb{R}^\mathbf{N}$. Hence, the operator in (7.1) takes the form

$$(\mathbf{L}v)(X) = \int_\Sigma \tilde{\mathcal{K}}(X, Y)v(Y)\,d\mu_\Sigma(Y),$$

where $\tilde{\mathcal{K}}$ is the integral kernel of the pseudodifferential operator with the principal symbol $\tilde{V}(X)|a_{-l}(X, \Xi)|^2$. We are now in the setting of the paper [15] (see also [14]). By [15, Theorem 6.4], the asymptotics of the eigenvalues of a weakly polar integral operator on a Lipschitz surface is given exactly by formula (2.4). We mention that the proof in those papers uses the Asymptotic Perturbation Lemma in [3, Lemma 1.5] in the analysis of operator convergence when a Lipschitz surface is approximated by smooth ones in a special way. Thus, this beautiful invention by Birman and Solomyak is used twice in our study, in quite different settings. ∎

Now we justify the eigenvalue asymptotic formula (2.11) for the case when there are several Lipschitz surfaces of possibly different dimensions, including, possibly, dimension $d = \mathbf{N}$, codimension $\mathfrak{d} = 0$, i.e., a domain in Ω with an absolutely continuous measure. In order to avoid excessive complications in the proof, we impose a geometric restriction.

Theorem 7.1. *Let Σ_j, $j = 1, \dots, J$, be compact Lipschitz surfaces of dimension d_j, $1 \le d_j \le \mathbf{N}$, with measures $P_j = V_j\mu_{\Sigma_j}$, where $V_j \in L^\Psi(\mu_{\Sigma_j})$. We assume that the* *surfaces of the same dimension are disjoint. Then the eigenvalues of the operator* $\mathbf{T}(P, \mathfrak{A}) = \sum \mathbf{T}(P_j, \mathfrak{A})$ *satisfy the asymptotic formula*

$$n_{\pm}(\lambda, \mathbf{T}(P, \mathfrak{A})) \sim \lambda^{-1} \sum_j C^{\pm}(V_j, \Sigma_j, \mathfrak{A}),$$

where $C^{\pm}(V_j, \Sigma_j, \mathfrak{A})$ *are given by* (2.4) *with* V_j, Σ_j *in place of* V, Σ.

Proof. If the surfaces Σ_j are disjoint so that all mutual distances are positive, the result follows by the inductive application of Lemma 3.1. Otherwise, we construct an approximation of the measure $P = \sum P_j$ by measures with disjoint surfaces Σ_j.

Relabeling the surfaces if necessary, we can assume that their dimensions increase with j, i.e., $d_1 \leq d_2 \leq \cdots \leq d_J$. We start with Σ_1. Consider the δ-neighborhood $U_1(\delta)$ of Σ_1 in $\mathbb{R}^{\mathbf{N}}$. If $d_j > d_1$, then the surface measure $\mu_{\Sigma_j}(\Sigma_j \cap U_1(\delta))$, $j > 1$, decays at least as $\delta^{d_j - d_1}$ for $\delta \to 0$. If $d_j = d_1$, then, according to our assumption, $\Sigma_j \cap U_1(\delta) = \varnothing$ for all sufficiently small $\delta > 0$. Either way, $\mu_{\Sigma_j}(\Sigma_j \cap U_1(\delta))$, $j > 1$, tends to zero as $\delta \to 0$. Therefore, for δ small enough, the averaged Ψ-norm of each V_j, $j > 1$, over $\Sigma_j \cap U_1(\delta)$ can be made arbitrarily small. We set

$$V_j^{(1)}(X) = V_j(X)(1 - \chi(X)), \quad j > 1,$$

where χ is the characteristic function of $U_1(\delta)$. Then the averaged norm of $V_j - V_j^{(1)}$ is small and Σ_1 is separated from the support of $V_j^{(1)}$, $j > 1$. Next we consider

$$\Sigma_2^{(1)} = \Sigma_2 \cap \operatorname{supp} V_2^{(1)}.$$

This piece of the surface Σ_2 is separated from Σ_1, but may cross Σ_j, $j > 2$. We repeat with $\Sigma_2^{(1)}$ the same procedure as above, considering its sufficiently small neighborhood in $\mathbb{R}^{\mathbf{N}}$ and then killing the remaining measures in this neighborhood. In this way, after a finite number of steps, we arrive at a system of separated measures, to which Lemma 3.1 can be applied, with further application of the already proved case of Theorem 2.4. In this construction, we introduce perturbations of V_j with small averaged Ψ-norms. By Theorem 2.1 and, again, [3, Lemma 1.5], the asymptotic eigenvalue formula is thus justified in full. ■

Acknowledgements. The research of Grigori Rozenblum was supported by the Russian Science Foundation (Project 20-11-20032).

Bibliography

[1] M. S. Agranovich, Potential-type operators and conjugation problems for second-order strongly elliptic systems in domains with a Lipschitz boundary. *Funct. Anal. Appl.* **43** (2009), 165–183

[2] M. S. Birman, A. A. Laptev and M. Z. Solomyak, The negative discrete spectrum of the operator $(-\Delta)^l - \alpha V$ in $\mathbb{R}^d$ for d even and $2l \geq d$. *Ark. Mat.* **35** (1997), 87–126

[3] M. S. Birman and M. Z. Solomjak, Spectral asymptotics of nonsmooth elliptic operators. I. *Trudy Moskov. Mat. Obšč.* **27** (1972), 3–52; English transl. *Trans. Moscow Math. Soc.* **27** (1972), 1–52

[4] M. S. Birman and M. Z. Solomjak, Quantitative analysis in Sobolev's imbedding theorems and applications to spectral theory. In *Tenth Mathematical School (Summer School, Kaciveli/Nalchik, 1972)*, pp. 5–189, Inst. Mat. Akad. Nauk Ukrain. SSR, Kiev, 1974 (in Russian); English transl. Amer. Math. Soc. Transl. Ser. 2 114, American Mathematical Society, Providence, 1980

[5] G. David and S. Semmes, *Fractured fractals and broken dreams*. Oxford Lect. Ser. Math. Appl. 7, Clarendon Press, New York, 1997

[6] G. Grubb, Spectral asymptotics for nonsmooth singular Green operators. *Comm. Partial Differential Equations* **39** (2014), 530–573

[7] M. de Guzmán, Differentiation of integrals in $\mathbf{R}^n$. Lecture Notes in Math. 481, Springer, Berlin, 1975

[8] M. Karuhanga and E. Shargorodsky, On negative eigenvalues of two-dimensional Schrödinger operators with singular potentials. *J. Math. Phys.* **61** (2020), Article ID 051509

[9] A. Kozhevnikov, The asymptotic behavior of the eigenvalues of an elliptic boundary value problem with λ in the equation and in the boundary condition. *Uspehi Mat. Nauk* **31** (1976), 265–266

[10] A. Kozhevnikov and S. Yakubov, On operators generated by elliptic boundary problems with a spectral parameter in boundary conditions. *Integral Equations Operator Theory* **23** (1995), 205–231

[11] S. Lord, F. Sukochev and D. Zanin, A last theorem of Kalton and finiteness of Connes' integral. *J. Funct. Anal.* **279** (2020), Article ID 108664

[12] V. Maz'ya, *Sobolev spaces with applications to elliptic partial differential equations*. Augmented edn., Grundlehren Math. Wiss. 342, Springer, Heidelberg, 2011

[13] G. Rozenblum and E. Shargorodsky, Eigenvalue asymptotics for weighted polyharmonic operator with a singular measure in the critical case. *Funct. Anal. Appl.*, submitted

[14] G. Rozenblum and G. Tashchiyan, Eigenvalue asymptotics for potential type operators on Lipschitz surfaces. *Russ. J. Math. Phys.* **13** (2006), 326–339

[15] G. Rozenblum and G. Tashchiyan, Eigenvalue asymptotics for potential type operators on Lipschitz surfaces of codimension greater than 1. *Opuscula Math.* **38** (2018), 733–758

[16] E. Shargorodsky, An estimate for the Morse index of a Stokes wave. *Arch. Ration. Mech. Anal.* **209** (2013), 41–59

[17] M. Solomyak, Piecewise-polynomial approximation of functions from $H^l((0,1)^d)$, $2l = d$, and applications to the spectral theory of the Schrödinger operator. *Israel J. Math.* **86** (1994), 253–275

[18] F. Sukochev and D. Zanin, Cwikel–Solomyak estimates on tori and Euclidean spaces. Preprint 2020, arXiv:2008.04494

[19] M. E. Taylor, *Partial differential equations. II*. Springer, New York, 1996

Relations between two parts of the spectrum of a Schrödinger operator and other remarks on the absolute continuity of the spectrum in a typical case

Oleg Safronov

To Ari Laptev on the occasion of his 70th birthday

We will discuss relations between different parts of spectra of differential operators. In particular, we will see that negative and positive spectra of Schrödinger operators are related to each other. However, there is a stipulation: one needs to consider two operators one of which is opened from the other by flipping the sign of the potential at each point x. If one knows only that the negative spectra of the two operators are discrete, then their positive spectra do not have gaps. If one knows more about the rate of accumulation of the discrete negative eigenvalues to zero, then one can say more about the absolutely continuous component of the positive spectrum. The second part of this article contains a discussion of spectral properties of a family of Schrödiger operators depending on a real parameter t. The results claim that the absolutely continuous spectrum of an operator of this family is essentially supported by the positive half-line for almost every t.

1 Introduction

Let us look at the set displayed in the picture below.

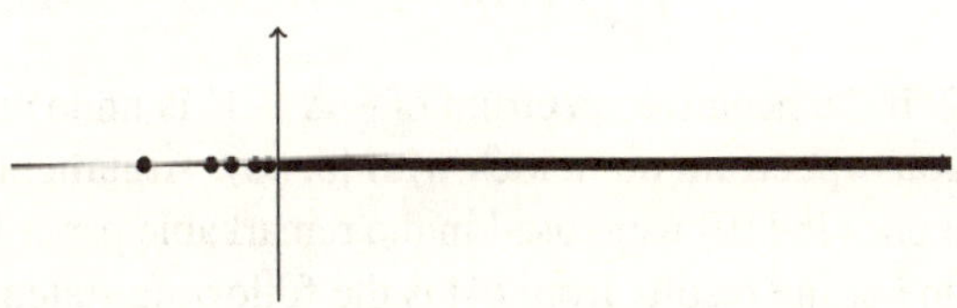

The vertical arrow divides this set into two parts: the discrete part and the continuous one. We see the dots on the left part, while the right part consists of the thick half-line.

Keywords: Schrödinger operators, absolutely continuous spectrum, discrete spectrum
2020 Mathematics Subject Classification: Primary 81Q10; secondary 47F10, 47A10

In general, there is no relation between one part and the other. That is no longer true if this picture represents the spectrum of a Schrödinger operator!

There is a relation between the two parts of the spectrum. It is particularly simple if the potential $V(x)$ in the Schrödinger equation is negative. In this situation, if we see only isolated dots to the left of the origin, then the right part of the spectrum consists of the positive half-line. In the general case, one has to consider two Schrödinger operators, one of which is obtained from the other by flipping the sign of the electric potential $V(x)$ at every point x.

Theorem 1.1. *Let $V \in L^\infty(\mathbb{R}^d)$. If the negative spectra of the two Schrödinger operators $-\Delta + V$ and $-\Delta - V$ consist of isolated eigenvalues of finite multiplicity possibly accumulating to the origin, then the spectra contain every point of the interval $[0, \infty)$.*

A complete proof of Theorem 1.1 can be found in our joint paper [11] with R. Killip and S. Molchanov (see also [4] for the case $d = 1$). The reason we do not discuss additional properties of the positive spectrum in this theorem is that there is no condition on the rate of accumulation of eigenvalues to zero. However, once we know that the eigenvalues tend to zero sufficiently fast, we can talk about absolute continuity.

Theorem 1.2. *Let $d = 1$, and assume that the negative spectra of both operators $-\frac{d^2}{dx^2} + V$ and $-\frac{d^2}{dx^2} - V$ are finite. Then the positive spectra of these operators are absolutely continuous.*

A proof of this striking result can be found the paper [2] by D. Damanik and R. Killip (see also [3]). The proof relies on the fact that if the negative spectra of the two operators are finite, then V is a function of the form $V = f' + W$, where f^2 and W are L^1-functions (actually, one can say more about f). Another example of a statement where conditions imposed on negative eigenvalues imply some properties of the potential is the Grigoryan–Netrusov–Yau theorem (see [10]):

Theorem 1.3. *Let $d = 2$. Let $V \geq 0$ be a bounded non-negative function on $\mathbb{R}^2$. Assume that the negative spectrum of the operator $-\Delta - V$ consist of N eigenvalues whose multiplicities are taken into account. Then $V \in L^1(\mathbb{R}^2)$ and there is a universal constant $C > 0$ such that*

$$\int_{\mathbb{R}^2} V(x)\, dx \leq CN.$$

Thus, in $d = 2$, if the negative spectrum of $-\Delta - V$ is finite and $V \geq 0$, then the absolutely continuous spectrum coincides with $[0, \infty)$. Arguments that are in some sense similar to the ones in [10] were used in the remarkable paper [4] by D. Damanik and C. Remling. One of the results from [4] is the following statement.

Theorem 1.4. *Let $d = 1$ and let $V \geq 0$ be a bounded function on $\mathbb{R}_+$. Assume that the negative spectrum of the operator $-\frac{d^2}{dx^2} - V$ on $\mathbb{R}_+$ with the Neumann condition at zero consists of eigenvalues $E_j < 0$. Then, for any $0 < p \leq \frac{1}{2}$, there is* *a constant* $C_p > 0$ *such that*

$$\int_0^\infty V^{p+\frac{1}{2}}(x)\,dx \le C_p \sum_j |E_j|^p.$$

As was mentioned in [15], this theorem implies the following statement.

Corollary 1.5. *Let* $d \ge 2$ *and let* $V \ge 0$ *be a bounded function on* $\mathbb{R}^d \setminus B_1$, *where* B_1 *is the unit ball. Assume that the negative spectrum of the operator* $-\Delta - V$ *on* $\mathbb{R}^d \setminus B_1$ *with the Neumann boundary conditions on the unit sphere consists of eigenvalues* $E_j < 0$. *Then, for any* $0 < p \le \frac{1}{2}$, *there is a constant* $C_{p,d} > 0$ *such that*

$$\int_{\mathbb{R}^d \setminus B_1} V^{p+\frac{1}{2}}(x) \frac{dx}{|x|^{d-1}} \le C_{p,d}\Big(1 + \sum_j |E_j|^p\Big).$$

Consequently, if $\sum_j |E_j|^{\frac{1}{2}} < \infty$ and $V \ge 0$, then the a.c. spectrum of $-\Delta - V$ coincides with $[0,\infty)$. It turns out that the condition that V is sign-definite can be omitted. However, if V is allowed to change its sign, then one has to consider two operators $-\Delta + V$ and $-\Delta - V$.

Theorem 1.6. *Let* $V \in L^\infty(\mathbb{R}^d)$. *Assume that the negative spectra of both operators* $-\Delta + V$ *and* $-\Delta - V$ *consist of isolated eigenvalues* $\{E_j^+\}_{j=1}^\infty$ *and* $\{E_j^-\}_{j=1}^\infty$ *satisfying the condition*

$$\sum_j |E_j^+|^{\frac{1}{2}} + \sum_j |E_j^-|^{\frac{1}{2}} < \infty.$$

Then the absolutely continuous spectrum of each of the two operators is essentially supported on the positive half-line $\mathbb{R}_+$.

This theorem was proved by Damanik and Remling in the one-dimensional case. The proof of this result in $d \ge 2$ can be found in my paper [18].

Another remarkable theorem from [4] gives an estimate of the Hausdorff dimension of a set S supporting the singular components of the spectral measures in the following sense:

$$\mu_{\text{sing}}^{\pm}(S) = \mu_{\text{sing}}^{\pm}(\mathbb{R}). \tag{1.1}$$

Note that in order to be able to talk about the spectral measures of $-\frac{d^2}{dx^2} \pm V$, one has to consider these operators on the half-line $\mathbb{R}_+$. The statement below does not depend on the boundary condition at $x = 0$.

Theorem 1.7. *Let* $d = 1$. *Assume that*

$$\sum_j |E_j^+|^p + \sum_j |E_j^-|^p < \infty \quad \textit{for } 0 \le p \le \frac{1}{4}.$$

Then there is a set $S \subset \mathbb{R}$ *satisfying* (1.1) *for which*

$$\dim S \le 4p.$$

One of the reasons two operators are needed in such theorems is that the continuous spectrum of a Schrödinger operator with a decaying potential has only one edge. There are other interesting operators whose continuous spectra have two edges. For instance, one can consider the so-called discrete Schrödinger operator defined as an operator on $\ell^2(\mathbb{Z}^d)$ by

$$[Hu](n) = \sum_{j:|n-j|=1} u(j) + V(n)u(n), \quad u \in \ell^2(\mathbb{Z}^d), \tag{1.2}$$

where V is a real-valued function on the lattice $\mathbb{Z}^d$. The spectrum of this operator might (but does not have to) look like the set in the picture below.

In this case, the role of the thick half-line is played by the line segment whose edges are $-2d$ and $2d$ (here, d is the dimension of the space). The following statement is proved in [11].

Theorem 1.8. *If the spectrum of the operator* (1.2) *outside of the interval* $[-2d, 2d]$ *is bounded and consists of isolated points possibly accumulating to the edges* $-2d$ *and* $2d$, *then the interval* $[-2d, 2d]$ *is contained in the spectrum of* H.

Note that for dimensions $d = 1$ and $d = 2$, this theorem was proved in the paper [1].

Let us now discuss spectral properties of the Dirac operator

$$\mathfrak{D} = \mathfrak{D}_0 + V(x) \tag{1.3}$$

acting in the space $L^2(\mathbb{R}^3; \mathbb{C}^4)$. The free Dirac operator in (1.3) is introduced by the formula

$$\mathfrak{D}_0 = \sum_{j=1}^{3} i\gamma_j \frac{\partial}{\partial x_j} + \gamma_0, \tag{1.4}$$

where γ_j are selfadjoint 4×4- matrices having very special properties, whose discussion will be omitted. The operator (1.3) is used to describe the motion of a charged quantum particle in the electric field whose potential is the function $V(x)$. Typically, the spectrum of a Dirac operator looks like the set displayed in the picture below.

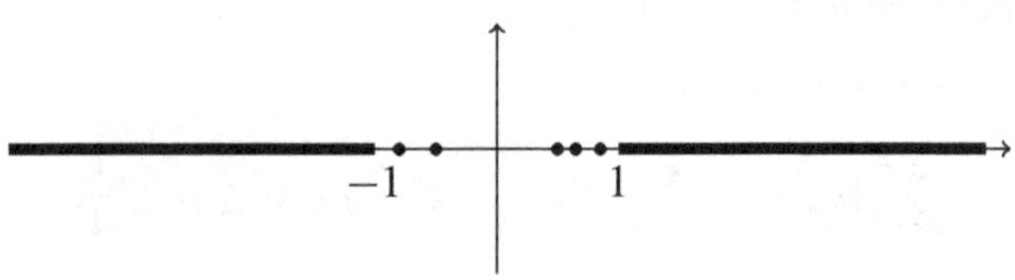

In particular, the operator (1.4) has purely absolutely continuous spectrum, which coincides with the set $(-\infty, 1] \cup [1, \infty)$.

In this picture, you see two thick half-lines and the dots in the gap between the points -1 and 1. Assume that it is the spectrum of a Dirac operator. Is it true that the rate of accumulation of the dots to the edges of the gap determines whether or not the absolutely continuous spectrum of this operator covers the two thick half-lines? A partial answer to this question is given in the paper [18], containing some results where besides the condition on the rate of accumulation of the eigenvalues to the edges of the spectrum one imposes a condition on the decay of V. In order to describe one of the results of [18], we need to define $\mathfrak{D}_V$ setting

$$\mathfrak{D}_V = \mathfrak{D}_0 + V\gamma_0. \tag{1.5}$$

Note that the operator (1.5) was studied by Denisov in [7].

Theorem 1.9. *Let $\mathfrak{D} = \mathfrak{D}_0 + V$ be the Dirac operator with a potential V satisfying the condition*

$$\int_{\mathbb{R}^3} \frac{|V|^2}{|x|^2}\,dx < \infty.$$

Suppose that the spectrum of $\mathfrak{D}$ in $(-1, 1)$ consists of eigenvalues E_j obeying the condition

$$\sum_j \sqrt{1 - E_j^2} < \infty. \tag{1.6}$$

Then the absolutely continuous spectrum of the operator $\mathfrak{D}_V$ is essentially supported on $(-\infty, -1] \cup [1, \infty)$

Theorem 1.9 makes one wonder: is there a Dirac operator whose spectrum does not have eigenvalues in the gap $(-1, 1)$ by some obvious reason? The answer to this question is affirmative. Let $A : \mathbb{R}^3 \to \mathbb{R}^3$ be a vector potential with components A_1, A_2 and A_3. Suppose that $A \in L^\infty(\mathbb{R}^3; \mathbb{R}^3)$ and consider the magnetic Dirac operator

$$\mathcal{D}_A = \sum_1^3 \gamma_j \Big(i \frac{\partial}{\partial x_j} + A_j(x) \Big) + \gamma_0.$$

If the vector-potential A is identically zero, then $\mathcal{D}_A = \mathfrak{D}_0$. Therefore, it is natural to view $\mathcal{D}_A$ as a perturbation of $\mathfrak{D}_0$. Our assumption, that $A(x)$ decays at the infinity in some integral sense, is a sufficient condition guaranteeing that the following sets coincide:

$$\sigma_{\text{a.c.}}(\mathcal{D}_A) = \sigma(\mathcal{D}_A) = \sigma(\mathfrak{D}_0) = \mathbb{R} \setminus (-1, 1).$$

In this case, the spectrum of $\mathcal{D}_A$ looks like the set in the picture below.

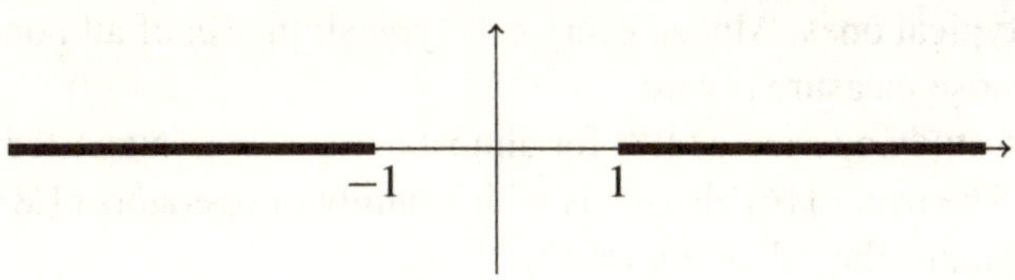

Since a condition of the form (1.6) is no longer needed, an analogue of Theorem 1.9 for the operator $\mathfrak{D}_A$ will be the following statement.

Theorem 1.10. *Let A be a bounded vector potential satisfying the conditions*

$$\int_{\mathbb{R}^3} |A|^2|x|^{-2}\,dx < \infty, \qquad (A(x), x) = 0 \quad \textit{for all } x \in \mathbb{R}^3.$$

Then the absolutely continuous spectrum of the operator $\mathfrak{D}_A$ is essentially supported on the set $(-\infty, -1] \cup [1, \infty)$.

One of the major challenges in the spectral theory of Schrödinger operators is the problem of proving the existence of the absolutely continuous spectrum under mild conditions on the decay of the potential V. The article [19] by Barry Simon contains a conjecture saying that, if $V \in L^\infty(\mathbb{R}^d)$ obeys

$$\int_{\mathbb{R}^d} \frac{|V(x)|^2}{|x|^{d-1}}\,dx < \infty, \tag{1.7}$$

then the a.c. spectrum covers the positive half-line $[0, \infty)$.

Among the results related to this conjecture are theorems that can not be obtained by the means of Scattering Theory. For instance, the paper [9] by Denisov contains a proof of the following remarkable statement.

Theorem 1.11. *Let $d = 3$. Assume that the support of the potential $V \in L^\infty(\mathbb{R}^3)$ is the union of the sparse spherical layers*

$$\operatorname{supp} V = \bigcup_{n\in\mathbb{N}\cup 0} \{x \in \mathbb{R}^3 : R_n < |x| < R_n + 1\},$$

where R_n satisfy the condition $R_{n+1} > e^{\beta R_n}$ with $\beta > 1$. Assume also that V obeys condition (1.7). *Then the absolutely continuous spectrum of the operator $-\Delta + V$ covers the interval* $[0, \infty)$.

Although first results related to Simon's conjecture appeared in the mathematical literature already fifteen years ago (see [12] and [14]), this problem is still open. A way to make this problem less challenging is to consider a collection of operators and then ask if most of the representatives of this collection have absolutely continuous spectra. Instead of a single operator, one considers a family of Schrödinger operators

$$-\Delta + \alpha V, \tag{1.8}$$

depending on a parameter $\alpha \in \mathbb{R}$. One studies only typical members of this family, ignoring the atypical ones. Almost every α is typical; the set of all non-typical values of α is a set whose measure is zero.

The idea to study operators (1.8) for almost every α was suggested by Denisov in the article [8]. The paper [16] also deals with a family of operators (1.8). Among other things, [16] contains the following result.

Theorem 1.12. *Suppose that V fulfills the so-called Simon condition*

$$\int \frac{V^2}{|x|^{d-1}}\,dx < \infty, \tag{1.9}$$

and the Fourier transform of $W = V|x|^{-\frac{d-1}{2}}$ satisfies

$$\int_{|\xi|<\delta} \frac{|\hat{W}(\xi)|^2}{|\xi|^2}\,d\xi < \infty, \quad \delta > 0, \tag{1.10}$$

then the absolutely continuous spectrum of the operator $H_\alpha = -\Delta + \alpha V$ covers the half-line $[0,\infty)$ for almost every α.

One should not think that potentials satisfying conditions of Theorem 1.12 are functions of a very special type, that never appear in physical examples. On the contrary, the way to prove the following theorem below is to show that the potential V satisfies conditions (1.9) and (1.10) (see [16]).

Theorem 1.13. *Let*

$$V = \sum_{n\in\mathbb{Z}^d} v_n\omega_n\chi(x-n), \quad x \in \mathbb{R}^d,\ d \ge 3,$$

where χ is the characteristic function of the unit cube $[0,1)^d$ and ω_n are bounded identically distributed independent random variables with zero expectations $\mathbb{E}[\omega_n] = 0$. Suppose that the real coefficients v_n satisfy the condition

$$\sum_{n\neq 0} \frac{v_n^2}{|n|^{d-1}} < \infty. \tag{1.11}$$

Then the absolutely continuous spectrum of $-\Delta + \alpha V$ is essentially supported on the interval $[0,\infty)$ almost surely for almost every $\alpha \in \mathbb{R}$.

Similar classes of potentials were considered by J. Bourgain and S. Denisov. In particular, the following theorem was proved in [6].

Theorem 1.14. *Let ϕ be a smooth function whose support is a subset of the unit ball in $\mathbb{R}^3$. Let*

$$V = \sum_{n\in\mathbb{Z}^3} v_n\omega_n\phi(x-2n), \quad x \in \mathbb{R}^3,$$

where and ω_n are bounded identically distributed independent random variables such that $\mathbb{E}[\omega_n^{2j+1}] = 0$ for all $j \in \mathbb{N}$. Suppose that the real coefficients v_n decay at infinity so that

$$|V(x)| \le C(1+|x|)^{-\frac{1}{2}-\delta}, \quad \delta > 0. \tag{1.12}$$

Then the absolutely continuous spectrum of $-\Delta + V$ covers the interval $[0,\infty)$ almost surely.

Obviously, assumption (1.12) is a much more restrictive condition than (1.11). Proving the claim of Theorem 1.13 for all α is still an open problem.

The method of the paper [16] was developed further in [17] . It was shown that a decay of the derivatives of the potential V can also imply the presence of the a.c. spectrum. Namely, let $\tilde V(r,\theta)$ be the function defined in polar coordinates by

$$\tilde V(r,\theta) = \int_0^r V(\rho,\theta)\,d\rho, \tag{1.13}$$

where r and ρ are the radial variables and $\theta = \frac{x}{|x|}$. Then the following statement holds true (see [17]).

Theorem 1.15. *Let $V \in L^\infty(\mathbb{R}^d)$. Let the function $\tilde V$ be defined by* (1.13)*. Suppose that*

$$\int_{\mathbb{R}^d} \frac{|\nabla \tilde V|^2}{|x|^{d-1}}\,dx < \infty. \tag{1.14}$$

Then the absolutely continuous spectrum of the operator $H_\alpha = -\Delta + \alpha V$ covers the half-line $[0,\infty)$ for almost every α.

Note that Theorem 1.15 nicely complements the main result of [13], where the conditions imposed on V imply that $|\nabla \tilde V(x)| \le C(1+|x|)^{-\frac12-\delta}$.

Let us give some examples showing that Theorem 1.15 is quite interesting.

Example 1. Let Φ be a smooth function on the unit sphere $\mathbb{S} = \{x \in \mathbb{R}^d : |x| = 1\}$. Then $\tilde V$ defined by (1.13) with

$$V = \frac{\Phi(\theta)}{(1+r)^{\frac12+\varepsilon}}, \qquad \theta = \frac{x}{|x|},\ r = |x|,\ \varepsilon > 0,$$

satisfies condition (1.14).

Example 2. Let Φ be a smooth function on the unit sphere $\mathbb{S} = \{x \in \mathbb{R}^d : |x| = 1\}$ and let $v \in L^2(0,\infty)$ be a function vanishing on $[0,\varepsilon)$ for some $\varepsilon > 0$. Then $\tilde V$ defined by (1.13) with

$$V = \Phi(\theta)v(r), \quad \theta = \frac{x}{|x|},\ r = |x|,\ \varepsilon > 0,$$

satisfies condition (1.14).

Note that in $d = 1$ condition (1.14) turns into

$$\int_{-\infty}^{\infty} V^2\,dx < \infty.$$

Therefore, Theorem 1.15 would be viewed as a direct generalization of a rather celebrated result of P. Deift and R. Killip [5], if it was proved for every α. Moreover, it gives new interesting examples of potentials for which the corresponding a.c. spectrum is essentially supported on the interval $[0,\infty)$. For instance, the positive function

$$V(r,\theta) = \frac{1}{(1+r)^{\frac12+\delta}}\left(2+\sin(r^\gamma\theta)\sin^n\left(\frac{\theta}{2}\right)\right) \quad \text{with } \gamma > \delta > 0,\ \theta \in [0,2\pi),$$

fulfills conditions of Theorem 1.15 in $d = 2$.

2 Proof of Theorem 1.1

While most of the proofs of theorems discussed in this survey are quite long, the proof of Theorem 1.1 is an exception. Therefore, we decided to present it again. Note that our paper [11] contains a more general statement that holds for some pseudo-differential operators. Here we focus only on Schrödinger operators.

Fix $\phi \in C_0^\infty(\mathbb{R}^d)$ having the property $\int |\phi|^2\,dx = 1$ and set

$$u_n(x) := n^{-\frac{d}{2}}\phi\Big(\frac{x}{n}\Big).$$

Let $E_\pm$ be the operator-valued spectral measures of $H_\pm$. Then $E_\pm((-\infty,0])H_\pm$ are compact operators, which transform a weakly convergent sequence u_n into two strongly convergent sequences. Therefore,

$$\|E_\pm((-\infty,0])H_\pm u_n\| \to 0 \quad \text{as } n\to\infty.$$

The latter implies that

$$\||H_+|^{\frac{1}{2}}u_n\|^2 + \||H_-|^{\frac{1}{2}}u_n\|^2 = (H_+u_n,u_n) + (H_-u_n,u_n) + o(1) \quad \text{as } n\to\infty.$$

Put differently,

$$\||H_+|^{\frac{1}{2}}u_n\|^2 + \||H_-|^{\frac{1}{2}}u_n\|^2 = 2\int_{\mathbb{R}^d}|\nabla u_n|^2dx + o(1) = o(1) \quad \text{as } n\to\infty.$$

Let $\eta\in\mathbb{R}^d$ be a vector such that $|\eta|^2=\lambda>0$. It is not difficult to show that for any $\psi\in C_0^\infty(\mathbb{R}^d)$,

$$\big|\big(e^{i\eta x}\psi,(H_\pm-\lambda)e^{i\eta x}u_n\big)\big| \le \|(H_\pm+2\gamma)^{\frac{1}{2}}\psi\|\cdot o(1) \quad \text{as } n\to\infty, \tag{2.1}$$

where $\gamma>0$ is a number such that $H_\pm+\gamma\ge 0$. The factor $o(1)$ on the right-hand side of (2.1) is independent of ψ. Indeed,

$$\begin{aligned}
&\big|\big(e^{i\eta x}\psi,(H_\pm-\lambda)e^{i\eta x}u_n\big) - (\psi, H_\pm u_n)\big| \\
&\quad\le \|\psi\|_{L^2}\Big(\int_{\mathbb{R}^d}(|\xi-\eta|^2-\lambda-|\xi|^2)^2|\hat u_n(\xi)|^2\,d\xi\Big)^{\frac{1}{2}} \\
&\quad= \|\psi\|_{L^2}\cdot o(1) \quad \text{as } n\to\infty.
\end{aligned} \tag{2.2}$$

Combining (2.2) with

$$|(\psi,H_\pm u_n)| \le \||H_\pm|^{\frac{1}{2}}\psi\|\cdot\||H_\pm|^{\frac{1}{2}}u_n\| = \||H_\pm|^{\frac{1}{2}}\psi\|\cdot o(1) \quad \text{as } n\to\infty,$$

we obtain (2.1).

On the other hand, since

$$\|(H_\pm+2\gamma)^{\frac{1}{2}}\psi\| \le C_{\gamma,\eta}\|(H_\pm+2\gamma)^{\frac{1}{2}}e^{i\eta x}\psi\|$$

with a constant $C_{\gamma,\eta}$ independent of ψ, we conclude that (2.1) can be written in the form

$$\left|\left(e^{i\eta x}\psi, (H_\pm - \lambda)e^{i\eta x}u_n\right)\right| \le \|(H_\pm + 2\gamma)^{\frac{1}{2}} e^{i\eta x}\psi\| \cdot o(1) \quad \text{as } n \to \infty. \tag{2.3}$$

Note that the factors $o(1)$ on the right-hand sides of (2.1)–(2.3) are also independent of ψ. Consequently, (2.3) implies

$$\|(H_\pm - \lambda)(H_\pm + 2\gamma)^{-\frac{1}{2}} e^{i\eta x}u_n\| \to 0 \quad \text{as } n \to \infty. \tag{2.4}$$

It follows from the relation (2.4) that 0 is in the essential spectra of the operators $(H_\pm - \lambda)(H_\pm + 2\gamma)^{-\frac{1}{2}}$, which can be true only if λ is in the essential spectra of $H_\pm$.

Bibliography

[1] D. Damanik, D. Hundertmark, R. Killip and B. Simon, Variational estimates for discrete Schrödinger operators with potentials of indefinite sign. *Comm. Math. Phys.* **238** (2003), 545–562

[2] D. Damanik and R. Killip, Half-line Schrödinger operators with no bound states. *Acta Math.* **193** (2004), 31–72

[3] D. Damanik, R. Killip and B. Simon, Schrödinger operators with few bound states. *Comm. Math. Phys.* **258** (2005), 741–750

[4] D. Damanik and C. Remling, Schrödinger operators with many bound states. *Duke Math. J.* **136** (2007), 51–80

[5] P. Deift and R. Killip, On the absolutely continuous spectrum of one-dimensional Schrödinger operators with square summable potentials. *Comm. Math. Phys.* **203** (1999), 341–347

[6] S. A. Denisov, Absolutely continuous spectrum of multidimensional Schrödinger operator. *Int. Math. Res. Not.* **2004** (2004), 3963–3982

[7] S. A. Denisov, On the absolutely continuous spectrum of Dirac operator. *Comm. Partial Differential Equations* **29** (2004), 1403–1428

[8] S. A. Denisov, Schrödinger operators and associated hyperbolic pencils. *J. Funct. Anal.* **254** (2008), 2186–2226

[9] S. A. Denisov, Wave propagation through sparse potential barriers. *Comm. Pure Appl. Math.* **61** (2008), 156–185

[10] A. Grigoryan, Y. Netrusov and S.-T. Yau, Eigenvalues of elliptic operators and geometric applications. In *Surveys in differential geometry. Vol. IX*, pp. 147–217, Surv. Differ. Geom. 9, International Press, Somerville, 2004

[11] R. Killip, S. Molchanov and O. Safronov, A relation between the positive and negative spectra of elliptic operators. *Lett. Math. Phys.* **107** (2017), 1799–1807

[12] A. Laptev, S. Naboko and O. Safronov, Absolutely continuous spectrum of Schrödinger operators with slowly decaying and oscillating potentials. *Comm. Math. Phys.* **253** (2005), 611–631

[13] G. Perelman, Stability of the absolutely continuous spectrum for multidimensional Schrödinger operators. *Int. Math. Res. Not.* **2005** (2005), 2289–2313

[14] O. Safronov, On the absolutely continuous spectrum of multi-dimensional Schrödinger operators with slowly decaying potentials. *Comm. Math. Phys.* **254** (2005), 361–366

[15] O. Safronov, Lower bounds on the eigenvalue sums of the Schrödinger operator and the spectral conservation law. *J. Math. Sci.* **166** (2010), 300–318

[16] O. Safronov, Absolutely continuous spectrum of a one-parameter family of Schrödinger operators. *St. Petersburg Math. J.* **24** (2013), 977–989

[17] O. Safronov, Absolutely continuous spectrum of a typical Schrödinger operator with a slowly decaying potential. *Proc. Amer. Math. Soc.* **142** (2014), 639–649

[18] O. Safronov, Absolutely continuous spectrum of a Dirac operator in the case of a positive mass. *Ann. Henri Poincaré* **18** (2017), 1385–1434

[19] B. Simon, Schrödinger operators in the twenty-first century. In *Mathematical physics 2000*, pp. 283–288, Imperial College Press, London, 2000

Bogoliubov theory for many-body quantum systems

Benjamin Schlein

Dedicated to Ari Laptev, on the occasion of his 70th birthday, in recognition of his contributions to mathematical physics and of his invaluable services to our community

We review some recent applications of rigorous Bogoliubov theory. We show how Bogoliubov theory can be used to approximate quantum fluctuations, both in the analysis of the energy spectrum and in the study of the dynamics of many-body quantum systems.

1 Introduction

In several physically interesting limits, the behavior of many-body quantum systems can be approximated by classical theories. In these regimes, classical theories often predict both energetic and dynamical properties of many-body Hamiltonians. Bogoliubov theory describes corrections to the classical limit, known as quantum fluctuations. In these short notes we are going to review recent progress in the mathematically rigorous understanding of the predictions of Bogoliubov theory, for some important examples of many-body quantum system.

In Section 2, we will consider Bose gases consisting of N particles trapped in a box or confined by external fields. In the mean-field regime, we will show how Bogoliubov theory predicts correction to Hartree theory and allows us to obtain norm approximations for the solution of the many-body Schrödinger equation and to determine the ground state energy and the low-energy excitation spectrum, up to errors vanishing as $N \to \infty$. We will also present recent developments which led to the understanding of thc low-energy spectrum in the more subtle Gross–Pitaevskii regime, where particles interact through a repulsive potential with scattering length of order N^{-1}.

Keywords: Many-body quantum mechanics, Bogoliubov theory, Bose–Einstein condensates, mean-field limit, Hartree–Fock theory, Fröhlich polaron
2020 Mathematics Subject Classification: 35P05, 35Q40, 35Q55, 47A75, 81Q10, 81V70, 82B10

In Section 3, we will then switch to Fermi gases, in the mean-field limit. For systems of N fermions, the mean-field regime is naturally linked with a semiclassical limit, with Planck's constant scaling as $N^{-\frac{1}{3}}$ and vanishing, as $N \to \infty$. Hartree–Fock theory, based on the restriction to Slater determinants, provides an approximation to the ground state energy. As we will explain in Section 3, the most important excitations of Slater determinants are particle-hole pairs of fermions, approximately behaving like bosons. As a consequence, it is possible to estimate the correlation energy of a mean-field Fermi gas, namely the difference between the true many-body ground state energy and the minimum of the Hartree–Fock functional, by means of bosonic Bogoliubov theory.

Finally, in Section 4 we will discuss the dynamics of a polaron in the strong coupling regime. A polaron is an electron, interacting with a quantized field describing the polarization of the crystal lattice. At strong coupling $\alpha \gg 1$, there is a separation of scale between the electron, which moves on times of order one and the radiation field, which instead moves slowly, on times of order α^2. The dynamics can be described, to leading order, by the Landau–Pekar equations. To obtain an accurate approximation, valid up to times of order α^2 and leading therefore to non-trivial changes in the radiation field, it is important to modify the coherent state, evolved through the Landau–Pekar equations, with a Bogoliubov dynamics (a unitary evolution with a generator quadratic in creation and annihilation operators).

2 Bose gases, energy and dynamics

To illustrate the main ideas of Bogoliubov theory and to explain how they can be rigorously implemented, let us consider a system of N bosons in the mean field regime. For simplicity, let us assume that particles move in the box $\Lambda = [0;1]^3$, with periodic boundary conditions. We are interested in the ground state energy and in the low-energy spectrum of the Hamilton operator

$$H_N^{\mathrm{mf}} = \sum_{j=1}^N -\Delta_{x_j} + \frac{1}{N}\sum_{i<j}^N V(x_i - x_j) \tag{2.1}$$

acting on $L^2_s(\Lambda^N)$, the subspace of $L^2(\Lambda^N)$ consisting of functions that are symmetric with respect to permutations of the N particles. To further simplify our discussion, let us also assume that the interaction potential is bounded and positive definite, i.e. $\widehat{V} \geq 0$. Then we immediately obtain

$$\begin{aligned} 0 &\leq \int_{\Lambda\times\Lambda} V(x-y)\Big[\sum_{j=1}^N \delta(x-x_j) - N\Big]\Big[\sum_{i=1}^N \delta(y-x_i) - N\Big]\, dx\, dy \\ &= \sum_{i,j=1}^N V(x_i - x_j) - N^2\widehat{V}(0) = 2\sum_{i<j}^N V(x_i-x_j) + NV(0) - N^2\widehat{V}(0). \end{aligned}$$

This implies that

$$H_N^{\mathrm{mf}} \geq \frac{N}{2}\widehat{V}(0) + \sum_{j=1}^{N} -\Delta_{x_j} - \frac{V(0)}{2} \tag{2.2}$$

and has two important consequences. First of all, when combined with the upper bound provided by the trivial trial state $\psi_N(x_1,\dots,x_N) = 1$ for all $(x_1,\dots,x_N) \in \Lambda^N$, it shows that the ground state energy E_N^{mf} of (2.1) is such that

$$\left| E_N^{\mathrm{mf}} - \frac{N}{2}\widehat{V}(0) \right| \leq C \tag{2.3}$$

uniformly in N. Furthermore, (2.2) also implies that low-energy states exhibit complete Bose–Einstein condensation in the zero-momentum mode $\varphi_0 \in L^2(\mathbb{R}^3)$, defined by $\varphi_0(x) = 1$ for all $x \in \Lambda$. In fact, for an arbitrary $\psi_N \in L^2_s(\Lambda^N)$ with

$$\langle \psi_N, H_N^{\mathrm{mf}} \psi_N \rangle \leq E_N^{\mathrm{mf}} + C \tag{2.4}$$

it follows from (2.2) that

$$\left\langle \psi_N, \sum_{j=1}^{N} -\Delta_{x_j} \psi_N \right\rangle \leq C$$

and therefore, with the notation $q_0 = 1 - |\varphi_0\rangle\langle\varphi_0|$ for the projection onto the orthogonal complement of φ_0, and with the bound $-\Delta \geq 2\pi q_0$ for the Laplacian on Λ, that

$$\langle \psi_N, d\Gamma(q_0)\psi_N \rangle \leq C, \tag{2.5}$$

where $d\Gamma(q_0) = \sum_{j=1}^N 1 \otimes \cdots \otimes q_0 \otimes \cdots \otimes 1$ measures the number of particles orthogonal to φ_0. Equation (2.5) shows that low-energy states ψ_N (satisfying (2.4)) exhibit condensation in the state φ_0, with only a bounded number of excitations (independent of N).

Given (2.3), it is interesting to investigate order-one corrections to the ground state energy of (2.1). In fact, determining the energy up to errors that vanish as $N \to \infty$ is also important because it allows us to resolve excited eigenvalues. To reach this goal, we are going to use a rigorous version of Bogoliubov theory, first proposed in [33].

While the leading-order contribution to the ground state energy of (2.1) is the energy of the condensate, the next order corrections are determined by the interaction between the excitations and the condensate. To estimate these contributions, it is convenient to factor out the condensate and to rewrite the Hamilton operator (2.1) in terms of its orthogonal excitations. To this end, we define the unitary map

$$U : L^2_s(\mathbb{R}^{3N}) \to \mathcal{F}^{\leq N}_{\perp\varphi_0} = \bigoplus_{n=0}^{N} L^2_{\perp\varphi_0}(\mathbb{R}^3)^{\otimes_s n}$$

so that $U\psi_N = \{\alpha_0, \alpha_1, \dots, \alpha_N\}$ if the N-particle wave function $\psi_N \in L^2_s(\Lambda^N)$ is given by

$$\psi_N = \alpha_0 \varphi_0^{\otimes N} + \alpha_1 \otimes_s \varphi_0^{\otimes(N-1)} + \cdots + \alpha_N \tag{2.6}$$

with $\otimes_s$ denoting symmetrized tensor product. Here $\alpha_j \in L^2_{\perp\varphi_0}(\Lambda)^{\otimes_s j}$, the symmetric tensor product of j copies of the orthogonal complement of φ_0. The vector $\{\alpha_0, \dots, \alpha_N\}$, in the truncated Fock space $\mathcal{F}^{\leq N}_{\perp\varphi_0}$, describes excitations of the condensate in the state ψ_N.

The action of the operator U can be easily expressed in the language of second quantization, in terms of creation and annihilation operators a_p^*, a_p satisfying canonical commutation relations

$$[a_p, a_q^*] = \delta_{p,q}, \quad [a_p, a_q] = [a_p^*, a_q^*] = 0 \tag{2.7}$$

for arbitrary momenta $p, q \in \Lambda^* = 2\pi\mathbb{Z}^3$. We find that

$$\begin{aligned} Ua_0^*a_0U^* &= N - \mathcal{N}_+, \quad & Ua_0^*a_pU^* &= \sqrt{N - \mathcal{N}_+}\,a_p, \\ Ua_q^*a_pU^* &= a_q^*a_p & Ua_q^*a_0U^* &= a_q^*\sqrt{N - \mathcal{N}_+} \end{aligned} \tag{2.8}$$

for all $p, q \in \Lambda_+^* = \Lambda^* \backslash \{0\}$. Here we introduced the operator

$$\mathcal{N}_+ = Ud\Gamma(q_0)U^* = \sum_{p\in\Lambda_+^*} a_p^*a_p,$$

measuring the number of excitations on $\mathcal{F}^{\leq N}_{\perp\varphi_0}$. In particular, (2.5) implies that low-energy states $\psi_N \in L^2_s(\Lambda^N)$ satisfying (2.4) are mapped by U into excitation vectors $\xi_N = U\psi_N$, with

$$\langle \xi_N, \mathcal{N}_+\xi_N\rangle \leq C. \tag{2.9}$$

Using U, we can now define the excitation Hamiltonian

$$\mathcal{L}_N^{\text{mf}} = UH_N^{\text{mf}}U^* : \mathcal{F}^{\leq N}_{\perp\varphi_0} \to \mathcal{F}^{\leq N}_{\perp\varphi_0}.$$

With (2.8), we find

$$\mathcal{L}_N^{\text{mf}} = UH_N^{\text{mf}}U^* = \mathcal{L}_N^{(0)} + \mathcal{L}_N^{(2)} + \mathcal{L}_N^{(3)} + \mathcal{L}_N^{(4)} \tag{2.10}$$

with

$$\begin{aligned} \mathcal{L}_N^{(0)} &= \frac{N-1}{2N}\widehat{V}(0)(N - \mathcal{N}_+) + \frac{\widehat{V}(0)}{2N}\mathcal{N}_+(N - \mathcal{N}_+) \\ \mathcal{L}_N^{(2)} &= \sum_{p\in\Lambda_+^*} p^2 a_p^*a_p + \sum_{p\in\Lambda_+^*} \widehat{V}(p)a_p^*\frac{N - \mathcal{N}_+ - 1}{N}a_p \\ &\quad + \frac{1}{2}\sum_{p\in\Lambda_+^*} \widehat{V}(p)(b_p^*b_{-p}^* + \text{h.c.}) \\ \mathcal{L}_N^{(3)} &= \frac{1}{\sqrt{N}} \sum_{p,q\in\Lambda_+^*,\, p+q\neq 0} \widehat{V}(p)\big[b_{p+q}^*a_{-p}^*a_q + a_q^*a_{-p}b_{p+q}\big] \\ \mathcal{L}_N^{(4)} &= \frac{1}{2N} \sum_{p,q\in\Lambda_+^*,\, r\neq -p,-q} \widehat{V}(r)a_{p+r}^*a_q^*a_pa_{q+r}, \end{aligned} \tag{2.11}$$

where we defined the modified creation and annihilation operators

$$\begin{aligned} b_p &= U \frac{a_0^*}{\sqrt{N}} a_p U^* = \sqrt{\frac{N - \mathcal{N}_+}{N}} a_p, \\ b_p^* &= U a_p^* \frac{a_0}{\sqrt{N}} U^* = a_p^* \sqrt{\frac{N - \mathcal{N}_+}{N}}. \end{aligned} \tag{2.12}$$

Notice that on states with only few excitations, we have $\mathcal{N}_+ \ll N$ and therefore $b_p^* \simeq a_p^*$, $b_p \simeq a_p$. From (2.8), we conclude that conjugation with the unitary map U can be seen as a rigorous version of the c-number substitution proposed by Bogoliubov, which consisted in replacing all a_0 and a_0^* operators, associated with the condensate wave function φ_0, by factors of $\sqrt{N}$.

Following Bogoliubov approach, we remark now that the terms $\mathcal{L}_N^{(3)}$ and $\mathcal{L}_N^{(4)}$ in (2.11) are formally of order $N^{-\frac{1}{2}}$ and, respectively, N^{-1}; thus, we expect them to be negligible, in the limit of large N. This simple observation can be made rigorous, at least for bounded potentials, using estimates of the form (2.9) on the number of excitations. To be more precise, since

$$\begin{aligned} \pm \mathcal{L}_N^{(3)} &\le \frac{C\|V\|_\infty}{\sqrt{N}} (\mathcal{N}_+ + 1)^{\frac{3}{2}}, \\ \pm \mathcal{L}_N^{(4)} &\le \frac{C\|V\|_\infty}{N} (\mathcal{N}_+ + 1)^2, \end{aligned} \tag{2.13}$$

one actually needs a bound for the second moment of $\mathcal{N}_+$, which can be obtained with a bit more work, combining (2.5) with a similar estimate for the commutator $[H_N^{\mathrm{mf}}, d\Gamma(q_0)]$ (in [33], this step is avoided by using localization in the number of excitations on the truncated Fock space $\mathcal{F}_{\perp\varphi_0}^{\le N}$).

Ignoring cubic and quartic terms on the right-hand side of (2.10), and recalling that $b_p \simeq a_p$, $b_p^* \simeq a_p^*$ on low-energy states, we conclude that the low-energy spectrum of $\mathcal{L}_N^{\mathrm{mf}}$ (and therefore the low-energy spectrum of (2.1)) are determined by the quadratic operator

$$\begin{aligned} \mathcal{L}_\infty^{\mathrm{mf}} = \frac{\widehat{V}(0)}{2}(N-1) &+ \sum_{p \in \Lambda_+^*} [p^2 + \widehat{V}(p)] a_p^* a_p \\ &+ \frac{1}{2} \sum_{p \in \Lambda_+^*} \widehat{V}(p) [a_p^* a_{-p}^* + a_p a_{-p}]. \end{aligned} \tag{2.14}$$

It should be noted that, while the excitation Hamiltonian (2.10) is an operator on the truncated Fock space $\mathcal{F}_{\perp\varphi_0}^{\le N}$, we think of the limiting Hamiltonian $\mathcal{L}_\infty^{\mathrm{mf}}$ as an operator on the full Fock space

$$\mathcal{F} = \bigoplus_{n \ge 0} L^2_{\perp\varphi_0}(\Lambda)^{\otimes_s n}.$$

To determine the spectrum of (2.14), we conjugate $\mathcal{L}_\infty^{\text{mf}}$ with the Bogoliubov transformation

$$T = \exp\Big[\frac{1}{2}\sum_{p\in\Lambda_+^*}(\tau_p a_p^* a_{-p}^* - \text{h.c.})\Big]. \tag{2.15}$$

As it easily follows from the canonical commutation relations (2.7), the action of T on creation and annihilation operators is given by

$$T^* a_p T = a_p \cosh\tau_p + a_{-p}^* \sinh\tau_p. \tag{2.16}$$

Applying (2.16) to (2.14), and choosing $\tau \in \ell^2(\Lambda_+^*)$ such that

$$\tanh(2\tau_p) = -\frac{\widehat{V}(p)}{p^2 + \widehat{V}(p)},$$

we get rid of all off-diagonal terms (of the form a^*a^* or aa) and arrive at

$$\begin{aligned} T^*\mathcal{L}_\infty^{\text{mf}} T = \frac{\widehat{V}(0)}{2}(N-1) - \frac{1}{2}\sum_{p\in\Lambda_+^*}\Big[p^2 + \widehat{V}(p) - \sqrt{|p|^4 + 2p^2\widehat{V}(p)}\Big] \\ + \sum_{p\in\Lambda_+^*}\sqrt{|p|^4 + 2p^2\widehat{V}(p)}\, a_p^* a_p. \end{aligned} \tag{2.17}$$

This expression allows us to conclude that the ground state energy of the Hamilton operator (2.1) is given by

$$E_N^{\text{mf}} = \frac{\widehat{V}(0)}{2}(N-1) - \frac{1}{2}\sum_{p\in\Lambda_+^*}\Big[p^2 + \widehat{V}(p) - \sqrt{|p|^4 + 2p^2\widehat{V}(p)}\Big] + o(1) \tag{2.18}$$

in the limit $N \to \infty$. Moreover, the spectrum of $H_N^{\text{mf}} - E_N$ below a fixed threshold $\zeta > 0$ consists of eigenvalues having the form

$$\sum_{p\in\Lambda_+^*} n_p\sqrt{|p|^4 + 2p^2\widehat{V}(p)} + o(1) \tag{2.19}$$

as $N \to \infty$. The first proof of (2.18) and (2.19) has been obtained in [43]. These results have been extended to systems of bosons on $\mathbb{R}^3$, trapped by an external potential in [22, 33]. Alternative approaches have been proposed also in [10, 42].

It turns out that Bogoliubov theory can also be used to study the time-evolution of bosons in the mean-field limit. From the point of view of physics, it is interesting to study the evolution of an initially trapped gas, resulting from a change of the external fields. Let us suppose, for example, that the external potential is turned off at time $t = 0$. In this case, we are led to the Schrödinger equation

$$i\partial_t\psi_{N,t} = H_N^{\text{mf}}\psi_{N,t} \tag{2.20}$$

for the evolution of the wave function $\psi_{N,t} \in L^2_s(\mathbb{R}^{3N})$. The Hamilton operator H_N^{mf} on the right-hand side of (2.20) has the same form as in (2.1), but now it acts on functions defined on the full space $\mathbb{R}^{3N}$ (rather than Λ^N). We are interested in the evolution of initial wave functions $\psi_{N,0}$ exhibiting complete Bose–Einstein condensation in a one-particle state φ. This assumption corresponds to the idea that at time $t = 0$, the trapped gas is prepared at equilibrium. At zero or at low temperature, this implies that $\psi_{N,0}$ exhibits complete condensation, with only a bounded number of excitations. We can therefore assume the initial data $\psi_{N,0}$ to satisfy (2.5), with q_0 replaced by $q = 1 - |\varphi\rangle\langle\varphi|$. Under general assumptions on the potential, one can then show that condensation is preserved by the time-evolution. For any fixed $t \in \mathbb{R}$, there exists namely a constant $C = C(t)$ such that

$$\langle \psi_{N,t}, d\Gamma(q_t)\psi_{N,t}\rangle \leq C(t), \tag{2.21}$$

where $q_t = 1 - |\varphi_t\rangle\langle\varphi_t|$ and φ_t is the solution of the Hartree equation

$$i\partial_t\varphi_t = -\Delta\varphi_t + (V * |\varphi_t|^2)\varphi_t \tag{2.22}$$

with the initial data $\varphi_{t=0} = \varphi$.

While (2.21) shows that $\psi_{N,t}$ can be approximated by the product state $\varphi_t^{\otimes N}$ when computing expectations of one-particle observables, we cannot expect convergence to hold in norm, i.e. $\|\psi_{N,t} - \varphi_t^{\otimes N}\| \not\to 0$. To obtain a norm approximation for the solution of the Schrödinger equation (2.20), we have instead to take into account excitations of the product state $\varphi_t^{\otimes N}$ that can be described, to leading order, by a time-dependent family of Bogoliubov transformations.

Let us explain this point in some more details. First of all, we remove the condensate from the solution $\psi_{N,t}$ of (2.20), so that we can focus on its orthogonal excitations. To this end, we define, for any fixed $t \in \mathbb{R}$, a unitary map $U_t : L^2_s(\mathbb{R}^{3N}) \to \mathcal{F}^{\leq N}_{\perp\varphi_t}$, similarly as we did around (2.6), but with the zero-momentum mode φ_0 replaced by the solution φ_t of the Hartree equation (2.22). We can then define the fluctuations dynamics

$$\mathcal{W}_N(t;s) = U_t e^{-iH_N^{\text{mf}}(t-s)} U_s^* : \mathcal{F}^{\leq N}_{\perp\varphi_s} \to \mathcal{F}^{\leq N}_{\perp\varphi_t}. \tag{2.23}$$

The excitation vector associated with the solution $\psi_{N,t}$ of equation (2.20) is determined by (2.23), since

$$\xi_{N,t} = U_t\psi_{N,t} = U_t e^{-iH_N t} U_0^* \xi_{N,0} = \mathcal{W}_N(t;0)\xi_{N,0}.$$

To find a good approximation for $\xi_{N,t}$, we observe that $\mathcal{W}_N(t;0)$ satisfies the equation

$$i\partial_t \mathcal{W}_N(t;0) = \mathcal{L}_N^{\text{mf}}(t)\mathcal{W}_N(t;0)$$

with the generator

$$\mathcal{L}_N^{\text{mf}}(t) = U_t H_N U_t^* + [i\partial_t U_t]U_t^*$$

which can be computed adapting the rules (2.8) to the present setting. Denoting by $\mathcal{N}_t$ the number of particles operator on $\mathcal{F}_{\perp\varphi_t}^{\leq N}$, from (2.21) we obtain the estimate $\langle \xi_{N,t}, \mathcal{N}_t \xi_{N,t} \rangle \leq C(t)$. Similarly to (2.13), this allows us to argue that cubic and quartic parts of $\mathcal{L}_N^{\text{mf}}(t)$ are negligible, and therefore that, in the limit of large N, $\mathcal{L}_N^{\text{mf}}(t)$ can be approximated (neglecting the constant part, which only produces an uninteresting phase), by its quadratic component

$$\begin{aligned}\mathcal{L}_\infty^{\text{mf}}(t) = &\int dx \nabla_x a_x^* \nabla_x a_x + \int dx (V * |\varphi_t|^2)(x) a_x^* a_x \\ &+ \int dx\, dy V(x-y) \varphi_t(x) \overline{\varphi}_t(y) a_x^* a_y \\ &+ \frac{1}{2} \int dx\, dy V(x-y) [\varphi_t(x)\varphi_t(y) a_x^* a_y^* + \overline{\varphi}_t(x)\overline{\varphi}_t(y) a_x a_y]\end{aligned} \tag{2.24}$$

generating, through

$$i\partial_t \mathcal{W}_\infty(t;s) = \mathcal{L}_\infty^{\text{mf}}(t) \mathcal{W}_\infty(t;s) \quad \text{with } \mathcal{W}_\infty(s;s) = 1$$

the Bogoliubov dynamics $\mathcal{W}_\infty$ (since (2.24) is quadratic, $\mathcal{W}_\infty$ acts on creation and annihilation operators as a time-dependent Bogoliubov transformation). The following theorem has been proven in [32].

Theorem 2.1. *Assume that $V \in L^\infty(\mathbb{R}^3)$ and $\varphi_0 \in H^1(\mathbb{R}^3)$ with $\|\varphi_0\|_2 = 1$. Let φ_t be the solution of (2.22). Consider $\xi_{N,0} \in \mathcal{F}_{\perp\varphi_0}^{\leq N}$, with $\langle \xi_{N,0}, \mathcal{N}_0 \xi_{N,0} \rangle \leq C$. Then, for a suitable phase $\omega_{N,t}$,*

$$\| \mathcal{W}_N(t;0)\xi_{N,0} - e^{i\omega_{N,t}} \mathcal{W}_\infty(t;0)\xi_{N,0} \| \leq \frac{C|t|}{\sqrt{N}}. \tag{2.25}$$

Since the limiting dynamics $\mathcal{W}_\infty$ (like its generator) acts on the full (rather than truncated) Fock space $\mathcal{F}$, the bound (2.25) should be understood in the norm of $\mathcal{F}$, after embedding $\mathcal{W}_N(t;0)\xi_{N,0}$ in the full Fock space. Equation (2.25) provides a norm approximation for the solution $\psi_{N,t} = U_t^* \mathcal{W}_N(t;0)\xi_{N,0}$ of the Schrödinger equation (2.20), in terms of the solution of the Hartree equation (2.22) (entering the map U_t) and of the Bogoliubov dynamics $\mathcal{W}_\infty(t;0)$. In fact, as proven in [32], Theorem 2.1 holds for more general potentials, requiring only $V^2 \leq C(1-\Delta)$, if one additionally assumes that the initial excitation vector $\xi_{N,0}$ has finite kinetic energy.

Quantum fluctuations around the Hartree dynamics for the evolution of coherent states have been first investigated in [28]. More recently, many other works have been devoted to the approximation of the many-body time-evolution in the mean-field regime [9, 20, 25, 38] and also in more singular limits [7, 8, 24, 29, 39].

To conclude this section, let us also briefly discuss more recent results on the Bose gas in the Gross–Pitaevskii regime, where N particles interact through a repulsive potential with scattering length of the order N^{-1}. We focus here on the analysis of the low-energy spectrum, and for simplicity we restrict our attention to systems defined on the box $\Lambda = [0;1]^3$, with periodic boundary conditions. The Hamilton operator has the form

$$H_N^{\mathrm{GP}} = \sum_{j=1}^{N} -\Delta_{x_j} + \sum_{i<j}^{N} N^2 V(N(x_i - x_j)) \tag{2.26}$$

where we assume $V \geq 0$ to have compact support.

To study the spectral properties of (2.26), we proceed as we did in the mean-field regime, factoring out the condensate by defining the excitation Hamiltonian $\mathcal{L}_N^{\mathrm{GP}} = U H_N^{\mathrm{GP}} U^*$ acting on the truncated Fock space $\mathcal{F}_{\perp\varphi_0}^{\leq N}$. The form of $\mathcal{L}_N^{\mathrm{GP}}$ is similar to (2.10), but with the potential $\widehat{V}(p)$ now replaced by $\widehat{V}(\frac{p}{N})$. The lack of decay of the potential in momentum space is the reason why, mathematically, the Gross–Pitaevskii limit is more subtle than the mean-field regime. It implies, first of all, that cubic and quartic terms in $\mathcal{L}_N^{\mathrm{GP}}$ are not negligible. In fact, they even contribute to the energy at leading order, effectively replacing $\widehat{V}(0)$ by $8\pi\mathfrak{a}_0$, with $\mathfrak{a}_0$ denoting the scattering length of the unscaled potential V.

The main difference with respect to the mean-field regime is that here correlations among particles, which were essentially negligible in the mean-field limit, become important. To model correlations and thus extract important contributions to the energy from cubic and quartic terms in $\mathcal{L}_N^{\mathrm{GP}}$, we can conjugate the excitation Hamiltonian with a generalized Bogoliubov transformation e^B, with the phase

$$B = \frac{1}{2} \sum_{p \in \Lambda_+^*} [\eta_p b_p^* b_{-p}^* - \text{h.c.}]. \tag{2.27}$$

To make sure that B maps $\mathcal{F}_{\perp\varphi_0}^{\leq N}$ back to itself, we use here the modified creation and annihilation operators defined in (2.12). Notice, however, that since $b_p \simeq a_p$ and $b_p^* \simeq a_p^*$ when $\mathcal{N}_+ \ll N$, one can prove that e^B acts similarly as a standard Bogoliubov transformation (2.15), satisfying an approximate version of (2.16), on states with only few excitations.

Let us now briefly (and heuristically) explain how conjugation with the Bogoliubov transformation $T = e^B$ helps to extract important contributions from cubic and quartic terms in $\mathcal{L}_N^{\mathrm{GP}}$. Expanding the action of $T = e^B$ to second order, we find

$$e^B \mathcal{L}_N^{\mathrm{GP}} e^{-B} = \mathcal{L}_N^{\mathrm{GP}} + [B, \mathcal{L}_N^{\mathrm{GP}}] + \frac{1}{2}[B,[B,\mathcal{L}_N^{\mathrm{GP}}]] + \cdots.$$

Decomposing (similarly as in (2.11)), $\mathcal{L}_N^{\mathrm{GP}} = \mathcal{L}_{\mathrm{GP}}^{(0)} + \mathcal{L}_{\mathrm{GP}}^{(2)} + \mathcal{L}_{\mathrm{GP}}^{(3)} + \mathcal{L}_{\mathrm{GP}}^{(4)}$, we observe that the off-diagonal part of $\mathcal{L}_{\mathrm{GP}}^{(2)}$, namely the operator

$$\frac{1}{2} \sum_{p \in \Lambda_+^*} \widehat{V}\left(\frac{p}{N}\right) [b_p^* b_{-p}^* + b_{-p} b_p], \tag{2.28}$$

can be combined with the commutator of B with the diagonal part of $\mathcal{L}_{\mathrm{GP}}^{(2)}$ (in particular, the main contribution arises from the commutator of B with the kinetic energy operator $\mathcal{K} = \sum_{p \in \Lambda_+^*} p^2 a_p^* a_p$) and with quadratic terms arising from the commutator $[B, \mathcal{L}_{\mathrm{GP}}^{(4)}]$. While $[B, \mathcal{K}]$ is quadratic and off-diagonal (since $[b_p, a_q^* a_q] = \delta_{pq} b_p$) and can therefore be directly compared with (2.28), the commutator $[B, \mathcal{L}_{\mathrm{GP}}^{(4)}]$ is, to leading order, quartic in creation and annihilation operators. Some of the quartic contributions, however, are not in normal order. When rearranged, they produced therefore quadratic off-diagonal terms that can also be added to (2.28). Through this procedure we essentially reconstruct the zero-energy scattering equation in front of the off-diagonal term $[b_p^* b_{-p}^* + b_{-p} b_p]$. With the correct choice of the coefficients η_p in (2.27), we can therefore make sure that the combination of these terms vanishes. At the same time, the commutator of B with the off-diagonal quadratic operator (2.28) produces, after normal ordering, a constant term that, combined with the factor $\frac{1}{2} N \widehat{V}(0)$ contained in $\mathcal{L}_{\mathrm{GP}}^{(0)}$ and with constant contributions arising from the second commutators $[B, [B, \mathcal{K}]]$ and $[B, [B, \mathcal{L}_N^{(4)}]]$, produces, to leading order, the correct ground state energy $4\pi \mathfrak{a}_0 N$, consistently with the results of [36].

Even after conjugation with e^B, it turns out that cubic and quartic terms in the new excitation Hamiltonian $\mathcal{G}_N^{\mathrm{GP}} = e^B \mathcal{L}_N^{\mathrm{GP}} e^{-B}$ are not negligible; they still contain order one contributions to the energy that do not vanish, as $N \to \infty$. To extract these contributions, we conjugate $\mathcal{G}_N^{\mathrm{GP}}$ with another unitary operator e^A, this time with A cubic in creation and annihilation operators. This leads us to a renormalized excitation Hamiltonian $\mathcal{J}_N^{\mathrm{GP}} = e^A \mathcal{G}_N^{\mathrm{GP}} e^{-A} = e^A e^B U^* H_N U e^B e^A$ which is now quadratic in creation and annihilation operators, up to errors that are negligible on states with few excitations. The low-energy spectrum of $\mathcal{J}_N^{\mathrm{GP}}$ can now be computed similarly as we did in (2.17) in the mean-field setting.

The strategy presented in the last two paragraphs is based on the crucial assumption that low-energy states of (2.26) exhibit complete Bose–Einstein condensation with only a bounded number of excitations. The argument (2.5),8 which proved optimal condensation in the mean-field setting, does not work here (the error $\frac{V(0)}{2}$ on the right-hand side of (2.2) would be of order N^2). While mathematical rigorous proofs of Bose–Einstein condensation in the Gross–Pitaevskii limit have been obtained in the last two decades, see in particular [34, 41], they do not provide sufficient control on the number of excitations to estimate errors arising in the construction of $\mathcal{J}_N^{\mathrm{GP}}$. It turns out, however, that proceeding similarly as outlined above, conjugating with quadratic and cubic phases, it is possible to map $\mathcal{L}_N^{\mathrm{GP}}$ into a new, renormalized Hamilton operator, where the singular interaction is replaced, effectively, by a mean-field type potential. After these transformations, optimal Bose–Einstein condensation can be shown similarly as in (2.2). This strategy was used in [4, 6]. Simplified proofs of optimal condensation have been later derived in [40], also for gases trapped by external potentials, and in [26]. Moreover, as pointed out in [13], the approach used in [14] to derive the Lee–Huang–Yang formula for the ground state energy of a dilute gas in the thermodynamic limit also leads to a proof of Bose–Einstein condensation, in and beyond the Gross–Pitaevskii regime.

Making use of the estimates on the number (and the energy) of excitations of the Bose–Einstein condensate derived in [6], the strategy discussed above provides the following theorem, established in [5].

Theorem 2.2. *Let $V \in L^3(\mathbb{R}^3)$ be non-negative, radial and compactly supported. Let $\mathfrak{a}_0 > 0$ denote the scattering length of V. Then the ground state energy E_N of (2.26) is given by*

$$E_N^{\mathrm{GP}} = 4\pi(N-1)\mathfrak{a}_0 + e_\Lambda \mathfrak{a}_0^2$$
$$- \frac{1}{2} \sum_{p \in \Lambda_+^*} \Big[p^2 + 8\pi\mathfrak{a}_0 - \sqrt{|p|^4 + 16\pi\mathfrak{a}_0 p^2} - \frac{(8\pi\mathfrak{a}_0)^2}{2p^2} \Big] + \mathcal{O}(N^{-\frac{1}{4}})$$

with an explicit constant $e_\Lambda > 0$ (the presence of the term e_Λ is a consequence of the fact that the scattering length $\mathfrak{a}_0$ is defined through a scattering equation on $\mathbb{R}^3$, while the system we are considering is restricted to Λ). Moreover, the spectrum of $H_N^{\mathrm{GP}} - E_N^{\mathrm{GP}}$ below a threshold ζ consists of eigenvalues having the form given

$$\sum_{p \in \Lambda_+^*} n_p \sqrt{|p|^4 + 16\pi\mathfrak{a}_0 p^2} + \mathcal{O}(N^{-\frac{1}{4}}(1+\zeta^3)),$$

where $n_p \in \mathbb{N}$ for all $p \in \Lambda_+^*$ and $n_p \neq 0$ for finitely many $p \in \Lambda_+^*$ only.

3 The correlation energy of a Fermi gas

In this section, we consider Fermi gases in the mean field limit. We focus here on a system of fermions moving in the box $\Lambda = [0;1]^3$, with periodic boundary conditions. Particles interact through a two-body potential, varying on distances of order one (thus comparable with the size of the box). Because of Pauli's exclusion principle, the kinetic energy of N fermions in Λ is of the order $N^{\frac{5}{3}}$, much larger than for bosons. To have a non-trivial competition between kinetic and potential energies, and to make sure they are both of order N, we consider the Hamilton operator

$$H_N^{\mathrm{F}} = \sum_{j=1}^{N} -\varepsilon^2 \Delta_{x_j} + \frac{1}{N} \sum_{i<j}^{N} V(x_i - x_j) \tag{3.1}$$

with the parameter $\varepsilon = N^{-\frac{1}{3}}$. As dictated by fermionic statistics, the operator (3.1) acts on $L_a^2(\Lambda^N)$, the subspace of $L^2(\Lambda^N)$ consisting of functions that are antisymmetric w.r.t. permutations of the N particles. From (3.1) we conclude that, in the fermionic setting, the mean-field regime is naturally linked with a semiclassical limit, with the small parameter $\varepsilon = N^{-\frac{1}{3}}$ playing the role of Planck's constant.

We are interested, in this section, in the ground state energy

$$E_N^{\mathrm{F}} = \inf_{\psi \in L_a^2(\Lambda^N):\, \|\psi\|=1} \langle \psi_N, H_N^{\mathrm{F}} \psi_N \rangle$$

of (3.1). An approximation to E_N^{F} is given by Hartree–Fock theory, where we restrict our attention to Slater determinants having the form

$$\psi_{\mathrm{sl}}(x_1, \dots, x_N) = C \det(\varphi_i(x_j))_{1 \le i,j \le N} \tag{3.2}$$

for an orthonormal family $\{\varphi_i\}_{i=1,\dots,N}$ in the one-particle space $L^2(\mathbb{R}^3)$, and for an appropriate normalization constant $C > 0$. The one-particle reduced density matrix $\omega = N\,\mathrm{tr}_{2,\dots,N}\,|\psi_{\mathrm{sl}}\rangle\langle\psi_{\mathrm{sl}}|$ of (3.2) is the orthogonal projection

$$\omega = \sum_{j=1}^{N} |\varphi_j\rangle\langle\varphi_j|$$

onto the N-dimensional subspace of $L^2(\mathbb{R}^3)$ spanned by the orbitals $\{\varphi_j\}_{j=1,\dots,N}$. The energy of (3.2) can then be expressed through the Hartree–Fock energy functional

$$\begin{aligned}\langle\psi_{\mathrm{s}}, H_N^{\mathrm{F}}\psi_s\rangle &= \mathrm{tr}\,\big(-\varepsilon^2\Delta\big)\omega + \frac{1}{2}\int_{\Lambda\times\Lambda} V(x-y)[\omega(x;x)\omega(y;y) - |\omega(x;y)|^2] \\ &=: \mathcal{E}_{\mathrm{HF}}(\omega). \end{aligned} \tag{3.3}$$

We define the Hartree–Fock energy

$$E_{\mathrm{HF}} = \inf \mathcal{E}_{\mathrm{HF}}(\omega),$$

where the infimum is taken over all rank N orthogonal projections on $L^2(\mathbb{R}^3)$. Obviously, $E_N^{\mathrm{F}} \leq E_{\mathrm{HF}}$. The difference between the true many-body ground state energy and the Hartree–Fock energy is defined as the correlation energy

$$E_{\mathrm{corr}} = E_N^F - E_{\mathrm{HF}} \leq 0. \tag{3.4}$$

The goal of this section is to present a recent estimate for the correlation energy E_{corr} of the Hamilton operator (3.1), obtained with the help of (bosonic) Bogoliubov theory.

We start by having a closer look at the Hartree–Fock energy E_{HF}. For every momentum $p \in \Lambda^* = 2\pi\mathbb{Z}^3$, consider the plane wave $\varphi_p(x) = e^{-ip\cdot x}$. The kinetic part of the Hartree–Fock functional is minimized by the Slater determinant, known as Fermi sea, with one particle in the state φ_p, for every p in the Fermi ball

$$B_F = \{p \in \Lambda^* : |p| \leq k_F N^{\frac{1}{3}}\}.$$

Here $k_F > 0$ is chosen so that $|B_F| = N$. We denote the reduced one-particle density matrix of the Fermi sea by

$$\omega_{\mathrm{sea}} = \sum_{p\in B_F} |\varphi_p\rangle\langle\varphi_p|.$$

While the Fermi sea minimizes the kinetic energy, if $V \neq 0$ it does not minimize the full Hartree–Fock functional (3.3). Nevertheless, the free Fermi sea provides a very good approximation for E_{HF}. In fact, it was recently shown in [21] that

$$E_{\mathrm{HF}} - \mathcal{E}_{\mathrm{HF}}(\omega_{\mathrm{sea}}) \leq Ce^{-CN^{\frac{1}{6}}}. \tag{3.5}$$

It follows from (3.5) that, up to sub-exponentially small corrections, the correlation energy of (3.1), as defined in (3.4), is given by

$$E_{\mathrm{corr}} \simeq E_N - \mathcal{E}_{\mathrm{HF}}(\omega_{\mathrm{sea}}).$$

To estimate this difference, it is convenient to factor out the Fermi sea and to study the Hamilton operator (3.1) from the point of view of its excitations. To reach this goal, we embed the system on the fermionic Fock space $\mathcal{F}_a = \bigoplus_{n\geq 0} L^2_a(\Lambda^n)$ and we write (3.1) in second quantized form as

$$H_N = \sum_{p\in\Lambda^*} \varepsilon^2 p^2 a_p^* a_p + \frac{1}{N} \sum_{p,q,r\in\Lambda^*} \widehat{V}(r) a_{p+r}^* a_q^* a_{q+r} a_p \tag{3.6}$$

using creation and annihilation operators satisfying canonical anti-commutation relations

$$\{a_p, a_q^*\} = a_p a_q^* + a_q^* a_p = \delta_{pq},$$
$$\{a_p, a_q\} = \{a_p^*, a_q^*\} = 0.$$

To factor out the Fermi sea, we introduce, following [3], the unitary particle-hole transformation $R_{\omega_{\text{sea}}} : \mathcal{F}_a \to \mathcal{F}_a$, mapping the vacuum $\Omega = \{1, 0, 0, \dots\}$ into the Fermi sea, i.e.

$$R_{\omega_{\text{sea}}} \Omega = \prod_{p\in B_F} a_p^* \Omega,$$

and satisfying the relation

$$R_{\omega_{\text{sea}}}^* a_p^* R_{\omega_{\text{sea}}} = \begin{cases} a_p^* & \text{if } p \notin B_F, \\ a_p & \text{if } p \in B_F \end{cases} \tag{3.7}$$

for all $p \in \Lambda^*$. After conjugation with $R_{\omega_{\text{sea}}}$, our reference state, the free Fermi sea, is described by the vacuum vector $\Omega \in \mathcal{F}_a$ and the operators a_p^* create excitations of the Fermi sea; if $p \in B_F^c$, they create new particles outside the Fermi sea, if instead $p \in B_F$, they create a hole in the Fermi sea.

Conjugating the operator (3.6) with the unitary operator $R_{\omega_{\text{sea}}}$, we define the excitation Hamiltonian $\mathcal{L}_N^F = R_{\omega_{\text{sea}}}^* H_N R_{\omega_{\text{sea}}}$. The precise form of $\mathcal{L}_N^F$ can easily be computed with the rules (3.7), distinguishing momenta in B_F and momenta in B_F^c. Focussing on the most important part of $\mathcal{L}_N^F$, we obtain

$$\mathcal{L}_N^F = \mathcal{E}_{\text{HF}}(\omega_{\text{sea}}) + \sum_{k\in\Lambda^*} |\varepsilon^2 k^2 - \mu| a_k^* a_k + \mathcal{Q}_N + \delta_N \tag{3.8}$$

with $\mu = k_F^2$ and

$$\mathcal{Q}_N = \frac{1}{N} \sum_{k\in\Lambda_+^*} \widehat{V}(k) \sum_{p,q\in B_F^c\cap B_F+k} \Big[a_p^* a_{p-k}^* a_{q-k} a_q + \frac{1}{2}\big(a_p^* a_{p-k}^* a_q^* a_{q+k}^* + a_{q+k} a_q a_{p-k} a_p\big)\Big] \tag{3.9}$$

and with an error operator δ_N, whose expectation is negligible on states with few excitations of the Fermi sea.

It is useful to notice that terms in (3.9) involve two momenta $p, q \in B_F^c$ and two momenta $p-k, q-k \in B_F$. They can all be expressed in terms of particle-hole creation operators

$$b_k^* = \sum_{p \in B_F^c \cap B_F + k} a_p^* a_{p-k}^* \tag{3.10}$$

and of their adjoint b_k. We find

$$\mathcal{Q}_N = \frac{1}{N} \sum_{k \in \Lambda^*} \widehat{V}(k) \Big[b_k^* b_k + \frac{1}{2}(b_k^* b_{-k}^* + b_k b_{-k}) \Big]. \tag{3.11}$$

When appropriately normalized, the operators b_k^*, b_k defined in (3.10) satisfy approximately bosonic commutation relations, up to errors that are small on states with few excitations. This suggests that one can estimate the ground state energy of the excitation Hamiltonian $\mathcal{L}_N^F$, and in particular the corrections to the energy of the free Fermi sea $\mathcal{E}_{\text{HF}}(\omega_{\text{sea}})$, by conjugating it with a bosonic Bogoliubov transformation, similarly as we did in (2.17), in the bosonic case.

Unfortunately, it is impossible to express the kinetic energy operator

$$\mathcal{K}^F = \sum_{k \in \Lambda^*} |\varepsilon^2 k^2 - \mu| a_k^* a_k$$

appearing on the right-hand side of (3.8) as a quadratic expression in b_k, b_k^*. In fact, the action of $\mathcal{K}^F$ on particle-hole excitations of the form $b_k^* \Omega$ is given by

$$\begin{aligned} \mathcal{K}^F b_k^* \Omega &= \sum_{q \in \Lambda^*} |\varepsilon^2 q^2 - \mu| a_q^* a_q b_k^* \Omega \\ &= \sum_{q \in \Lambda^*} \sum_{p \in B_F^c \cap B_F + k} |\varepsilon^2 q^2 - \mu| a_q^* a_q a_p^* a_{p-k}^* \Omega \\ &= \sum_{p \in B_F^c \cap B_F + k} \big[|\varepsilon^2 p^2 - \mu| + |\varepsilon^2 (p-k)^2 - \mu| \big] a_p^* a_{p-k}^* \Omega \\ &= \varepsilon^2 k^2 b_k^* \Omega + \sum_{p \in B_F^c \cap B_F + k} \varepsilon^2 k \cdot p a_p^* a_{p-k}^* \Omega. \end{aligned} \tag{3.12}$$

The last term is again a particle-hole excitation, but it is not created by an operator of the form (3.10). This computation suggests, however, that this problem can be solved by localizing the bosonized fields in momentum space. Let us restrict our attention to interactions having compact support in Fourier space, and let us assume that $\widehat{V}(k) = 0$ if $|k| > K$. For $|k| \leq K$, particle-hole fields in (3.10) only involve momenta in a shell of size $2K$ around the Fermi sphere ∂B_F. We decompose this shell in M patches $\{B_\alpha\}_{\alpha=1,\dots,M}$, each covering approximately the same area on the Fermi sphere. For any $|k| < K$ and $\alpha = 1, \dots, M$, we can then define localized particle-hole operators

$$b_{k,\alpha}^* = \frac{1}{n_{k,\alpha}} \sum_{p \in (B_\alpha \cap B_F^c) \cap [(B_\alpha \cap B_F) + k]} a_{p-k}^* a_p^*$$

with the normalization constant $n_{k,\alpha} = |(B_\alpha \cap B_F^c) \cap [(B_\alpha \cap B_F) + k]|^{\frac{1}{2}}$ (to be more precise, in this construction we have to remove patches which are almost orthogonal to the vector k, to make sure that $n_{k,\alpha} \neq 0$). Like the operators b_k^*, b_k, also the new localized fields $b_{k,\alpha}^*, b_{k,\alpha}$ satisfy approximately bosonic commutation relations (up to mistakes that are negligible, on states with few excitations). Moreover, since

$$b_k = \sum_\alpha n_{k,\alpha} b_{k,\alpha}, \quad b_k^* = \sum_\alpha n_{k,\alpha} b_{k,\alpha}^*,$$

we conclude from (3.11) that $\mathcal{Q}_N$ can be written as a quadratic expression in the new localized fields $b_{k,\alpha}, b_{k,\alpha}^*$. The advantage of $b_{k,\alpha}, b_{k,\alpha}^*$, compared with (3.10), is that now, denoting by ω_α the central point of the patch B_α on the Fermi sphere, and proceeding as in (3.12), we can approximate

$$\mathcal{K}^F b_{k,\alpha}^* \Omega \simeq [\varepsilon^2 k^2 + \varepsilon^2 k \cdot \omega_\alpha] b_{k,\alpha}^* \Omega$$

if M is large enough, so that momenta in the patch B_α are not too far from its middle point ω_α. Thus, also the kinetic energy operator $\mathcal{K}^F$ can be approximately written as a quadratic form in the new fields. From (3.8), it follows that the correction to the energy of the Fermi sea can be determined using a Bogoliubov transformation to diagonalize an operator that is quadratic in creation and annihilation operators that satisfy approximately bosonic commutation relations. This approach has been recently developed in [1, 2]; the main result of these works is the following theorem.

Theorem 3.1. *Assume $\widehat{V} : \Lambda^* \to \mathbb{R}$ is compactly supported, with $\widehat{V}(k) = \widehat{V}(-k)$ for all $k \in \Lambda^*$ and with $\|\widehat{V}\|_{\ell^1}$ small enough. Then the correlation energy of the Hamilton operator* (3.1)*, as defined in* (3.4)*, is given, with the notation $\kappa = (6\pi^2)^{\frac{1}{3}}$ and for an arbitrarily small $\delta > 0$, by*

$$\begin{aligned} E_{\text{corr}} = \varepsilon\kappa \sum_{k \in \Lambda^*} |k| \Big[\frac{1}{\pi} \int_0^\infty \log\Big(1 + 2\pi\kappa \widehat{V}(k)\Big(1 - \lambda \arctan\Big(\frac{1}{\lambda}\Big)\Big)\Big)\, d\lambda \\ - \frac{\pi}{2}\kappa \widehat{V}(k)\Big] + \mathcal{O}(\varepsilon^{\frac{17}{16}-\delta}). \end{aligned} \tag{3.13}$$

Estimate (3.13) follows from an upper bound, obtained in [2] (with no assumption on the size of the interaction) and a lower bound, proven in [1]. By expanding in the potential, (3.13) implies that

$$\lim_{N\to\infty} \frac{E_{\text{corr}}}{\varepsilon} = -4\pi^4 (1 - \log 2) \sum_{k \in \Lambda^*} |k| |\widehat{V}(k)|^2 (1 + \mathcal{O}(\widehat{V}(k))),$$

a result that has been previously established in [27], through rigorous second-order perturbation theory. If we extrapolate (3.13) to the Coulomb potential $\widehat{V}(k) = |k|^{-2}$, we find that it agrees with the formula predicted by Gell-Mann and Brueckner in [19], using the random phase approximation.

4 Dynamics of a Fröhlich polaron

A polaron is an electron in a ionic crystal, coupled with the polarization field that it creates. If the wavelength of the electron is large compared with the lattice spacing, the polaron can be described by the Fröhlich model, with the electron moving in a continuous medium and with the polarization of the crystal described by a quantized radiation field, whose excitations are known as phonons. In the limit of strong coupling $\alpha \gg 1$, it is convenient to express the Hamilton operator in rescaled coordinates, defining

$$H_P = -\Delta + \mathcal{N} + \int \frac{dk}{|k|}\big[e^{-ik\cdot x}a_k^* + e^{ik\cdot x}a_k\big] \tag{4.1}$$

as acting on the Hilbert space $\mathcal{H} = L^2(\mathbb{R}^3) \otimes \mathcal{F}$, with the bosonic Fock space

$$\mathcal{F} = \bigoplus_{n\geq 0} L^2(\mathbb{R}^3)^{\otimes_s n}.$$

Here, a_k^*, a_k are creation and annihilation operators (in momentum space), satisfying rescaled canonical commutation relations

$$[a_k, a_{k'}^*] = \alpha^{-2}\delta(k-k'), \quad [a_k, a_{k'}] = [a_k^*, a_{k'}^*] = 0. \tag{4.2}$$

Moreover,

$$\mathcal{N} = \int dk a_k^* a_k$$

is the operator measuring the number of phonons; it reflects the fact that the electron mainly interacts with optical phonons, whose dispersion is approximately constant.

Despite the fact that the form factor $\kappa(k) = |k|^{-1}$ is not square integrable, and therefore the interaction cannot be bounded in terms of the field energy $\mathcal{N}$, using the kinetic energy of the electron one can show that the Hamilton operator (4.1) is well-defined and bounded below, since, for every $\varepsilon > 0$, there is $C_\varepsilon > 0$ such that

$$\pm\big[a(e^{ikx}\kappa) + a^*(e^{ikx}\kappa)\big] \leq -\varepsilon\Delta + C_\varepsilon \|(1+|k|)^{-1}\kappa\|_2^2(\mathcal{N} + \alpha^{-2}).$$

Observe that the coupling constant α only enters (4.1) through the commutation relations for the operators a_k, a_k^*. In fact, (4.2) shows that, in the strong coupling regime $\alpha \gg 1$, the quantized polarization field approaches a classical limit. This suggests that the minimal energy of (4.1) can be estimated with product states of the form $\Psi = \psi \otimes W(\alpha^2\varphi)\Omega$, with the coherent state $W(\alpha^2\varphi)\Omega$ on $\mathcal{F}$, generated by the Weyl operator $W(\alpha^2\varphi) = \exp(a^*(\alpha^2\varphi) - a(\alpha^2\varphi))$. With the relation

$$W(\alpha^2\varphi)^* a_k W(\alpha^2\varphi) = a_k + \varphi(k)$$

we obtain the Pekar energy functional

$$\begin{aligned}
&\langle \psi \otimes W(\alpha^2\varphi)\Omega, H_P \psi \otimes W(\alpha^2\varphi)\Omega\rangle \\
&= \int |\nabla\psi(x)|^2\, dx + \int |\varphi(k)|^2\, dk + 2\,\mathrm{Re} \int |\psi(x)|^2 \frac{1}{|x-y|^2}\check{\varphi}(y)\, dx\, dy \\
&=: \mathcal{E}_{\text{pekar}}(\psi, \varphi).
\end{aligned} \tag{4.3}$$

Completing the square, we find that (4.3) can be minimized choosing first

$$\varphi(k) = -(2\pi)^{\frac{3}{2}}|k|^{-1}\widehat{|\psi|^2}(k),$$

and then ψ to minimize the functional

$$\mathcal{F}(\psi) = \inf_{\varphi} \mathcal{E}_{\text{pekar}}(\psi,\varphi) = \int |\nabla\psi(x)|^2\,dx - \int \frac{1}{|x-y|}|\psi(x)|^2|\psi(y)|^2\,dx\,dy$$

among all normalized $\psi \in L^2(\mathbb{R}^3)$. As proven in [11,35], the resulting energy captures, to leading order, the true ground state energy of (4.1) in the limit of large α, i.e.

$$\lim_{\alpha\to\infty} \inf_{\Psi\in\mathcal{H}:\,\|\Psi\|=1} \langle\Psi, H_P\Psi\rangle = \inf_{\psi\in L^2(\mathbb{R}^3):\,\|\psi\|_2=1} \mathcal{F}(\psi). \tag{4.4}$$

Here, we are going to consider the time-evolution generated by the Fröhlich Hamiltonian (4.1). In particular, in view of (4.4), we are interested in the dynamics of initial Pekar product states of the form $\Psi = \psi \otimes W(\alpha^2\varphi)\Omega$. Taking variations of the Pekar functional (4.3), we may expect that the evolution of Ψ can be approximated by evolved Pekar states $e^{-iH_P t}\Psi \simeq \psi_t \otimes W(\alpha^2\varphi_t)\Omega$, with (ψ_t,φ_t) solving the Landau–Pekar equations

$$\begin{cases} i\partial_t\psi_t = (-\Delta + V_{\varphi_t})\psi_t, \\ i\alpha^2\partial_t\varphi_t = \varphi_t + \sigma_{\psi_t} \end{cases} \tag{4.5}$$

with

$$V_\varphi(x) = \int \frac{dk}{|k|}\big[e^{ik\cdot x}\varphi(k) + e^{-ik\cdot x}\overline{\varphi}(k)\big], \quad \sigma_\psi(k) = (2\pi)^{\frac{3}{2}}\frac{1}{|k|}\widehat{|\psi|^2}(k). \tag{4.6}$$

The convergence (4.4) of the energy and the emergence of the Landau–Pekar equations (4.5) confirm that, for strong coupling, the quantized polarization field approaches a classical limit and can be described by a classical field. It is important to stress, however, that, while the phonon field becomes classical, as $\alpha\to\infty$, the electron remains a quantum particle (the first equation in (4.5) is the usual Schrödinger equation, for a particle moving in the potential V_{φ_t}). This creates a separation of time-scales between the two components of the system. While the electron moves on times of order one, the phonon field, and its classical approximation φ_t, move much slower, on times of order α^2, as indicated by the second equation in (4.5).

The first rigorous proofs of the validity of the Landau–Pekar equations (4.5) were obtained in [15, 17], for times $|t| \lesssim \alpha$, for which the phonon field remains essentially constant. To reach longer times, one can consider an initial Pekar state with the electron wave function minimizing the energy in the potential generated by the applied polarization field. In other words, we choose an initial polarization field $\varphi_0 \in L^2(\mathbb{R}^3)$ such that the operator $h_{\varphi_0} = -\Delta + V_{\varphi_0}$ has a simple and isolated eigenvalue $e(\varphi_0)$ at the bottom of its spectrum, with a unique positive and normalized eigenvector ψ_{φ_0}, and then we take ψ_{φ_0} as the initial electron wave function. For such initial data, it was proven in [31] that

$$\left\| e^{-iH_P t}(\psi_{\varphi_0} \otimes W(\alpha^2\varphi_0)\Omega) - e^{i\omega_t}\psi_t \otimes W(\alpha^2\varphi_t)\Omega \right\|^2 \leq C\frac{|t|}{\alpha^2} \tag{4.7}$$

for an appropriate phase ω_t, where (ψ_t, φ_t) denote the solution of the Landau–Pekar equations (4.5) with initial data $(\psi_{\varphi_0}, \varphi_0)$. The bound (4.7) was first established in [23], for the special initial data $\psi_0 \otimes W(\alpha^2\varphi_0)\Omega$, with (ψ_0, φ_0) minimizing the Pekar energy functional (4.3) (this implies that ψ_0 is the ground state vector of h_{φ_0}, but also that $\varphi_0 = -\sigma_{\psi_0}$, as defined in equation (4.6)). In this case, the solution of the Landau–Pekar equations is stationary, up to an irrelevant oscillating phase (but of course, the many-body dynamics is not stationary). The result of [23] relies on the remark that the spectral gap above the ground state of the Schrödinger operator h_{φ_0} produces an oscillatory phase, which can be used to show that the electron does not leave the ground state for longer times. The same idea has also been used in [31] to handle the more general family of initial data, combined with a new adiabatic theorem showing that, for times $|t| \leq C\alpha^2$, ψ_t can be approximated by the normalized ground state ψ_{φ_t} of the Schrödinger operator h_{φ_t}, if this holds true at time $t = 0$ (the idea of making use of an adiabatic theorem in this setting and the proof of an adiabatic theorem for a one-dimensional version of the Landau–Pekar equations previously appeared in [16]).

The bound (4.7) shows that the Landau–Pekar equations (4.5) provide a good approximation to the polaron dynamics, in the limit of large α, for all $|t| \ll \alpha^2$. Although this time scale is longer than those considered in previous works, the polarization field still remains essentially unchanged. It turns out that, in order to cover times over which the polarization field φ_t experiences non-trivial changes, we have to modify the Pekar ansatz to take into account excitations of the coherent state describing the polarization field. This goal can be achieved by means of a Bogoliubov dynamics, with a time-dependent generator, quadratic in creation and annihilation operators. Let us denote by $\Upsilon_t \in \mathcal{F}$ the solution of the equation

$$i\partial_t \Upsilon_t = (\mathcal{N} - \mathcal{A}_t)\Upsilon_t \tag{4.8}$$

with the initial data $\Upsilon_0 = \Omega$. Here we defined

$$\mathcal{A}_t = \int \frac{dk\, dk'}{|k|\,|k'|} \langle \psi_{\varphi_t}, e^{-ik\cdot} R_t e^{ik'\cdot} \psi_{\varphi_t} \rangle_{L^2(\mathbb{R}^3)} (a_k^* + a_{-k})(a_{-k}^* + a_{k'})$$

where ψ_{φ_t} is, as above, the normalized ground state of the operator h_{φ_t} and

$$R_t = q_t(h_{\varphi_t} - e(\varphi_t))^{-1} q_t$$

is its resolvant, restricted by $q_t = 1 - |\psi_{\varphi_t}\rangle\langle\psi_{\varphi_t}|$ onto the orthogonal complement of ψ_{φ_t}.

To heuristically understand the emergence of equation (4.8), observe that the fluctuation dynamics $\mathcal{U}(t) = e^{-i\omega_t} W^*(\alpha^2\varphi_t) e^{-iH_P t} W(\alpha^2\varphi_0)$, associated with the approximation (4.7), is such that

$$i\partial_t \mathcal{U}(t) = \mathcal{G}(t)\mathcal{U}(t)$$

with the generator

$$\mathcal{G}(t) = h_{\varphi_t} + \mathcal{N} + \int \frac{dk}{|k|} \{[e^{-ik\cdot x} - \langle \psi_t, e^{-ik\cdot x}\psi_t\rangle] a_k^* + \text{h.c.}\}. \tag{4.9}$$

Expanding (4.9) (or the unitary evolution it defines) around the effective electron Hamiltonian h_{φ_t} to second order in perturbation theory, we arrive at the generator on the right-hand side of (4.8) (more precisely, the number of particles operator $\mathcal{N}$ already appears as a first-order perturbation, while the term $\mathcal{A}_t$ emerges in second order).

By using the Bogoliubov dynamics (4.8), the following theorem has been established in [30] (and in [37], for the special initial data minimizing the Pekar functional, leading to a stationary solution of (4.5)).

Theorem 4.1. *Assume $\varphi_0 \in L^2(\mathbb{R}^3)$ is such that $e(\varphi_0) = \inf \sigma(h_{\varphi_0}) < 0$ (so that h_{φ_0} has a unique positive ground state ψ_{φ_0}, with eigenvalue $e(\varphi_0)$ separated from the rest of the spectrum). Let $(\psi_t, \varphi_t) \in H^1(\mathbb{R}^3) \times L^2(\mathbb{R}^3)$ denote the solution of the Landau–Pekar equations* (4.5)*, with initial data $(\psi_{\varphi_0}, \varphi_0)$. With Υ_t defined as in* (4.8) *and with a suitable phase $\omega_t \in \mathbb{R}$, we find constants $C, T > 0$ such that*

$$\big\| e^{-iH_P t}(\psi_{\varphi_0} \otimes W(\alpha^2\varphi_0)\Omega) - e^{i\omega_t}\psi_t \otimes W(\alpha^2\varphi_t)\Upsilon_t \big\| \le C\alpha^{-1} \tag{4.10}$$

for all times $|t| \le T\alpha^2$.

The condition $|t| \le T\alpha^2$, that was already mentioned above in relation with the adiabatic theorem, guarantees that the gap above the isolated ground state energy of h_{φ_t} remains open; for a certain family of initial data, it has been recently shown in [12] that the gap remains open for all times. For such data, estimate (4.10) can be extended to all $|t| \ll \alpha^4$.

Note that the addition of the Bogoliubov dynamics is crucial. It is impossible to approximate the evolution of initial Pekar states $\Psi_0 = \psi_{\varphi_0} \otimes W(\alpha^2\varphi_0)\Omega$ by states of the same form, for times of order $t \simeq \alpha^2$ or larger. On the other hand, the bound (4.10) implies that the electron reduced density matrix $\gamma_t^{\text{el}} = \text{tr}_{\mathcal{F}} |e^{-iH_P t}\Psi_0\rangle\langle e^{-iH_P t}\Psi_0|$ and the one-phonon density matrix

$$\gamma_t^{\text{ph}}(k;k') = \langle e^{-iH_P t}\Psi_0, a_{k'}^* a_k e^{-iH_P t}\Psi_0\rangle$$

can both be approximated by the solution of (4.5), in the sense that

$$\|\gamma_t^{\text{el}} - |\psi_t\rangle\langle\psi_t|\|_{\text{tr}} \le C\alpha^{-1},$$
$$\|\gamma_t^{\text{ph}} - |\varphi_t\rangle\langle\varphi_t|\|_{\text{tr}} \le C\alpha^{-\frac{1}{4}}$$

for large α (see [30, Corollary I.5]). These bounds establish, at the level of the one-particle reduced densities, the validity of the Landau–Pekar equations up to times of order α^2, which allow the polarization field φ_t to undergo substantial changes.

Observe, finally, that Bogoliubov theory can also be applied to determine corrections to the ground state energy (4.4), in the limit of large α; results in this directions have been recently obtained in [18].

Acknowledgements. The author gratefully acknowledges partial support from NCCR SwissMAP, from the Swiss National Science Foundation through the Grant "Dynamical and energetic properties of Bose–Einstein condensates" and from the European Research Council through the ERC-AdG CLaQS.

Bibliography

[1] N. Benedikter, P. T. Nam, M. Porta, B. Schlein and R. Seiringer, Correlation energy of a weakly interacting fermi gas. Preprint 2020, arXiv:2005.08933v1; to appear in *Ivent. Math.*

[2] N. Benedikter, P. T. Nam, M. Porta, B. Schlein and R. Seiringer, Optimal upper bound for the correlation energy of a Fermi gas in the mean-field regime. *Comm. Math. Phys.* **374** (2020), 2097–2150

[3] N. Benedikter, M. Porta and B. Schlein, Mean-field evolution of fermionic systems. *Comm. Math. Phys.* **331** (2014), 1087–1131

[4] C. Boccato, C. Brennecke, S. Cenatiempo and B. Schlein, Complete Bose–Einstein condensation in the Gross–Pitaevskii regime. *Comm. Math. Phys.* **359** (2018), 975–1026

[5] C. Boccato, C. Brennecke, S. Cenatiempo and B. Schlein, Bogoliubov theory in the Gross–Pitaevskii limit. *Acta Math.* **222** (2019), 219–335

[6] C. Boccato, C. Brennecke, S. Cenatiempo and B. Schlein, Optimal rate for Bose–Einstein condensation in the Gross–Pitaevskii regime. *Comm. Math. Phys.* **376** (2020), 1311–1395

[7] C. Boccato, S. Cenatiempo and B. Schlein, Quantum many-body fluctuations around nonlinear Schrödinger dynamics. *Ann. Henri Poincaré* **18** (2017), 113–191

[8] C. Brennecke, P. T. Nam, M. Napiórkowski and B. Schlein, Fluctuations of N-particle quantum dynamics around the nonlinear Schrödinger equation. *Ann. Inst. H. Poincaré Anal. Non Linéaire* **36** (2019), 1201–1235

[9] X. Chen, Second order corrections to mean field evolution for weakly interacting bosons in the case of three-body interactions. *Arch. Ration. Mech. Anal.* **203** (2012), 455–497

[10] J. Dereziński and M. Napiórkowski, Excitation spectrum of interacting bosons in the mean-field infinite-volume limit. *Ann. Henri Poincaré* **15** (2014), 2409–2439

[11] M. D. Donsker and S. R. S. Varadhan, Asymptotics for the polaron. *Comm. Pure Appl. Math.* **36** (1983), 505–528

[12] D. Feliciangeli, S. Rademacher and R. Seiringer, Persistence of the spectral gap for the Landau–Pekar equations. *Lett. Math. Phys.* **111** (2021), Paper No. 19

[13] S. Fournais, Length scales for BEC in the dilute Bose gas. Preprint 2020, arXiv: 2011.00309v2

[14] S. Fournais and J. P. Solovej, The energy of dilute Bose gases. *Ann. of Math. (2)* **192** (2020), 893–976

[15] R. L. Frank and Z. Gang, Derivation of an effective evolution equation for a strongly coupled polaron. *Anal. PDE* **10** (2017), 379–422

[16] R. L. Frank and Z. Gang, A non-linear adiabatic theorem for the one-dimensional Landau–Pekar equations. *J. Funct. Anal.* **279** (2020), Article ID 108631

[17] R. L. Frank and B. Schlein, Dynamics of a strongly coupled polaron. *Lett. Math. Phys.* **104** (2014), 911–929

[18] R. L. Frank and R. Seiringer, Quantum corrections to the Pekar asymptotics of a strongly coupled polaron. *Comm. Pure Appl. Math.* **74** (2021), 544–588

[19] M. Gell-Mann and K. A. Brueckner, Correlation energy of an electron gas at high density. *Phys. Rev. (2)* **106** (1957), 364–368

[20] J. Ginibre and G. Velo, The classical field limit of scattering theory for nonrelativistic many-boson systems. I and II. *Comm. Math. Phys.* **66** (1979), 37–76, and **68** (1979), 45–68

[21] D. Gontier, C. Hainzl and M. Lewin, Lower bound on the Hartree–Fock energy of the electron gas. *Phys. Rev. A* **99** (2019), no. 5, Article ID 052501

[22] P. Grech and R. Seiringer, The excitation spectrum for weakly interacting bosons in a trap. *Comm. Math. Phys.* **322** (2013), 559–591

[23] M. Griesemer, On the dynamics of polarons in the strong-coupling limit. *Rev. Math. Phys.* **29** (2017), Article ID 1750030

[24] M. Grillakis and M. Machedon, Pair excitations and the mean field approximation of interacting bosons, I. *Comm. Math. Phys.* **324** (2013), 601–636

[25] M. G. Grillakis, M. Machedon and D. Margetis, Second-order corrections to mean field evolution of weakly interacting bosons. I. *Comm. Math. Phys.* **294** (2010), 273–301

[26] C. Hainzl, Another proof of BEC in the GP-limit. Preprint 2020, arXiv:2011.09450

[27] C. Hainzl, M. Porta and F. Rexze, On the correlation energy of interacting fermionic systems in the mean-field regime. *Comm. Math. Phys.* **374** (2020), 485–524

[28] K. Hepp, The classical limit for quantum mechanical correlation functions. *Comm. Math. Phys.* **35** (1974), 265–277

[29] E. Kuz, Exact evolution versus mean field with second-order correction for bosons interacting via short-range two-body potential. *Differential Integral Equations* **30** (2017), 587–630

[30] N. Leopold, D. Mitrouskas, S. Rademacher, B. Schlein and R. Seiringer, Landau–Pekar equations and quantum fluctuations for the dynamics of a strongly coupled polaron. Preprint 2020, arXiv:2005.02098v1

[31] N. Leopold, S. Rademacher, B. Schlein and R. Seiringer, The Landau–Pekar equations: Adiabatic theorem and accuracy. *Anal. PDE*, to appear; arXiv:1904.12532v2

[32] M. Lewin, P. T. Nam and B. Schlein, Fluctuations around Hartree states in the mean-field regime. *Amer. J. Math.* **137** (2015), 1613–1650

[33] M. Lewin, P. T. Nam, S. Serfaty and J. P. Solovej, Bogoliubov spectrum of interacting Bose gases. *Comm. Pure Appl. Math.* **68** (2015), 413–471

[34] E. H. Lieb and R. Seiringer, Proof of Bose–Einstein condensation for dilute trapped gases. *Phys. Rev. Lett.* **88** (2002), Article ID 170409

[35] E. H. Lieb and L. E. Thomas, Exact ground state energy of the strong-coupling polaron. *Comm. Math. Phys.* **183** (1997), 511–519; erratum, *Comm. Math. Phys.* **188** (1997), 499–500

[36] E. H. Lieb, J. Yngvason, Ground State Energy of the low density Bose gas. *Phys. Rev. Lett.* **80** (1998), 2504—2507

[37] D. Mitrouskas, A note on the Fröhlich dynamics in the strong coupling limit. Preprint 2020, arXiv:2003.11448

[38] D. Mitrouskas, S. Petrat and P. Pickl, Bogoliubov corrections and trace norm convergence for the Hartree dynamics. *Rev. Math. Phys.* **31** (2019), Article ID 1950024

[39] P. T. Nam and M. Napiórkowski, Bogoliubov correction to the mean-field dynamics of interacting bosons. *Adv. Theor. Math. Phys.* **21** (2017), 683–738

[40] P. T. Nam, M. Napiórkowski, J. Ricaud and A. Triay, Optimal rate of condensation for trapped bosons in the Gross–Pitaevskii regime. Preprint 2020, arXiv:2001.04364v1

[41] P. T. Nam, N. Rougerie and R. Seiringer, Ground states of large bosonic systems: The Gross–Pitaevskii limit revisited. *Anal. PDE* **9** (2016), 459–485

[42] A. Pizzo, Bose particles in a box III. A convergent expansion of the ground state of the Hamiltonian in the mean field limiting regime. Preprint 2015, arXiv:1511.07026v1

[43] R. Seiringer, The excitation spectrum for weakly interacting bosons. *Comm. Math. Phys.* **306** (2011), 565–578

A statistical theory of heavy atoms: Asymptotic behavior of the energy and stability of matter

Heinz Siedentop

Dedicated to Ari Laptev on the occasion of his septuagesimal birthday. His ideas in analysis have inspired many

We give the asymptotic behavior of the ground state energy of Engel's and Dreizler's relativistic Thomas–Fermi–Weizsäcker–Dirac functional for heavy atoms for fixed ratio of the atomic number and the velocity of light. Using a variation of the lower bound, we show stability of matter.

1 Introduction

Heavy atoms require a relativistic description because of the extremely fast moving inner electrons. However, a statistical theory of the atom in the spirit of Thomas [29] and Fermi [12,13] yields a functional which is unbounded from below because the semi-classical relativistic Fermi energy is too weak to prevent mass from collapsing into the nucleus. (See Gombas [17, Section 14] and [18, Chapter III, Section 16]. Gombas also suggested that Weizsäcker's (non-relativistic) inhomogeneity correction would solve this problem. Tomishia [30] carried this suggestion through.) Because of the same reason the relativistic generalization of the Lieb–Thirring inequality by Daubechis is not directly applicable to Chandrasekhar operators with Coulomb potentials but requires a separate treatment of the singularity. Frank and Ekholm [10] found a way circumventing this problem treating critical potentials of Schrödinger operators; later Frank, Lieb and Seiringer [14] accomplished the same for Chandrasekhar operators with Coulomb potentials. It amounts to a Thomas–Fermi functional with a potential whose critical singularity has been extracted. However, there is a drawback, namely the Thomas–Fermi constant of this functional is smaller than the classical one, i.e., we cannot expect asymptotically correct results.

Keywords: Heavy atom, asymptotic behavior of the ground energy, Engel–Dreizler functional
2020 Mathematics Subject Classification: Primary 81V45; secondary 81V74

Here we discuss an alternative relativistic density functional which can handle Coulomb potentials of arbitrary strength: Engel and Dreizler [11] derived a functional $\mathcal{E}_{c,Z}^{\mathrm{TFWD}}$ of the electron density ρ from quantum electrodynamics (in contrast to Gombas' ad hoc procedure of adding the non-relativistic Weizsäcker correction). It is – in a certain sense – a generalization of the non-relativistic Thomas–Fermi–Weizsäcker–Dirac functional to the relativistic setting, a feature that it shares with the functional investigated by Lieb et al. [22]. However, it does not suffer from the problem that it becomes unbounded from below for heavy atoms. We will show here, that it has – unlike the functional which can be obtained from [14] – the same asymptotic behavior as the atomic quantum energy. The price to pay is the absence of a known inequality relating it to the full quantum problem. One could speculate that it might be an upper bound on the ground state energy. The way we prove the upper bound of the asymptotics might nourishes such thoughts. However, even in the non-relativistic context this is open despite numerical evidence and claims to the contrary, e.g., by March and Young [25]: the arguments given contain a gap. In other words, such a claim would have – even in the non-relativistic context – at best the status of a conjecture.

Engel's and Dreizler's functional is the relativistic Thomas–Fermi functional (see Chandrasekhar [7] [in the ultrarelativistic limit] and Gombas [17, Section 14] for the general case) with an inhomogeneity and exchange correction different from the non-relativistic terms but with an integrand tending pointwise to their non-relativistic analogue for large velocity of light c. In Hartree units it reads for atoms of atomic number Z and electron density ρ

$$\mathcal{E}_{c,Z}^{\mathrm{TFWD}}(\rho) := \mathcal{T}^{\mathrm{W}}(\rho) + \mathcal{T}^{\mathrm{TF}}(\rho) - \mathcal{X}(\rho) + \mathcal{V}(\rho). \tag{1.1}$$

The first summand on the right is an inhomogeneity correction of the kinetic energy generalizing the Weizsäcker correction. Using the Fermi momentum $p(x) := (3\pi^2\rho(x))^{\frac{1}{3}}$ it is

$$\mathcal{T}^{\mathrm{W}}(\rho) := \int_{\mathbb{R}^3} \mathrm{d}x \frac{3\lambda}{8\pi^2} (\nabla p(x))^2 c\, f\left(\frac{p(x)}{c}\right)^2 \tag{1.2}$$

with

$$f(t)^2 := t(t^2+1)^{-\frac{1}{2}} + 2t^2(t^2+1)^{-1}\mathfrak{Arsin}(t), \tag{1.3}$$

where $\mathfrak{Arsin}$ is the inverse function of the hyperbolic sine. The parameter $\lambda \in \mathbb{R}_+$ is given by the gradient expansion as $\frac{1}{9}$ but is in the non-relativistic analogue sometimes taken as an adjustable parameter (Weizsäcker [31], Yonei and Tomishima [32] and Lieb [20,21]).

The second summand is the relativistic generalization of the Thomas–Fermi kinetic energy, namely

$$\mathcal{T}^{\mathrm{TF}}(\rho) := \int_{\mathbb{R}^3} \mathrm{d}x \frac{c^5}{8\pi^2} T^{\mathrm{TF}}\left(\frac{p(x)}{c}\right) \tag{1.4}$$

with

$$T^{\mathrm{TF}}(t) := t(t^2+1)^{\frac{3}{2}} + t^3(t^2+1)^{\frac{1}{2}} - \mathfrak{Arsin}(t) - \frac{8}{3}t^3. \tag{1.5}$$

The third summand is a relativistic generalization of the exchange energy. It is

$$\mathcal{X}(\rho) := \int_{\mathbb{R}^3} \mathrm{d}x \frac{c^4}{8\pi^3} X\left(\frac{p(x)}{c}\right) \tag{1.6}$$

with

$$X(t) := 2t^4 - 3[t(t^2+1)^{\frac{1}{2}} - \mathfrak{Arsin}(t)]^2. \tag{1.7}$$

Eventually, the last summand is the potential energy, namely the sum of the electron-nucleus and the electron-electron interaction. It is

$$\mathcal{V}(\rho) := -Z \int_{\mathbb{R}^3} \mathrm{d}x \rho(x)|x|^{-1} + \underbrace{\frac{1}{2}\int_{\mathbb{R}^3} \mathrm{d}x \int_{\mathbb{R}^3} \mathrm{d}y \rho(x)\rho(y)|x-y|^{-1}}_{=:\mathcal{D}[\rho]}. \tag{1.8}$$

Using $F(t) := \int_0^t \mathrm{d}s f(s)$, the functional $\mathcal{E}_{c,Z}^{\mathrm{TFWD}}$ is naturally defined on

$$P := \{\rho \in L^{\frac{4}{3}}(\mathbb{R}^3) : \rho \geq 0,\ \mathcal{D}[\rho] < \infty,\ F \circ p \in D^1(\mathbb{R}^3)\} \tag{1.9}$$

and bounded from below [9] for all c and Z. In fact, Chen et al. [8] obtained a Thomas–Fermi type lower bound for fixed ratio $\kappa := \frac{Z}{c}$. Unfortunately, its Thomas–Fermi constant γ_e is less than $\gamma_{\mathrm{TF}} := (3\pi^2)^{\frac{2}{3}}$, the correct physical value.

For comparison we need the non-relativistic Thomas–Fermi functional

$$\mathcal{E}_Z^{\mathrm{TF}}(\rho) := \frac{3}{10}\gamma_{\mathrm{TF}} \int_{\mathbb{R}^3} \mathrm{d}x \rho(x)^{\frac{5}{3}} + \mathcal{V}(\rho) \tag{1.10}$$

defined on $I := \{\rho \in L^{\frac{5}{3}}(\mathbb{R}^3) : \rho \geq 0,\ \mathcal{D}[\rho] < \infty\}$. The functional is bounded from below (Simon [27]) and its infimum fulfills the scaling relation

$$E^{\mathrm{TF}}(Z) := \inf \mathcal{E}_Z^{\mathrm{TF}}(I) = -e^{\mathrm{TF}} Z^{\frac{7}{3}}, \tag{1.11}$$

where $e^{\mathrm{TF}} = -E^{\mathrm{TF}}(1)$ (Gombas [17], Lieb and Simon [23]). There exists a unique minimizer of $\mathcal{E}_Z^{\mathrm{TF}}$ which we denote by σ.

Our first result is

Theorem 1.1. *Assume $\kappa := \frac{Z}{c} \in \mathbb{R}_+$ fixed. Then, as $Z \to \infty$,*

$$\inf \mathcal{E}_{c,Z}^{\mathrm{TFWD}}(P) = -e^{\mathrm{TF}} Z^{\frac{7}{3}} + O(Z^2). \tag{1.12}$$

Our second result is the stability of the second kind of the functional which we address in Section 3.

From a mathematical perspective it might come as a surprise that Engel's and Dreizler's density functional – derived by purely formal methods from a quantum theory which is still lacking a full mathematical understanding – yields a fundamental feature like the ground state energy of heavy atoms to leading order quantitatively correct, in full agreement with the N-particle descriptions of heavy atoms like the Chandrasekhar Hamiltonian and no-pair Hamiltonians (Sørensen [26], [6], Solovej [28], Frank et al. [15, 16], [19]). It remains to be seen whether this is also true for other quantities like the density or whether the functional can be used as a tool to investigate relativistic many particle systems – like Thomas–Fermi theory in non-relativistic many body quantum mechanics – or, whether it even can shed light on a deeper understanding of quantum electrodynamics.

2 Bounds on the energy

2.1 Upper bound on the energy

We begin with an innocent lemma.

Lemma 2.1. *Assume $\rho : \mathbb{R}^3 \to \mathbb{R}_+$ such that $p := (3\pi^2\rho)^{\frac{1}{3}}$ has partial derivatives with respect to all variables at $x \in \mathbb{R}^3$. Then*

$$\frac{3}{8\pi^2}|\nabla p(x)|^2 c f\left(\frac{(p(x)}{c}\right) \leq |(\nabla\sqrt{\rho})(x)|^2. \tag{2.1}$$

Thus every nonnegative ρ with $\nabla\sqrt{\rho} \in L^2(\mathbb{R}^3)$ fulfills

$$\mathcal{T}^{\mathrm{W}}(\rho) \leq \lambda \int_{\mathbb{R}^3} |\nabla\sqrt{\rho}|^2. \tag{2.2}$$

Proof. We set $\psi := \sqrt{\rho}$ and compute

$$\begin{aligned}
&\frac{3}{8\pi^2}|\nabla p(x)|^2 c f\left(\frac{(p(x)}{c}\right)\\
&= \frac{3^2}{8}|\nabla \sqrt[3]{\rho}(x)|^2\left(\frac{\psi^{\frac{2}{3}}(x)}{\sqrt{1+\left(\frac{(p(x)}{c}\right)^2}} + 2\frac{\psi^{\frac{2}{3}}(x)\left(\frac{(p(x)}{c}\right)\mathfrak{Arsin}\left(\frac{(p(x)}{c}\right)}{1+\left(\frac{(p(x)}{c}\right)^2}\right) \qquad (2.3)\\
&\leq \frac{1}{2}|\nabla\psi(x)|^2 \max\left\{\frac{\sqrt{1+t^2}+2t\,\mathfrak{Arsin}(t)}{1+t^2} : t \in \mathbb{R}_+\right\} \leq |\nabla\psi(x)|^2
\end{aligned}$$

which gives the desired result. ∎

Of course the illuminati are hardly impressed by (2.2), since dominating relativistic energies by non-relativistic ones is common place for them. Presumably not even use of the numerically correct value 1.658290113 of the maximum in the proof instead of the estimate 2 would change that.

Now we turn to the upper bound on the left side of (1.12). It will be practical to use the non-relativistic Thomas–Fermi–Weizsäcker functional

$$\mathcal{E}_Z^{\mathrm{nrTFW}}(\rho) := \frac{\beta}{2}\int_{\mathbb{R}^3}|\nabla\sqrt{\rho}|^2 + \mathcal{E}_Z^{\mathrm{TF}}(\rho), \tag{2.4}$$

where $\beta \in \mathbb{R}_+$. It is defined on

$$J := \{\rho \in L^{\frac{5}{3}}(\mathbb{R}^3) : \rho \geq 0,\ \sqrt{\rho} \in D^1(\mathbb{R}^3),\ \mathcal{D}[\rho] < \infty\}, \tag{2.5}$$

has a unique minimizer ρ_W with

$$\int_{\mathbb{R}^3} \rho_W \leq Z + C \quad \text{and} \quad \int_{\mathbb{R}^3} \rho_W^{\frac{5}{3}} = O(Z^{\frac{7}{3}}) \tag{2.6}$$

(see [1–3]). Moreover,

$$E^{\mathrm{nrTFW}}(Z) = \mathcal{E}_Z^{\mathrm{nrTFW}}(\rho_W) = E^{\mathrm{TF}}(Z) + D_\beta Z^2 + o(Z^2) \tag{2.7}$$

for some β-dependent constant $D_\beta \in \mathbb{R}_+$ (Lieb and Liberman [21] and Lieb [20, Formula (1.6)]).

In the following we pick $\beta = 2$ and use the minimizer ρ_W of the non-relativistic Thomas–Fermi–Weizsäcker functional as a test function.

We estimate the exchange term first. Since $-X(t) \leq t^4$, we get

$$-\mathcal{X}(\rho_W) \leq \frac{(3\pi^2)^{\frac{4}{3}}}{8\pi^3} \int_{\mathbb{R}^3} \rho_w^{\frac{4}{3}} \leq C \sqrt{\int_{\mathbb{R}^3} \rho_W^{\frac{5}{3}} \int_{\mathbb{R}^3} \rho_W} = O(Z^{\frac{5}{3}}). \tag{2.8}$$

Thus, since $T^{\mathrm{TF}}(t) \leq \frac{3}{5} t^5$,

$$\inf \mathcal{E}_{c,Z}^{\mathrm{TFWD}}(P) \leq \mathcal{E}_{c,Z}^{\mathrm{TFWD}}(\rho_W) \leq \mathcal{E}_Z^{\mathrm{nrTFW}}(\rho_W) + O(Z^{\frac{5}{3}}) = E^{\mathrm{TF}}(Z) + O(Z^2) \tag{2.9}$$

which concludes the proof of the upper bound.

2.2 Lower bound on the energy

We set

$$đ\xi := \frac{\mathrm{d}\xi}{h^3} = \frac{\mathrm{d}\xi}{(2\pi)^3}. \tag{2.10}$$

(Note that the rationalized Planck constant $\hbar$ equals one in Hartree units and, therefore, $h = 2\pi$.)

We introduce the notation $(a)_- := \min\{0, a\}$ and write

$$\varphi_\sigma := \frac{Z}{|\cdot|} - \sigma * |\cdot|^{-1} \tag{2.11}$$

for the Thomas–Fermi potential of the minimizer σ. We start again with a little lemma.

Lemma 2.2. *Assume $\kappa = \frac{Z}{c}$ fixed. Then, as $Z \to \infty$,*

$$\int_{|x|>\frac{1}{Z}} \mathrm{d}x \int_{\mathbb{R}^3} đ\xi \Big(\frac{\xi^2}{2} - \varphi_\sigma(x)\Big)_- - \int_{|x|>\frac{1}{Z}} \mathrm{d}x \int_{\mathbb{R}^3} đ\xi \big(\sqrt{c^2\xi^2 + c^4} - c^2 - \varphi_\sigma(x)\big)_- = O(Z^2). \tag{2.12}$$

Again, it does not come as a surprise to the physicist that relativistic and non-relativistic theory give the same result up to errors, if the innermost electrons, i.e., in particular the fast moving, are disregarded.

Proof. Since

$$\frac{\xi^2}{2} \geq \sqrt{c^2\xi^2 + c^4} - c^2, \tag{2.13}$$

the left side of the claimed inequality cannot be negative. Thus, we merely need an upper bound:

$$\begin{aligned}
&\int_{|x|>\frac{1}{Z}} \mathrm{d}x \int_{\mathbb{R}^3} đ\xi \left(\frac{\xi^2}{2} - \varphi_\sigma(x)\right)_- \\
&\qquad - \int_{|x|>\frac{1}{Z}} \mathrm{d}x \int_{\mathbb{R}^3} đ\xi \left(\sqrt{c^2\xi^2 + c^4} - c^2 - \varphi_\sigma(x)\right)_- \\
&\leq \int_{|x|>\frac{1}{Z}} \mathrm{d}x \left\{ c^5 \int_{\frac{\xi^2}{2} < \frac{\varphi_\sigma(x)}{c^2}} đ\xi \left[\frac{1}{2}\xi^2 - \left(\sqrt{\xi^2+1} - 1\right)\right] \right. \\
&\qquad \left. - \int_{\sqrt{c^2\xi^2+c^4} - c^2 < \varphi_\sigma(x) \leq \frac{\xi^2}{2}} đ\xi \left(\sqrt{c^2\xi^2 + c^4} - c^2 - \varphi_\sigma(x)\right) \right\} \\
&\leq c^5 \int_{|x|>\frac{1}{Z}} \mathrm{d}x \left(\int_{\frac{\xi^2}{2} < \frac{\varphi_\sigma(x)}{c^2}} đ\xi + \int_{\sqrt{\xi^2+1}-1 < \frac{\varphi_\sigma(x)}{c^2} \leq \frac{\xi^2}{2}} đ\xi \right) \\
&\qquad \times \left[\frac{1}{2}\xi^2 - \left(\sqrt{\xi^2+1} - 1\right)\right] \\
&\leq \frac{c^5}{8} \int_{|x|>\frac{1}{Z}} \mathrm{d}x \left(\int_{\frac{\xi^2}{2} < \frac{\varphi_\sigma(x)}{c^2}} đ\xi + \int_{\sqrt{\xi^2+1}-1 < \frac{\varphi_\sigma(x)}{c^2} \leq \frac{\xi^2}{2}} đ\xi \right) |\xi|^4 \\
&\leq \frac{c^5}{8} \int_{|x|>\frac{1}{Z}} \mathrm{d}x \int_{\sqrt{\xi^2+1}-1 < \frac{\varphi_\sigma(x)}{c^2}} đ\xi |\xi|^4 \\
&\leq \frac{c^2}{8\kappa^3} \int_{|x|>1} \mathrm{d}x \int_{\sqrt{\xi^2+1}-1 < \frac{\kappa^2}{|x|}} đ\xi |\xi|^4
\end{aligned} \tag{2.14}$$

where we used

$$\varphi_\sigma(x) \leq \frac{Z}{|x|} \tag{2.15}$$

in the last inequality. Moreover, the resulting last integral obviously exists and is independent of Z. Thus the left side of the claimed inequality is bounded from above by a constant depending only on κ times Z^2 quod erat demonstrandum. ∎

We turn to the lower bound on the left side of (1.12) and follow initially [8]. In fact, apart from minor modifications, we copy the high density part and focus on the low density part. We pick any $\rho \in P$ and address the parts of the energy separately.

The Weizsäcker energy. Since $F(t) \geq \frac{t}{2}\sqrt{\mathfrak{Arsin}(t)}$ (see [9, Formula (90)]), Hardy's inequality gives the lower bound

$$
\begin{aligned}
\mathcal{T}^{\mathrm{W}}(\rho) &\geq \frac{3\lambda c}{2^7\pi^2}\int_{\mathbb{R}^3}\mathrm{d}x\,\frac{p(x)^2\mathfrak{Arsin}(\frac{p(x)}{c})}{|x|^2}\\
&= \frac{3^{\frac{5}{3}}\lambda c}{2^7\pi^{\frac{2}{3}}}\underbrace{\int_{\mathbb{R}^3}\mathrm{d}x\,\frac{\rho(x)^{\frac{2}{3}}\mathfrak{Arsin}(\frac{p(x)}{c})}{|x|^2}}_{=:\,\mathcal{H}(\rho)}.
\end{aligned}
\tag{2.16}
$$

The Potential energy. Since σ is positive, we have $\varphi_\sigma(x) \leq \frac{Z}{|x|}$. Then

$$
\begin{aligned}
\mathcal{V}(\rho) &= -\int_{\mathbb{R}^3}\mathrm{d}x\,\varphi_\sigma(x)\rho(x) - 2\mathcal{D}(\sigma,\rho) + \mathcal{D}[\rho]\\
&\geq -\int_{\mathbb{R}^3}\mathrm{d}x\,\varphi_\sigma(x)\rho(x) - \mathcal{D}[\sigma].
\end{aligned}
\tag{2.17}
$$

Splitting the integrals at s, using (2.17), and Schwarz's inequality yields

$$
\begin{aligned}
\mathcal{V}(\rho) &\geq -\int_{\frac{p(x)}{c}<s}\mathrm{d}x\,\varphi_\sigma(x)\rho(x)\\
&\quad - Z\int_{\frac{p(x)}{c}\geq s}\mathrm{d}x\,\frac{\rho(x)^{\frac{1}{3}}}{|x|}\mathfrak{Arsin}\Big(\frac{p(x)}{c}\Big)^{\frac{1}{2}}\frac{\rho(x)^{\frac{2}{3}}}{\mathfrak{Arsin}(\frac{p(x)}{c})^{\frac{1}{2}}} - \mathcal{D}[\sigma]\\
&\geq -\frac{Z}{\mathfrak{Arsin}(s)^{\frac{1}{2}}}\mathcal{H}(\rho)^{\frac{1}{2}}\mathcal{T}_{>}(\rho)^{\frac{1}{2}} - \int_{\frac{p(x)}{c}<s}\mathrm{d}x\,\varphi_\sigma(x)\rho(x) - \mathcal{D}[\sigma]
\end{aligned}
\tag{2.18}
$$

with

$$
\mathcal{T}_{>}(\rho) := \int_{\frac{p(x)}{c}>s}\mathrm{d}x\,\rho(x)^{\frac{4}{3}}.
\tag{2.19}
$$

The Thomas–Fermi term. First, we note that

$$
\mathbb{R}_+ \to \mathbb{R}_+,\ t \mapsto \frac{T^{\mathrm{TF}}(t)}{t^4}
\tag{2.20}
$$

is monotone increasing from 0 to 2. Thus

$$
\begin{aligned}
\mathcal{T}^{\mathrm{TF}}(\rho) &= \int_{\frac{p(x)}{c}<s}\mathrm{d}x\,\frac{c^5}{8\pi^2}T^{\mathrm{TF}}\Big(\frac{p(x)}{c}\Big) + \int_{\frac{p(x)}{c}\geq s}\mathrm{d}x\,\frac{c^5}{8\pi^2}T^{\mathrm{TF}}\Big(\frac{p(x)}{c}\Big)\\
&\geq \int_{\frac{p(x)}{c}<s}\mathrm{d}x\,\frac{c^5}{8\pi^2}T^{\mathrm{TF}}\Big(\frac{p(x)}{c}\Big) + \int_{\frac{p(x)}{c}\geq s}\mathrm{d}x\,\frac{T^{\mathrm{TF}}(s)}{s^4}\frac{3}{8}(3\pi^2)^{\frac{1}{3}}c\rho(x)^{\frac{4}{3}}\\
&= \int_{\frac{p(x)}{c}<s}\mathrm{d}x\,\frac{c^5}{8\pi^2}T^{\mathrm{TF}}\Big(\frac{p(x)}{c}\Big) + \frac{3}{8}\frac{T^{\mathrm{TF}}(s)}{s^4}\gamma_{\mathrm{TF}}^{\frac{1}{2}}c\mathcal{T}_{>}(\rho).
\end{aligned}
\tag{2.21}
$$

Exchange energy. Since X is bounded from above and $X(t) = O(t^4)$ at $t = 0$, we have that for every $\alpha \in [0, 4]$ there is an η_0 such that $X(t) \leq \eta_0 t^\alpha$. We pick $\alpha = 3$ in which case $\xi_0 \approx 1.15$. Thus, with $\eta := \frac{\eta_0}{4\pi} \approx 0.0914$, we have

$$\mathcal{X}(\rho) \leq \frac{c\eta_0}{4\pi} N = \eta c N. \tag{2.22}$$

The total energy. Adding everything up yields

$$\begin{aligned}\mathcal{E}_{c,Z}^{\mathrm{TFWD}}(\rho) \geq{}& \frac{3^{\frac{5}{3}}\lambda c}{2^7\pi^{\frac{2}{3}}}\mathcal{H}(\rho) + \frac{3}{8}\frac{T^{\mathrm{TF}}(s)}{s^4}\gamma_{\mathrm{TF}}{}^{\frac{1}{2}} c\mathcal{T}_>(\rho) - \frac{Z}{\mathfrak{Arsin}(s)^{\frac{1}{2}}}\mathcal{H}(\rho)^{\frac{1}{2}}\mathcal{T}_>(\rho)^{\frac{1}{2}} \\ &+ \int_{\frac{p(x)}{c}<s} \mathrm{d}x \frac{c^5}{8\pi^2} T^{\mathrm{TF}}\left(\frac{p(x)}{c}\right) - \varphi_\sigma(x)\rho(x)) - \mathcal{D}[\sigma] - \xi c N. \end{aligned}\tag{2.23}$$

We pick $s \in \mathbb{R}_+$ such that the sum of the first three summands on the right of (2.23) is a complete square, i.e., fulfilling

$$\sqrt{\frac{3^{\frac{5}{3}}}{2^7\pi^{\frac{2}{3}}}\frac{3T^{\mathrm{TF}}(s)(3\pi^2)^{\frac{1}{3}}}{8s^4}} = \frac{Z}{c\sqrt{\lambda}}\frac{1}{2\mathfrak{Arsin}(s)^{\frac{1}{2}}}. \tag{2.24}$$

The solution is uniquely determined, because of (2.20) (and the line below) and $\mathfrak{Arsin}(s)$ is also monotone increasing from 0 to ∞. Denote the corresponding s by s_0. Obviously, s_0 does not depend on c and Z independently but only on the ratio $\kappa := \frac{Z}{c}$ and is strictly monotone increasing from 0 to ∞.

We set

$$\begin{aligned} I_{s,Z} &:= \left\{x \in \mathbb{R}^3 : \frac{p(x)}{c} < s,\ |x| < \frac{1}{Z}\right\}, \\ A_{s,Z} &:= \left\{x \in \mathbb{R}^3 : \frac{p(x)}{c} < s,\ |x| \geq \frac{1}{Z}\right\}. \end{aligned}\tag{2.25}$$

Then

$$\mathcal{E}_{c,Z}^{\mathrm{TFWD}}(\rho) \geq I + A - \mathcal{D}[\sigma] - \xi c N \tag{2.26}$$

with

$$I := \int_{I_{s,Z}} \mathrm{d}x \left(\frac{c^5}{8\pi^2} T^{\mathrm{TF}}\left(\frac{p(x)}{c}\right) - \varphi_\sigma(x)\rho(x)\right), \tag{2.27}$$

$$A := \int_{A_{s,Z}} \mathrm{d}x \left(\frac{c^5}{8\pi^2} T^{\mathrm{TF}}\left(\frac{p(x)}{c}\right) - \varphi_\sigma(x)\rho(x)\right). \tag{2.28}$$

We estimate I from below by dropping the TF-term, using $\varphi_\sigma(x) \leq \frac{Z}{|x|}$, and observing that $x \in I_{s,Z}$. We get

$$I \geq -C_\kappa Z c^3 \int_0^{Z^{-1}} \mathrm{d}r r = -C_\kappa Z^2, \tag{2.29}$$

where C_κ is a generic constant depending on κ only. In other words, we can pull the Coulomb tooth paying a negligible price.

Next we estimate A from below by keeping $A_{s,Z}$ fixed and minimizing the integrand at each point $x \in A_{s,Z}$ by varying the values $p(x) \in \mathbb{R}_+$. We get

$$\begin{aligned} A &\geq 2 \int_{A_{s,Z}} \mathrm{d}x \int_{\mathbb{R}^3} đ\xi \big(\sqrt{c^2\xi^2 + c^4} - c^2 - \varphi_\sigma(x)\big)_- \\ &\geq 2 \int_{|x| \geq \frac{1}{Z}} \mathrm{d}x \int_{\mathbb{R}^3} đ\xi \big(\sqrt{c^2\xi^2 + c^4} - c^2 - \varphi_\sigma(x)\big)_-. \end{aligned} \tag{2.30}$$

Although at first glance the first inequality might seem abrupt, it is easily checked that the Thomas–Fermi functional (relativistic or non-relativistic, restricted to some region in space M) with kinetic energy $T(\xi)$ and external potential φ is merely the marginal functional (integrating out the momentum variable ξ) of the phase space variational principle

$$\mathcal{E}_\Gamma(\gamma) := \int_M \mathrm{d}x \int_{\mathbb{R}^3} đ\xi (T(\xi) - \varphi(x))\gamma(x,\xi) \tag{2.31}$$

with $\gamma(M, \mathbb{R}^3) \subset [0, 2]$ and the choice

$$\gamma(x,\xi) := \chi_{\{(x,\xi) \in M \times \mathbb{R}^3 : |\xi| < p(x)\}} \tag{2.32}$$

for given Fermi momentum p. Eventually, since γ has only values between 0 and 2, (2.31) is obviously minimized by the characteristic function of the support of the negative part of $T(\xi) - \varphi(x)$ times 2.

Metaphorically speaking the second inequality of (2.30) states, that we can bound the energy from below by a toothless relativistic Thomas–Fermi energy.

Now, by Lemma 2.2 this equals the corresponding non-relativistic expression up to errors of order $O(Z^2)$, i.e., we have

$$\begin{aligned} \mathcal{E}_{c,Z}^{\mathrm{TFWD}}(\rho) &\geq 2 \int_{|x| \geq \frac{1}{Z}} \mathrm{d}x \int_{\mathbb{R}^3} đ\xi \Big(\frac{1}{2}\xi^2 - \varphi_\sigma(x)\Big)_- - \mathcal{D}[\sigma] - C_\kappa Z^2 \\ &\geq 2 \int_{\mathbb{R}^3} \mathrm{d}x \int_{\mathbb{R}^3} đ\xi \Big(\frac{1}{2}\xi^2 - \varphi_\sigma(x)\Big)_- - \mathcal{D}[\sigma] - C_\kappa Z^2 \\ &= \mathcal{E}_Z^{\mathrm{TF}}(\sigma) - C_\kappa Z^2 = -e^{\mathrm{TF}} Z^{\frac{7}{3}} - C_\kappa Z^2 \end{aligned} \tag{2.33}$$

which concludes the proof of the desired lower bound and therefore also the proof of (1.12).

3 Stability of matter

For several atoms the potential term $\mathcal{V}$ in (1.8) is replaced by

$$\mathcal{V}(\rho) := -\sum_{k=1}^{K} \int_{\mathbb{R}^3} \mathrm{d}x \frac{Z_k \rho(x)}{|x - R_k|} + \mathcal{D}[\rho] + \sum_{1 \leq k < l \leq K} \frac{Z_k Z_l}{|R_k - R_l|} \tag{3.1}$$

with pairwise different positions of the nuclei $\mathfrak{R} := (R_1, \ldots, R_K) \in \mathbb{R}^{3K}$, and atomic numbers $\mathfrak{Z} := (Z_1, \ldots, Z_K) \in \mathbb{R}_+^K$. The first term is the attraction potential between the electrons and the nuclei, the second term, the electron-electron interaction, is unchanged, and the third is the repulsion between the nuclei. We write $\mathcal{E}_{\mathfrak{R},\mathfrak{Z}}^{\mathrm{TFWD}}$ for the otherwise unchanged Engel–Dreizler functional. We are interested in finding a lower bound to $\mathcal{E}_{\mathfrak{R},\mathfrak{Z}}^{\mathrm{TFWD}}$ which is uniform in the density ρ as long as the constraint $\int \rho = N$ is respected, is uniform in $R_1, \ldots, R_K$ and linear in N and K, i.e., we wish to show that the energy per particle is bounded from below by the same constant irrespective of the electron density and the positions of the nuclei. This property is also known as stability of the second kind, thermodynamic stability, and stability of matter.

Benguria et al. [4] showed stability of matter for the massless version of this functional without the regularizing $\mathfrak{Arsin}$-term in $\mathcal{T}^{\mathrm{W}}$ and without the exchange correction.

Initially we restrict to the case of equal atomic numbers Z. We begin by extracting all Coulomb teeth which is conveniently done using [24, Formula (5.2)]: given disjoint balls $B_1, \ldots, B_K$ with centers at $R_1, \ldots, R_K$ and radii $D_1, \ldots, D_K$, we have

$$-\Delta \geq \sum_{\kappa=1}^{K} \left(\frac{1}{4|\cdot - R_\kappa|^2} - D_k^{-2} Y\left(\frac{|\cdot - R_\kappa|}{D_\kappa} \right) \right)_+ \chi_{B_\kappa} \tag{3.2}$$

with

$$Y(r) = 1 + \frac{r^2}{4}. \tag{3.3}$$

We also set

$$H_k(x) := \frac{2}{D_k} \sqrt{Y\left(\frac{|x - R_k|}{D_K} \right)}. \tag{3.4}$$

We pick the radii maximal, namely D_k as half the distance of the k-th nucleus to its nearest neighbor.

We also use Lieb and Yau's electrostatic inequality [24, Formula (4.5)]: we write $\Gamma_1, \ldots, \Gamma_K$ for the Voronoi cells of $R_1, \ldots, R_K$ and set

$$\Phi(x) := \sum_{l=1}^{K} \Gamma_l(x) \sum_{k=1, k \neq l}^{K} \frac{Z}{|x - R_k|}, \tag{3.5}$$

i.e., the nuclear potential at point x of all nuclei except the one from the cell in which x lies. With this notation the electrostatic inequality reads

$$\frac{1}{2} D[\nu] - Z \int_{\mathbb{R}^3} \Phi(x) \mathrm{d}\nu(x) + Z^2 \sum_{1 \leq k < l \leq K} \frac{1}{|R_k - R_l|} \geq \frac{Z^2}{8} \sum_{k=1}^{K} \frac{1}{D_k} \tag{3.6}$$

for any bounded measure and $Z \in \mathbb{R}_+$.

With these tools we modify the atomic lower bounds term by term.

3.1 The Weizsäcker energy

Instead of (2.16) we get

$$\mathcal{T}^{\mathrm{W}}(\rho) \geq \frac{3^{\frac{5}{3}}\lambda c}{2^7\pi^{\frac{2}{3}}} \sum_{k=1}^{K} \underbrace{\int_{B_k} \mathrm{d}x \rho(x)^{\frac{2}{3}} \mathfrak{Arsin}\Big(\frac{p(x)}{c}\Big)\big(|x-R_k|^{-2} - H_k(x)^2\big)_+}_{=:\,\mathcal{H}_k(\rho)}, \tag{3.7}$$

where we use (3.2) and $F(t) \geq \frac{t}{2}\sqrt{\mathfrak{Arsin}(s)}$ (see [9, Formula (90)]).

3.2 The potential energy

Using (3.6) and $\sqrt{(a^2-b^2)_+} \geq a-b$ for $a, b \in \mathbb{R}_+$, we get

$$\begin{aligned}
\mathcal{V}(\rho) \geq & -\sum_{k=1}^{K}\Big(\int_{B_K}\mathrm{d}x + \int_{\Gamma_k\setminus B_k}\mathrm{d}x\Big)\frac{Z\rho(x)}{|x-R_k|} + \frac{Z^2}{8}\sum_{k=1}^{K}\frac{1}{D_k} \\
\geq & -\sum_{k=1}^{K} Z\Big\{\int_{B_k,\, \frac{p(x)}{c}\geq s}\mathrm{d}x\rho(x)\Big[\sqrt{\Big(\frac{1}{|x-R_k|^2} - H_k(x)^2\Big)_+} + H_k(x)\Big] \\
& + \int_{B_k,\, \frac{p(x)}{c}\leq s}\mathrm{d}x\frac{\rho(x)}{|x-R_k|} + \int_{\Gamma_k\setminus B_k}\frac{\rho(x)}{|x-R_k|} - \frac{Z}{8}\frac{1}{D_k}\Big\} \\
\geq & -\sum_{k=1}^{K}\Big(\frac{Z}{\sqrt{\mathfrak{Arsin}(s)}}\sqrt{\mathcal{H}_{R_k}(\rho)\mathcal{T}_{R_k}(\rho)} + Z\int_{B_k,\, \frac{p(x)}{c}\geq s}\mathrm{d}x H_k(x)\rho(x) \\
& + \int_{B_k,\, \frac{p(x)}{c}\leq s}\mathrm{d}x\frac{Z\rho(x)}{|x-R_k|} + \int_{\Gamma_k\setminus B_k}\mathrm{d}x\frac{Z\rho(x)}{|x-R_k|}\Big) + \frac{Z^2}{8}\sum_{k=1}^{K}\frac{1}{D_k}
\end{aligned} \tag{3.8}$$

with

$$\mathcal{T}_{R_k}(\rho) := \int_{B_k,\, \frac{p(x)}{c}\geq s}\mathrm{d}x\rho(x)^{\frac{4}{3}}. \tag{3.9}$$

3.3 The combined Thomas–Fermi and exchange terms

We use

$$T^{\mathrm{TF}}(t) \geq 2t^4 - \frac{8t^3}{3}, \tag{3.10}$$

estimate (2.22), and set $\delta := (3\pi^2)^{\frac{1}{3}}$. This yields

$$\begin{aligned}
\mathcal{T}^{\mathrm{TF}}(\rho) - \mathcal{X}(\rho) &= \int_{\mathbb{R}^3}\mathrm{d}x\frac{c^5}{8\pi^2}T^{\mathrm{TF}}\Big(\frac{p(x)}{c}\Big) - \eta c N \\
&\geq \frac{3}{4}\delta c\int_{\mathbb{R}^3}\mathrm{d}x\rho(x)^{\frac{4}{3}} - C_c N.
\end{aligned} \tag{3.11}$$

3.4 The total energy

Adding again all up yields

$$\mathcal{E}_{c,Z}^{\mathrm{TFWD}}(\rho) \tag{3.12}$$

$$\geq \sum_{k=1}^{K} c\Bigg[\frac{3^{\frac{5}{3}}\lambda}{2^7\pi^{\frac{2}{3}}}\mathcal{H}_{R_k}(\rho) + \frac{3}{8}\delta\mathcal{T}_{R_k}(\rho) - \frac{\kappa}{\sqrt{\mathfrak{Arsin}(s)}}\sqrt{\mathcal{H}_{R_k}(\rho)\mathcal{T}_{R_k}(\rho)} \tag{3.13}$$

$$+ \int_{B_k,\ \frac{p(x)}{c}<s} \mathrm{d}x\left(\frac{3}{4}\delta\rho(x)^{\frac{4}{3}} - \frac{\kappa\rho(x)}{|x-R_k|}\right) \tag{3.14}$$

$$+ \int_{B_k,\ \frac{p(x)}{c}>s} \mathrm{d}x\left(\frac{3}{8}\delta\rho(x)^{\frac{4}{3}} - \kappa H_k(x)\rho(x)\right) \tag{3.15}$$

$$+ \int_{\Gamma_k\setminus B_k} \mathrm{d}x\left(\frac{3}{4}\delta\rho(x)^{\frac{4}{3}} - \frac{\kappa\rho(x)}{|x-R_k|}\right)\Bigg] \tag{3.16}$$

$$+ \sum_{k=1}^{K}\frac{Z^2}{8D_k} - C_c N. \tag{3.17}$$

We pick s such that

$$2\sqrt{\frac{3^{\frac{5}{3}}\lambda}{2^7\pi^{\frac{2}{3}}}\frac{3}{8}\delta} = \frac{\kappa}{\sqrt{\mathfrak{Arsin}(s)}}, \tag{3.18}$$

which makes (3.13) a sum of complete squares. Next

$$\begin{aligned}(3.14) &\geq \delta(cs)^4 \inf\left\{\int_{p(x)<1}\mathrm{d}x\left(\frac{3}{4}\rho(x)^{\frac{4}{3}} - \frac{\kappa\rho(x)}{cs\delta|x|}\right) : \rho\in P\right\}\\ &\geq \frac{cs\kappa^3}{\delta^2}\inf\left\{\int_{p(x)<1}\mathrm{d}x\left(\frac{3}{4}\rho(x)^{\frac{4}{3}} - \frac{\rho(x)}{|x|}\right) : \rho\in P\right\} \geq -C\frac{cs\kappa^3}{\delta^2},\end{aligned} \tag{3.19}$$

where we replaced p by csp in the first step and x by $\frac{\kappa}{cs\delta}x$ in the second step. Thus (3.19) yields, after summation, K times a constant which is irrelevant for stability. Furthermore,

$$(3.15) \geq -\frac{2^4\kappa^4\cdot 4\pi}{2\cdot 4\delta^3 D_k}\int_0^1 \mathrm{d}r\, r^2 2^4\left(1+\frac{r^2}{4}\right)^2 = -\frac{5944\pi\kappa^4}{105\delta^3 D_k} \tag{3.20}$$

and using $\int_{\Gamma_k\setminus B_k}\mathrm{d}x|x-R_k|^{-4} \leq \frac{3\pi}{D_k}$ (Lieb et al. [22, Formula (4.6)]), we get

$$(3.16) \geq -\frac{\kappa^4}{4\delta^3}\int_{\Gamma_k\setminus B_k}\frac{1}{|x-R_k|^4} \geq -\frac{3\pi\kappa^4}{4\delta^3 D_k}. \tag{3.21}$$

Thus, the energy per particle is bounded from below uniformly in ρ, K, and N, if

$$c\left(\frac{2972\pi\kappa^4}{105\delta^3} + \frac{3\pi\kappa^4}{4\delta^3}\right) \leq \frac{Z^2}{8}, \tag{3.22}$$

i.e.,

$$Z \le Z_{\max} := 3\frac{\sqrt{1686370\pi}}{48182} c^{\frac{3}{2}}. \tag{3.23}$$

Numerically, using the physical value of the velocity of light $c = 137.037$ (Bethe and Salpeter [5, p. 84]), we get

$$Z_{\max} \approx 229.9029615 \tag{3.24}$$

covering liberally all known elements. The result can be condensed into

Theorem 3.1. *There exists a constant C such that for all $\rho \in P$ and all pairwise different $R_1, \dots, R_K \in \mathbb{R}^3$ and $Z_1 = \dots = Z_K \in [0, Z_{\max}]$,*

$$\mathcal{E}^{\mathrm{TFWD}}_{\mathfrak{R},\mathfrak{Z}}(\rho) \ge C \cdot (K + N). \tag{3.25}$$

We conclude with two remarks:

(1) Theorem 3.1 holds actually for all $\mathfrak{Z} \in [0, Z_{\max}]^K$. Our proof obviously generalizes to that case, since the potential estimate (3.6) also generalizes in the obvious way.

(2) It might be surprising that there is no requirement on a minimal velocity of light which is independent of the value of Z. This is different from other unrenormalized relativistic models like the Thomas–Fermi functional with inhomogeneity correction $(\sqrt{\rho}, |\nabla|\sqrt{\rho})$ investigated by Lieb et al. [22, Formula (2.7)]. There we were forced to control the exchange energy by the Thomas–Fermi term. Here, this is no longer necessary: due to the renormalization in Engel's and Dreizler's derivation the exchange energy is bounded from below by a multiple of the particle number.

Acknowledgements. Special thanks go to Rupert Frank for many inspiring discussions, in particular for pointing out that in addition to separation in high and low density regimes, a localization in space near the nucleus could be useful, and for critical reading of a substantial part of the manuscript. Thanks go also to Hongshuo Chen for critical reading of the manuscript.

Partial support by the Deutsche Forschungsgemeinschaft (DFG, German Research Foundation) through Germany's Excellence Strategy EXC – 2111 – 390814868 is gratefully acknowledged.

Bibliography

[1] R. Benguria, *The von Weizsäcker and exchange corrections in the Thomas–Fermi theory*. Ph.D. thesis, Department of Physics, Princeton, 1979

[2] R. Benguria, H. Brézis and E. H. Lieb, The Thomas–Fermi–von Weizsäcker theory of atoms and molecules. *Comm. Math. Phys.* **79** (1981), 167–180

[3] R. Benguria and E. H. Lieb, The most negative ion in the Thomas–Fermi–von Weizsäcker theory of atoms and molecules. *J. Phys. B* **18** (1985), 1045–1059

[4] R. Benguria, M. Loss and H. Siedentop, Stability of atoms and molecules in an ultrarelativistic Thomas–Fermi–Weizsäcker model. *J. Math. Phys.* **49** (2008), Article ID 012302

[5] H. A. Bethe and E. E. Salpeter, *Quantum mechanics of one- and two-electron atoms*. Springer, Berlin, 1957

[6] R. Cassanas and H. Siedentop, The ground-state energy of heavy atoms according to Brown and Ravenhall: absence of relativistic effects in leading order. *J. Phys. A* **39** (2006), 10405–10414

[7] S. Chandrasekhar, The maximum mass of ideal white dwarfs. *Astrophys. J.* **74** (1931), 81–82

[8] H. Chen, R. L. Frank and H. Siedentop, A statistical theory of heavy atoms: Energy and excess charge. Preprint 2020, arXiv:2010.12074

[9] H. Chen and H. Siedentop, On the excess charge of a relativistic statistical model of molecules with an inhomogeneity correction. *J. Phys. A* **53** (2020), Article ID 395201

[10] T. Ekholm and R. L. Frank, On Lieb–Thirring inequalities for Schrödinger operators with virtual level. *Comm. Math. Phys.* **264** (2006), 725–740

[11] E. Engel and R. M. Dreizler, Field-theoretical approach to a relativistic Thomas–Fermi–Dirac–Weizsäcker model. *Phys. Rev. A* **35** (1987), 3607–3618

[12] E. Fermi, Un metodo statistico per la determinazione di alcune proprietá dell'atomo. *Atti Reale Accad. Naz. Lincei Rend. Classe Sci. Fis. Mat. Nat.* **6** (1927), 602–607

[13] E. Fermi, Eine statistische Methode zur Bestimmung einiger Eigenschaften des Atoms und ihre Anwendung auf die Theorie des periodischen Systems der Elemente. *Z. Phys.* **48** (1928), 73–79

[14] R. L. Frank, E. H. Lieb and R. Seiringer, Stability of relativistic matter with magnetic fields for nuclear charges up to the critical value. *Comm. Math. Phys.* **275** (2007), 479–489

[15] R. L. Frank, H. Siedentop and S. Warzel, The ground state energy of heavy atoms: relativistic lowering of the leading energy correction. *Comm. Math. Phys.* **278** (2008), 549–566

[16] R. L. Frank, H. Siedentop and S. Warzel, The energy of heavy atoms according to Brown and Ravenhall: the Scott correction. *Doc. Math.* **14** (2009), 463–516

[17] P. Gombás, *Die statistische Theorie des Atoms und ihre Anwendungen*. Springer, Wien, 1949

[18] P. Gombás, Statistische Behandlung des Atoms. In *Handbuch der Physik. Atome II*, pp. 109–231, Springer, Berlin, 1956

[19] M. Handrek and H. Siedentop, On the maximal excess charge of the Chandrasekhar–Coulomb Hamiltonian in two dimension. *Lett. Math. Phys.* **103** (2013), 843–849

[20] E. H. Lieb, Analysis of the Thomas–Fermi–von Weizsäcker equation for an infinite atom without electron repulsion. *Comm. Math. Phys.* **85** (1982), 15–25

[21] E. H. Lieb and D. A. Liberman, Numerical calculation of the Thomas–Fermi–von Weizsäcker function for an infinite atom without electron repulsion. Technical Report LA-9186-MS, Los Alamos National Laboratory, Los Alamos, 1982

[22] E. H. Lieb, M. Loss and H. Siedentop, Stability of relativistic matter via Thomas–Fermi theory. *Helv. Phys. Acta* **69** (1996), 974–984

[23] E. H. Lieb and B. Simon, The Thomas–Fermi theory of atoms, molecules and solids. *Advances in Math.* **23** (1977), 22–116

[24] E. H. Lieb and H.-T. Yau, The stability and instability of relativistic matter. *Comm. Math. Phys.* **118** (1988), 177–213

[25] N. H. March and W. H. Young, Variational methods based on the density matrix, *Proc. Phys. Soc.* **72** (1958), 182–192

[26] T. Østergaard Sørensen, The large-Z behavior of pseudorelativistic atoms. *J. Math. Phys.* **46** (2005), Article ID 052307

[27] B. Simon, *Functional integration and quantum physics*. Pure Appl. Math. 86, Academic Press, New York, 1979

[28] J. P. Solovej, T. O. Sørensen and W. L. Spitzer, Relativistic Scott correction for atoms and molecules. *Comm. Pure Appl. Math.* **63** (2010), 39–118

[29] L. H. Thomas, The calculation of atomic fields, *Proc. Camb. Phil. Soc.* **23** (1927), 542–548

[30] Y. Tomishima, A relativistic Thomas–Fermi theory. *Progr. Theoret. Phys.* **42** (1969), 437–447

[31] C. F. von Weizsäcker, Zur Theorie der Kernmassen, *Z. Phys.* **96** (1935), 431–458

[32] K. Yonei and Y. Tomishima, On the Weizsäcker correction to the Thomas–Fermi theory of the atom. *J. Phys. Soc. Japan* **20** (1965), 1051–1057

Homogenization of the higher-order Schrödinger-type equations with periodic coefficients

Tatiana Suslina

To Ari Laptev on the occasion of his 70th birthday

In $L_2(\mathbb{R}^d;\mathbb{C}^n)$, we consider a matrix strongly elliptic differential operator A_ε of order $2p$, $p \geqslant 2$. The operator A_ε is given by $A_\varepsilon = b(\mathbf{D})^* g(\frac{\mathbf{x}}{\varepsilon}) b(\mathbf{D})$, $\varepsilon > 0$, where $g(\mathbf{x})$ is a periodic, bounded, and positive definite matrix-valued function, and $b(\mathbf{D})$ is a homogeneous differential operator of order p. We prove that, for fixed $\tau \in \mathbb{R}$ and $\varepsilon \to 0$, the operator exponential $e^{-i\tau A_\varepsilon}$ converges to $e^{-i\tau A^0}$ in the norm of operators acting from the Sobolev space $H^s(\mathbb{R}^d;\mathbb{C}^n)$ (with a suitable s) into $L_2(\mathbb{R}^d;\mathbb{C}^n)$. Here A^0 is the effective operator. Sharp-order error estimate is obtained. The results are applied to homogenization of the Cauchy problem for the Schrödinger-type equation $i\partial_\tau \mathbf{u}_\varepsilon = A_\varepsilon \mathbf{u}_\varepsilon + \mathbf{F}$, $\mathbf{u}_\varepsilon|_{\tau=0} = \boldsymbol{\phi}$.

1 Introduction

The paper concerns homogenization theory of periodic differential operators (DOs). First of all, we mention the books [1, 20].

1.1 Operator error estimates

In a series of papers [2–4] by Birman and Suslina, an operator-theoretic (spectral) approach to homogenization problems was developed. In $L_2(\mathbb{R}^d;\mathbb{C}^n)$, a wide class of matrix strongly elliptic second-order DOs A_ε was studied. The operator A_ε is given by $A_\varepsilon = b(\mathbf{D})^* g(\frac{\mathbf{x}}{\varepsilon}) b(\mathbf{D})$, $\varepsilon > 0$, where $g(\mathbf{x})$ is a bounded and positive definite $(m \times m)$-matrix-valued function periodic with respect to some lattice $\Gamma \subset \mathbb{R}^d$, and $b(\mathbf{D}) = \sum_{l=1}^d b_l D_l$ is a first-order DO. Here b_l are constant $(m \times n)$-matrices. It is assumed that $m \geqslant n$ and the symbol $b(\boldsymbol{\xi})$ has maximal rank.

Keywords: Homogenization, periodic differential operators, operator error estimates
2020 Mathematics Subject Classification: 35B27

In [2], it was shown that the resolvent $(A_\varepsilon + I)^{-1}$ converges in the operator norm in $L_2(\mathbb{R}^d; \mathbb{C}^n)$ to the resolvent of an effective operator A^0, and

$$\|(A_\varepsilon + I)^{-1} - (A^0 + I)^{-1}\|_{L_2(\mathbb{R}^d) \to L_2(\mathbb{R}^d)} \leqslant C\varepsilon. \tag{1.1}$$

The effective operator is given by $A^0 = b(\mathbf{D})^* g^0 b(\mathbf{D})$, where g^0 is a constant positive matrix called the *effective* matrix. In [16], a similar result was obtained for the parabolic semigroup

$$\|e^{-\tau A_\varepsilon} - e^{-\tau A^0}\|_{L_2(\mathbb{R}^d) \to L_2(\mathbb{R}^d)} \leqslant C(\tau)\varepsilon, \quad \tau > 0. \tag{1.2}$$

Estimates (1.1) and (1.2) are order-sharp. Such inequalities are called *operator error estimates* in homogenization theory.

A different approach to operator error estimates (the shift method) was developed by Zhikov and Pastukhova. In [19, 21, 22], estimates (1.1) and (1.2) were obtained for the operators of acoustics and elasticity. Further results were discussed in a survey [23].

The operator error estimates for the Schrödinger-type and hyperbolic equations were studied in [5] and in the recent works [6–8, 11, 17]. In operator terms, the behavior of the operator-valued functions $e^{-i\tau A_\varepsilon}$, $\cos(\tau A_\varepsilon^{1/2})$, $A_\varepsilon^{-1/2}\sin(\tau A_\varepsilon^{1/2})$, $\tau \in \mathbb{R}$, was investigated. It turned out that the nature of the results differed from the case of elliptic and parabolic equations: the type of the operator norm must be changed.

Let us dwell on the case of the operator exponential $e^{-i\tau A_\varepsilon}$. In [5], the following sharp-order estimate was proved:

$$\|e^{-i\tau A_\varepsilon} - e^{-i\tau A^0}\|_{H^3(\mathbb{R}^d) \to L_2(\mathbb{R}^d)} \leqslant C(1 + |\tau|)\varepsilon. \tag{1.3}$$

In [6, 17], it was shown that in the general case the result (1.3) is sharp both regarding the type of the operator norm and regarding the dependence of the estimate on τ (it is impossible to replace $(1 + |\tau|)$ on the right by $(1 + |\tau|)^\alpha$ with $\alpha < 1$). On the other hand, under some additional assumptions the result admits improvement:

$$\|e^{-i\tau A_\varepsilon} - e^{-i\tau A^0}\|_{H^2(\mathbb{R}^d) \to L_2(\mathbb{R}^d)} \leqslant C(1 + |\tau|)^{\frac{1}{2}}\varepsilon. \tag{1.4}$$

The operator-theoretic approach was applied to the higher-order operators A_ε in [10, 18]. It was assumed that the operator A_ε is given by

$$A_\varepsilon = b(\mathbf{D})^* g\left(\frac{\mathbf{x}}{\varepsilon}\right) b(\mathbf{D}), \quad \operatorname{ord} b(\mathbf{D}) = p \geqslant 2, \quad \varepsilon > 0, \tag{1.5}$$

where $g(\mathbf{x})$ is a periodic, bounded, and positive definite $(m \times m)$-matrix-valued function, and $b(\mathbf{D}) = \sum_{|\beta|=p} b_\beta D^\beta$. Here b_β are constant $(m \times n)$-matrices. It is assumed that $m \geqslant n$ and the symbol $b(\boldsymbol{\xi})$ has maximal rank. In [10, 18], an estimate of the form (1.1) for such operators A_ε was obtained. A more accurate approximation for the resolvent of A_ε was found recently in [14, 15]. The shift method was applied to homogenization of the elliptic higher-order operators in the papers [12, 13] by Pastukhova.

1.2 Main results

In the present paper, the behavior of the operator exponential $e^{-i\tau A_\varepsilon}$ for the operator A_ε of order $2p$ given by (1.5) is studied. Our main result is the following estimate:

$$\big\| e^{-i\tau A_\varepsilon} - e^{-i\tau A^0} \big\|_{H^{2p+1}(\mathbb{R}^d)\to L_2(\mathbb{R}^d)} \leqslant C(1+|\tau|)\varepsilon. \tag{1.6}$$

Here $A^0 = b(\mathbf{D})^* g^0 b(\mathbf{D})$ is the effective operator. By the interpolation with the obvious estimate $\| e^{-i\tau A_\varepsilon} - e^{-i\tau A^0} \|_{L_2\to L_2} \leqslant 2$, we also obtain "intermediate" results:

$$\big\| e^{-i\tau A_\varepsilon} - e^{-i\tau A^0} \big\|_{H^s(\mathbb{R}^d)\to L_2(\mathbb{R}^d)} \leqslant C(s)(1+|\tau|)^{\frac{s}{2p+1}} \varepsilon^{\frac{s}{2p+1}}, \quad 0 \leqslant s \leqslant 2p+1.$$

Under some additional assumptions formulated in terms of the spectral characteristics of $A = b(\mathbf{D})^* g(\mathbf{x}) b(\mathbf{D})$ near the bottom of the spectrum, it is proved that

$$\big\| e^{-i\tau A_\varepsilon} - e^{-i\tau A^0} \big\|_{H^{2p+2}(\mathbb{R}^d)\to L_2(\mathbb{R}^d)} \leqslant C(1+|\tau|)\varepsilon^2. \tag{1.7}$$

This means that the difference $e^{-i\tau A_\varepsilon} - e^{-i\tau A^0}$ is of order $O(\varepsilon^2)$ in a suitable norm. It should be noted that the imposed additional assumptions are valid automatically for a scalar operator A_ε (i.e., $n = 1$) with real-valued coefficients. By the interpolation, we deduce

$$\big\| e^{-i\tau A_\varepsilon} - e^{-i\tau A^0} \big\|_{H^s(\mathbb{R}^d)\to L_2(\mathbb{R}^d)} \leqslant C(s)(1+|\tau|)^{\frac{s}{2p+2}} \varepsilon^{s/(p+1)}, \quad 0 \leqslant s \leqslant 2p+2.$$

In particular, for $s = p + 1$ we have

$$\big\| e^{-i\tau A_\varepsilon} - e^{-i\tau A^0} \big\|_{H^{p+1}(\mathbb{R}^d)\to L_2(\mathbb{R}^d)} \leqslant C(1+|\tau|)^{\frac{1}{2}} \varepsilon. \tag{1.8}$$

This improves (1.6) regarding both the type of the norm and the dependence on τ.

We stress that, for the second-order operators A_ε, there is an analog of (1.8) (cf. (1.4)), but there is no analog of (1.7).

The above results are applied to homogenization of the Cauchy problem for the Schrödinger-type equation

$$i\,\partial_\tau \mathbf{u}_\varepsilon(\mathbf{x},\tau) = A_\varepsilon \mathbf{u}_\varepsilon(\mathbf{x},\tau) + \mathbf{F}(\mathbf{x},\tau), \quad \mathbf{u}_\varepsilon(\mathbf{x},0) = \boldsymbol{\phi}(\mathbf{x}).$$

1.3 Method

We rely on the operator-theoretic approach. By the scaling transformation, the problem is reduced to the study of the operator $e^{-i\tau\varepsilon^{-2p}A}$. Next, using the Floquet–Bloch theory, we expand A in the direct integral of the operators $A(\mathbf{k})$ acting in $L_2(\Omega;\mathbb{C}^n)$ and given by $b(\mathbf{D}+\mathbf{k})^* g(\mathbf{x}) b(\mathbf{D}+\mathbf{k})$ with periodic boundary conditions; here Ω is the cell of the lattice Γ. Since $A(\mathbf{k})$ is an analytic operator family with compact resolvent, it can be studied by means of the analytic perturbation theory with respect to the one-dimensional parameter $t = |\mathbf{k}|$. It turns out that only the spectral characteristics of $A(\mathbf{k})$ near the bottom of the spectrum are responsible for homogenization. It is convenient to study the family $A(\mathbf{k})$ in the framework of an abstract operator-theoretic scheme.

1.4 Plan of the paper

The paper consists of five sections. In Section 2, the abstract operator-theoretic method is developed. In Section 3, we introduce the class of operators A acting in $L_2(\mathbb{R}^d; \mathbb{C}^n)$ and describe the direct integral expansion for A. Section 4 is devoted to application of the abstract results to the operator family $A(\mathbf{k})$. In Section 5, using the results of Section 4, we obtain approximations for the operator exponential of A. Section 6 is devoted to homogenization problems: we deduce approximations for the exponential $e^{-i\tau A_\varepsilon}$ from the results of Section 5 and apply them to find approximations for the solutions of the Cauchy problem.

1.5 Notation

Let $\mathfrak{H}$ and $\mathfrak{H}_*$ be complex separable Hilbert spaces. By $(\cdot,\cdot)_\mathfrak{H}$ and $\|\cdot\|_\mathfrak{H}$ we denote the inner product and the norm in $\mathfrak{H}$, respectively; the symbol $\|\cdot\|_{\mathfrak{H}\to\mathfrak{H}_*}$ denotes the norm of a linear continuous operator acting from $\mathfrak{H}$ to $\mathfrak{H}_*$. Sometimes, we omit the indices. If $\mathfrak{N}$ is a subspace of $\mathfrak{H}$, then $\mathfrak{N}^\perp$ denotes its orthogonal complement. If A is a closed linear operator in $\mathfrak{H}$, its domain and kernel are denoted by $\operatorname{Dom} A$ and $\operatorname{Ker} A$, respectively; $\sigma(A)$ stands for the spectrum of A.

The inner product and the norm in $\mathbb{C}^n$ are denoted by $\langle\cdot,\cdot\rangle$ and $|\cdot|$, respectively, $\mathbf{1}_n = \mathbf{1}$ is the unit $(n\times n)$-matrix. We denote $\mathbf{x} = (x_1,\dots,x_d) \in \mathbb{R}^d$, $iD_j = \frac{\partial}{\partial x_j}$, $j = 1,\dots,d$, $\mathbf{D} = -i\nabla = (D_1,\dots,D_d)$.

The class L_2 of $\mathbb{C}^n$-valued functions in a domain $\mathcal{O}\subset\mathbb{R}^d$ is denoted by $L_2(\mathcal{O};\mathbb{C}^n)$. The Sobolev classes of $\mathbb{C}^n$-valued functions in a domain $\mathcal{O}$ are denoted by $H^s(\mathcal{O};\mathbb{C}^n)$. For $n = 1$, we write simply $L_2(\mathcal{O})$, $H^s(\mathcal{O})$, but sometimes we use such simple notation also for the spaces of vector-valued or matrix-valued functions.

2 Abstract operator-theoretic scheme

2.1 Polynomial nonnegative operator pencils

Let $\mathfrak{H}$ and $\mathfrak{H}_*$ be complex separable Hilbert spaces. Let $X(t)$ be a family of operators (a polynomial pencil) of the form

$$X(t) = \sum_{j=0}^{p} X_j t^j, \quad t\in\mathbb{R},\ p\in\mathbb{N},\ p\geqslant 2.$$

The operators $X(t)$ and X_j act from $\mathfrak{H}$ into $\mathfrak{H}_*$. It is assumed that the operator X_0 is densely defined and closed, while the operator X_p is defined on the whole $\mathfrak{H}$ and bounded. In addition, we impose the following conditions.

Condition 2.1. For any $j = 0,\dots,p$ and $t\in\mathbb{R}$, we have

$$\operatorname{Dom} X(t) = \operatorname{Dom} X_0 \subset \operatorname{Dom} X_j \subset \operatorname{Dom} X_p = \mathfrak{H}.$$

Condition 2.2. For any $j = 0, \dots, p-1$ and $u \in \operatorname{Dom} X_0$ we have

$$\|X_j u\|_{\mathfrak{H}_*} \leqslant C_0 \|X_0 u\|_{\mathfrak{H}_*},$$

where a constant $C_0 \geqslant 1$ is independent of j and u.

Under the above assumptions, the operator $X(t)$ is closed for $|t| \leqslant (2(p-1)C_0)^{-1}$. *Our main object* is the following family of nonnegative selfadjoint operators in $\mathfrak{H}$:

$$A(t) = X(t)^* X(t), \quad t \in \mathbb{R}, \ |t| \leqslant (2(p-1)C_0)^{-1}.$$

Denote $A(0) = X_0^* X_0 =: A_0$, $\mathfrak{N} := \operatorname{Ker} A_0 = \operatorname{Ker} X_0$, and $\mathfrak{N}_* := \operatorname{Ker} X_0^*$. Let P be the orthogonal projection of $\mathfrak{H}$ onto the subspace $\mathfrak{N}$, and let P_* be the orthogonal projection of $\mathfrak{H}_*$ onto the subspace $\mathfrak{N}_*$.

Condition 2.3. Suppose that the point $\lambda_0 = 0$ is an isolated point of the spectrum of A_0, and $n := \dim \mathfrak{N} < \infty$, $n \leqslant n_* := \dim \mathfrak{N}_* \leqslant \infty$.

By d^0 we denote the distance from the point $\lambda_0 = 0$ to $\sigma(A_0) \setminus \{\lambda_0\}$. Let $F(t,h)$ be the spectral projection of the operator $A(t)$ corresponding to the interval $[0,h]$. We fix a positive number $\delta \leqslant \min\{\frac{d^0}{36}, \frac{1}{4}\}$ and choose a number $t_0 > 0$ such that

$$t_0 \leqslant \delta^{\frac{1}{2}} C_1^{-1}, \quad \text{where } C_1 = \max\{(p-1)C_0, \|X_p\|\}. \tag{2.1}$$

Note that $t_0 \leqslant \frac{1}{2}$. The operator $X(t)$ is automatically closed for $|t| \leqslant t_0$, because $t_0 \leqslant (2(p-1)C_0)^{-1}$. According to [18, Proposition 3.10], for $|t| \leqslant t_0$ we have

$$F(t,\delta) = F(t,3\delta), \quad \operatorname{rank} F(t,\delta) = n.$$

This means that, for $|t| \leqslant t_0$, the operator $A(t)$ has exactly n eigenvalues counting their multiplicities on the interval $[0,\delta]$, and the interval $(\delta, 3\delta)$ is free of the spectrum. We write $F(t) := F(t,\delta)$.

2.2 Operators Z, R, and S

Let $\mathcal{D} = \operatorname{Dom} X_0 \cap \mathfrak{N}^\perp$. Obviously, $\mathcal{D}$ is a Hilbert space with the inner product $(f_1, f_2)_{\mathcal{D}} = (X_0 f_1, X_0 f_2)_{\mathfrak{H}_*}$, $f_1, f_2 \in \mathcal{D}$.

Let $u \in \mathfrak{H}_*$. Consider the equation $X_0^*(X_0 \psi - u) = 0$ for $\psi \in \mathcal{D}$, which is understood in the weak sense:

$$(X_0 \psi, X_0 \zeta)_{\mathfrak{H}_*} = (u, X_0 \zeta)_{\mathfrak{H}_*}, \quad \text{for all } \zeta \in \mathcal{D}. \tag{2.2}$$

The right-hand side of (2.2) is an antilinear continuous functional of $\zeta \in \mathcal{D}$. Hence, by the Riesz theorem, there exists a unique solution $\psi \in \mathcal{D}$, and $\|X_0 \psi\|_{\mathfrak{H}_*} \leqslant \|u\|_{\mathfrak{H}_*}$.

Now, let $\omega \in \mathfrak{N}$ and $u = -X_p \omega$. In this case, the solution of equation (2.2) is denoted by $\psi(\omega)$. We define a bounded linear operator $Z : \mathfrak{H} \to \mathcal{D}$ putting

$$Z\omega = \psi(\omega), \quad \omega \in \mathfrak{N}, \qquad Zv = 0, \quad v \in \mathfrak{N}^\perp.$$

Next, we define the operator $R : \mathfrak{N} \to \mathfrak{N}_*$ by the relation $R\omega = X_0\psi(\omega) + X_p\omega$, $\omega \in \mathfrak{N}$. Another representation for R is given by $R = P_* X_p|_{\mathfrak{N}}$. The selfadjoint operator $S = R^*R : \mathfrak{N} \to \mathfrak{N}$ is called the *spectral germ* of the operator family $A(t)$ at $t = 0$. The germ S is called *non-degenerate* if $\operatorname{Ker} S = \{0\}$.

2.3 Analytic branches of eigenvalues and eigenvectors of $A(t)$

According to the analytic perturbation theory (see [9] and also [10, 18]), for $|t| \leqslant t_0$ there exist real-analytic functions $\lambda_j(t)$ (the branches of eigenvalues) and real-analytic $\mathfrak{H}$-valued functions $\varphi_j(t)$ (the branches of eigenvectors) such that

$$A(t)\varphi_j(t) = \lambda_j(t)\varphi_j(t), \quad j = 1, \dots, n, \ |t| \leqslant t_0,$$

and the set $\{\varphi_j(t)\}_{j=1}^n$ forms an orthonormal basis in the space $F(t)\mathfrak{H}$ for $|t| \leqslant t_0$. For sufficiently small $t_* \in (0, t_0]$ we have the following convergent power series expansions (see [18, Theorem 3.15]):

$$\lambda_j(t) = \gamma_j t^{2p} + \mu_j t^{2p+1} + \cdots, \quad j = 1, \dots, n, \ |t| \leqslant t_*; \tag{2.3}$$

$$\varphi_j(t) = \omega_j + t\varphi_j^{(1)} + \cdots, \quad j = 1, \dots, n, \ |t| \leqslant t_*. \tag{2.4}$$

We have $\gamma_j \geqslant 0$, $\mu_j \in \mathbb{R}$. The set $\omega_1, \dots, \omega_n$ forms an orthonormal basis in $\mathfrak{N}$. The numbers γ_j and the vectors ω_j are eigenvalues and eigenvectors of the spectral germ: $S\omega_j = \gamma_j\omega_j$, $j = 1, \dots, n$.

2.4 Threshold approximations

The following statement was proved in [10, 18]. Below different constants depending only on p are denoted by $C(p)$.

Proposition 2.4. *Suppose that Conditions* 2.1, 2.2, *and* 2.3 *are satisfied. Then for* $|t| \leqslant t_0$ *we have*

$$\|F(t) - P\| \leqslant C_2|t|, \qquad C_2 = C(p)C_T, \tag{2.5}$$

$$\|A(t)F(t) - t^{2p}SP\| \leqslant C_3|t|^{2p+1}, \quad C_3 = C(p)C_T^{2p+1}. \tag{2.6}$$

Here $C_T = p\,C_0^2 + \|X_p\|^2\delta^{-1}$.

More accurate threshold approximations were found in [15, Theorem 3.2].

Proposition 2.5. *Suppose that Conditions* 2.1, 2.2, *and* 2.3 *are satisfied. Let*

$$G := (RP)^* X_1 Z + (X_1 Z)^* RP. \tag{2.7}$$

In terms of the expansions (2.3) *and* (2.4),

$$G = \sum_{j=1}^n \mu_j (\cdot, \omega_j)_{\mathfrak{H}}\omega_j + \sum_{j=1}^n \gamma_j \big((\cdot, \varphi_j^{(1)})_{\mathfrak{H}}\omega_j + (\cdot, \omega_j)_{\mathfrak{H}}\varphi_j^{(1)}\big).$$

Then for $|t| \leqslant t_0$ *we have*

$$\|F(t) - P\| \leqslant C_4 |t|^p, \qquad C_4 = C(p) C_T^p, \tag{2.8}$$

$$\|A(t)F(t) - t^{2p} SP - t^{2p+1} G\| \leqslant C_5 t^{2p+2}, \quad C_5 = C(p) C_T^{2p+2}. \tag{2.9}$$

2.5 Approximation for $e^{-i\tau A(t)}$

Proposition 2.6. *Denote*

$$J(t,\tau) := \big(e^{-i\tau A(t)} - e^{-i\tau t^{2p} SP}\big) P. \tag{2.10}$$

For $\tau \in \mathbb{R}$ *and* $|t| \leqslant t_0$ *we have*

$$\|J(t,\tau)\| \leqslant 2C_2|t| + C_3 |\tau| |t|^{2p+1}. \tag{2.11}$$

Proof. We put $E(t,\tau) := e^{-i\tau A(t)} F(t) - e^{-i\tau t^{2p} SP} P$,

$$\Sigma(t,\tau) := e^{i\tau t^{2p} SP} E(t,\tau) = e^{i\tau t^{2p} SP} F(t) e^{-i\tau A(t)} - P.$$

Obviously,

$$\|J(t,\tau)\| \leqslant \|E(t,\tau)\| + \|F(t) - P\|. \tag{2.12}$$

We have $\Sigma(t,0) = F(t) - P$ and

$$\Sigma'(t,\tau) := \frac{d\Sigma(t,\tau)}{d\tau} = i e^{i\tau t^{2p} SP} \big(t^{2p} SP - A(t)F(t)\big) F(t) e^{-i\tau A(t)}.$$

Since $\Sigma(t,\tau) = \Sigma(t,0) + \int_0^\tau \Sigma'(t,\rho)\, d\rho$, it follows that

$$\|E(t,\tau)\| = \|\Sigma(t,\tau)\| \leqslant \|F(t) - P\| + |\tau| \|t^{2p} SP - A(t)F(t)\|. \tag{2.13}$$

Combining this with the estimates in (2.5), (2.6) and (2.12), we arrive at the required estimate (2.11). ∎

In the case where $G = 0$, the result can be improved.

Proposition 2.7. *Let* G *be the operator* (2.7). *Suppose that* $G = 0$. *Let* $J(t,\tau)$ *be the operator* (2.10). *Then for* $\tau \in \mathbb{R}$ *and* $|t| \leqslant t_0$ *we have*

$$\|J(t,\tau)\| \leqslant 2C_4 |t|^p + C_5 |\tau| t^{2p+2}. \tag{2.14}$$

Proof. Estimate (2.14) follows from (2.8), (2.9), (2.12), (2.13) and the condition $G = 0$. ∎

2.6 Approximation for the operator $\exp(-i\tau\varepsilon^{-2p} A(t))$

Let $\varepsilon > 0$. We study the behavior of the operator $\exp(-i\tau\varepsilon^{-2p} A(t))$ for $\tau \in \mathbb{R}$ and small ε. Let us estimate the operator $J(t, \tau\varepsilon^{-2p})$ multiplied by the "smoothing factor" $\varepsilon^s (t^2 + \varepsilon^2)^{-\frac{s}{2}}$ with $s = 2p + 1$. (In applications to DOs, such multiplying turns into smoothing.)

Theorem 2.8. *Let $J(t,\tau)$ be the operator* (2.10). *For $\tau \in \mathbb{R}$, $\varepsilon > 0$, and $|t| \leqslant t_0$ we have*

$$\|J(t,\tau\varepsilon^{-2p})\|\frac{\varepsilon^{2p+1}}{(t^2+\varepsilon^2)^{p+\frac{1}{2}}} \leqslant (C_2 + C_3|\tau|)\varepsilon. \tag{2.15}$$

Proof. From (2.11) with τ replaced by $\tau\varepsilon^{-2p}$ it follows that

$$\begin{aligned}\|J(t,\tau\varepsilon^{-2p})\|\frac{\varepsilon^{2p+1}}{(t^2+\varepsilon^2)^{p+\frac{1}{2}}} &\leqslant \big(2C_2|t| + C_3|\tau|\varepsilon^{-2p}|t|^{2p+1}\big)\frac{\varepsilon^{2p+1}}{(t^2+\varepsilon^2)^{p+\frac{1}{2}}}\\ &\leqslant (C_2 + C_3|\tau|)\varepsilon,\end{aligned}$$

as desired. ∎

In the case where $G = 0$, this result can be improved.

Theorem 2.9. *Let $J(t,\tau)$ be the operator* (2.10). *Let G be the operator* (2.7). *Suppose that $G = 0$. Then for $\tau \in \mathbb{R}$, $\varepsilon > 0$, and $|t| \leqslant t_0$ we have*

$$\|J(t,\tau\varepsilon^{-2p})\|\frac{\varepsilon^{2p+2}}{(t^2+\varepsilon^2)^{p+1}} \leqslant \big(2C_4 t_0^{p-2} + C_5|\tau|\big)\varepsilon^2. \tag{2.16}$$

Proof. Estimate (2.16) follows from (2.14) with τ replaced by $\tau\varepsilon^{-2p}$:

$$\begin{aligned}\|J(t,\tau\varepsilon^{-2p})\|\frac{\varepsilon^{2p+2}}{(t^2+\varepsilon^2)^{p+1}} &\leqslant \big(2C_4|t|^p + C_5|\tau|\varepsilon^{-2p}t^{2p+2}\big)\frac{\varepsilon^{2p+2}}{(t^2+\varepsilon^2)^{p+1}}\\ &\leqslant 2C_4 t_0^{p-2}\varepsilon^2 + C_5|\tau|\varepsilon^2.\end{aligned}$$

We took into account that $p \geqslant 2$ and $|t| \leqslant t_0$. ∎

3 Periodic differential operators in $L_2(\mathbb{R}^d;\mathbb{C}^n)$

3.1 Lattices. The Gelfand transformation

Let $\mathbf{a}_1,\dots,\mathbf{a}_d$ be a basis in $\mathbb{R}^d$ generating the lattice Γ:

$$\Gamma = \left\{\mathbf{a} \in \mathbb{R}^d : \mathbf{a} = \sum_{j=1}^{d} l_j\mathbf{a}_j,\ l_j \in \mathbb{Z}\right\},$$

and let $\Omega \subset \mathbb{R}^d$ be the elementary cell of Γ:

$$\Omega = \left\{\mathbf{x} \in \mathbb{R}^d : \mathbf{x} = \sum_{j=1}^{d} \xi_j\mathbf{a}_j,\ 0 < \xi_j < 1\right\}.$$

The basis $\mathbf{b}_1,\dots,\mathbf{b}_d$ in $\mathbb{R}^d$ dual to $\mathbf{a}_1,\dots,\mathbf{a}_d$ is defined by the relations

$$\langle\mathbf{b}_i,\mathbf{a}_j\rangle = 2\pi\delta_{ij}.$$

This basis generates the lattice $\widetilde{\Gamma}$ dual to Γ. Let $\widetilde{\Omega}$ be the central Brillouin zone of $\widetilde{\Gamma}$ given by

$$\widetilde{\Omega} = \{\mathbf{k} \in \mathbb{R}^d : |\mathbf{k}| < |\mathbf{k} - \mathbf{b}|,\ 0 \neq \mathbf{b} \in \widetilde{\Gamma}\}.$$

We use the notation $|\Omega| = \operatorname{meas} \Omega$, $|\widetilde{\Omega}| = \operatorname{meas} \widetilde{\Omega}$. Note that $|\Omega||\widetilde{\Omega}| = (2\pi)^d$.

Let r_0 be the radius of the ball inscribed in $\operatorname{clos} \widetilde{\Omega}$. We have $2r_0 = \min_{0 \neq \mathbf{b} \in \widetilde{\Gamma}} |\mathbf{b}|$.

Below, $\widetilde{H}^s(\Omega)$ *stands for the subspace of all functions* $f \in H^s(\Omega)$ *such that the* Γ*-periodic extension of* f *to* $\mathbb{R}^d$ *belongs to* $H^s_{\mathrm{loc}}(\mathbb{R}^d)$.

Initially, the Gelfand transformation $\mathcal{U}$ is defined on the functions $\mathbf{v}$ belonging to the Schwartz class $\mathcal{S}(\mathbb{R}^d; \mathbb{C}^n)$ by the formula

$$\widetilde{\mathbf{v}}(\mathbf{k}, \mathbf{x}) = (\mathcal{U}\mathbf{v})(\mathbf{k}, \mathbf{x}) = |\widetilde{\Omega}|^{-\frac{1}{2}} \sum_{\mathbf{a} \in \Gamma} e^{-i\langle \mathbf{k}, \mathbf{x}+\mathbf{a}\rangle} \mathbf{v}(\mathbf{x} + \mathbf{a}), \quad \mathbf{x} \in \Omega,\ \mathbf{k} \in \widetilde{\Omega}.$$

Then $\mathcal{U}$ extends by continuity up to a *unitary mapping*

$$\mathcal{U} : L_2(\mathbb{R}^d; \mathbb{C}^n) \to \int_{\widetilde{\Omega}} \oplus L_2(\Omega; \mathbb{C}^n)\, d\mathbf{k} =: \mathcal{K}. \tag{3.1}$$

The relation $\mathbf{v} \in H^p(\mathbb{R}^d; \mathbb{C}^n)$ is equivalent to $\widetilde{\mathbf{v}} \in L_2(\widetilde{\Omega}; \widetilde{H}^p(\Omega; \mathbb{C}^n))$. Under the transformation $\mathcal{U}$, the operator of multiplication by a bounded Γ-periodic function in $L_2(\mathbb{R}^d; \mathbb{C}^n)$ turns into multiplication by the same function on the fibers of the direct integral $\mathcal{K}$ (see (3.1)). The linear DO $b(\mathbf{D})$ of order p applied to $\mathbf{v} \in H^p(\mathbb{R}^d; \mathbb{C}^n)$ turns into the operator $b(\mathbf{D} + \mathbf{k})$ applied to $\widetilde{\mathbf{v}}(\mathbf{k}, \cdot) \in \widetilde{H}^p(\Omega; \mathbb{C}^n)$.

3.2 Factorized operators of order $2p$

In $L_2(\mathbb{R}^d; \mathbb{C}^n)$, we consider an operator A formally given by the differential expression

$$A = b(\mathbf{D})^* g(\mathbf{x}) b(\mathbf{D}). \tag{3.2}$$

Here $g(\mathbf{x})$ is a Hermitian $(m \times m)$-matrix-valued function, in general, with complex entries. It is assumed that $g(\mathbf{x})$ is Γ-periodic, bounded, and positive definite:

$$g, g^{-1} \in L_\infty(\mathbb{R}^d), \quad g(\mathbf{x}) > 0. \tag{3.3}$$

The operator $b(\mathbf{D})$ is given by

$$b(\mathbf{D}) = \sum_{|\beta|=p} b_\beta \mathbf{D}^\beta,$$

where b_β are constant $(m \times n)$-matrices, in general, with complex entries. It is assumed that $m \geqslant n$ and that the symbol $b(\boldsymbol{\xi}) = \sum_{|\beta|=p} b_\beta \boldsymbol{\xi}^\beta$ satisfies $\operatorname{rank} b(\boldsymbol{\xi}) = n$ for $0 \neq \boldsymbol{\xi} \in \mathbb{R}^d$. This condition is equivalent to the estimates

$$\alpha_0 \mathbf{1}_n \leqslant b(\boldsymbol{\theta})^* b(\boldsymbol{\theta}) \leqslant \alpha_1 \mathbf{1}_n, \quad \boldsymbol{\theta} \in \mathbb{S}^{d-1},\ 0 < \alpha_0 \leqslant \alpha_1 < \infty, \tag{3.4}$$

with some constants α_0, α_1.

The precise definition of the operator A is given in terms of the quadratic form

$$\mathfrak{a}[\mathbf{u},\mathbf{u}] = \int_{\mathbb{R}^d} \langle g(\mathbf{x}) b(\mathbf{D})\mathbf{u}(\mathbf{x}), b(\mathbf{D})\mathbf{u}(\mathbf{x})\rangle \, d\mathbf{x}, \quad \mathbf{u} \in H^p(\mathbb{R}^d;\mathbb{C}^n). \tag{3.5}$$

By using the Fourier transform and (3.3)–(3.4), it is easy to check that

$$c_0 \int_{\mathbb{R}^d} |\mathbf{D}^p \mathbf{u}(\mathbf{x})|^2 \, d\mathbf{x} \leqslant \mathfrak{a}[\mathbf{u},\mathbf{u}] \leqslant c_1 \int_{\mathbb{R}^d} |\mathbf{D}^p \mathbf{u}(\mathbf{x})|^2 \, d\mathbf{x}, \quad \mathbf{u} \in H^p(\mathbb{R}^d;\mathbb{C}^n). \tag{3.6}$$

Here $|\mathbf{D}^p \mathbf{u}(\mathbf{x})|^2 := \sum_{|\beta|=p} |\mathbf{D}^\beta \mathbf{u}(\mathbf{x})|^2$. The constants c_0, c_1 are given by

$$c_0 = C(p)\alpha_0 \|g^{-1}\|^{-1}_{L_\infty}, \quad c_1 = C(p)\alpha_1 \|g\|_{L_\infty}. \tag{3.7}$$

Hence, the form (3.5) is closed and nonnegative. By definition, A is a selfadjoint operator in $L_2(\mathbb{R}^d;\mathbb{C}^n)$ generated by this form.

Note that the operator A can be written as

$$A = X^* X,$$

where $X : L_2(\mathbb{R}^d;\mathbb{C}^n) \to L_2(\mathbb{R}^d;\mathbb{C}^m)$ is a closed operator defined by

$$X = g^{\frac{1}{2}} b(\mathbf{D}), \quad \operatorname{Dom} X = H^p(\mathbb{R}^d;\mathbb{C}^n).$$

3.3 Operators $A(\mathbf{k})$ in $L_2(\Omega;\mathbb{C}^n)$

Let $\mathbf{k} \in \mathbb{R}^d$. In $L_2(\Omega;\mathbb{C}^n)$, we consider the quadratic form

$$\mathfrak{a}(\mathbf{k})[\mathbf{u},\mathbf{u}] = \int_{\Omega} \langle g(\mathbf{x}) b(\mathbf{D}+\mathbf{k})\mathbf{u}(\mathbf{x}), b(\mathbf{D}+\mathbf{k})\mathbf{u}(\mathbf{x})\rangle \, d\mathbf{x}, \quad \mathbf{u} \in \widetilde{H}^p(\Omega;\mathbb{C}^n). \tag{3.8}$$

By using the discrete Fourier transform and (3.3)–(3.4), it is easy to check that

$$\begin{aligned} c_0 \int_{\Omega} |(\mathbf{D}+\mathbf{k})^p \mathbf{u}(\mathbf{x})|^2 \, d\mathbf{x} &\leqslant \mathfrak{a}(\mathbf{k})[\mathbf{u},\mathbf{u}] \\ &\leqslant c_1 \int_{\Omega} |(\mathbf{D}+\mathbf{k})^p \mathbf{u}(\mathbf{x})|^2 \, d\mathbf{x}, \quad \mathbf{u} \in \widetilde{H}^p(\Omega;\mathbb{C}^n). \end{aligned}$$

Here c_0, c_1 are the same constants as in (3.6); see (3.7). Hence, the form (3.8) is closed and nonnegative. A selfadjoint operator in $L_2(\Omega;\mathbb{C}^n)$ corresponding to this form is denoted by $A(\mathbf{k})$. Formally, we have $A(\mathbf{k}) = b(\mathbf{D}+\mathbf{k})^* g(\mathbf{x}) b(\mathbf{D}+\mathbf{k})$.

Note that the operator $A(\mathbf{k})$ can be written as

$$A(\mathbf{k}) = X(\mathbf{k})^* X(\mathbf{k}),$$

where $X(\mathbf{k}) : L_2(\Omega;\mathbb{C}^n) \to L_2(\Omega;\mathbb{C}^m)$ is a closed operator defined by

$$X(\mathbf{k}) = g^{\frac{1}{2}} b(\mathbf{D}+\mathbf{k}), \quad \operatorname{Dom} X(\mathbf{k}) = \widetilde{H}^p(\Omega;\mathbb{C}^n).$$

3.4 Direct integral expansion for the operator A

Using the Gelfand transform $\mathcal{U}$ defined in Section 3.1, we expand the operator A in the direct integral of the operators $A(\mathbf{k})$. Let $\mathbf{v} \in H^p(\mathbb{R}^d; \mathbb{C}^n)$ and $\widetilde{\mathbf{v}}(\mathbf{k}, \mathbf{x}) = (\mathcal{U}\mathbf{v})(\mathbf{k}, \mathbf{x})$. Then $\widetilde{\mathbf{v}} \in L_2(\widetilde{\Omega}; \widetilde{H}^p(\Omega; \mathbb{C}^n))$ and

$$\mathfrak{a}[\mathbf{v}, \mathbf{v}] = \int_{\widetilde{\Omega}} \mathfrak{a}(\mathbf{k})[\widetilde{\mathbf{v}}(\mathbf{k}, \cdot), \widetilde{\mathbf{v}}(\mathbf{k}, \cdot)]\, d\mathbf{k}. \tag{3.9}$$

Conversely, if $\widetilde{\mathbf{v}} \in L_2(\widetilde{\Omega}; \widetilde{H}^p(\Omega; \mathbb{C}^n))$, then $\mathbf{v} = \mathcal{U}^{-1}\widetilde{\mathbf{v}} \in H^p(\mathbb{R}^d; \mathbb{C}^n)$ and identity (3.9) is fulfilled. This means that

$$\mathcal{U} A \mathcal{U}^{-1} = \int_{\widetilde{\Omega}} \oplus A(\mathbf{k})\, d\mathbf{k}. \tag{3.10}$$

4 Application of the abstract results to $A(\mathbf{k})$

4.1 Incorporation of the operators $A(\mathbf{k})$ in the abstract scheme

We apply the scheme of Section 2, putting $\mathfrak{H} = L_2(\Omega; \mathbb{C}^n)$ and $\mathfrak{H}_* = L_2(\Omega; \mathbb{C}^m)$.

We write $\mathbf{k}$ as $\mathbf{k} = t\boldsymbol{\theta}$, where $t = |\mathbf{k}|$ and $\boldsymbol{\theta} \in \mathbb{S}^{d-1}$. The roles of $X(t)$ and $A(t)$ are played by the operators $X(\mathbf{k}) =: X(t, \boldsymbol{\theta})$ and $A(\mathbf{k}) =: A(t, \boldsymbol{\theta})$. They depend on the one-dimensional parameter t and the additional parameter $\boldsymbol{\theta}$, which was absent in the abstract scheme. He have to take care about this and to prove estimates uniform in $\boldsymbol{\theta}$.

Let us check that all the assumptions of Section 2 are fulfilled. We have

$$X(\mathbf{k}) = g^{\frac{1}{2}} \sum_{|\beta|=p} b_\beta (\mathbf{D} + \mathbf{k})^\beta = g^{\frac{1}{2}} \sum_{|\beta|=p} b_\beta \sum_{\gamma \leqslant \beta} C_\beta^\gamma t^{|\beta-\gamma|} \boldsymbol{\theta}^{\beta-\gamma} \mathbf{D}^\gamma.$$

Hence, the operator $X(\mathbf{k}) =: X(t, \boldsymbol{\theta})$ can be written as

$$X(t, \boldsymbol{\theta}) = X_0 + \sum_{j=1}^{p} t^j X_j(\boldsymbol{\theta}),$$

where the operator $X_0 = g^{\frac{1}{2}} b(\mathbf{D})$, $\operatorname{Dom} X_0 = \widetilde{H}^p(\Omega; \mathbb{C}^n)$, is closed, the operators $X_1(\boldsymbol{\theta}), \dots, X_{p-1}(\boldsymbol{\theta})$ are given by

$$X_j(\boldsymbol{\theta}) = g^{\frac{1}{2}} \sum_{|\beta|=p} b_\beta \sum_{\substack{\gamma \leqslant \beta \\ |\gamma| = p-j}} C_\beta^\gamma \boldsymbol{\theta}^{\beta-\gamma} \mathbf{D}^\gamma, \quad \operatorname{Dom} X_j(\boldsymbol{\theta}) = \widetilde{H}^{p-j}(\Omega; \mathbb{C}^n), \tag{4.1}$$

and the operator $X_p(\boldsymbol{\theta}) = g^{\frac{1}{2}} b(\boldsymbol{\theta})$ is bounded from $\mathfrak{H}$ to $\mathfrak{H}_*$.

Obviously, Condition 2.1 is satisfied. Condition 2.2 is also satisfied with

$$C_0 = C(d, p) \alpha_1^{\frac{1}{2}} \alpha_0^{-\frac{1}{2}} \|g\|_{L_\infty}^{\frac{1}{2}} \|g^{-1}\|_{L_\infty}^{\frac{1}{2}} (1 + r_0^{-1})^{p-1}, \tag{4.2}$$

where $C(d, p)$ depends only on d and p; see [10, Proposition 5.2].

By (3.4), we obtain the uniform bound for the norm of $X_p(\boldsymbol{\theta})$:

$$\|X_p(\boldsymbol{\theta})\| \leqslant \alpha_1^{\frac{1}{2}} \|g\|_{L_\infty}^{\frac{1}{2}}, \quad \boldsymbol{\theta} \in \mathbb{S}^{d-1}. \tag{4.3}$$

Let $\mathfrak{N} = \operatorname{Ker} A(0) = \operatorname{Ker} X_0$. It is easy to check that $\mathfrak{N}$ consists of constant vector-valued functions (see [10, Proposition 5.1]):

$$\mathfrak{N} = \{\mathbf{u} \in L_2(\Omega; \mathbb{C}^n) : \mathbf{u}(\mathbf{x}) = \mathbf{c} \in \mathbb{C}^n\}. \tag{4.4}$$

So, $\dim \mathfrak{N} = n$. The orthogonal projection of $L_2(\Omega; \mathbb{C}^n)$ onto $\mathfrak{N}$ is the operator of the averaging over the cell:

$$P\mathbf{u} = |\Omega|^{-1} \int_{\Omega} \mathbf{u}(\mathbf{x})\, d\mathbf{x}, \quad \mathbf{u} \in L_2(\Omega; \mathbb{C}^n). \tag{4.5}$$

Let $\mathfrak{N}_* = \operatorname{Ker} X_0^*$ and $n_* = \dim \mathfrak{N}_*$. The condition $m \geqslant n$ ensures that $n \leqslant n_*$. Moreover, either $n_* = \infty$ (if $m > n$), or $n_* = n$ (if $m = n$). See [10, Section 5.1] for details.

Since the embedding of $\widetilde{H}^p(\Omega; \mathbb{C}^n)$ into $L_2(\Omega; \mathbb{C}^n)$ is compact, the spectrum of the operator $A(0)$ is discrete. The point $\lambda_0 = 0$ is an isolated eigenvalue of $A(0)$ of multiplicity n; the corresponding eigenspace $\mathfrak{N}$ is given by (4.4). Thus, Condition 2.3 is satisfied.

Let d^0 be the distance from the point $\lambda_0 = 0$ to the rest of the spectrum of $A(0)$. According to [10, (5.17)],

$$d^0 \geqslant \alpha_0 \|g^{-1}\|_{L_\infty}^{-1} (2r_0)^{2p}. \tag{4.6}$$

In Section 2.1 it was required to fix a positive number $\delta \leqslant \min\{\frac{d^0}{36}, \frac{1}{4}\}$. Using (4.6), we choose δ as follows:

$$\delta = \min\left\{\alpha_0 \|g^{-1}\|_{L_\infty}^{-1} \frac{(2r_0)^{2p}}{36}, \frac{1}{4}\right\}. \tag{4.7}$$

Next, the constant $C_1(\boldsymbol{\theta}) = \max\{(p-1)C_0, \|X_p(\boldsymbol{\theta})\|\}$ now depends on $\boldsymbol{\theta}$ (see the second relation in (2.1)). Using (4.3), we see that $C_1(\boldsymbol{\theta}) \leqslant C_1$, where

$$C_1 = \max\{(p-1)C_0, \alpha_1^{\frac{1}{2}} \|g\|_{L_\infty}^{\frac{1}{2}}\}. \tag{4.8}$$

Here C_0 is given by (4.2). According to (2.1), we fix a number $t_0 \leqslant \delta^{\frac{1}{2}} C_1(\boldsymbol{\theta})^{-1}$ as follows:

$$t_0 = \delta^{\frac{1}{2}} C_1^{-1}, \tag{4.9}$$

where δ and C_1 are defined by (4.7) and (4.8), respectively.

4.2 The operators $Z(\boldsymbol{\theta})$, $R(\boldsymbol{\theta})$, and $S(\boldsymbol{\theta})$

For the operator family $A(t, \boldsymbol{\theta})$, the operators Z, R, and S defined in Section 2.2 in the abstract setting depend on the parameter $\boldsymbol{\theta}$.

To describe these operators, we introduce the $(n \times m)$-matrix-valued function $\Lambda(\mathbf{x})$ which is a Γ-periodic solution of the following problem:

$$b(\mathbf{D})^* g(\mathbf{x})(b(\mathbf{D})\Lambda(\mathbf{x}) + \mathbf{1}_m) = 0, \quad \int_\Omega \Lambda(\mathbf{x})\, d\mathbf{x} = 0. \tag{4.10}$$

The equation is understood in the weak sense: for each $\mathbf{C} \in \mathbb{C}^m$ we have

$$\Lambda \mathbf{C} \in \widetilde{H}^p(\Omega; \mathbb{C}^n)$$

and

$$\int_\Omega \langle g(\mathbf{x})(b(\mathbf{D})\Lambda(\mathbf{x})\mathbf{C} + \mathbf{C}), b(\mathbf{D})\eta(\mathbf{x})\rangle\, d\mathbf{x} = 0, \quad \eta \in \widetilde{H}^p(\Omega; \mathbb{C}^n).$$

Then (cf. [10, Section 5.3])

$$Z(\boldsymbol{\theta}) = [\Lambda] b(\boldsymbol{\theta}) P, \tag{4.11}$$

where $[\Lambda]$ denotes the operator of multiplication by the matrix-valued function $\Lambda(\mathbf{x})$. The operator $R(\boldsymbol{\theta})$ is given by

$$R(\boldsymbol{\theta}) = [g^{\frac{1}{2}}(b(\mathbf{D})\Lambda + \mathbf{1}_m)] b(\boldsymbol{\theta})|_{\mathfrak{N}}. \tag{4.12}$$

Then (cf. [10, Section 5.3]), the spectral germ $S(\boldsymbol{\theta}) = R(\boldsymbol{\theta})^* R(\boldsymbol{\theta})$ acts in the subspace $\mathfrak{N}$ (see (4.4)) and is represented as

$$S(\boldsymbol{\theta}) = b(\boldsymbol{\theta})^* g^0 b(\boldsymbol{\theta}), \quad \boldsymbol{\theta} \in \mathbb{S}^{d-1}. \tag{4.13}$$

Here g^0 is the so-called *effective matrix* (of size $m \times m$) given by

$$g^0 = |\Omega|^{-1} \int_\Omega \widetilde{g}(\mathbf{x})\, d\mathbf{x}, \quad \widetilde{g}(\mathbf{x}) := g(\mathbf{x})(b(\mathbf{D})\Lambda(\mathbf{x}) + \mathbf{1}_m).$$

It turns out that the effective matrix g^0 is positive definite. So, the germ $S(\boldsymbol{\theta})$ is non-degenerate. We mention some properties of g^0; see [10, Propositions 5.3, 5.4, 5.5].

Proposition 4.1. *Denote*

$$\overline{g} = |\Omega|^{-1} \int_\Omega g(\mathbf{x})\, d\mathbf{x}, \quad \underline{g} = \left(|\Omega|^{-1} \int_\Omega g(\mathbf{x})^{-1}\, d\mathbf{x} \right)^{-1}.$$

The effective matrix g^0 satisfies the following estimates (the Voigt–Reuss bracketing):

$$\underline{g} \leqslant g^0 \leqslant \overline{g}.$$

In the case where $m = n$, we have $g^0 = \underline{g}$.

Proposition 4.2. *The following statements hold.*

(1) *Let $\mathbf{g}_k(\mathbf{x})$, $k = 1, \dots, m$, be the columns of the matrix $g(\mathbf{x})$. The relation $g^0 = \overline{g}$ is equivalent to the identities*

$$b(\mathbf{D})^* \mathbf{g}_k(\mathbf{x}) = 0, \quad k = 1, \dots, m. \tag{4.14}$$

(2) *Let* $\mathbf{l}_k(\mathbf{x})$, $k = 1, \dots, m$, *be the columns of the matrix* $g(\mathbf{x})^{-1}$. *The relation* $g^0 = \underline{g}$ *is equivalent to the representations*

$$\mathbf{l}_k(\mathbf{x}) = \mathbf{l}_k^0 + b(\mathbf{D})\mathbf{v}_k(\mathbf{x}), \quad \mathbf{l}_k^0 \in \mathbb{C}^m, \ \mathbf{v}_k \in \widetilde{H}^p(\Omega; \mathbb{C}^n), \quad k = 1, \dots, m. \tag{4.15}$$

Remark. In the case where $g^0 = \underline{g}$, the matrix $\widetilde{g}(\mathbf{x})$ is constant: $\widetilde{g}(\mathbf{x}) = g^0 = \underline{g}$.

4.3 The effective operator

By (4.13) and the homogeneity of the symbol $b(\mathbf{k})$, we have

$$S(\mathbf{k}) := t^{2p} S(\boldsymbol{\theta}) = b(\mathbf{k})^* g^0 b(\mathbf{k}), \quad \mathbf{k} \in \mathbb{R}^d. \tag{4.16}$$

Expression (4.16) is the symbol of the DO

$$A^0 = b(\mathbf{D})^* g^0 b(\mathbf{D}), \quad \operatorname{Dom} A^0 = H^{2p}(\mathbb{R}^d; \mathbb{C}^n), \tag{4.17}$$

which is called the *effective operator* for A.

Let $A^0(\mathbf{k})$ be the operator family in $L_2(\Omega; \mathbb{C}^n)$ corresponding to the operator A^0. Then $A^0(\mathbf{k})$ is given by the differential expression $b(\mathbf{D} + \mathbf{k})^* g^0 b(\mathbf{D} + \mathbf{k})$ on the domain $\widetilde{H}^{2p}(\Omega; \mathbb{C}^n)$.

By (4.5) and (4.16), we have

$$S(\mathbf{k}) P = A^0(\mathbf{k}) P. \tag{4.18}$$

4.4 The operator $G(\boldsymbol{\theta})$

For $A(t, \boldsymbol{\theta})$, the operator G defined by (2.7) in the abstract setting depends on $\boldsymbol{\theta}$:

$$G(\boldsymbol{\theta}) = (R(\boldsymbol{\theta})P)^* X_1(\boldsymbol{\theta}) Z(\boldsymbol{\theta}) + (X_1(\boldsymbol{\theta}) Z(\boldsymbol{\theta}))^* R(\boldsymbol{\theta}) P.$$

Let $B_1(\boldsymbol{\theta}; \mathbf{D})$ be the DO of order $p - 1$ such that $X_1(\boldsymbol{\theta}) = g^{\frac{1}{2}} B_1(\boldsymbol{\theta}; \mathbf{D})$ (see (4.1)). Then

$$B_1(\boldsymbol{\theta}; \mathbf{D}) = \sum_{|\beta| = p} b_\beta \sum_{\substack{\gamma \leqslant \beta \\ |\gamma| = p-1}} C_\beta^\gamma \boldsymbol{\theta}^{\beta - \gamma} \mathbf{D}^\gamma.$$

Using (4.11) and (4.12), we obtain

$$G(\boldsymbol{\theta}) = b(\boldsymbol{\theta})^* g^{(1)}(\boldsymbol{\theta}) b(\boldsymbol{\theta}) P, \tag{4.19}$$

where $g^{(1)}(\boldsymbol{\theta})$ is a Hermitian $(m \times m)$-matrix given by

$$g^{(1)}(\boldsymbol{\theta}) = |\Omega|^{-1} \int_\Omega \big(\widetilde{g}(\mathbf{x})^* B_1(\boldsymbol{\theta}; \mathbf{D}) \Lambda(\mathbf{x}) + (B_1(\boldsymbol{\theta}; \mathbf{D}) \Lambda(\mathbf{x}))^* \widetilde{g}(\mathbf{x})\big) \, d\mathbf{x}. \tag{4.20}$$

We distinguish some cases where the operator (4.19) is equal to zero.

Proposition 4.3. *The following statements hold.*

(1) *Suppose that relations* (4.14) *are satisfied. Then* $\Lambda(\mathbf{x}) = 0$, *whence* $g^{(1)}(\boldsymbol{\theta}) = 0$ *and* $G(\boldsymbol{\theta}) = 0$.

(2) *Suppose that relations* (4.15) *are satisfied. Then* $g^{(1)}(\boldsymbol{\theta}) = 0$ *and* $G(\boldsymbol{\theta}) = 0$.

(3) *Suppose that* $n = 1$ *and the matrices* $g(\mathbf{x})$, b_β, $|\beta| = p$, *have real entries. Then* $G(\boldsymbol{\theta}) = 0$ *for any* $\boldsymbol{\theta} \in \mathbb{S}^{d-1}$.

Proof. Obviously, if relations (4.14) are satisfied, then the solution $\Lambda(\mathbf{x})$ of problem (4.10) is equal to zero. From (4.20) it follows that $g^{(1)}(\boldsymbol{\theta}) = 0$. Then, by (4.19), we have $G(\boldsymbol{\theta}) = 0$.

If relations (4.15) are satisfied, then $\widetilde{g}(\mathbf{x}) = g^0 = \underline{g}$. Since the integral over the cell of the derivatives of a periodic function is equal to zero, it follows that

$$\int_\Omega B_1(\boldsymbol{\theta};\mathbf{D})\Lambda(\mathbf{x})\,d\mathbf{x} = 0$$

and hence, we have $g^{(1)}(\boldsymbol{\theta}) = 0$. Consequently, by (4.19), $G(\boldsymbol{\theta}) = 0$.

Now, suppose that $n = 1$ and the matrices $g(\mathbf{x})$, b_β, $|\beta| = p$, have real entries. Then, for p even, the solution $\Lambda(\mathbf{x})$ of problem (4.10) is a $(1 \times m)$-matrix with real entries. Hence, $\widetilde{g}(\mathbf{x}) = g(\mathbf{x})(b(\mathbf{D})\Lambda(\mathbf{x}) + \mathbf{1}_m)$ is an $(m \times m)$-matrix with real entries. Next, $B_1(\boldsymbol{\theta};\mathbf{D})\Lambda(\mathbf{x})$ is an $(m \times m)$-matrix with imaginary entries. By (4.20), $g^{(1)}(\boldsymbol{\theta})$ is a Hermitian $(m \times m)$-matrix with imaginary entries. Consequently, $b(\boldsymbol{\theta})^* g^{(1)}(\boldsymbol{\theta}) b(\boldsymbol{\theta})$ is equal to zero, as a Hermitian imaginary (1×1)-matrix.

For p odd, the solution $\Lambda(\mathbf{x})$ of problem (4.10) is a $(1 \times m)$-matrix with imaginary entries. The matrix $\widetilde{g}(\mathbf{x})$ has real entries. Next, $B_1(\boldsymbol{\theta};\mathbf{D})\Lambda(\mathbf{x})$ is an $(m \times m)$-matrix with imaginary entries. Hence, $g^{(1)}(\boldsymbol{\theta})$ is a Hermitian $(m \times m)$-matrix with imaginary entries. Again, $b(\boldsymbol{\theta})^* g^{(1)}(\boldsymbol{\theta}) b(\boldsymbol{\theta})$ is equal to zero, as a Hermitian imaginary (1×1)-matrix. ∎

Remark. In the general case, the operator $G(\boldsymbol{\theta})$ may be nonzero. In particular, it is easy to give examples of the scalar operator $A = b(\mathbf{D})^* g(\mathbf{x}) b(\mathbf{D})$ (i.e., $n = 1$), where $g(\mathbf{x})$ is a Hermitian matrix with complex entries, such that the corresponding operator $G(\boldsymbol{\theta})$ is not zero.

4.5 Approximation for the operator exponential of $A(\mathbf{k})$

In $L_2(\mathbb{R}^d;\mathbb{C}^n)$, consider the operator $H_0 := -\Delta$. Let $H_0(\mathbf{k})$ be the operator family in $L_2(\Omega;\mathbb{C}^n)$ corresponding to the operator H_0. Then $H_0(\mathbf{k})$ is given by the differential expression $|\mathbf{D} + \mathbf{k}|^2$ on the domain $\widetilde{H}^2(\Omega;\mathbb{C}^n)$. Denote

$$R_0(\mathbf{k},\varepsilon) := \varepsilon^2 (H_0(\mathbf{k}) + \varepsilon^2 I)^{-1}. \tag{4.21}$$

Clearly, we have

$$R_0(\mathbf{k},\varepsilon)^{\frac{s}{2}} P := \varepsilon^s (t^2 + \varepsilon^2)^{-\frac{s}{2}} P, \quad t = |\mathbf{k}|,\ s > 0. \tag{4.22}$$

We apply Theorem 2.8 to the operator family $A(t,\boldsymbol{\theta}) = A(\mathbf{k})$. Note that, by (4.16) and (4.18), $e^{-i\tau t^{2p} S(\boldsymbol{\theta})P}\,P = e^{-i\tau A^0(\mathbf{k})}\,P$. Thus, the operator (2.10) turns into

$$J(\mathbf{k},\tau) := \big(e^{-i\tau A(\mathbf{k})} - e^{-i\tau A^0(\mathbf{k})}\big)P.$$

It remains to implement the values of the constants in estimates. The constants δ and t_0 are given by (4.7) and (4.9), respectively; they do not depend on $\boldsymbol{\theta}$. The constant C_0 is given by (4.2). Using (4.3), we can replace the constant

$$C_T(\boldsymbol{\theta}) = pC_0^2 + \|X_p(\boldsymbol{\theta})\|^2\delta^{-1}$$

depending now on $\boldsymbol{\theta}$ by $C_T = pC_0^2 + \alpha_1\|g\|_{L_\infty}\delta^{-1}$. According to (2.5) and (2.6), we put $C_2 = C(p)C_T$, $C_3 = C(p)C_T^{2p+1}$.

Now, applying (2.15) and taking (4.22) into account, we obtain for $\tau \in \mathbb{R}$, $\varepsilon > 0$, $|\mathbf{k}| \leqslant t_0$,

$$\big\|J(\mathbf{k},\tau\varepsilon^{-2p})R_0(\mathbf{k},\varepsilon)^{p+\frac{1}{2}}P\big\|_{L_2(\Omega)\to L_2(\Omega)} \leqslant (C_2 + C_3|\tau|)\varepsilon. \tag{4.23}$$

For $\mathbf{k} \in \widetilde{\Omega}$, $|\mathbf{k}| > t_0$, the estimates are trivial. Obviously, we have $\|J(\mathbf{k},\tau)\| \leqslant 2$ and $\|R_0(\mathbf{k},\varepsilon)\| \leqslant 1$. By (4.22),

$$\big\|R_0(\mathbf{k},\varepsilon)^{\frac{1}{2}}P\big\|_{L_2(\Omega)\to L_2(\Omega)} \leqslant t_0^{-1}\varepsilon, \quad \mathbf{k}\in\widetilde{\Omega},\ \ |\mathbf{k}| > t_0,\ \ \varepsilon > 0.$$

Hence, for $\tau \in \mathbb{R}$, $\mathbf{k} \in \widetilde{\Omega}$, $|\mathbf{k}| > t_0$, $\varepsilon > 0$,

$$\big\|J(\mathbf{k},\tau\varepsilon^{-2p})R_0(\mathbf{k},\varepsilon)^{p+\frac{1}{2}}P\big\|_{L_2(\Omega)\to L_2(\Omega)} \leqslant 2\,t_0^{-1}\varepsilon. \tag{4.24}$$

Finally, let us show that, within the margin of error, the projection P in estimates (4.23) and (4.24) can be removed. Indeed, using the discrete Fourier transform, we have for $\mathbf{k} \in \widetilde{\Omega}$ and $\varepsilon > 0$,

$$\big\|R_0(\mathbf{k},\varepsilon)^{\frac{1}{2}}(I-P)\big\|_{L_2(\Omega)\to L_2(\Omega)} = \max_{0\neq\mathbf{b}\in\widetilde{\Gamma}} \varepsilon(|\mathbf{b}+\mathbf{k}|^2+\varepsilon^2)^{-\frac{1}{2}} \leqslant r_0^{-1}\varepsilon. \tag{4.25}$$

Consequently, for $\tau \in \mathbb{R}$, $\mathbf{k} \in \widetilde{\Omega}$, $\varepsilon > 0$,

$$\big\|\big(e^{-i\tau\varepsilon^{-2p}A(\mathbf{k})} - e^{-i\tau\varepsilon^{-2p}A^0(\mathbf{k})}\big)R_0(\mathbf{k},\varepsilon)^{p+\frac{1}{2}}(I-P)\big\|_{L_2(\Omega)\to L_2(\Omega)} \leqslant 2r_0^{-1}\varepsilon. \tag{4.26}$$

Combining (4.23), (4.24) and (4.26), we arrive at the following result.

Theorem 4.4. *For $\tau \in \mathbb{R}$, $\varepsilon > 0$, and $\mathbf{k} \in \widetilde{\Omega}$ we have*

$$\big\|\big(e^{-i\tau\varepsilon^{-2p}A(\mathbf{k})} - e^{-i\tau\varepsilon^{-2p}A^0(\mathbf{k})}\big)R_0(\mathbf{k},\varepsilon)^{p+\frac{1}{2}}\big\|_{L_2(\Omega)\to L_2(\Omega)} \leqslant \mathfrak{C}_1(1+|\tau|)\varepsilon.$$

The constant $\mathfrak{C}_1$ depends only on d, p, α_0, α_1, $\|g\|_{L_\infty}$, $\|g^{-1}\|_{L_\infty}$, and the parameters of the lattice Γ.

Note that, by (4.22),

$$\|R_0(\mathbf{k},\varepsilon)P\| \leqslant t_0^{-2}\varepsilon^2, \quad \mathbf{k}\in\widetilde{\Omega},\ |\mathbf{k}|>t_0,\ \varepsilon>0. \tag{4.27}$$

Similarly to (4.25), we have

$$\|R_0(\mathbf{k},\varepsilon)(I-P)\| = \max_{0\neq\mathbf{b}\in\widetilde{\Gamma}} \varepsilon^2(|\mathbf{b}+\mathbf{k}|^2+\varepsilon^2)^{-1} \leqslant r_0^{-2}\varepsilon^2, \quad \mathbf{k}\in\widetilde{\Omega},\ \varepsilon>0. \tag{4.28}$$

Applying Theorem 2.9 and using (4.27)–(4.28), we arrive at the following result.

Theorem 4.5. *Let $G(\boldsymbol{\theta})$ be the operator given by* (4.19) *and* (4.20). *Suppose that $G(\boldsymbol{\theta})=0$ for any $\boldsymbol{\theta}\in\mathbb{S}^{d-1}$. Then for $\tau\in\mathbb{R}$, $\varepsilon>0$ and $\mathbf{k}\in\widetilde{\Omega}$ we have*

$$\big\|\big(e^{-i\tau\varepsilon^{-2p}A(\mathbf{k})}-e^{-i\tau\varepsilon^{-2p}A^0(\mathbf{k})}\big)R_0(\mathbf{k},\varepsilon)^{p+1}\big\|_{L_2(\Omega)\to L_2(\Omega)} \leqslant \mathfrak{C}_2(1+|\tau|)\varepsilon^2.$$

The constant $\mathfrak{C}_2$ depends only on d, p, α_0, α_1, $\|g\|_{L_\infty}$, $\|g^{-1}\|_{L_\infty}$, and the parameters of the lattice Γ.

5 Approximation for the operator exponential of A

Let A be the operator in $L_2(\mathbb{R}^d;\mathbb{C}^n)$ given by $A=b(\mathbf{D})^*g(\mathbf{x})b(\mathbf{D})$; see (3.2). Let $A^0=b(\mathbf{D})^*g^0b(\mathbf{D})$ be the effective operator (4.17). Recall the notation $H_0=-\Delta$ and denote

$$R_0(\varepsilon) := \varepsilon^2(H_0+\varepsilon^2 I)^{-1}. \tag{5.1}$$

From expansion (3.10) it follows that

$$e^{-i\tau\varepsilon^{-2p}A} = \mathcal{U}^{-1}\Big(\int_{\widetilde{\Omega}}\oplus e^{-i\tau\varepsilon^{-2p}A(\mathbf{k})}\,d\mathbf{k}\Big)\mathcal{U}.$$

The operator $e^{-i\tau\varepsilon^{-2p}A^0}$ admits a similar expansion. The operator (5.1) is decomposed into the direct integral of the operators (4.21):

$$R_0(\varepsilon) = \mathcal{U}^{-1}\Big(\int_{\widetilde{\Omega}}\oplus R_0(\mathbf{k},\varepsilon)\,d\mathbf{k}\Big)\mathcal{U}.$$

From these direct integral expansions, taking into account that $\mathcal{U}$ is unitary, we obtain

$$\begin{aligned}&\big\|\big(e^{-i\tau\varepsilon^{-2p}A}-e^{-i\tau\varepsilon^{-2p}A^0}\big)R_0(\varepsilon)^{\frac{s}{2}}\big\|_{L_2(\mathbb{R}^d)\to L_2(\mathbb{R}^d)}\\ &\quad= \sup_{\mathbf{k}\in\widetilde{\Omega}}\big\|\big(e^{-i\tau\varepsilon^{-2p}A(\mathbf{k})}-e^{-i\tau\varepsilon^{-2p}A^0(\mathbf{k})}\big)R_0(\mathbf{k},\varepsilon)^{\frac{s}{2}}\big\|_{L_2(\Omega)\to L_2(\Omega)}.\end{aligned}$$

Combining this with Theorem 4.4, we arrive at the following result.

Theorem 5.1. *For $\tau\in\mathbb{R}$ and $\varepsilon>0$ we have*

$$\big\|\big(e^{-i\tau\varepsilon^{-2p}A}-e^{-i\tau\varepsilon^{-2p}A^0}\big)R_0(\varepsilon)^{p+\frac{1}{2}}\big\|_{L_2(\mathbb{R}^d)\to L_2(\mathbb{R}^d)} \leqslant \mathfrak{C}_1(1+|\tau|)\varepsilon. \tag{5.2}$$

Similarly, Theorem 4.5 implies the following result.

Theorem 5.2. *Let $G(\boldsymbol{\theta})$ be the operator given by* (4.19) *and* (4.20). *Suppose that $G(\boldsymbol{\theta}) = 0$ for any $\boldsymbol{\theta} \in \mathbb{S}^{d-1}$. Then for $\tau \in \mathbb{R}$ and $\varepsilon > 0$ we have*

$$\left\| \left(e^{-i\tau\varepsilon^{-2p}A} - e^{-i\tau\varepsilon^{-2p}A^0}\right) R_0(\varepsilon)^{p+1} \right\|_{L_2(\mathbb{R}^d)\to L_2(\mathbb{R}^d)} \leqslant \mathfrak{C}_2(1+|\tau|)\varepsilon^2.$$

6 Homogenization of the Schrödinger-type equation

6.1 The operator A_ε. The scaling transformation

We use the notation $g^\varepsilon(\mathbf{x}) := g(\varepsilon^{-1}\mathbf{x})$, $\varepsilon > 0$. Our main object is the operator A_ε acting in $L_2(\mathbb{R}^d;\mathbb{C}^n)$ and formally given by

$$A_\varepsilon = b(\mathbf{D})^* g^\varepsilon(\mathbf{x}) b(\mathbf{D}).$$

The precise definition of A_ε is given in terms of the corresponding quadratic form; cf. Section 3.2. Our goal is to approximate the operator exponential $e^{-i\tau A_\varepsilon}$ for small ε.

Let T_ε be the scaling transformation defined by $(T_\varepsilon \mathbf{u})(\mathbf{x}) = \varepsilon^{\frac{d}{2}}\mathbf{u}(\varepsilon\mathbf{x})$. Then T_ε is unitary in $L_2(\mathbb{R}^d;\mathbb{C}^n)$. We have $A_\varepsilon = \varepsilon^{-2p} T_\varepsilon^* A T_\varepsilon$. Hence,

$$e^{-i\tau A_\varepsilon} = T_\varepsilon^* e^{-i\tau\varepsilon^{-2p}A} T_\varepsilon.$$

A similar relation holds for the effective operator A^0:

$$e^{-i\tau A^0} = T_\varepsilon^* e^{-i\tau\varepsilon^{-2p}A^0} T_\varepsilon.$$

Applying the scaling transformation to the resolvent of $H_0 = -\Delta$, we obtain

$$(H_0 + I)^{-1} = \varepsilon^2 T_\varepsilon^* (H_0 + \varepsilon^2 I)^{-1} T_\varepsilon = T_\varepsilon^* R_0(\varepsilon) T_\varepsilon.$$

Using these relations and taking into account that T_ε is unitary, we have

$$\begin{aligned} &\left\| \left(e^{-i\tau A_\varepsilon} - e^{-i\tau A^0}\right)(H_0+I)^{-\frac{s}{2}} \right\|_{L_2(\mathbb{R}^d)\to L_2(\mathbb{R}^d)} \\ &\quad = \left\| \left(e^{-i\tau\varepsilon^{-2p}A} - e^{-i\tau\varepsilon^{-2p}A^0}\right) R_0(\varepsilon)^{\frac{s}{2}} \right\|_{L_2(\mathbb{R}^d)\to L_2(\mathbb{R}^d)}, \quad s > 0. \end{aligned} \tag{6.1}$$

6.2 Approximation for the operator exponential $e^{-i\tau A_\varepsilon}$

Combining (6.1) and (5.2), we see that for $\tau \in \mathbb{R}$, $\varepsilon > 0$,

$$\left\| \left(e^{-i\tau A_\varepsilon} - e^{-i\tau A^0}\right)(H_0+I)^{-(p+\frac{1}{2})} \right\|_{L_2(\mathbb{R}^d)\to L_2(\mathbb{R}^d)} \leqslant \mathfrak{C}_1(1+|\tau|)\varepsilon.$$

As $(H_0+I)^{p+\frac{1}{2}}$ is an isometric isomorphism of the Sobolev space $H^{2p+1}(\mathbb{R}^d;\mathbb{C}^n)$ onto $L_2(\mathbb{R}^d;\mathbb{C}^n)$, this yields the following result.

Theorem 6.1. *For $\tau \in \mathbb{R}$ and $\varepsilon > 0$ we have*

$$\left\| e^{-i\tau A_\varepsilon} - e^{-i\tau A^0} \right\|_{H^{2p+1}(\mathbb{R}^d) \to L_2(\mathbb{R}^d)} \leqslant \mathfrak{C}_1 (1 + |\tau|) \varepsilon. \tag{6.2}$$

The constant $\mathfrak{C}_1$ depends only on d, p, α_0, α_1, $\|g\|_{L_\infty}$, $\|g^{-1}\|_{L_\infty}$, and the parameters of the lattice Γ.

Interpolating between the obvious estimate

$$\left\| e^{-i\tau A_\varepsilon} - e^{-i\tau A^0} \right\|_{L_2(\mathbb{R}^d) \to L_2(\mathbb{R}^d)} \leqslant 2$$

and (6.2), we obtain the following corollary.

Corollary 6.2. *Let $0 \leqslant s \leqslant 2p + 1$. For $\tau \in \mathbb{R}$ and $\varepsilon > 0$ we have*

$$\left\| e^{-i\tau A_\varepsilon} - e^{-i\tau A^0} \right\|_{H^s(\mathbb{R}^d) \to L_2(\mathbb{R}^d)} \leqslant C(s) (1 + |\tau|)^{\frac{s}{2p+1}} \varepsilon^{\frac{s}{2p+1}}.$$

Here $C(s) = 2^{1 - \frac{s}{2p+1}} \mathfrak{C}_1^{\frac{s}{2p+1}}$.

Similarly, by the scaling transformation, we deduce the following result from Theorem 5.2.

Theorem 6.3. *Let $G(\boldsymbol{\theta})$ be the operator given by* (4.19) *and* (4.20). *Suppose that $G(\boldsymbol{\theta}) = 0$ for any $\boldsymbol{\theta} \in \mathbb{S}^{d-1}$. Then for $\tau \in \mathbb{R}$ and $\varepsilon > 0$ we have*

$$\left\| e^{-i\tau A_\varepsilon} - e^{-i\tau A^0} \right\|_{H^{2p+2}(\mathbb{R}^d) \to L_2(\mathbb{R}^d)} \leqslant \mathfrak{C}_2 (1 + |\tau|) \varepsilon^2.$$

The constant $\mathfrak{C}_2$ depends only on d, p, α_0, α_1, $\|g\|_{L_\infty}$, $\|g^{-1}\|_{L_\infty}$, and the parameters of the lattice Γ.

Corollary 6.4. *Under the assumptions of Theorem* 6.3, *let $0 \leqslant s \leqslant 2p + 2$. For $\tau \in \mathbb{R}$ and $\varepsilon > 0$ we have*

$$\left\| e^{-i\tau A_\varepsilon} - e^{-i\tau A^0} \right\|_{H^s(\mathbb{R}^d) \to L_2(\mathbb{R}^d)} \leqslant C'(s) (1 + |\tau|)^{\frac{s}{2p+2}} \varepsilon^{\frac{s}{p+1}}.$$

Here $C'(s) = 2^{1 - \frac{s}{2p+2}} \mathfrak{C}_2^{\frac{s}{2p+2}}$. In particular, for $s = p + 1$

$$\left\| e^{-i\tau A_\varepsilon} - e^{-i\tau A^0} \right\|_{H^{p+1}(\mathbb{R}^d) \to L_2(\mathbb{R}^d)} \leqslant C'(p + 1) (1 + |\tau|)^{\frac{1}{2}} \varepsilon. \tag{6.3}$$

Recall that some sufficient conditions ensuring that $G(\boldsymbol{\theta}) \equiv 0$ are given in Proposition 4.3.

Remark. Theorem 6.3 shows that, if $G(\boldsymbol{\theta}) \equiv 0$, the difference $e^{-i\tau A_\varepsilon} - e^{-i\tau A^0}$ is of order $O(\varepsilon^2)$ in a suitable norm. Estimate (6.3) improves (6.2) regarding both the norm type and the dependence of the estimate on τ. We note that for the second-order operators there is analog of estimate (6.3) (see [6]), but there is no analog of Theorem 6.3.

6.3 Homogenization of the Cauchy problem for the Schrödinger-type equation

Let $\mathbf{u}_\varepsilon(\mathbf{x}, \tau)$ be the solution of the following Cauchy problem:

$$\begin{aligned} i\partial_\tau \mathbf{u}_\varepsilon(\mathbf{x},\tau) &= b(\mathbf{D})^* g^\varepsilon(\mathbf{x}) b(\mathbf{D}) \mathbf{u}_\varepsilon(\mathbf{x},\tau) + \mathbf{F}(\mathbf{x},\tau), \quad && \mathbf{x}\in\mathbb{R}^d,\ \tau\in\mathbb{R}, \\ \mathbf{u}_\varepsilon(\mathbf{x},0) &= \boldsymbol{\phi}(\mathbf{x}), && \mathbf{x}\in\mathbb{R}^d, \end{aligned} \tag{6.4}$$

where $\boldsymbol{\phi} \in L_2(\mathbb{R}^d;\mathbb{C}^n)$ and $\mathbf{F} \in L_{1,\mathrm{loc}}(\mathbb{R}; L_2(\mathbb{R}^d;\mathbb{C}^n))$. The solution of problem (6.4) admits the following representation:

$$\mathbf{u}_\varepsilon(\cdot,\tau) = e^{-i\tau A_\varepsilon}\boldsymbol{\phi} - i\int_0^\tau e^{-i(\tau-\widetilde{\tau})A_\varepsilon}\mathbf{F}(\cdot,\widetilde{\tau})\,d\widetilde{\tau}. \tag{6.5}$$

Let $\mathbf{u}_0(\mathbf{x}, \tau)$ be the solution of the homogenized Cauchy problem:

$$\begin{aligned} i\partial_\tau \mathbf{u}_0(\mathbf{x},\tau) &= b(\mathbf{D})^* g^0 b(\mathbf{D}) \mathbf{u}_0(\mathbf{x},\tau) + \mathbf{F}(\mathbf{x},\tau), \quad && \mathbf{x}\in\mathbb{R}^d,\ \tau\in\mathbb{R}, \\ \mathbf{u}_0(\mathbf{x},0) &= \boldsymbol{\phi}(\mathbf{x}), && \mathbf{x}\in\mathbb{R}^d. \end{aligned} \tag{6.6}$$

The solution of problem (6.6) can be represented as

$$\mathbf{u}_0(\cdot,\tau) = e^{-i\tau A^0}\boldsymbol{\phi} - i\int_0^\tau e^{-i(\tau-\widetilde{\tau})A^0}\mathbf{F}(\cdot,\widetilde{\tau})\,d\widetilde{\tau}. \tag{6.7}$$

Theorem 6.5. *Let $\mathbf{u}_\varepsilon(\mathbf{x}, \tau)$ be the solution of the Cauchy problem* (6.4). *Let $\mathbf{u}_0(\mathbf{x}, \tau)$ be the solution of the homogenized problem* (6.6).

(1) *Let $0 \leqslant s \leqslant 2p + 1$. If $\boldsymbol{\phi} \in H^s(\mathbb{R}^d;\mathbb{C}^n)$ and $\mathbf{F} \in L_{1,\mathrm{loc}}(\mathbb{R}; H^s(\mathbb{R}^d;\mathbb{C}^n))$, then for $\tau \in \mathbb{R}$ and $\varepsilon > 0$ we have*

$$\begin{aligned} \|\mathbf{u}_\varepsilon(\cdot,\tau) - \mathbf{u}_0(\cdot,\tau)\|_{L_2(\mathbb{R}^d)} &\leqslant C(s)(1+|\tau|)^{\frac{s}{2p+1}}\varepsilon^{\frac{s}{2p+1}} \\ &\quad\times \big(\|\boldsymbol{\phi}\|_{H^s(\mathbb{R}^d)} + \|\mathbf{F}\|_{L_1((0,\tau);H^s(\mathbb{R}^d))}\big). \end{aligned} \tag{6.8}$$

(2) *If $\boldsymbol{\phi} \in L_2(\mathbb{R}^d;\mathbb{C}^n)$ and $\mathbf{F} \in L_{1,\mathrm{loc}}(\mathbb{R}; L_2(\mathbb{R}^d;\mathbb{C}^n))$, then for $\tau \in \mathbb{R}$ we have*

$$\lim_{\varepsilon\to 0}\|\mathbf{u}_\varepsilon(\cdot,\tau) - \mathbf{u}_0(\cdot,\tau)\|_{L_2(\mathbb{R}^d)} = 0.$$

Proof. Statement (1) follows from Corollary 6.2 and representations (6.5), (6.7).

Estimate (6.8) with $s = 0$ means that the norm $\|\mathbf{u}_\varepsilon(\cdot,\tau) - \mathbf{u}_0(\cdot,\tau)\|_{L_2(\mathbb{R}^d)}$ is uniformly bounded provided $\boldsymbol{\phi} \in L_2(\mathbb{R}^d;\mathbb{C}^n)$ and $\mathbf{F} \in L_{1,\mathrm{loc}}(\mathbb{R}; L_2(\mathbb{R}^d;\mathbb{C}^n))$. Hence, using statement (1) with $s = 2p + 1$ and applying the Banach–Steinhaus theorem, we obtain statement (2). ■

Similarly, from Corollary 6.4 we deduce the following result.

Theorem 6.6. *Let $G(\boldsymbol{\theta})$ be the operator given by* (4.19) *and* (4.20). *Suppose that $G(\boldsymbol{\theta}) = 0$ for any $\boldsymbol{\theta} \in \mathbb{S}^{d-1}$. Let $\mathbf{u}_\varepsilon(\mathbf{x}, \tau)$ be the solution of the Cauchy problem* (6.4). *Let $\mathbf{u}_0(\mathbf{x}, \tau)$ be the solution of the homogenized problem* (6.6). *Let $0 \leqslant s \leqslant 2p + 2$.*

If $\boldsymbol{\phi} \in H^s(\mathbb{R}^d;\mathbb{C}^n)$ *and* $\mathbf{F} \in L_{1,\mathrm{loc}}(\mathbb{R};H^s(\mathbb{R}^d;\mathbb{C}^n))$*, then for* $\tau \in \mathbb{R}$ *and* $\varepsilon > 0$ *we have*

$$\|\mathbf{u}_\varepsilon(\cdot,\tau) - \mathbf{u}_0(\cdot,\tau)\|_{L_2(\mathbb{R}^d)} \leqslant C'(s)(1+|\tau|)^{\frac{s}{2p+2}}\varepsilon^{s/(p+1)} \times \big(\|\boldsymbol{\phi}\|_{H^s(\mathbb{R}^d)} + \|\mathbf{F}\|_{L_1((0,\tau);H^s(\mathbb{R}^d))}\big).$$

Acknowledgements. This research was supported by the Russian Science Foundation (grant no. 17-11-01069).

Bibliography

[1] A. Bensoussan, J.-L. Lions and G. Papanicolaou, *Asymptotic analysis for periodic structures*. Stud. Math. Appl. 5, North-Holland Publishing, Amsterdam, 1978

[2] M. S. Birman and T. A. Suslina, Second order periodic differential operators. Threshold properties and homogenization. *Algebra i Analiz* **15** (2003), no. 5, 1–108 (in Russian); English transl. *St. Petersburg Math. J.* **15** (2004), no. 5, 639–714

[3] M. S. Birman and T. A. Suslina, Homogenization with corrector term for periodic elliptic differential operators. *Algebra i Analiz* **17** (2005), no. 6, 1–104 (in Russian); English transl. *St. Petersburg Math. J.* **17** (2006), no. 6, 897–973

[4] M. S. Birman and T. A. Suslina, Homogenization with corrector for periodic differential operators. Approximation of solutions in the Sobolev class $H^1(\mathbb{R}^d)$. *Algebra i Analiz* **18** (2006), no. 6, 1–130 (in Russian); English transl. *St. Petersburg Math. J.* **18** (2007), no. 6, 857–955

[5] M. S. Birman and T. A. Suslina, Operator error estimates in the homogenization problem for nonstationary periodic equations. *Algebra i Analiz* **20** (2008), no. 6, 30–107 (in Russian); English transl. *St. Petersburg Math. J.* **20** (2009), no. 6, 873–928

[6] M. A. Dorodnyi, Operator error estimates for homogenization of the nonstationary Schrödinger-type equations: Sharpness of the results. Appl. Anal. (2021), DOI 10.1080/00036811.2021.1901886

[7] M. A. Dorodnyi and T. A. Suslina, Spectral approach to homogenization of hyperbolic equations with periodic coefficients. *J. Differential Equations* **264** (2018), 7463–7522

[8] M. A. Dorodnyi and T. A. Suslina, Homogenization of hyperbolic equations with periodic coefficients in $\mathbb{R}^d$: Sharpness of the results. *Algebra i Analiz* **32** (2020), no. 4, 3–136 (in Russian); English transl. *St. Petersburg Math. J.* **32** (2021), no. 4, to appear

[9] T. Kato, *Perturbation theory for linear operators*. Class. Math., Springer, Berlin, 1995

[10] A. A. Kukushkin and T. A. Suslina, Homogenization of high-order elliptic operators with periodic coefficients. *Algebra i Analiz* **28** (2016), no. 1, 89–149 (in Russian); English transl. *St. Petersburg Math. J.* **28** (2017), no. 1, 65–108

[11] Y. M. Meshkova, On operator error estimates for homogenization of hyperbolic systems with periodic coefficients. *J. Spectr. Theory*, to appear; arXiv:1705.02531v4

[12] S. E. Pastukhova, Estimates in homogenization of higher-order elliptic operators. *Appl. Anal.* **95** (2016), 1449–1466

[13] S. E. Pastukhova, Operator error estimates for homogenization of fourth order elliptic equations. *Algebra i Analiz* **28** (2016), no. 2, 204–226 (in Russian); English transl. *St. Petersburg Math. J.* **28** (2017), no. 2, 273–289

[14] V. A. Sloushch and T. A. Suslina, Homogenization of the fourth-order elliptic operator with periodic coefficients with correctors taken into account. *Funktsional. Anal. i Prilozhen.* **54** (2020), no. 3, 94–99 (in Russian); English transl. *Funct. Anal. Appl.* **54** (2020), no. 3, 224–228

[15] V. A. Sloushch and T. A. Suslina, Threshold approximations for the resolvent of a polynomial nonnegative operator pencil. *Algebra i Analiz* **33** (2021), to appear (in Russian); English transl. *St. Petersburg Math. J.* **33** (2022), to appear

[16] T. A. Suslina, On homogenization of periodic parabolic systems. *Funktsional. Anal. i Prilozhen.* **38** (2004), no. 4, 86–90 (in Russian); English transl. *Funct. Anal. Appl.* **38** (2004), no. 4, 309–312

[17] T. A. Suslina, Spectral approach to homogenization of nonstationary Schrödinger-type equations. *J. Math. Anal. Appl.* **446** (2017), 1466–1523

[18] N. A. Veniaminov, Homogenization of higher-order periodic differential operators. *Algebra i Analiz* **22** (2010), no. 5, 69–103 (in Russian); English transl. *St. Petersburg Math. J.* **22** (2011), no. 5, 751–775

[19] V. V. Zhikov, On operator estimates in homogenization theory. *Dokl. Akad. Nauk* **403** (2005), no. 3, 305–308 (in Russian); English transl. *Dokl. Math.* **72** (2005), no. 1, 534–538

[20] V. V. Zhikov, S. M. Kozlov and O. A. Oleinik, *Homogenization of Differential Operators*. Springer, Berlin, 1994

[21] V. V. Zhikov and S. E. Pastukhova, On operator estimates for some problems in homogenization theory. *Russ. J. Math. Phys.* **12** (2005), no. 4, 515–524

[22] V. V. Zhikov and S. E. Pastukhova, Estimates of homogenization for a parabolic equation with periodic coefficients. *Russ. J. Math. Phys.* **13** (2006), no. 2, 224–237

[23] V. V. Zhikov and S. E. Pastukhova, Operator estimates in homogenization theory. *Uspekhi Mat. Nauk* **71** (2016), no. 3, 27–122 (in Russian); English transl. *Russian Math. Surveys* **71** (2016), no. 3, 417–511

Trace formulas for the modified Mathieu equation

Leon A. Takhtajan

To my friend Ari Laptev on the occasion of his 70th birthday

For the radial and one-dimensional Schrödinger operator H with growing potential $q(x)$ we outline a method of obtaining the trace identities – an asymptotic expansion of the Fredholm determinant $\det_F(H - \lambda I)$ as $\lambda \to -\infty$. As an illustrating example, we consider Schrödinger operator with the potential $q(x) = 2\cosh 2x$, associated with the modified Mathieu equation.

1 Introduction

Mathieu functions, solutions of the Mathieu equation and modified Mathieu equation, have many applications in mathematics and theoretical physics. In particular, the modified Mathieu differential equation

$$\frac{d^2\psi}{dx^2} + (\lambda - 2\cosh 2x)\psi = 0$$

on the real axis $-\infty < x < \infty$ is a Schrödinger equation

$$H\psi = \lambda\psi$$

with the Hamiltonian

$$H = -\frac{d^2}{dx^2} + 2\cosh 2x \tag{1.1}$$

in the Hilbert space $L^2(\mathbb{R})$.

A positive, self-adjoint operator H given by (1.1) is a Hamiltonian of the quantum two-particle periodic Toda chain (after separation of the center of mass), and has a purely discrete spectrum. As its classical analog, the quantum N-particle periodic

Keywords: Fredholm determinant, Green–Liouville method, Mathieu equation, radial and one-dimensional Schrödinger operator, Riccati equation, trace identities
2020 Mathematics Subject Classification: Primary 47E05; secondary 34E05

Toda chain is also completely integrable, as was shown by Gutzwiller [9] for $N = 2, 3$ and by Sklyanin [20] for general N. For $N = 2$ the exact quantization condition in [9] was obtained by applying Floquet theory to the potential $q(x) = 2\cosh 2x$ with pure imaginary period, and is formulated in terms of infinite Hill determinants (see [25, Chapter XIX]). Sklyanin's method of quantum separation of variables was developed further by Pasquier and Gaudin [17], and by Kharchev and Lebedev [10].

Recently, Nekrasov and Shatashvili [15] found a remarkable connection between quantum integrable systems and $\mathcal{N} = 2$ four-dimensional supersymmetric Yang–Mills theories in the special Ω-background. Among other interesting results they showed, at the physical level of rigor, that exact quantization conditions for the quantum N-particle periodic Toda chain can be written in terms of the so-called effective twisted superpotential for the low energy effective two-dimensional gauge theory. In particular, Nekrasov and Shatashvili gave a new interpretation of Gutzwiller and Kharchev and Lebedev quantization conditions.

Replacing the kinetic term $\boldsymbol{P}^2$ in the Hamiltonian H by $2\cosh \boldsymbol{P}$, where

$$\boldsymbol{P} = -i\frac{d}{dx}$$

is the quantum-mechanical momentum operator, one gets the following functional-difference operator on $L^2(\mathbb{R})$:

$$\hat{H} = 2\cosh \boldsymbol{P} + 2\cosh 2\boldsymbol{Q}, \tag{1.2}$$

where $\boldsymbol{Q}$ is the quantum-mechanical position operator. The operator $\hat{H}$ is a positive self-adjoint operator on $L^2(\mathbb{R})$ with a purely discrete spectrum, and $\hat{H}^{-1}$ is a trace class operator. Operators of such type naturally appear in the theory of topological strings as a quantization of mirror curves of non-compact Calabi–Yau threefolds. These operators have a purely discrete spectrum, and their Fredholm determinants

$$\det_F(\hat{H} - \lambda I) = \frac{\det(\hat{H} - \lambda I)}{\det \hat{H}} = \prod_{n=1}^{\infty}\left(1 - \frac{\lambda}{\lambda_n}\right),$$

where λ_n are the eigenvalues, are believed to encode deep relations in the enumerative geometry of the corresponding Calabi–Yau manifolds, called "topological strings/spectral theory (TS/ST) duality". We refer the reader to Mariño's comprehensive survey [14] for the precise description of the method, formulation of the main conjectures and references. Developing this method, Grassi, Gu and Mariño in [7] conjectured an exact formula for the Fredholm determinant of the modified Mathieu operator in terms of the so-called "exact quantum periods". Though in certain special cases TS/ST conjectures can be confirmed by numerical computations, their mathematical derivation seems to be out of reach.

Nevertheless, it is quite remarkable that operators of the type (1.2) can be rigorously studied by the method developed by Ari Laptev more then twenty years ago in [11]. Namely, it was proved in [12] (see also [13]) that these functional-difference operators have a purely discrete spectrum and their eigenvalue counting function satisfies the Weyl's law.

The logarithmic derivative of the Fredholm determinant $a(\lambda) = \det_F(\hat{H} - \lambda I)$ is

$$-\frac{d}{d\lambda}\log a(\lambda) = \operatorname{Tr}(\hat{H} - \lambda I)^{-1} = \sum_{n=1}^{\infty}\frac{1}{\lambda - \lambda_n}.$$

As $\lambda \to -\infty$, it admits the asymptotic expansion

$$\frac{d}{d\lambda}\log a(\lambda) = \alpha_0(\nu) + \sum_{n=1}^{\infty}\frac{c_n}{\nu^n} + O(|\nu|^{-\infty}),$$

where $\nu = \sqrt{-\lambda}$ and the function $\alpha_0(\nu)$ is determined by the asymptotic of the eigenvalues (cf. [3, Theorem 3.1]). The coefficients c_{2k} represent regularized divergent series $\sum_{n=1}^{\infty}\lambda_n^k$ and it is expected that they are expressed in terms of the potential of the functional-difference operator $\hat{H}$. Relations of such type are called *trace identities*.

The trace identities for the Sturm–Liouville operators were derived in the classic papers by Gelfand and Levitan [5] and Dikii [3], for the radial Schrödinger operator with rapidly decaying potential $q(x)$ – by Buslaev and Faddeev in [1], and for the one-dimensional Schrödinger operator with rapidly decaying potential $q(x)$ – by Faddeev and Zakharov in [26] (see also [21]).

Though the trace identities for growing as $|x| \to \infty$ potential $q(x)$ have not been explicitly considered in the literature, a closed topic – semiclassical expansion in a complex domain, illustrated by the case of the homogeneous quartic harmonic oscillator – was investigated by Voros in [23, 24]. One should also mention the relative trace identities – the asymptotic expansion of

$$\operatorname{Tr}\left\{\left(-\frac{d^2}{dx^2} + q(x) + r(x) - \lambda I\right)^{-1} - \left(-\frac{d^2}{dx^2} + q(x) - \lambda I\right)^{-1}\right\}$$

as $\lambda \to -\infty$, where potential $q(x)$ grows as $|x| \to \infty$, and function $r(x)$ is rapidly decaying. The interesting example $q(x) = x^2$ was considered by Pushnitski and Sorrell in [19].

In the present paper we outline a simple method for the derivation of the trace identities for the Schrödinger operator with the potential $q(x)$ that rapidly grows to infinity as $|x| \to \infty$, and illustrate it by explicit formulas for the case $q(x) = 2\cosh 2x$, the modified Mathieu operator (1.1). In Section 2.1 we briefly review the Liouville–Green method, which is used to obtain the trace identities for the radial Schrödinger operator in Section 2.2.1, and for the one-dimensional Schrödinger operator in Section 2.2.2. In Section 3 we illustrate our method on two examples: a well-known radial Schrödinger operator with potential $q(x) = e^{2x}$ in Section 3.1, and modified Mathieu operator (1.1) in Section 3.2. Our main result – the trace identities for the modified Mathieu operator – is presented in Theorem 3.1. Finally, in Section 4 we prove that modified Mathieu differential equation has a solution that behaves like the modified Bessel function of the second kind as $x \to \infty$.

2 General case

Here we consider the differential equation

$$-\psi'' + q(x)\psi = \lambda\psi \tag{2.1}$$

on the half-line $0 < x < \infty$ and on the real line $\infty < x < \infty$, where $q(x)$ is a positive smooth function that grows to infinity as $x \to \infty$ (or $|x| \to \infty$).

2.1 The Liouville–Green method

When $\lambda < 0$, decaying as $x \to \infty$ solution of (2.1) has the following double asymptotic:

$$\psi_1(x,\lambda) = \frac{C_1(\lambda)}{\sqrt[4]{q(x)-\lambda}} e^{-\int_0^x \sqrt{q(s)-\lambda}\,ds}(1+\varepsilon_1(x,\lambda)), \tag{2.2}$$

where in may cases

$$|\varepsilon_1(x,\lambda)| \le \frac{\tau(x)}{|\lambda|}, \quad \tau(x) \to 0 \quad \text{as } x \to \infty.$$

Similarly, solution of (2.1) that grows as $x \to \infty$ has the asymptotic

$$\psi_2(x,\lambda) = \frac{C_2(\lambda)}{\sqrt[4]{q(x)-\lambda}} e^{\int_0^x \sqrt{q(s)-\lambda}\,ds}(1+\varepsilon_2(x,\lambda)). \tag{2.3}$$

This follows from the classical Liouville–Green method, developed by Olver [16] (see also [4]).

Put $Q(x,\lambda) = q(x) - \lambda$ and introduce

$$\chi(x,\lambda) = \log\psi_1(x,\lambda) + \frac{1}{4}\log Q(x,\lambda) + \int_0^x \sqrt{Q(s,\lambda)}\,ds - \log C_1(\lambda) \tag{2.4}$$

and

$$\sigma(x,\lambda) = \chi'(x,\lambda) = \frac{\psi_1'(x,\lambda)}{\psi_1(x,\lambda)} + \frac{Q'(x,\lambda)}{4Q(x,\lambda)} + \sqrt{Q(x,\lambda)}, \tag{2.5}$$

where prime stands for the x-derivative. Denoting for brevity

$$\sigma = \sigma(x,\lambda) \quad \text{and} \quad Q = Q(x,\lambda),$$

we see that the function σ satisfies the Riccati equation

$$\begin{aligned}\sigma' &= \frac{\psi_1''}{\psi_1} - \left(\frac{\psi_1'}{\psi_1}\right)^2 + \frac{Q''}{4Q} - \frac{1}{4}\left(\frac{Q'}{Q}\right)^2 + \frac{Q'}{2\sqrt{Q}} \\ &= Q - \left(\sigma - \frac{Q'}{4Q} - \sqrt{Q}\right)^2 + \frac{Q''}{4Q} - \frac{1}{4}\left(\frac{Q'}{Q}\right)^2 + \frac{Q'}{2\sqrt{Q}}\end{aligned}$$

or

$$\sigma' = -\sigma^2 + 2\sqrt{Q}\sigma + \frac{Q'}{2Q}\sigma + \frac{Q''}{4Q} - \frac{5}{16}\left(\frac{Q'}{Q}\right)^2. \tag{2.6}$$

2.2 Trace identities

The Riccati equation (2.6) can be used to obtain an asymptotic expansion of $\sigma(x,\lambda)$ as $\lambda \to -\infty$ and get the trace identities for the Schrödinger operator with growing as $x \to \infty$ potential $q(x)$ on $L^2(0,\infty)$ and on $L^2(\mathbb{R})$. In the latter case potential $q(x)$ is assumed to be an even function.

2.2.1 Radial Schrödinger operator. Here we consider the Schrödinger operator H on $L^2(0,\infty)$ with a smooth positive potential $q(x)$ satisfying

$$\int_0^\infty \frac{dx}{\sqrt{q(x)}} < \infty, \tag{2.7}$$

supplemented with the boundary condition $\psi(0) = 0$. Restricting further, we consider potentials satisfying the inequality $q(x) \geq Cx^{2+\varepsilon}$ for some $\varepsilon > 0$. It follows from the Weyl's law that H^{-1} is a trace class operator, and its Fredholm determinant is an entire function of order $\frac{1}{2}$.

For every λ equation (2.1) for $0 < x < \infty$ has a unique solution $\psi(x,\lambda)$ with the asymptotic

$$\psi(x,\lambda) = \frac{1}{\sqrt[4]{q(x)}} e^{-\int_0^x \sqrt{q(s)}\,ds}(1 + o(1)) \quad \text{as } x \to \infty. \tag{2.8}$$

For fixed x solution $\psi(x,\lambda)$ is an entire function of λ of order $\frac{1}{2}$ and $\psi(0,\lambda)$ is proportional to the Fredholm determinant $a(\lambda)$ of the operator H,

$$a(\lambda) = \frac{\psi(0,\lambda)}{\psi(0,0)}.$$

Comparing equations (2.2) and (2.8) for $\lambda < 0$, we see that for such λ solution $\psi(x,\lambda)$ has asymptotic (2.2) with

$$C_1(\lambda) = e^{\int_0^\infty (\sqrt{Q(x,\lambda)} - \sqrt{q(x)})\,dx}.$$

It follows from (2.7) that the integral in this formula is convergent, so $C_1(\lambda)$ is well defined.

Putting $\lambda = -\nu^2$ in formula (2.4), we have

$$\begin{aligned}\chi(x,\lambda) = \log\psi(x,\lambda) + \frac{1}{4}\log(q(x)+\nu^2) + \int_0^x \sqrt{q(s)+\nu^2}\,ds \\ - \int_0^\infty \left(\sqrt{q(s)+\nu^2} - \sqrt{q(s)}\right) ds.\end{aligned}$$

From here we obtain

$$\begin{aligned}\chi(0,\lambda) &= \log a(\lambda) - a_0(\lambda), \\ \lim_{x\to\infty} \chi(x,\lambda) &= 0,\end{aligned}$$

where

$$a_0(\lambda) = -\frac{1}{4}\log(q(0)+\nu^2) + \int_0^\infty \left(\sqrt{q(x)+\nu^2} - \sqrt{q(x)}\right) dx - \log\psi(0,0).$$

It is not difficult to show that $a_0(\lambda)$ admits an asymptotic expansion as $\lambda = -\nu^2 \to \infty$,

$$a_0(\lambda) = \alpha(\nu) + \sum_{n=1}^{\infty} \frac{\alpha_n}{\nu^n} + O(|\nu|^{-\infty}), \tag{2.9}$$

where the leading term – a function $\alpha(\nu)$ – is determined by the asymptotic of the eigenvalues and can be computed explicitly (see examples in Section 3).

The function $\sigma(x,\lambda) = \chi'(x,\lambda)$ satisfies the Riccati equation (2.6), which admits the asymptotic solution

$$\sigma(x,\lambda) = \sum_{n=1}^{\infty} \frac{c_n(x)}{\nu^n} + O(|\nu|^{-\infty}), \tag{2.10}$$

where the coefficients $c_n(x)$ are determined recursively in terms of the potential $q(x)$ and its derivatives. Since

$$\log\chi(0,\lambda) = -\int_0^\infty \sigma(x,\lambda)\,dx,$$

we have

$$\log a(\lambda) = \alpha_0(\lambda) - \int_0^\infty \sigma(x,\lambda)\,dx. \tag{2.11}$$

Using the asymptotic expansion (2.10), we obtain the trace identities

$$\log a(\lambda) = \alpha(\nu) + \sum_{n=1}^{\infty} \frac{c_n}{\nu^n} + O(|\nu|^{-\infty}),$$

where

$$c_n = \alpha_n - \int_0^\infty c_n(x)\,dx.$$

2.2.2 One-dimensional Schrödinger operator. Here we assume that the potential $q(x)$ is a smooth even function satisfying (2.7). A fundamental system of solutions of the differential equation (2.1) is given by solutions $\psi_1(x,\lambda)$ and $\psi_2(x,\lambda)$ with the following asymptotic as $x \to \infty$:

$$\psi_1(x,\lambda) = \frac{1}{\sqrt[4]{q(x)}} e^{-\int_0^x \sqrt{q(s)}\,ds}(1+o(1)), \tag{2.12}$$

$$\psi_2(x,\lambda) = \frac{1}{\sqrt[4]{q(x)}} e^{\int_0^x \sqrt{q(s)}\,ds}(1+o(1)). \tag{2.13}$$

Assuming that

$$\lim_{x\to\infty} \frac{q'(x)}{\left(\sqrt{q(x)}\right)^3} = 0,$$

we get from (2.12) and (2.13) that

$$W(\psi_1, \psi_2) = 2,$$

where $W(f, g) = fg' - f'g$ is the Wronskian of two functions. The functions $\psi_1(x, \lambda)$ and $\psi_2(x, \lambda)$ satisfy asymptotic formulas (2.2)–(2.3), where

$$C_1(\lambda) = e^{\int_0^\infty (\sqrt{Q(x,\lambda)} - \sqrt{q(x)})\, dx} \quad \text{and} \quad C_2(\lambda) = e^{-\int_0^\infty (\sqrt{Q(x,\lambda)} - \sqrt{q(x)})\, dx}.$$

Another fundamental system of solutions is given by the functions $\psi_1(-x, \lambda)$ and $\psi_2(-x, \lambda)$, and we have

$$\psi_1(x, \lambda) = t_{11}(\lambda)\psi_1(-x, \lambda) + t_{12}(\lambda)\psi_2(-x, \lambda), \tag{2.14}$$

where

$$t_{12}(\lambda) = \frac{1}{2} W(\psi_1(x, \lambda), \psi_1(-x, \lambda)).$$

For fixed x the functions $\psi_1(x, \lambda)$ and $\psi_2(x, \lambda)$ are entire functions of order $\frac{1}{2}$, as it can be shown using condition (2.7). Therefore, $t_{12}(\lambda)$ is an entire function of order $\frac{1}{2}$ with zeros at the eigenvalues of the operator H. As in the previous section, for the Fredholm determinant $a(\lambda)$ of the operator H we obtain

$$a(\lambda) = \frac{t_{12}(\lambda)}{t_{12}(0)}.$$

It follows from (2.2) that the function $\chi(x, \lambda)$, defined by (2.4), satisfies

$$\lim_{x\to\infty} \chi(x, \lambda) = 0.$$

To investigate its behavior as $x \to -\infty$, we observe that it follows from (2.12)–(2.13) that as $x \to -\infty$ the first term in (2.14) is exponentially small with respect to the second term, so

$$\lim_{x\to-\infty} (\log \psi_1(x, \lambda) - \log \psi_2(-x, \lambda)) = \log t_{12}(\lambda).$$

Since the antiderivative $\int_0^x \sqrt{Q(s, \lambda)}\, ds$ is an odd function of x, we obtain

$$\lim_{x\to-\infty} \chi(x, \lambda) = \log a(\lambda) - a_0(\lambda),$$

where now

$$a_0(\lambda) = \int_{-\infty}^{\infty} \left(\sqrt{q(x) + \nu^2} - \sqrt{q(x)}\right) dx - \log t_{12}(0).$$

As in Section 2.2.1, the function $a_0(\lambda)$ admits an asymptotic expansion of type (2.9). The corresponding Riccati equation (2.6) has an asymptotic solution

$$\sigma(x,\lambda) = \sum_{n=1}^{\infty} \frac{c_n(x)}{\nu^n} + O(|\nu|^{-\infty}),$$

where the coefficients $c_n(x)$ are determined recursively and are expressed in terms of the potential $q(x)$ and its derivatives. Thus we obtain the trace identities

$$\log a(\lambda) = \alpha(\nu) + \sum_{n=1}^{\infty} \frac{c_n}{\nu^n} + O(|\nu|^{-\infty}),$$

where

$$c_n = \alpha_n - \int_{-\infty}^{\infty} c_n(x)\,dx.$$

Remark. The interesting example of the harmonic oscillator, namely the potential $q(x) = x^2$, does not satisfy condition (2.7). However, one can modify the outlined here method by exploiting the substitution $x = \sqrt{-\lambda}t$.[1] In this case the Fredholm determinant is

$$a(\lambda) = \frac{2^{-\frac{\lambda}{2}}\sqrt{\pi}}{\Gamma(\frac{1-\lambda}{2})}$$

and corresponding trace identities give an alternate derivation of the Stirling asymptotic expansion for the Euler gamma-function (for an equivalent approach, see [23, Part IV]).

3 Examples

3.1 Potential $q(x) = e^{2x}$ on the half-line

It is well known that the differential equation

$$-\psi'' + e^{2x}\psi = \lambda\psi$$

has two linearly independent solutions, the modified Bessel function of the first kind $I_\nu(e^x)$, and the modified Bessel function of the second kind $K_\nu(e^x)$, where $\nu = \sqrt{-\lambda}$. For fixed ν and real $y \to \infty$ we have the asymptotic

$$K_\nu(y) = \sqrt{\frac{\pi}{2y}}e^{-y}\big(1 + O(y^{-1})\big),$$

so the eigenvalues determined by the boundary condition $\psi(0) = 0$ are the zeros $\{\lambda_n\}_{n=1}^{\infty}$ of an entire function $K_\nu(1)$, where $\lambda = -\nu^2$. The eigenvalues λ_n are positive,

[1]This is the substitution used in [16] to obtain uniform asymptotic for Weber functions (parabolic cylinder functions).

simple and accumulate to infinity, and

$$a(\lambda) = \frac{K_\nu(1)}{K_0(1)}.$$

Remark. It is well known (see, e.g., [18]) that the total number of zeros of $K_{ik}(1)$ with $0 < k < T$ is

$$\frac{T}{\pi} \log \frac{2T}{e} + O(1),$$

so for the eigenvalue counting function $N(\lambda)$ we obtain

$$N(\lambda) = \frac{\sqrt{\lambda}}{\pi} \log \frac{2\sqrt{\lambda}}{e} + O(1).$$

This also easily follows from the Weyl's law. Thus as $n \to \infty$ for the n-th eigenvalue λ_n we have

$$\lambda_n \simeq \left(\frac{2\pi n}{eW(\frac{2\pi n}{e})}\right)^2,$$

where $W(x)$ is the Lambert function, $W(x)e^{W(x)} = x$ for $x > 0$.

The asymptotic expansion of $\log a(\lambda)$ as $\lambda = -\nu^2 \to -\infty$ can be easily obtained from the well-known asymptotic of the modified Bessel function of the second kind.[2] However, it can also be derived directly, using the method outlined in Section 2.2.1.

Namely, we have

$$\begin{aligned}\int_0^x \sqrt{e^{2s} + \nu^2}\, ds = \sqrt{\nu^2 + e^{2x}} + \nu x - \nu \log\big(\nu + \sqrt{\nu^2 + e^{2x}}\big) \\ - \sqrt{\nu^2 + 1} + \nu \log\big(\nu + \sqrt{\nu^2 + 1}\big).\end{aligned}$$

Since

$$\sqrt{\nu^2 + e^{2x}} + \nu x - \nu \log\big(\nu + \sqrt{\nu^2 + e^{2x}}\big) = e^x + O(e^{-x}) \quad \text{as } x \to \infty,$$

comparison with (2.2) gives $K_\nu(e^x) = \psi_1(x, \lambda)$, where

$$C_1(\lambda) = \sqrt{\frac{\pi}{2}}\, e^{\sqrt{\nu^2+1} - \nu \log(\nu - \sqrt{\nu^2+1})}.$$

Of course, this is a well-known asymptotic of the modified Bessel function of the second kind,

$$K_\nu(e^x) = \sqrt{\frac{\pi}{2}} \frac{e^{-\left\{\sqrt{\nu^2+e^{2x}} + \nu x - \nu \log(\nu + \sqrt{\nu^2+e^{2x}})\right\}}}{\sqrt[4]{\nu^2 + e^{2x}}} (1 + O(|\nu|^{-1})),$$

which is uniform for $-\infty < x < \infty$ (see [16, Chapter 10, Section 7]).

---

[2] One needs to put $z = \frac{1}{\nu}$ and $p = \frac{\nu}{\sqrt{1+\nu^2}}$ in [16, Chapter 10, Section 7, formula (7.17)].

Now the function $\chi(x,\lambda)$, defined in (2.4), satisfies

$$\lim_{x\to\infty}\chi(x,\lambda)=0,$$
$$\chi(0,\lambda)=\log a(\lambda)-a_0(\lambda),$$

where

$$a_0(\lambda)=\sqrt{\nu^2+1}-\nu\log\bigl(\nu+\sqrt{\nu^2+1}\bigr)-\frac{1}{4}\log(\nu^2+1)+\frac{1}{2}\log\frac{\pi}{2}-\log K_0(1).$$

The function $a_0(\lambda)$ admits an asymptotic expansion (2.9), where

$$\alpha_0(\lambda)=\nu\log(2\nu)-\nu-\frac{1}{2}\log\nu+\frac{1}{2}\log\frac{\pi}{2}-\log K_0(1).$$

This formula agrees with the asymptotic

$$K_\nu(e^x)=\sqrt{\frac{\pi}{2\nu}}\left(\frac{e^{x+1}}{2\nu}\right)^{-\nu}\bigl(1+O(|\nu|^{-1})\bigr)$$

for fixed x and $\nu\to\infty$, and with the Weyl's law.

The function $\sigma(x,\lambda)$, defined in (2.5), satisfies the Riccati equation (2.6). Its asymptotic expansion (2.10) can be easily obtained by using the substitution

$$\tau(t,\lambda)=\sigma(t+\log\nu,\lambda),$$

which transforms (2.6) into the equation

$$\tau'=-\tau^2+2\nu\sqrt{1+e^{2t}}\,\tau+\frac{e^{2t}}{1+e^{2t}}\tau+\frac{4e^{2t}-e^{4t}}{4(1+e^{2t})^2}.$$

We have

$$\tau(t,\nu)=\sum_{n=1}^{\infty}\frac{c_n(t)}{\nu^n},$$

where

$$c_1(t)=\frac{e^{4t}-4e^{2t}}{8(1+e^{2t})^{\frac{5}{2}}}$$

and recursively

$$2\sqrt{1+e^{2t}}\,c_{n+1}(t)=c_n'(t)-\frac{e^{2t}}{1+e^{2t}}c_n(t)+\sum_{k=1}^{n}c_k(t)c_{n-k}(t).$$

Plugging this into (2.11), we obtain

$$\log a(\lambda)=\nu\log(2\nu)-\nu-\frac{1}{2}\log\nu+\frac{1}{2}\log\frac{\pi}{2}+\sum_{n=1}^{\infty}\frac{\alpha_n-c_n(\nu)}{\nu^n}+O(|\nu|^{-\infty}),$$

where

$$c_n(\nu)=-\int_{-\log\nu}^{\infty}\sigma_n(t)\,dt.$$

Moreover, it can be shown (cf. [16, Chapter 10, Section 7]) that the coefficients $c_n(\nu)$ are polynomials in $\frac{\nu}{\sqrt{\nu^2+1}}$, which gives the asymptotic expansion of $\log a(\lambda)$.

3.2 Potential $q(x) = 2\cosh 2x$ on the line

Here we consider the modified Mathieu differential equation

$$-\psi'' + 2\cosh 2x\,\psi = \lambda\psi, \quad -\infty < x < \infty. \tag{3.1}$$

It is well known that the antiderivative of the function $\sqrt{2\cosh 2x - \lambda}$ is expressed in terms of elliptic integrals. Namely, let

$$F(\varphi, k) = \int_0^{\varphi} \frac{d\theta}{\sqrt{1-k^2\sin^2\theta}} \quad \text{and} \quad E(\varphi, k) = \int_0^{\varphi} \sqrt{1-k^2\sin^2\theta}\,d\theta$$

be, correspondingly, elliptic integrals of the first and second kinds, where $0 < k^2 < 1$ and $-\frac{\pi}{2} < \varphi < \frac{\pi}{2}$. Putting $\lambda = 2 - \nu^2$, where $\nu > 0$, we have for $\nu > 2$ (see [6, formula (22) on p. 136])

$$\begin{aligned}\int_0^x \sqrt{\nu^2 - 2 + 2\cosh 2s}\,ds = \nu(F(\varphi, k) - E(\varphi, k))\\ + \tanh\frac{x}{2}\sqrt{\nu^2 - 2 + 2\cosh 2x},\end{aligned} \tag{3.2}$$

where

$$\varphi = \sin^{-1}\left(\tanh\frac{x}{2}\right) \quad \text{and} \quad k^2 = \frac{\nu^2-4}{\nu^2}. \tag{3.3}$$

The corresponding solutions $\psi_1(x,\lambda)$ and $\psi_2(x,\lambda)$ of the differential equation (3.1) with asymptotics (2.2)–(2.3) are

$$\psi_1(x,\lambda) \simeq \frac{C_1(\lambda)}{\sqrt[4]{2\cosh 2x + \nu^2}} e^{\nu(F(\varphi,k)-E(\varphi,k)) + \tanh\frac{x}{2}\sqrt{\nu^2-2+2\cosh 2x}}, \tag{3.4}$$

$$\psi_2(x,\lambda) \simeq \frac{C_2(\lambda)}{\sqrt[4]{2\cosh 2x + \nu^2}} e^{-\nu(F(\varphi,k)-E(\varphi,k)) - \tanh\frac{x}{2}\sqrt{\nu^2-2+2\cosh 2x}}, \tag{3.5}$$

where $\lambda = 2 - \nu^2 \to -\infty$ (cf. [8]).

Since $\varphi \to \frac{\pi}{2}$ as $x \to \infty$, we have

$$\int_0^x \sqrt{\nu^2 - 2 + 2\cosh 2s}\,ds = \nu(\boldsymbol{K}(k) - \boldsymbol{E}(k))) + e^x + O(e^{-x}),$$

where $\boldsymbol{K}(k)$ and $\boldsymbol{E}(k)$ are, respectively, complete elliptic integrals of the first and second kinds. Choosing the constants in (3.4)–(3.5) as

$$C_1(\lambda) = \sqrt{\frac{\pi}{2}}\, e^{-\nu(\boldsymbol{K}(k)-\boldsymbol{E}(k))} \quad \text{and} \quad C_2(\lambda) = \sqrt{\frac{1}{2\pi}}\, e^{\nu(\boldsymbol{K}(k)-\boldsymbol{E}(k))},$$

we get a solution $\psi_1(x,\lambda)$ with the same asymptotic as $x\to\infty$ as the modified Bessel function of the second kind $K_{i\sqrt{\lambda}}(e^x)$,

$$\psi_1(x,\lambda)=\sqrt{\frac{\pi}{2e^x}}\,e^{-e^x}\big(1+O(e^{-x})\big)\quad\text{as } x\to\infty,$$

and a solution $\psi_2(x,\lambda)$ with the same asymptotic as $x\to\infty$ as the modified Bessel function of the first kind $I_{i\sqrt{\lambda}}(e^x)$,

$$\psi_2(x,\lambda)=\sqrt{\frac{1}{2\pi e^x}}\,e^{e^x}\big(1+O(e^{-x})\big)\quad\text{as } x\to\infty.$$

We have $W(\psi_1,\psi_2)=1$ and

$$\psi_1(x,\lambda)=t_{11}(\lambda)\psi_1(-x,\lambda)+t_{12}(\lambda)\psi_2(-x,\lambda),$$

where

$$t_{12}(\lambda)=W(\psi_1(x,\lambda),\psi_1(-x,\lambda))$$

is an entire function of order $\frac{1}{2}$ with zeros – the eigenvalues λ_n of the Schrödinger operator (1.1). The corresponding Fredholm determinant is

$$a(\lambda)=\frac{t_{12}(\lambda)}{t_{12}(0)}.$$

Let $\chi(x,\lambda)$ be the function defined in (2.4). Since the antiderivative (3.2)–(3.3) is an odd function of x, as in Section 2.2.2 we get

$$\lim_{x\to\infty}\chi(x,\lambda)=0, \tag{3.6}$$

$$\lim_{x\to-\infty}\chi(x,\lambda)=\log a(\lambda)+\log\pi+\log t_{12}(0)-2\nu(\boldsymbol{K}(k)-\boldsymbol{E}(k)). \tag{3.7}$$

We state the main result of this section.

Theorem 3.1. *The Fredholm determinant $a(\lambda)$ of the Schrödinger operator* (1.1) *admits the following asymptotic expansion as $\lambda=2-\nu^2\to-\infty$:*

$$\log a(\lambda)=2\nu(\boldsymbol{K}(k)-\boldsymbol{E}(k))-\log\pi-\log t_{12}(0)+\sum_{n=1}^{\infty}\frac{c_n}{\nu^n}+O(|\nu|^{-\infty}), \tag{3.8}$$

where $k^2=1-4\nu^{-2}$ and the coefficients c_n are determined explicitly.

Proof. It follows from (3.6)–(3.7) that

$$\log a(\lambda)=2\nu(\boldsymbol{K}(k)-\boldsymbol{E}(k))-\log\pi-\log t_{12}(0)-\int_{-\infty}^{\infty}\sigma(x,\lambda)\,dx, \tag{3.9}$$

where $\sigma(x,\lambda)$ is defined in (2.5) and satisfies the Riccati equation (2.6). Using

$$2\cosh 2x-\lambda=\nu^2+2\cosh 2x-2=\nu^2+4\sinh^2 x,$$

we can rewrite the equation (2.6) as

$$\begin{aligned}\frac{d\sigma}{dx} &= -\sigma^2 + 2\sqrt{\nu^2 + 4\sinh^2 x}\,\sigma + \frac{2\sinh 2x}{\nu^2 + 4\sinh^2 x}\sigma \\ &\quad + \frac{8\cosh 2x}{\nu^2 + 4\sinh^2 x} - 5\Big(\frac{\sinh 2x}{\nu^2 + 4\sinh^2 x}\Big)^2.\end{aligned}$$

It is convenient to change variables by $\sinh x = \frac{\nu}{2}\sinh y$, so

$$\nu^2 + 4\sinh^2 x = \nu^2\cosh^2 y$$

and

$$\frac{dx}{dy} = \frac{\nu\cosh y}{2\cosh x} = \frac{\nu\cosh y}{\sqrt{4 + \nu^2\sinh^2 y}} = \Big(1 - \frac{k^2}{\cosh^2 y}\Big)^{-\frac{1}{2}}.$$

Next, introduce the function $\tau(y,\lambda) = \sigma(\sinh^{-1}(\frac{\nu}{2}\sinh y), \lambda)$, so

$$\int_{-\infty}^{\infty} \sigma(x,\lambda)\,dx = \int_{-\infty}^{\infty} \tau(y,\lambda)\Big(1 - \frac{k^2}{\cosh^2 y}\Big)^{-\frac{1}{2}}\,dy. \tag{3.10}$$

The Riccati equation for $\tau(y,\lambda)$ takes the form

$$\begin{aligned}\Big(1 - \frac{k^2}{\cosh^2 y}\Big)^{\frac{1}{2}}\frac{d\tau}{dy} &= -\tau^2 + 2\nu\cosh y\,\tau + \tanh y\Big(1 - \frac{k^2}{\cosh^2 y}\Big)^{\frac{1}{2}}\tau \\ &\quad + 4\tanh^2 y - \frac{5}{4}\tanh^4 y \\ &\quad + \frac{1}{\nu^2\cosh^2 y}\Big(8 - \frac{5}{4}\tanh^2 y\Big).\end{aligned} \tag{3.11}$$

Now it is straightforward to show that equation (3.11) admits an asymptotic solution

$$\tau(y,\lambda) = \sum_{n=1}^{\infty}\frac{\tau_n(y)}{\nu^n},$$

where

$$\tau_1(y) = -\frac{1}{2\cosh y}\Big(4\tanh^2 y - \frac{5}{4}\tanh^4 y\Big).$$

The coefficients $\tau_n(y)$ are obtained recursively using the expansion

$$\Big(1 - \frac{k^2}{\cosh^2 y}\Big)^{\frac{1}{2}} = \sum_{m=1}^{\infty}(-1)^m\binom{\frac{1}{2}}{m}\frac{k^{2m}}{\cosh^{2m} y}$$

and the binomial expansion for $k^{2m} = (1 - 4\nu^{-2})^m$; only finitely many terms from these expansions are needed in order to get the n-th term. Substituting the asymptotic expansion for $\tau(y,\lambda)$ into (3.10) and (3.9) and using

$$\left(1-\frac{k^2}{\cosh^2 y}\right)^{-\frac{1}{2}}=\sum_{m=1}^{\infty}(-1)^m\binom{-\frac{1}{2}}{m}\frac{k^{2m}}{\cosh^{2m} y},$$

we get the desired asymptotic expansion (3.8). ∎

Remark. Since

$$\boldsymbol{K}(k)=\log\frac{4}{\sqrt{1-k^2}}+o(1) \quad \text{and} \quad \boldsymbol{E}(k)=1+o(1) \quad \text{as } k\to 1,$$

we have as $\nu\to\infty$,

$$\nu(\boldsymbol{K}(k)-\boldsymbol{E}(k))=\nu\log(2\nu)-2\nu+o(1).$$

Here the $o(1)$ term – the remainder – is of the form $f_1(\nu^{-1})+f_2(\nu^{-1})\log\nu$, where $f_1(x)$ and $f_2(x)$ are convergent for $|x|<1$ power series in x^2 that vanish at $x=0$ (see [2, p. 54]).

4 Existence of the solution $\psi_1(x,\lambda)$

Here for the convenience of the reader we prove that differential equation (3.1) has a solution $\psi_1(x,\lambda)$ which asymptotically as $x\to\infty$ behaves like the modified Bessel function of the second kind. Namely, we have the following result.

Theorem 4.1. *The modified Mathieu equation* (3.1) *has a solution* $\psi_1(x,\lambda)$ *with the following asymptotic:*

$$\psi_+(x,\lambda)=K_{ik}(e^x)(1+o(1)) \quad \textit{as } x\to\infty, \quad \textit{where } \lambda=k^2.$$

For fixed x the function $\psi_1(x,\lambda)$ is entire of order $\frac{1}{2}$.

Proof. We consider the Schrödinger operator $H=-\frac{d^2}{dx^2}+2\cosh 2x$ as a perturbation of the operator $H_0=-\frac{d^2}{dx^2}+e^{2x}$ by a small as $x\to\infty$ potential e^{-2x}. The operator H_0 has a simple absolutely continuous spectrum $[0,\infty)$, and for $\lambda\in\mathbb{C}\setminus[0,\infty)$ its resolvent $R^0_\lambda=(H_0-\lambda I)^{-1}$ is an integral operator in $L^2(\mathbb{R})$ with the integral kernel (see, e.g., [22, Section 4.15])

$$R^0_\lambda(x,y)=\begin{cases} I_{-ik}(e^x)K_{ik}(e^y), & x\le y,\\ K_{ik}(e^x)I_{-ik}(e^y), & x\ge y.\end{cases}$$

Here $k=\sqrt{\lambda}$ and $\operatorname{Im}k>0$. The operator H_0 has a Volterra-type Green's function

$$G(x,y,k)=\begin{cases} I_{-ik}(e^x)K_{ik}(e^y)-K_{ik}(e^x)I_{-ik}(e^y), & x\le y,\\ 0, & x>y.\end{cases}$$

Since

$$K_\nu(z) = \frac{\pi}{2\sin\pi\nu}(I_{-\nu}(z) - I_\nu(z)),$$

it follows that $G(x, y, k)$ is an even function of k and, therefore, is an entire function of λ of order $\frac{1}{2}$.

The function $\psi_1(x, \lambda)$ satisfies the integral equation

$$\psi_1(x, \lambda) = K_{ik}(e^x) + \int_x^\infty G(x, y, k)e^{-2y}\psi_1(y, \lambda)\,dy,$$

which can be solved by successive approximations. Indeed, put $f_0(x, k) = K_{ik}(e^x)$ and

$$f_n(x, k) = \int_x^\infty G(x, y, k)e^{-2y} f_{n-1}(y, k)\,dy.$$

Using the estimates

$$|K_{ik}(e^x)| \le Ce^{-e^x}e^{-\frac{1}{2}x} \quad \text{and} \quad |I_{ik}(e^x)| \le Ce^{e^x}e^{-\frac{1}{2}x} \tag{4.1}$$

for $a \le x < \infty$ (the constant C depends on a), we have $f_n(x, -k) = f_n(x, k)$ and

$$|f_n(x, k)| \le \frac{2^n C^{2n+1}}{3^n n!} e^{-e^x} e^{-\frac{6n+1}{2}x}, \tag{4.2}$$

which can easily be proved by induction. Namely, (4.2) holds for $n = 0$, and using estimates (4.1), we get

$$\begin{aligned}
|f_{n+1}(x, k)| &\le |I_{-ik}(e^x)| \int_x^\infty e^{-2y}|K_{ik}(e^y) f_n(y, k)|\,dy \\
&\quad + |K_{ik}(e^x)| \int_x^\infty e^{-2y}|I_{-ik}(e^y) f_n(y, k)|\,dy \\
&\le \frac{2^n C^{2n+3}}{3^n n!}\Big[e^{e^x}e^{-\frac{1}{2}x}\int_x^\infty e^{-2y}e^{-2e^y}e^{-\frac{1}{2}y}e^{-\frac{6n+1}{2}y}\,dy \\
&\quad + e^{-e^x}e^{-\frac{1}{2}x}\int_x^\infty e^{e^y}e^{-\frac{1}{2}y}e^{-2y}e^{-e^y}e^{-\frac{6n+1}{2}y}\,dy\Big] \\
&\le \frac{2^{n+1}C^{2n+3}}{3^{n+1}(n+1)!}e^{-e^x}e^{-\frac{6n+7}{2}x}.
\end{aligned}$$

Thus

$$\psi_1(x, \lambda) = \sum_{n=0}^\infty f_n(x, k)$$

is given by the absolutely convergent series and

$$|\psi_1(x, \lambda) - K_{ik}(e^x)| \le Ce^{-e^x}e^{-\frac{7}{2}x}. \qquad \blacksquare$$

As in Section 3.2, we have

$$t_{12}(\lambda) = W(\psi_1(x,\lambda), \psi_1(-x,\lambda)) = -2\psi_1(0,\lambda)\psi_1'(0,\lambda),$$

and the eigenvalue problem (3.1) is equivalent to two radial eigenvalue problems

$$\begin{aligned} -\psi'' + 2\cosh 2x\,\psi &= k^2\psi, \quad 0 < x < \infty, \qquad \psi(0) = 0,\\ -\psi'' + 2\cosh 2x\,\psi &= k^2\psi, \quad 0 < x < \infty, \qquad \psi'(0) = 0. \end{aligned}$$

Acknowledgements. It is a pleasure to thank Ari Laptev for the discussion of the trace identities and for drawing my attention to the reference [19]. I am also grateful to Marcos Mariño for the useful comments and for pointing to the references [7] and [23, 24].

Bibliography

[1] V. S. Buslaev and L. D. Faddeev, Formulas for traces for a singular Sturm–Liouville differential operator. *Dokl. Akad. Nauk SSSR* **132** (1960), 13–16 (in Russian); English transl. *Soviet Math. Dokl.* **1** (1960), 451–454

[2] A. Cayley, *An elementary treatise on elliptic functions*. George Bell and Sons, London, 1895; reprinted by Dover Publications, New York, 1961

[3] L. A. Dikii, Trace formulas for Sturm–Liouville differential operators. *Uspehi Mat. Nauk (N.S.)* **13** (1958), 111–143 (in Russian); English transl. *Transl. Amer. Math. Soc. (2)* **18** (1958), 81–115

[4] M. V. Fedoryuk, *Asymptotic analysis: Linear ordinary differential equations*. Springer, Berlin, 1993

[5] I. M. Gelfand and B. M. Levitan, On a simple identity for the eigenvalues of a second-order differential operator. *Dokl. Akad. Nauk. USSR* **88** (1953), 953–956 (in Russian)

[6] I. S. Gradshtein and I. M. Ryzhik, *Tables Of Integrals, Series And Products*, 7th edn., Academic Press, New York, 2007

[7] A. Grassi, J. Gu and M. Mariño, Non-perturbative approaches to the quantum Seiberg–Witten curve. *J. High Energy Phys.* **2020** (2020), Paper No. 106

[8] N. S. Grigoreva, Uniform asymptotic expansions of the solutions of the Mathieu and modified Mathieu equations. *Zap. Naučn. Sem. Leningrad. Otdel Mat. Inst. Steklov. (LOMI)* **62** (1976), 60–91 (in Russian); English transl. *J. Soviet Math.* **11** (1979), 700–721

[9] M. C. Gutzwiller, The quantum mechanical Toda lattice. *Ann. Physics* **124** (1980), 347–381; The quantum mechanical Toda lattice II, *Ann. Physics* **133** (1981), 304–331

[10] S. Kharchev and D. Lebedev, Integral representation for the eigenfunctions of a quantum periodic Toda chain. *Lett. Math. Phys.* **50** (1999), 53–77

[11] A. Laptev, Dirichlet and Neumann eigenvalue problems on domains in Euclidean spaces. *J. Funct. Anal.* **151** (1997), 531–545

[12] A. Laptev, L. Schimmer and L. A. Takhtajan, Weyl type asymptotics and bounds for the eigenvalues of functional-difference operators for mirror curves. *Geom. Funct. Anal.* **26** (2016), 288–305

[13] A. Laptev, L. Schimmer and L. A. Takhtajan, Weyl asymptotics for perturbed functional difference operators. *J. Math. Phys.* **60** (2019), Article ID 103505

[14] M. Mariño, Spectral theory and mirror symmetry. In *String-Math 2016*, pp. 259–294, Proc. Sympos. Pure Math. 98, American Mathematical Society, Providence, 2018

[15] N. A. Nekrasov and S. L. Shatashvili, Quantization of integrable systems and four dimensional gauge theories. In *Proceedings of the 16th International Congress on Mathematical Physics (Prague, 2009)*, pp. 265–289, World Scientific Publishing, Hackensack, 2010

[16] F. W. J. Olver, *Asymptotics and special functions*. Academic Press, New York, 1974

[17] V. Pasquier and M. Gaudin, The periodic Toda chain and a matrix generalization of the Bessel function recursion relations. *J. Phys. A* **25** (1992), 5243–5252

[18] G. Pólya, Bemerkung über die Integraldarstellung der Riemannschen ξ-Funktion. *Acta Math.* **48** (1926), 305–317

[19] A. Pushnitski and I. Sorrell, High energy asymptotics and trace formulas for the perturbed harmonic oscillator. *Ann. Henri Poincaré* **7** (2006), 381–396

[20] E. K. Sklyanin, The quantum Toda chain. In *Nonlinear equations in classical and quantum field theory (Meudon/Paris, 1983/1984)*, pp. 196–233, Lecture Notes in Phys. 226, Springer, Berlin, 1985

[21] L. A. Takhtajan, Etudes on the resolvent. *Uspekhi Mat. Nauk* **75** (2020), 155–194 (in Russian); English transl. *Russian Math. Surveys* **75** (2020), 147–186

[22] E. C. Titchmarsh, *Eigenfunction expansions associated with second-order differential equations. Part I*. 2nd edn., Clarendon Press, Oxford, 1962

[23] A. Voros, *Spectre de l'équation de Schrödinger et méthode BKW*. Publ. Math. Orsay 81, Université de Paris-Sud, Orsay, 1981

[24] A. Voros, The return of the quartic oscillator: The complex WKB method. *Ann. Inst. H. Poincaré Sect. A (N.S.)* **39** (1983), 211–338

[25] E. T. Whittaker and G. N. Watson, *A course of modern analysis*. Cambridge University Press, Cambridge, 1927

[26] V. E. Zaharov and L. D. Faddeev, Korteweg–de Vries equation: A completely integrable Hamiltonian system. *Funkcional. Anal. i Priložen.* **5** (1971), 18–27 (in Russian); English transl. *Funct. Anal. Appl.* **5** (1971), 280–287

Eigenvalue accumulation and bounds for non-selfadjoint matrix differential operators related to NLS

Christiane Tretter

To Ari Laptev on the occasion of his 70th birthday with great appreciation

We establish results on the accumulation and location of the non-real spectrum of non-selfadjoint matrix differential operators arising in the study of non-linear Schrödinger equations (NLS) in $\mathbb{R}^d$. In particular, without restrictions on the decay rate of the potentials to 0 at ∞, we show that the non-real spectrum cannot accumulate anywhere on the real axis. Under some weak assumptions satisfied, e.g., by L_p-potentials with $p > \frac{d}{2}$, $p \geq 2$, we prove that there are only finitely many non-real eigenvalues and that the non-real eigenvalues are located in a bounded lens-shaped region centered at the origin. Our key tool to prove this is a recent result on the existence of $\mathcal{J}$-semi-definite invariant subspaces for $\mathcal{J}$-selfadjoint operators in Krein spaces as well as abstract operator matrix methods.

1 Introduction

In the study of stability/instability of solutions of non-linear Schrödinger equations (NLS) $\mathrm{i}\partial_t\psi = -\Delta\psi - F(|\psi|^2)\psi$ in $\mathbb{R}^d$ with a real-valued non-negative function F, matrix differential operators of the form

$$\mathcal{H} = \begin{pmatrix} -\Delta + \mu + U & W \\ -W & -(-\Delta + \mu + U) \end{pmatrix}, \tag{1.1}$$

where

$$\operatorname{dom} \mathcal{H} = H^2(\mathbb{R}^d, \mathbb{C}) \oplus H^2(\mathbb{R}^d, \mathbb{C}),$$

in the Hilbert space $L^2(\mathbb{R}^d, \mathbb{C}) \oplus L^2(\mathbb{R}^d, \mathbb{C})$ with $\mu > 0$ and real-valued functions U, W play an important role. In NLS particular non-linearities F and linearizations

---

Keywords: Eigenvalue bounds, eigenvalue accumulation, non-linear Schrödinger operator, NLS, operator matrix, non-selfadjoint operator, Krein space

2020 Mathematics Subject Classification: Primary 35P15; secondary 35Q55, 47A10, 47B50, 47A15

around a ground state $\phi > 0$ give rise to coefficients U, W of the form

$$U = -F(|\phi|^2) - F'(|\phi|^2)|\phi|^2, \quad W = -F'(|\phi|^2)|\phi|^2$$

with certain decay properties to 0 at ∞ such as exponential or polynomial decay, see, e.g., [4, 10, 18, 19].

Here we follow the spirit of [12] where no restrictions on this decay were made. The main result in [12] concerns the exponential decay of eigenfunctions corresponding to eigenvalues in the strip $\{z \in \mathbb{C} : -\mu < \operatorname{Re} z < \mu\}$ under the assumptions that U, W are Δ-bounded with Δ-bound 0 and that U, W tend to 0 at ∞. It is easy to see that, even under these weak assumptions, the essential spectrum of $\mathcal{H}$ equals $(-\infty, -\mu] \cup [\mu, \infty)$, but nothing seems to be known about the accumulation properties and location of the discrete non-real spectrum.

The present paper contributes to fill this gap. We prove two results, the first one under the same weak assumptions as in [12], the second, stronger one under the additional assumption of "polynomially growing constant" in the relative bound 0 condition, see (2.1) below. This assumption allows, in particular, for $U, W \in L^p(\mathbb{R}^d, \mathbb{R})$ with $p > \frac{d}{2}$, $p \geq 2$, and hence for U, W satisfying a decay property

$$|U(x)|, |W(x)| \leq C(1 + \|x\|^2)^{-1-\epsilon}, \quad x \in \mathbb{R}^d,$$

for $\epsilon > 0$, as considered in NLS, see, e.g., [10, A3)] and [17, (4.14)].

More precisely, our first result shows that, under the sole assumption that U, W are Δ-bounded with Δ-bound 0 and U, W tend to 0 at ∞ as in [12], *the non-real eigenvalues of $\mathcal{H}$ cannot accumulate anywhere at its essential spectrum* $(-\infty, -\mu] \cup [\mu, \infty)$. Our second, main result shows, e.g., in the case $U, W \in L^p(\mathbb{R}^d)$ with $p > \frac{d}{2}$, $p \geq 2$, that the *non-real eigenvalues of $\mathcal{H}$ are confined to a bounded lens-shaped region around the origin*, see Figure 1.1 below, and *$\mathcal{H}$ has at most finitely many non-real eigenvalues*. Moreover, the enclosing region for the non-real eigenvalues is described explicitly, including an upper bound for the maximal value of the imaginary part of the non-real eigenvalues. The key ingredient in the proof of these results is a recent joint result with H. Langer on the existence of $\mathcal{J}$-semi-definite invariant subspaces for $\mathcal{J}$-selfadjoint operators in Krein spaces with unbounded "imaginary part", see [13].

Both results of this paper exhibit some interesting features. The first result on the non-accumulation of eigenvalues of $\mathcal{H}$ contrasts results for Schrödinger operators with complex potentials in [2]. Therein a potential in $L^p(\mathbb{R}^d) \cap L^\infty(\mathbb{R}^d)$ with $p > d$, having arbitrarily small norms in both spaces, was constructed so that the non-real eigenvalues accumulate *everywhere* on the real axis. This solved a conjecture by A. Laptev which originated in work of B. Pavlov dating back to 1966, see [16].

In our second result the *shape* of the bounded region enclosing the non-real eigenvalues of the matrix operator $\mathcal{H}$ is unexpected, even when viewed from two different perspectives. Bounded enclosures for non-real eigenvalues have been derived for many different operators, including Schrödinger, Dirac, Jacobi, fractional Schrödinger, discrete Schrödinger or bilayer graphene operators with complex potentials. For Schrödinger operators, the enclosures consist of one disc, see, e.g., [1, 5, 11, 14] and, since $\mathcal{H}$ has two Schrödinger operators with different signs on the diagonal, one would expect the union of two discs as an enclosure. For Dirac operators the enclosures consist of two discs, see, e.g., [6, 8], and, since the essential spectrum of $\mathcal{H}$ is of the same form as for Dirac operators, one would also expect the union of two discs. However, the enclosure obtained in this paper is the *intersection* of two discs and hence much smaller than what one would have expected.

The above discussion only touches upon the very active research area of eigenvalue enclosures for non-selfadjoint operators in mathematical physics, sparked by a paper of Abramov, Aslanyan and Davies twenty years ago, see [1]; many more papers and authors deserve to be mentioned. Moreover, it seems to be impossible to keep any account of this thriving area up-to-date since no references can be added after the publication of a paper. My solution for this problem is to recommend to click on the link https://mathscinet.ams.org/mathscinet/search/publications.html?refcit=2540070&loc=refcit listing all citations of the influential paper [14] by Laptev and Safronov which inspired so many colleagues around Ari and beyond to work in this area, including me. Thank you very much indeed, Ari, for this and for being such a motivating postdoctoral host for my two former PhD students, J.-C. Cuenin and S. Bögli.

2 Auxiliary spectral bounds of independent interest

In this section we only assume that U, W are Δ-bounded with Δ-bound 0. In this case an abstract perturbation result for non-symmetric perturbations S of selfadjoint, not necessarily semi-bounded operators H_0, see [7, Corollary 2.6], applies. It shows that for arbitrarily small angle $\varphi \in (0, \frac{\pi}{2})$ there exists a radius $R(\varphi) > 0$ such that the spectrum of the (closed) operator $H_0 + S$ satisfies

$$\sigma(H_0 + S) \subset \{z \in \mathbb{C} : |\arg z| \leq \varphi\} \cup \big(-\{z \in \mathbb{C} : |\arg z| \leq \varphi\}\big) \cup K_{R(\varphi)}(0),$$

i.e., $\sigma(H_0 + S)$ lies in the union of two sectors with semi-angle φ around the real axis and a ball $K_{R(\varphi)}(0)$ of radius $R(\varphi)$ centered at 0. Typically, the price for making the semi-angle φ smaller is that the radius $R(\varphi)$ increases and, vice versa, achieving good bounds on the imaginary axis may drive the semi-angle towards $\frac{\pi}{2}$.

In the following we will study this interplay more precisely. To this end, we use a finer characterization of the relative bound 0 property often encountered, e.g., in applications to differential operators. This characterization takes into account the growth of the constant $a \geq 0$ when $b \searrow 0$ in the relative bound 0 inequality $\|Sx\| \leq a\|x\| + b\|H_0x\|$ for $x \in \operatorname{dom} H_0 \subset \operatorname{dom} S$. Examples for at most polynomial growth of a when $b \searrow 0$ include Schrödinger operators with L_p-potentials for $p > \frac{d}{2}$, $p \geq 2$, see Corollary 3.3.

The following results will not only play an important role in proving our main result on the operator $\mathcal{H}$ originating in the NLS in the next section, but they may be of independent interest. We begin with the semi-bounded case.

Proposition 2.1. *Let H_0 be a selfadjoint operator in a Hilbert space $\mathscr{H}$ satisfying $\min\sigma(H_0) \geq 0$, and let S be an H_0-bounded operator in $\mathscr{H}$ with H_0-bound 0 and at most polynomially growing constant, i.e., there exist constants a, $b \geq 0$ and $\kappa > 0$ such that, for arbitrarily small $\varepsilon > 0$,*

$$\|Sx\| \leq a\,\varepsilon^{-\kappa}\|x\| + b\,\varepsilon\|H_0 x\|, \quad x \in \operatorname{dom} H_0 \subset \operatorname{dom} S, \tag{2.1}$$

and let $\lambda \in \mathbb{C} \setminus [\min\sigma(H_0), \infty)$. Then $\lambda \in \sigma(H_0 + S)$ implies that

$$\begin{cases} |\lambda - \min\sigma(H_0)| \leq ab^\kappa \dfrac{(\kappa+1)^{\kappa+1}}{\kappa^\kappa} & \text{if } \operatorname{Re}\lambda \leq 0, \\ \dfrac{|\operatorname{Im}\lambda|^{\kappa+1}}{|\lambda|^\kappa} \leq ab^\kappa \dfrac{(\kappa+1)^{\kappa+1}}{\kappa^\kappa} & \text{if } \begin{cases} 0 \leq \operatorname{Re}\lambda \leq \min\sigma(H_0), \\ |\lambda - \frac{1}{2}\min\sigma(H_0)| \geq \frac{1}{2}\min\sigma(H_0), \end{cases} \\ \dfrac{|\lambda - \min\sigma(H_0)|^{\kappa+1}}{(\min\sigma(H_0))^\kappa} \leq ab^\kappa \dfrac{(\kappa+1)^{\kappa+1}}{\kappa^\kappa} & \text{if } |\lambda - \frac{1}{2}\min\sigma(H_0)| \leq \frac{1}{2}\min\sigma(H_0), \\ |\lambda - \min\sigma(H_0)| \dfrac{|\operatorname{Im}\lambda|^\kappa}{|\lambda|^\kappa} \leq ab^\kappa \dfrac{(\kappa+1)^{\kappa+1}}{\kappa^\kappa} & \text{if } \operatorname{Re}\lambda \geq \min\sigma(H_0). \end{cases}$$

Proof. Let $\lambda \notin [\min\sigma(H_0), \infty)$. Then, by assumption and (2.1), $S(H_0-\lambda)^{-1}$ is bounded with

$$\begin{aligned} \|S(H_0-\lambda)^{-1}\| &\leq a\varepsilon^{-\kappa}\|(H_0-\lambda)^{-1}\| + b\varepsilon\|H_0(H_0-\lambda)^{-1}\| \\ &= a\varepsilon^{-\kappa} \sup_{t\in\sigma(H_0)} \frac{1}{|t-\lambda|} + b\varepsilon \sup_{t\in\sigma(H_0)} \frac{|t|}{|t-\lambda|} \\ &=: \alpha_\lambda\varepsilon^{-\kappa} + \beta_\lambda\varepsilon =: g_\lambda(\varepsilon) \end{aligned}$$

for every $\varepsilon \in (0,\infty)$. Clearly, by a Neumann series argument, $\lambda \in \sigma(H_0+S)$ necessitates that

$$1 \leq \|S(H_0-\lambda)^{-1}\| \leq g_\lambda(\varepsilon)$$

for every $\varepsilon \in (0,\infty)$. It is easy to see that g_λ attains a global minimum at $\varepsilon_0 \in (0,\infty)$ with

$$\varepsilon_0 = \left(\kappa\frac{\alpha_\lambda}{\beta_\lambda}\right)^{\frac{1}{\kappa+1}}, \quad g_\lambda(\varepsilon_0) = (\kappa+1)\kappa^{\frac{1}{\kappa+1}-1}\alpha_\lambda^{\frac{1}{\kappa+1}}\beta_\lambda^{1-\frac{1}{\kappa+1}}.$$

Therefore, $\lambda \in \sigma(H_0+S)$ implies that

$$g_\lambda(\varepsilon_0) \geq 1$$

which is equivalent to

$$\alpha_\lambda\beta_\lambda^\kappa \geq \frac{\kappa^\kappa}{(\kappa+1)^{\kappa+1}}.$$

A careful, but elementary analysis shows that

$$\sup_{t\in\sigma(H_0)} \frac{1}{|t-\lambda|} \begin{cases} = \dfrac{1}{|\lambda - \min\sigma(H_0)|} & \text{if } \operatorname{Re}\lambda \le \min\sigma(H_0), \\ \le \dfrac{1}{|\operatorname{Im}\lambda|} & \text{if } \operatorname{Re}\lambda \ge \min\sigma(H_0), \end{cases} \tag{2.2}$$

and

$$\sup_{t\in\sigma(H_0)} \frac{|t|}{|t-\lambda|} \begin{cases} = 1 & \text{if } \operatorname{Re}\lambda \le 0, \\ \le \dfrac{|\lambda|}{|\operatorname{Im}\lambda|} & \text{if } \begin{cases} \operatorname{Re}\lambda \ge 0, \\ |\lambda - \frac{1}{2}\min\sigma(H_0)| \ge \frac{1}{2}\min\sigma(H_0), \end{cases} \\ = \dfrac{\min\sigma(H_0)}{|\lambda - \min\sigma(H_0)|} & \text{if } |\lambda - \frac{1}{2}\min\sigma(H_0)| \le \frac{1}{2}\min\sigma(H_0). \end{cases} \tag{2.3}$$

Combining the above estimates in the condition $g_\lambda(\varepsilon_0) \ge 1$, we obtain the claimed necessary inequalities in the four different regions of $\mathbb{C} \setminus [\min\sigma(H_0), \infty)$. ■

Remark. (i) If H_0 in Proposition 2.1 is bounded from below, but $\min\sigma(H_0) < 0$, then the inequalities in (2.3) become

$$\sup_{t\in\sigma(H_0)} \frac{|t|}{|t-\lambda|} \begin{cases} = 1 & \text{if } \begin{cases} \operatorname{Re}\lambda \le 0, \\ |\lambda - \min\sigma(H_0)| \ge \min\sigma(H_0), \end{cases} \\ \le \dfrac{|\lambda|}{|\operatorname{Im}\lambda|} & \text{if } \operatorname{Re}\lambda \ge 0, \\ = \dfrac{\min\sigma(H_0)}{|\lambda - \min\sigma(H_0)|} & \text{if } |\lambda - \min\sigma(H_0)| \le |\min\sigma(H_0)|. \end{cases} \tag{2.4}$$

Hence similar results as in Proposition 2.1 can be derived also in this case, now in four different regions obtained by combining the estimates (2.4) with those in (2.2) in the proof of Proposition 2.1.

(ii) If $\sigma(H_0) = [\min\sigma(H_0), \infty)$, then the inequalities for the suprema in (2.2), (2.3), (2.4) are, in fact, equalities.

It is not difficult to check that analogous arguments as in the proof of Proposition 2.1 yield the following result for non-semibounded operators, with or without symmetric spectral gap $(-m, m)$ around 0.

Corollary 2.2. *Let H_0 be a selfadjoint operator in a Hilbert space $\mathscr{H}$, let S be an H_0-bounded operator in $\mathscr{H}$ with H_0-bound 0 and at most polynomially growing constant as in* (2.1)*, and let $\lambda \in \mathbb{C} \setminus \mathbb{R}$. Then $\lambda \in \sigma(H_0 + S)$ implies that*

$$|\lambda| \frac{|\operatorname{Im}\lambda|^\kappa}{|\lambda|^\kappa} \le ab^\kappa \frac{(\kappa+1)^{\kappa+1}}{\kappa^\kappa}.$$

If there exists $m > 0$ with

$$\sigma(H_0) \subset (-\infty, -m] \cup [m, \infty)$$

and if $\lambda \in \mathbb{C} \setminus ((-\infty, -m] \cup [m, \infty))$, then

$$\begin{cases} \dfrac{|\mathrm{Im}\,\lambda|^{\kappa+1}}{|\lambda|^{\kappa}} \le ab^{\kappa} \dfrac{(\kappa+1)^{\kappa+1}}{\kappa^{\kappa}} & \text{if } |\mathrm{Re}\,\lambda| \le m,\ |\lambda \mp \tfrac{1}{2}m| \ge \tfrac{1}{2}m, \\ \dfrac{|\lambda \mp m|^{\kappa+1}}{m^{\kappa}} \le ab^{\kappa} \dfrac{(\kappa+1)^{\kappa+1}}{\kappa^{\kappa}} & \text{if } |\lambda \mp \tfrac{1}{2}m| \le \tfrac{1}{2}m, \\ |\lambda \mp m| \dfrac{|\mathrm{Im}\,\lambda|^{\kappa}}{|\lambda|^{\kappa}} \le ab^{\kappa} \dfrac{(\kappa+1)^{\kappa+1}}{\kappa^{\kappa}} & \text{if } |\mathrm{Re}\,\lambda| \ge m. \end{cases}$$

In particular, $\mathbb{C}_{\pm} \cap \varrho(H_0 + S) \neq \emptyset$ for $\mathbb{C}_{\pm} := \{z \in \mathbb{C} : \mathrm{Im}\, z \gtrless 0\}$.

The last result applies directly to the operator $\mathcal{H}$ in (1.1) viewed as a relative bound 0 perturbation of the diagonal operator $H_0 = \mathrm{diag}(-\Delta + \mu, \Delta - \mu)$ with symmetric spectral gap $(-\mu, \mu)$. However, the enclosing region obtained from it is unbounded. Our main theorem in the next section will provide a much stronger result under the additional assumption that U, W tend to 0 at ∞.

3 Invariant subspaces and non-real spectrum

While in the previous section we only assumed that U, W are Δ-bounded with Δ-bound 0, we will now also suppose that U, W tend to 0 at ∞. These two assumptions together ensure that U, W are Δ-compact. This enables us to employ a recent result on the existence of $\mathcal{J}$-semi-definite invariant subspaces of $\mathcal{J}$-selfadjoint operators in Krein spaces, see [13].

The assumption that U, W are Δ-bounded with Δ-bound 0 implies that the off-diagonal entries $\pm W$ are $\pm(-\Delta + \mu + U)$-bounded with relative bound 0, see [21, Lemma 2.1.6], and hence the operator matrix $\mathcal{H}$ in (1.1) is diagonally dominant of order 0, see [20] or [21, Definition 2.2.3]. Since U, W tend to 0 at ∞ and hence U, W are $-\Delta$-compact, it also readily follows from [15, Theorem 2.2], see also [21, Theorem 2.4.1], that

$$\sigma_{\mathrm{ess}}(\mathcal{H}) = \sigma_{\mathrm{e}i}(\mathcal{H}) = (-\infty, -\mu] \cup [\mu, \infty), \quad i = 1, 2, 3, 4, 5,$$

where

$$\sigma_{\mathrm{ess}}(\mathcal{H}) := \{\lambda \in \mathbb{C} : \mathcal{H} - \lambda \text{ is not Fredholm}\}$$

is the essential spectrum $\sigma_{\mathrm{e}3}(\mathcal{H})$ in [9, Section IX.1] for non-selfadjoint operators. Note that $\sigma_{\mathrm{ess}}(\mathcal{H}) = \sigma_{\mathrm{e}i}(\mathcal{H})$, $i = 1, 2, 3, 4$, therein since U, W are real-valued and hence $\mathcal{H}$ is symmetric with respect to complex conjugation, see [9, Theorem IX.1.6], and $\sigma_{\mathrm{ess}}(\mathcal{H}) = \sigma_{\mathrm{e}5}(\mathcal{H})$ since $\mathbb{C}_{\pm} \cap \varrho(\mathcal{H}) \neq \emptyset$ by Corollary 2.2.

The key tool to prove our main result relies on the $\mathcal{J}$-selfadjointness of the operator $\mathcal{H}$ with respect to the indefinite inner product $[\cdot,\cdot]$ on $\mathcal{K}$ given by

$$[x,y] := (\mathcal{J}x, y), \quad \mathcal{J} := \begin{pmatrix} I_{L^2(\mathbb{R}^d,\mathbb{C})} & 0 \\ 0 & -I_{L^2(\mathbb{R}^d,\mathbb{C})} \end{pmatrix}, \quad x, y \in \mathcal{K},$$

rendering the product Hilbert space $\mathcal{K} := L^2(\mathbb{R}^d,\mathbb{C}) \oplus L^2(\mathbb{R}^d,\mathbb{C})$ a Krein space, see [21, Theorem 2.6.6]. Here $\mathcal{H}$ is called $\mathcal{J}$-selfadjoint if $\mathcal{J}\mathcal{H}$ is selfadjoint in $L^2(\mathbb{R}^d,\mathbb{C}) \oplus L^2(\mathbb{R}^d,\mathbb{C})$ with the Hilbert space inner product.

The spectrum of every $\mathcal{J}$-selfadjoint operator in a Krein space is symmetric to the real axis. Since U, W are real-valued, $\mathcal{H}$ is not only $\mathcal{J}$-selfadjoint, but $\mathrm{i}\mathcal{H}$ is $\widetilde{\mathcal{J}}$-selfadjoint with respect to the indefinite inner product $\langle\cdot,\cdot\rangle$ on $\mathcal{K}$ given by

$$[x,y] := (\widetilde{\mathcal{J}}x, y), \quad \widetilde{\mathcal{J}} := \begin{pmatrix} 0 & \mathrm{i}I_{L^2(\mathbb{R}^d,\mathbb{C})} \\ -\mathrm{i}I_{L^2(\mathbb{R}^d,\mathbb{C})} & 0 \end{pmatrix}, \quad x, y \in \mathcal{K}.$$

This two-fold indefinite symmetry is the abstract reason for the well-known property that the spectrum of $\mathcal{H}$ is symmetric to both $\mathbb{R}$ and $\mathrm{i}\mathbb{R}$.

The following notions for subspaces of Krein spaces will be needed in the sequel, see, e.g., [3] and also [13, Section 1]. A subspace $L \subset \mathcal{K}$ is called $\mathcal{J}$-non-negative if $[x,x] \geq 0$ for all $x \in L$ and maximal $\mathcal{J}$-non-negative if it is not properly contained in any other $\mathcal{J}$-non-negative subspace of $\mathcal{K}$; analogously, (maximal) $\mathcal{J}$-non-positive subspaces are defined. A subspace L is called $\mathcal{J}$-neutral if $[x,x] = 0$ for all $x \in L$. Finally, the $\mathcal{J}$-orthogonal companion of L is defined as $L^{[\perp]} := \{x \in \mathcal{K} : [x, L] = 0\}$.

The following theorem is the first main result of this paper. It shows that the non-real eigenvalues of the operator $\mathcal{H}$ in (1.1) associated to NLS cannot accumulate anywhere at the essential spectrum $(-\infty,-\mu] \cup [\mu,\infty)$ of $\mathcal{H}$.

Theorem 3.1. *Suppose that U, W are Δ-bounded with Δ-bound 0 and that U, W tend to 0 at ∞. Then $\mathcal{H}$ possesses invariant subspaces $\mathcal{L}_+$ and $\mathcal{L}_- = \mathcal{L}_+^{[\perp]}$, i.e.,*

$$\mathcal{H}(\mathcal{L}_\pm \cap \operatorname{dom}\mathcal{H}) \subset \mathcal{L}_\pm, \tag{3.1}$$

such that $\mathcal{L}_+$ is maximal $\mathcal{J}$-non-negative, $\mathcal{L}_-$ is maximal $\mathcal{J}$-non-positive,

$$\overline{\mathcal{L}_\pm \cap (H^2(\mathbb{R}^d,\mathbb{C}) \oplus H^2(\mathbb{R}^d,\mathbb{C}))} = \mathcal{L}_\pm,$$
$$P_\pm\big(\mathcal{L}_\pm \cap (H^2(\mathbb{R}^d,\mathbb{C}) \oplus H^2(\mathbb{R}^d,\mathbb{C}))\big) = H^2(\mathbb{R}^d,\mathbb{C}).$$

The restrictions $\mathcal{H}|_{\mathcal{L}_+}$ and $\mathcal{H}|_{\mathcal{L}_-}$ are isomorphic to the operators in $L^2(\mathbb{R}^d,\mathbb{C})$ given by

$$\mathcal{H}|_{\mathcal{L}_+} \cong -\Delta + \mu + U + WK, \quad \mathcal{H}|_{\mathcal{L}_-} \cong -(-\Delta + \mu + U) + WK^*, \tag{3.2}$$

where $K : L^2(\mathbb{R}^d,\mathbb{C}) \to L^2(\mathbb{R}^d,\mathbb{C})$ is a contraction, i.e., $\|K\| \leq 1$, satisfying

$$\mathcal{L}_+ = \left\{\begin{pmatrix} x \\ Kx \end{pmatrix} : x \in L^2(\mathbb{R}^d,\mathbb{C})\right\}, \quad \mathcal{L}_- = \left\{\begin{pmatrix} -K^*y \\ y \end{pmatrix} : y \in L^2(\mathbb{R}^d,\mathbb{C})\right\}.$$

Moreover, for every $z \in \sigma(\mathcal{H}) \cap \mathbb{C}_+ = \sigma_p(\mathcal{H}) \cap \mathbb{C}_+$ *in the open upper half-plane* $\mathbb{C}_+$ *with corresponding algebraic eigenspace* $\mathcal{S}_z$*, we have*

$$z \in \sigma(\mathcal{H}) \cap \mathbb{C}_+ \implies z \in \sigma(\mathcal{H}|_{\mathcal{L}_+}) \cap \sigma(\mathcal{H}|_{\mathcal{L}_-}), \qquad \mathcal{S}_z \subset \mathcal{L}_+ \cap \mathcal{L}_-, \tag{3.3}$$

and $\sigma(\mathcal{H}) \setminus \mathbb{R}$ *does not have any finite accumulation point, in particular, it does not accumulate anywhere at* $\sigma_{\text{ess}}(\mathcal{H}) = (-\infty, -\mu] \cup [\mu, \infty)$.

Proof. The assumptions on U, W ensure that $\mathcal{H}$ is diagonally dominant of order 0 and the off-diagonal entry W is $(-\Delta + \mu + U)$-compact. Therefore $\mathcal{H}$ satisfies the conditions of [13, Theorem 4.2]. The latter implies the existence of maximal $\mathcal{J}$-semi-definite invariant subspaces $\mathcal{L}_\pm$ with the required properties and of the contraction $K : L^2(\mathbb{R}^d, \mathbb{C}) \to L^2(\mathbb{R}^d, \mathbb{C})$ representing them.

To prove (3.2), we define $Q_+ : \mathcal{L}_+ \to L^2(\mathbb{R}^d, \mathbb{C})$ by

$$Q_+ \begin{pmatrix} x \\ Kx \end{pmatrix} := x, \quad x \in L^2(\mathbb{R}^d, \mathbb{C}).$$

Then Q_+ is bijective and, together with the invariance $\mathcal{H}(\mathcal{L}_\pm \cap \operatorname{dom} \mathcal{H}) \subset \mathcal{L}_\pm$ of $\mathcal{L}_\pm$ for $\mathcal{H}$ by (3.1), it follows that

$$Q_+ \mathcal{H}|_{\mathcal{L}_+} Q_+^{-1} = -\Delta + \mu + U + WK$$

in $L^2(\mathbb{R}^d, \mathbb{C})$. The proof for $\mathcal{H}|_{\mathcal{L}_-}$ is analogous.

The implication (3.3) follows from [13, Theorems 4.2, 3.1] if, therein, we choose $\widehat{\sigma} = \sigma(\mathcal{H}) \cap \mathbb{C}_+$. By [13, Theorem 4.2] it also follows that non-real eigenvalues of $\mathcal{H}$ may accumulate at most at the intersection of the essential spectra of the diagonal elements of $\mathcal{H}$, i.e., at

$$\begin{aligned} &\sigma_{\text{ess}}(-\Delta + \mu + U) \cap \sigma_{\text{ess}}(-(-\Delta + \mu + U)) \\ &= \sigma_{\text{ess}}(-\Delta + \mu) \cap \sigma_{\text{ess}}(-(-\Delta + \mu)) \\ &= (-\infty, -\mu] \cap [\mu, \infty) = \emptyset, \end{aligned}$$

and hence they cannot have any finite accumulation point. ∎

The previous theorem will now be used to establish the main result of this paper concerning the location of the non-real spectrum, i.e., of the non-real eigenvalues of $\mathcal{H}$. It shows that the non-real eigenvalues of $\mathcal{H}$ are confined to a bounded lens-shaped region around the origin, whose extreme points on the imaginary axis and on the real axis are quantified in terms of the coefficients U, W of $\mathcal{H}$.

Theorem 3.2. *Suppose that* U, W *are* Δ*-bounded with* Δ*-bound* 0 *and at most polynomially growing constant, i.e., there exist constants* $a_U, a_W, b_U, b_W \geq 0$ *and* $\kappa > 0$ *such that, for arbitrarily small* $\varepsilon > 0$,

$$\|Ux\| \leq a_U \varepsilon^{-\kappa} \|x\| + b_U \varepsilon \|(-\Delta + \mu)x\|, \quad x \in H^2(\mathbb{R}^d, \mathbb{C}), \tag{3.4}$$

$$\|Wx\| \leq a_W \varepsilon^{-\kappa} \|x\| + b_W \varepsilon \|(-\Delta + \mu)x\|, \quad x \in H^2(\mathbb{R}^d, \mathbb{C}), \tag{3.5}$$

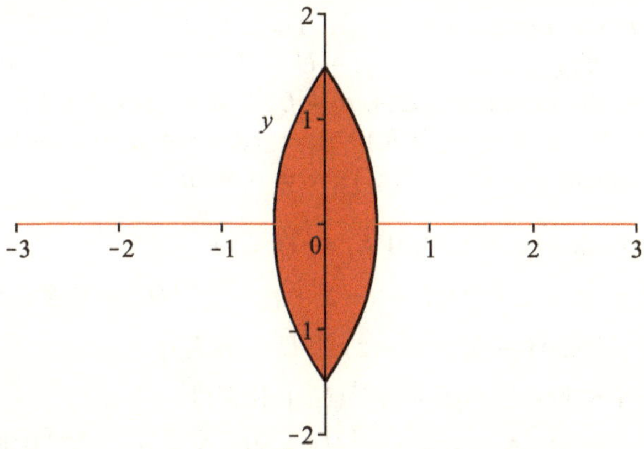

Figure 1.1. The lens-shaped spectral enclosure for $\sigma(\mathcal{H}) \setminus \mathbb{R}$ in Theorem 3.2, (3.6), in the case $\sigma_{\mathrm{ess}}(\mathcal{H}) = (-\infty, -2] \cup [2, \infty)$, i.e., $\mu = 2$, and $r_{U,W} = 2.5$.

and assume that U, W tend to 0 at ∞. Then the non-real spectrum

$$\sigma(\mathcal{H}) \setminus \mathbb{R} = \sigma_{\mathrm{p}}(\mathcal{H}) \setminus \mathbb{R}$$

of $\mathcal{H}$ in (1.1) *is confined to the bounded lens-shaped region*

$$\begin{aligned} \sigma(\mathcal{H}) \setminus \mathbb{R} \subset \{z \in \mathbb{C} : \operatorname{Re} z \le 0, |z - \mu| \le r_{U,W}\} \\ \cap \{z \in \mathbb{C} : \operatorname{Re} z \ge 0, |z + \mu| \le r_{U,W}\}, \end{aligned} \tag{3.6}$$

see Figure 1.1*, where $r_{U,W}$ is given by*

$$r_{U,W} := (a_U b_U^\kappa + a_W b_W^\kappa) \frac{(\kappa + 1)^{\kappa+1}}{\kappa^\kappa}, \tag{3.7}$$

and $\mathcal{H}$ has at most finitely many non-real eigenvalues. In particular, every non-real eigenvalue $z \in \sigma(\mathcal{H}) \setminus \mathbb{R}$ satisfies

$$|\operatorname{Re} z - \mu| \le r_{U,W}, \quad |\operatorname{Re} z + \mu| \le r_{U,W}, \quad |\operatorname{Im} z| \le \sqrt{r_{U,W}^2 - \mu^2}.$$

Remark. Theorem 3.2 also yields a criterion for the absence of non-real spectrum of $\mathcal{H}$, namely $r_{U,W} \le \mu$, i.e.,

$$(a_U b_U^\kappa + a_W b_W^\kappa) \frac{(\kappa + 1)^{\kappa+1}}{\kappa^\kappa} \le \mu \implies \sigma(\mathcal{H}) \setminus \mathbb{R} = \emptyset; \tag{3.8}$$

if the inequality on the left-hand side is strict, then

$$\sigma(\mathcal{H}) \subset (-\infty, -\mu + r_{U,W}] \,\dot{\cup}\, [\mu - r_{U,W}, \infty)$$

exhibits a spectral gap around 0, compare [7]. In the context of operators $\mathcal{H}$ arising from NLS, it is usually known that $0 \in \sigma(\mathcal{H})$, therefore the interesting case is that $r_{U,W} \ge \mu$ or even $r_{U,W} > \mu$, i.e., the case that (3.8) is violated.

Proof of Theorem 3.2. Let $z \in \sigma(\mathcal{H}) \setminus \mathbb{R}$. Then, by (3.3) and (3.2) in Theorem 3.1, $z \in \sigma(\mathcal{H}|_{\mathcal{L}_+}) \cap \sigma(\mathcal{H}|_{\mathcal{L}_-}) = \sigma(-\Delta + \mu + U + WK) \cap \sigma(-(-\Delta + \mu + U) + WK^*)$. First we consider the operator $-\Delta + \mu + U + WK$. Since $z \notin \mathbb{R}$ by assumption, we have $z \in \sigma(-\Delta + \mu + U + WK) \setminus \mathbb{R} = \sigma_{\mathrm{p}}(-\Delta + \mu + U + WK) \setminus \mathbb{R}$. Hence there exists an element $x \in H^2(\mathbb{R}^d, \mathbb{C}))$, $x \neq 0$, with

$$
\begin{aligned}
&(-\Delta + \mu + U + WK - z)x = 0 \\
\iff &(-\Delta + \mu - z)\big(I + (-\Delta + \mu - z)^{-1}(U + WK)\big)x = 0 \\
\iff &-1 \in \sigma\big((-\Delta + \mu - z)^{-1}(U + WK)\big) \\
\implies &1 \leq \|(-\Delta + \mu - z)^{-1}(U + WK)\| \\
&\quad\leq \|(-\Delta + \mu - z)^{-1}U\| + \|(-\Delta + \mu - z)^{-1}W\| \\
\implies &1 \leq \|U(-\Delta + \mu - \overline{z})^{-1}\| + \|W(-\Delta + \mu - \overline{z})^{-1}\|;
\end{aligned}
$$

here in the last two lines we have used that K is a contraction by Theorem 3.1, that $(-\Delta + \mu - z)^{-1}$ is bounded and hence, e.g.,

$$((-\Delta + \mu - z)^{-1}U)^* = U^*((-\Delta + \mu - z)^{-1})^* = U(-\Delta + \mu - \overline{z})^{-1}$$

because U, W are real-valued. Using the estimates in the proof of Proposition 2.1 with $H_0 = -\Delta + \mu$, $m = \mu$ and with $S = U$ and $S = W$, we conclude that

$$
\begin{aligned}
|z - \mu| &= |\overline{z} - \mu| \\
&\leq a_U b_U^{\kappa} \frac{(\kappa + 1)^{\kappa+1}}{\kappa^{\kappa}} + a_W b_W^{\kappa} \frac{(\kappa + 1)^{\kappa+1}}{\kappa^{\kappa}} \\
&= r_{U,W} \quad \text{if } \operatorname{Re} z \leq 0.
\end{aligned}
$$

Similarly, considering the operator $-(-\Delta + \mu + U) + WK^*$, we arrive at

$$
\begin{aligned}
|z + \mu| &= |\overline{z} + \mu| \\
&\leq a_U b_U^{\kappa} \frac{(\kappa + 1)^{\kappa+1}}{\kappa^{\kappa}} + a_W b_W^{\kappa} \frac{(\kappa + 1)^{\kappa+1}}{\kappa^{\kappa}} \\
&= r_{U,W} \quad \text{if } \operatorname{Re} z \geq 0.
\end{aligned}
$$

This completes the proof of (3.6). The last two estimates are immediate consequences of (3.6). Since the spectral enclosure on the right-hand side of (3.6) is bounded and, by Theorem 3.1, the non-real eigenvalues of $\mathcal{H}$ have no finite accumulation point, there can only be finitely many. ■

Corollary 3.3. *Suppose that U, $W \in L_p(\mathbb{R}^d, \mathbb{R})$ for some $p > \frac{d}{2}$, $p \geq 2$. Then the non-real spectrum $\sigma(\mathcal{H}) \setminus \mathbb{R} = \sigma_{\mathrm{p}}(\mathcal{H}) \setminus \mathbb{R}$ of $\mathcal{H}$ in* (1.1) *is confined to the bounded lens-shaped region*

$$\sigma(\mathcal{H}) \setminus \mathbb{R} \subset \{z \in \mathbb{C} : \operatorname{Re} z \leq 0,\ |z - \mu| \leq r_p\} \cap \{z \in \mathbb{C} : \operatorname{Re} z \geq 0,\ |z + \mu| \leq r_p\}$$

with r_p given by

$$r_p := C_p^{\frac{2p}{2p-d}} \Big(\|U\|_p^{\frac{2p}{2p-d}} + \|W\|_p^{\frac{2p}{2p-d}} \Big) \frac{(2p)^{\frac{2p}{2p-d}}}{(2p-d)d^{\frac{d}{2p-d}}},$$

where

$$C_p := (2\pi)^{-\frac{d}{p}} 2 \Big(\pi^{\frac{d}{2}} \frac{\Gamma(p-\frac{d}{2})}{\Gamma(p)} \Big)^{\frac{1}{p}},$$

and $\mathcal{H}$ has at most finitely many non-real eigenvalues.

Proof. In the following, we denote the norm in $L_p(\mathbb{R}^d, \mathbb{C})$ by $\|\cdot\|_p$, also in the case $p = 2$ instead of $\|\cdot\|$ in $L_2(\mathbb{R}^d, \mathbb{C})$. Under the given assumptions on p, the estimate

$$\|Uf\|_2 \le C_p \|U\|_p \Big(t^{-\frac{d}{p}} \|f\|_2 + t^{-\frac{d}{p}+2} \|(-\Delta + \mu) f\|_2 \Big), \quad f \in H^2(\mathbb{R}^d, \mathbb{C})),$$

and the analogous estimate for W can be proved following the lines of the proof of [22, Satz 17.7].[1] If we set $\varepsilon := t^{-\frac{d}{p}+2}$, then (3.4) and (3.5) hold with

$$\kappa = \frac{d}{2p-d}, \quad a_U = b_U = C_p \|U\|_p, \quad a_W = b_W = C_p \|W\|_p.$$

Inserting this into (3.7), we find that

$$\begin{aligned} r_{U,W} &= \Big((C_p \|U\|_p)^{\frac{2p}{2p-d}} + (C_p \|W\|_p)^{\frac{2p}{2p-d}} \Big) \frac{(\frac{2p}{2p-d})^{\frac{2p}{2p-d}}}{(\frac{d}{2p-d})^{\frac{d}{2p-d}}} \\ &= C_p^{\frac{2p}{2p-d}} \Big(\|U\|_p^{\frac{2p}{2p-d}} + \|W\|_p^{\frac{2p}{2p-d}} \Big) \frac{(2p)^{\frac{2p}{2p-d}}}{(2p-d)d^{\frac{d}{2p-d}}}. \end{aligned}$$

Now the claims follow from Theorem 3.2 if we set $r_p := r_{U,W}$. ∎

Acknowledgements. The author gratefully acknowledges the support of the Swiss National Science Foundation (SNF), under grant no. 169104. The author would like to thank the referee for the careful reading of this paper and for pointing out some typos.

Bibliography

[1] A. A. Abramov, A. Aslanyan and E. B. Davies, Bounds on complex eigenvalues and resonances. *J. Phys. A* **34** (2001), 57–72

[2] S. Bögli, Schrödinger operator with non-zero accumulation points of complex eigenvalues. *Comm. Math. Phys.* **352** (2017), 629–639

[3] J. Bognár, *Indefinite inner product spaces*. Springer, New York, 1974

[1] Note that in [7, formula (5.12)] the factors $\frac{\pi^{\frac{d}{2}}}{\Gamma(\frac{d}{2})} B(\frac{d}{2}, p - \frac{d}{2}) = \pi^{\frac{d}{2}} \frac{\Gamma(p-\frac{d}{2})}{\Gamma(p)}$ are missing the exponent $\frac{1}{p}$.

[4] S. Cuccagna, D. Pelinovsky and V. Vougalter, Spectra of positive and negative energies in the linearized NLS problem. *Comm. Pure Appl. Math.* **58** (2005), 1–29

[5] J.-C. Cuenin, Eigenvalue bounds for Dirac and fractional Schrödinger operators with complex potentials. *J. Funct. Anal.* **272** (2017), 2987–3018

[6] J.-C. Cuenin, A. Laptev and C. Tretter, Eigenvalue estimates for non-selfadjoint Dirac operators on the real line. *Ann. Henri Poincaré* **15** (2014), 707–736

[7] J.-C. Cuenin and C. Tretter, Non-symmetric perturbations of self-adjoint operators. *J. Math. Anal. Appl.* **441** (2016), 235–258

[8] P. D'Ancona, L. Fanelli and N. M. Schiavone, Eigenvalue bounds for non-selfadjoint Dirac operators. *Math. Ann.* (2020), DOI 10.1007/s00208-021-02158-x

[9] D. E. Edmunds and W. D. Evans, *Spectral theory and differential operators*. Oxford Math. Monogr., Clarendon Press, New York, 1987

[10] M. B. Erdoğan and W. Schlag, Dispersive estimates for Schrödinger operators in the presence of a resonance and/or an eigenvalue at zero energy in dimension three. II. *J. Anal. Math.* **99** (2006), 199–248

[11] R. L. Frank, Eigenvalue bounds for Schrödinger operators with complex potentials. *Bull. Lond. Math. Soc.* **43** (2011), 745–750

[12] D. Hundertmark and Y.-R. Lee, Exponential decay of eigenfunctions and generalized eigenfunctions of a non-self-adjoint matrix Schrödinger operator related to NLS. *Bull. Lond. Math. Soc.* **39** (2007), 709–720

[13] H. Langer and C. Tretter, Maximal $\mathcal{J}$-semi-definite invariant subspaces of unbounded $\mathcal{J}$-selfadjoint operators in Krein spaces. *J. Math. Anal. Appl.* **494** (2021), Article ID 124597

[14] A. Laptev and O. Safronov, Eigenvalue estimates for Schrödinger operators with complex potentials. *Comm. Math. Phys.* **292** (2009), 29–54

[15] M. Marletta and C. Tretter, Essential spectra of coupled systems of differential equations and applications in hydrodynamics. *J. Differential Equations* **243** (2007), 36–69

[16] B. S. Pavlov, On a non-selfadjoint Schrödinger operator. In *Probl. math. phys., No. I: Spectral theory and wave processes*, pp. 102–132, Izdat. Leningrad. Univ., Leningrad, 1966 (in Russian)

[17] I. Rodnianski and W. Schlag, Time decay for solutions of Schrödinger equations with rough and time-dependent potentials. *Invent. Math.* **155** (2004), 451–513

[18] I. Rodnianski, W. Schlag and A. Soffer, Asymptotic stability of N-soliton states of NLS. Preprint 2003, arXiv:math/0309114v1

[19] W. Schlag, Stable manifolds for an orbitally unstable nonlinear Schrödinger equation. *Ann. of Math. (2)* **169** (2009), 139–227

[20] C. Tretter, Spectral issues for block operator matrices. In *Differential equations and mathematical physics (Birmingham, AL, 1999)*, pp. 407–423, AMS/IP Stud. Adv. Math. 16, American Mathematical Society, Providence, 2000

[21] C. Tretter, *Spectral theory of block operator matrices and applications*. Imperial College Press, London, 2008

[22] J. Weidmann, *Lineare Operatoren in Hilberträumen. Teil II*. Math. Leitf., B. G. Teubner, Stuttgart, 2003

Scattering theory for Laguerre operators

Dimitri R. Yafaev

To Ari Laptev on the occasion of his 70th birthday

We study Jacobi operators J_p, $p > -1$, whose eigenfunctions are Laguerre polynomials. All operators J_p have absolutely continuous simple spectra coinciding with the positive half-axis. This fact, however, by no means imply that the wave operators for the pairs J_p, J_q where $p \neq q$ exist. Our goal is to show that, nevertheless, this is true and to find explicit expressions for these wave operators. We also study the time evolution of $(e^{-Jt} f)_n$ as $|t| \to \infty$ for Jacobi operators J whose eigenfunctions are different classical polynomials. For Laguerre polynomials, it turns out that the evolution $(e^{-J_p t} f)_n$ is concentrated in the region where $n \sim t^2$ instead of $n \sim |t|$ as happens in standard situations. As a by-product of our considerations, we obtain universal relations between amplitudes and phases in asymptotic formulas for general orthogonal polynomials.

1 Introduction

1.1 Jacobi operators

Jacobi operators J are discrete analogues of one-dimensional differential operators. They are defined in the space $\ell^2(\mathbb{Z}_+)$ by the formula

$$(Jf)_n = a_{n-1} f_{n-1} + b_n f_n + a_n f_{n+1}, \quad n \in \mathbb{Z}_+ = \{0, 1, 2, \ldots\}, \quad a_{-1} = 0. \quad (1.1)$$

We always suppose that $a_n > 0$ and $b_n = \bar{b}_n$. Of course spectral properties of Jacobi operators depend crucially on a behavior of the coefficients a_n and b_n as $n \to \infty$.

In the simplest case $a_n = a_\infty > 0$, $b_n = 0$, the operator J (known as the "free" discrete Schrödinger operator) has the absolutely continuous spectrum $[-2a_\infty, 2a_\infty]$, and its eigenfunctions are expressed via Chebyshev polynomials of second kind. Operators J whose eigenfunctions are Jacobi polynomials are natural generalizations of this operator.

Keywords: Jacobi operators, Laguerre polynomials, asymptotic formulas
2020 Mathematics Subject Classification: 33C45, 39A70, 47A40, 47B39

The situation $a_n \to \infty$ as $n \to \infty$ is also quite common in applications. We discuss only the case where the Carleman condition

$$\sum_{n=0}^{\infty} a_n^{-1} = \infty \tag{1.2}$$

is satisfied. Suppose that $\frac{b_n}{2a_n} \to \gamma$ as $n \to \infty$. If $|\gamma| < 1$, then the spectrum of the operator J is purely absolutely continuous and coincides with the whole axis $\mathbb{R}$; see [2,5]. A famous example is

$$a_n = \sqrt{\frac{n+1}{2}}, \qquad b_n = 0. \tag{1.3}$$

Eigenfunctions of the corresponding operator J are the Hermite polynomials. If $|\gamma| > 1$ ($|\gamma| = \infty$ is admitted), then the spectrum of J is discrete. The case $|\gamma| = 1$ is critical, and the spectral properties of J depend on details of the behavior of $\frac{b_n}{2a_n} - \gamma$ as $n \to \infty$. The results pertaining to this situation are scarce. We mention only the papers [6, 7] and the references therein.

Here we study an important particular case

$$a_n = a_n^{(p)} = \sqrt{(n+1)(n+1+p)}, \quad b_n = b_n^{(p)} = 2n + p + 1, \quad p > -1. \tag{1.4}$$

Thus we have $\gamma = 1$. For all p, the Jacobi operators $J = J_p$ (we call them the Laguerre operators) with the recurrence coefficients (1.4) have absolutely continuous spectra coinciding with $[0, \infty)$. Eigenfunctions of these operators are the Laguerre polynomials $L_n^{(p)}(\lambda)$. Our goal is to investigate an asymptotic behavior of the unitary group e^{-iJ_pt} as $t \to \pm\infty$. We show that, for different p, these asymptotics are essentially the same although the operators $J_q - J_p$ are not even compact unless $q = p$.

Another goal of the paper is to obtain detailed asymptotic formulas for $(e^{-iJt}f)_n$ as $|t| \to \infty$ for sufficiently arbitrary Jacobi operators J. Here we suppose that an asymptotic behavior of the corresponding orthogonal polynomials $P_n(\lambda)$ as $n \to \infty$ is known. These general results are illustrated on examples of the classical polynomials. In particular, for Laguerre operators J_p, we show that the evolution $(e^{-J_pt}f)_n$ is concentrated in the region where $n \sim t^2$ instead of $n \sim |t|$ as happens in standard situations.

1.2 Scattering theory

We work in the framework of scattering theory. Let us briefly recall its basic notions. We refer, e.g., to the book [10] for a detailed presentation. Consider a couple of self-adjoint operators A and B in some Hilbert space $\mathcal{H}$. In view of our applications, we suppose that both of these operators are absolutely continuous. Scattering theory studies the strong limits

$$W_\pm = W_\pm(B, A) = \operatorname*{s-lim}_{t \to \pm\infty} e^{iBt} e^{-iAt} \tag{1.5}$$

known as the wave operators for the pair A, B. If the limits (1.5) exist, then the wave operators possess several useful features. In particular, they are isometric and enjoy the intertwining property $BW_\pm = W_\pm A$. It follows that the restriction of the operator B on the image $\operatorname{Ran} W_\pm$ of $W_\pm$ is unitary equivalent to the operator A. If both wave operators $W_\pm(B, A)$ and $W_\pm(A, B)$ exist, then the operators A and B are unitarily equivalent. In this case the spectra of the operators A and B coincide.

The existence of the limits (1.5) is a non-trivial problem. We emphasize that the unitary equivalence of operators A and B does not imply the existence of the wave operators (1.5). A notorious example is given by the pair of multiplication A, $(Au)(x) = xu(x)$, and differential $B = -\frac{id}{dx}$ operators in the space $L^2(\mathbb{R})$. Even simpler example is given by $B = A + cI$, where I the identity operator and $c \neq 0$ is a real number. Generally speaking, the limits (1.5) (or their appropriate modifications) exist if the perturbation $B - A$ is in some sense small. This might mean different things. For example, it suffices to assume that the operator $B - A$ is trace class or that it acts as an integral operator with smooth kernel in the diagonal representation of the operator A (or B).

Our aim is to develop scattering theory for pairs of the operators J_p, J_q. According to (1.4) we have

$$a_n^{(q)} - a_n^{(p)} = \frac{q-p}{2} + O(n^{-1}), \quad n \to \infty, \qquad b_n^{(q)} - b_n^{(p)} = q - p, \tag{1.6}$$

so that the operator $J_q - J_p$ is not even compact. Therefore standard methods of scattering theory do not work in this case. It turns out however that, from the viewpoint of scattering theory, the diagonal and off-diagonal terms in (1.6) compensate each other. Note that if only one of the coefficients a_n or b_n is changed, then the spectrum of the operator J_p is shifted. In this case the wave operators cannot exist.

Although the difference $J_q - J_p$ is by no means small, there exists a natural one-to-one correspondence between eigenfunctions of the operators J_p for different p. Their asymptotics as $n \to \infty$ differ by a phase shift only. This allows us to show that the wave operators $W_\pm(J_q, J_p)$ exist for all $p, q > -1$. This result is obtained by a direct calculation which yields also explicit expressions for the wave operators.

1.3 Structure of the paper

Jacobi operators and associated orthogonal polynomials, in particular, Laguerre polynomials, are discussed in Section 2.

In Section 3, we prove the existence of the wave operators $W_\pm(J_q, J_p)$ (Theorem 3.1). We also construct scattering theory for pairs $J, \widetilde{J}$ where $J = J_p$ for some $p > -1$ and the coefficients of a Jacobi operator $\widetilde{J}$ are sufficiently close to those of J (Theorem 3.8).

In Section 4.1, we exhibit a link between a large time behavior of $(e^{-iJt} f)_n$ and asymptotics of associated orthogonal polynomials $P_n(\lambda)$ as $n \to \infty$. This leads to universal relations between amplitudes and phases in asymptotic formulas for $P_n(\lambda)$.

These results are illustrated in Sections 4.2–4.4 on examples of classical polynomials. The results for Laguerre, Jacobi and Hermite polynomials are stated as Theorems 4.5, 4.9 and 4.10, respectively. The Hermite operator J is somewhat exceptional since the evolution $e^{-iJt} f$ is dispersionless in this case.

Below $\|\cdot\|$ is the norm in the space $\ell^2(\mathbb{Z}_+)$; C are different positive constants whose precise values are of no importance.

2 Jacobi operators and orthogonal polynomials

2.1 Orthogonal polynomials

Here we collect necessary information about the Jacobi operators J given by formula (1.1) and associated orthogonal polynomials $P_n(z)$; see the books [1, 8] for a comprehensive presentation. Given coefficients $a_n > 0$, $b_n = \bar{b}_n$, $n \in \mathbb{Z}_+$, one constructs $P_n(z)$ by the recurrence relation

$$a_{n-1} P_{n-1}(z) + b_n P_n(z) + a_n P_{n+1}(z) = z P_n(z), \quad n \in \mathbb{Z}_+, \ z \in \mathbb{C}, \tag{2.1}$$

and the boundary conditions $P_{-1}(z) = 0$, $P_0(z) = 1$. Then $P_n(z)$ is a polynomial of degree n. Obviously, $P(z) = \{P_n(z)\}_{n=0}^\infty$ satisfies the equation $JP(z) = zP(z)$, that is, it is an "eigenvector" of the operator J.

We consider the operator J in the space $\ell^2(\mathbb{Z}_+)$. Let us denote by J_0 the minimal operator defined by formula (1.1) on a set $\mathcal{D}$ of vectors $f = \{f_n\}_{n=0}^\infty$ with only a finite number of non-zero components. This operator is symmetric; moreover, it is essentially self-adjoint if the Carleman condition (1.2) is satisfied. In particular, condition (1.2) holds true for the coefficients (1.4). For essentially self-adjoint operators, the domain $\mathcal{D}(J)$ of the closure J of the operator J_0 consists of all vectors $f \in \ell^2(\mathbb{Z}_+)$ such that $Jf \in \ell^2(\mathbb{Z}_+)$. The spectrum of the self-adjoint operator J is simple with $e_0 = (1, 0, 0, \dots)^\top$ being a generating vector. Therefore it is natural to define the spectral measure of J by the relation $d\rho(\lambda) = d(E(\lambda)e_0, e_0)$, where $E(\lambda)$ is the spectral family of the operator J. The polynomials $P_n(\lambda)$ (we call them orthonormal) are orthogonal and normalized in the spaces $L^2(\mathbb{R}; d\rho)$:

$$\int_{-\infty}^{\infty} P_n(\lambda) P_m(\lambda)\, d\rho(\lambda) = \delta_{n,m}; \tag{2.2}$$

as usual, $\delta_{n,n} = 1$ and $\delta_{n,m} = 0$ for $n \neq m$.

Alternatively, given a probability measure $d\rho(\lambda)$, the polynomials $P_0(\lambda)$, $P_1(\lambda)$, $\dots$, $P_n(\lambda), \dots$ can be obtained by the Gram–Schmidt orthonormalization of the monomials $1, \lambda, \dots, \lambda^n, \dots$ in the space $L^2(\mathbb{R}_+; d\rho)$. It is easy to see that $P_n(\lambda)$ is a polynomial of degree n, that is, $P_n(\lambda) = k_n(\lambda^n + r_n \lambda^{n-1} + \cdots)$ with $k_n \neq 0$. One usually requires $k_n > 0$. The recurrence coefficients a_n, b_n can be recovered by the formulas $a_n = \frac{k_n}{k_{n+1}}$, $b_n = r_n - r_{n+1}$.

One defines a mapping $\mathbf{\Phi} : \ell^2(\mathbb{Z}_+) \to L^2(\mathbb{R}; d\rho)$ by the formula

$$(\mathbf{\Phi} f)(\lambda) = \sum_{n=0}^{\infty} P_n(\lambda) f_n, \quad f = \{f_n\}_{n=0}^{\infty} \in \mathcal{D}. \tag{2.3}$$

This mapping is isometric according to (2.2). It is also unitary if the set of all polynomials $P_n(\lambda)$, $n \in \mathbb{Z}_+$, is dense in $L^2(\mathbb{R}; d\rho)$. This condition is satisfied if the operator J_0 is essentially self-adjoint. By putting together definitions (1.1), (2.1) and (2.3), it is easy to check the intertwining property $(\mathbf{\Phi} J f)(\lambda) = \lambda(\mathbf{\Phi} f)(\lambda)$.

Let us now discuss, at a heuristic level, asymptotic formulas for $P_n(\lambda)$ an $n \to \infty$ (and λ is fixed). Typically, on the absolutely continuous spectrum of a Jacobi operator J when $d\rho(\lambda) = \tau(\lambda) d\lambda$, the orthonormal polynomials $P_n(\lambda)$ have oscillating asymptotics

$$P_n(\lambda) = 2\kappa(\lambda) n^{-r} \cos \Omega_n(\lambda) + o(n^{-r}), \tag{2.4}$$

$$\Omega_n'(\lambda) = \omega(\lambda) n^s + o(n^s), \tag{2.5}$$

where $\kappa(\lambda) > 0$, $s \in (0, 1]$ and we can suppose that $\omega(\lambda) > 0$. The exponents r, s, the amplitude $\kappa(\lambda)$ and the phase $\omega(\lambda)$ are determined by the recurrence coefficients a_n and b_n. Our considerations (see Section 4.1) show that these quantities are necessarily linked by universal relations:

$$2r + s = 1, \tag{2.6}$$

$$2\pi \tau(\lambda) \kappa^2(\lambda) = s\omega(\lambda). \tag{2.7}$$

In particular, relations (2.4), (2.5) and hence (2.6), (2.7) are satisfied for the classical polynomials.

2.2 Laguerre operators

Suppose now that the recurrence coefficients a_n, b_n are given by formulas (1.4). In this case the orthogonal polynomials $L_n^{(p)}(z)$ defined by relations (2.1) with the boundary conditions $L_{-1}^{(p)}(z) = 0$, $L_0^{(p)}(z) = 1$ are known as the Laguerre polynomials. Note that the normalized polynomials $L_n^{(p)}(z)$ we consider here are related to the Laguerre polynomials $\mathbf{L}_n^{(p)}(z)$ defined in [3, Section 10.12] or in [9, Section 5.1] by the equality

$$L_n^{(p)}(z) = (-1)^n \sqrt{\frac{\Gamma(1+n)\Gamma(1+p)}{\Gamma(1+n+p)}} \mathbf{L}_n^{(p)}(z). \tag{2.8}$$

According to [3, asymptotic formula (10.15.1)] for positive λ, we have

$$\begin{aligned} L_n^{(p)}(\lambda) = (-1)^n \sqrt{\frac{\Gamma(1+p)}{\pi}} \lambda^{-\frac{p}{2}-\frac{1}{4}} e^{\frac{\lambda}{2}} n^{-\frac{1}{4}} \cos\Big(2\sqrt{n\lambda} - \frac{2p+1}{4}\pi\Big) \\ + O(n^{-\frac{3}{4}}) \end{aligned} \tag{2.9}$$

as $n \to \infty$. This asymptotics is uniform in $\lambda \in [\lambda_0, \lambda_1]$ if $0 < \lambda_0 < \lambda_1 < \infty$.

Let us now consider the Laguerre operators J_p defined by formula (1.1), where a_n, b_n are given by (1.4). The spectral measures of the operators J_p are supported on the half-axis $[0, \infty)$, they are absolutely continuous and are given by the relation (see, e.g., [3, formula (10.12.1)])

$$d\rho_p(\lambda) = \tau_p(\lambda) d\lambda, \quad \text{where } \tau_p(\lambda) = \Gamma(p+1)^{-1}\lambda^p e^{-\lambda}, \quad \lambda \in \mathbb{R}_+. \tag{2.10}$$

Since the measure $d\rho_p(\lambda)$ is absolutely continuous, it is convenient to reduce the Jacobi operator J_p to the operator A of multiplication by λ in the space $L^2(\mathbb{R}_+)$. To that end, we put

$$\varphi_n^{(p)}(\lambda) = \sqrt{\tau_p(\lambda)} L_n^{(p)}(\lambda), \quad \lambda \in \mathbb{R}_+, \tag{2.11}$$

and introduce a mapping $\Phi_p : \ell^2(\mathbb{Z}_+) \to L^2(\mathbb{R}_+)$ by the formula (cf. (2.3))

$$(\Phi_p f)(\lambda) = \sum_{n=0}^{\infty} \varphi_n^{(p)}(\lambda) f_n, \quad f = \{f_n\}_{n=0}^{\infty} \in \mathcal{D}, \quad \lambda \in \mathbb{R}_+. \tag{2.12}$$

The operator $\Phi_p^* : L^2(\mathbb{R}_+) \to \ell^2(\mathbb{Z}_+)$ adjoint to Φ_p is given by the equality

$$(\Phi_p^* g)_n = \int_0^{\infty} \varphi_n^{(p)}(\lambda) g(\lambda)\, d\lambda, \quad n \in \mathbb{Z}_+.$$

The operator Φ_p is unitary, that is,

$$\Phi_p^* \Phi_p = I, \quad \Phi_p \Phi_p^* = I, \tag{2.13}$$

and enjoys the intertwining property

$$\Phi_p J_p = \mathsf{A} \Phi_p. \tag{2.14}$$

3 Wave operators

3.1 Two Laguerre operators

One of our main results is stated as follows.

Theorem 3.1. *Let J_p be a Jacobi operator with matrix elements* (1.4) *in the space $\ell^2(\mathbb{Z}_+)$. Define the unitary operators $\Phi_p : \ell^2(\mathbb{Z}_+) \to L^2(\mathbb{R}_+)$ by formulas* (2.11) *and* (2.12)*. Then, for all $p, q > -1$, the wave operators $W_\pm(J_q, J_p)$ exist and*

$$W_\pm(J_q, J_p) = e^{\pm \frac{i(q-p)\pi}{2}} \Phi_q^* \Phi_p.$$

We start a proof with a simple standard statement.

Lemma 3.2. *The claim of Theorem* 3.1 *is equivalent to the relation*

$$\operatorname*{s-lim}_{t \to \pm\infty} (\Phi_p^* - \mu_\pm \Phi_q^*) e^{-i\mathsf{A}t} = 0, \quad \mu_\pm = e^{\pm \frac{i(q-p)\pi}{2}}. \tag{3.1}$$

Proof. Let f be an arbitrary element of the space $\ell^2(\mathbb{Z}_+)$ and $g = \Phi_p f$. In view of the properties (2.13) and (2.14), we have

$$\begin{aligned}\|e^{iJ_q t}e^{-iJ_p t} f - \mu_\pm \Phi_q^* \Phi_p f\| &= \|e^{-iJ_p t}\Phi_p^* g - \mu_\pm e^{-iJ_q t}\Phi_q^* g\| \\ &= \|(\Phi_p^* - \mu_\pm \Phi_q^*)e^{-i\mathsf{A}t} g\|.\end{aligned}$$

Since the left- and right-hand sides here tend to zero at the same time, this concludes the proof. ∎

It suffices to check (3.1) on a set $C_0^\infty(\mathbb{R}_+)$ dense in $L^2(\mathbb{R}_+)$. Let the function $\varphi_n^{(p)}(\lambda)$ be defined by equalities (2.10) and (2.11). It follows from asymptotic formula (2.9) that

$$\varphi_n^{(p)}(\lambda) = (-1)^n 2^{-1}\pi^{-\frac12}\lambda^{-\frac14}(n+1)^{-\frac14}\big(\nu_p e^{2i\sqrt{n\lambda}} + \bar\nu_p e^{-2i\sqrt{n\lambda}}\big) + r_n^{(p)}(\lambda) \tag{3.2}$$

where $\nu_p = e^{-\frac{i\pi(2p+1)}{4}}$ and

$$|r_n^{(p)}(\lambda)| \le C(n+1)^{-\frac34} \tag{3.3}$$

uniformly on every compact subinterval of $\mathbb{R}_+$.

Let us define mappings $V_\pm : C_0^\infty(\mathbb{R}_+) \to \ell^2(\mathbb{Z}_+)$ by the formula

$$(V_\pm g)_n = (-1)^n 2^{-1}\pi^{-\frac12}(n+1)^{-\frac14}\int_0^\infty e^{\pm 2i\sqrt{n\lambda}}\lambda^{-\frac14} g(\lambda)\, d\lambda. \tag{3.4}$$

Equality (3.2) implies that

$$\Phi_p^* e^{-i\mathsf{A}t} g = \nu_p V_+ e^{-i\mathsf{A}t} g + \bar\nu_p V_- e^{-i\mathsf{A}t} g + R_p(t) g, \tag{3.5}$$

where

$$(R_p(t)g)_n = \int_0^\infty r_n^{(p)}(\lambda) e^{-i\lambda t} g(\lambda)\, d\lambda.$$

First, we check that the remainder in (3.5) is negligible.

Lemma 3.3. *Let $g \in C_0^\infty(\mathbb{R}_+)$. Then*

$$\lim_{|t|\to\infty} \|R_p(t) g\| = 0. \tag{3.6}$$

Proof. By the Riemann–Lebesgue lemma, every integral in the sum

$$\|R_p(t)g\|^2 = \sum_{n=0}^\infty \Big| \int_0^\infty r_n^{(p)}(\lambda) e^{-i\lambda t} g(\lambda)\, d\lambda \Big|^2 \tag{3.7}$$

tends to zero as $|t| \to \infty$. Estimate (3.3) allows us to use the dominated convergence theorem. Therefore the sum (3.7) tends to zero as $|t| \to \infty$. ∎

We need also the following elementary assertion.

Lemma 3.4. *For an arbitrary $G \in C_0^\infty(\mathbb{R}_+)$, an estimate*

$$\left| \int_0^\infty e^{\pm 2i\sqrt{n\lambda} - i\lambda t} G(\lambda)\, d\lambda \right| \leq C_k(\sqrt{n} + |t|)^{-k}, \quad \mp t > 0, \tag{3.8}$$

is true for all $k \in \mathbb{Z}_+$.

Proof. Let us use a formula

$$\int_0^\infty e^{\pm 2i\sqrt{n\lambda} - i\lambda t} G(\lambda)\, d\lambda = i \int_0^\infty e^{\pm 2i\sqrt{n\lambda} - i\lambda t} \left(\frac{G(\lambda)}{\pm\sqrt{n}\lambda^{-\frac{1}{2}} - t} \right)' d\lambda \tag{3.9}$$

which can be verified by a direct integration by parts. Suppose that $\operatorname{supp} G \subset [\lambda_1, \lambda_2]$. Then

$$\sqrt{n}\lambda^{-\frac{1}{2}} + |t| \geq \sqrt{n}\lambda_2^{-\frac{1}{2}} + |t|$$

and $\lambda^{-\frac{3}{2}} \leq \lambda_1^{-\frac{3}{2}}$. Therefore the right-hand side of (3.9) is estimated by $C_1(\sqrt{n} + |t|)^{-1}$ which proves (3.8) for $k = 1$. Further integrations by parts in (3.9), yield (3.8) for an arbitrary k. ■

Corollary 3.5. *For the operators* (3.4) *and all $g \in C_0^\infty(\mathbb{R}_+)$, we have*

$$\lim_{t \to \pm\infty} \| V_\mp e^{-i\mathrm{A}t} g \| = 0. \tag{3.10}$$

Using now relation (3.5) and taking into account Lemma 3.3, we arrive at the following result.

Lemma 3.6. *Let $g \in C_0^\infty(\mathbb{R}_+)$. Then*

$$\lim_{t \to \pm\infty} \| \Phi_p^* e^{-i\mathrm{A}t} g - \nu_p^{\pm 1} V_\pm e^{-i\mathrm{A}t} g \| = 0.$$

The same result is of course true for $\Phi_q^* e^{-i\mathrm{A}t} g$. This yields relation (3.1) with $\mu_\pm = (\nu_p \bar{\nu}_q)^{\pm 1}$. Using Lemma 3.2, we conclude the *proof of Theorem* 3.1. ■

For the scattering operator $S = W_+^* W_-$, we obtain a very simple expression.

Proposition 3.7. *Under the assumptions of Theorem* 3.1, *we have $S = e^{i(p-q)\pi} I$.*

Recall that the wave operators $W_\pm(B, A)$ for a couple of self-adjoint operators A and B can be represented as products of an appropriate Fourier transform for the operator A and the inverse transform corresponding to B. For Schrödinger operators in the space $L^2(\mathbb{R}_+)$, this is discussed, for example, in [11, Section 4.2 (see formula (2.30))]. Normally, there are two natural sets of eigenfunctions of the operators A and B. This leads to two wave operators. In our case these sets of eigenfunctions almost coincide so that the wave operators $W_\pm(J_q, J_p)$ differ by a phase factor only whence the scattering operator is almost trivial.

3.2 Perturbation theory

Here we choose some $p > -1$ and construct scattering theory for the pair $J = J_p$, $\widetilde{J} = J + V$, where the operator V is in some sense small. We do not assume that V is a Jacobi operator, but, in particular, our results apply to Jacobi operators. We denote by $\widetilde{\mathcal{H}}_{\mathrm{ac}}$ the absolutely continuous subspace of the operator $\widetilde{J}$.

Let us define an operator N in the space $\ell^2(\mathbb{Z}_+)$ by the formula

$$(Nf)_n = (n+1) f_n.$$

Our goal is to prove the following result.

Theorem 3.8. *Let $J = J_p$, where $p > -1$, be the Laguerre operator with matrix elements* (1.4), *and let $\widetilde{J} = J + V$ where V is a symmetric operator such that*

$$V = N^{-r} T N^{-r_0} \tag{3.11}$$

for some bounded operator T.

(1) *If $r_0 > \frac{1}{4}$, $r > \frac{1}{4}$, then the wave operators $W_\pm(\widetilde{J}, J)$ exist and are complete, that is,* $\operatorname{Ran} W_\pm(\widetilde{J}, J) = \widetilde{\mathcal{H}}_{\mathrm{ac}}$.

(2) *If $r_0 > \frac{1}{4}$, $r > \frac{1}{2}$, then the singular spectrum of the operator $\widetilde{J}$ consists of eigenvalues of finite multiplicities that may accumulate to the point* 0 *only.*

Theorem 3.8 will be proven in the next subsection. Note that under its assumptions the operator $V = \widetilde{J} - J$ belongs to the Hilbert–Schmidt but not to the trace class. Therefore the assertion about the wave operators $W_\pm(\widetilde{J}, J)$ does not follow from the classical Kato–Rosenblum theorem.

Corollary 3.9. *Let a_n, b_n be defined by formulas* (1.4), *and let $\widetilde{J}$ be a self-adjoint Jacobi operator with matrix elements $\tilde{a}_n$, $\tilde{b}_n$ such that*

$$\tilde{a}_n - a_n = O(n^{-\rho}), \quad \tilde{b}_n - b_n = O(n^{-\rho}). \tag{3.12}$$

(1) *If $\rho > \frac{1}{2}$, then the wave operators $W_\pm(\widetilde{J}, J)$ exist and are complete.*

(2) *If $\rho > \frac{3}{4}$, then the singular spectrum of the operator $\widetilde{J}$ consists of eigenvalues that may accumulate to the point* 0 *only.*

Proof. Let us introduce diagonal matrices A and B with the elements a_n and b_n and the shift S: $(\mathsf{S}f)_n = f_{n+1}$, $n \in \mathbb{Z}_+$. Then $J = A\mathsf{S} + \mathsf{S}^* A + B$ and with obvious notation, we have

$$\widetilde{J} - J = (\widetilde{A} - A)\mathsf{S} + \mathsf{S}^*(\widetilde{A} - A) + (\widetilde{B} - B).$$

The operators $N^r(\widetilde{A} - A)N^{r_0}$ and $N^r(\widetilde{B} - B)N^{r_0}$ are bounded if $r_0 + r = \rho$. Since the operator $N^{-r}\mathsf{S}N^r$ is also bounded, we see that $N^r(\widetilde{J} - J)N^{r_0}$ is bounded as long as $r_0 + r = \rho$. Let us now use Theorem 3.8. For the proof of part (1), we set $r_0 = r = \frac{\rho}{2}$. Part (2) follows if $r_0 = \frac{1}{2}(\rho - \frac{1}{4})$ and $r = \frac{1}{2}(\rho + \frac{1}{4})$. ■

Remark. Under the assumptions of Corollary 3.9 one can find asymptotics of the associated orthogonal polynomials. Essentially, it is the same as for the Laguerre polynomials, that is, given (up to a phase shift) by formula (2.9).

Example 3.10. Consider the Jacobi operator $\widetilde{J}$ with the coefficients

$$\tilde{a}_n = n + \alpha, \quad \tilde{b}_n = 2n + 2\alpha - 1, \quad \alpha > \frac{1}{2}.$$

Up to a shift by $2\alpha - 1$, this operator is related to the birth and death processes (see [4, Section 5.2]). Now conditions (3.12) with $\rho = 1$ are satisfied for $a_n = a_n^{(p)}$, $b_n = b_n^{(p)}$, where $p = 2(\alpha - 1)$.

3.3 Strong smoothness

Our proof of Theorem 3.8 relies on a notion of strong smoothness (see [10, Definition 4.4.5]). We will check strong J-smoothness of the operator N^{-r} for $r > \frac{1}{4}$. Recall that the operator $\Phi = \Phi_p$ is defined by formula (2.12), where the functions $\varphi_n(\lambda) = \varphi_n^{(p)}(\lambda)$ are linked to the Laguerre polynomials $L_n(\lambda) = L_n^{(p)}(\lambda)$ by equalities (2.10), (2.11).

Lemma 3.11. *Let Λ be a compact subinterval of $\mathbb{R}_+$ and $r > \frac{1}{4}$. Then*

$$|(\Phi N^{-r} f)(\lambda)| \leq C \|f\|. \tag{3.13}$$

Moreover,

$$|(\Phi N^{-r} f)(\mu) - (\Phi N^{-r} f)(\lambda)| \leq C |\mu - \lambda|^s \|f\| \tag{3.14}$$

if $s < 2r - \frac{1}{2}$ and $s \leq 1$. The constants C in (3.13) *and* (3.14) *do not depend on $f \in \ell^2(\mathbb{Z}_+)$ and $\lambda, \mu \in \Lambda$.*

Proof. Asymptotics (2.9) of $L_n^{(p)}(\lambda)$ imply that

$$|\varphi_n(\lambda)| \leq C(1+n)^{-\frac{1}{4}}, \tag{3.15}$$

where the constant C does not depend on $n \in \mathbb{Z}_+$ and on $\lambda \in \Lambda$. By the Schwarz inequality, it now follows from definition (2.12) that

$$|(\Phi N^{-r} f)(\lambda)| \leq C \sum_{n=0}^{\infty} (1+n)^{-\frac{1}{4}-r} |f_n| \leq C \sqrt{\sum_{n=0}^{\infty} (1+n)^{-\frac{1}{2}-2r}} \; \|f\|$$

and the series is convergent if $2r > \frac{1}{2}$. This proves (3.13).

For a proof of (3.14), we need an estimate on derivatives of $\frac{dL_n^{(p)}(\lambda)}{d\lambda}$ of Laguerre polynomials for large n. Let us use [9, formula (5.1.14)] for Laguerre polynomials $\mathbf{L}_n^{(p)}(\lambda)$ linked to $L_n^{(p)}(\lambda)$ by equality (2.8):

$$\frac{d}{d\lambda} \mathbf{L}_n^{(p)}(\lambda) = -\mathbf{L}_{n-1}^{(p+1)}(\lambda)$$

whence

$$\frac{d}{d\lambda} L_n^{(p)}(\lambda) = \sqrt{\frac{n}{p+1}} L_{n-1}^{(p+1)}(\lambda).$$

It now follows from estimate (3.15) that

$$|\varphi_n'(\lambda)| \le C(1+n)^{\frac{1}{4}}. \tag{3.16}$$

Since

$$|\varphi_n(\mu) - \varphi_n(\lambda)| \le 2 \sup_{x\in\Lambda} |\varphi_n(x)|$$

and

$$|\varphi_n(\mu) - \varphi_n(\lambda)| \le \sup_{x\in\Lambda} |\varphi_n'(x)||\mu - \lambda|,$$

we have an estimate

$$|\varphi_n(\mu) - \varphi_n(\lambda)| \le \Big(2 \sup_{x\in\Lambda} |\varphi_n(x)|\Big)^{1-s} \sup_{x\in\Lambda} |\varphi_n'(x)|^s |\mu - \lambda|^s$$

for any $s \in [0,1]$. Therefore using (3.15) and (3.16), we see that

$$|\varphi_n(\mu) - \varphi_n(\lambda)| \le C(1+n)^{-\frac{1}{4}+\frac{s}{2}} |\mu - \lambda|^s.$$

This yields an estimate for the operator (2.12):

$$\begin{aligned} |(\Phi N^{-r} f)(\mu) - (\Phi N^{-r} f)(\lambda)| &\le \sum_{n=0}^{\infty} (1+n)^{-r} |\varphi_n(\mu) - \varphi_n(\lambda)||f_n| \\ &\le \|f\| \sqrt{\sum_{n=0}^{\infty} (1+n)^{-2r} |\varphi_n(\mu) - \varphi_n(\lambda)|^2} \\ &\le C\|f\||\mu - \lambda|^s \end{aligned}$$

provided $s < 2r - \frac{1}{2}$. Thus we get (3.14). ■

The operator N^{-r} satisfying estimates (3.13)–(3.14) is called strongly J-smooth with exponent $s \in (0,1]$ on the interval Λ.

Theorem 4.6.4 of [10] states that if a perturbation V admits representation (3.11) with the operators N^{-r_0} and N^{-r} strongly J-smooth (with some exponents $s_0, s > 0$) on all compact subintervals Λ of $\mathbb{R}_+$, then the wave operators $W_\pm(\widetilde{J}, J)$ exist and are complete. This is part (1) of Theorem 3.8.

Theorems 4.7.9 and 4.7.10 of [10] state that all spectral results enumerated in part (2) of Theorem 3.8 are true provided $s > \frac{1}{2}$ (and $s_0 > 0$). According to Lemma 3.11 we can choose $s > \frac{1}{2}$ if $r > \frac{1}{2}$. This concludes the proof of part (2) of Theorem 3.8. ■

Finally, we note that an unusually weak assumption $\rho > \frac{1}{2}$ (instead of the standard $\rho > 1$) in (3.12)) is explained by a decay (3.15) of eigenfunctions of J.

4 Time evolution

It is a common wisdom that an asymptotic behavior of $e^{-iJt}f$ as $t \to \infty$ is determined by spectral properties of the Jacobi operator J and by asymptotics of the corresponding orthonormal polynomials $P_n(\lambda)$ as $n \to \infty$. We first discuss this general idea at a heuristic level and derive new relations between amplitudes and phases in asymptotic formulas for $P_n(\lambda)$. Then we illustrate the formulas obtained on examples of the classical polynomials.

4.1 Universal asymptotic relations

Assume that the spectrum of a Jacobi operator J is absolutely continuous on an interval Λ and the corresponding measure $d\rho(\lambda) = \tau(\lambda)d\lambda$ has a smooth weight $\tau(\lambda)$ for $\lambda \in \Lambda$. Let $E(\Lambda)$ be the spectral projection of the operator J corresponding to Λ, and let the operator $\mathbf{\Phi}$ diagonalizing J be defined by formula (2.3). Choose f such that $f = E(\Lambda)f$ and set $g(\lambda) = \sqrt{\tau(\lambda)}(\mathbf{\Phi} f)(\lambda)$. Clearly, $\|g\|_{L^2(\Lambda)} = \|f\|$ because the operator $\mathbf{\Phi}$ is isometric. If $P_n(\lambda)$ satisfy relation (2.4), then, asymptotically,

$$(e^{-iJt}f)_n = n^{-r}\int_\Lambda \varkappa(\lambda)\big(e^{i\Omega_n(\lambda)-i\lambda t} + e^{-i\Omega_n(\lambda)-i\lambda t}\big)g(\lambda)\,d\lambda, \tag{4.1}$$

where $\varkappa(\lambda) = \sqrt{\tau(\lambda)}\kappa(\lambda)$. Here and below we keep track only of leading terms in asymptotic formulas as $t \to \infty$ and $n \to \infty$; *lower-order terms are neglected.* We suppose that the phase $\Omega_n(\lambda)$ obeys condition (2.5) where $\omega'(\lambda) \neq 0$ for $\lambda \in \Lambda$ and that $g \in C_0^\infty(\Lambda)$.

Stationary points of the integrals (4.1) are determined by the equations

$$\pm\,\omega(\lambda)n^s = t. \tag{4.2}$$

Suppose, for definiteness, that $t \to +\infty$. Then equation (4.2) may have a solution (necessary unique) for the sign "+" only so that the second term (with $e^{-i\Omega_n(\lambda)}$) in (4.1) can be neglected. Let $\sigma = s^{-1}$, $\xi = \frac{n}{t^\sigma}$, and let $\lambda = \lambda(\xi)$ be the solution of the equation

$$\omega(\lambda) = \xi^{-s}. \tag{4.3}$$

Applying the stationary phase method to the first integral in (4.1), we see that

$$(e^{-iJt}f)_n = (2\pi)^{\frac{1}{2}} n^{-r-\frac{s}{2}} e^{i\phi(\xi,t)} |\omega'(\lambda(\xi))|^{-\frac{1}{2}} \varkappa(\lambda(\xi)) g(\lambda(\xi)), \tag{4.4}$$

where

$$\phi(\xi,t) = \pm\frac{\pi}{4} + \Omega_{\xi t^\sigma}(\lambda(\xi))t - \lambda(\xi)t \quad \text{if } \pm\,\omega'(\lambda) > 0.$$

Let

$$h(\xi) = \xi^{-r-\frac{s}{2}} |\omega'(\lambda(\xi))|^{-\frac{1}{2}} \varkappa(\lambda(\xi)) g(\lambda(\xi)) \tag{4.5}$$

so that (4.4) reads as

$$(e^{-iJt}f)_n = (2\pi)^{\frac{1}{2}} t^{-\frac{(2r+s)\sigma}{2}} e^{i\phi(\frac{n}{t^\sigma},t)} h\Big(\frac{n}{t^\sigma}\Big). \tag{4.6}$$

Since the operators e^{-iJt} are unitary, it follows from (4.6) that

$$2\pi \lim_{t\to\infty}\left(t^{-(2r+s)\sigma}\sum_{n=0}^{\infty}\left|h\left(\frac{n}{t^{\sigma}}\right)\right|^{2}\right)=\|f\|^{2}. \tag{4.7}$$

Observe that the integral sums

$$N^{-1}\sum_{n=0}^{\infty}\left|h\left(\frac{n}{N}\right)\right|^{2}\to\int_{\lambda^{-1}(\Lambda)}|h(\xi)|^{2}\,d\xi \tag{4.8}$$

as $N\to\infty$, by the definition of the integral. Let us here set $N=t^{\sigma}$ and compare (4.8) with (4.7). First, we obtain relation (2.6) since the powers of t in the left-hand sides of (4.7) and (4.8) should be the same. It follows that $\xi^{-r-\frac{s}{2}}$ in (4.5) can be replaced by $\xi^{-\frac{1}{2}}$. Second, comparing the right-hand sides, using definition (4.5) and taking into account that $\|f\|=\|g\|_{L^2(\Lambda)}$, we see that

$$2\pi\int_{\lambda^{-1}(\Lambda)}\xi^{-1}|\omega'(\lambda(\xi))|^{-1}\varkappa^{2}(\lambda(\xi))|g(\lambda(\xi))|^{2}\,d\xi=\int_{\Lambda}|g(\lambda)|^{2}\,d\lambda. \tag{4.9}$$

Differentiating relation (4.3), we find that

$$\omega'(\lambda(\xi))\lambda'(\xi)=-s\xi^{-s-1}=-s\xi^{-1}\omega(\lambda(\xi)).$$

Substituting this expression for $\omega'(\lambda(\xi))$ into the left-hand side of (4.9), we rewrite equality (4.9) as

$$2\pi s^{-1}\int_{\Lambda}\varkappa^{2}(\lambda)\omega(\lambda)^{-1}|g(\lambda)|^{2}\,d\lambda=\int_{\Lambda}|g(\lambda)|^{2}\,d\lambda.$$

Since $g\in C_0^{\infty}(\Lambda)$ is arbitrary, this yields relation (2.7) between asymptotic coefficients in (2.4), (2.5) and the spectral measure.

According to formula (4.4) the functions $(e^{-iJt}f)_n$ "live" in the region where $n\sim|t|^{\sigma}$. For example, for the Laguerre polynomials, it follows from (2.9) that $\sigma=2$. This is fairly unusual since scattering states are normally concentrated in the region where $n\sim|t|$. Similarly, for continuous operators of Schrödinger type we have the relation $x\sim|t|$. Indeed, consider, for example, the operator $H=-\frac{d^2}{dx^2}$ in the space $L^2(\mathbb{R})$. In this case, we have

$$(e^{-iHt}f)(x)=e^{\mp\frac{\pi i}{4}}(2|t|)^{-\frac{1}{2}}e^{i\frac{x^2}{4t}}\hat{f}\left(\frac{x}{2t}\right)+\varepsilon(x,t),$$

where $\hat{f}(\xi)=(2\pi)^{-\frac{1}{2}}\int_{-\infty}^{\infty}e^{-ix\xi}f(x)\,dx$ is the Fourier transform of $f(x)$ and the norm in $L^2(\mathbb{R})$ of the term $\varepsilon(\cdot,t)$ tends to zero as $|t|\to\infty$.

The above arguments relied on the stationary phase method and strongly used the assumption $\omega'(\lambda)\neq 0$. Let us now consider, at a very heuristic level, the dispersionless case $\omega'(\lambda) = 0$. We suppose that condition (2.4) is satisfied with $\Omega_n(\lambda) = \nu_n\lambda + \delta_n$, where $\nu_n = \omega n^s + o(n^{s-1})$, $s \in (0, 1]$ and δ_n do not depend on λ. Then it follows from (4.1) that, up to a term which tends to zero in $\ell^2(\mathbb{Z}_+)$ as $|t| \to \infty$,

$$(e^{-iJt} f)_n = (2\pi)^{\frac{1}{2}} (n+1)^{-r} \big(e^{i\delta_n} \widehat{G}(t - \nu_n) + e^{-i\delta_n} \widehat{G}(t + \nu_n)\big), \tag{4.10}$$

where $\widehat{G}(t)$ is the Fourier transform of $G(\lambda) = \varkappa(\lambda) g(\lambda)$. If $\omega > 0$ and $t \to +\infty$, the second term in the right-hand side is negligible. The operators e^{-iJt} being unitary, it follows from (4.10) that

$$2\pi \sum_{n=0}^{\infty} (n+1)^{-2r} |\widehat{G}(t - \nu_n)|^2 \to \|f\|^2 = \int_{-\infty}^{\infty} |g(\lambda)|^2 \, d\lambda \quad \text{as } t \to +\infty. \tag{4.11}$$

It is natural to expect that the limit of the left-hand side here is determined by n such that $\omega n^s \sim t$ whence $n^{-2r} \sim (\frac{t}{\omega})^{-2\sigma r}$. Let us set $m = n - (\frac{t}{\omega})^{\sigma}$ and use that $t - \nu_n \sim -st(\frac{\omega}{t})^{\sigma} m$. Then (4.11) implies that

$$(2\pi t^{-2\sigma r + \sigma - 1} \omega^{2\sigma r - \sigma} s^{-1}) \Big(N^{-1} \sum_{m=-\infty}^{\infty} \Big| \widehat{G}\Big(-\frac{m}{N}\Big)\Big|^2 \Big) \to \int_{-\infty}^{\infty} |g(\lambda)|^2 \, d\lambda, \tag{4.12}$$

where $N = s^{-1} t^{\sigma - 1} \omega^{-\sigma} \to \infty$. According to (4.8) and the Parseval identity the second factor on the left has a finite limit $\int_{-\infty}^{\infty} |G(\lambda)|^2 \, d\lambda$. Therefore the power of t in the first factor should be zero which yields equality (2.6). Now relation (4.12) shows that

$$2\pi \int_{-\infty}^{\infty} |G(\lambda)|^2 \, d\lambda = \omega s \int_{-\infty}^{\infty} |g(\lambda)|^2 \, d\lambda.$$

Since $G(\lambda) = \varkappa(\lambda) g(\lambda)$ and $g \in C_0^{\infty}(\Lambda)$ is arbitrary, we again arrive at equality (2.7) where $\omega(\lambda) = \omega$ does not depend on λ.

Let us summarize the results obtained. Suppose that the orthonormal polynomials $P_n(\lambda)$ have asymptotic behavior (2.4) with the phase $\Omega_n(\lambda)$ satisfying (2.5). Then necessarily relations (2.6) and (2.7) hold true. Precise conditions guaranteeing (2.6) and (2.7) and proofs of these relations will be published elsewhere.

4.2 Laguerre polynomials

Let the operators J_p be defined by formula (1.1) with the coefficients (1.4). Theorem 3.1 shows that, for all $p, q > -1$,

$$\lim_{t \to \pm\infty} \|e^{-iJ_p t} f - e^{-iJ_q t} \tilde{f}_{\pm}\| = 0 \quad \text{if } \tilde{f}_{\pm} = W_{\pm}(J_q, J_p) f,$$

that is, the time evolution of $e^{-iJ_p t} f$ is the same for all $p > -1$; only the initial data are changed.

Our goal here is to obtain explicit asymptotic formulas for $(e^{-iJ_p t} f)_n$ as $t \to \pm\infty$. It turns out that the asymptotics of $(e^{-iJ_p t} f)_n$ depends crucially on the ratio $\frac{n}{t^2}$. Below we omit the index p. Since $e^{-iJt} f = \Phi^* e^{-iAt} \Phi f$, Lemma 3.6 can be reformulated as follows.

Lemma 4.1. *Let $g = \Phi f \in C_0^\infty(\mathbb{R}_+)$, and let the operators $V_\pm$ be defined by formula* (3.4). *Then*

$$e^{-iJt} f = \nu^{\pm 1} V_\pm e^{-i\mathsf{A}t} g + \varepsilon_\pm(t),$$

where $\|\varepsilon_\pm(t)\| \to 0$ as $t \to \pm\infty$.

Thus, we have to find the asymptotics of

$$(V_\pm e^{-i\mathsf{A}t} g)_n = (-1)^n 2^{-1} \pi^{-\frac{1}{2}} (n+1)^{-\frac{1}{4}} \int_0^\infty e^{\pm 2i\sqrt{n\lambda} - i\lambda t} G(\lambda)\, d\lambda, \tag{4.13}$$

where $G(\lambda) = \lambda^{-\frac{1}{4}} g(\lambda)$ as $t \to \pm\infty$.

The first assertion shows that these functions are small as $|t| \to \infty$ both for relatively "small" and "large" n.

Lemma 4.2. *Let the operators $V_\pm$ be defined by formula* (3.4), *and let $g \in C_0^\infty(\mathbb{R}_+)$. Suppose that* $\operatorname{supp} g \subset [\lambda_1, \lambda_2]$ *and choose $\mu_1 < \lambda_1$, $\mu_2 > \lambda_2$. Then, for all $n \le \mu_1 t^2$ and $n \ge \mu_2 t^2$, all $k \in \mathbb{Z}_+$ and some constants C_k, we have the estimates*

$$\big|(V_\pm e^{-i\mathsf{A}t} g)_n\big| \le C_k(\sqrt{n} + |t|)^{-k} \quad \textit{for all } t \in \mathbb{R}. \tag{4.14}$$

Proof. We proceed from formula (3.9) and estimate its right-hand side. If $n \le \mu_1 t^2$, then

$$|\sqrt{n}\lambda^{-\frac{1}{2}} - t| \ge |t| - \sqrt{n}\lambda_1^{-\frac{1}{2}} \ge |t|\Big(1 - \Big(\frac{\mu_1}{\lambda_1}\Big)^{\frac{1}{2}}\Big) \ge c(\sqrt{n} + |t|),$$

where

$$c(\sqrt{\mu_1} + 1) = 1 - \sqrt{\frac{\mu_1}{\lambda_1}} > 0.$$

Quite similarly, if $n \ge \mu_2 t^2$, then

$$|\sqrt{n}\lambda^{-\frac{1}{2}} - t| \ge \sqrt{n}\lambda_2^{-\frac{1}{2}} - |t| \ge \sqrt{n}(\lambda_2^{-\frac{1}{2}} - \mu_2^{-\frac{1}{2}}) \ge c(\sqrt{n} + |t|),$$

where

$$c(\sqrt{\mu_2} + 1) = \sqrt{\frac{\mu_2}{\lambda_2}} - 1 > 0.$$

According to (3.9), this proves (4.14) for $k = 1$.

Integrating by parts k times in (3.9), we obtain (4.14). ∎

To find asymptotics of the integral in (4.13), we use the stationary phase method. Put

$$\xi = \sqrt{n}|t|^{-1} \quad \text{and} \quad \theta(\lambda, \xi) = -2\xi\sqrt{\lambda} + \lambda.$$

Then

$$\int_0^\infty e^{\pm 2i\sqrt{n\lambda} - i\lambda t} G(\lambda)\, d\lambda = \int_0^\infty e^{-i\theta(\lambda,\xi)t} G(\lambda)\, d\lambda =: \mathcal{I}(t, \xi). \tag{4.15}$$

Differentiating $\theta(\lambda,\xi)$ in λ, we find that

$$\theta'(\lambda,\xi) = -\xi\lambda^{-\frac{1}{2}} + 1 \quad \text{and} \quad \theta''(\lambda,\xi) = 2^{-1}\xi\lambda^{-\frac{3}{2}}.$$

The stationary point $\lambda_0 = \lambda_0(\xi)$ of the phase $\theta(\lambda,\xi)$ is determined by the equation $\theta'(\lambda_0,\xi) = 0$ whence $\lambda_0 = \xi^2$. Therefore the stationary phase method yields

$$\mathcal{J}(t,\xi) = e^{\mp\frac{i\pi}{4}} e^{-i\theta(\lambda_0,\xi)t} \sqrt{\frac{2\pi}{\theta''(\lambda_0,\xi)|t|}}\, G(\lambda_0) + o(|t|^{-\frac{1}{2}})$$

as $t \to \pm\infty$. Since $\theta(\lambda_0,\xi) = -\xi^2$ and $\theta''(\lambda_0,\xi) = 2^{-1}\xi^{-2}$, we arrive at the following intermediate result.

Lemma 4.3. *Let $G \in C_0^\infty(\mathbb{R}_+)$. Then the integral* (4.15) *has asymptotics*

$$\mathcal{J}(t,\xi) = 2\pi^{\frac{1}{2}} e^{\mp\frac{i\pi}{4}} e^{i\xi^2 t}|t|^{-1}|\xi| G(\xi^2) + o(|t|^{-1}), \quad t \to \pm\infty, \tag{4.16}$$

with the estimate of the remainder uniform in ξ from compact subintervals of $\mathbb{R}_+$.

Let us come back to formula (4.13). Set

$$(U(t)g)_n = (-1)^n e^{\frac{in}{t}} |t|^{-1} g\Big(\frac{n}{t^2}\Big), \quad t \neq 0. \tag{4.17}$$

Lemma 4.4. *Let $G \in C_0^\infty(\mathbb{R}_+)$ and $0 < \mu_1 < \mu_2 < \infty$. Then*

$$\sup_{n \in (\mu_1 t^2, \mu_2 t^2)} \big|(V_\pm e^{-i\mathsf{A}t} g)_n - e^{\mp\pi i/4}(U(t)g)_n\big| = o(|t|^{-1}), \quad t \to \pm\infty. \tag{4.18}$$

Proof. Let us set in (4.16) $G(\lambda) = \lambda^{-\frac{1}{4}} g(\lambda)$ and $\xi = \sqrt{n} t^{-1}$ so that

$$|\xi| G(\xi^2) = |\xi|^{\frac{1}{2}} g(\xi^2) = n^{\frac{1}{4}} |t|^{-\frac{1}{2}} g(n/t^2).$$

Thus, it follows from Lemma 4.3 that, as $t \to \pm\infty$,

$$\int_0^\infty e^{\pm 2i\sqrt{n\lambda} - i\lambda t} \lambda^{-\frac{1}{4}} g(\lambda)\, d\lambda = 2\pi^{\frac{1}{2}} e^{\mp\frac{i\pi}{4}} n^{\frac{1}{4}} e^{\frac{in}{t}} |t|^{-1} g\Big(\frac{n}{t^2}\Big) + o(|t|^{-1}),$$

as long as $n \in (\mu_1 t^2, \mu_2 t^2)$. Putting together this relation with (4.13), we arrive at (4.18). ■

Now we are in a position to obtain an asymptotic formula for $e^{-iJ_p t} f$ as $t \to \pm\infty$.

Theorem 4.5. *Let J_p, $p > -1$, be a Jacobi operator with matrix elements* (1.4), *and let the operator $U(t)$ be given by formula* (4.17). *Define the operators Φ_p by formulas* (2.11) *and* (2.12) *and suppose that $\Phi_p f \in C_0^\infty(\mathbb{R}_+)$. Then*

$$\lim_{t \to \pm\infty} \|e^{-iJ_p t} f - e^{\mp\frac{i(p+1)\pi}{2}} U(t)\Phi_p f\| = 0. \tag{4.19}$$

Proof. According to Lemma 4.1, we can replace here $e^{-iJt}f$ by $\nu^{\pm 1}V_\pm e^{-i\mathsf{A}t}g$, where $g = \Phi f$. Suppose that $\operatorname{supp} g \subset [\lambda_1, \lambda_2]$ and choose $\mu_1 < \lambda_1$, $\mu_2 > \lambda_2$. It follows from Lemma 4.2 that

$$\Big(\sum_{n\leq\mu_1 t^2} + \sum_{n\geq\mu_2 t^2}\Big)|(V_\pm e^{-i\mathsf{A}t}g)_n|^2 \to 0 \tag{4.20}$$

as $t \to \pm\infty$. According to (4.18), we also have

$$\sum_{\mu_1 t^2\leq n\leq\mu_2 t^2} |(V_\pm e^{-i\mathsf{A}t}g)_n - e^{\mp\frac{\pi i}{4}}(U(t)g)_n|^2 = o(1).$$

Combined with (4.20) this yields relation (4.19). ∎

According to (2.9) for the Laguerre polynomials, we have $\Lambda = \mathbb{R}_+$, $r = \frac{1}{4}$, $s = \frac{1}{2}$ and

$$\begin{aligned}\kappa(\lambda) &= \frac{1}{2}\sqrt{\frac{\Gamma(1+p)}{\pi}}\lambda^{-\frac{p}{2}-\frac{1}{4}}e^{\frac{\lambda}{2}},\\ \Omega_n(\lambda) &= \pi n + 2\sqrt{n\lambda} - \frac{2p+1}{4}\pi,\\ \omega(\lambda) &= \lambda^{-\frac{1}{2}}.\end{aligned}$$

Since $\tau(\lambda)$ is given by (2.10), the identity in (2.7) is satisfied.

4.3 Jacobi polynomials

In this subsection we define a Jacobi operator by its spectral measure $d\rho(\lambda)$. We suppose that this measure is supported on the interval $[-1, 1]$ and

$$d\rho(\lambda) = \tau(\lambda)d\lambda, \quad \lambda \in (-1, 1), \tag{4.21}$$

where

$$\tau(\lambda) = k(1-\lambda)^\alpha(1+\lambda)^\beta, \quad \alpha, \beta > -1. \tag{4.22}$$

The weight function $\tau(\lambda) = \tau_{\alpha,\beta}(\lambda)$ (as well as all other objects discussed below) depends on α and β, but these parameters are often omitted in notation. The constant $k = k_{\alpha,\beta}$ is chosen in such a way that the measure (4.21) is normalized, i.e.,

$$\rho(\mathbb{R}) = \rho((-1, 1)) = 1.$$

The orthonormal polynomials $\mathsf{G}_n(\lambda) = \mathsf{G}_n^{(\alpha,\beta)}(\lambda)$ determined by the measure defined in (4.21)–(4.22) are known as the Jacobi polynomials.

Let $J = J_{\alpha,\beta}$ be the Jacobi operator with the spectral measure $d\rho(\lambda) = d\rho_{\alpha,\beta}(\lambda)$. Explicit expressions for its matrix elements a_n, b_n can be found, for example, in the books [3,9], but we do not need them. We here note only asymptotic formulas

$$\begin{aligned}a_n &= \frac{1}{2} + 2^{-4}(1-2\alpha^2-2\beta^2)n^{-2} + O(n^{-3}),\\ b_n &= 2^{-2}(\beta^2-\alpha^2)n^{-2} + O(n^{-3})\end{aligned} \tag{4.23}$$

for the matrix elements and (see [9, formula (8.21.10)])

$$\begin{aligned} \mathsf{G}_n(\lambda) = 2^{\frac{1}{2}}(\pi k)^{-\frac{1}{2}}(1-\lambda)^{-\frac{1+2\alpha}{4}}(1+\lambda)^{-\frac{1+2\beta}{4}} \\ \times \cos\Big((n+\gamma)\arcsin\lambda - \frac{\pi(2n+\beta-\alpha)}{4}\Big) + O(n^{-1}), \end{aligned} \tag{4.24}$$

where $\gamma = \frac{\alpha+\beta+1}{2}$, for the orthonormal polynomials. Estimate of the remainder in (4.24) is uniform in λ from compact subsets of $(-1,1)$.

Similarly to the cases of the Laguerre polynomials, we set $\varphi_n(\lambda) = \sqrt{\tau(\lambda)}\mathsf{G}_n(\lambda)$ and define the mapping $\Phi : \ell^2(\mathbb{Z}_+) \to L^2(-1,1)$ by (2.12), where $\lambda \in (-1,1)$. Using (4.22) and (4.24), we obtain the representation

$$e^{-iJt} f = \Phi^* e^{-i\mathsf{A}t} g = V_+ e^{-i\mathsf{A}t} g + V_- e^{-i\mathsf{A}t} g + R(t)g \tag{4.25}$$

with $g = \Phi f \in C_0^\infty(-1,1)$, where

$$(V_\pm g)_n = (2\pi)^{-\frac{1}{2}} i^{\mp n} e^{\pm\frac{i(\alpha-\beta)\pi}{4}} \int_{-1}^{1} e^{\pm i(n+\gamma)\arcsin\lambda}(1-\lambda^2)^{-\frac{1}{4}} g(\lambda)\, d\lambda,$$

and the remainder $R(t)g$ satisfies condition (3.6).

Let us state analogues of Lemmas 3.4 and 4.2.

Lemma 4.6. *For an arbitrary $G \in C_0^\infty(-1,1)$, an estimate*

$$\Big|\int_{-1}^{1} e^{\pm in\arcsin\lambda - i\lambda t} G(\lambda)\, d\lambda\Big| \le C_k(n+|t|)^{-k} \tag{4.26}$$

is true for all $k \in \mathbb{Z}_+$ if $\mp t > 0$.

Proof. Integrating by parts, we see that

$$\int_{-1}^{1} e^{\pm in\arcsin\lambda - i\lambda t} G(\lambda)\, d\lambda = \pm i \int_{-1}^{1} e^{\pm in\arcsin\lambda - i\lambda t} \Big(\frac{G(\lambda)}{n(1-\lambda^2)^{-\frac{1}{2}} + |t|}\Big)' d\lambda.$$

This yields (4.26) for $k = 1$ because $|\lambda| \le c < 1$ on the support of G. Further integrations by parts lead to estimates (4.26) for all k. ■

Quite similarly, we obtain also the following result.

Lemma 4.7. *Let $G \in C_0^\infty(-1,1)$. Then estimates* (4.26) *hold for all $t \in \mathbb{R}$ and all $k \in \mathbb{Z}_+$ if either $n \le \delta|t|$ or $n \ge (1-\delta)|t|$ for a sufficiently small number δ depending on* supp G.

We use these lemmas with

$$G_\pm(\lambda) = e^{\pm i\gamma\arcsin\lambda}(1-\lambda^2)^{-\frac{1}{4}} g(\lambda) \tag{4.27}$$

and take equality (4.25) into account. According to Lemma 4.6, relation (3.10) is satisfied so that it suffices to consider $V_\pm e^{-i\mathsf{A}t} g$ as $t \to \pm\infty$. According to Lemma 4.7, we only have to study

$$(V_\pm e^{-i\mathrm{A}t} g)_n = (2\pi)^{-\frac{1}{2}} i^{\mp n} e^{\pm\frac{i(\alpha-\beta)\pi}{4}} \int_{-1}^{1} e^{\pm i n \arcsin\lambda - i\lambda t} G_\pm(\lambda)\, d\lambda$$

for $\delta|t| \le n \le (1-\delta)|t|$, where $\delta > 0$.

Let us now set

$$\xi = n|t|^{-1}, \quad \theta(\lambda,\xi) = \xi \arcsin\lambda - \lambda \tag{4.28}$$

and consider an integral

$$\mathcal{J}(t,\xi) = \int_{-1}^{1} e^{i\theta(\lambda,\xi)t} G(\lambda)\, d\lambda, \quad \xi \in (0,1), \tag{4.29}$$

where $G \in C_0^\infty((-1,1)\setminus\{0\})$. Differentiating $\theta(\lambda,\xi)$ in λ, we see that

$$\theta'(\lambda,\xi) = \xi(1-\lambda^2)^{-\frac{1}{2}} - 1 \quad \text{and} \quad \theta''(\lambda,\xi) = \xi\lambda(1-\lambda^2)^{-\frac{3}{2}}.$$

The stationary points $\lambda_\pm = \lambda_\pm(\xi)$ of the phase $\theta(\lambda,\xi)$ are determined by the equation $\theta'(\lambda_\pm,\xi) = 0$ whence

$$\lambda_\pm = \pm\lambda_0, \quad \lambda_0 = \sqrt{1-\xi^2}$$

and

$$\theta''(\lambda_\pm,\xi) = \pm\xi^{-2}\sqrt{1-\xi^2}.$$

Note also that

$$\theta(\lambda_0,\xi) = \xi \arccos\xi - \sqrt{1-\xi^2} =: \psi(\xi). \tag{4.30}$$

Therefore the stationary phase method yields the following intermediate result.

Lemma 4.8. *Let the phases $\theta(\lambda,\xi)$ and $\psi(\xi)$ be given by formulas* (4.28) *and* (4.30), *respectively. Then the integral* (4.29) *has asymptotics*

$$\begin{aligned} \mathcal{J}(t,\xi) = (2\pi)^{\frac{1}{2}} |t|^{-\frac{1}{2}} \xi (1-\xi^2)^{-\frac{1}{4}} \big(& e^{\pm\frac{i\pi}{4}} e^{i\psi(\xi)t} G\big(\sqrt{1-\xi^2}\big) \\ & + e^{\mp\frac{i\pi}{4}} e^{-i\psi(\xi)t} G\big(-\sqrt{1-\xi^2}\big)\big) + o(|t|^{-\frac{1}{2}}) \end{aligned} \tag{4.31}$$

as $t \to \pm\infty$. The estimate of the remainder in (4.31) *is uniform in ξ from compact subintervals of* (0, 1).

Let us now apply Lemma 4.8 to functions (4.27). We set

$$h_\pm(\xi) = e^{\pm\frac{i\pi}{4}} e^{\pm i\gamma \arccos\xi} \xi^{\frac{1}{2}} (1-\xi^2)^{-\frac{1}{4}} g\big(\pm\sqrt{1-\xi^2}\big) \tag{4.32}$$

and

$$(U(t)g)_n = i^{\mp n} e^{\pm\frac{i(\alpha-\beta)\pi}{4}} |t|^{-\frac{1}{2}} \big(e^{i\psi(\xi)t} h_+(\xi) + e^{-i\psi(\xi)t} h_-(\xi)\big), \tag{4.33}$$

where $\xi = n|t|^{-1} \in (0,1)$, $\pm t > 0$ and the function $\psi(\xi)$ is defined by formula (4.30). For $n \ge |t|$, we set $(U(t)g)_n = 0$.

Putting the results obtained together, we state our final result.

Theorem 4.9. *Let $J = J_{\alpha,\beta}$ be the Jacobi operator corresponding to the weight function* (4.21). *Define the operator $\Phi : \ell^2(\mathbb{Z}_+) \to L^2(-1,1)$ by formula* (2.12)*, where $\lambda \in (-1,1)$, and suppose that $g = \Phi f \in C_0^\infty((-1,1)\setminus\{0\})$. Let the operator $U(t)$ be given by equalities* (4.32) *and* (4.33). *Then*

$$\lim_{t\to\pm\infty} \|e^{-iJt} f - U(t)\Phi f\| = 0.$$

The operator $J_{1/2,1/2} =: J^{(0)}$ is particularly simple. It is known as the free discrete Schrödinger operator. In this case, we have $a_n = \frac{1}{2}$, $b_n = 0$ for all n. The corresponding orthonormal polynomials $\mathsf{G}_n^{(1/2,1/2)}(\lambda)$ are the Chebyshev polynomials of second kind. According to (4.23), the difference $J - J^{(0)}$ is trace class for all α and β, so that the wave operators $W_\pm(J, J^{(0)})$ as well as $W_\pm(J^{(0)}, J)$ exist. Similarly to Theorem 3.1, this fact can also be checked by a direct calculation using relations (4.25) and (3.10); see [12] for details. This yields also explicit expressions for the wave operators: $W_\pm(J, J^{(0)}) = \Phi^* \Sigma_\pm \Phi^{(0)}$, where $\Sigma_\pm$ is the multiplication operator in the space $L^2(-1,1)$ by the function

$$e^{\mp i \frac{\pi(\alpha-\beta)}{4}} e^{\mp i 2^{-1}(\alpha+\beta-1)\arcsin\lambda}.$$

For Jacobi polynomials, we have $\Lambda = (-1,0)$ or $\Lambda = (0,1)$, $r = 0$, $s = 1$ and

$$\omega(\lambda) = (1-\lambda^2)^{-\frac{1}{2}}.$$

Comparing (4.22) and (4.24), we also see that

$$\varkappa(\lambda) = (2\pi)^{-\frac{1}{2}}(1-\lambda^2)^{-\frac{1}{4}}$$

whence $2\pi\tau(\lambda)\varkappa^2(\lambda) = \omega(\lambda)$ which is consistent with (2.7).

We finally note that the asymptotic behavior of the evolution $e^{-iJt} f$ looks similar to the corresponding result for the Klein–Gordon operator $\sqrt{-d^2/dx^2+1}$ in the space $L^2(\mathbb{R})$.

4.4 Hermite polynomials

The Hermite polynomials $H_n(z)$ are determined by the recurrence coefficients (1.3). As usual, relations (2.1) for $H_n(z)$ are complemented by the boundary conditions $H_{-1}(z) = 0$, $H_0(z) = 1$. According to [3, asymptotic formula (10.15.18)], we have

$$H_n(\lambda) = 2^{\frac{1}{2}} \pi^{-\frac{1}{4}} e^{\frac{\lambda^2}{2}} (2n+1)^{-\frac{1}{4}} \cos\Big(\sqrt{2n+1}\,\lambda - \frac{\pi n}{2}\Big) + O(n^{-\frac{3}{4}}) \tag{4.34}$$

as $n \to \infty$. This asymptotics is uniform in $\lambda \in \mathbb{R}$ on compact subintervals.

Let us consider the Jacobi operators J defined by formula (1.1), where a_n, b_n are given by (1.3). The spectral measure of J equals $d\rho(\lambda) = \pi^{-\frac{1}{2}} e^{-\lambda^2} d\lambda$, where $\lambda \in \mathbb{R}$ (see, e.g., [3, formula (10.13.1)]). Thus, $d\rho(\lambda)$ is absolutely continuous and its support is the whole axis $\mathbb{R}$.

Following the scheme exposed in Section 2.2, we reduce the Jacobi operator J to the operator A of multiplication by λ in the space $L^2(\mathbb{R})$. To that end, we introduce a mapping $\Phi : \ell^2(\mathbb{Z}_+) \to L^2(\mathbb{R})$ by the formula

$$(\Phi f)(\lambda) = \pi^{-\frac{1}{4}} \sum_{n=0}^{\infty} e^{-\frac{\lambda^2}{2}} H_n(\lambda) f_n, \quad f = \{f_n\}_{n=0}^{\infty} \in \mathcal{D}, \quad \lambda \in \mathbb{R}. \tag{4.35}$$

The operator Φ is unitary and enjoys the intertwining property $\Phi J = \mathsf{A}\Phi$.

Putting together formulas (4.34) and (4.35), we obtain representation (4.25) where

$$(V_\pm g)_n = i^{\mp n} (2\pi)^{-\frac{1}{2}} (2n+1)^{-\frac{1}{4}} \int_{-\infty}^{\infty} e^{\pm i\sqrt{2n+1}\lambda} g(\lambda)\, d\lambda$$

and the remainder $R(t)g$ satisfies condition (3.6). Then

$$(V_\pm e^{-i\mathsf{A}t} g)_n = i^{\mp n} (2n+1)^{-\frac{1}{4}} \hat{g}\big(t \mp \sqrt{2n+1}\big),$$

where $\hat{g}(x)$ is the Fourier transform of g. Since $\hat{g}(x) = O(x|^{-k})$ as $|x| \to \infty$ for all $k > 0$, we see that

$$\lim_{t\to\pm\infty} \sum_{n=0}^{\infty} (2n+1)^{-\frac{1}{2}} \big|\hat{g}\big(t \pm \sqrt{2n+1}\big)\big|^2 = 0,$$

whence relation (3.10) follows. Thus, representation (4.25) implies the result below.

Theorem 4.10. *Let J be a Jacobi operator with matrix elements* (1.3), *and let the operator $U(t)$ be given by the formula*

$$(U(t)g)_n = i^{\mp n} (2n+1)^{-\frac{1}{4}} \hat{g}\big(t \mp \sqrt{2n+1}\big), \quad \pm t > 0. \tag{4.36}$$

Define the operator Φ by formula (4.35) *and suppose that $\Phi f \in C_0^\infty(\mathbb{R})$. Then*

$$\lim_{t\to\pm\infty} \|e^{-iJt} f - U(t)\Phi f\| = 0.$$

Corollary 4.11. *For all $g \in C_0^\infty(\mathbb{R})$, we have*

$$\lim_{t\to\pm\infty} \sum_{n=0}^{\infty} (2n+1)^{-\frac{1}{2}} \big|\hat{g}\big(t \mp \sqrt{2n+1}\big)\big|^2 = \|\hat{g}\|^2_{L^2(\mathbb{R})}. \tag{4.37}$$

According to (4.36) the evolution $e^{-iJt} f$ is dispersionless. Clearly, it is similar to time evolutions for first-order differential operators.

Finally, we note that for the Hermite polynomials, $\Lambda = \mathbb{R}$, $r = \frac{1}{4}$, $s = \frac{1}{2}$, and

$$\kappa(\lambda) = 2^{-\frac{3}{4}} \pi^{-\frac{1}{4}} e^{\frac{\lambda^2}{2}}, \quad \Omega_n(\lambda) = \sqrt{2n+1}\,\lambda - \frac{\pi n}{2}, \quad \omega(\lambda) = \sqrt{2}.$$

Thus, the identity in (2.7) remains true, and relation (4.37) is a particular case of (4.11).

Acknowledgements. The research was supported by the Russian Foundation for Basic Research, Grant No. 20-01-00451 A.

Bibliography

[1] N. I. Akhiezer, *The classical moment problem and some related questions in analysis*. Translated by N. Kemmer, Oliver and Boyd, Edinburgh, 1965

[2] A. I. Aptekarev and J. S. Geronimo, Measures for orthogonal polynomials with unbounded recurrence coefficients. *J. Approx. Theory* **207** (2016), 339–347

[3] A. Erdélyi, W. Magnus, F. Oberhettinger and F. G. Tricomi, *Higher transcendental functions. Vol. 1–2*. McGraw–Hill, New York, 1953

[4] M. E. H. Ismail, *Classical and quantum orthogonal polynomials in one variable*. Encyclopedia Math. Appl. 98, Cambridge University Press, Cambridge, 2005

[5] J. Janas and S. Naboko, Jacobi matrices with power-like weights—grouping in blocks approach. *J. Funct. Anal.* **166** (1999), 218–243

[6] J. Janas, S. Naboko and E. Sheronova, Asymptotic behavior of generalized eigenvectors of Jacobi matrices in the critical ("double root") case. *Z. Anal. Anwend.* **28** (2009), 411–430

[7] S. Naboko and S. Simonov, Titchmarsh–Weyl formula for the spectral density of a class of Jacobi matrices in the critical case. Preprint 2019, arXiv:1911.10282

[8] P. G. Nevai, Orthogonal polynomials. *Mem. Amer. Math. Soc.* **18** (1979), no. 213

[9] G. Szegö, *Orthogonal polynomials*. American Mathematical Society, Providence, 1978

[10] D. R. Yafaev, *Mathematical scattering theory*. Transl. Math. Monogr. 105, American Mathematical Society, Providence, 1992

[11] D. R. Yafaev, *Mathematical scattering theory*. Math. Surveys Monogr. 158, American Mathematical Society, Providence, 2010

[12] D. R. Yafaev, Analytic scattering theory for Jacobi operators and Bernstein–Szegö asymptotics of orthogonal polynomials. *Rev. Math. Phys.* **30** (2018), Article ID 1840019

List of contributors

Rafael D. Benguria
Pontificia Universidad Católica de Chile
Santiago, Chile
rbenguri@fis.puc.cl

Soledad Benguria
University of Wisconsin – Madison
USA
benguria@math.wisc.edu

Andrea Cianchi
Università di Firenze, Italy
andrea.cianchi@unifi.it

E. Brian Davies
King's College London
United Kingdom
edward.brian.davies@gmail.com

Jean Dolbeault
Université Paris-Dauphine, France
dolbeaul@ceremade.dauphine.fr

Geneviève Dusson
Université Bourgogne Franche-Comté
Besançon, France
genevieve.dusson@math.cnrs.fr

Maria J. Esteban
Université Paris-Dauphine, France
esteban@ceremade.dauphine.fr

Clotilde Fermanian Kammerer
Université Paris Est Créteil, France;
Université Gustave Eiffel
Marne-la-Vallée, France
clotilde.fermanian@u-pec.fr

Søren Fournais
Aarhus University, Denmark;
Institute for Advanced Study
Princeton, USA
fournais@math.au.dk, fournais@ias.edu

Rupert L. Frank
Ludwig-Maximilans Universität München
Germany;
Munich Center for Quantum Science and Technology (MCQST)
Germany;
California Institute of Technology
Pasadena, USA
r.frank@lmu.de, rlfrank@caltech.edu

David Gontier
University of Paris-Dauphine, France
gontier@ceremade.dauphine.fr

Denis S. Grebenkov
CNRS – Ecole Polytechnique
Palaiseau, France
denis.grebenkov@polytechnique.edu

Bernard Helffer
Université de Nantes, France
Bernard.Helffer@univ-nantes.fr

Dirk Hundertmark
Karlsruhe Institute of Technology
Germany;
University of Illinois at Urbana-Champaign
USA
dirk.hundertmark@kit.edu

Alexei Ilyin
Keldysh Institute of Applied Mathematics
Moscow, Russia
ilyin@keldysh.ru

Victor Ivrii
University of Toronto, Canada
ivrii@math.toronto.edu

Thomas F. Kieffer
Georgia Institute of Technology, Atlanta, USA
tkieffer3@gatech.edu

Christian Klein
Université de Bourgogne-Franche-Comté
Dijon, France
christian.klein@u-bourgogne.fr

Hynek Kovařík
Università degli studi di Brescia, Italy
hynek.kovarik@unibs.it

Peer Kunstmann
Karlsruhe Institute of Technology
Germany
peer.kunstmann@kit.edu

Stanislas Kupin
Université de Bordeaux
Talence, France
skupin@math.u-bordeaux.fr

Simon Larson
California Institute of Technology
Pasadena, USA
larson@caltech.edu

Caroline Lasser
Technische Universität München
Germany
classer@ma.tum.de

Mathieu Lewin
University of Paris-Dauphine, France
mathieu.lewin@math.cnrs.fr

Elliott H. Lieb
Princeton University, USA
lieb@princeton.edu

Michael Loss
Georgia Institute of Technology
Atlanta, USA
loss@math.gatech.edu

Vladimir Maz'ya
Linköping University, Sweden;
University of Liverpool, United Kingdom;
RUDN University, Moscow, Russia
vladimir.mazya@liu.se

Evgeny Mikhalkin
Siberian Federal University
Krasnoyarsk, Russia
mikhalkin@bk.ru

Nicolas Moutal
Ecole Polytechnique, Palaiseau, France
nicolas.moutal@gmail.com

Sergey Naboko†
St. Petersburg State University, Russia

Tobias Ried
Max Planck Institute for Mathematics in the Sciences (MiS)
Leipzig, Germany
tobias.ried@mis.mpg.de

Didier Robert
Université de Nantes, France
didier.robert@univ-nantes.fr

Grigori Rozenblum
Chalmers University of Technology
Gothenburg, Sweden;
St. Petersburg State University
Russia
grigori@chalmers.se

Oleg Safronov
University of North Carolina at Charlotte
USA
osafrono@uncc.edu

Benjamin Schlein
University of Zurich, Switzerland
benjamin.schlein@math.uzh.ch

Eugene Shargorodsky
King's College London
United Kingdom;
Technische Universität Dresden
Germany
eugene.shargorodsky@kcl.ac.uk

Heinz Siedentop
Ludwig-Maximilans Universität München
Germany;
Munich Center for Quantum Science and Technology (MCQST)
Germany
h.s@lmu.de

Israel Michael Sigal
University of Toronto, Canada
im.sigal@utoronto.ca

Johannes Sjöstrand
Université de Bourgogne-Franche-Comté
Dijon, France
johannes.sjostrand@u-bourgogne.fr

Jan Philip Solovej
University of Copenhagen, Denmark
solovej@math.ku.dk

Benjamin Stamm
RWTH Aachen University, Germany
best@mathcces.rwth-aachen.de

Vitaly Stepanenko
Siberian Federal University
Krasnoyarsk, Russia
v-stepanen@mail.ru

Nikola Stoilov
Université de Bourgogne-Franche-Comté
Dijon, France
nikola.stoilov@u-bourgogne.fr

Tatiana A. Suslina
St. Petersburg State University, Russia
t.suslina@spbu.ru

Leon A. Takhtajan
Stony Brook University
USA;
Euler International Mathematical Institute
St. Petersburg, Russia
leontak@math.stonybrook.edu

Christiane Tretter
Universität Bern, Switzerland
tretter@math.unibe.ch

Avgust Tsikh
Siberian Federal University
Krasnoyarsk, Russia
atsikh@sfu-kras.ru

Semjon Vugalter
Karlsruhe Institute of Technology
Germany
semjon.wugalter@kit.edu

Dimitri R. Yafaev
Université Rennes
France;
St. Petersburg State University
Russia
yafaev@univ-rennes1.fr

Sergey Zelik
Keldysh Institute of Applied Mathematics
Moscow, Russia;
Lanzhou University
China;
University of Surrey
Guildford, United Kingdom
s.zelik@surrey.ac.uk